RICE: PRODUCTION AND UTILIZATION

RICE: PRODUCTION AND UTILIZATION

Bor S. Luh, Ph.D.

Food Technologist
Department of Food Science and Technology
University of California, Davis

AVI PUBLISHING COMPANY, INC.
Westport, Connecticut

Frontispiece Courtesy of Dr. D.S. Mikkelsen

©Copyright 1980 by
THE AVI PUBLISHING COMPANY, INC.
Westport, Connecticut

Library of Congress Cataloging in Publication Data

Main entry under title:

Rice—production and utilization.

  Includes index.
  1. Rice. 2. Rice processing. I. Luh, Bor Shiun,
1916—
SB191.R5R497   633'.18   79–13864
ISBN 0-87055-332-1

Printed in the United States of America

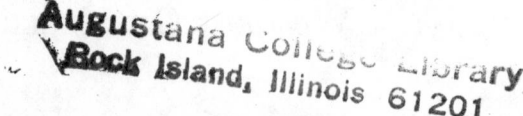

Preface

Rice is one of the most important grains for the world's inhabitants. The hope for better nourishment of the world's population will depend on the development of better rice varieties and improved methods for rice production and processing. During the past 3 decades, interest in rice research has increased greatly in many countries. The development of several new and better varieties by the International Rice Research Institute in the Philippines has stimulated many research institutes and agricultural experiment stations to test the performance of these varieties in other countries where the climate, soil properties, cultural practices and environments are different. The methods of harvesting, handling, drying and milling rough rice have improved as a result of the research development efforts of many agricultural and mechanical engineers, and the rice milling industry.

This book is written in an attempt to assemble the new information on rice production and utilization. It is hoped that the book will be helpful and useful to rice growers, research workers and processors. Students studying the agronomy of rice plants and the technology of rice utilization will also find the book important in their search for new knowledge.

It is a great pleasure to acknowledge the excellent cooperation of the contributors of this book. They have given their knowledge, experience and time generously to the readers. The manuscripts of the various chapters were prepared carefully with creative originality. Many friends have provided information and offered assistance in preparing and editing the content. I wish to thank them for their kindness and assistance in completing this work. The California Rice Research Board, Yuba City, California, and the California Rice Growers' Association, Sacramento, California, have helped greatly in our research work on rice utilization at the University of California, Davis. Many graduate students and work-study students have contributed to the knowledge of rice utilization through research and experiments. I appreciate the assistance of Dr. Y.K.

Liu, Dominic Wong, Irene Ross and Mrs. Clara Robison of the Department of Food Science and Technology, University of California, Davis, who have worked tirelessly in proofreading and retyping the manuscripts.

The encouragements and advice of Dr. B.S. Schweigert and Dr. T.A. Nickerson, Department of Food Science and Technology, University of California, Davis; Mr. Robert R. Mickus, California Rice Growers' Association, Sacramento, California; Dr. Robin M. Saunders, Western Regional Research Laboratory, Albany, California; the U.S. Department of Agriculture; and Dr. A.M. Izadi, Director of the Agricultural Experiment Station of Pahlavi University, Shiraz, Iran, are all gratefully acknowledged.

I wish to thank Dr. M. Ejlali, University of Tehran, Karadj, Iran; Prof. T.T. Fang, National Taiwan University, Taipei, Taiwan; Dr. Homero Fonseca, University of São Paulo, ESALQ, Piracicaba São Paulo, Brasil; Dr. J.R. Ramirez, C.I.E.P.E., San Felipe, Yaracuy, Venezuela; Dr. Hugo C. Silva, University of São Paulo, Rib Preto, São Paulo, Brasil; Dr. Farrokh Sherkat, University of Tabriz, Tabriz, Iran; and Dr. M.M. Tabekhia, University of Mansoura, Mansoura, Egypt, for their valuable contributions when they spent their sabbatical year at Davis.

Thanks are also expressed to Khadijeh Siavosh Haghighi, Librarian of the College of Agriculture, Pahlavi University, Shiraz, Iran, who has helped me in many ways to obtain the literature for the rice book.

The staff members of the Department of Food Technology and Nutrition, Pahlavi University, have reviewed the chapters on rice utilization. I thank Drs. Ara Nahapetian, J. Jamalian, Ahmad Karbassi, Prof. M. Ghavami, and R. Aramian for their constructive review and comments of the chapters on rice flour, rice enrichment, parboiled rice, quick cooking rice, canning, freezing and freeze-drying, breakfast cereals and baby foods, snack foods, rice vinegar, rice hulls and rice oil.

The assistance and contributions made by S.Y. Chen, Annie Chin, Cesar Delgado Rendon, Debora Drust, Sara de Guzman, Linda Lethem, S. Karbassi, Chat Myron Kwan, Kenneth Lau, Peter Lo, Debora Niaps, Ikam Onykwelu, Ann Przybyla, Teresa Tsang, Angela Vargas, K.H. Yang, Martin M.T. Yan, and Betty B. Yih on various aspects of rice research and editing work are most valuable. I thank them for their interest and effort in achieving the goal of this task.

The author would also like to express his appreciation to Dr. D.K. Tressler, Dr. N.W. Desrosier and Ms. Lisa E. Melilli of the AVI Publishing Company for their encouragement and assistance in bringing this book into being.

B.S. LUH

May 1979   Davis, California

Contributors

SALVADOR BARBER, Ph.D., Director, Institute de Agroquimica Y Tecnologia de Alimentos, Jaime Roig 11, Valencia-10 Spain.

CARMEN BENEDITO DE BARBER, Ph.D., Cereal and Proteaginous Laboratory, Institute de Agroquimica Y Tecnologia de Alimentos, Jaime Roig 11, Valencia-10, Spain.

AMARA BHUMIRATANA, Ph.D., Professor and Director, Institute of Food Research and Product Development, Kasetsart University, Bangkok, Thailand.

CLARENCE C. BOWLING, M.S., Associate Professor of Entomology, Research and Extension Center, Texas A & M University, Beaumont, Texas.

SHIU-CHAN CHANG, M.S., Associate Research Fellow, Department of Food Chemistry, Food Industry Research and Development Institute, Hsinchu, Taiwan, R.O.C.

TE-TZU CHANG, Ph.D., Geneticist and Leader, Genetic Resources Program, the International Rice Research Institute, Los Baños, Laguna, Philippines.

WILLIAM TIEN HUNG CHANG, Ph.D., Senior Research Fellow, Department of Food Microbiology, Food Industry Research and Development Institute, Hsinchu, Taiwan, R.O.C.

ROBERT R. COGBURN, M.S., Research Entomologist, Market Quality Research Division, South Region, U.S. Department of Agriculture, Beaumont, Texas.

SURAJIT K. DEDATTA, Ph.D., Agronomist and Head, Agronomy Department, the International Rice Research Institute, Los Baños, Laguna, Philippines.

WEN-HWEI HSU, M.S., Associate Microbiologist, Department of Food Microbiology, Food Industry Research and Development Institute, Hsinchu, Taiwan, R.O.C.

BIENVENIDO O. JULIANO, Ph.D., Chemist and Head, Chemistry Department, the International Rice Research Institute, Los Baños, Laguna, Philippines.

BARBARA M. KENNEDY, Ph.D., Associate Professor of Nutrition, Department of Nutritional Sciences, University of California, Berkeley, California.

MIN-NAM LAI, M.S., Food Microbiologist, Department of Food Microbiology, Food Industry Research and Development Institute, Hsinchu, Taiwan, R.O.C.

CHENG-CHANG LI, Ph.D., Professor, Department of Agronomy, National Chung-Hsing University, Taichung, Taiwan, R.O.C.

CHIN-FUNG LI, Ph.D., Senior Research Fellow and Head, Department of Food Technology and Department of Food Engineering, Food Industry Research and Development Institute, Hsinchu, Taiwan, R.O.C.

YUAN-KUANG LIU, Ph.D., Research Enzymologist, Del Monte Corporation Research Center, Walnut Creek, California.

JONATHAN J. LU, Ph.D., Associate Professor of Geography, Department of Geography, University of Northern Iowa, Cedar Falls, Iowa.

BOR S. LUH, Ph.D., Food Technologist, Department of Food Science and Technology, University of California, Davis, California.

JACK MATTHEWS, Ph.D., Biochemist, Food Product Research, Engineering and Development Laboratory, Southern Regional Research Center, U.S. Department of Agriculture, New Orleans, Louisiana.

ROBERT R. MICKUS, Director, Research and Product Development, California Rice Growers' Association, Sacramento, California.

DUANE S. MIKKELSEN, Ph.D., Professor of Agronomy, Department of Agronomy and Range Science, University of California, Davis, California.

GEORGE E. MILLER, Jr., B.S., Extension Agricultural Engineer, Agricultural Engineering Extension, University of California, Davis, California.

SHU-HUANG OU, Ph.D., Principal Plant Pathologist, Plant Pathology Department, the International Rice Research Institute. Los Baños, Laguna, Philippines.

ROBERT L. ROBERTS, Research Food Technologist, Cereal Product Research Unit, Western Regional Research Center, U.S. Department of Agriculture, Berkeley, California.

ROBIN M. SAUNDERS, Ph.D., Research Leader, Cereal Product Research Unit, Western Regional Research Center, U.S. Department of Agriculture, Berkeley, California.

R. PAUL SINGH, Ph.D., Assistant Professor of Food Engineering, Department of Agricultural Engineering and Department of Food Science and Technology, University of California, Davis, California.

JAMES J. SPADARO, B.S., Research Leader, Food Products Research, Engineering and Development Laboratory, Southern Regional Research Center, U.S. Department of Agriculture, New Orleans, Louisiana.

JAMES F. STEFFE, M.S., Research Assistant, Department of Agricultural Engineering, University of California, Davis, California.

BENITO S. VERGARA, Ph.D., Plant Physiologist, Plant Physiology Department, the International Rice Research Institute, Los Baños, Laguna, Philippines.

JAMES I. WADSWORTH, M.S., Chemical Engineer, Food Products Research, Engineering and Development Laboratory, U.S. Department of Agriculture, New Orleans, Louisiana.

HSI-HUA WANG, Ph.D., Professor, Department of Agricultural Chemistry, National Taiwan University, Taipei, Taiwan, R.O.C.

BILL D. WEBB, Ph.D., Research Chemist, Rice Quality Laboratory, South Region, U.S. Department of Agriculture, Rice Research Center, Texas A & M University, Beaumont, Texas.

Contents

Rice in its Temporal and Spatial Perspectives

Jonathan J. Lu
Te-Tzu Chang

RICE IN RETROSPECT

Rice is generally considered a semiaquatic annual plant, although it could survive as a perennial in the tropics or subtropics. Cultivars of the 2 cultivated species, *Oryza sativa* L. and *O. glaberrima* Steud., can grow in a wide range of water-soil regimes, from a prolonged period of flooding in deep water to dryland on hilly slopes.

Remarkable diversity exists in the crop because of its long history of cultivation and selection under diverse climatic, edaphic and biotic environments, frequently in geographically separated areas. Dispersal by people to many geographic areas has further intensified and accelerated the process of ecogenetic diversification.

Today rice is grown in more than 100 countries on every continent of the world (except Antarctica), extending from 53° latitude N to 40° latitude S.

Antiquity and Importance

A Plant of Antiquity.—As a member of the grass family, the progenitors of the *Oryza* species could have differentiated themselves as early as the Cretaceous period (Stebbins 1971). The above postulate draws support from the pantropical distribution of 20 species in the genus which indicates a close relationship between the components of the Gondwanaland continents and the genomic composition of 13 species (Chang 1976D). The fracture and subsequent drift of the Gondwanaland components could have begun between early Cretaceous and the Tertiary (Hallam 1974). Landbridges could have existed in the Pacific region south of the equator until the early Tertiary (Van Steenis 1962).

A Crop of Early Importance.—Although archaeological evidence from south and Southeast Asia is far from complete, rice nevertheless is most likely one of the major cereals that the prehistoric people in this region gathered as a food supplement to game and fish they hunted. The gathering of wild-growing rices, largely of the perennial species, preceded the beginning of agriculture in the humid tropics of Asia. Such seed gathering operations could develop into cultivation practices following the Neothermal age (10,000 to 15,000 BP) when annual forms began to appear at the periphery of the habitats of the wild rices. Alternating periods of drought and large variations in temperature during the Neothermal accelerated the development of the annual forms in east India, northern Southeast Asia and southwest China (Whyte 1972; Chang 1976B).

Rice cultivation in Asia likely began independently and concurrently at many sites within or bordering a broad belt (see Fig. 1.1) that extends from the Ganges plains below the eastern foothills of the Himalayas, through upper Burma, northern Thailand, Laos and Vietnam, to southwest and south China (Roschevicz 1931; Ramiah 1937; Chatterjee 1951; Chang 1964, 1976B; Morinaga 1968).

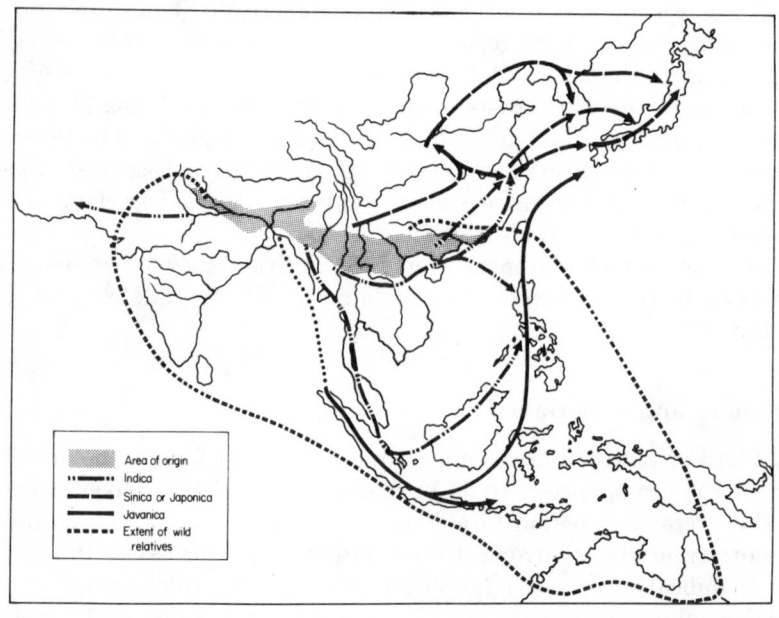

Adapted from Chang (1976C)

FIG. 1.1. EXTENT OF WILD RELATIVES AND SPREAD OF ECOGEOGRAPHIC RACES OF *O. SATIVA* IN ASIA AND OCEANIA

Archaeological digs conducted in 1973—74 near Homutu village in Yu-yao county of Chekiang province, China, uncovered rices dated more than 6960 years old (CPAM 1976; Anon 1977). Further north, domestication of rice seemed to have occurred around 3200—2500 B.C. in the "nuclear area" of northern China as evidenced by the finding of rice glume remains at the Yang-shao site (K.C. Chang 1978).

Cultivated rice excavated from the Gangetic plain in India dated back to about 4530 B.C. (Vishnu-Mittre 1976). The rice remains excavated from Non Nok Tha site in northern Thailand dated back to about 4000 B.C. (Solheim 1972), but the sample may represent a noncultivated form (Chang 1976B).

The above findings indicate that between the 4th and 5th millennia B.C. a substantial portion of the people who inhabited the lowlands of South and Southeast Asia, and the southern tier of China (including southwest and east China) were already depending on rice as a staple food. Once adopted, this tasty and versatile food grain quickly replaced other cereals and root crops as the principal food (Chang 1976B,C).

The cultivation of African rices is more recent (Porteres 1950, 1956). The spread of the African rice from the central Niger delta to Senegal and Gambia between 1500 and 800 B.C. (Porteres 1976) suggests that rice was rapidly becoming one of the staple foods in those areas of West Africa during that period.

The wide dispersal of rice cultivars partly resulted from the role that rice had played as an international commodity during the early days of trade between countries or even continents. The Romans did not grow rice. They imported it ever since the early days of their empire for making an emollient drink (Wickizer and Bennett 1941). Traders brought rice from tropical Asia and China to north Africa, Europe, Australia and the Americas. The Asian traders also depended on the calorie-rich rice as one of their main staples as they traveled afar.

Rice in the World Food Economy

Rice as a Mainstay.—Although cultural geographers have identified rice as the staple food for the Asians, as maize (corn) for the Americans, millet for the Africans, and wheat for the Europeans, the importance of rice on a global basis cannot be overemphasized. It was said in 1966, that "every day more than one-third of the world's population rises to face a future dominated by the single word, rice," and by the end of the current century, "the number of people dependent on rice for their staple food will be greater than the present world population" (Anon 1966).

Rice is very important in the world food economy in that (1) it is the main staple food for ½ to ⅔ of the world's population, and (2) it accounts for ⅓ to ½ of the daily caloric intake in many Asian countries, including Japan. Indeed, it is also the major source of protein for the masses of Asian people. Furthermore, rice is becoming a main staple food in many of the African and Latin American countries. It also plays an important role in the economies of many countries: Thailand, Burma, Vietnam, Egypt, and to some extent the United States.

Rice Versus Other Cereal Crops.—In terms of area and production rice ranks second to wheat in 1977 among the world's cereal crops (FAO 1978). In yields, however, rice is second to maize only by 13% but 54% higher than wheat (Table 1.1).

TABLE 1.1. WORLD CEREALS: 1977 AND 1978[1]

	Area[2]		Yield[3]		Production[4]	
	1977	1978	1977	1978	1977	1978
Rice	142,842	144,727	2,566	2,600	366,505	376,996
Wheat	232,382	226,108	1,664	1,880	386,596	425,217
Barley	91,368	84,995	1,894	2,150	173,094	183,021
Maize	118,453	120,900	2,952	2,980	349,676	360,033
Rye	13,904	15,020	1,709	1,770	23,767	26,570
Oats	30,640	29,046	1,713	1,760	52,472	51,154
Millet	65,453	49,029	655	750	42,886	36,718
Sorghum	43,650	47,869	1,269	1,440	55,413	69,099

Source: FAO (1978) and USDA (1979).

[1]Data for these 2 years are from FAO and USDA, respectively.
[2]Harvested, in 1000 ha.
[3]In kg per ha.
[4]In 1000 MT.

The importance of rice in the world food economy, however, should also be viewed from different perspectives. Table 1.2 shows the nutrient value of rice compared to wheat, maize, barley, oats and millet. From the table, it is clear that rice produces more calories and carbohydrates per hectare than any other cereal. Its per unit land yield of protein is second only to oats.

Assuming a daily calorie intake of 3200 calories per person, 1 ha of rice (1977 world average yield) can sustain 2055 persons per day or 5.63 persons per year. This is to compare with the 5.06 and 3.67 persons per year sustained by 1 ha of maize (corn) and wheat, respectively (Table 1.3).

On a protein requirement basis, Table 1.3 also shows that 1 ha of rice can sustain 2132 persons per day, as compared with 2167 and 2090 persons per day sustained by oats and wheat, respectively, or 1921 persons per day by maize. The protein content in oats is highest in proportion among these cereals but it is used today mostly to feed animals (especially horses) rather than as a human food.

TABLE 1.2. NUTRIENT VALUE OF RICE COMPARISON WITH OTHER SELECTED CEREALS

Cereals	Rice Oryza sativa		Wheat Triticum aestivum		Maize Zea mays		Barley Hordeum sativum		Oats Avena sativa		Millet Eleusine coracana	
Nutrient Per	100g[1]	ha	100g[2]	ha	100g[1]	ha	100g[1]	ha	100g[1]	ha	100g[1]	ha
Yield[4]	NA[5]	2,566	NA[5]	1,664	NA[5]	2,952	NA[5]	1,894	NA[5]	1,713	NA[5]	655
Adjusted yield[6]	White rice[7]	1,847.5	White flour[8]	1,214.7	Corn meal[9]	1,664.9	Pearl barley[10]	1,041.7	Rolled oats[11]	991.8	Whole grain[12]	655
Protein[13]	7.5	138,563	11.2	136,046	7.5	124,868	8.2	85,419	14.2	140,836	5.6	36,680
Fat[13]	1.7	31,408	1.0	12,147	1.1	18,314	1.0	10,417	7.0	69,426	1.5	9,825
Carbohydrate (incl. fibre)[13]	77.7	1,435,508	74.7	907,381	78.8	1,311,941	78.3	815,651	68.2	676,408	78.0	510,900
Calories	356	6,577,100	353	4,287,891	355	5,910,395	357	3,718,869	396	3,927,528	336	2,200,800

[1]Source: Bradley (1956). [2]Source: Waller (1949). [3]Source: Aykroyd and Doughty (1970). [4]Source: FAO 1977 yield data in kg per ha. Adjust from the 1977 FAO data by respective conversion factor according to United States standard as detailed in following footnotes. [5]Not applicable. [6]In kg per ha. Adjust from the FAO (1978). [7]Converted to white (milled) rice at 72% of the paddy. [8]Converted to white flour at 73% of the whole grain. [9]Converted to degermed corn meal at 56.4% of the shelled corn. [10]Converted to pearl at 55% of the unprocessed barley. [11]Converted to rolled oats at 57.9% of the harvested oats. [12]Harvested grain. [13]In g.

TABLE 1.3. SUSTAINING CAPABILITY[1] OF RICE AND SELECTED CEREAL CROPS

| | On Caloric Basis[2] | | On Protein Basis[3] | |
	person/day	person/year	person/day	person/year
Rice[4]	2055.3	5.63	2131.7	5.84
Wheat[5]	1340.0	3.67	2090.0	5.73
Maize[6]	1847.0	5.06	1921.1	5.26
Barley[7]	1162.2	3.18	1314.1	3.60
Oats[8]	1227.4	3.36	2166.7	5.94
Millet[9]	687.8	1.88	564.3	1.55

[1] In number of people per ha. Based on FAO 1977 (crop) yield data and Table 1.2.
[2] Assuming 3200 calories per person per day and 1,168,000 calories per person per 365-day year.
[3] Assuming 65 per person per day and 23,725 g per person per 365-day year.
[4] Converted to white rice.
[5] Converted to white flour.
[6] Converted to degermed corn meal.
[7] Converted to pearl barley.
[8] Converted to rolled oats.
[9] Harvested grain.

RICE IN ITS TEMPORAL PERSPECTIVE

The Origin and Evolution of Rice

The Genus *Oryza.*—The genus *Oryza* includes 20 valid species. Table 1.4 shows the species' names (and some former designations), their chromosome numbers, genomes and geographic distribution. A working key to the 20 species and their geographic races has been provided by Chang (1976E). Various synonyms of the *Oryza* species under several classifications and nomenclature systems proposed by different botanists were listed by Chang (1964)[1], Nayar (1973) and Oka (1975).

Roschevicz (1931) grouped the *Oryza* species into 4 sections, 1 of which (section *rynchoryza*) has been moved to the genus *Rynchoryza*. Ghose, Ghatge and Subrahmanyan (1956) divided the genus into 3 sections: *sativa, officinalis* and *granulata*. Tateoka (1964) also recognized 3 sections: *sativa, granulatae* and *ridleyianae*. In all cases, the 2 cultivated species (cultigens) and their immediate wild relatives belong to section *sativa*.

A biosystematic treatment of species in the genus has been made by Sampath (1962), Chang (1964), Oka (1964), Morishima (1969) and Sharma and Shastry (1973). Studies on interspecific hybrids with respect to

[1] Nine taxa, either of doubtful validity or having names of uncertain application, have been removed from the lists enumerated by Roschevicz (1931), by Chevalier (1932) and by others. Five other species have been transferred to other genera (see Chang 1970, 1976E). Tateoka (1964) has enumerated the above 20 species along with 2 others that were later removed from the genus. For further information, see also Nayar (1973) and Oka (1975).

TABLE 1.4. *ORYZA*: SPECIES, CHROMOSOME NUMBERS, GENOME SYMBOLS, AND DISTRIBUTION OF ENDEMIC HABITATS[1]

Species Name (Synonym)	Chromosome No. (X = 12; 2n =)	Genome Group	Graphic Distribution
O. alta	48	CCDD	Central and South America
O. australiensis	24	EE	Australia
O. barthii (O. breviligulata)	24	A^gA^g	West Tropical Africa
O. brachyantha	24	FF	West Tropical and Central Africa
O. eichingeri	24, 48	CC, BBCC	East and Central Africa
O. glaberrima	24	A^gA^g	West Tropical Africa
O. grandiglumis	24,48	CCDD	South America
O. granulata	24	—	South and Southeast Asia
O. latifolia	48	CCDD	Central and South America and West Indies
O. longiglumis	48	—	New Guinea (Irian Jaya)
O. longistaminata (O. barthii)	24	A^lA^l	Africa
meyeriana	24	—	Southeast Asia and southern China
O. minuta	48	BBCC	Southeast Asia and New Guinea
O. nivara, (O. fatua, O. rufipogon)	24	AA	South and Southeast Asia, southern China and Australia
O. officinalis	24	CC	South and Southeast Asia, southern China and New Guinea
O. punctata	48, 24	BBCC	East Africa
O. ridleyi	48	—	Southeast Asia and New Guinea
O. rufipogon (O. perennis, O. fatua, O. perennis subsp. balunga)	24	AA	South and Southeast Asia, southern China and New Guinea
(O. perennis subsp. cubensis)		$A^{cu}A^{cu}$	South America
O. sativa	24	AA	Asia
O. schlechteri	—	—	New Guinea (Papua New Guinea)

[1]Source: Adapted from Chang (1970, 1976A,C).

F_1 fertility and meiotic behavior have been summarized by Chang (1964) and Nayar (1973). The determination of the genomic composition of 5 species in the genus remains to be made. We need also to make biosystematic studies of those wild species outside section *sativa,* especially those having the CC or BBCC genomic constitution, because the B and C genomes appear to have partial homology with genome A (Chang 1976A).

Genera rather closely related to *Oryza* in the tribe Oryzae are *Leersia* (x = 12) and *Zizania* (x = 15). The wild rice of North America is *Z. aquatica.* Intergeneric hybridization between rice as one parent and "kaoliang" (sorghum), "gu," broad-sheath bamboo, corn, wheat and a number of grasses as the other parent have been attempted but no viable or useful progeny was obtained (Wu *et al.* 1961; Hsia *et al.* 1964; Yieh 1964; Rugsachat and Claridad 1971). Further attempts by Chinese botanists in the early 1970s led to the production of F_1 and F_2 progenies from crosses between rice and reed (Ni-feng Inst. Agric. Sci. 1977) and rice and bamboo (Hai-feng Inst. Agric. Sci. 1977). However, the usefulness of progenies from such wide crosses needs to be elucidated by further research.

The Asian and African Cultivated Rices.—Morphologically *O. glaberrima* differs from *O. sativa* in having shorter ligules, absence of secondary branching on the panicles, glabrous glumes and glabrous to slightly scabrid leaves, although the geographic distribution of the 2 cultivated species appears distinct. Each cultigen forms a species-group, which consists of 1 wild perennial form, and 1 annual wild form, the annual cultivated species and 1 annual weed form (see Fig 1.2). The weed form results from hybridization among the above.

Therefore, several rice researchers believe that the 2 cultigens have had a common ancestor in the distinct past—"*O. perennis* Moench," an ambiguous designation used differently by various researchers and which may also include the annual forms (Chevalier 1932; Chatterjee 1948; Sampath and Rao 1951; Richharia 1960; Oka 1964). Porteres (1956) recognized that the common progenitor was a rhizomatous and floating form but he did not name it. Part of the controversies concerning the evolutionary processes of the 2 cultivated species stemmed from differences in naming and characterizing the putative progenitors. Contrasting views on this subject have been reviewed and discussed by Chang (1964, 1976C) and Nayar (1973).

Although the parallel variation from the more primitive form to the cultivated species in each species-group was described earlier by Porteres (1956) and Morishima *et al.* (1963). Chang (1976A,B,C) recognized that the Gondwanaland supercontinents were the original habitats of the genus. Therefore, the evolutionary process of the 2 cultigens in Fig. 1.2 could satisfactorily explain the independent and parallel evolutionary pathway of the 2 species-groups (Chang 1976A,B). The general process is: wild perennial → wild (prototype) annual → cultivated annual. In addition, an annual weed race that is a conglomerate of natural hybrids of all intergrades among the above 3 exists alongside of the 3 species. This scheme can also resolve many earlier controversies surrounding this subject (Chang 1976C). Interpretation of the terms "perennial" vs "annual" with respect to growth habit and mode of reproduction has been given by Chang (1976C,D); the weed race, by Harlan (1965).

Based on recent paleo-geological and -meteorological studies, Chang (1976B) postulated that the prototype of *O. sativa* evolved primarily along the foothills of the Himalayas in South Asia and its associated mountain ranges in the mainland of Southeast Asia and southwest China. The widespread distribution of the wild races in insular South-

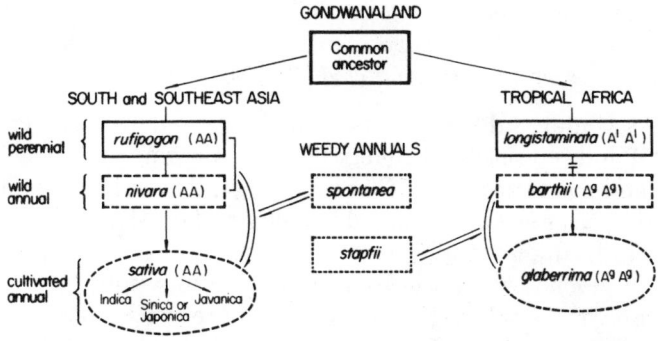

Adapted from Chang (1976C)

FIG. 1.2. EVOLUTIONARY PATHWAY OF 2 CULTIVATED SPECIES OF RICE

Taxa boxed by solid lines are *wild perennials*. Taxa boxed by broken lines are *annuals*. Arrow with solid line indicates *direct descent*. Arrow with broken line indicates *indirect descent*. Double arrows indicate introgressive hybridization.

east Asia, Sri Lanka and Taiwan was most probably facilitated by land-bridges which remained as recent as the Pleistocene (Van Steenis 1962). The continuous and varied movements of people in Asia during prehistoric times was an important factor in the broad distribution of early *O. sativa* forms which led to the formation of 3 ecogeographic races: Sinica (formerly known as Japonica), Indica and Javanica (Chang 1976C).

Porteres (1950) postulated that the African cultigen, *O. glaberrima*, originated in the Niger River delta. The primary center of diversity for *O. glaberrima* is in the swampy basin of the upper Niger River and 2 secondary centers to the southwest near the Guinean coast. The primary center was probably formed around 1500 B.C. while the secondary centers were formed 500 to 700 years later (Porteres 1956, 1976).

Judging from the history of domestication and the extent of diversity within the species, it appears quite possible that the genetic differentiation of *O. sativa* in Asia predated that of *O. glaberrima* in West Africa (Chang 1976B,C).

Much of the divergence in views concerning the ancestral species of *O. sativa* resulted from (1) uncertainties or confusions in characterizing and naming the 2 closely related wild species, (2) scarcity of finding in nature the typical specimens of the 2 wild taxa, and (3) continuous hybridization among the 2 wild taxa, the weed race and the cultigen (Chang 1976C). A similar situation exists in Africa. In many parts of West Africa, recently collected samples of *O. glaberrima* show that it has lost some of its distinctive features, partly because of mixed planting with *O. barthii* and *O. sativa* in the same field (Chang *et al.* 1977).

Changes Under Cultivation and Domestication.—A number of morphological and physiological changes occurred during the process of cultivation and domestication of *O. sativa*. Larger leaves, longer and thicker culms, and longer panicles resulted in a larger plant size. There were also increases in the number of leaves and in their rate of development, in the number of secondary panicle branches, in grain weight, in the rate of seedling growth, in tillering capacity, and in the synchronization of tiller development and panicle formation. The period of grain filling became longer. Concurrently, there were decreases in (or losses of) pigmentation, rhizome formation, ability to float in deep water, awning, shattering, duration of grain dormancy and intensity of photoperiodicity. The frequency of cross-pollination also declined so that the crop became more inbred than the wild rices (Chang 1976A).

The rice plant underwent further changes when it was widely disseminated by cultivators from the humid tropics to subtropical zones, where great diversity is found today. Further dispersal, hybridization and selection led it into temperate zones as far north as 53° latitude North (Ting 1961).

Morphological and physiological changes associated with cultivation have been discussed by Oka (1975) and Chang (1976A,B). The combined forces of natural selection and cultural practices led to the great ecologic diversity now found in *O. sativa* cultivars (Oka 1975; Chang 1976A). Cultivators' preferences and socioreligious traditions have added much morphologic diversity to the rice cultivars (Chang 1976B).

Differentiation of the Ecogeographic Races.—The tropical race of *O. sativa*, the Indica race, is most likely the prototype of 2 other ecogeographic races, Javanica and Sinica (also known as "Keng" or "Japonica"). Initially, the prototype was grown in shallowly flooded or hydromorphic sites of the humid tropics, mainly in South Asia and the Southeast Asia mainland.

As the Indica race was brought into the islands of Indonesia, natural and human selection, along with accidental hybridization among different varieties, led to the differentiation of the tall, thick culmed, large and bold grained Javanica race (bulu or gundil).

When the tropical race was carried across mountain ranges into southwest China, a cool tolerant, low amylose, and short and thick grained race (keng) developed and found its way into central China and the coastal regions south of the Yangtze River. The keng race was grown in Honan Province of north China at least 4500 years BP. The cultivation in Chekiang and Kiangsu Provinces of the hsien (Indica) race dates back to at least 6700 years BP (Inst. of Archaeol. 1972; CPAM 1976). Even now in the mountainous areas of Yunnan Province, the Indica type, mixtures of the Indica and keng types, and the keng type could be found at the lower, middle and higher elevations, respectively (Chen 1957; Kwang-tung Agric.-Forestry Coll. and Yunnan Univ. 1974). Because the cultivation of the keng race in China preceded that in Japan by more than 4000 years, this temperate region race should be designated as Sinica instead of Japonica (Chang 1976B,C).

Table 1.5 compares the morphoagronomic and grain features of the 3 ecogeographic races.

TABLE 1.5. ECOGEOGRAPHIC RACES OF *ORYZA SATIVA*: COMPARISON OF THEIR MORPHOLOGIC AND PHYSIOLOGIC CHARACTERISTICS

Indica	Sinica (or Japonica)	Javanica
Broad to narrow, light green leaves	Narrow, dark green leaves	Broad, stiff, light green leaves
Long to short, slender, somewhat flat grains	Short, roundish grains	Long, broad, thick grains
Profuse tillering	Medium tillering	Low tillering
Tall to intermediate plant stature	Short to intermediate plant stature	Tall plant stature
Mostly awnless	Awnless to long awned	Long awned or awnless
Thin and short hairs on lemma and palea	Dense and long hairs on lemma and palea	Long hairs on lemma and palea
Easy shattering	Low shattering	Low shattering
Soft plant tissues	Hard plant tissues	Hard plant tissues
Varying sensitivity to photoperiod	None to low sensitivity to photoperiod	Low sensitivity to photoperiod
23–31% amylose	10–24% amylose	20–25% amylose
Variable gelatinization temperatures (low or intermediate)	Low gelatinization temperature	Low gelatinization temperature

Source: Adapted from Chang and Bardenas (1965).

Ecoedaphic and Genetic Differentiation.—The deep water rices of Asia, which have remarkable ability for internode elongation, may have acquired the floating habit from the wild perennial race (*O. rufipogon*) as rice cultivators moved into areas where the water was deep. These rices branched at the higher nodes, and had rapid internode elongation during rises in water depth, adventitious roots at the upper nodes, and had location- and maturity-specific photoperiod sensitivity (Chang 1976B). The development of the deep water rice could have involved the differentiation hybridization cycles described by Harlan (1970).

When rice was introduced into higher grounds where the soil had poor moisture retention, the upland type with early maturity, low tillering capacity and deep and thick roots evolved (Chang *et al.* 1972).

In the Ganges River delta, rice crops for the 3 seasons *bora, aus* and *aman* (winter, summer and autumn), have evolved to fit into varying water and temperature regimes.

Deliberate human selection in a disruptive manner along with contributions from the weed race, led to enormous diversity among rice cultivars in adaptiveness to climatic factors (mainly temperature and photoperiod), water regimes (deep, shallow, dryland and various intermediaries), and edaphic conditions (saline, alkaline, acid, etc.). The trend in ecologic diversification following dispersal and selection is shown in Fig. 1.3. Some of the ecoedaphic differentiation in rice developed parallel to the evolution in cultural practices—from shifting cultivation, to broadcasting or dibbling in permanent sites, to transplanting in bunded fields (Chang 1976B). The present extent of variousness in climatic adaptiveness among *O. sativa* cultivars is indeed remarkable for a tropical species (Chang and Oka 1976).

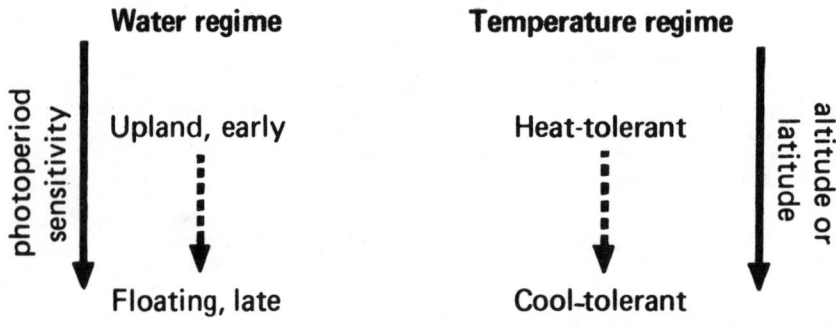

Adapted from Chang (1976B)

FIG. 1.3. ECOLOGIC DIVERSIFICATION IN RICE FOLLOWING DISPERSAL AND SELECTION

As the tropics-based cultivars were brought northward and cultivated under more restricted growing seasons, the loss in photoperiod- and thermo-sensitivity was accompanied by reduced plant height, more determinate tillering habits, decreased competitiveness with weeds, more synchronous heading among tillers, a longer grain ripening period and weaker ratooning ability (Chang and Oka 1976). A more thrifty and productive plant type evolved in the temperate regions—this applies to both the keng (Sinica) and the hsien (Indica) types.

Genetic differentiation has developed parallel to the ecogeographic diversification process. The occurrence of sterility and restriction on recombination in crosses between any 2 geographic races of *O. sativa*, especially the Indica and Sinica (Japonica) races, has been well documented (IRRI 1964; Chang 1964). Recent cytogenetic analyses of hybrids of Indica × Indica and Indica × Javanica crosses at the International Rice Research Institute (IRRI) reveal a suprisingly large spectrum of both chromosomal and nonchromosomal meiotic aberrations. Genic imbalance could also be involved (Engle *et al.* 1969; Dolores *et al.* 1975). Similarly, restricted segregation and aberrant recombination showed up when upland varieties were crossed with semidwarf Indica varieties (Hung and Chang 1977).

The contrast between the restricted flow of genes among different ecogeographic races of *O. sativa* and the relatively free flow among *O. sativa* and its wild relatives indicate not only the large number of sites where cultivation and domestication took place, but also the antiquity of the process of genetic differentiation (Chang 1976C).

In West Africa, *O. glaberrima* is a dominant crop grown in flooded areas of the Niger and Sokoto River basins where it is broadcast on hoed fields. On shallowly flooded land, a rainfed lowland crop is directly sown either by broadcasting or dibbling or transplanting. About 75% of the land planted to rice in Africa belongs to the upland culture, largely under the bush fallow or grass fallow system. *O. glaberrima* seeds are broadcast or dibbled after the ground has been hoed. Some African farmers still use axes, hoes and bush knives in land preparation (Chang 1976B). In hydromorphic soils, the *glaberrima* behaves like a self-perpetuated weed.

The African cultigen and its annual wild rice (*O. barthii*) are less diverse than their Asian counterparts. While Porteres (1956) recognized 2 subspecies (*vulgaris* and *humilis*) in *O. glaberrima*, Chu and Oka (1972) considered that the *barthii-glaberrima* complex could be differentiated into 2 ecotypes: deep water and upland (see also Oka 1974). For areas where the water is deep, the African cultivars are inferior to the Asian cultivars in the ability of internode elongation (Morishima *et al.* 1962).

Geographic Dispersal of *O. sativa*

Rice seeds were dispersed around the world initially as a portion of the foodstuff carried by travelers or traders. Commercial cargoes of rice have also served as sources of seed. Rice cultivation techniques were undoubtedly introduced or affected by those early settlers or traders who brought rice to a foreign land. In several instances, repeated and independent introductions of rice seeds took place before certain varieties became established as commercial types. Grist (1975) and Chang (1976B) have described examples in which cultural practices for growing lowland rice in various Asian countries were greatly influenced by those developed in China.

Spreading Throughout Asia.—The aggregate of centers and noncenters in north and northeast India, Bangladesh, the northern portion of mainland Southeast Asia, and southwest and south China forms a belt of primary diversity of *O. sativa* (see Fig. 1.1). From the Indian subcontinent and mainland Southeast Asia, the Indica race spread southward into Sri Lanka where rice was a rainfed (nonirrigated) crop before 543 B.C., eastward into the Malay Archipelago (date unknown), and northward into central and south China where it was known as the hsien type and has been grown since 5005 ± 130 B.C. (CPAM 1976). Rice cultivation began in Indonesia between 1600 B.C. (Grist 1975) and 1084 B.C. (Gushchin 1930). From Indonesia, the Javanica race spread to the Philippines, Taiwan, Ryukyus and Japan (Chang 1976B,C).

From north and east China, rice was introduced into Korea before the 6th to 5th century B.C. (Hyo 1977) and into Japan in the 3rd century B.C. (Ando 1951; Morinaga 1957, 1967). Several routes for the dispersal of rice from China to Japan could have existed: (1) from the lower Yangtze Valley to Kyushu Island, (2) from north China to Honshu Island, (3) via Korea to northern Kyushu, or (4) a combination of any of the above.

The Chinese introduced the terraced rice cultivation to the mountainous Luzon Island of the Philippines during the 2nd millenium B.C. (Erygin and Natal'in 1968), but the rice seeds could have come from several sources.

It is thought that rice and its culture spread to upper South Asia prior to the later movements of the Aryan migration because of the identical name in Zend and Sanskrit and which is also similar to the Old High Persian (Copeland 1924). The Middle East acquired rice from the Indian subcontinent probably as early as 1000 B.C. Persia loomed large as the principal stepping stone from South Asia toward points west of the Persian Empire. Rice cultivation could have begun between the 3rd century B.C. in Persia (Bertin *et al.* 1971) and sometime after 400 B.C. in Syria (Burkill 1935). Central Asia probably obtained its rice from China

(Gushchin 1930). The probable routes of dispersal of the 3 geographic races of *O. sativa* within Asia are shown in Fig. 1.1.

On Reaching Europe.—The early records of Egypt as well as the Bible had no mention of rice (Gushchin 1930). During the early Roman Empire, the Romans learned about rice because of the expedition of Alexander the Great to India (*c* 327–324 B.C.). The Romans did not cultivate the crop but imported it to make an emollient drink (Wickizer and Bennett 1941). The Greeks learned to use rice from the Persians (Grist 1975).

Different routes by which rice was introduced into Europe have been suggested. One probable route was that the Arabs brought rice from Persia to Egypt between the 4th and the 1st centuries B.C. (Porteres 1950; Grist 1975). From Greece or Egypt or Ethiopia, rice went to Spain and Sicily around 883 A.D. A second theory was that rice was introduced into Naples in the 9th and 10th centuries A.D.; it spread to central and northern Italy between 1468 and 1475 (Angelini 1936; Carrasco-Garcia 1952). Another alternate route was that the Moors took rice from Persia to Spain (*c* 711 A.D.), and later to Italy between the 13th and 16th centuries (Pinolini 1929; Gushchin 1930; Grist 1959; Ciferri 1960; Erygin and Natal'in 1968). Another probable route was directly from the Far East to Italy (Piacco 1959). Commercial production appeared to expand after 43 varieties from the Philippines and China, including several upland varieties of the Philippines, were introduced around 1840 (Bortoloni 1844; Piacco 1959; Ciferri 1960).

Sevilla of Spain seems to have served as a focal point from which rice spread throughout southern Spain and Levant; thence to Portugal (8th century) and Italy (G. Lopez Campos 1978). The Moors in Spain (711–1492 A.D.) brought rice to Portugal which, in turn, introduced it to Brazil (Silva 1955). Spain could also have introduced rices from the Philippines, Malaya, China and Japan (Mora y Llorens 1939). Portugal obtained a number of commercial types from Italy and China (Silva 1978). France obtained commercial types from Italy during the 18th and 19th centuries (R. Marie 1978).

The large-scale cultivation of rice in southern Europe was delayed frequently by local opposition stemming from fears of spreading malaria (Wickizer and Bennett 1941; Carrasco-Garcia 1952).

The Turks brought rice from Asia Minor to the Balkan Peninsula. From Italy rice went to Bulgaria, Yugoslavia and Romania around 1468 A.D. (Gushchin 1930).

In areas around the Caspian Sea of the U.S.S.R., the Russians first obtained rice from East Asia and also from Iran at the time of Peter I in the early 1700s (Erygin and Natal'in 1968). The commercial varieties of the 1920s probably came from Japan, while those grown after World

War II were either introduced from Japan, Korea, China, or European varieties or bred from them.

Spreading Through Africa.—Madagascar received the Asian rices (*O. sativa*) probably as early as 1000 B.C., when the first settlers arrived in the Tuliar region in the southwest. The second group of settlers came from Indonesia in the Christian era. The third group came from south India from 10 to 12 centuries later. Thus, the Javanica race was planted on the central plateau while the other areas were planted with the Indica type (M. Arraudeau 1978). The Arabs also brought rice to northwest Madagascar and to Mozambique between the 8th and the 10th centuries (Porteres 1950).

Madagascar has likely served as an intermediary for the rices later introduced into East Africa, largely of the Indica type. On the other hand, East Africa probably obtained many varieties directly from the Indian subcontinent.

The Arabs took Asian rice from Syria to Egypt and later to the west coast of North Africa during the 10th century (Porteres 1950), although trade between Egypt and India began in the 1st century B.C. (Burkill 1935). Rice was cultivated in the Nile Valley around 639 A.D. (Erygin and Natal'in 1968).

The Portuguese introduced the Asian rices into West Africa between the 15th century (Porteres 1950) and the 17th century (Grist 1959). Undoubtedly, many varieties were also brought in by the colonizers from Spain, France and England during a period from the 18th to the 19th century. Rice also traveled from Mozambique to the Congo during the 19th century (Porteres 1950).

Across the Ocean.—The Caribbean Islands obtained their rice from Europe in the 15th century. Countries in Central and South America introduced their rices during the latter part of the 15th century through the 18th century (Bertin *et al.* 1971). Brazil received rices from Portugal or other European countries (Benton 1970), while most of the other Latin American countries obtained rices largely from Spain (Grist 1975). Mexico received its first lot of rice around 1522 A.D. in a cargo mixed with wheat (Icasbalceta 1866). Introductions from tropical and temperate Asia took place in the 19th century (Bertin *et al.* 1971). There was much exchange of cultivars among countries in Central America, South America and North America.

Rice cultivation was initially made in the United States as a trial planting in Virginia around 1609. Other plantings were also made along the south Atlantic coast from that time on. Some of the early introductions came from Madagascar. Rice production was well established in South Carolina about 1690. Production spread to southwest Louisiana, to the adjoining area in Texas and to central Arkansas. California began to grow rice from 1909 to 1912 (Gray and Thompson 1958; Adair *et al.* 1962). Rice was introduced into Hawaii between 1853 (Grist 1975) and 1862 by the Chinese (Lu 1971).

Trial planting of rice in New South Wales of Australia began in 1891, although introduction into Queensland and the northern territories could have been earlier (Grist 1965). Commercial production started in 1923 (Grist 1975). New Guinea began to grow rice in the 19th century (Bertin *et al.* 1971). The Germans brought rice into New Guinea while the French introduced rice into Caledonia (Grist 1965).

Geographic Dispersal of *O. glaberrima*

African rice (*O. glaberrrima*) had its primary center of diversity in the central delta of the Niger River. The predominant varieties are the floating and semifloating types (Porteres 1950). The center began about 1500 B.C. A secondary center formed later on both sides of the coastal Gambia River around 800 B.C. A third center existed in the mountainous region of Guinea (Porteres 1956, 1976). Porteres (1965) has mapped the dispersal patterns of both *O. glaberrima* and *O. sativa* in West Africa. Harlan (1973) and Chang (1975) have mapped the geographic distribution of *O. glaberrima* and its wild relatives in Africa.

The African rices have been found in French Guiana, El Salvador and Panama. The African species were most likely introduced from Guinea of West Africa (Porteres 1976) during the time of the slave trade (Bertin *et al.* 1971).

RICE IN ITS SPATIAL CONTEXT

As a result of long years of diffusion and selection, *Oryza sativa,* Linn. *grammineae* is grown today in a great variety of physical and cultural environments. These complex environments have intricately affected *O. sativa*, the common rice of today, in many profound ways. For example, the proportion of available cropland devoted to rice (or to a certain type of rice) in a given country, in addition to the climoedaphic, hygrobiotic and socioeconomic factors, is also determined by the cultural dietary preferences. There exists, therefore, in the world rice economy a spatial inequality of rice hectarage, yield, supply and demand situations, as well as market capabilities.

World Rice Hectarage

As noted earlier, the world rice regions straddle a wide band of latitudes between 53°N and 35°S. Within this broad band, a great variety of rice matures in different time lengths ranging from a little over 3 months to more than 11 months. Thus, in many areas, especially those in the tropics, double and even triple cropping is often being practiced. It is therefore very difficult to tally a precise total world rice hectarage. Published data from 1975, for instance, range from 140,239,000 ha (Palacpac 1977), to 140,880,000 ha reported by the Food and Agriculture Organization of the United Nations (FAO 1976), and to 142,342,000 ha (USDA 1977A). The difference of 2.1 million ha, even at the average yield of 2.441 MT per hectare as reported by FAO (1976), gives a magnitude of 5.1 MMT of paddy—equal to ½ of the rice that entered the international trade market.

Distribution Pattern.—Over the last 25 years, the world's rice hectarage has increased from 108.8 million ha in 1954 to 144.7 million ha in 1978 (Table 1.6), an increase of 33% (Table 1.7). Change in rice hectarage, however, has not been uniform throughout the world. Table 1.7 shows both temporal and spatial changes of world rice hectarage between 1954 and 1978. As noted, regional changes vary greatly.

TABLE 1.6. WORLD RICE: AREA, YIELD AND PRODUCTION, 1954–1978

Year	Area (1000 ha)	Yield (kg/ha)	Production (1000 MT)
1954	108,783	1,798	195,640
1955	110,833	1,907	211,359
1956	117,027	1,891	211,275
1957	115,271	1,883	217,035
1958	117,671	2,019	237,553
1959	116,804	1,960	228,899
1960	120,172	1,950	234,749
1961	120,239	2,000	241,062
1962	121,864	2,000	244,097
1963	122,229	2,090	255,611
1964	125,234	2,180	272,655
1965	123,902	2,070	256,480
1966	125,318	2,120	265,125
1967	127,073	2,250	286,526
1968	128,286	2,250	288,957
1969	131,623	2,290	301,247
1970	131,174	2,380	311,830
1971	132,048	2,410	317,742
1972	131,484	2,340	308,152
1973	135,757	2,440	330,872
1974	138,007	2,440	336,770
1975	143,056	2,520	360,575
1976[1]	141,659	2,460	348,900
1977[1]	143,461	2,550	366,071
1978[1]	144,727	2,600	376,996

Source: Palacpac (1977); USDA (1978, 1979).
[1] USDA, preliminary.

TABLE 1.7. WORLD RICES HECTARAGE: PERCENTAGE OF WORLD TOTAL, BY CONTINENT AND REGION; PERCENTAGE INCREASE, 1978 OVER 1954

Continent/Region	1978[1]	1954[1]	Percentage Change	Percentage of World Total 1978	1954
Asia	130,252	100,957	29.0	90.01	92.81
East Asia	41,720	34,019	22.6	28.83	31.27
Southeast Asia	34,090	24,486	39.2	23.56	22.51
South Asia	53,784	41,858	28.5	37.16	38.48
West Asia and others	658	594	10.8	0.46	0.55
Africa	4,760	2,805	69.7	3.29	2.58
Europe	1,007	544	85.1	0.70	0.50
U.S.S.R. (including Soviet Asia)	600	136	341.2	0.41	0.12
Others	407	408	(0.3)[2]	0.28	0.38
North America	2,053	1,490	37.8	1.41	1.37
United States	1,238	1,032	20.0	0.85	0.95
Others	815	458	78.0	0.56	0.42
South America	6,563	2,971	121.0	4.53	2.73
Brazil	5,400	2,425	122.7	3.73	2,23
Others	1,163	546	113.0	0.80	0.50
Oceania	92	16	475.0	0.06	0.01
World Total	144,727	108,783	33.0	100.00	100.00

Source: Palacpac (1978); USDA (1979).
[1]In 1000 hectares.
[2]( )denotes decrease.

Asia as a whole has not kept pace with the world increase. Within Asia, only Southeast Asia increased at a greater rate than the world (39.2 versus 33.0). While South Asia nearly kept pace with Asia as a whole, both East and West Asia lagged behind Asia as well as the world.

Africa and Europe, on the other hand, have kept pace with each other and at more than twice the world rate. South America and Oceania, by far, have the largest percentage of increase. North America, however, increased only slightly ahead of the world. It is interesting to note that each of these continents is dominated by 1 country: Malagasy in Africa, Soviet Union in Europe, the United States in North America, Brazil in South America, and Australia in Oceania.

As a result of a greater percentage increase in these continents, especially in South America, Asia's share of the world's rice hectarage in 1978 has declined to 90.01% from the 92.81% in 1954.

The world's rice hectarage, nonetheless, is still greatly concentrated in Asia. This great concentration of rice land in Asia could be viewed from several standpoints. First, greater population pressure leads to greater intensive use of the land (Carter 1975). Crop competition in terms of greater nutrient output naturally made the Asians in favor of rice. As we have demonstrated before, rice has a greater supportive capability both in terms of caloric and protein outputs (Table 1.1, 1.2; Univ. of Calif. Food Task Force 1974).

Related to this is the fact that rice is the staple for many Asians—their only available source of energy and protein. The per capita need of rice therefore is much higher in Asia than in any other continent. Following this, there is the Asians' passion for rice, a dietary preference and cultural bias they inherited at birth. Many Asians feel that they have not eaten a meal if there is no rice served. Furthermore, in many of the waterlogged areas of tropical Asia (as well as in Africa and America), rice is the only food crop that can be grown during the rainy season.

Figure 1.4 is a cartogram showing the world rice hectarage in 1976 based on U.S. Department of Agriculture data (USDA 1977A).[2] Here India and China appear as the world's rice giants: each devoted 38.0 and 34.5 million ha, respectively, of their cropland to rice.

The next group of countries, although dropping far behind the giants, each harvested at least 5 million ha of rice in 1976. These include Bangladesh (10.1), Indonesia (8.8), Thailand (8.5), Brazil (5.4), North and South Vietnam (5.3), and Burma (5.2).

The Philippines, Japan, Pakistan, Cambodia, Nepal, South Korea and Malagasy belong to the next group, having devoted at least 1 million ha to rice. The United States could also be included in this group. Its 985,000 ha of rice harvested in 1976 actually represented a 13% cutback from the 1,134,000 ha harvested in 1975.

Limitations on Expansion.—During the last 25 years, the world's rice land has been steadily expanding at an average annual rate of 1.32%. As the world's rice eating population continues to expand at a faster rate (2.5% per year) than the world as a whole (1.7%), the world's need for rice must also expand at an increasing rate. Will the world be able to continue its expansion of rice land at the rate it has been? What are the prospects?

It has been estimated that the land area of the world is roughly 13.149 billion ha (derived from the President's Sci. Advisory Comm. 1967). Of this amount, only ¼ or 3.287 billion ha are potentially arable (Natl. Acad. Sci. 1969). In 1970, the world's cultivated land was estimated at 1.461 billion ha (Univ. Calif. Food Task Force 1974). This gives the remaining 1.826 billion ha as the uncultivated arable land the world possessed in 1970.

Assuming that between 1970 and 1976 the annual growth rate of the world's cultivated land is the same as the expansion of the world's rice

---

[2]World rice hectarage (Table 1.6) has increased by 3.28% from 1976 to 144.7 million ha in 1978; the relative importance of individual countries remains pretty much the same.

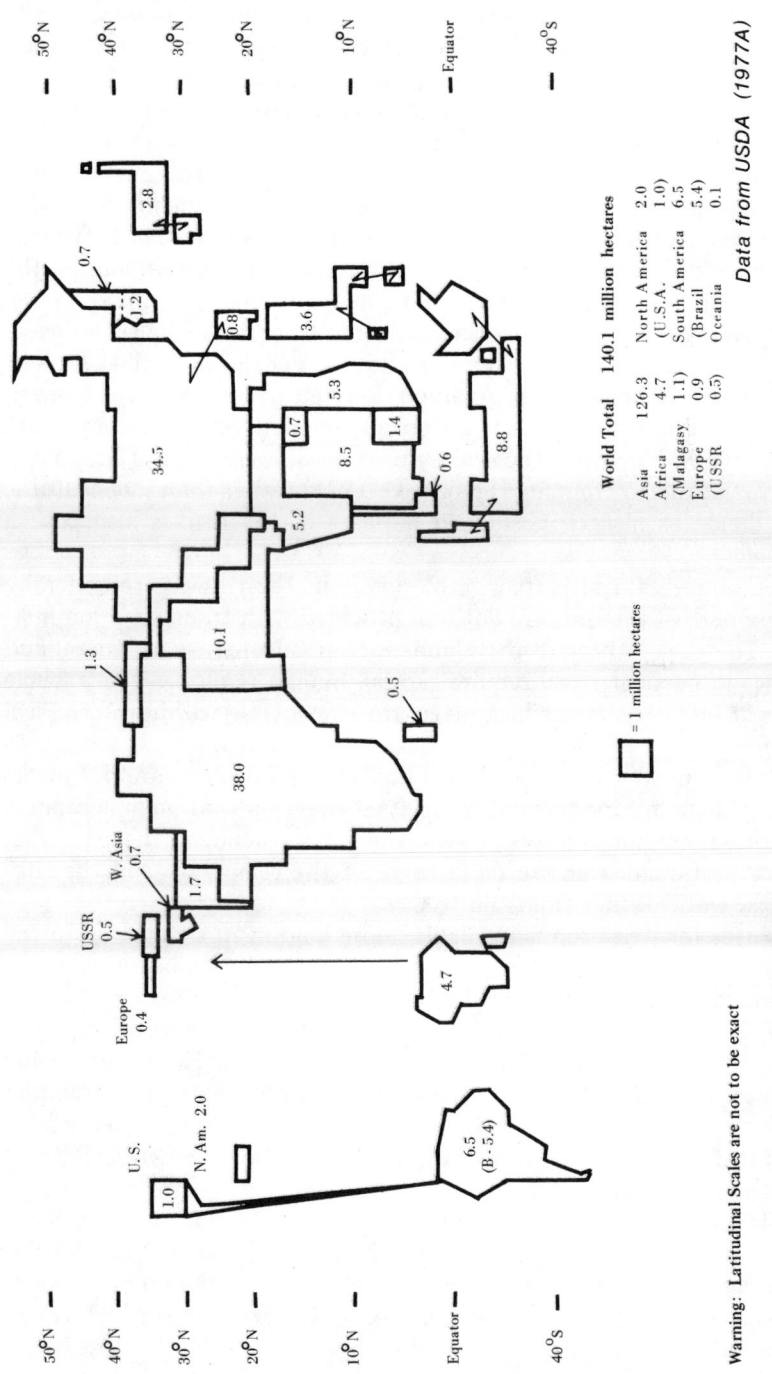

World Total	140.1 million hectares		
Asia	126.3	North America	2.0
Africa	4.7	(U.S.A.	1.0)
(Malagasy	1.1)	South America	6.5
Europe	0.9	(Brazil	5.4)
(USSR	0.5)	Oceania	0.1

Data from USDA (1977A)

= 1 million hectares

Warning: Latitudinal Scales are not to be exact

FIG. 1.4.   WORLD RICE HECTARAGE: 1976

land (1.09% per year), we then must subtract 0.096 billion ha from the 1970 figure to get 1.73 billion ha of uncultivated arable land for the world for 1977.

Since about 60% of the world's uncultivated arable land is in the tropics (Miller 1975), this gives us a maximum total of 1.038 billion ha for possible future expansion of rice land in the tropics. Of this amount, only about 300 million ha are in the Amazon and Congo basins where rainfall is abundant for rice growing (Univ. Calif. Food Task Force 1974). To this, may be added another 100—150 million ha of well watered land in the humid tropics of Sumatra and other Pacific islands.

However, the lush tropical vegetation (rainforest and jungles) is deceptive (Miller 1975). The soils have low inherent chemical fertility (Askew et al. 1970), their nutrient reservoir is often of low capacity (Janzen 1975), and the true laterite is a tropical phenomenon (Carter and Pendleton 1956). Moreover, the ecology of tropical agriculture is one that is in favor of mixed cropping (Igbozurike 1971) rather than monoculture. Abundance of insects, pests and diseases in the tropics also pose a problem.

Furthermore, uncultivated arable lands in the tropics are generally remote and, at present, are not well provided with transportation linkages. To develop these tropical lands for agricultural use in general and for rice in particular will require capital, human resources and a whole series of infrastructure which often are unavailable within the tropical region.

Physical constraints and cultural limitations, however, are not insurmountable. If future demand warrants its development, man is capable to tackle these limitations at a cost. But for the moment, the expansion of rice hectarage seems to be in favor of the temperate and the subtropical world rather than the tropics.

If water for irrigation is available, more land in Australia and in the Soviet Union could be brought under paddy cultivation, especially in Krasnodar Kray and Kazakhstan around the Caspian Sea area. In the rice belt of southern United States, croplands that are suitable to rice amount to 3.5 times its 1975 harvested rice hectarage. Even after taking into account the limitations imposed by available water for irrigation and/or agronomic factors associated with rotation, the United States could easily double its 1975 rice hectarage just within the rice belt (Grant and Holder 1975).

In Japan, there are roughly half a million ha of land formerly devoted to rice but currently not under rice cultivation (Off. of the Prime Minist. 1976). Due to increased productivity of farmers and the problems associated with rice over production, the government of Japan had to pay the nation's farmers not to grow rice on some of their land. These lands, retired from rice production, could easily be replanted to rice when the demand for rice in Japan and/or the world market price for rice is sufficiently high to warrant their retrieval (Lu 1972A).

Spatial Variation in Rice Yield

Anthony Baily once said that the world food problem lies in the fact that God ceased to create land long ago but He is still creating people. Even though there are still lands available for possible expansion of world rice areas in the near future, the land, nevertheless, is a *fixed* resource. Conversion of marginal land for rice cultivation futhermore, is costly. Increases in future world rice supply rests, therefore, heavily upon increases in per unit land yield.

Comparative Rice Yields.—Table 1.8 compares world rice yields per hectare among world countries or geographic areas between 1966 and 1978. Over this 12 year period, the world had increased its rice yields from 18.2 quintals per hectare (Q/ha) in 1966 to 26.0 Q/ha in 1978, an increase of 42.6% or an average of 3.55% per year (as compared with 1.20% area expansion).

TABLE 1.8. WORLD RICE: COMPARATIVE YIELD AND CHANGE BY COUNTRY OR GEOGRAPHIC AREA, 1966 AND 1978

Country or Geographic Area	1978 Yield[1]	1978 Percentage[2]	1966 Yield[1]	1966 Percentage[2]	Change[3] Quantity[1]	Change[3] Percentage
North America						
Costa Rica	23.1	88.9	14.3	78.6	8.8	61.5
Cuba	24.5	94.2	21.3	117.0	3.2	15.0
Dominican Republic[5]	26.4	101.5	23.4	128.6	3.0	12.8
El Salvador	38.5	148.1	25.0	137.4	13.5	54.0
Guatemala[10]	32.2	123.9	18.8	103.3	13.4	71.3
Haiti	11.7	45.0	10.9	59.9	0.8	7.3
Honduras	22.2	85.4	10.0	55.0	12.2	122.0
Jamaica and Dep.[10]	62.5	240.4	20.0	110.0	42.5	212.5
Mexico	26.0	100.0	24.3	133.5	1.7	7.0
Nicaragua	31.9	122.7	24.2	133.0	7.7	31.8
Panama	17.7	68.1	10.6	58.2	7.1	67.0
Trinidad-Tobago	27.0	103.9	25.0	137.4	2.0	8.0
United States	50.4	193.9	48.4	265.9	2.0	4.1
South America						
Argentina	34.2	131.5	35.0	192.3	−0.8	−0.2
Bolivia	15.7	60.4	15.3	84.1	0.4	2.6
Brazil	15.6	60.0	15.8	86.8	−0.2	−1.3
Chile[5]	33.2	127.7	25.5	140.1	8.0	31.8
Colombia[5]	43.0	165.4	19.4	106.6	23.6	121.5
Ecuador	21.5	82.7	18.3	100.6	3.2	17.5
Guyana	24.0	92.3	19.1	105.0	4.9	25.7
Paraguay	19.7	75.8	25.7	141.2	−6.0	−23.4

TABLE 1.8. *(CONTINUED)*

Peru	38.1	146.5	39.0	214.3	0.9	−2.3
Surinam	37.3	143.5	29.7	163.2	7.6	25.6
Uruguay	40.0	153.9	34.1	187.4	5.9	17.3
Venezuela	32.1	123.5	17.7	97.3	14.4	81.4
Europe						
Bulgaria	38.9	149.6	34.3	188.5	4.6	13.4
France	32.8	126.2	35.7	196.2	−2.9	−8.1
Greece	50.0	192.3	47.1	258.8	2.9	6.2
Hungary	18.9	72.7	16.3	89.6	2.6	16.0
Italy	51.0	196.2	47.0	258.2	4.0	8.5
Portugal	33.0	126.9	44.0	241.8	−11.0	−25.0
Romania	17.5	67.3	28.0	153.8	−10.5	−37.5
Spain	60.3	231.9	63.6	349.5	− 3.3	− 5.2
Yugoslavia	52.5	201.9	46.0	252.8	6.5	14.1
U.S.S.R. (including Soviet Asia)	50.0	192.3	28.7	157.7	21.3	74.2
Africa						
Algeria	20.0	76.9	30.0	164.8	−10.0	−33.3
Angola	12.5	48.1	13.6	74.7	− 1.1	− 8.1
Chad[7]	8.0	30.8	13.2	72.5	− 5.2	−39.4
Egypt	52.0	200.0	47.3	259.9	4.7	9.9
Gambia	13.3	51.2	13.7	75.3	−0.4	− 2.9
Ghana[5]	17.0	65.4	11.5	63.2	5.5	47.8
Guinea[4]	5.6	21.5	8.7	47.8	− 3.1	−35.6
Guinea-Bissau	9.2	35.4	NA[8]	NA[8]	NA[8]	NA[8]
Ivory Coast	11.4	43.8	10.6	58.2	0.8	7.5
Liberia	13.1	50.4	7.3	40.1	5.8	79.5
Malagasy Republic[7]	16.4	63.1	20.2	111.0	−3.8	−18.8
Mali	10.0	38.5	7.8	42.9	2.2	28.2
Morocco	37.0	142.3	41.7	229.1	−4.7	−11.2
Mozambique[4]	10.8	41.5	15.0	82.4	−4.2	−28.0
Nigeria	20.4	78.5	16.8	92.3	3.6	21.4
Senegal	13.3	51.2	14.2	78.0	−0.9	−6.3
Sierra Leone	14.8	56.9	11.9	65.4	2.9	24.4
Tanzania	15.4	59.2	11.0	60.4	4.4	40.0
Upper Volta	9.5	36.5	9.7	53.3	− 0.2	− 2.1
Zaire	6.8	26.2	8.0	44.0	−1.2	−15.0
Asia						
Afghanistan	21.9	84.2	15.3	84.1	6.6	43.1
Bangladesh	20.3	78.1	15.8	86.8	4.5	28.5
Burma	19.8	76.2	13.9	76.4	5.9	42.5
Cambodia	10.6	40.8	12.2	67.0	−1.6	−13.1
China (mainland)	35.6	136.9	31.8	174.7	3.8	12.0
China (Taiwan)	44.5	171.2	40.9	224.7	3.6	8.8
Hong Kong	20.0	76.9	18.9	103.8	1.1	5.8
India	19.7	75.8	13.0	71.4	6.7	51.5
Indonesia[5]	29.6	113.9	17.9	98.4	11.7	65.4
Iran	40.9	157.3	37.5	206.0	3.4	9.1
Iraq	30.2	116.2	16.4	90.1	13.8	84.2
Japan[10]	64.0	246.2	49.0	269.2	15.0	30.6
Korea (North)[5]	51.3	197.3	37.9	208.2	13.4	35.4
Korea (South)[9]	62.1	238.9	44.2	242.9	17.9	40.5
Laos	10.3	39.6	8.1	44.5	2.2	27.2
Malaysia (West)[5]	28.7	110.4	25.6	140.7	3.1	12.1
Nepal	19.8	76.2	18.2	100.0	1.6	8.8
Pakistan	24.0	92.3	14.5	79.7	9.5	65.5
Philippines	19.0	73.1	13.2	72.5	5.8	43.9
Sabah (E. Malaysia)	22.3	85.8	20.9	114.8	1.4	6.7
Sarawak (E. Malaysia)	13.3	51.2	10.8	59.3	2.5	23.2
Saudi Arabia	30.0	115.4	30.0	164.8	0	0

TABLE 1.8. (*CONTINUED*)

Sri Lanka (Ceylon)	23.9	91.9	15.6	85.7	8.3	53.2
Syria[6]	20.0	76.9	20.0	109.9	0	0
Thailand	18.0	69.2	19.4	106.6	−1.4	−7.2
Turkey	37.9	145.8	35.5	195.1	2.4	6.3
Vietnam (North)	19.8	76.2	17.7	97.3	2.1	11.9
Vietnam (South)	*[11]	*[11]	18.9	103.8	*[11]	*[11]
Oceania						
Australia[7]	62.0	238.5	71.3	391.6	−9.3	−13.0
World Average	26.0	100	18.2	100	7.8	42.6

Source: USDA (1977A, 1979).

[1]In quintals per hectare (Q/ha).
[2]Percentage of world average.
[3]1978 over 1966.
[4]Yield fluctuated along an improving trend but declined sharply in 1978 for various reasons.
[5]Yield has fluctuated along a strong upward trend especially since mid 1970s.
[6]Yield has consistently followed an improving trend but declined sharply in 1978.
[7]Yield peaked around mid 1960s and early 1970s but has since been declining.
[8]Not available or not applicable.
[9]Yield has consistently followed a strong upward trend but declined slightly in 1978.
[10]Yield consistent or fluctuated upward with sharp increase in 1978.
[11]Included in North Vietnam.

From Table 1.8, it is obvious that in either time period, rice yield per hectare varies widely throughout the world: ranging from 40.1% (Liberia) to 3.9 times (Australia) the world average of 18.2 Q/ha in 1966, and from 21.5% (Guinea) to 2.46 times (Japan) the world average of 26.0 Q/ha in 1978. During this period (1966−1978) changes in rice yields among world nations and geographic areas also varied greatly: ranging from an increase of 212% (Jamaica) to a decrease of 39.4% (Chad).

The wide range of world rice yield and the varying degrees of change thereof mean that the world rice supply could possibly be increased even just by bringing those low yielding countries to the world average. This, of course, is easier said than done; there are a great many factors affecting world rice yields.

Factors Affecting World Rice Yields.—From a geographic viewpoint, quality of the land (which varies from one part of the world to another) and the latitudinal location are very important. The former, because of its inherent nature (terrain, soil fertility and hydrological conditions, etc.), has an immediate and apparent effect on rice yields. The latter, because it determines seasonality, day length, isolation and the overall climatic regime (especially temperature), also affects rice yields in a profound way.

In this conjunction, it is interesting to note (Fig. 1.5) that countries or geographic areas which had, in 1976, a rice yield that is twice as much as the world average of 24.5 Q/ha, are all located at the latitude beyond the Tropic of Cancer and Tropic of Capricorn, on the poleward limits of rice cultivation. With the exceptions of Afghanistan, Pakistan and Iraq, countries of lower than the world average rice yield are all located

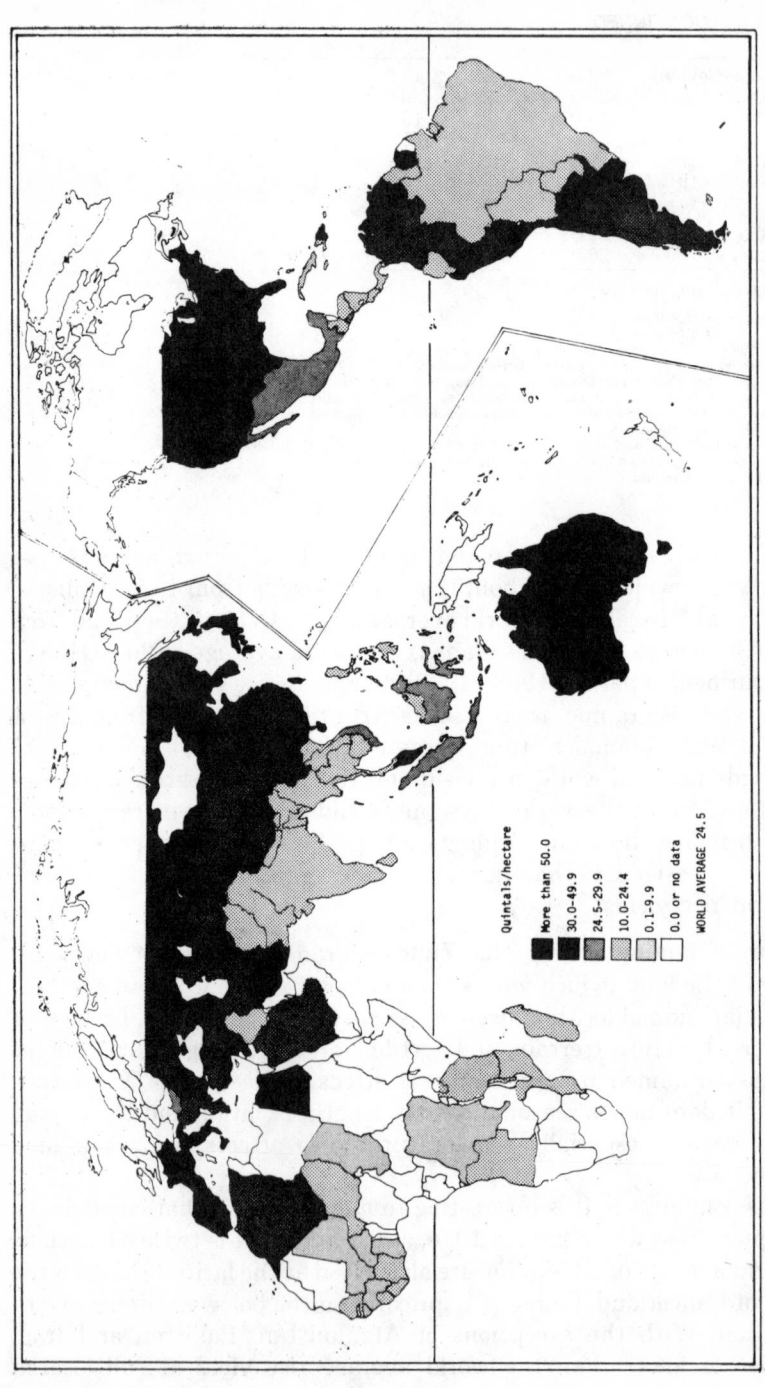

Data from USDA (1977A)

FIG. 1.5.   WORLD RICE YIELD: 1976

between the 2 tropics. Futhermore, most of the countries that have a yield that is higher than the world average have a midlatitudinal location (see Chapter 3).

The wide range of available technology is another factor affecting the world rice yield. This includes varietal selection of seeds specifically bred for a given geographic region as seen in the case of the Tong-il rice of Korea (IRRI 1976B). Moreover, technique of irrigation controls, timing of fertilization, correct application of herbicides and insecticides or genetic improvement in pest resistance, as well as effective usage of machinery—all have gradually become or are becoming an intricate part of the modern rice culture toward better yields. To these may also be added the factor of spatial adjustment leading toward greater degree of efficiency in land use and maximum yields as exemplified by the successful story of United States' rice (Lu 1976).

There is also an institutional factor affecting world rice yield as seen in the role of decision making. Through price-incentive policies, rice farmers in Japan, for example, have been encouraged and indeed were enabled to take advantage of the available technologies to boost their rice yields to the highest possibility (Lu 1972A).

World Supply and Demand for Rice

In 1960, the world produced 240.1 million metric tons (MMT) of wheat and 234.8 MMT of rice (rough or paddy). In 1975, the production was 348.5 MMT for wheat and 360.6 MMT for rice (USDA 1976B; Table 1.6). This gives an increase of 53.58% for rice but only 45.17% for wheat. The greater and faster pace of increase in rice production was due to greater world demand for rice during the same period.[3]

World Demand for Rice.—Between 1960 and 1962, the world had an average midyear population of 3,046,595,600. Between 1974 and 1976, the average midyear population of the world stood at 3,968,764,000. The increase in world population between these 2 time periods was 30.27%. On the other hand, the world's consumption of rice during the same time reference had increased faster at 42.03%: from an annual average of 161.3 MMT (milled) between 1960 and 1962 to an annual average of 229.1 MMT between 1974 and 1976.

It looks as though people are eating more rice, but why? During these 1½ decades, the real income of the world has increased. As people's incomes increase many of them shifted their consumption behavior in

---

[3]Wheat production, however, had leaped to 415.6 MMT in 1976 (USDA 1978). Despite cutbacks in 1977, world wheat production in 1978 stood at an all time high of 425.2 MMT (USDA 1979).

preference of finer food such as the versatile rice. Among the important starchy foods in West Africa, for instance, the demand for rice has surpassed maize, yams and cassava.

On a per capita term, the average world rice consumption between 1960 and 1962 was 52.95 kg (milled) per person per year. This amount has been increased to an average of 57.73 kg per person per year between 1964 and 1976. To be sure, people of the world as a whole are consuming more rice today than they were before. But who are they?

Table 1.9 answers this question in a highly generalized way. Asia, wherein a great bulk of the world's rice is consumed, has not been keeping up with the world trend of increasing rice consumption. In fact, it is about 15% below the trend. Both North and South America have increased their rice consumption much faster than the world average; Africa and Europe increased nearly twice as fast as the world. Especially worth noting, however, is the Soviet Union which has increased its rice consumption 366% during the same time framework.

Asia, nevertheless, remains the world's biggest rice consumer. In any given year, Asia accounts for 85–90% of the world's rice consumed. Asia not only has the world's largest rice consuming population but also has the world's highest per capita rice consumers. As Fig. 1.6 reveals, nearly all of the world's high to very high per capita rice consumers are located in Asia. Outside Asia, only Malagasy, Surinam, Sierra Leone, Guyana and Liberia are comparable.

In 1975, every Asian consumed an average 94.70 kg of milled rice. This compares with the average of 58.42 for the world, 33.54 for South America, 20.13 for Africa, 15.55 for Oceania, and 8.77 for North and Central America. The 247.46 kg of milled rice consumed in 1975 by each South Vietnamese is unmatched anywhere in the world, not even by the Thais who followed next with 212.54 kg per capita. For comparative per capita rice consumption, see Table 1.11.

World Supply of Rice.—Despite the slight decline in the demand for rice in some parts of the world since the early 1970s and in the projected demand for 1977/1978, the total world demand for rice, as noted before (Table 1.9), did increase at a faster rate than its population increases. Here one may raise a question: Has the world been able to supply its need for rice? The answer is fortunately yes. World production of 377.0 MMT of rough (paddy) rice in 1978 represents an increase of 60.6 over its production of 234.8 MMT in 1960 or 124.0% over its 1950 production of 168.3 MMT (Table 1.10).

The pace of increase, however, has not been uniform throughout the world (Table 1.10). Asia, where roughly 90% of the world's rice is produced, has the smallest percentage increase among all world continents except Africa. Its increase of 119.17% between 1950 and 1978, indeed, is

TABLE 1.9. WORLD RICE CONSUMPTION[1] BY CONTINENTS AND SUBREGIONS: SELECTED YEARS, 1960–1977[2]

Continent and Region	1960/61–1962/63		1969/70–1971/72		1974/75–1976/77		1977/1978		Change[4]	
	Quantity	Share[3]	Quantity	Share[3]	Quantity	Share[3]	Quantity	Share[3]	Quantity	Percent[5]
Africa	4.31	2.67	6.19	2.98	7.71	3.36	8.51	3.62	3.40	78.89
North Africa[6]	1.76	1.09	2.70	1.30	3.54	1.54	4.00	1.70	1.78	101.14
Central Africa	2.36	1.46	3.22	1.55	3.73	1.63	4.05	1.72	1.37	58.05
East Africa	0.15	0.09	0.20	0.10	0.36	0.16	0.38	0.16	0.21	140.00
South Africa	0.04	0.03	0.07	0.03	0.08	0.03	0.08	0.04	0.04	100.00
Asia[7]	114.79	89.75	185.94	89.43	197.78	86.32	201.03	85.58	52.99	36.60
East Asia	83.51	51.77	108.52	52.20	120.44	52.56	120.79	51.42	36.93	44.22
Southeast Asia	13.01	8.06	18.79	9.03	15.70	6.86	16.04	6.83	2.69	20.68
South Asia	48.27	29.92	58.63	28.20	61.64	26.90	64.20	27.33	13.37	27.70
South America	4.46	2.76	5.57	2.68	7.00	3.06	7.91	3.37	2.54	56.95
Brazil	3.45	2.14	4.00	1.92	4.87	2.13	5.40	2.30	1.42	41.16
Others	1.01	0.62	1.57	0.76	2.13	0.93	2.51	1.07	1.12	110.89
North America	1.53	0.95	2.52	1.21	2.50	1.09	2.14	0.91	0.97	63.40
United States	0.95	0.59	1.66	0.80	1.41	0.61	1.45	0.62	0.46	48.42
Others[8]	0.58	0.36	0.86	0.41	1.09	0.48	0.69	0.29	0.51	87.93
Europe	1.76	1.09	2.68	1.29	3.22	1.41	3.25	1.38	1.46	82.96
Western Europe	1.18	0.73	1.21	0.58	1.30	0.57	1.19	0.51	0.12	10.17
Eastern Europe	0.25	0.16	0.37	0.18	0.38	0.17	0.50	0.21	0.13	52.00
U.S.S.R.	0.33	0.20	1.10	0.53	1.54	0.61	1.56	0.66	1.21	366.67
Oceania	0.04	0.03	0.07	0.03	0.08	0.03	0.07	0.03	0.04	100.00
Others	4.43	2.75	4.94	2.38	10.84	4.73	12.00	5.11	6.41	147.70
World total	161.32	100.00	207.91	100.00	229.13	100.00	234.91	100.00	67.81	42.03

Source: USDA (1977B).
[1]Milled, in MMT.
[2]Crop years: 3 year average for 1960–1962, 1969–1971, and 1974–1976, and projection for 1977–78.
[3]Percentage of world total.
[4]1974/75–1976/77 over 1960/61–1962/63.
[5]Percentage increase over the above 2 time periods.
[6]Includes Middle East.
[7]Southwest Asia not included.
[8]Includes Middle America.

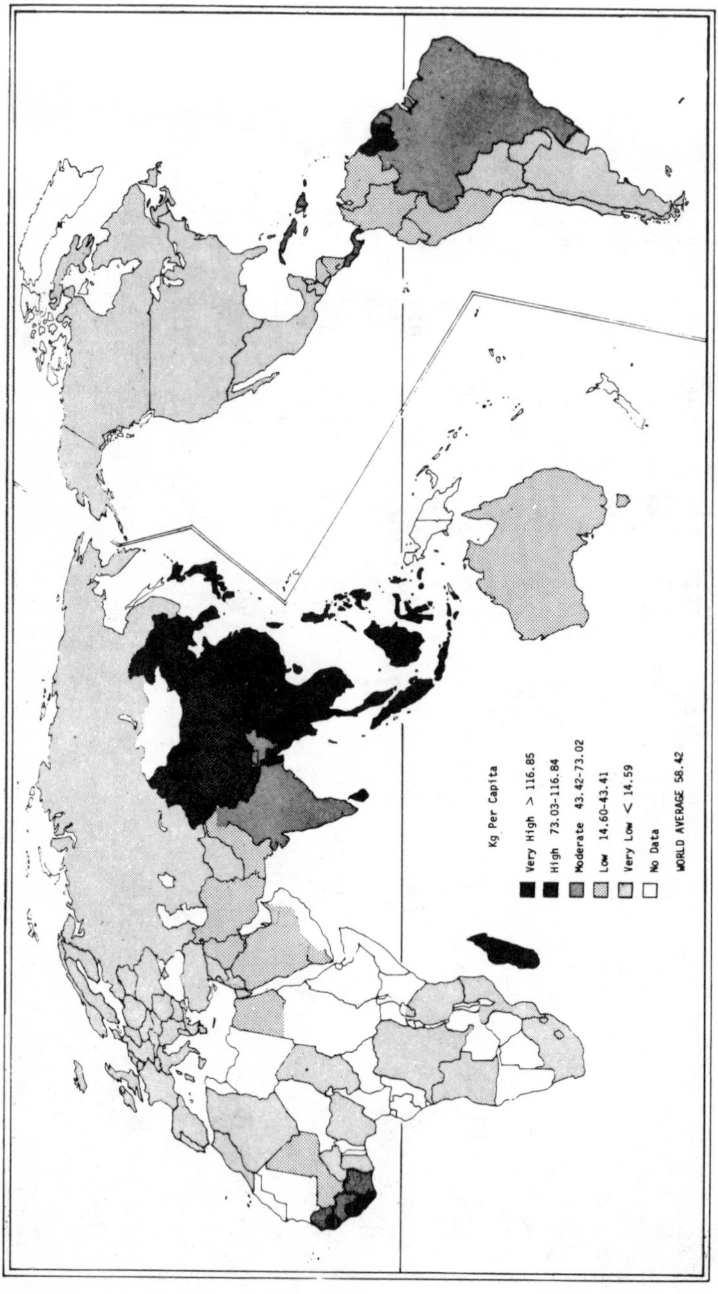

Data from USDA (1976A)

FIG. 1.6. WORLD RICE PER CAPITA CONSUMPTION: 1975

TABLE 1.10. WORLD RICE: PRODUCTION AND CHANGE¹ SELECTED YEARS, 1950–1978

Continent and Region	1950 Quantity	1950 Share²	1960 Quantity	1960 Share²	1970 Quantity	1970 Share²	1978 Quantity	1978 Share²	Change Quantity	Change Percent	Change Share²
Africa	3,621	2.15	4,730	2.04	7,156	2.30	7,788	2.07	4,167	115.08	2.00
North America	2,398	1.43	3,551	1.53	5,195	1.67	8,148	2.16	5,750	239.78	2.75
South America	4,142	2.46	6,501	2.80	9,523	3.06	12,385	3.28	8,243	199.01	3.95
Europe	1,496	0.89	1,627	0.70	3,069	0.99	4,853	1.29	3,357	224.40	1.61
Oceania	72	0.04	128	0.06	247	0.08	571	0.15	499	693.06	0.24
Asia	156,614	93.03	215,427	92.87	285,598	91.90	343,251	91.05	186,637	119.17	89.45
East	74,160	44.05	97,191	41.90	134,469	43.27	160,699	42.63	86,539	116.69	41.48
Southeast	33,601	19.96	45,911	19.79	61,628	19.83	72,691	19.28	39,090	116.34	18.73
South	47,861	28.43	71,017	30.62	87,359	28.11	107,664	28.56	59,803	124.95	28.66
Southwest	992	0.59	1,308	0.56	2,142	0.69	2,197	0.58	1,205	121.47	0.58
World³	168,343	100.00	234,749	100.00	311,830	100.00	376,996	100.00	208,653	123.95	100.00

Source: Derived from Palacpac (1978) and USDA (1979).
¹Production, rough (paddy), in 1000 MT. Change, 1978 over 1950: quantity in 1000 MT.
²Regional share of the world total.
³May not add up due to rounding and from different sources.

below the world average. One may rightly suspect the 693% increase in rice production in Oceania (mainly Australia) to be deceptive because of its late start as a rice producer—with only 72,000 MT produced in 1950. But considering its high yield of 62.0 Q/ha (third highest in the world next only to Japan's 64.0 and S. Korea's 62.1), and its latitudinal location, Australia has real potential to increase its rice production if more water, as noted before, can be diverted for irrigation.

Over the last 29 years, production of rice in South America (mainly Brazil) has attained the highest absolute increase outside Asia. North America (mainly the United States), on the other hand, has excelled in percentage increase. While Europe (dominated by the Asiatic Soviet Union) has more than tripled its rice production, Africa (mainly Egypt with Malagasy following suit) only little more than doubled its share—trailing behind the world average increase by 9%. Thus, variations in change of rice production among the world continents were considerable.

On a per capita basis, the world production of rice varies even more. In 1975, the per capita availability of milled rice for the world as a whole stood at 59.96 kg, but as for the world nations, it varies from a low of 0 in non-rice producing countries to a high of over 271 kg in Guyana and Surinam (Table 1.11). It is apparent that there exists a spatial inequality in both world supply and world demand for rice.

Degree of Self-sufficiency.—Based on the ratio between the production and consumption of rice, an attempt has been made to show the degree of self-sufficiency among the world nations (Fig. 1.7). As seen in Table 1.11, the world as a whole has a ratio of 1.03. This means that in 1975, the world's production of milled rice was slightly ahead of its consumption by a small margin of about 3%. It should be noted that in any given year, mill production of rice may be greater (or smaller) than its equivalent of farm production. This is due to carryovers (or stocks) from previous year(s).

In Fig. 1.7, countries (or geographic areas) with a ratio of 0.81 to 1.02 are treated as "self-sufficient," or nearly so. Countries with a ratio between 1.03 and 1.50 are termed as "adequate." Those with a ratio of greater than 1.5 could be called "abundant" because their production is at least 50% higher than their domestic demand. Countries whose ratio falls between 0.80 and 0.50 are "deficit" countries. They must import rice from time to time in order to satisfy their domestic needs. A ratio that lies below 0.49 could mean "severe deficit." Countries that normally produce less than ½ their domestic demand of milled rice naturally depend heavily on imports to meet their needs.

Among the abundant countries in 1975 one finds, perhaps with some degree of amazement, that Uruguay ranked first with a self-sufficient

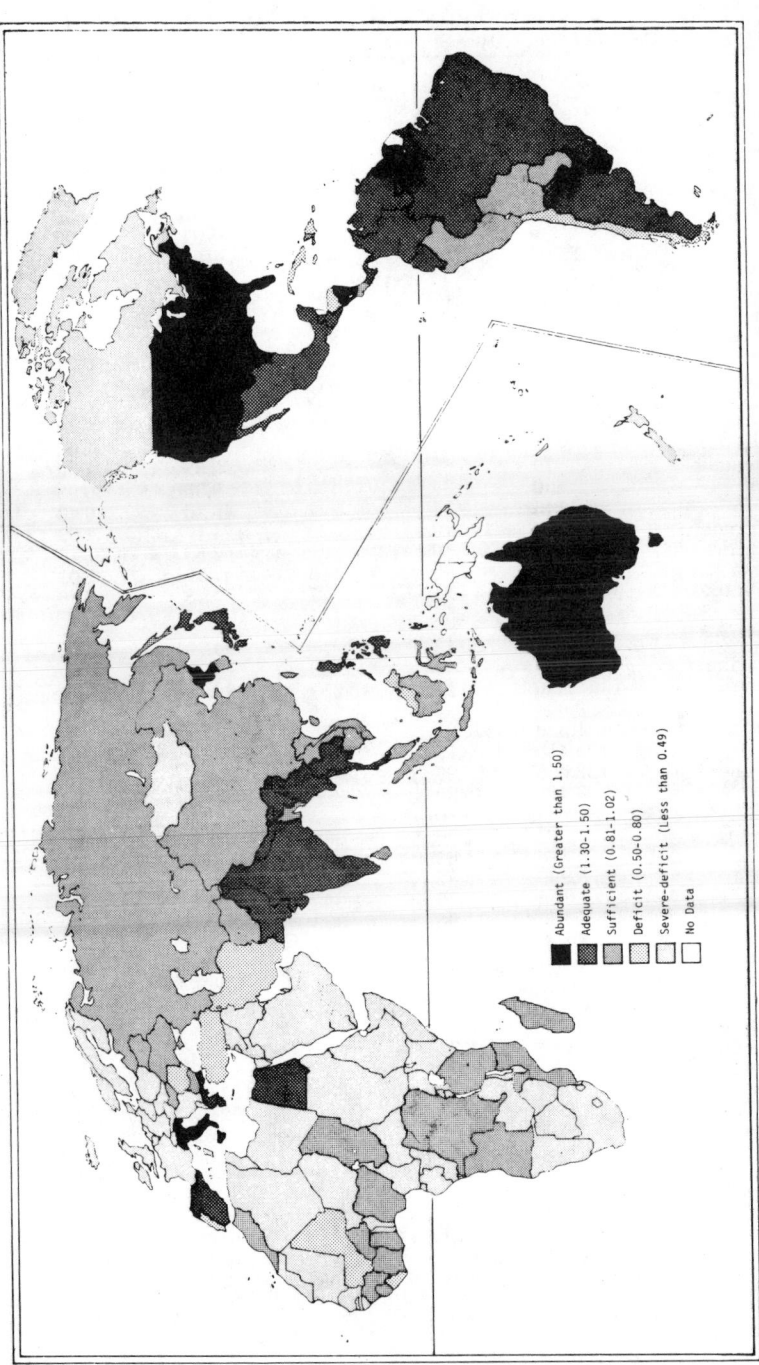

FIG. 1.7. WORLD RICE: DEGREE OF SUFFICIENCY (DOMESTIC SUPPLY / DEMAND RATIO)

Calculated from data given in USDA (1976A)

TABLE 1.11. WORLD RICE: PRODUCTION AND CONSUMPTION BY MAJOR
COUNTRIES OR GEOGRAPHIC AREAS: 1975[1]

Country or Geographic Area	Production		Consumption		Degree of Sufficiency[6]
	Total[2]	Per Capita[3]	Total[4]	Per Capita[5]	
Africa		18.51		20.13	0.92
Algeria	6	0.24	14	0.84	0.29
Angola	20	1.95	13	1.95	1.00
Chad	30	5.21	21	5.21	1.00
Egypt	2,500	44.99	1,475	39.62	1.14
Gambia	30	38.17	31	59.16	0.65
Ghana	79	5.55	53	5.55	1.00
Guinea	380	55.93	277	62.73	0.89
Ivory Coast	460	44.81	321	48.10	0.93
Liberia	183	77.63	154	100.46	0.77
Malagasy	2,000	159.13	1,475	183.37	0.87
Mali	110	12.81	93	16.32	0.79
Mauritius	0	0	85	96.37	0
Morocco	24	0.93	16	0.93	1.00
Mozambique	79	5.63	52	5.63	1.00
Nigeria	600	6.34	409	6.50	0.98
Senegal	146	19.55	227	45.76	0.43
Sierra Leone	530	121.16	339	121.16	1.00
South Africa	0	0	72	2.83	0
Tanzania	170	7.65	115	7.59	1.01
Upper Volta	40	4.31	27	4.48	0.96
Zaire	250	6.55	184	7.39	0.87
Asia		96.13		94.70	1.02
Afghanistan	400	13.49	260	13.49	1.00
Bangladesh	19,820	180.58	13,357	182.72	0.99
Burma	9,000	191.87	5,544	177.47	1.08
Cambodia	1,500	120.22	1,175	144.88	0.83
China[7]	125,415	98.16	84,066	96.73	1.02
Mainland	122,000	97.26	81,760	95.85	1.02
Taiwan	3,415	145.99	2,306	143.69	1.02
Cyprus	0	0	2	3.13	0
Hong Kong	10	1.60	362	82.89	0.02
India	68,318	75.83	43,720	72.87	1.04
Indonesia	23,900	123.82	17,092	130.22	0.95
Iran	1,155	23.36	1,029	31.62	0.75
Iraq	200	11.96	383	34.43	0.35
Israel	0	0	38	11.01	0
Japan	16,456	107.97	11,385	102.61	1.05
Jordan	0	0	25	9.25	0
Korea, North	3,700	158.72	2,266	142.95	1.11
Korea, South	6,485	134.52	4,719	135.96	0.99
Laos	911	179.23	672	203.45	0.88
Lebanon	0	0	21	7.32	0
Malaysia[8]	2,125	115.29	1,602	134.62	0.86
Nepal	2,200	116.53	1,165	92.67	1.26
Pakistan	3,658	34.67	1,943	27.65	1.25
Phillipines	6,515	101,24	3,850	92.04	1.10
Saudi Arabia	3	0.28	132	18.36	0.02
Singapore	0	0	210	93.33	0
Sri Lanka	1,100	53.48	1,148	82.08	0.65
Syria	5	0.41	76	10.35	0.04
Thailand	15,000	239.51	8,785	212.54	1.13
Turkey	235	3.81	208	5.17	0.74

TABLE 1.11. (CONTINUED)

Vietnam[9]	11,900	171.09	7,735	171.09	1.00
North Vietnam	4,400	120.17	2,860	120.17	1.00
South Vietnam	7,500	247.46	4,875	247.46	1.00
Europe		2.80		3.33	0.84
Austria	0	0	30	3.99	0
Belux[10]	0	0	33	3.25	0
Bulgaria	70	3.25	46	5.27	1.00
Denmark	0	0	10	1.98	0
France	43	0.55	140	2.65	0.21
Germany, West	0	0	123	1.99	0
Greece	103	7.19	56	6.19	1.16
Hungary	75	4.46	55	5.22	0.85
Italy	931	11.68	291	5.21	2.24
Netherlands	0	0	35	2.56	0
Norway	0	0	5	1.25	0
Poland	0	0	71	2.09	0
Portugal	122	9.02	125	14.27	0.63
Rumania	53	1.60	64	3.01	0.53
Spain	378	7.36	196	5.53	1.33
Sweden	0	0	14	1.71	0
Switzerland	0	0	16	2.50	0
United Kingdom	0	0	115	2.05	0
Yugoslavia	33	0.94	41	1.92	0.49
North America		16.16		8.77	1.84
Canada	0	0	57	2.50	0
Costa Rica	143	45.22	89	45.22	1.00
Cuba	420	29.46	493	53.21	0.55
Dominican Republic	233	32.79	204	43.43	0.75
El Salvador	57	9.23	33	8.24	1.12
Guatemala	64	8.12	40	7.23	1.05
Haiti	82	11.56	53	11.56	1.00
Honduras	26	5.27	22	7.24	0.73
Jamaica	3	0.99	42	20.70	0.05
Mexico	600	6.65	350	5.82	1.14
Nicaragua	116	33.88	45	20.88	1.62
Panama	175	68.35	104	62.35	1.10
Trinidad-Tobago	18	11.17	52	48.42	0.23
United States	5,788	19.51	1,365	6.39	3.05
South America		39.22		33.54	1.17
Argentina	325	8.31	149	5.87	1.42
Bolivia	75	9.41	53	9.41	1.00
Brazil	8,500	53.95	5,000	46.67	1.16
Chile	75	4.78	67	6.54	0.73
Colombia	1,614	44.56	963	40.91	1.09
Ecuador	322	28.67	167	24.80	1.16
Guyana	326	271.81	83	104.93	2.59
Paraguay	50	12.47	33	12.47	1.00
Peru	456	19.60	344	22.03	0.89
Surinam	151	271.43	59	168.57	1.61
Uruguay	208	44.06	33	10.77	4.09
Venezuela	369	20.01	198	16.51	1.21
Oceania					
Australia	450	23.85	210	15.55	1.53
U.S.S.R.	2,000	5.11	1,550	6.09	0.84
World Total	353,300	59.96	231,747	58.42	1.03

Source: USDA (1976A).
[1]Crop year or marketing year
[2]Rough (paddy) production in 1000 MT.
[3]Milled rice in kg per person per year. Derived from dividing the total milled production in 1975 (USDA 1976A) by the mid year population of 1975 (Encycl. Britannica 1977).
[4]Milled, in 1000 MT.

TABLE 1.11. *(CONTINUED)*

[5]Milled, in kg per person per year; derived from dividing total consumption by the midyear population.
[6]Defined as the ratio between per capita production (supply) and per capita consumption (demand).
[7]Total of Mainland and Taiwan.
[8]Includes Malaya, Sabah and Sarawak.
[9]Sum of North and South Vietnam.
[10]Belgium and Luxembourg.

ratio of 4.09. It was followed by the United States (3.05), Guyana (2.59), Italy (2.24), Nicaragua (1.62), Surinam (1.61) and Australia (1.53).

Projections on supply and demand for rice were carried out by FAO (e.g., 1965, 1966, 1971). The projection, however, could only be said to be nearly satisfactory. World per capita demand for rice, for instance, is found to be underestimated by about 14%. The projections were given as 49.7 (trend) and 50.1 (high), comparing with the actual of 58.4 kg per person for 1975 (compare Table 1.11).

Rice as a Commodity of Trade

The existence of spatial inequalities in world rice production and consumption gives rise to a situation, common to any commodity, in which a certain geographic region(s) or country has extra supplies of given goods and another region(s) has a definite need for it. This situation was termed by Ullman (1956) as complementarity which, together with transferability and lack of intervening opportunity form the bases for spatial interaction such as intra- or international trade. Despite its relatively small quantity entering the international market, rice as a commodity of trade is subject to the same analytical framework (Lu 1975).

Factors Affecting World Rice Trade.—Many interrelated factors are at work to form the market forces which in turn determine the mutually related volume and price of trade. An analytical tool called spatial price equilibrium was employed to study the trade aspect of the world rice industry by El-Amir (1967). Spatial price equilibrium treats regional supply and demand in terms of market price differences and transportation costs between regions.

However, in the international markets, besides the available quantity for and the price of trade, the ability to pay with "hard" currencies (or foreign exchange reserves), the ability to transfer the goods, and the absence of an intervening "distance" are very important. The latter has been expounded further to include physical, economic and cultural as well as political distances (Lu 1975).

The political distance,[4] although often overlooked by theoretical economists, is one of the most important factors affecting the world trade of

---

[4]Some call it "institutional factor."

rice. This may be seen in the dramatic decline in rice exports from the United States to Cuba in the early 1960s, and the loss of the United States rice market in Vietnam in the late 1960s (Lu 1975).

As the hard currency shortage and the need for importing more rice by many of the Third and Fourth World countries are unfortunate coincidences, government policies then definitely play another important role affecting world rice trade. For instance, the passage by the United States Congress of the Public Law 83-480 (PL 480), the so-called Agricultural Trade Development and Assistance Act of 1954 and its subsequent amendments, have tremendous impacts on the exports of United States rice (and other agricultural commodities) as well as on the rice importing countries.

The United States government does not intend to indefinitely prolong the PL 480 exports. In fact, United States rice growers have often been challenged to expand commercial rice markets (Yeutter 1975). But as the stocks of United States rice mounted again in the mid 1970s, and the need for rice in the Third and Fourth World continues to exist, the Congress of the United States must continue its appropriations for the PL 480 programs. At its height, the PL 480 exports accounted for 78.7% of the total export of United States rice. Only until very recently has the PL 480 sale of United States rice dropped consecutively slightly below the ½ mark of the total United States rice exports in 1974 and 1975 (USDA 1977A).

Concessional sales offered by China (mainland) and by Japan have similar effects. Faced with the rice surplus and storage problems in the late 1960s, the Japanese government began to offer long-term concessional sales of rice to Korea, Indonesia and Pakistan. This was the main reason that Japan was able to divert its rice surplus problem and peaked its export of milled rice to nearly 1 MMT in 1970–71 (Lu 1972A).

Two other factors, both geographic, also helped the Japanese to cut into the United States market in Korea. The first is the cultural preference of the Koreans who, like the Japanese, prefer the "pearl" type of low-amylose short grain rice produced both in Japan and in California. The second, perhaps a less known one, is the factor of geographic propinquity which the United States suffers by a very large margin of sheer physical distance and consequently, higher transportation costs (Lu 1971).

The *type* of rice mentioned above, because of its inherent cooking quality, is another factor affecting regional-cultural preferences for rice— thus its international trade.

Based on the ratio of grain width to this length, rice can be classified into short, medium, and long and extra-long grain rices. In general, the medium and short grain rices have lower amylose contents (less than 25%), have a sticky texture and are relatively firm; they tend to stick together when cooked (Williams *et al.* 1958). The long grain rice, on the other hand, has higher amylose content (greater than 25%), with an inherent flaky texture and minimum splitting; it does not tend to stick together when cooked.

The short grain type of rice, called pearl, is preferred by the Japanese, the Koreans and the Hawaiians. The great majority of the world's people, however, prefer the medium to long grain types of rice. Thus, world markets for these types of rice are more or less defined by this regional-cultural preference. It is, therefore, relatively difficult, for instance, for the Californian or Japanese pearl rice to cut into the markets for the long grain rice produced in the southern United States or in Thailand. In fact, this is one of the important reasons that California has been gradually shifting its rice acreage from the traditional short grain pearl rice to the medium and even long grain rices after it had lost its market in Japan in the late 1960s and recently in Korea.

Pattern of World Rice Trade.—From an historical perspective, World War II has exerted a depressing effect on world rice trades. The total world exports of milled rice in the pre-war years of 1937–1939 stood at an annual average of 7.7 MMT (Grist 1975). It dropped nearly 40% to an annual average of 4.7 MMT in the 1950–54 period (Table 1.12). In fact, only until very recently has the world fully regained its pre-war strength in rice exports: the peak total export of 8.367 MMT in 1977 was 633,000 MT above the average level of 1937–39.

With its tremendous volume of production and enormous demand for rice, Asia naturally dominates the scene of world rice trade. In the importing sector, Asia's share has been fluctuating between 60 and 80% of the world total—with an average of about 65% between 1960 and 1977. In exporting, Asia's share has been declining—from an overwhelming average of 85% of the world total in 1937–39 to only around 56% in 1977.

Meanwhile, the share of other continents in the total world rice exports has been enlarging. Especially worth noting is North America which has increased its share from 1.6% in 1937–39 to slightly over 27% in 1977. The export of rice from North America, however, is monopolized by a single country—the United States—which in 1975 exported 1,743,954 MT of milled rice to 104 countries or geographic areas (USDA 1977C).[5]

From a geographic perspective, the pattern of world rice trade is shaped, among other factors, by the spatial inequalities of world rice supply and demand as noted before. The best way to show a clear pattern of world rice trade perhaps is through the "balance sheet" approach. This will require information from various possible sources, including

---

[5]The total exports of U.S. rice in 1977 amounted to more than 2.2 MMT (Table 1.12).

TABLE 1.12. WORLD RICE EXPORTS:[1] 1937-1977[2] BY CONTINENTS AND MAJOR EXPORTING COUNTRIES

Continent and County	1937−39[3]	1950−54	1955−59	1960−64	1965−69	1970−74	1977[4]
Africa	116	149.6	288.0	374.2	431.0	373.6	217.0
Egypt	108	111.9	228.0	142.6	142.0	330.4	200.0
Others	8	37.7	60.0	231.6	289.0	43.2	17.0
North America	122	651.2	695.0	1,101.6	1,696.2	1,750.6	2,238.0
United States	122	630.6	686.8	1,081.6	1,678.2	1,731.4	2,220.0
Others	*5	20.6	8.2	20.0	18.0	19.2	18.0
South America	87	207.0	149.0	195.2	393.4	436.4	494.0
Argentina	0	8.2	23.8	6.4	44.8	67.8	100.0
Brazil	54	87.0	33.3	41.4	152.0	103.0	102.0
Guyana	15	27.6	31.2	80.6	95.4	68.8	100.0
Surinam	6	6.6	14.2	19.8	19.6	33.2	50.0
Uruguay	2	13.0	12.2	18.2	38.0	66.4	115.0
Others	10	64.6	34.3	28.8	43.6	97.2	27.0
Asia	6,557	3,361.0	4,527.2	4,759.8	3,750.0	4,097.8	4,709.0
Burma	2,966	1,247.4	1,687.6	1,624.6	748.6	434.2	600.0
China, Mainland	9	157.4	834.6	748.8	932.8	1,501.8	1,000.0
China, Taiwan	638	*5	ND[6]	55.0[7]	126.6[7]	0	200.0
Japan	0	2.0	*8	*8	55.0	509.6	100.0
Pakistan	279	28.0	73.2	124.6	137.2	394.2	711.0
Thailand	1,386	1,399.8	1,266.2	1,484.0	1,373.6	1,089.4	1,500.0
Others	1,279[9]	526.4	665.6	722.8	376.2	168.6	598.0
Europe	136	325.4	403.8	330.4	305.8	527.4	469.0
Italy	136	232.8	200.6	147.4	139.0	377.0	244.0
Spain	0	38.6	66.0	55.8	78.4	48.6	50.0
Others	0	54.0	137.2	127.2	88.4	101.8	175.0
Australia	13	31.8	40.8	74.2	91.0	163.2	300.0
World Total	7,734	4,726.0	6,103.8	6,835.2	6,667.4	7,349.0	8,367.0

Source: Palacpac (1977); USDA (1978); Grist (1975).
[1]Milled, in 1000 MT.
[2]Three-year average: 1937−39; 5 year average: 1950−1974; yearly: 1977.
[3]Source: Grist (1975). Grist put North and South America together under western hemisphere.
[4]Source: USDA (1978), preliminary.
[5]Included in the one above.
[6]No separate data.
[7]Derived from USDA (1976A)
[8]Less than 500 MT.
[9]Including 54,000 MT from Iran and 1,225,000 MT from Indochina.

FAO Trade Intelligence, USDA Rice Market News, USDA Foreign Agriculture Circular, as well as official trade statistics from almost all nations. Unfortunately, data from these sources are not always available and/or detailed enough or in agreement with each other for accurate presentation. Furthermore, available data for a complete up-to-date balance sheet often are several years behind. At best, one can only derive a gross pattern.

Table 1.13, admittedly, is far from being complete and not detailed enough. Nevertheless, it gives a gross account for world rice trade in 1974.[6] Several points, which may also be applicable to other recent years,

[6]USDA has compiled tables for each of the following years from 1963 to 1970 (see USDA 1972B).

are apparent: (1) There exists in the export sector of rice trade a quasi-monopoly situation. The 5 major exporting countries of the United States, Thailand, China, Pakistan and Burma account for nearly ¾ of the total supply. (2) There is an enormous intra-Asian flow of rice. (3) Rice trade within Southeast Asia is also considerable. (4) While the United States is the main supplier of rice to Europe and Africa, a greater portion (more than ⅘) of its exports landed in Asia in 1974.

An idealized flow of world rice trade based on minimizing the total transportation cost was suggested for 1975 by El-Amir (1967). Political realities and other current situations, however, have offset the idealized economic consideration.

Trend.—Before 1957, there were only 2 "millionaires" in world rice trade in terms of quantity exported. Of the 2, Burma led Thailand until 1964. Then China joined the club in 1958 and the United States in 1962. Burma dropped out from the club in 1967. In 1969 the United States, which had just become the world's leading rice exporter 2 years earlier, was the only millionaire. Since 1971, the leading role in world rice exports has been played by China, Thailand and the United States. Between 1971 and 1975, the United States led the world with an average export of 1.78 MMT of milled rice, followed by China (1.51) and Thailand (1.14)—while the world average stood at 7.44 MMT.

With the erratic monsoons, frequent typhoons and the greater increases in rice consuming populations as sure facts in life, Asia's dependence on other continents, especially North America, to supply its subregional shortages of rice is expected to continue. This means that the United States, the smallest rice producer among world exporters, which has recently become a major supplier to Asia's need for rice, is likely to continue its leading role in the world rice trade. This is because the rice industry of the United States is both enterprising and dynamic (Lu 1971, 1976), and it has sufficient productive resources (Godwin and Jones 1970) to meet the increasing demand at least into the foreseeable future.

An area of increasing importance in world rice trade is that of Southwest Asia and the Middle East. Despite its increases in rice production in recent years, these regions have also stepped up their imports of rice from 405,000 MT in 1964 to 1,242,000 MT in 1977 (USDA 1978). The increased oil revenue which improved the living standards of these people has also raised the demand for rice.

Coming new onto the scene of world rice trade is the exportation of rice from North Korea. The increasing export activities, from 44,000 MT in

TABLE 1.13. WORLD RICE TRADE: 1974[1] BY MAJOR COUNTRIES OF ORIGIN AND DESTINATION

| Destination | Origin | | | | | | | Total |
	U.S.	Thailand	China[2]	Pakistan	Japan	Burma	Other[3]	Import[4]
Indonesia	69.2	127.5	427.0	111.6	134.8	68.8	209.1	1148.0
South Korea	498.6	*[5]	*[6]	*[6]	44.7	*[6]	[6]	543.3[7]
Iran	450.8	1.1	*[6]	62.5	*[6]	*[6]	*[6]	513.9[7]
Malaysia[8]	0	68.7	221.0	*[6]	*[6]	4.0	72.3	366.0
Hong Kong	5.5	113.7	154.0	*[6]	*[6]	1.0	69.8	344.0
Sri Lanka	0	0	113.7	98.2	*[6]	102.9	15.2	330.0
Bangladesh	295.1	3.8	*[6]	*[6]	*[6]	*[6]	*[6]	298.9[7]
Philippines	*[5]	47.1	52.2	*[6]	68.7	*[6]	41.0	209.0
Iraq	110.0	82.2	*[6]	*[6]	*[6]	*[6]	8.8	200.0
Saudi Arabia	72.1	69.7	*[6]	31.1	*[6]	*[6]	*[6]	172.9[7]
India[9]	0.5	0	*[6]	*[6]	*[6]	*[6]	154.5	155.0
Singapore	1.6	88.8	14.0	5.5	*[6]	*[6]	34.1	144.0
Japan	19.3	23.6	29.0	*[6]	*[6]	*[6]	*[6]	71.9[7]
Other-Asia[10]	278.4	220.9	*[9]	*[9]	*[9]	*[9]	*[9]	861.0
Europe[10],[11]	182.0	40.7	*[9]	*[9]	*[9]	*[9]	*[9]	1165.0
Africa[10]	153.9	126.9	*[9]	*[9]	*[9]	*[9]	*[9]	721.0
Other-World[12]	69.8	32.3	429.1	163.3	34.5	37.3	*[9]	583.0
Total Export	2206.8	1046.0	1440.0	477.7	282.7	214.0		*[13]

Source: Palacpac (1978); USDA (1977C, J, 1978).

[1]Crop year except Thailand which is a calendar year; in 1000 MT.
[2]Mainland only
[3]Differences between the total import and the sum total of the countries of origin as listed.
[4]Total according to USDA (1978) except where indicated otherwise.
[5]Less than 500 MT.
[6]Assume to be 0.
[7]Sum total of the countries of origin as listed.
[8]Includes Sabah and Sarawak.
[9]Details are not available.
[10]Tallied from USDA (1978) except the United States (USDA 1977C) and Thailand (USDA 1977J).
[11]Includes the USSR.
[12]Differences between the total export and the sum total of the countries of destination as listed.
[13]Total world export of rice according to Palacpac (1978) was 7,367,000 MT; according to USDA (1978) it was 7,389,000 MT. But the total world import was 7,912,000 MT (marketing year) and 7,908,000 MT (calendar year), respectively.

1964 to peak at 340,000 MT in 1974, have helped North Korean rice find its way into markets as far as Iran and the Middle East (USDA 1978).

A trend in world rice trade that is both unique and interesting is found in Japan. As late as 1966, Japan still had to import nearly 500,000 MT of milled rice to satisfy its own needs. But importation of rice to Japan has been declining due to increasing production there which eventually led to a surplus of rice. Japan not only became a net exporter of rice in 1968, but also climbed up the export ladder to almost join the millionaire club in 1970 (USDA 1978).

Nearly all of the exports of rice from Japan during these last few years were made through concessional sales. Since farm prices of rice in Japan have been supported by the government at very high levels—often 2 to 3 times the world market prices—Japan thus cannot be expected to stay in the world rice export business for long. Japan's position in world rice trade, therefore, has been marginal (USDA 1976C).

During the last 10 years, no significant amounts of rice were imported to Japan. On the other hand, the average annual export of 468,000 MT of rice from Japan between 1968 and 1973 reflects only a short-term surplus situation and Japan's export of rice has been declining since its peak year in 1970 (USDA 1978).

But as the economic growth in Japan has slowed down in recent years, a shift in consumer behavior toward more demand for rice is expected. Indeed, the government of Japan is anticipating a return to yester-years' higher per capita rice consumption in the days ahead. How to divert the retired land back to rice cultivation without creating an undue surplus is truly a very interesting situation. The trend of world rice trade in Japan, therefore, is one that moves toward neither import nor export.

High Yielding Varieties (HYVs)

During the 15 years between 1960 and 1975, the world had increased its rice production by more than 50%. Many factors had contributed to this although opinions differ as to which was the most decisive factor (Efferson 1972). Among the many contributing factors was the increasing use of the high yielding varieties (HYVs) and the associated technology.

The HYVs of rice generally refer to the recently developed short stiff strawed, nitrogen responsive, photoperiod sensitive widely adapted and potentially high yielding rice varieties. This definition of HYVs, however, includes not only the semidwarfs but also the non-semidwarfs such as IR5, C4-63, Jagannath, Pankaj, Pelita I/1 and the RD series of Thailand (Dalrymple 1978).

Since the large-scale introduction of Taichung Native 1 (TN1) in India in 1965 and the release of IR8 rice in late 1966 by the International

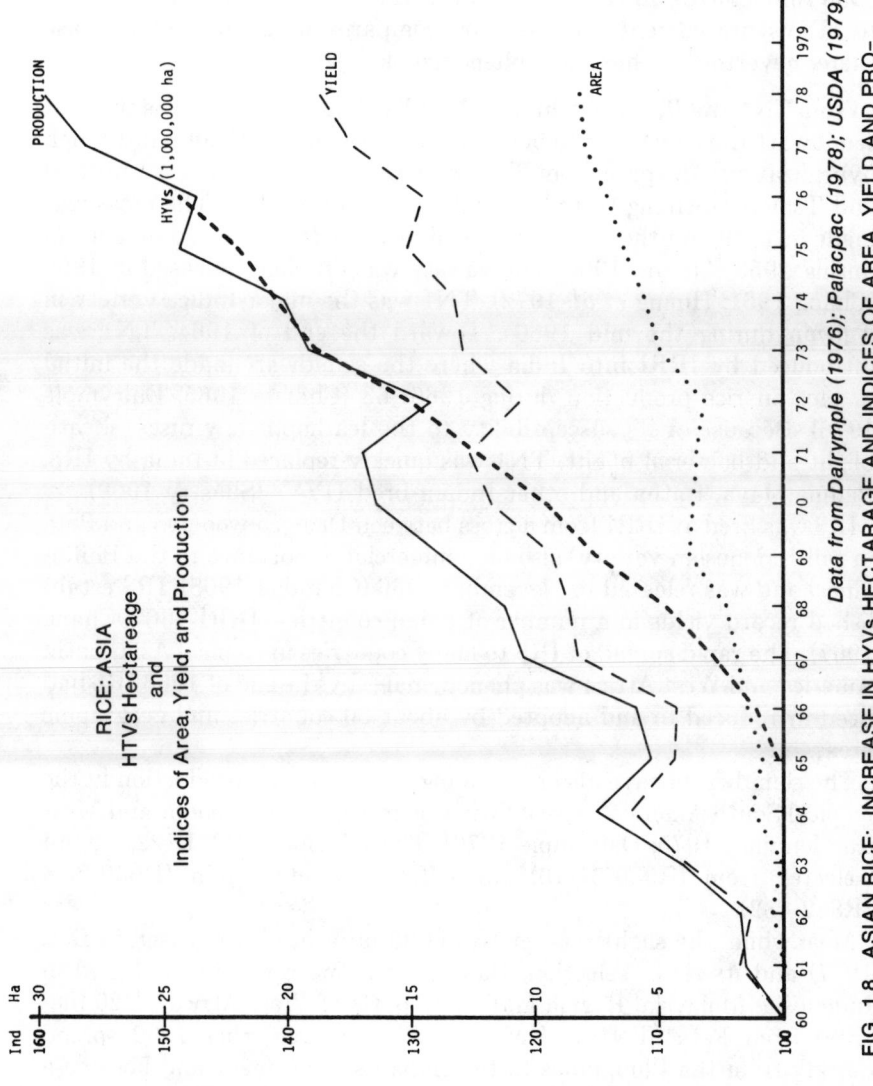

RICE: ASIA
HTVs Hectareage
and
Indices of Area, Yield, and Production

Data from Dalrymple (1976); Palacpac (1978); USDA (1979)

FIG. 1.8. ASIAN RICE: INCREASE IN HYVs HECTARAGE AND INDICES OF AREA, YIELD AND PRO-
DUCTION

Rice Research Institute (IRRI), areas planted with HYVs have been expanding rapidly. In Asia, where the first introduction of HYVs took place, areas planted with these new varieties expanded from 49,000 ha in 1965—66 to more than 24 million ha in 1976—77 as seen in Fig. 1.8 (Dalrymple 1978). In Africa and Latin America, although areas planted to HYVs are currently far less by comparison, the rate of increase there, nevertheless, has been phenomenal.

From TN1 and IR8 to Yushin.—TN1 (Taichung Native 1) was the first semidwarf Indica variety to better respond to fertilization bred through hybridization. The parents of TN1 were Dee-geo-woo-gen (a semidwarf) and Tsai-yuan-chung (a tall, drought resistant Indica). The cross was made in 1949 and the selection named in 1956. It was tested in Taiwan during 1953—55 and 1960. The variety was officially released in 1960 (Chang 1961; Huang et al. 1972). TN1 was the major Indica variety in Taiwan during the mid 1960s. Toward the end of 1964, TN1 was introduced by IRRI into India where the semidwarf made the initial impact on rice production during 1966—68 (Chalam 1965; Dalrymple 1976). Because of its susceptibility to the leafhoppers, viruses, sheath blight and bacterial blight, TN1 was quickly replaced in India by IR8, Padma, Jaya, Ratna and other Indian-bred HYVs (Shastry 1969).

IR8 was bred at IRRI from a cross between Dee-geo-woo-gen and Peta (a tall Indonesian variety, also of commercial importance in the Philippines) and was released in November 1966 (Chandler 1968). IR8 established record yields in a number of Asian countries (IRRI 1967; Chang 1967). The rapid spread of IR8 to many countries in tropical Asia, Latin America and West Africa was phenomenal—by the end of 1966 IR8 had been introduced to and adopted by about 60 countries and geographic areas.

The semidwarfs have also made a big impact on rice production in the tropical South America, especially in Colombia, Cuba, Ecuador and Mexico (Jennings 1975; Dalrymple 1976). These include IR8, IR22, CICA4 (selected from IR930-31-10) and CICA6 (selected from IR930-2 × IR822-432).

Meanwhile, the slightly taller IR5 (1.35 m in height, released in Dec. 1967) and its sister selection, Bahagia, became rapidly established in Indonesia, Malaysia, Burma and many parts of West Africa. IR20 (derived from Peta[3]/TN1//TKM6) released in December 1969 spread quickly from the Philippines to the Indian subcontinent and became a major variety in the Philippines, Bangladesh, the Tamil Nadu state of India and South Vietnam. Likewise, IR26 diffused quickly in the Philippines and in Indonesia during 1974—76, but it also was replaced by subsequent IRRI releases because of dynamic changes in the brown planthopper populations (Chang 1977).

Meanwhile, a large number of IRRI breeding lines were tested by researchers under different national programs. Those lines that were named as varieties at national and state levels numbered 60 at the end of 1976. Such releases were enumerated in the *IRRI Annual Reports* (1971–1977) and by Dalrymple (1976, 1978).

The rapid spread of the HYVs has been documented in detail by Dalrymple (1969, 1976). In 1976–77, the 25.27 million ha HYVs planted in Asia (South and East), Near East, Africa (excluding N. Africa) and Latin America, accounted for 26% of the total rice hectarage (Dalrymple 1976).

On mainland China, the first semidwarf mutant, Ai-chio-nan-te, appeared around 1960. It was soon followed by a score of semidwarfs of hybrid origin that quickly spread since 1959 (Kwangtung Prov. Acad. Agric. Sci. 1966). Other commercially important semidwarfs were Kwang-chang-ai, Cheng-chu-ai, Nung-ken 58, Chiang-nan-ai and Ai-nan-tsao 1 (Stavis 1974; Dalrymple 1976).

The single recessive gene for semidwarfism in Dee-geo-woo-gen was widely used by Asian rice breeders in developing other HYVs. By 1975, 39% of the improved varieties in India were derivatives of TN1 and 55% derived from IR8 (Hargrove 1976). Most of the semidwarfs on the China mainland also carried the same recessive gene, although the sources were different (Chang 1977).

The semidwarfing gene also found its way into the temperate zone as seen in the recent releases of several varieties in the area. Tong-il of South Korea was selected from IR667-98-1-2 (IR8//Yukara/TN1). It quickly became a major HYV there. More recent releases include Yushin (IR667-98-2-3/IR1317-392-1) and Milyang 23 (Suweon 232/IR24). In California, Calrose 76 derived its short stature from induced mutant D7 which has the same semidwarfing gene as IR8 (Rutger and Peterson 1976).

Up-to-date listings of HYVs may be found in the frequently revised work of Dalrymple (latest edition, Sept. 1978) and in issues of the *IRRI Annual Report.*

In recent years, some of the HYVs related to Dee-geo-woo-gen have slightly taller plant stature than the typical semidwarfs (c 100 cm). Examples are IR34, RD9 and Masria.

Extension into New Areas.—Most of the HYVs of rice are photoperiod insensitive and thus are early maturing (105–135 days). Shorter growing period facilitates intensive multiple cropping in the lower latitudes or, in the case of the temperate zone, permits possible extension of HYVs into new frontiers in the higher latitudes where the growing season may be as short as 4 months. This, of course, depends on the specific variety, location and crop season (Chang and Vergara 1971).

In a physiological sense, the HYVs do not necessarily require more water than the traditional local varieties. In fact, the total amount of water needed may be less because of the short growing period. But to offer their full potential, the HYVs require an assured water supply. For this reason, they naturally have their greatest impacts in irrigation areas. In fact, most of the extensions of the HYVs so far have been in areas with newly constructed irrigation projects. The more notable examples are the Yamassukro area of the Ivory Coast, Nigeria, the Senegal River banks of Senegal and Mauritania, the lake areas of East Africa, and various parts of Brazil, Mexico and Malaysia. The dry season rice crop of Thailand is a unique example in which the high yielding potential of the semidwarf RD1, irrigation water and high solar radiation in the dry season combine their impacts to produce high yields approximating 5 MT/ha.

In Punjab of India the HYVs of rice quickly replaced other dryland cereals grown in the monsoon season. Here hectarage devoted to HYVs has increased from 1% of the total cropped area in 1966 to more than 90% in 1975 (Saini 1976).

Impact on Cropping Intensity.—Because of their early maturing nature and photoperiod insensitivity, the HYVs have definitely contributed more to rice production in many areas through the intensification of double and multiple cropping than the sheer fact of their enormous yield potential.

The classic example may be gleaned from the impact of the large-scale introduction of early maturing and drought resistant Champa varieties into the lower Yangtze area of China during the 11th century which greatly increased China's rice producing capacity and eventually, led to a population boom (Ho 1956). A more recent example is the effect of "ponlai" varieties on the multiple cropping system in Taiwan (Iso 1954; Huang *et al.* 1972).

The impact of HYVs on cropping intensity may be viewed in 3 similar practices with different degrees of intensity. First, there is a *simple multiple cropping* system under continuous monoculture of rice—2 or more crops of rice per year. This system is likely to occur in areas where the demand for rice is most acute, where water is available year-round and the temperatures warm enough throughout the year. If carefully managed, this system is likely to give the greatest production of rice from a given unit of land in a given crop year.

A subtype of this simple multiple cropping system is the *ratooning*, a system in which the stubbles, after the first harvest is completed, are left in the field with additional fertilizer, water and other inputs to encourage the new shoots to grow back. There is considerable savings in time and costs per unit of land under this system (Grist 1965, 1975).

The system, practiced in central China and Taiwan in former years for intensive use of land, is now gaining wider acceptance in the southern United States since the early 1960s to gain additional income (Anon. 1963). Ratooning, however, has its drawbacks. First, although the production costs per unit of land of a ratooned crop are lower, its yields are lower. Thus, the costs per unit weight of a ratooned crop is higher than that of the sown crop. Furthermore, the incidence of insect pest is also higher than that of the sown crop (Grist 1975). Rice cultivar for ratooning crops, therefore, should have high levels of resistance to diseases and insects as well as high yield potentials (Bahar and De Datta 1977).

There is also a *relay multiple cropping* system of combined multiple culture of rice with other crop(s). As a greater yield of the HYVs is assured and surplus of rice is a likely result, 1 (or more) season of rice under the simple multiple cropping system may be diverted to other crop(s), either to diversify the agricultural economy or for the purpose of crop rotation. In practice, the relay multiple cropping differs only slightly from the simple multiple cropping in that a different crop, rather than another crop of rice, is planted *after* the previous rice crop has been harvested.[7] The relay type of multiple cropping rice with other crop(s) has been practiced in south China and Taiwan.

A third system, also reaping the benefit of early maturity of the HYVs, is the *intensive* or *maximum multiple cropping* system of interculture. The system of planting a different crop on the same plot of land *before* a crop is harvested has been practiced by farmers in southern China for millennia. However, in some areas of tropical Asia, 2 crops of rice of different varietal types grown together for a short period of time in the same field has also been practiced. Recent tests under rigid scientific management at the IRRI have proven that the interculture system of maximum multiple cropping is economically and ecologically sound (Bradfield 1972). Hopefully, this research will pave the way for a wider application and adoption of this system.

RICE IN PROSPECTS

Assessment of the Current Situation

The world in the 1970s has had a rather healthy rice economy for 3

---

[7]Some, however, are of the opinion that it is *before* a previous crop has been harvested (Beets 1977); but two crops growing together in the same field is known as *interculture* or *intercropping*.

reasons: (1) There is in the world today a deeper awareness of its own strengths as well as weaknesses in its battle to grow enough rice for the world's inhabitants. (2) There have been wider commitments by world nations (rice deficit and surplus countries alike) to produce more rice in conjunction with their effort to combat hunger. (3) There has been greater international cooperation for rice improvements (Umali 1972). By pooling together the available financial, technological and human resources, the world is now able to deal more effectively with the world's rice problem.

This does not mean, however, that the world's rice economy is immune to any setbacks. In fact, despite an increase of 20.9% in rice production between 1970 and 1978 (Table 1.6), the world has already experienced 2 setbacks. The first setback, occurring in 1972, was due primarily to bad weather conditions hitting various parts of the world and the reduction of the year's rice area. The result was a 3% reduction in world rice output. But the rice economy of the world quickly recovered. Indeed, it was because the economy was healthy that the world immediately witnessed a 7.4% rebound to 331 MMT of rough rice produced in 1973.

This was followed by 2 more consecutive record harvests of 337 MMT in 1974 and 361 MMT in 1975. Fears of low rice prices and concerns of high costs of carryover then caused world rice areas to reduce from its 1975 record of 143 million ha to 141.7 million ha in 1976. The desire to avoid the consequences of unusually large stocks although was met happily also by unseasonal weather in some rice producing areas of Asia in 1976, the year's output was reduced only by 3.2%, from its record high of 361 MMT in 1975, to 349 MMT in 1976 (USDA 1977H).

Weather conditions in monsoon Asia were generally fair in 1977, although several areas in Southeast Asia suffered from drought earlier in the main season. Thus, world rice production in 1977 climbed up again from its 1976 setback, to reach a new record of 366 MMT. Coupled with area expansion and a higher yield in 1978, world production of rice then reached a level it had never had before at 377 MMT. But, characteristically, the monsoon is neither dependable nor reliable. And the areal extent of the monsoonal influence in Asia is so vast that in a given year, chances for any part of what Huke (1976) called the rice "core area" and, especially the "transitional belt" to be hit by floods or droughts or both remain very great.

So long as the monsoon remains unpredictable, a bounteous crop of rice can never be assured unless measures are taken to ensure an adequate water supply. Indeed, more irrigation facilities have been extended in recent years to cover a larger rice hectarage even within the rice core area of Southeast Asia. The irrigated rice fields in Thailand, for instance, have increased 42% from 1,287,000 ha in 1958 to 1,829,000 ha in 1974. Burma, on the other hand, nearly doubled its irrigated rice area in 15 years: from 472,700 ha in 1961 it fluctuated upward to 850,000 ha in 1976. Bangladesh has extended its irrigated rice area the most: from 460,000 ha in 1965 to 1,162,000 ha in 1974 (Palacpac 1978), more than a 1.5 time increase in a decade.

Technology and Constraints

Technology.—The choice of alternatives available to farmers for growing a crop of rice under the traditional cultural practice is rather limited. Modern technology, however, has made available a number of alternatives to farmers for growing a successful rice crop. These include the development of water control schemes; the application of chemical fertilizer, insecticides and pesticides; the introduction of new rice varieties; as well as the improvement of cultural practices. Indeed, in recent years technology and time have modified and enhanced the resource bases of rice production.

Water Controls.—Rice is not an aquatic plant, yet it grows better under a flooded condition. One reason for this is that root porosities, which affect plant growth, are highest under completely flooded conditions (Das and Jat 1977). Furthermore, flooding in rice fields checks weed growth. Water controls (flooding and draining) in paddy culture, therefore, are necessary and irrigation, even under a humid environment, is deemed to be desirable. It has been well documented (Yamada 1975) that rice grown under irrigation consistently gives higher yields than those grown under nonirrigation conditions.

However, irrigation in rice growing is not simply a matter of securing sufficient water; it should also be viewed from the standpoint of efficiency and optimality (Lu 1971). It has been reported that too much water can be detrimental to a rice crop as well as an insufficient amount (Smith 1966).

Furthermore, the traditional method of irrigation has been by way of open canals which may incur losses through seepage and/or evaporation. Recent developments to conserve water in rice growing include the use of underground concrete or lucite irrigation pipelines, as well as hoses above ground. This is especially so in regions where precipitation and/or water supply is not overly abundant. The use of pipelines or hoses in irrigation, in addition to preventing water losses, also conserve space. It has been estimated that for each 304.8 m (1000 ft) of open canal replaced with a pipeline or hose, 1.13 ha (2.8 acres) of land can be returned to rice production (Johnston and Miller 1973).

Chemical Applications.—Modern technology also made chemicals available to help rice farmers grow a better crop. Chemicals are used in 2 specific ways, namely, to feed the rice plant and to control rice pests, diseases and weeds. The importance of feeding the rice plant may be exemplified by a study conducted by Mikkelsen who presented a cost-benefit analysis before a group of California rice growers (Lu 1971).

According to Mikkelsen, when no fertilizer was used, the yield of rice was 28.81 Q/ha. But the yields increased to 41.81, 56.16, 64.79 and 71.07 Q/ha when fertilizer was applied respectively in the amount of 34, 67, 100 and 135 kg per ha. Thus, the profit per hectare increased from a deficit of $61.78 when no fertilizer was used, to some 54.36, 190.27, 266.87 and 321.23 dollars per ha (as quoted in Lu 1971 with original figures in pounds and acres). However, it is very difficult to give generalizations on the optimum amount of fertilizer to be applied for rice is grown today under various climatic conditions, in diverse edaphic environments and with different cultural practices. Furthermore, different rice varieties have inherent characteristics that respond to varying levels of fertilization (Grist 1969).

Moreover, the chemistry of the flooded soil works in such a way that, for example, when ammonium (NH_4) fertilizer is applied to the surface of the flooded rice field, it will quickly be converted through the process of denitrification into nitrate (NO_3), nitrite (NO_2), and then into gaseous nitrogen (N_2) which escapes readily from the soil. It has been estimated that as much as 80% of the nitrogen fertilizer applied directly to the surface of the paddy soil is lost in this way. To correct this situation, Pearsall (1969) has suggested that the ammonium fertilizer be placed in the lower (reduction) layer of the soil before it is flooded (Wrigley 1969).

Not only the placement, but also the timing of fertilizer application is of great importance (Evatt 1966). Recently, a technique of employing the DD-50 concept for timing the midseason fertilizer application toward maximum yield (instead of measuring the internodal elongation) has been tested. It was found to be useful (Downey and Wells 1975; Downey *et al.* 1976) and may eventually gain a wider application.

The feeding of the rice plant, therefore, requires knowledge of the nature of the soils, the varietal characteristics of the rice planted, methods of weed control, and water and other managerial information affecting the use of fertilizer, as well as the nature of the fertilizer itself. All of these lead to the conclusion that while modern technology has made available chemicals to aid in rice production, technical information and scientific procedure in fertilizer in addition to cultivation skill, are very important to successful rice growing.

Plant Breeding and Genetic Improvement.—Another technological contribution to modern rice production is found in the area of plant breeding through genetic manipulation (Brown 1970; Jennings 1975). Plant breeding produces improved genotypes that can efficiently produce high yields or good qualities or both under a given set of environments, especially favorable conditions. The improved varieties, such as the IR rices resulting from 2 basic varietal changes, have so dramatically increased the yield potential that they have won for themselves the fame of the *green revolution*. These changes include a shortening of the plant height to reduce lodging and to increase the grain/straw growth ratio, and a broadening of plant adaptability over latitude and crop seasons because of their low degree of photoperiod sensitivity.

Thus, the purpose of rice breeding is to improve seeds with specific varietal characteristics, making it available to farmers and adaptable to their particular environmental conditions. However, most of the breeding efforts devoted so far, have been to the HYVs intended for the irrigated areas. In fact, enhancing yield potential was the most frequent objective given by plant breeders, according to an IRRI report (1976A). But the productive potential of the HYVs of rice could only be realized on about 25 to 30% of the world's rice land where the hydrological environment is relatively favorable (Brady and Athwal 1975; IRRI 1976A).

Rice lands, whose hydrological environments are less than favorable, remain very large. It has been estimated that there are 15.3 million ha of the world's rice land that are flat or sloping without bund, prepared and seeded under dry conditions, and that are entirely dependent on rainfall for the required moisture. This is what is known commonly as the "upland rice" (though somewhat misnomered), the "providence rice" in Louisiana (Lee 1960), dryland rice, or the "K'an-t'ien-t'ien" in China (Lu 1971). Furthermore, in the lowlands of tropical Asia, there are between 50 and 56% of the paddy fields that are entirely rainfed without irrigation. Such fields are subject to a very high frequency of drought or flood, or suffer one stress after another.

Moreover, about 10% of the rice land in tropical Asia (and some more in West Africa) have water levels exceeding 100 cm for 3 to 4 months during the monsoon seasons, in which deep water and floating rices can better adapt although with low yields (IRRI 1975B; Vergara 1976). In addition, there are large tracts of rice land in South and Southeast Asia where, due to relatively high elevation, only rices tolerant to cool temperatures produce acceptable yields.

Due to inherent limitations, yields from these rices usually are very low. The yield of upland rice, for instance, is about 0.5 MT/ha in Africa, from 0.5 to 1.5 MT/ha in Asia and from 1.0 to 4.0 MT/ha in Latin America. The constraints imposed by nature, however, are not final. Substantially higher yields of upland rice are possible through further development of improved varieties and of cultural practices (De Datta 1975).

Most of the world's land that is suitable for wetland rice culture has already been developed. Future increase in world rice production will depend on the expansion of irrigation facilities for the rainfed wetland areas and the improvement of rices for all rainfed areas.

Constraints.—But the application of these technologies in rice production is not entirely without limitations or consequences. In irrigation, for instance, the traditional methods usually involved only gravity flow in what may be called the "field-to-field overflow" system. This system allows the water to carry away at least part of the fertilizer applied, and thus makes the application of fertilizer less effective. In modern irrigation, this problem could be corrected through delivery of water directly to the field where it is needed. However, this system requires energy input and so availability and cost become limiting factors, especially for smaller farms. The application of chemicals in rice growing, for another example, increases the chance of polluting the soil and the environment, whether it be for nourishing the plant or for controlling the pests.

The adoption of the HYVs of rice also has its limitations and consequences. The wider adoption of the HYVs increases the use of fertilizer and the demand for water and its controls. Increased application of fertilizer, it has been found, induces serious crop damage by insect pests and diseases. Flood control, on the other hand, tends to interrupt the natural balance of supply and removal of soil nutrients.

The short season HYVs enables the wider practices of double or multiple cropping which may lead to treble or quadruple per hectare rice yields. But the HYVs, in order to attain such a high yield, require "from four to seven times as much water" as the traditional varieties (Miller 1975). The availability of the needed water then becomes a limiting factor to its wider adoption. Moreover, increases in double or multiple cropping are apt to disrupt the existing soil nutrient equilibrium. Furthermore, in the process of intense selection for HYVs, the plant tends to reduce its defense ability in order to balance its internal resource and thus becomes highly susceptible to insects and diseases for internal reasons as well as for genetic uniformity (Janzen 1975). It is true that some new varieties were bred for resistance to certain pests and diseases, but their vulnerability to outbreak of new pests and diseases remains relatively high (NAS 1972).

There is also the constraint of not fully realizing the genetic potential of the improved seeds. Two yield gaps were identified: (1) between the experiment station yield and the potential farm yield, and (2) between the potential and the actual farm yields (IRRI 1977; Herdt and Barker

1977). Various constraining factors have been identified and elucidated by the IRRI and other researchers. These have been reported in the *IRRI Annual Reports* (IRRI 1975A and 1977) and are discussed by Chang (1977). One may categorize these factors under biophysical and socioeconomic constraints. The biophysical constraints include vagaries of weathers, inequalities of water supply and control, differences in soil fertilities and problems, varietal adaptability, problems of insects, weeds and diseases, as well as differences in cultural practices, etc. The socioeconomic constraints include attitudes and beliefs, lack of information and knowledge, tradition and institutions, credit and economic behavior, price and markets, as well as limited available inputs and unwillingness to take risks, etc.

As long as these constraints exist, the yield gap will remain. Fortunately, these and other constraints have gradually been brought to light and are being identified. Researchers are now paying more attention to tackling these problems. These efforts will eventually narrow the yield gaps, and the genetic potential and expectations of these improved seeds will gradually be realized.

Taking into consideration these constraints, problems and consequences, genetic improvements of rice certainly open up a field to which modern technology can increase its contribution. Indeed, the IRRI and other national rice research programs have already devoted their efforts not only to improve the yields of deep water, floating (Vergara 1976) and upland rices (Chang *et al.* 1975), but also to enhance rice tolerances to low temperature (Vergara and Coffman 1976), extreme temperature (Chang *et al.* 1977) and various soil conditions. Examples for the latter include tolerance to bog soil (Sri Lanka Dep. Agric. 1977), soils of various mineral toxicities and deficiences (Ponnamperuma and Ikehashi 1977).

Future Supply, Demand and Trade

In the foreseeable future, the spatial pattern of world rice supply and demand is likely to remain about the same as it has always been—with the great bulk of rice being produced and consumed on a large number of subsistence farms in Asia. However, with continuing efforts, greater committments and wider cooperation among world nations, the stage of self-sufficiency in many of the rice deficit countries is enhancing. Thus in the long-run, the volume of rice trade on the international market may be reduced. Short-term fluctuations in world rice trade, due to unseasonal weather conditions, nonetheless, will remain as a dominating force.

Selected Importing Countries.—During the years between 1970 and 1974, Indonesia topped the world with an annual average import of

1,308,400 MT of milled rice, more than 18% of the average world export of 7.1 MMT. It was followed by South Korea (564,000 MT), Singapore and Malaysia (449,400 MT), Hong Kong (363,200 MT), Sri Lanka (359, 000 MT and the Philippines (239,000 MT). Other importing countries with a relatively large quantity include Cuba, Malagasy, India, Saudi Arabia and other Middle East countries. Among these, Korea and likely the Philippines could become self-sufficient in the near future; the rest are likely to remain importers of rice for some time.

In South Korea, the recently engineered and released (in 1972) Tong-il variety is believed to have "triggered a Korean rice revolution" (IRRI 1976B). The Tong-il variety gives a 30% increase in yield over the traditional Korean rice varieties. Furthermore, the Korean government has designed a tremendous campaign for its expansion. Together with the release of a yet higher yield variety, Yushin, South Korean's dependence on imports of rice has been greatly lessened. In fact, South Korea started to export some rice in 1977 (USDA 1978).

The Philippines, one of the 2 original centers of the green revolution and the home of the IR rice varieties, once achieved self-sufficiency from 1968 to 1970. With continuing expansion of the HYVs hectarage and the aggressive "Masagana 99" program sponsored by the government, rice production in the Philippines has recovered from its setbacks in 1971 and 1972. In 1977, the Philippines had produced sufficient rice also for export—totaling 29,000 MT (USDA 1978), with about 8000 MT exported to Indonesia.

The situation in Indonesia is somewhat different. The HYVs hectarage is expanding, the double cropped rice areas have been extending, more marshlands have been drained for rice cultivation, and indeed rice production in Indonesia has steadily been increasing in recent years which culminated in a drastic curtailment of milled rice imports in 1975 by nearly ½ its record import in 1972. The situation, however, was rather short-lived. For reasons not clearly known, Indonesia's imports of rice bounced back to 1.5 MMT in 1976 and 2.5 MMT in 1977 (USDA 1978).

As a "new rich" among the petroleum exporting countries, perhaps Indonesia may still find importing rice easier and cheaper than becoming self-sufficient. In the foreseeable future, Indonesia is likely to remain a relatively important rice importer.

Among the Exporting Countries.—The amount of rice trading on the international market each year, indeed, is very small by comparison with wheat or with the total quantity of rice produced. Nevertheless, analysis of major rice exporters reveal some interesting facts.

The world in the latter 1970s has had 4 major rice exporters—the United States, China, Thailand and Pakistan. Between fiscal years 1974 and 1976, these 4 countries exported an annual average of 5.3 MMT of milled rice which accounted for 68% of the world average total of 7.8 MMT. Pakistan is a late comer to these ranks and still is a relatively small exporter with an annual average of only 0.6 MMT for the same time period. Burma, Indochina and Egypt have also been rice exporters of some importance in past years.

*The United States.—*The United States is a small producer of world rice. Between the crop years of 1969 and 1978, the United States harvested annually an average crop of 4.7 MMT of rice (rough) from an annual average of 925,300 ha of land. This is 1.41% and 0.67%, respectively, of the world total (derived from USDA 1978B). But small domestic consumption helped to make the United States a very important rice supplier (Lu 1977). Between the marketing years of 1974 and 1978, the United States exported an annual average of more than 2 MMT of milled rice, which contributed about ¼ of the annual average total of 8.7 MMT traded on the world market during the same 5 year period (USDA 1978B).

The importance of the United States as a supplier of world rice may further be seen in the following aspects (Lu 1971). First, the United States is the world's *leading* supplier of rice. Since its assumption as the world's leading rice exporter in 1967, the export of United States rice has increased from 1.89 MMT in 1967 to 2.2 MMT in 1977 (preliminary). During these 12 marketing years, the United States was exceeded in its rice export 3 times by Thailand (1971, 1975 and 1976) and twice (1972, 1973) by China (USDA 1978). But in all these 5 years, it was not because of the inability of the United States to meet the world demand. Rather, it was because of the anticipated large world stocks and the declining rice prices that area devoted to rice in the United States, under a government price support program, was cut by 10% in 1969 and by another 15% in 1970. In spite of these cuts and their resulting production, the United States export of rice still leads the world's annual average (of these 11 years from 1967 to 1977) by a very large margin: 1.87 MMT as compared with Thailand's 1.45 MMT and China's 1.16 MMT (USDA 1978).

Second, the United States is also the world's *most reliable* supplier of rice. In most rice exporting countries, rice is a staple diet and their rice exports are marginal activities; rice is available for export only after their domestic demand has been taken care of. In the United States, on the other hand, rice is not a staple food and exports of rice are a commercial enterprise. In fact, the exportation of rice, after it exceeded domestic sales in 1961, has been the main activity of the United States rice industry.

In most years, while other countries may have an inadequate rice supply, owing to bad weather, political unrest or mismanagement, etc., the United States has always had more than enough to spare. The erratic and inadequate supply of rice from other countries during the last 25 years has created an opportunity for the United States rice industry to respond dynamically with its available resources for an increasingly wider world market.

The key strength of the United States as a reliable supplier of rice lies in the efficiency of its rice economy. The production of rice as a crop at the farm level, the process of rice as a product at the mill level, and the export of rice as a commodity at various marketing levels, all are efficiently geared together not only to make the United States one of the world's largest and the most stable and reliable supplier of rice, but also the provider of the world's most consistently high quality rice.

Third, the United States is a *residual* supplier of world rice, meaning that the United States does not have a traditional market for its rice. Efforts must be made in order to open up a new market. Furthermore, the United States is bound by a domestic price support and acreage allotment farm program and controls. Thus, in practice, the United States is not free to increase its rice production and, in this sense, is not free to supply the world with a product which it can produce more cheaply than many other world rice producers.

When world stocks of rice are mounting and before the world prices for rice are to drastically decline, the program would enforce a reduction in United States rice acreage, leading to a reduction in rice output. This mechanism, in essence, makes the United States a residual supplier of rice as seen in the 25% reduction of the United States rice acreage in 1969 and 1970, and its subsequent curtailment of rice production from 1970 to 1972.

On the other hand, when a world shortage of rice is in sight, the program would allow rice farmers of the United States to go for full production. With land resources (Godwin and Jones 1970), capital, equipment and technology readily available, it takes only a crop season for the United States to quickly replenish its rice supply as seen in the cases of 1974 and 1975. The role played by the United States as a residual supplier helps to prevent deterioration of both world price and supply of rice.

The rice industry of the United States is dynamic and enterprising. The current trend is moving toward more efficient use of labor and other resources. The rice economy of the United States is sound. It can respond quickly toward any foreseeable demand. The main problem facing the United States rice industry is not in supply. It has sufficient resources, natural and otherwise, and managerial skill to produce an efficient and competitive rice crop with high quality. The only problem which the

United States cannot deal with is the lack of demand. As long as there is a demand, there will always be abundant United States rice to supply such a demand, at least in the foreseeable future (Lu 1971).

China.—China, the unchallengeable champion of world rice producers, became a major world rice exporter relatively late. In fact, before World War II, China was a net importer of rice with considerable quantity. It was only since 1951 that China, for the first time in this century, became a net exporter of rice (USDA 1972A) and the quantity of its export rose quickly to join the millionaire club in 1958. But before 1970, China's rice exports fluctuated from a low of slightly over 0.5 MMT to 1.6 MMT, reflecting a period of consolidating the unsecured foundation as a world exporter of rice.

In the early 1970s, China appeared to have wanted to become a leader in world rice exports. In fact, the 2.14 MMT of milled rice exported by China in 1972 surpassed the United States by 409,000 MT, and was over 2.5 times of what Thailand exported in the same year. Although China's rice export has since declined, its export of 1.98 MMT in 1973, however, again led the world, and its export of 1.44 MMT in 1974 was 0.5 MMT higher than Thailand (as compared with 2.2 MMT exported by the U.S.). The Chinese export of rice reached a low ebb of 0.7 MMT in 1976—only ⅓ its peak in 1972. China's export of rice, however, again reached the 1 million mark in 1977 (USDA 1978).

The leading position of China as a rice exporter may not be as secure as the above statistics seem to indicate. It depends to a very large degree on the trend of population growth and the policies of the government. It has been suggested that in the immediate past, China had practiced a policy of selling the higher priced rice in exchange for the lower priced wheat and/or other foodstuffs—a wise and handsome trade-off. Indeed, there is a positive correlation between China's exports of rice and its imports of wheat although the causal relation is not quite clear. Furthermore, in 1975 China had, for the second time since the 1960s, imported some 140,000 MT of milled rice despite its exports of nearly 1 MMT in the same year (USDA 1978).

Furthermore, it also seems to have been a practice of China to arrange shipment of rice from abroad to a third country. In 1976, for example, about ½ of the 200,000 MT of Thai rice traded for China's crude oil was shipped from Thailand directly to Sri Lanka (Schatz 1977). If the above practices are consistently in line with China's trade policy, then China belongs to a different class among the world's major rice exporters who normally export the surplus from their own stocks.

What is the prospect of China as a major rice exporter? It depends on several things: (1) China's current production and, to a lesser extent, its previous stocks; (2) the price ratio between rice and wheat on the international markets; (3) China's political ties with other Asian and the Third and Fourth World countries; and (4) Barter trade agreements on other commodities between China and the countries concerned.

With China's population as large as it is today, an increase of 1 kg per capita rice consumption will quickly deplete nearly 1 MMT of exportable rice in no time. On the other hand, a slight tightening of the belt with a "volunteer" reduction of 0.5 kg which is but a small fraction of its 96.73 kg per capita consumption, will result in nearly 0.5 MMT exportable rice. Given the strict government control over food distribution and rationing in China, an "eat 1 kg less rice a year" campaign would raise no protest at all. Thus, the planned amount for export and the policy and implementation of such a plan is another factor determining the quantity of rice export of China.

Thailand.—The rise of Thailand as a rice exporter was fostered almost entirely by the trade agreement with Great Britain in the heyday of the colonial expansion in Southeast Asia (Brown 1963). As the only country in the rice bowl region which still remains an important rice exporter, Thailand climbed its way to top the world with a record of 1.9 MMT exported in 1964.

This deltaic country of Chao Phraya is part of what is known as the rice bowl of Southeast Asia because of its ability to supply other rice deficit countries of Asia with its surpluses. But the surplus had not been the result of the unusual fertility of the land or productivity of the farmers. In fact, Thailand's average rice yield between 1972 and 1978 was only 17.8 Q/ha as compared with 26.8 in Indonesia, 35.0 in China (Mainland), not to mention 50.5 in the United States and 58.1 in Japan (derived from USDA 1979). Rather, it was due primarily to the region's relative underpopulation.

The country's population, however, is rapidly increasing. Thailand which enumerated 26 million inhabitants in 1960, had an estimated midyear population of 45.1 million in 1978 (Kane and Myers 1978) and is currently growing at a rate of about 1 million a year. Assuming the 1975 level of per capita consumption which was at 212.5 kg (Table 1.11), up from 166.9 in 1969 (Work 1969), Thailand needs 9.58 MMT of milled rice just to feed its people in 1978.

In the 1975—76 crop year, Thailand produced 15.2 MMT of rough rice. In 1976—77, Thailand planted 8.5 million ha of rice (USDA 1977A) which, at an estimated seeding rate of 100 kg/ha, requires 0.85 MMT of rough rice for seed. This gives Thailand 14.4 MMT of rough rice convertible, at a rate of 0.66 (Palacpac 1977), to 9.5 MMT of milled rice. Subtracting 8.78 MMT for the year's domestic consumption, there were only 0.72 MMT of milled rice from the 1975—76 production available for export which, when previous stocks were added, was very close to the actual export of 0.93 MMT for the marketing year of 1975 (USDA 1978). But for 1978, with a domestic consumption estimated at 9.58 MMT and farm production at 15.5 MMT (USDA 1979), Thailand would have a hard time finding enough exportable rice.

It is therefore no surprise to see that Thailand's export of rice has been declining since its 1964 climax. In fact, Thailand's immediate concerns and efforts have not been so much on regaining its position as the world's leading rice exporter as on reaching an export goal of 1.5 MMT annually and to maintain it (Work 1969). This goal was once achieved in 1971–72 when exports rose to 1.5 MMT and exceeded by more than 0.5 MMT in 1972–73 (Palacpac 1978).

But the official, controlled rice export figures are likely to be understated because of the illegal exports which are not included in the government statistics. Due to the relatively low government export procurement price, considerable amounts of Thai rice have been smuggled into Laos, Cambodia and Malaysia along its borders. Fortunately, the RDI HYV grown in the dry season has at times helped to counterbalance the domestic short-supply drained by these uncontrolled exports and thus helped to stabilize domestic prices of rice.

In order to fully attain its goal, Thailand must: (1) increase its per hectare rice yield, (2) take proper price-incentive measures to increase production, (3) practice multiple cropping wherever possible, (4) expand its rice area to come close to the nation's potential which Thailand has been doing since 1971 (see Palacpac 1977), and (5) reduce its per capita rice consumption through substitution of other cereals.

However, there are internal as a well as external handicaps. Internal factors include the limited irrigation availability and flood control facilities, a high degree of tenancy, and the traditional cultural values and attitudes of Thai farmers (Lu 1971). Externally, competition and low prices are most serious. In addition to competition from Pakistan and the United States, the recent end of arms conflict in Indochina no doubt will permit the reemergence of Vietnam (South) and Cambodia as rice exporters and thus further stiffen the competition (Wibulseth 1976). The short run effect of stiff competition is low export market prices which may result in increasing domestic stocks, depressing farm prices and/or smuggling activities.

Under normal circumstances, low market and especially farm prices have a dampening effect on production. But the Thai government is determined to reach its export goal and, therefore, is willing to pay its farmers a guaranteed price (Wibulseth 1976). Nevertheless, Thailand's rice export potential will be determined by the intricate relationship between its capability to increase production, its adaptability to reduce per capita rice consumption, its ability to compete successfully in the world rice market and most of all, its determination to slow down the population growth.

Twelve years ago, in looking into the future developments of the rice industry of Thailand, J. Norman Efferson felt that Thailand would not be able to reach its export goal of 1.5 MMT. In all likelihood, he said, it would be in the vicinity of 1 MMT annually (Efferson 1967).[8] Indeed, Thailand's annual average export of rice between 1968 and 1977 was 1.49 MMT (USDA 1978).

Preliminary reports indicate that Thailand has greatly boosted its rice export to 2.9 MMT in the 1977 marketing year but slipped back again to 1.5 MMT in 1978 (USDA 1978). Thus, positive measures, such as those mentioned above, must be taken seriously. Otherwise, Thailand may not even be able to reach its reduced export target of 1 MMT by the 1990s (Wibulseth 1976).

Pakistan.—Pakistan only relatively recently became the world's fourth rice exporter. This is due partly to West Pakistan's being a small rice producer before 1968, and partly to the large "domestic" demand in its former "east wing." However, having taken advantage of the HYVs since their large-scale planting in 1968, Pakistan has boosted its rice production. Furthermore, since the secession of East Pakistan (Bangladesh) in 1971, Pakistan has enhanced its rice export ability: the average annual export since 1972 has been at the 0.67 MMT level.

At present, the market for Pakistani rice is primarily in the Middle East countries. This is due to : (1) the increased demand for rice in these countries in recent years, (2) the close resemblance of Pakistan's Basmati rices to the preferred locally grown rices in Iran and other nearby countries, (3) the geographic propinquity to these countries which Pakistan enjoys, and (4) Pakistan's determination to catch these markets with competitive prices. Whether a wider market will be available to Pakistan or not depends not only on its capacity to produce, but also on its ability to improve the quality of its coarse rice for export.

Burma.—Rice became a commercial enterprise in Burma more than a century ago due to increased external demand, especially from western Europe (Cheng 1968). The export of Burmese rice reached the million metric ton mark in 1882 and peaked with 3.1 MMT (rough) just before World War II; but its postwar exports fluctuated around 1.5 to 1.7 MMT (milled). While rice production in Burma has been maintained at an annual level of 8.3 MMT (rough) between 1964 and 1978, rice exports

---

[8]But in the light of Thailand's latest achievement and his recent visit in 1977, Mr. Efferson may have a different opinion on the matter.

fluctuated downward from a high of 1.8 MMT in 1961 to reach the lowest ebb of 199,000 MT in 1973–1974 marketing year (USDA 1978).

Several factors have contributed to this decline (Lu 1971). The drastic change in political atmosphere has manifested in (1) radical transformation of the market structure and the marketing system as the result of nationalization, and (2) the low agricultural prices determined by the new government's policies. As a consequence, farmers have been keeping more rice on the farms resulting in (1) a general increase in per capita rural rice consumption, and (2) more black market activities when the normal marketing system became worsened. Coupled with a relatively high rate of population increase and slow adoption of HYVs,[9] Burma then found its rice exporting capacity greatly reduced since 1966.

Good weather generally prevailed in Burma during the last two years. It helped Burma to regain partially, its export strength. But future success depends largely on Burma's ability and willingness to revitalize an industry "now stifled by excessive government regulations, inadequate infrastructure and lack of production incentive" (Lederer 1977). Unless the sociopolitical and socioeconomic environment in Burma can be improved successfully within a reasonable time, it seems unlikely that Burma will be able to compete favorably in the world rice market in the foreseeable future.

Egypt.—The rice export of Egypt, although very small in world proportion, is very important to its own national economy. Rice ranks second in Egypt's agricultural exports only to raw cotton. Because the market value of rice usually is relatively high, it has been a policy of Egypt, like that of China, to sell its rice to finance the purchase of wheat and other foodstuffs.

However, the price for Egyptian rice has been closely related to the fluctuating supply and demand situation of rice in Japan because of the similar type of rice (Japonica or Sinica) produced in both countries (Ishida 1975). In recent years, the supply and demand situation of rice in Japan has become stabilized. In the long run, it would seem to be better for Egypt "to convert some of its rice production from the Japonica to the Indica type" (Ishida 1975).

The per hectare yield of rice in Egypt is among the highest in the world. Indeed, its 1972–78 average yield of 52.4 Q/ha, although 5.3 and 5.7 Q behind Australia and Japan, trails closely behind South Korea to become the fourth highest in the world (USDA 1979). However, Egypt's potential as one of the world's major rice exporters is very difficult to realize

---

[9]Even as recent as in 1976–77, the HYVs accounted for only about 7% of Burma's total rice hectarage (derived form Dalrymple 1978 and USDA 1978).

because of Egypt's relatively high rate of population growth and increasing per capita rice consumption.

Perspectives on Government Policy

In the foregoing discussion, we have mentioned briefly the farm program of price support and acreage allotment of the United States, the trade-off practices of China and Egypt, the price incentive policies in Japan and Thailand's policy of taking care of its own domestic needs first. All of these reflect clearly the impact on the world rice situation by the respective governments' policies. As we look into the future, what framework of policies and course of action should governments of the world formulate and take on? This perhaps, could be looked at from the following mutually related factors.

First, there is the perspective on policies towards increasing world rice production and the increasing world population. The world's population is currently growing at about 72 million persons per year. At the current rate of per capita consumption (Table 1.11), the world will have to provide an additional 4.2 MMT of milled rice per year just to keep pace with this growth in population. Taking into account the rising level of world affluence and the resulting income effect on rice, the annual increase of demand for milled rice will definitely be more than 5 MMT. This is equivalent to about 7.6 MMT of rough rice a year. So far the world has been able to provide for such an increase. But in the long-run, the impact of HYVs will eventually wear out. Besides, breakthroughs like the HYVs occur so infrequently (Efferson 1972). Policies leading not only to increasing rice supply, but also toward reducing population growth appear to be very urgent.

Furthermore, the HYVs have been introduced for over a decade. Their impact, at present, seems yet to be fully realized. For in tropical Asia, where the HYVs have their greatest impact, the yield of rice has increased only about 20% between 1966 and 1975. Yield gaps are found to exist between the experiment station yield and the actual farm yield and between the potential farm yields and the actual farm yield (Herdt and Baker 1977). As land resources in Asia are rather limited, effort must be made to reduce or even to close these yield gaps. Policies are needed not only to realize the full potential of the HYVs, but also to deal with the related environmental questions.

Second, there is the perspective on policies of achieving self-sufficiency or depending on imports. A question may be raised here: Is self-sufficiency necessarily a commendable goal to pursue? Given appropriate incentives, self-sufficiency certainly is possible to attain, but, of course, at a cost. This is clearly demonstrated by the successful achievement in

Japan recently.

Lester R. Brown called the price support policies of Japan to maintain its rice self-sufficiency "the most flagrant violator of the laws of comparative advantage of any cereal-producing country" (Brown 1970). How many governments in the world, especially in Asia, can afford the luxury of self-sufficiency? Would self-sufficiency be any better for countries such as Malaysia and Indonesia who find it difficult to meet their rice demand but have other natural resources to supplement their rice demand through imports? If the answer to this question is positive, then policies concerning quantity, trade-off and strategy are needed.

Finally, there is the perspective on policies leading to international cooperation and national development programs. As was mentioned, one sign that shows the world's rice economy is healthy is the fact that there exists now in the world a greater degree of international cooperation in rice research. The results of this cooperation, as seen in the achievements of IRRI, are very encouraging and gratifying.

One can never emphasize enough the importance of international cooperation. It has been said that the selection of the whole line of IR rices would not have been successfully achieved with such a rapid pace if it were not for the already existing Taichung Native 1 variety in Taiwan. Robert F. Chandler, Jr., first director of the IRRI, rightfully said (1972) that "IRRI's advances in changing the architecture of the tropical rice plant *originated* (italics ours) in Taiwan ..." and that the fast progress which the IRRI has made "would not have been possible without the availability of the dwarfing gene from Dee-geo-woo-gen ..."

The Chinese rice researchers should be proud of their contribution of these parental varieties to the successful improvement of the tropical rices. But the semidwarfing gene from Taiwan alone could not have contributed that much. The point is this: International cooperation on agricultural research has achieved what national efforts could not have done. "Although the creation of Taichung Native 1 in Taiwan was a single advance," said Dr. Chandler, Jr., "it requires an international organization such as the IRRI to implement a massive crossing program resulting in an extensive series of semidwarf genetic lines ..."

It has been a very gratifying experience to see that in recent years many *national* rice research centers have vastly expanded and strengthened their research and development programs. The long list of recently released improved rice varieties enumerated by Dalrymple (1978) is an indication of marked progress. With increased international collaboration on genetic conservation, evaluation and utilization, and on rice testing through the network initiated by IRRI, we can anticipate greater strides in varietal and other technological improvements (IRRI 1974,

[1976]; Chang 1977).

As we look into the future, the results of international cooperation must be continually disseminated to countries where improving rice yield is an urgent matter. In the process, the IRRI probably can continue to provide assistance on technological know-how, but it is the country concerned that must plan the national program and provide the means of support for the research program and the development of a viable rice industry. The responsibility thus rests upon the national government and its policy makers.

REFERENCES

ADAIR, C.R., MILLER, M.D. and BEACHELL, H. M. 1962. Rice improvement and culture in the United States. Adv. Agron. *14*, 61-108.

ANDO, H. 1951. Miscellaneous Records on the Ancient History of Rice Crops in Japan. Chiku Publishing Co., Tokyo. (Japanese)

ANGELINI, F. 1936. Introduction of rice in Italy. Il Riso. Tecnica Ed Economia della Coltivazione Soc. Anonima "Arte della Stampa," Rome. (Italian)

ANON. 1963. Second crop rice—a significant factor in total rice production in Texas. Rice J. *66* (5) 10-12, 43.

ANON. 1966. Rice: Mainstay and hope for 1.1 billion people. For. Agric. *4* (43) 6.

ANON. 1977. A Neolithic village nearly 7000 years old. China Reconstr. *26* (6) 22-23.

ARRAUDEAU, M. 1978. Personal communication.

ASKEW, G. P. *et al.* 1970. Soil landscapes in northeastern Mato Grosso. Georg. J. *136* (2) 211-227.

AYKROYD, W.R. and DOUGHTY, J. 1970. Wheat in Human Nutrition. Food Agric. Organ., Rome.

BAHAR, F.A. and DE DATTA, S.K. 1977. Prospects of increasing tropical rice production through ratooning. Agron. J. *69* (4) 536-540.

BEETS, W.C. 1977. Multiple cropping. World Crops and Livestock *29* (1) 25-27.

BENTON, W. 1970. Enciclopedia Barsa *2*, 190-194. Encyclopedia Britannica Ed. Etda., Rio de Janeiro, Brazil. (Portuguese)

BERTIN, J., HERMARDINQUER, J., KEUL, M. and RANDLES, W.G.L. 1971. Atlas of Food Crops. Ecole Pratique des Hautes Etudes. Mouton, Paris and The Hague. (French and English)

BORTOLONI, G. 1844. Approximate relation of 43 varieties of rice. Mem. Soc. Agraria *1*, 73-78. Bologna. (Italian)

BRADFIELD, R. 1972. Maximizing food production through multiple cropping systems centered on rice. *In* Rice, Science and Man. Int. Rice Res. Inst., Los Baños, Philippines.

BRADLEY, A.V. 1956. Tables of Food Values. Charles A. Bennett Co., Peoria, Ill.

BRADY, N.C. and ATHWAL, D.S. 1975. Future emphasis on upland rice. *In* Major Research in Upland Rice. Int. Rice Res. Inst., Los Baños, Philippines.

BROWN, L.R. 1963. Agricultural diversification and economic development in Thailand: a case study. U.S. Dep. of Agric. Econ. Rep. *(8),* Washington, D.C.

BROWN, L.R. 1970. Seeds of Change: The Green Revolution and Development in the 1970s. Praeger Publishers, New York.

BURKILL, I.H. 1935. A Dictionary of the Economic Products of the Malay Peninsula, Vol. 22. Ministry of Agriculture and Cooperatives, Kuala Lumpur.

CARRASCO GARCIA, J.M. 1952. Summary of Rice Grower. Editorial Guerri S.A., Valencia, Spain. (Spanish)

CARTER, G.F. 1975. Man and the Land: A Cultural Geography. Holt, Rinehart and Winston, New York.

CARTER, G.F. and PENDLETON, R.R. 1956. The humid soil: process and time. Geogr. R. *46* (4) 488-507.

CHALAM, G.V. 1965. Taichung Native 1 promises rice in plenty. Indian Farming *15* (7) 34-35.

CHANDLER, R.F. 1968. Dwarf rice—a giant in tropical Asia. *In* Science for Better Living. The Yearbook of Agriculture, U. S. Dep. Agric., Washington, D.C.

CHANDLER, R.F. 1972. IRRI—the first decade. *In* Rice, Science and Man. Int. Rice Res. Inst., Los Baños, Philippines.

CHANG, K.C. 1978. Personal communication.

CHANG, T.T. 1961. Recent advances in rice breeding in Taiwan. *In* Crop and Seed Improvement in Taiwan, Republic of China, May 1959—January 1961. Joint Comm. on Rural Reconstr. Pl. Ind. Ser. 22, Taipei.

CHANG, T.T. 1964. Present knowledge of rice genetics and cytogenetics. Int. Rice Res. Inst. Tech. Bull. *1,* Los Baños, Philippines.

CHANG, T.T. 1967. The genetic basis of wide adaptability and yielding ability of rice varieties in the tropics. Int. Rice Comm. Newsl. *16* (4) 4-12.

CHANG, T.T. 1970. Rice. *In* Genetic Resources in Plants—Their Exploration and Conservation. O.H. Frankel and E. Bennett (Editors). Blackwell Scientific Publications, Oxford and Edinburgh.

CHANG, T.T. 1975. Species of rice. *In* Rice, 5th Edition. D.H. Grist (Editor). Longman, London.

CHANG, T.T. 1976A. Rice (*Oryza sativa* and *Oryza glaberrima*). *In* Evolution of Crop Plants. N.W. Simmonds (Editor). Longman, London.

CHANG, T.T. 1976B. The rice cultures. Philos. Trans. R. Soc. (London) B275, London, 143-157.

CHANG, T.T. 1976C. The origin, evolution, cultivation, dissemination, and diversification of Asian and African rices. Euphytica *25*, 425-441.

CHANG, T.T. 1976D. Paleogeographic origin of the wild taxa in the genus *Oryza* and their genomic relationships. Int. Rice Res. Newsl. 2, 4.

CHANG, T.T. 1976E. Manual on Genetic Conservation of Rice Germ Plasm for Evaluation and Utilization. Int. Rice Res. Inst., Los Baños, Philippines.

CHANG, T.T. 1977. Genetics and evolution of the Green Revolution. Paper presented at the UNESCO-CSIC Symp. on Genetics and Ethics, Oct. 10-14. Madrid.

CHANG, T.T. and BARDENAS, E.A. 1965. The morphology and varietal characteristics of the rice plant. Int. Rice Res. Inst. Tech. Bull. 4. IRRI, Los Baños, Philippines.

CHANG, T.T., LORESTO, G.C. and TAGUMPAY, O. 1972. Agronomic and growth characteristics of upland and lowland varieties. *In* Rice Breeding. Int. Rice Res. Inst., Los Baños, Philippines.

CHANG, T.T., MARCIANO, A.P. and LORESTO, G.C. 1977. Morpho-agronomic variousness and economic potentials of *Oryza glaberrima* and wild species in the genus *Oryza*. Paper presented at the IRAT-ORSTOM Meeting on African rice species, Jan. 25-26. Paris.

CHANG, T.T. and OKA, H.I. 1976. Genetic variousness in the climatic adaptation of rice cultivars. *In* Climatic and Rice. Int. Rice Res. Inst., Los Baños, Philippines.

CHANG, T.T. and VERGARA, B.S. 1971. Ecological and genetic aspects of photoperiod-sensitivity and thermo-sensitivity in relation to the regional adaptability of rice varieties. Int. Rice Comm. Newsl. 20 (2) 1-10.

CHANG, T.T. *et al.* 1975. Varietal improvement of upland rice. *In* Major Research in Upland Rice. Int. Rice Res. Inst., Los Baños, Philippines.

CHANG, W.L., SU, H.P. and LIN, F.H. 1977. Rice response to extreme temperatures in Saudi Arabia. Int. Rice Res. Newsl. 1, 4.

CHATTERJEE, D. 1948. A modified key and enumeration of the species of *Oryza* Linn. Indian J. Agric. Sci. 18 (3) 185-192.

CHATTERJEE, D. 1951. Note on the origin and distribution of wild and cultivated rices. Indian J. Genet. 11, 18-22.

CHEN, K.S. 1957. The implications of processing the lowland rice varieties of Yunnan on the evolution of rice. Southwest Agric. Sci. 1957 (2) 107-110. (Chinese)

CHENG, S.H. 1968. The Rice Industry of Burma, 1852−1940. The University of Malaya Press, Kuala Lumpur, Malaysia.

CHEVALIER, A. 1932. New contribution on the systematic studies of rice. Rev. Bot. Appl. Agric. Trop. 12, 1014-1032. (French)

CHU, Y.E. and OKA, H.I. 1972. The distribution and effects of genes causing F_1 weakness in *Oryza brevilgulata* and *O. glaberrima*. Genet. 70, 163-173.

CIFERRI, R. 1960. Features of the History of Rice in Italy. Ente Nazionale Risi Utticio Studio, Quaderno 8. Milan, Italy. (Italian)

COPELAND, E.B. 1924. Rice. Macmillan, London.

CPAM. 1976. Reconnaissance of the neolithic site at Ho-mu-tu in Yuyao County, Chekiang Province. Wen Wu 8, 6-14. (Chinese)

DALRYMPLE, D.G. 1969. Imports and Plantings of High-yielding Varieties of Wheat and Rice in Less Developed Nations. U.S. Dep. Agric. and U.S. Agency Int. Dev., Washington, D.C.

DALRYMPLE, D.G. 1976. Development and spread of high-yielding varieties of wheat and rice in the less developed nations. Foreign Agric. Econ. Rep. 95. U.S. Dep. Agric. and U.S. Agency Int. Dev., Washington, D.C.

DALRYMPLE, D.G. 1978. Development and spread of high-yielding varieties of wheat and rice in the less developed nations. Foreign Agric. Econ. Rep. 95, 6th ed. U.S. Dep. Agric. and U.S. Agency Int. Dev., Washington, D.C.

DAS, D.K. and JAT, R.L. 1977. Influence of three soil-water regimes on root porosity and growth of four rice varieties. Agron. J. 69 (2) 197-200.

DE DATTA, S.K. 1975. Upland rice around the world. In Major Research in Upland Rice. Int. Rice Res. Inst., Los Baños, Philippines.

DOLORES, R.C., CHANG, T.T. and RAMIREZ, D.A. 1975. Cytogenetics of sterility in F_2 hybrids of indica × indica and indica × javanica varieties of rice (Oryza sativa L.). Cytologia 40, 639-647.

DOWNEY, D.A. and WELLS, B.R. 1975. Air temperatures in the Starbonnet rice canopy and their relationship to nitrogen timing, grain yield, and water temperature. Arkansas Agric. Exp. Sta. Bull. 796.

DOWNEY, D.A., HUEY, B.A. and WELLS, B.R. 1976. Using growing degree days to determine timing of midseason N application to rice. Arkansas Farm Res. 25 (2) 2.

EFFERSON, J.N. 1967. Recent developments in the rice industry of Thailand, Part II, marketing and future developments. Rice J. 70 (4) 33-37.

EFFERSON, J.N. 1972. Outlook for world rice production and trade. In Rice, Science and Man. Int. Rice Res. Inst., Los Baños, Philippines.

EL-AMIR, M.R.A. EL-FATTAH. 1967. Location models for the world rice industry. Ph.D. Dissertation. University of California, Berkeley.

ENCYCL. BRITANNICA. 1977. World population and areas. In Britannica Book of the Year. Encyclopedia Britannica, Chicago.

ENGLE, L.M., CHANG, T.T. and RAMIREZ, D.A. 1969. The cytogenetics of sterility in F_1 hybrids of indica × indica and indica × javanica varieties of rice (Oryza sativa L.). Phil. Agric. 53, 289-307.

ERYGIN, P.S. and NATAL'IN, N.B. 1968. Rice. Moscow. (Russian)

EVATT, N.W. 1966. Texas. In Rice fertilization recommendation. Rice J. 69 (1) 28-29.

FAO (FOOD AGRIC. ORGAN. U.N.). 1965. Long Term Trends: Development Plans and Production Projections for 1975. FAO, Rome.

FAO. 1966. Agriculture Commodities—Projection for 1975 and 1985. FAO, Rome.

FAO. 1971. Agriculture Commodity Projections, 1970–1980. FAO, Rome.

FAO. 1975. Production Yearbook 1974. FAO, Rome.

FAO. 1976. Production Yearbook 1975. FAO, Rome.

FAO. 1978. Production Yearbook 1977. FAO, Rome.

GHOSE, R.L.M., GHATGE, M.B. and SUBRAHMANYAN, V. 1956. Rice in India. Indian Counc. of Agric. Res., New Delhi.

GODWIN, M.R. and JONES, L.L. 1970. The Southern Rice Industry. Texas A & M Univ. Press, College Station, Texas.

GRANT, W.R. and HOLDER, S.H. 1975. Recent changes and the potential for U.S. rice acreage. *In* Rice Situation. U.S. Dep. Agric. *RS-26.* Washington, D.C.

GRAY, L.C. and THOMPSON, E.K. 1958. History of Agriculture in the Southern United States to 1860. Peter Smith, Gloucester, Mass.

GRIST, D.H. 1959. Rice, 3rd Edition. Longman, Green and Co., London.

GRIST, D.H. 1965. Rice, 4th Edition. Longman, London.

GRIST, D.H. 1969. Cultivation of rice. World Crops Livestock *21* (1) 17-21.

GRIST, D.H. 1975. Rice, 5th Edition. Longman, London.

GUSICHIN, G.G. 1930. Rice. Sel'khozgiz, Moscow. (Russian)

HAI-FENG INST. AGRIC. SCI. (KWANGTUNG). 977. Distinct hybridization between rice and bamboo. Acta Bot. Sinica *19* (1) 90-92. (Chinese)

HALLAM, A. 1974. A Revolution in the Earth Sciences: From Continental Drift to Plate Tectonics. Oxford Univ. Press, London.

HARGROVE, T.R. 1976. The diffusion of genetic materials among rice breeders in India. Int. Rice Res. Newsl. *1*, 5-6.

HARLAN, J.R. 1965. The possible role of weed races in the evolution of cultivated plants. Euphytica *14*, 173-176.

HARLAN, J.R. 1970. Evolution of cultivated plants. *In* Genetic Resources in Plants—Their Exploration and Conservation. O.H. Frankel and E. Bennett (Editors). Blackwell Scientific Publications, Oxford and Edinburgh.

HARLAN, J.R. 1973. Genetic resources of some major field crops in Africa. *In* Survey of Crop Genetic Resources in their centres of Diversity. O.H. Frankel, (Editor). FAO/Int. Biol. Prog., Rome.

HERDT, R.W. and BARKER, R. 1977. The status and potential of food production in Asia. Paper presented to the Int. Rice Res. Inst.-U.N. Univ. Workshop on an Interface Between Agric., Nutr. and Food Sci., Feb. 28-March 3. Los Baños, Philippines.

HO, P.T. 1956. Early-ripening rices in Chinese history. Econ. Hist. Rev. *9*, 200-218.

HYO, J.L. 1977. The oldest rice in Korea. *In* Korea Daily, Feb. 18. Seoul. (Korean)

HSIA, Y.S., KAO, Y.C. and CHIANG, Y.C. 1964. A preliminary report on the intergeneric crosses between paddy rice and gu. Genet. Sinica *5* (5), 41-54, pl. 1-4. (Chinese with English summary)

HUANG, C.H., CHANG, W.L. and CHANG, T.T. 1972. Ponlai varieties and

Taichung Native 1. *In* Rice Breeding. Int. Rice Res. Inst., Los Baños, Philippines.

HUKE, R. 1976. Geography and climate of rice. *In* Climate and Rice. Int. Rice Res. Inst., Los Baños, Philippines.

HUNG, H.H.and CHANG, T.T. 1977. Aberrant segregation of three marker-genes in crosses between upland and semidwarf-lowland varieties of rice. SABRAO J. *8* (2) 127-134.

ICASBALCETA, J.G. 1866. Report made by Andres de Tapia. *In* Collection of Documents for the History of Mexico, Vol. 2. Antigua Libreria, Mexico. (Spanish)

IGBOZURIKE, M.U. 1971. Ecological balance in tropical agriculture. Geogr. R. *61* (4) 519-529.

INST. OF ARCHAEOL. (PEKING). 1972. Report on radio carbon-determined dates. II. K'ao-ku *5*, 52-58. (Chinese)

IRRI (INT. RICE RES. INST.). 1964. Rice: Genetics and Cytogenetics. Elsevier, Amsterdam.

IRRI. 1967. Annual Report for 1966. IRRI, Los Baños, Philippines.

IRRI. 1971. Annual Report for 1970. IRRI, Los Baños, Philippines.

IRRI. 1972. Annual Report for 1971. IRRI, Los Baños, Philippines.

IRRI. 1973. Annual Report for 1972. IRRI, Los Baños, Philippines.

IRRI. 1974. Annual Report for 1973. IRRI, Los Baños, Philippines.

IRRI. 1975A. Annual Report for 1974. IRRI, Los Baños, Philippines.

IRRI. 1975B. Improved rice varieties needed for world's deep water regions. IRRI Rept. *3.*

IRRI. 1976A. Annual Report for 1975. Los Baños, Philippines.

IRRI. 1976B. How Tongil triggered a Korean rice revolution. IRRI Reptr. *3.*

IRRI. [1976]. The International Rice Testing Program. IRRI, Los Baños, Philippines. (unpublished)

IRRI. 1977. Annual Report for 1976. IRRI, Los Baños, Philippines. (forthcoming)

IRRI. 1978. Annual Report for 1977. IRRI, Los Baños, Philippines. (in process)

ISHIDA, S. 1975. Production and trade of rice in Egypt. *In* Rice in Asia. Assoc. Jpn. Agric. Sci. Soc., Univer. of Tokyo Press, Tokyo.

ISO, E. 1954. Rice and Crops in Its Rotation in Subtropical Zones. Japan FAO Assoc., Tokyo.

JANZEN, D.H. 1975. Tropical agroecosystems. *In* Food: Politics, Economics, Nutrition and Research. P.H. Abelson (Editor). Amer. Assoc. Adv. Sci., Washington, D.C.

JENNINGS, P.R. 1975. Rice breeding and world food production. *In* Food: Politics, Economics, Nutrition, and Research. Amer. Assoc. Adv. Sci., Washington, D.C.

JOHNSTON, T.H. and MILLER, M.D. 1973. Culture. *In* Rice in the United States: Varieties and Production. Agric. Handbk. *289.* U.S. Dep. Agric., Washington, D.C.

KANE, T.T. and MYERS, P.F. 1978. World Population Data Sheet. Population Reference Bureau, Inc., Washington, D.C.

KWANGTUNG AGRIC.–FORESTRY COLL. and YUNNAN UNIV. 1974. A report on the vertical distribution of the rice varieties in Szemao, Yunnan. Acta Bot. Sinica *16,* 208-222. (Chinese with English summary)

KWANGTUNG PROV. ACAD. AGRIC. SCI. 1966. Preliminary report on breeding rice in Kwangtung. Zuowu Xuebao (Crop Sci. J.) *5,* 33-40. (Chinese)

LEDERER, T.H. 1977. Two good crops in a row boost Burma's rice exports. Foreign Agric. *15* (22) 8–9.

LEE, C. 1960. A culture history of rice with special reference to Louisiana. Ph.D. Dissertation. Louisiana State University.

LOPEZ, G. 1978. Personal Communication.

LU, J.J. 1971. The demand for United States rice: an economic-geographic analysis. Ph.D. Dissertation. University of Washington.

LU, J.J. 1972A. Japan: mizuho no kuni, land of bounteous rice and plenteous problems. Chung Chi J. *11* (2) 66-72.

LU, J.J. 1972B. The declining export demand for U.S. rice: a production dilemma. *In* International Geography 1972. W.P. Adam and F.M. Helleiner (Editors). University of Toronto Press, Montreal.

LU, J.J. 1975. International situations and the United States rice exports: selected examples. Ecumene *7* (1) 35-41.

LU, J.J. 1976. Spatial adjustments in U.S. rice production: a historical-geographical appraisal. *In* Proc. of the Sixteenth Rice Technical Working Group. Texas A & M Univ. Press, College Station, Texas.

LU, J.J. 1977. Spatial pattern of rice consumption in the United States: some observations. Paper presented at the annual meeting of the Southwest Soc. Sci. Assoc., April 18-21. Denver.

MARIE, R. 1978. Personal Communication.

MIKKELSEN, D.S. 1971. *Cited by* J.J. Lu 1971. *In* The demand for United States rice: An economic-geographic analysis. Ph.D. Dissertation. University of Washington.

MILLER, G.T. 1975. Living in the Environment: Concepts, Problems, and Alternatives. Wadsworth Publishing Co., Belmont, Calif.

MORA Y LLORENS, R.F. DE. 1939. The Rice: its Cultivation, Milling and Commerce. Salvat Editores S. A. Barcelone, Spain. (Spanish)

MORINAGA, T. 1957. Rice of Japan. Yokendo, Tokyo. (Japanese)

MORINAGA, T. 1967. Rice of Japan. J. Agric. Soc. Japan *988,* 1-10. (Japanese)

MORINAGA, T. 1968. Origin and geographical distribution of Japanese rice. JARQ (Jpn. Agric. Res. Q.) *3* (2) 1-5.

MORISHIMA, H. 1969. Phenetic similarity and phylogenetic relationships among strains of *Oryza perennis*, estimated by methods of numerical taxonomy. Evolution *23*, 429-443.

MORISHIMA, H., HINATA, K. and OKA, H.I. 1962. Floating ability and drought resistance in wild and cultivated species of rice. Indian J. Genet. Pl. Breed. *22*, 1-11.

MORISHIMA, H., HINATA, K. and OKA, H.I. 1963. Comparison of modes of evolution of cultivated forms from two wild rice species, *Oryza breviligulata* and *O. perennis*. Evolution *17*, 170-181.

NATL. ACAD. SCI. (NAS). 1969. Resources and Man: A Study and Recomendations. W.H. Freeman & Co., San Francisco.

NATL. ACAD. SCI. (NAS). 1972. Genetic Vulnerability of Major Crops. Washington, D.C.

NAYAR, N.M. 1973. Origin and cytogenetics of rice. Adv. Genet. *17*, 153-292.

NI-FENG INST. AGRIC. SCI. (KIANGSI). 1977. Achievement in the hybridization between rice and reed. Acta Bot. Sinica *19* (1) 90-92. (Chinese)

OFF. PRIME MINIST. 1976. Japan Statistical Yearbook. Office of the Prime Minister, Tokyo.

OKA, H.I. 1964. Pattern of interspecific relationships and evolutionary dynamics in *Oryza. In* Rice Genetics and Cytogenetics. Elsevier, Amsterdam.

OKA, H.I. 1974. Experimental studies on the origin of cultivated rice. Genet. *78*, 475-486.

OKA, H.I. 1975. The origin of cultivated rice and its adaptive evolution. *In* Rice in Asia. Assoc. Jpn. Agric. Sci. Soc., Univ. of Tokyo Press, Tokyo.

PALACPAC, A.C. 1977. World Rice Statistics. Int. Rice Res. Inst., Los Baños, Philippines. (mimeo)

PALACPAC, A.C. 1978. World Rice Statistics. Int. Rice Res. Inst., Los Baños, Philippines. (mimeo)

PARKER, J.B. 1974. Poor crops, short stocks spark West Asia's rice imports. Foreign Agric. *12* (9) 8-9, 16.

PEARSALL, W.H. 1969. *Cited by* G. Wrigley 1969. *In* Tropical Agriculture. Frederick A. Praeger, N.Y.

PIACCO, R. 1959. The first rice variety grown in Italy. Riso *8* (12) 12-14. (Italian)

PINOLINI, D. 1929. Rice and Its Cultivation. Vallardi, Milan. (Italian)

PONNAMPERUMA, F.N. and IKEHASHI, H. 1977. Breeding for adverse-soils tolerance. Int. Rice Res. Newsl. 2, 4-5.

PORTERES, R. 1950. Old agriculture of Africa intertropical. Agron. Trop. *5*, 489-507. (French)

PORTERES, R. 1956. Taxonomic botany of the cultivated rices of *Oryza sativa* and *Oryza glaberrima*. J. Agric. Trop. Bot. Appl. *3*, 341-384, 541-580, 627-700, 721-856. (French)

PORTERES, R. 1976. African cereals: *Eleusine,* fonio, black fonio, teff, *Brachiaria, Paspallum,* Pennisetum and African rice. *In* Origins of African Plant Domestication. J.B. Harlan, J.M.J. de Wet and A.B.L. Stemler (Editors). Mouton Publishers, Hague and Paris.

PRESIDENT'S SCI. ADVISORY COMM. 1967. Report of the Panel on the World Food Supply, Vol. 2. The World Food Problem. The White House, Washington, D.C.

RAMIAH, K. 1937. Rice in Madras—A Popular Handbook. Government Press, Madras.

RICHHARIA, R.H. 1960. Origins of cultivated rice. Indian J. Genet. *20,* 1-4.

ROSCHEVICZ, R.J. 1931. A contribution to the knowledge of rice. Bull. Appl. Bot. Genet. Plant Breed. (Leningrad) *27* (4) 1-133. (Russian with English summary)

RUGSACHAT, S. and CLARIDAD, F.B. 1971. Experiments on intergeneric crosses of rice with some genera of the family Gramineae. Araneta J. Agr. *18,* 147-163.

RUTGER, J.N. and PETERSON, M.L. 1976. Improved short stature rice. Calif. Agric. *30* (6) 406.

SAINI, S.S. 1976. Impact of semidwarfs on the area planted, production, and yield of rice in the Punjab of India. Int. Rice Res. Newsl. *1,* 3-4.

SAMPATH, S. 1962. The genus *Oryza:* its taxonomy and species interrelationships. Oryza *1,* 1-29.

SAMPATH, S. and RAO, M.B.V.N. 1951. Interrelationships between species in the genus *Oryza.* Indian J. Genet. *11,* 14-17.

SCHATZ, H.L. 1977. U.S. rice exports up, competition running strong. Foreign Agric. *15* (24) 6-7, 12.

SHARMA, S.D. and SHASTRY, S.V.S. 1973. Evolution in genus *Oryza. In* Advancing Frontiers in Cytogenetics. Hindustan Publishing Co., New Delhi.

SHASTRY, S.V.S. 1969. New high-yielding varieties of rice: Jaya and Padma. Indian Farming *19,* 5-13.

SILVA, M.V.E. 1955. Essentials for the History of Rice in Portugal. Bol. Fed. Gremios Lavoura Beira Litoral 7 (Coimbra). (Portuguese)

SILVA, M.V.E. 1978. Personal communication.

SOLHEIM, W.G., II. 1972. An earlier agricultural revolution. Sci. Am. *226* (4) 34-41.

SMITH, T. 1966. Too much water can reduce yield. Rice J. *69 Dec.,* 13-14.

SRI LANKA DEP. AGRIC. 1977. BW78—a new variety for low country wet zone. Agric. Newsl. *4.*

STAVIS, B. 1974. Making green revolution—the politics of agricultural development in China. Rural Development Monograph No. *1.* Rural Dev. Comm., Cornell Univ., Ithaca, New York.

STEBBINS, G.L. 1971. Chromosomal Evolution in Higher Plants. Edward Arnold, London.

TATEOKA, T. 1963. Taxonomic studies in *Oryza*. III. Key to the species and their enumeration. Bot. Mag. (Tokyo) *76*, 165-173.

TATEOKA, T. 1964. Taxonomic studies of the genus *Oryza*. *In* Rice Genetics and Cytogenetics. Elsevier, Amsterdam.

TING, Y. 1961. Chinese Culture of Lowland Rice. Agricultural Publishing Society, Peking. (Chinese)

ULLMAN, E.L. 1956. The role of transportation and the bases for interaction. *In* Man's Role in Changing the Face of the Earth. W.L. Thomas (Editor). Univ. of Chicago Press, Chicago.

UMALI, D.L. 1972. Rice improvement through international cooperation. *In* Rice, Science and Man. Int. Rice Res. Inst., Los Baños, Philippines.

UNIV. CALIF. FOOD TASK FORCE. 1974. A Hungry World: The Challenge to Agriculture. Univ. of California Press, Berkeley.

USDA (U.S. DEP. AGRIC.). 1972A. Agricultural trade of the People's Republic of China, 1935–1969. USDA Foreign Agric. Econ. Rep. *83*.

USDA. 1972B. Review of world rice markets and major suppliers. U.S. Dep of Agric. FAS-M-246.

USDA. 1976A. Foreign agricultural circular, rice. U.S. Dep. of Agric. FR-1-76.

USDA. 1976B. 26 Years of World Cereal Statistics: Area, Yield, Production, 1950–75, by Country and Region. U.S. Dep. Agric., Washington, D.C.

USDA. 1976C. Foreign agricultural circular, rice. U.S. Dep. Agric. FR-2-76.

USDA. 1977A. Rice situation. U.S. Dep. Agric. RS-29.

USDA. 1977B. World agriculture situation. U.S. Dep. Agric., WAS-13.

USDA. 1977C. Foreign agriculture circular, rice. U.S. Dep. Agric. FR-2-77.

USDA. 1977D. Foreign agriculture circular, rice. U.S. Dep. Agric. FR-1-77.

USDA. 1977E. Foreign agriculture circular, grains. U.S. Dep. Agric. FG-9-77.

USDA. 1977F. Foreign agriculture circular, grains. U.S. Dep. Agric. FG-13-77.

USDA. 1977G. Foreign agriculture circular, rice. U.S. Dep. Agric. FR-3-77.

USDA. 1977H. Foreign agriculture circular, grains. U.S. Dep.Agric. FG-24-77.

USDA. 1977I. People's Republic of China agriculture situation: review of 1976 and outlook for 1977. USDA Foreign Agric. Econ. Rep. *137*.

USDA. 1977J. Foreign Agriculture circular, rice. U.S. Dep. Agric. FR-5-77.

USDA. 1978. Foreign agriculture circular, grains. U.S. Dep. Agric., FG-4-78.

USDA.1978A. Foreign agriculture circular, grains. U.S. Dep. Agric., FG-8-78.

USDA. 1978B. Foreign agriculture circular, grains. U.S. Dep. Agric., FG-18-78.

USDA. 1979. Foreign agriculture circular, grains. U.S. Dep. Agric., FG-2-79.

VAN STEENIS, C.G.G.J. 1962. The land-bridge theory in botany with particular reference to tropical plants. Blumea *11*, 235-542.

VERGARA, B.S. 1976. Culture of floating rice. *In* Proceedings of the Sixteenth Rice Technical Working Group. Texas A & M Univ. Press, College Station, Texas.

VERGARA, B.S. and COFFMAN, W.R. 1976. Low temperature tolerance program at IRRI. *In* proceedings of the Sixteenth Rice Technical Working Group. Texas A & M. Univ. Press, College Station, Texas.

VISHNU-MITTRE. 1976. Discussion (in India: local and introduced crops). *In* The Early History of Agriculture. Philos. Trans. R. Soc., London, B275, 141.

WALLER, D.S. 1949. Nutritive Value of Foods (G. Wahr, 1944. Ann Arbor, Mich.). Revised by Univ. of Michigan Hospital, Ann Arbor, Mich.

WHYTE, R.O. 1972. The Gramineae, wild and cultivated of monsoonal and equatorial Asia. I. Southeast Asia. Asian Persp. *15*, 127−151.

WIBULSETH, S. 1976. Thailand's future as rice exporter questioned. Foreign Agric. *14* (14) 10-11.

WICKIZER, V.D. and BENNETT, M.K. 1941. The Rice Economy of Monsoon Asia. Stanford Univ. Press, Palo Alto, Calif.

WILLIAMS, V.R. *et al.* 1958. Varietal differences in amylose content of rice starch. Agric. Food Chem. *6* (1) 47-48.

WORK, S.H. 1969. Thailand striving to build rice production. Foreign Agric. *79* (17) 2-5.

WRIGLEY, G. 1969. Tropical Agriculture. Frederick A. Praeger, New York.

WU, S.H., TSAI, C.K., WANG, C.C and WANG, H.M. 1961. The hybrids of rice and kaoliang and cytogenetical studies on the progenies of certain hybrids. Acta Bot. Sin. *9*, 191-218, pl. 1-8. (Chinese with English summary)

YAMADA, N. 1975. Technical problems of rice production in tropical Asia. *In* Rice in Asia. Assoc. Jpn. Agric. Sci. Soc. Univ. of Tokyo Press, Tokyo.

YEUTTER, C.K. 1975. *In* U.S. growers urged to expand commercial market. Foreign Agric. *13* (22) 5-6.

YIEH, M.F. 1964. The distant hybridization between broad sheath bamboo (*Arundinaria latiofolia* keng) and wheat, barley and rice. Genet. Sinica *5*, 145-149. (Chinese)

2

Rice Plant Growth and Development

Benito S. Vergara

The life cycle of the rice plant is generally completed within the range of 100 to 210 days, with the mode falling between 110 to 150 days. In the temperate countries, the average duration from sowing to harvest is about 130 to 150 days. Varieties with growth duration of 150 to 210 days are usually photoperiod sensitive and planted in the deep water areas. Temperature and day length are the 2 environmental factors that affect the development of the rice plant.

Growth of the rice plant can be divided into 3 main phases (Vergara 1970):

Vegetative phase—from seed germination to panicle initiation
Reproductive phase—from panicle initiation to flowering
Ripening phase—from flowering to full maturity

These main phases overlap each other within a rice hill or a rice crop and the ripening phase physiologically does not start until 3 weeks after fertilization.

Figure 2.1 illustrates the growth behavior in the tropics of the variety maturing in 120 days (IR8) compared with one maturing in 150 days (Peta) (Vergara and Chang 1976). Duration of the reproductive (35 days) and ripening phases (25–35 days) are fairly constant for the 2 varieties grown under tropical conditions (Vergara and Chang 1976). The difference in the total growth duration is in the vegetative phase. The length of the vegetative phase depends on the sensitivity of the variety to day length and temperature and its inherent late maturity or long basic vegetative phase (Vergara 1970; Vergara and Chang 1976).

If IR8 is grown in the temperate areas or high altitude areas in the tropics, the growth duration may be as long as 180 to 200 days compared to 130 days. The delay in growth duration is the effect of low temperature mainly on the vegetative and ripening phases. The ripening phase is prolonged from 25 to 30 days under tropical conditions to as long as 60 days in temperate conditions.

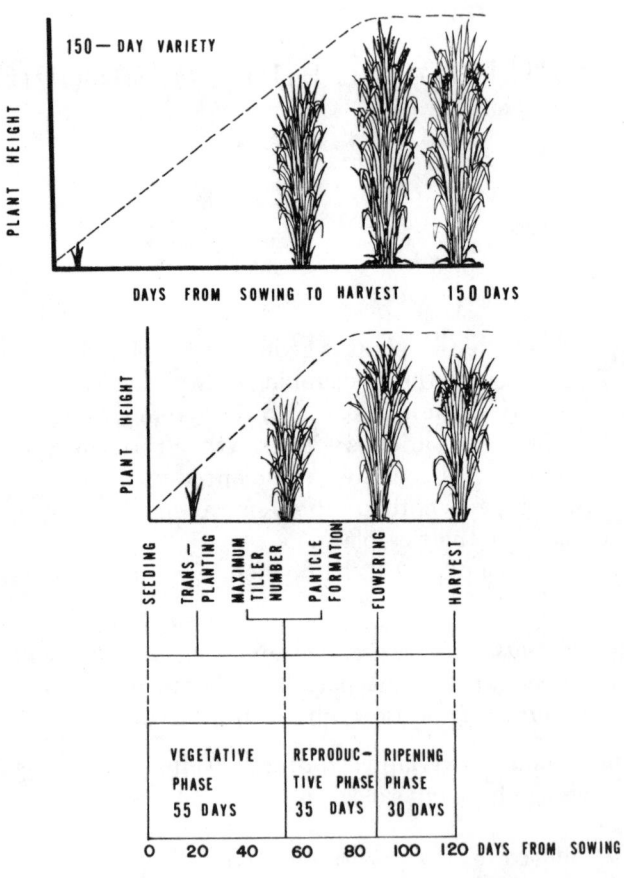

Courtesy of Dr. Benito S. Vergara

FIG. 2.1. GROWTH STAGES OF A 120-DAY AND 150-DAY RICE CULTIVAR GROWN UNDER TROPICAL CONDITIONS

VEGETATIVE PHASE

Seedling growth during the early stages will vary depending upon the method of sowing. Rice can be sown directly in dry land, broadcast in a puddled soil or broadcast in a soil prepared dry but flooded with 10 cm of water. As well, rice can be sown on a dry or wet bed nursery and then transplanted 20 to 70 days after sowing, sown using a special method called *dapog* and transplanted after 11 days, or sown on a wet bed nursery and transplanted to a bigger wet bed nursery after 1 month and retransplanted to the main field after another month (double transplanting). As such, the nursery period can be 0 in the case of direct seeding, 10 to 20 days in wet bed nursery and as long as 90 days in a double transplanting rice culture.

Transplanting is still a common practice in Southeast Asia and the growth of the plant is affected by transplanting. It takes at least 4 days before the rice plant can recover from the uprooting and the roots that develop tend to be shallower than the direct seeded plant.

Tillering

Tillering and the production of leaves are the main visible activity during the vegetative phase. In the tropics, maximum tiller number occurs 40 to 60 days after transplanting, depending upon the tillering capacity of the variety, the spacing used and the fertility level. The high yielding tropical varieties will produce 25 to 30 tillers when grown as isolated plants while the temperate varieties produce 8 to 12 tillers. In temperate areas such as the United States where seeds are drilled or broadcast densely, tillering ability is not as important and the maximum tiller number stage is reached within 30 days when the seedlings have emerged and produced 1 or 3 tillers. In Japan and Korea, transplanting is still being practiced and maximum tiller number is attained around 30 to 50 days after transplanting.

For transplanted rice, panicle initiation should coincide with maximum tiller number stage for uniformity in the flowering and maturity of the grains. In long growth duration varieties and in densely broadcast crops, panicle initiation occurs after the maximum tiller number stage. This lag period between maximum tiller number and panicle initiation is known as *lag-vegetative growth.* The tillers develop in size and weight during the lag-vegetative growth. If the lag-vegetative growth is long, internode elongation might occur even before panicle initiation. This increase in internode length may make the plants more susceptible to lodging as in the case of the traditional tropical cultivars.

Maximum tiller number stage is usually followed by a decrease in tiller number per unit area. This reduction may be as large as 60% in some cultivars planted in the tropics and as low as 10% for the high yielding cultivars both in the temperate and tropical areas. Rice cultivars and rice culture is geared in maintaining a high tiller number per unit area, minimizing the decrease in tiller number after the maximum tiller number stage.

Tiller number and the resulting panicle number per unit area is the main component of grain yield.

Leafing

Leaves are produced on the main culm at an average of 1 per week but can be modified by environmental factors. The interval between leaf production is shorter during early stages of growth (4—5 days) and longer at later stages (8—9 days). Cultivars differ in the number of leaves produced on the main culm (Fig. 2.2). The new high yielding cultivars in the tropics have approximately 14 to 18 leaves, similar to most of the cultivars from the temperate regions. Low temperature and long days increase the number of leaves produced before the panicle comes out.

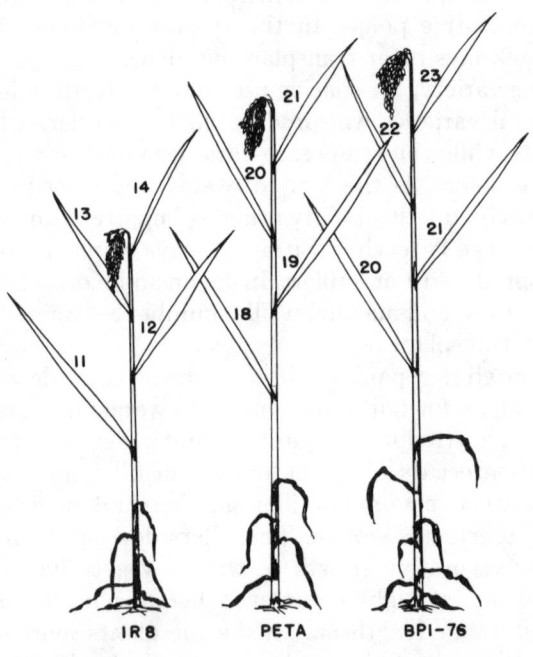

Courtesy of Dr. Benito S. Vergara

FIG. 2.2. NUMBER OF LEAVES ON THE MAIN STEM
OF 3 TROPICAL CULTIVARS GROWN DURING THE
RAINY SEASON

In a cultivar with 14 leaves on the main culm, the longest leaf is the fourth leaf from the flag leaf (Fig 2.3). This is the leaf before panicle formation. The succeeding leaves are generally smaller, probably a competition for substrate between the leaves and the developing panicle. The life duration of a leaf differs according to its position on the culm. The upper leaves have longer life duration than the lower leaves.

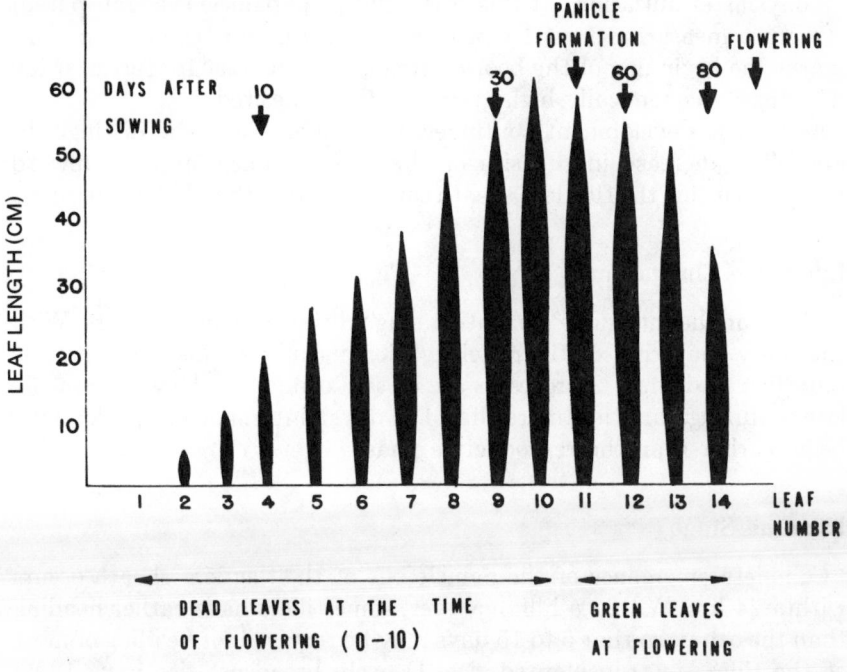

Courtesy of Dr. Benito S. Vergara

FIG. 2.3. LENGTH OF LEAVES ON THE MAIN STEM

REPRODUCTIVE PHASE

Panicle Initiation

The reproductive phase may begin before the maximum tiller number is reached, at about the time of the highest tillering activities, or thereafter. The reproductive phase is marked by the initiation of the panicle primordium and its development.

In the tropics, panicle initiation occurs approximately 70 to 75 days before the expected date of maturity of any cultivar. This period is more variable in the temperate areas where low temperature can lengthen the ripening period.

For photoperiod sensitive cultivars planted during the regular season, panicle initiation, the eventual panicle exsertion and anthesis occurs uniformly. Within 5 days after the first heading, all of the panicles have exserted. In photoperiod insensitive cultivars, it takes a longer time for 100% heading.

As the young panicle develops, it becomes visible to the naked eye 10 to

13 days after initiation. At this stage, the small panicle is a transparent structure measuring 1 to 2 mm long with a fuzzy or spongy tip. This marks the beginning of the booting stage, the last stage for fertilizer top dressing if economically high grain yields are desired.

As panicle development continues, the spikelets become distinguishable. The increase in the size of the young panicle and its upward extension inside the flag leaf sheath causes the sheath to bulge (booting).

Internode Elongation

Time for the internode elongation stage differs among cultivars. With the early maturing cultivars, elongation usually begins after panicle initiation and these 2 processes are closely connected. However, in the late maturing varieties (more than 150 days), internode elongation may begin earlier than the reproductive phase.

Heading Stage

Complete emergence of the panicle out of the flag leaf sheath occurs within 24 hrs. Within a hill, however, some tillers have earlier heading than the others so that 5 to 15 days may be required for heading of most of the tillers in transplanted rice. Densely broadcast rice have 100% heading within a shorter range.

The exsertion of the panicle and especially the length of the peduncle is greatly affected by temperature. Low temperature may result in incomplete exsertion of the panicle especially in the tropical lowland cultivars. This is the result of the nonelongation of the last internode.

Flowering Stage

Anthesis (blooming or flowering) begins with the protrusion of the first dehiscing anthers in the terminal spikelets on the panicle branches. Anthesis of all the spikelets on a panicle may take 7 days to be completed (Fig. 2.4). Anthesis within a panicle usually starts from the top moving downwards. Cultivars with large panicles show the highest percentage of empty spikelets on the lower ⅓ portion of the panicle.

On a clear day, anthesis starts around 1000 to 1400 hrs. It might be earlier on warm days or delayed several days on cold or humid days. Rice is generally self-pollinated.

RIPENING PHASE

In the tropics it takes 25 to 35 days from heading to maturity, while it takes 30 to 60 days in the temperate areas. Low temperature prolongs the ripening period. The rice grain develops after pollination and fertilization (Fig. 2.5). Grain development is a continuous process and the grain undergoes distinct changes before it fully matures.

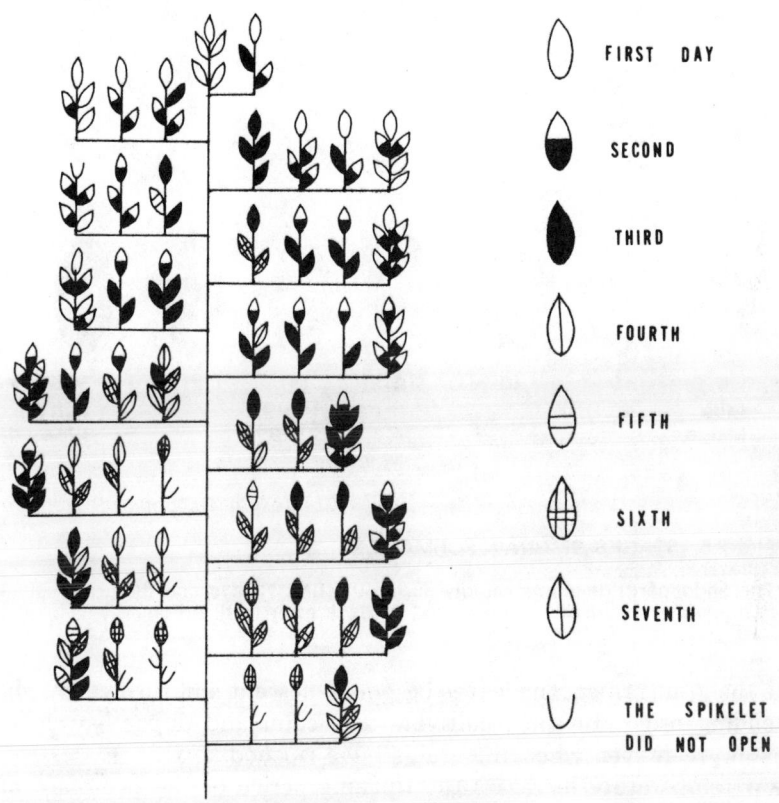

FIRST   DAY

SECOND

THIRD

FOURTH

FIFTH

SIXTH

SEVENTH

THE   SPIKELET
DID   NOT   OPEN

Courtesy of Dr. Benito S. Vergara

FIG. 2.4. FLOWERING ORDER OF A LARGE PANICLE

The contents of the caryopsis (the starchy portion of the grain) are first watery in consistency but later turn milky. Then the milky caryopsis turns into soft dough and later turns into hard dough.

At 21 days after fertilization, the spikelet has reached its maximum weight. The individual grain is matured when the caryopsis is fully developed in size and is hard, clear and free from greenish tint. The crop is considered ripe when more than 90% of the grains in the panicles are fully ripened.

Since it takes 7 days for all the spikelets in a panicle to open, full maturity for the whole panicle does not occur until 30 days after flowering. A few extra days are further needed to ripen all the grains since the panicles do not come out at the same time.

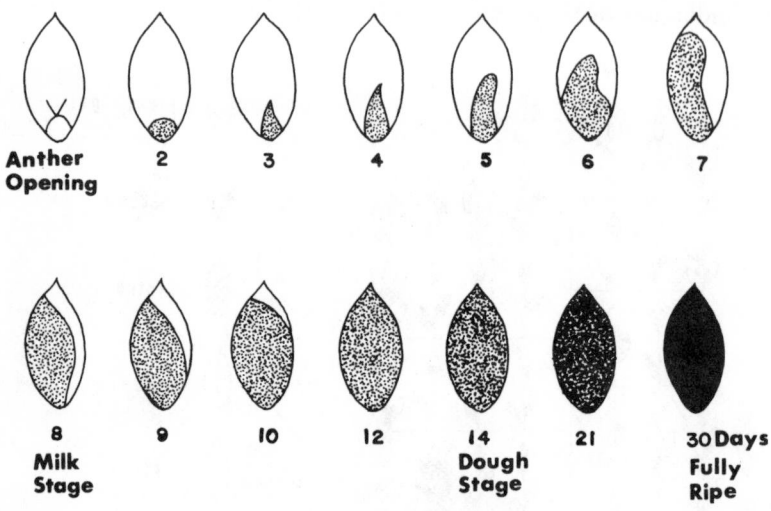

Courtesy of Dr. Benito S. Vergara

FIG. 2.5. STAGES OF GRAIN FORMATION

The endosperm develops rapidly during the first 7 days and the grain usually reaches its maximum weight around 21 days after fertilization.

As the grains ripen, the leaves become senescent and turn yellowish in ascending order. In some cultivars, the culm and upper leaves may remain green even when the grains have ripened.

Low temperature increases the ripening period of the rice plant and this increase has been considered to be partially responsible for the heavier grains and more fully filled grains of rice crops grown in the temperate areas.

Under favorable conditions of moisture and fertility, new tillers may grow from stubble of harvested plants. These new tillers constitute the ratoon crop. Very marked differences are noted in the ratooning ability of cultivars. In the tropics, ratooning is not popular since virus diseases often show up in the ratoon.

EFFECT OF TEMPERATURE ON GROWTH AND DEVELOPMENT

Rice can be classified into Indica, Japonica and Javanica types. The Japonica types are generally resistant to low temperatures—cultivars from California, the United States, Korea, Japan, Spain and other European countries.

In the mountainous regions of the tropics, Indica types are also planted and some degree of cold tolerance is noted. However, unlike Japonica types, very little improvement has been made on Indica types.

A cultivar may be tolerant to low temperature at seedling stage but not at anthesis, making breeding work more difficult.

Figure 2.6 shows the effect of low temperature at different stages of growth (Kaneda 1972). Poor germination, leaf discoloration, stunting, incomplete exsertion of panicle, increased degenerated spikelets, failure in anthesis, increased spikelet sterility, increased grain shattering and delayed heading are some of the adverse effects of low temperature.

Low temperature is most damaging to grain yields if it occurs at panicle initiation stage and at anthesis. The low temperatures generally reported in the literature range from $13°-21°C$ ($55.4°-69.8°F$). This is above the usual concept of low temperature in temperate regions.

High day temperatures ($35°-40°C$, $95°-104°F$) like the rice areas in Iran, Egypt and India, can also be detrimental to the growth of the rice plant. During the vegetative stage, high temperatures can result in reduced tillering and degeneration of the tips of young leaves. During panicle initiation and formation, high temperature can result in reduced spikelet number and degeneration of the spikelets formed. It is at flowering, however, that high temperature is most damaging to rice; it causes high spikelet sterility.

Optimum temperatures for germination, seedling growth, tillering, panicle initiation and development, anther dehiscence and ripening have been determined by a number of workers. Those optimum temperatures, however, vary not only with the cultivar but also the plant status prior to the temperature regime. Having the optimum temperature for certain growth phases of the rice plant probably is not necessary for obtaining high grain yields. In many cases, optimum values are determined on the basis of maximum vegetative growth rather than on direct relevance to grain yield, i.e., having the optimum temperature for leaf area production may result in heavy mutual shading at an early stage of growth resulting in low grain yields.

Grain development is prolonged at low temperatures so that the grains are fully filled. Although the translocation of carbohydrates is slowed down at low temperatures, the prolonged ripening phase results in longer time to fill up the spikelets. Low night temperatures at ripening is favorable since the longevity of the leaves is prolonged and consequently dry matter production is increased. Also, the lower temperature during ripening is considered effective in reducing the respiratory consumption of carbohydrates. Temperatures lower than the critical $20°C$ ($68°F$),

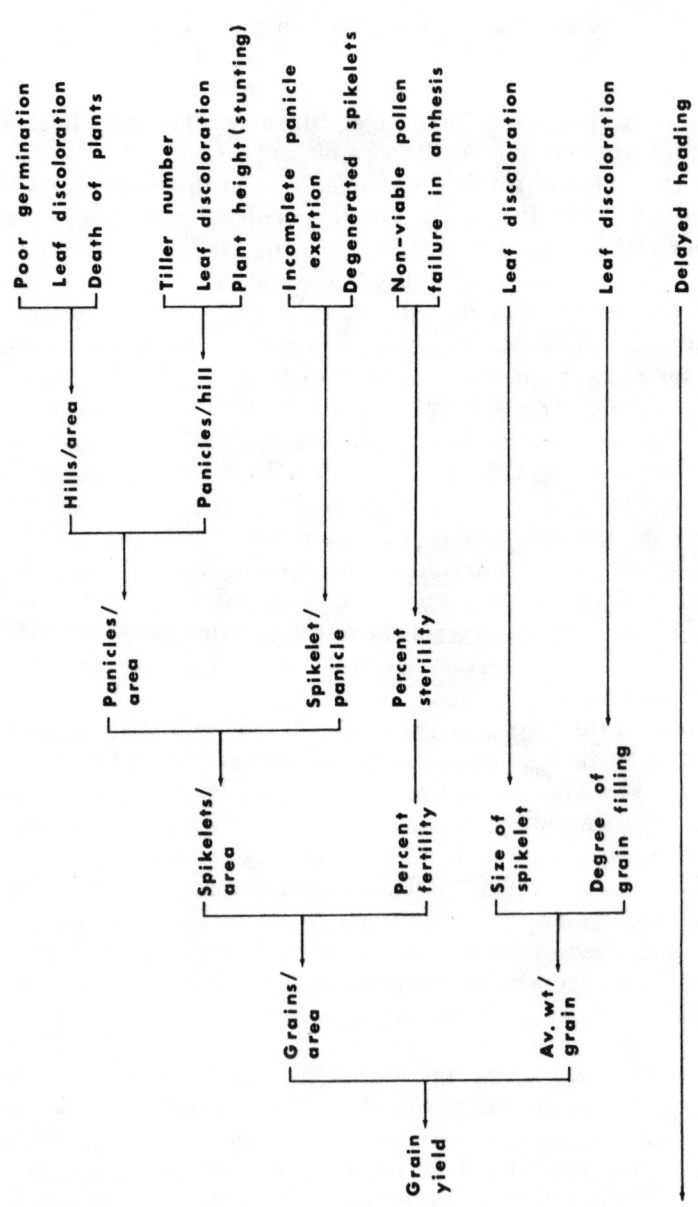

FIG. 2.6. MODIFIED PATH DIAGRAM OF DIFFERENT TYPES OF COLD DAMAGE AFFECTING RICE FIELDS

From Kaneda (1972)

however, may result in imperfect grains; a "notched belly" develops (Nagato and Kobayashi 1957).

Grain ripening is accelerated by high temperatures resulting in prematurity. The impediment in ripening at high temperature is mainly the result of the inability of the spikelet to serve as a "sink" (Aimi et al. 1959; Nagato et al. 1966; Bhattacharya 1970), i.e., the plant may still be able to produce the carbohydrates but the opening to the spikelet is already closed. This prematurity may result in partially chalky kernels and milk-white kernels and increases the thickness of the bran and aleurone layers (Nagato and Ebata 1960; Nagato et al. 1961, 1966).

Low temperature may change the length of the kernel. Generally, rice grains shatter easily from the panicle and dormancy of the grain is shorter when the temperatures during ripening are low (Vergara and Visperas 1971).

It is generally known that at lower temperatures, the harvest index (grain yield/total dry matter produced) is greater. Food partitioning is quite apparent in the rice plant and has been given as one of the reasons for the higher grain yields in the temperate areas. In Japan, the yields and harvest index are higher in the northern than southern regions. In the tropical countries, the rice plant produces more straw than grains. The better rice cultivars have higher harvest index.

EFFECT OF PHOTOPERIOD ON DEVELOPMENT

The main effect of day length on the rice plant is on panicle initiation and development. The traditional cultivars are generally photoperiod sensitive and classified as short day plants. Most of the recent cultivars, however, are insensitive to photoperiod. They have been purposely bred so that there is very little change in growth duration regardless of the season or month during which they are planted.

Insensitivity also increases adaptability. These cultivars can be planted at different latitudes without much change in their growth duration unless temperature becomes a limiting factor.

Photoperiod sensitive cultivars are still being used in areas where standing water remains on the field for long periods of time and there is no means of draining it. Because of photoperiod sensitivity, the rice plant does not initiate the panicle primordium until the high water level has definitely started to recede and insures harvest at a time when the water level is low or the field is almost dry. These cultivars need growth durations from 150 to 200 days and photoperiod can make it possible for these cultivars to have such long growth duration.

REFERENCES

AIMI, R., SAWAMURA, H. and KONNO, S. 1959. Physiological studies on the mechanism of crop plants. The effect of the temperature upon the behavior of carbohydrates and some related enzymes during the ripening of the rice plant. Proc. Crop Sci. Soc. Jpn. 27, 405-407. (Japanese with English summary)

BHATTACHARYA, B. 1970. Effects of various ranges of day and night temperatures at the ripening period on the grain production in rice plants. J. Fac. Agric. Kyushu Univ. 16, 85-140.

KANEDA, C. 1972. Terminal report on studies on the breeding for cold resistance. Int. Rice Res. Inst., Los Baños, Philippines. (mimeo)

NAGATO, K. and EBATA, M. 1960. Effects of temperature in the ripening periods upon the development and qualities of lowland rice kernels. Proc. Crop Sci. Jpn. 28, 275-278. (Japanese with English summary)

NAGATO, K., EBATA, M. and KISHI, Y. 1966. Effects of high temperature during ripening period on the qualities of Indica rice. Proc. Crop Sci. Soc. Jpn. 35, 239-244. (Japanese with English summary)

NAGATO, K., EBATA, M. and KONO, Y. 1961. On the adaptability of rice varieties to high temperature in the ripening periods. Proc. Crop Sci. Soc. Jpn. 29, 337-340. (Japanese with English summary)

NAGATO, K. and KOBAYASHI, Y. 1957. Studies on the occurrence of notched-belly kernels (Dogire-mai) in rice plants. Proc. Crop Sci. Soc. Jpn. 26, 13-14.

VERGARA, B.S. 1970. Plant growth and development. In Rice Production Manual. Univ. of the Philippines, Laguna.

VERGARA, B.S. and CHANG, T.T. 1976. The flowering response of the rice plant to photoperiod: a review of the literature. Int. Rice Res. Inst. Tech. Bull. 8. Los Baños, Philippines.

VERGARA, B.S. and VISPERAS, R.M. 1971. Effect of Temperature on the Physiology and Morphology of the Rice Plant. Int. Rice Res. Inst., Los Baños, Philippines. (mimeo)

3

Genetics and Breeding

Te-Tzu Chang
Cheng-Chang Li

Much of the increases in rice yield after World War II stemmed from expanded irrigation works, increased use of chemical fertilizers, effective control of pests and widespread adoption of improved varieties. The combined use of high yielding varieties (HYVs) and nitrogen fertilizers in irrigated areas has contributed more than any other factor to increased rice yields in South, Southeast and East Asia since the late 1960s. Varietal improvement triggered the beginning of the Green Revolution among Asian rice farmers which began in 1967—68.

GENETICS

Genetic diversity and genetic information provide the foundation for all sound and efficient plant breeding programs. The first genetical study on rice was probably made by Van der Stok (*c* 1908) in Java (Ramiah and Rao 1953). Since then several hundred publications on inheritance studies have appeared in rice literature but a relatively small proportion of the papers were related to traits of economic importance (Chang 1966). It remained for the simply inherited recessive gene in the Chinese semidwarfs to provide rice breeders with the impetus to significant progress in rice breeding.

Morphologic Traits

The most frequently studied subject has been the inheritance of anthocyanin pigments on different plant parts, although the color manifestations are of secondary economic importance. Rice cultivars vary greatly in pigmentation, distribution and intensity thus providing a fascinating array of materials for taxonomic or genetical studies. Anthocyanin pigments can be found in all of the vegetative organs and several floral parts but not in the embryo or endosperm.

Detailed studies by Indian, Japanese and United States workers have elucidated and simplified the genetic control of anthocyanin pigments by a complementary gene system of 3 basic genes (*C, A* and *P-*) and the occasional participation of an inhibitor gene. In tropical varieties other modifying genes may also operate. In the *C-A-P-* gene system, *C* is the basic gene for chromogen production, *A* controls the conversion of chromogen into anthocyanin and *P-* (to be specified as *P, Pau, Pg, Pin, Pl, Pla, Plg, Pr, Prp, Ps, Psh, Pu, Px* or *Pw*) determines the site or sites at which the pigments will appear. Multiple alleles and different dominance levels have been found in the *C, A* and *Pl* loci. Inhibitors for *P, Pl* and *Pla* have been identified. The gene for apiculus color (*P*) has been helpful in analyzing the inheritance of pigmentation in other vegetative organs (Nagao 1951), although exceptions have been found (Jones 1929; Ramiah 1945).

TABLE 3.1. MENDELIAN GENES OF RICE (*O. Sativa*)[1,2]

Genes or alleles	Phenotype and gene system
A^s, A^E, A, A^d, A^m, a	Anthocyanin activator; a member of the $C, A, P-$ complementary gene system
al	Albino; multiple loci
am	Amyloseless or glutinous; synonym of *wx* (waxy)
An˙	Awned, multiple loci
as	Asynaptic sterile
au	Rudimentary auricles; multiple loci
Ban	Brown awn
bc	Brittle culm
Bd	Beaked hull (tip of lemma recurved over palea)
Bf	Brown furrows (dark pigments in furrows of lemma and palea); inhibited by *I-Bf*
Bh	Black hull; complementary genes
bl	Physiologic discoloration of leaf showing dark brown or blackish mottling; multiple loci
bn	Bent node (culm forms angle at a node)
Bph or *bph*	Brown planthopper resistance; multiple loci
Br	Branching at nodes
$C^{Bs}, C, C^B, C^{Bp}, C^{Bt}, C^{Br},$ $C^{Bd}, C^{Bk}, C^{Bc}, C^{Bm}, c^*$	Chromogen production; a member of the $C, A, P-$ complementary system; higher alleles express in several plant parts
Ce	*Cercospora* resistance
chl	Chlorina (chlorophyll deficiency)
Cl*	Clustered spikelets, also super-cluster (*En-Cl*)
cls	Cleistogamous spikelets
clw	Claw-shaped spikelets
d	Dwarf, multiple loci
da	Double awn
Dn	Dense or compact panicles; epistatic to *Ur*
Dn_2	Dense vs lax
Dn_3	Normal vs lax
Dp	Depressed, underdeveloped palea

TABLE 3.1. (*CONTINUED*)

Genes or alleles	Phenotype and gene system
dp	Depressed, underdeveloped palea (in Jpn. testers)
ds	Desynapsis
dw	Deep-water tolerance or floating ability; complementary genes
Ef	Early flowering (low photoperiod sensitivity; polymeric series)
En	Enhancer (intensifier), e.g., *En-C* and *En-Cl*
er	Erect culms vs spreading or procumbent
Ex	Exserted vs enclosed panicle
Fg	Fuzzy hull vs normal pubescence
Fgr	Fragrant flower
fs	Fine stripe
g^*	Long sterile lemmas; multiple loci
ga	Gametophyte genes; multiple loci
Gd	Grain dormancy, anisomeric series
gf	Gold furrows on palea and lemma
gh^*	Gold hull vs straw
gl^*	Glabrous leaves, glumes; multiple loci
Glh or glh	Green leafhopper resistance; multiple loci
Gm	Extremely long sterile lemma or lemmas; epistatic to *g*
Gs	Grassy stunt (virus) resistance
gw	Green, white stripes
H^m, H^i, H^g, H^f	Non-anthocyanin hull colors; expresses in presence of *Gh*
He	*Helminthosporium* resistance
Hg	Hairy glume
Hl	Hairy leaf
hsp	Hull spot
I or i	Inhibitor gene, e.g. *I-P1*, *i-Se*
1−	Lethal character
1a	Lazy or ageotropic
Lf^*	Late flowering; multiple loci
lg^*	Liguleless (auricles, collar also absent); synonomous to juncturaless
Lh	Long, dense hairs on leaf blade vs normal pubescence
lmx	Extra lemma (polyhusk)
lu	Lutescent (seedling loses chlorophyll, dies)
lx	Lax vs normal panicle (in Jpn. varieties)
Lx	Lax vs normal panicle
Lx_2	Lax vs compact panicle
me	Multiple embryos (polyembryonic)
mp	Multiple pistils
ms	Male sterile
nal	Narrow leaf; multiple loci
nk	Notched kernel, or *I-Nk*
nl	Neck leaf
o	Open hull (parted lemma, palea)
P, P^k, P^c, p (or $Ap, ap)^*$	Purple apiculus in the presence of *C, A*; multiple alleles in *P* series; multiple loci in *Ap* series
Pau	Purple auricles in presence of *C, A*
Pc	Purple coleoptile
Pg	Purple sterile lemmas in presence of *C, A*
Ph	Phenol staining of hull, bran
Pi	*Piricularia* resistance; multiple loci; multiple alleles
Pin	Purple internode in presence of *C, A*
Pl^*	Purple leaf in presence of *C, A;* purple leaf

TABLE 3.1. *(CONTINUED)*

Genes or alleles	Phenotype and gene system
Pla	Purple leaf apex in presence of *C, A*
Plg	Purple ligule in presence of *C, A*
Plm	Purple leaf apex, margin
Pn*	Purple node in presence of *C, A*
Pr	Purple lemma, palea in presence of *C, A*; multiple loci
Prp	Purple pericarp; multiple loci
*Ps**	Purple stigma in presence of *C, A*; multiple loci
Psh	Purple leaf sheath in presence of *C, A*
Pu	Purple pulvinus in presence of *C, A*
Pw	Purple wash (faded purple leaves) in presence of *C, A*
Px	Purple leaf axil in presence of *C, A*
*Rc**	Brown pericarp with reddish speckles, basic to *Rd*
Rd	Red pericarp in presence of *Rc*
ri	Verticillate (whorled) arrangement of rachises
rk	Round spikelet in Jpn. varieties
Rk	Short, round spikelet in Indian varieties
rl	Rolled leaf
Rsv	Stripe virus resistance; complementary genes
s	Sterility (general); multiple loci
Sc	*Sclerotium oryzae* resistance
sd	Semidwarf; multiple loci
*Se**	Strong photoperiod sensitivity; one or duplicate loci; suppressed by *i-Se*; epistatic to genes for short BVP (*Ef* series)
Sg	Permeability of testa to water
sh	Shattering in Jpn. varieties
Sh	Shattering in Indian varieties
Sk	Scented rice
sn	Sinuous neck
spr	Spreading panicle branches
T	Tallness (general); *I-T* or *i-T*, short non-dwarf plant
Tf	Difficult (tough) dehulling
th	Easy threshing
tk	Twisted kernel
tl	Twisted leaves (no midrib)
tri	Triangular spikelet
Ur	Undulate rachis
*v**	Virescent seedling or green, white stripe; multiple loci
vr	Variegated seedling
wb	White belly in endosperm
wc	White core or center; synonomous to wb_2
*Wh**	White hull
ws	White stripes on leaves, culm, spikelets
*wx**	Waxy or glutinous endosperm
Xa or *xa*	*Xanthomonas oryzae* resistance; multiple loci
xt	*Xanthomonas translucens* resistance; multiple loci
y	Yellow leaf of seedlings, non-lethal; *l-y* xantha, lethal
Yd	Yellow dwarf (virus) resistance
z	Zebra stripes on seedling leaves

TABLE 3.1. *(CONTINUED)*

Source: Adapted and modified from Kinoshita (1976).
[1]Symbols recommended by FAO Rice Comm. (IRC 1960). The list includes revisions and additions.
[2]Letter superscript for multiple alleles; letter subscripts for complementary genes; numeral subscripts for duplicate loci
or polymeric genes.
*Linkage group known in both Japanese and Indian testers.

Table 3.1 enumerates the important genes controlling pigmentation and other morphological characteristics. The complexity of pigmentation patterns in Japanese testers may be illustrated as follows:

(1) The orders of dominance among the multiple alleles of the 3 basic genes are:

$$C^{Bs} > C^B > C^{Bp} > C^{Bt} > C^{Br} > C^{Bm} > c$$
$$A^s > A^E > A > A^d > a$$
$$Pl > Pl^{\bar{w}} > pl$$

(2) Color development is affected by light intensity, growth stage, and fading and leaching. The interaction of genes in Japanese varieties and their differential expression at 2 growth stages may be illustrated by the following genotypes (Takahashi 1957; Nagao *et al.* 1962; Kinoshita 1976):

Genotype	Apiculus Color at Flowering	Apiculus Color at Maturity
$C^B A P$	Blackish red purple	Faded purple
$C^{Bp} A P$	Pansy purple	Faded red purple
$C^{Bt} A P$	Tyrian rose	Faded pink
$C^{Bm} A P$	Rose red	Straw white
$C^{Bm} A^E P$	Faint red	Almost white
$C^B A^d P$	Amaranth purple	Tawny
$C^{Bp} A^d P$	Pomegranate purple	Light tawny
$c A P$	White	Straw white
$C^{Bp} a P$	White	Tawny
$c a P$	White	Straw white

(3) Pleiotropic effect is shown by the P and Pl genes. For instance, Pl also affects pigment distribution in the leaf sheath, pulvinus, auricles, ligules, internode, node and rachis.

(4) Color of the different layers of seedcoat can be affected by different genes or sets of genes: Rc and Rd in complementary fashion; C, A and Prp; $Pl^{\bar{w}}$ and C.

The probable number of genes controlling the pigmentation expression in the various organs may vary from 2 to 7 (see Chang 1964). Among the morphologically diverse varieties of India, as many as 11 genes may be involved in the color segregation of 3 crosses (Kadam, in Jodon 1955).

Several nonanthocyanin pigments control hull colors. Gold hull is inherited as a simple recessive (gh); white hull, a single dominant (Wh). In the presence of Gh, the H^m, H^i, H^g and H^f alleles confer other dark hull colors. Brown furrows on glumes are controlled by a single dominant, Bf; brown spots on glumes are due to bf; black hull by complementary genes, Bh_a and Bh_b (Anon. 1963).

Pubescence on the surface of leaf blades and hull is dominant to glabrous (gl) (Ramiah and Rao 1953; Jodon 1955). However, some varieties may have hairs on the blade surface, while the glume surfaces are smooth. In Japanese varieties, 2 complementary genes, Hl_a and Hl_b, in addition to Gl, confer long hairs to leaves. Another Hg gene controls long pubescence on floral glumes and also affects pubescence on leaf margins, auricles and rachises (Nagao et al. 1960). Smooth hull is a desirable trait in mechanically harvested and processed rice. It has been incorporated into all varieties of southern United States.

Dwarfs characterized by shortened internodes and undersized grains are generally controlled by single gene recessives (d), but independent genes (d_1, d_2, etc.) govern different types (Jones 1933; Nagao 1951). Double dwarfs or intermediate types may be obtained from dwarf × dwarf crosses (Akemine 1925; Hsieh 1962). In one case, a single dominant controls the dwarf character (Sugimoto 1923). Most dwarfs respond to gibberellic acid. The discrete monogenic inheritance of the Japanese dwarfs was expressed when they were crossed with Japanese cultivars of intermediate height, but such segregation may not be found in crosses with tall varieties of the tropics (Loresto and Chang 1977). From crosses of tall Indica × Japanese dwarfs and of tall Indica × Japanese cultivars, it appears that the dwarfs carried another recessive gene for short height (i-T) which was also present in the background of the Japanese cultivars (Heu and Suh 1974; Loresto and Chang 1977). On the other hand, an intermediate statured tropical variety such as Baok may also carry a dominant inhibitor to tall height, I-T (Balakrishna Rao et al. 1972).

Intermediate statured dwarfs with normal panicles and grains are frequently found in irradiated progenies, frequently under monogenic control (Shah et al. 1961; Huang 1961). The genetic control of agronomically useful semidwarfs will be discussed in the section on Quantitative Traits in this chapter.

Different genetic postulates have been made to interpret the inheritance of awning when parents not only differ in awn length but on the presence of awns on different parts of the panicle. In Japanese varieties from 1 to 3 genes (An_1, An_2 and An_3) with cumulative effect could be involved in varying degrees of awning (Nagao and Takashi 1942).

Long or full awn: $An_1 An_2 An_3$ or $An_1 An_2 an_3$
Medium awn: $An_1 an_2 An_3$ or $An_1 an_2 an_3$
Short awn: $an_1 An_2 An_3$ or $an_1 An_2 an_3$
Tip awn: $an_1 an_2 An_3$
Awnless: $an_1 an_2 an_3$

In Chinese varieties, F_2 ratios of 3:1, 9:7 and 15:1 have been reported (Kuang et al. 1946).

Compact or dense panicles behaved as a single dominant (Dn) over lax or normal panicles, while in other studies lax panicles (Lx) were dominant over normal panicles (Anon. 1963). In other studies, F_2 ratios of 9:7 with the dense panicle either dominant or recessive, have been reported (Ghose et al. 1960). Spreading panicle branches (spr) behaved as a recessive to the nonspreading type (Anon. 1963).

A round or extremely short spikelet (Rk) has been reported to be dominant over the slightly longer oval type (Ramiah and Rao 1953) involving 1 gene (Chao 1928) or as many as 4 complementary genes (Kadam and D'Cruz 1960). A large grained mutant behaved as a simple recessive to the normal (Ramiah and Rao 1953). Another large grain, semisterile mutant was also inherited as a single recessive (Nagai 1958). Wide and continuous variations in the grain dimensions of commercial varieties suggest that grain size and shape are complex quantitative traits.

Genetic information on many other morphologic mutants or variants of largely academic interest is also summarized in Table 3.1. Descriptions of such mutant traits are given by Chang and Bardenas (1965).

Quantitative Traits

Evidence of quantitative inheritance may be detected in nearly all of the traits that have economic importance. However, traits of a quantitative nature have not received sufficient attention from rice researchers. On the other hand, several truly quantitative traits have been oversimplified as qualitative or oligogenic cases of inheritance.

Ikeno (1919, see Ramiah 1933) was probably the first rice worker to recognize the quantitative nature of plant height in one of his crosses. Ramiah (1933) also found both oligogenic and continuous segregations for plant height from different crosses. Genetical studies after 1950 were aided by biometrical techniques in the interpretation of quantitative data.

Plant stature is an important trait that is related to the harvest index, growth duration, nitrogen responsiveness and lodging resistance. The short culm length of the Chinese semidwarfs $(c\ 1\ m$ in the tropics) is primarily controlled by 1 pair of recessive alleles (sd) which expresses potently in crosses with tall tropical varieties such as Peta (Chang *et al.* 1965; Shen *et al.* 1965; Aquino and Jennings 1966; Heu *et al.* 1968). The quantitative and rather complex nature of semidwarfism is indicated by:

(1) a dominance of tallness which is incomplete,
(2) a varying number of modifiers (minor genes of extremely unequal or oppositional effect) affecting the expression of the major recessive gene in sister selections homozygous for the *sd* alleles, and
(3) some other crosses involving moderately tall and semidwarf parents which have failed to produce the F_2 ratio of 3 tall:1 semidwarf (Chang *et al.* 1965; Shen *et al.* 1965; IRRI 1967A,B) indicating the interaction between *sd* and other genes in the background of the tall parents.

The recessive gene in the Peta/I-geo-tze cross, however, showed a high heritability estimate of 71–84% (Chang *et al.* 1965). The recessive nature of the semidwarfing gene not only facilitates the selection for true breeding F_2 plants, but also aids in recovering the desired combination of semidwarfism with erect leaves, high tillering and early maturity. The semidwarf stature in several Assam rices of India, several induced mutants of Indica origin, a spontaneous mutant of Burma, and an induced mutant from Calrose 76 can be traced to the same recessive gene (Hu *et al.* 1970: IRRI 1974A; Rutger and Peterson 1976), which is apparently a readily mutable locus.

Another useful source of short stature once used by the United States and the International Rice Research Institute (IRRI) breeders was from selections derived from Century Patna 231/SLO-17. The 1 m height was controlled by polymeric genes which were nonallelic to the Dee-geo-woo-gen recessive gene (IRRI 1967A; Chang 1967C; Loresto and Chang 1977). A semidwarf from France, Fanny, carried the same set of genes as in CP231/SLO-17. In a slightly taller intermediate dwarf from the United States, 2 recessive genes control the short stature (Loresto and Chang 1978).

Between parents that do not markedly differ in height, polymeric genes or polygenes determine the differences in plant height (Sakai and Niles 1957; Mohamed and Hanna 1964; IRRI 1967A; Wu 1968; Kawano and Takahashi 1969; Li and Chang 1970; Li 1975B). In such crosses tall height generally showed partial dominance to short stature (Wu 1968; Li and Chang 1970). Plant height and growth duration are positively correlated in intermediate to tall genotypes (Chang and Tagumpay 1970; Chang *et al.* 1973).

In a half diallel cross involving 2 dwarfs, several semidwarfs, an intermediate dwarf and 2 tall tropical varieties, a semidwarf such as Fanny showed an excess of dominant alleles, while a dwarf showed an excess of recessive alleles. The other parents have nearly equal proportions of dominant and recessive alleles. Significant genotype × season interactions also added complexity to such genetic analysis of culm length (Loresto and Chang 1977). Differences in F_1 plant height between reciprocal crosses of *O. nivara* × Indica varieties have been observed (R.C. Dolores, unpublished thesis).

Path analysis of lodging resistance in different varieties and crosses showed the significant contributions from plant height, leafsheath wrapping of the lower internodes, the length of the 2 basal internodes, and the cross sectional area of culm tissue at the basal internodes to straw strength (Chang and Liu 1967; Chang 1967B). Plant height ranks as the primary causal factor. In a tall × dwarf cross, height was negatively correlated with the lodging resistance factor (cLr) of the culms (r = −0.71); however, in a tall × intermediate-tall cross, the correlation between height and cLr decreased to a r value of −0.58 (Chang 1967B). Information on character association in the above crosses was summarized by Chang and Vergara (1972).

Angle of the lower leaves in crosses involving parents of contrasting plant types turned out to a dynamic trait that the angles of F_1 plants were predominantly droopy at the juvenile stage, but the leaf angles of F_2 plants later turned into a nearly normal distribution at heading with slightly more erect leaf progenies (see Chang and Vergara 1972). In several crosses, the horizontal flag leaves showed partial dominance to erect ones (Chang and Vergara 1972; Kramer 1973). In the progenies of a tall × semidwarf cross, erect flag leaves in semidwarf lines showed the largest correlation value with grain yield; in intermediate and tall lines, erect leaves below the flag leaf were associated with higher yield levels (Chang and Tagumpay 1970).

Leaf length, leaf width and total leaf area showed nearly normal distribution in some crosses but also indicated complex gene interactions in other crosses (Kawano and Takahashi 1969; Chang and Vergara 1972; Kramer 1973).

Tiller number and panicle number are positively correlated, although the ratio of tillers to panicles may vary among cultivars and is subject to environmental influence. The quantitative nature of tillering ability was indicated by the early studies of Ramiah and Rao (1953). Panicle number per plant has been shown to be largely controlled by additive gene action. High panicle number is partially dominant to a low count (Wu 1968; Li and Chang 1970; Li 1975B). Maternal effect on panicle number has been reported by Wu (1968).

In diallel crosses panicle number was controlled by both additive and dominance effects. High panicle number was partially dominant to a low count. But different parental arrays varied in the order of dominance and that different parents carried unequal proportions of dominant and recessive alleles (Wu 1968; Li and Chang 1970; Li 1975B).

Differences in panicle length are largely controlled by additive gene action with dominance at some loci. Long panicles are partially dominant to short ones (Wu 1968; Li and Chang 1970; Li 1975B). Panicle length shows positive correlation with plant height and growth duration (Chandraratna 1964; Chang et al. 1973).

The number of primary branches on a panicle appears to be partly controlled by additive genes but is also affected by complex gene interactions (Wu 1968).

The number of spikelets on a panicle partly determines the density of the panicle. Additive effect and partial dominance control spikelet number in diallel crosses (Wu 1968; Li and Chang 1970; Li 1975B). The direction of dominance varies from cross to cross (Li and Chang 1970). In some Indian varieties, lax panicles (lx) were monogenically dominant to compact ones (Parnell et al. 1922; Ramiah 1930), while in other crosses F_2 ratios of 9 lax: 7 compact and 7 lax: 9 compact were also obtained (Ghose et al. 1960). In Japanese varieties, lax panicles appeared as a recessive trait under the control of 1 or 2 duplicate genes (Takezaki 1932; Morinaga and Kuriyama 1948).

Panicle exsertion varies from completely enclosed to partly enclosed and from exserted to fully exserted. The extreme types of poor exsertion may be controlled by 3 recessive genes (Sethi et al. 1937). Continuous variation due to polygenes appears more plausible (Ramiah 1932).

Grain dimensions are traits of economic importance, but genetical studies related to size and shape were rather limited in number and scope. Genetic postulates on grain length vary from monogenic (Chao 1928; Ramiah et al. 1931), digenic (Bollich 1957), trigenic (Ramiah and Parthasarathy 1933), to essentially polygenic inheritance (Jones et al. 1935; Morinaga et al. 1943; Chakravarti 1948; Mitra 1962; Nakata and Jackson 1973; T.M. Chang 1974; Somrith 1974). Dave (1939) obtained the following order of dominance when discrete segregation patterns were observed: long > medium > short > very short, although in other cases, short was dominant to long (Anon. 1963). In the case of polygenic inheritance, both additive and dominance effects were detected (Somrith 1974), although the direction of dominance varied from one cross to another (Jones et al. 1935; Poonai 1966; Somrith 1974).

Grain width showed polygenic (Ramiah and Parthasarathy 1933; Jones et al. 1935; T.M. Chang 1974) or polymeric (3 to 5 genes) inheritance with narrow spikelet partially dominant over broad spikelet (Bollich 1957).

Grain shape is expressed by the ratio of length/width. The ratio showed essentially normal distribution in 3 crosses (Nakata and Jackson 1973;

Somrith 1974). Long grain and slender shape are frequently correlated (Ramiah and Parthasarathy 1933; Somrith 1974), although exceptions are also known (Sakai and Pinto 1959).

Grain thickness also appeared to be under polygenic control (Nakata and Jackson 1973).

Grain weight (estimated by the total weight of 100 grains) showed as an essentially normal distribution (Chandraratna and Sakai 1960; T.M. Chang 1971; Somrith 1974), although some degree of nonallelic interaction effect was also detected (Somrith 1974). The maternal influence on grain weight has been reported (Chandraratna and Sakai 1960; T.M. Chang 1971; Somrith 1974).

Heritability estimates of economic traits based on F_2 or subsequent generations show the following decreasing order in magnitude: number of days from seeding to heading (range: 53—95%), spikelet length (7—91%), plant height (32—93%), panicle length (25—84%), spikelet width (8—69%), grain weight (40—60%), number of grains per panicle (24—33%), panicle weight (18—36%), panicles per plant (9—72%), tillers per plant (10—35%), and grain yield per plant (8—22%) (Bollich 1957; Sakai and Niles 1957; Nei 1960; Chandraratna 1964; Chang et al. 1965; Yang 1970; Sakai and Pinto 1959; Li and Chang 1970; Ikehashi and Ito 1971; Somrith 1974).

Growth Characteristics and Ecological Adaptation

Rice cultivars vary greatly in their growth duration and reactions to changes in day length and temperature. The growth duration of a cultivar not only determines its suitability in a particular crop season at a given location but also affects its range of adaptiveness at different locations. The description of earliness or lateness is not as biologically meaningful as the characterization of the components of growth duration: the basic vegetative phase (BVP), the photoperiod sensitive phase (PSP) and the optimum photoperiod at which the shortest growth duration in a photoperiod sensitive genotype is obtained (see Chandraratna 1964; Vergara and Chang 1976). By studying diverse germplasm under controlled photoperiods, varieties may be categorized as completely insensitive, essentially insensitive, weakly sensitive and strongly sensitive (see Chang and Vergara 1971; Vergara and Chang 1976).

In crosses between insensitive parents that differ markedly in BVP, the F_1 plants frequently show dominance or overdominance of earliness because a series of dominant Ef genes control the BVP. The polymeric Ef series may range from 1 gene to several genes and are cumulative but unequal in effect (Chang et al. 1969). They belong to the anisomeric class of Grant (1964). The *Ef* genes may explain the monogenic (Hector 1922; Ramiah 1933) or multigenic control heading date with earliness partially dominant to lateness (Jones 1928; Kudo 1968; Khaleque and Eunus 1975).

The *Ef* genes may also account for some of the continuous distribution in the F_2 population which is markedly skewed towards the early-maturing parent (Jones 1928; Kudo 1968; Chang *et al.* 1965; Chang *et al.* 1969; Heu *et al.* 1968). In crosses involving parents that differ little in growth duration, the F_2 distributions appear essentially normal indicating additive gene action of many minor genes or polygenes (Chandraratna 1964; Shen *et al.* 1965; Chang *et al.* 1969; Li and Chang 1970).

In other crosses where photoperiodic reactions of the parents were not determined, lateness is dominant to earliness. The F_2 distribution may be 3 late: 1 early or 9 late: 7 early (Ramiah 1933; Jones 1933; Jones *et al.* 1935; Nandi and Ganguli 1941; Kawase 1961; Sen *et al.* 1964). Although the late flowering (*Lf*) gene or genes have been postulated to explain such F_2 segregation, the probable role of photoperiod sensitive gene (*Se*) or genes cannot be entirely ignored.

Strong photoperiod sensitivity is controlled by 1 (*Se*) or 2 (*Se₁* and *Se₂*) dominant genes (Chandraratna 1955; Yu and Yao 1957; Sampath and Seshu 1961; Kuriyama 1965; Chang *et al.* 1969). The *Se* genes are potent in expression in that panicle initiation is completely suppressed under long day length in spite of a short BVP as long as the prevailing day length has not reached the critical photoperiod (the lower limit of photoperiod at which panicle initiation will be triggered); i.e., *Se* is epistatic to the *Ef* genes under a long photoperiod. Moreover, an inhibitor (*i-Se*) may suppress the action of *Se* and produce a F_2 ratio of 9 sensitive: 7 insensitive (Chang *et al.* 1969). In crosses between sensitive parents, a short critical photoperiod is dominant to a long one. Likewise, a short optimum photoperiod appears to be dominant to a long one (Li 1970). An association between a long PSP and a short BVP and between a short optimum and a short critical photoperiod showed up in several F_2 populations (Chang *et al.* 1969; Li 1970). In other crosses involving distantly related parents, the F_2 ratios were reversed as 1 (sensitive): 3 (insensitive) or 1:15 in several crosses, suggesting the recessive nature of photoperiod sensitivity. The complicating effects of temperature response may have been involved in these crosses (Sampath and Seshu 1961). The difficulty of studying photoperiod reaction under a changing natural day length has been pointed out by Chang *et al.* (1969).

In strongly sensitive × weakly sensitive crosses, the strong photoperiod response appeared to be dominant to the weak reaction in the F_1 and F_2 plants. In the F_2 populations, the strong photoperiodicity behaved as a single dominant to the weak response. A few insensitive F_2 plants were also found, however (IRRI 1975A).

Variable F_2 segregation patterns appeared in weakly sensitive × insensitive crosses. Although the distribution was continuous and the short PSP plants predominated in all 4 crosses, transgressive segregation for long PSP was obtained in all crosses. A few strongly sensitive progenies were recovered in 1 cross (Lin 1972; Chang and Oka 1976). On the other hand, the segregation for BVP essentially followed the anisomeric type of Ef genes (Lin 1972).

While the partitioning of the growth duration into BVP and PSP under controlled photoperiods can account for the diverse array of previous findings on growth duration (Chang et al. 1969), the complicating effect of temperature responses cannot be ignored. It is known that tropical varieties can be retarded during the vegetative growth phase or during panicle initiation if cool temperatures prevail. Varieties of the temperate zone may show a greatly shortened vegetative growth phase and a shorter panicle development phase if high temperatures occur at the juvenile stage.

A complex of 6 genes was proposed by Fuke (1955) to explain the segregation for photoperiod and temperature responses in Japanese varieties. Late-maturing varieties have all 6 pairs of dominant alleles, medium-late ones have 5 and early ones have 4. One of the photoperiod-sensitive genes appeared to be linked with a gene for red apiculus. However, Fuke did not demonstrate the specific effect of the individual genes. Controlled environments are needed to elucidate photoperiod × temperature interactions (Chang and Oka 1976).

From an early maturing variety of northern China, a modifier gene m_a was found to promote flower initiation under low temperatures in the first crop season in Taiwan. Under higher temperatures in the second crop season, this gene interacted with a basic gene for earliness (E) in further reducing the vegetative growth period. Another modifier m_b (at the same locus as m_a) found in an early maturing variety of northern Japan showed almost the same effect on flower initiation. That m_a and m_b were isoalleles with slightly different effects was demonstrated by comparing isogenic lines carrying those genes (Tsai and Oka 1966, 1970). The E locus could be a compound locus located on the 8th linkage group of the Japanese testers (Tsai 1976).

The responses of rice varieties to low temperatures at specific growth stages and the adverse effects are:

(1) the period from germination to seedling establishment—poor germination and stunted seedlings.

(2) the period from seedling establishment to panicle initiation—slow seedling growth or reduced vigor and leaf discoloration, and

(3) the period from panicle initiation to ripening—stunting, poor tillering, partial panicle exsertion, degenerated panicle tips and sterility;
or delayed heading with irregular maturity (IRRI 1972).

Germinability under cool temperature (15°C, 59°F) is governed by 4 or
more dominant genes. These genes are linked with wx, d_2, d_6 and I-Bf
(Sasaki et al. 1973). In Japanese varieties, 2 or more dominant genes
control tolerance to low air temperatures (19°C, 66.2°F, or slightly below)
at the meiotic phase (Sakai and Shimazaki 1948). Tolerance to cool water
temperatures (18°C, 64.4°F, or slightly below) is controlled by 4 or more
dominant genes that have unequal potencies and unequal additive effect
(Toriyama and Futsuhara 1960; Li 1975B). Genetic correlation of seedling vigor under cool temperatures with earlinesss and large seed size has
been indicated, but these traits were not correlated with adult plant
height (Li 1975B). Heritability estimates of cool tolerance both at seedling stage (Erickson 1968; Li 1975B) and at panicle initiation period (Futsuhara and Toriyama 1971) were high.

The lazy or ageotropic growth habit was inherited as a single recessive,
la (Jones and Adair 1938; Ramiah and Rao 1953). In other crosses, the
erect habit (er) was recessive to spreading (Anon. 1963). Again, a continuous range of variation in the culm angle and the growth phase at
which the prostration of the lower internodes begins to occur may be
found among cultivars.

Shattering of grains from the panicle is more pronounced in the wild
species, the Indica race of $O.$ $sativa$ and the African rices ($O.$ $glaberrima$)
than in the upland varieties, Javanica and Sinica (or Japonica) races of
$O.$ $sativa$. In crosses between a wild strain and a cultivar shattering was
controlled by 1 or 2 dominant genes, Sh (Ramiah and Hanumantharao
1936; Kadam 1936). Because the degree of shattering varies continuously among cultivars, it is not surprising to find that one report by
Jones (1933) indicated the monogenic and dominant nature of shattering
while in another cross, difficult threshing (Th) in Japanese varieties was
dominant over easy shattering (Jones 1933). A multigenic interpretation
appears more plausible (Sethi et al. 1937; Sakai and Niles 1957; Takahashi 1964).

The shedding character is associated with the relative degree of development of the abscission layer between the spikelet and the pedicel.
Shedding appeared to be inversely associated with the thickness of cell
walls in the abscission layer (Nagai 1958). Shattering is generally more
serious in the dry season than in the wet season or when lowland varieties
are grown in upland culture.

Grain dormancy is essential to cultivars that mature during the rainy
season. Dormancy in Japanese varieties was reported to be controlled by

2 complementary genes; high permeability of testa to water by another gene, Sg (Takahashi 1962). In other studies, seed dormancy appeared to be a dominant trait governed by a number of genes (Shanmugasundaram 1953; Narayanan Namboodiri and Ponnaiya 1963). Determination of grain dormancy at several intervals following harvest led to the finding that a varying number of anisomeric genes (Gd) controlled the strength and duration of dormancy (Chang and Yen 1969; Chang and Tagumpay 1973). Grain dormancy is not genetically correlated with growth duration although superficial character association in diverse cultivars has led to the postulate that early maturing rices have no dormancy (Chang and Tagumpay 1973). Diffusible inhibitors from the hull and bran of tropical varieties that confer dormancy have been identified.

The wide adaptive range and stable high yield of the semidwarfs and "Ponlai" varieties under year-round cultivation in the irrigated areas of the tropics has been well established (Chang 1967A; Ram et al. 1970; Evelyn et al. 1971; Seshu et al. 1974; Kikuchi et al. 1975; Suzuki and Kikuchi 1975). Varietal characteristics that contribute to such wide adaptiveness are a moderately long BVP, low sensitivity to photoperiod, moderately high tillering ability, low response to variations in temperature and high panicle fertility (Chang and Oka 1976). Moreover, both TN1 and IR8 were found to have higher levels of heat tolerance than C4-64 (Osada et al. 1973); IR8 could also tolerate higher water temperatures than Japanese varieties (IRRI, undated). On the other hand, for adaptiveness to adverse environments such as a limited growing season, poor soil damage or an adverse soil factor, the reintroduction of photoperiod sensitivity, a moderately tall plant stature along with a specific tolerance or resistance trait may be necessary (Chang and Oka 1976).

Adaptation to deep water or floating ability is essentially the ability for the submerged internodes of a cultivar to elongate and to produce tillers at the higher nodes or both. Earlier studies in India led to the postulate that 2 complementary recessive genes, dw_1 and dw_2, conferred floating ability (Ramiah and Ramaswami 1941). Recent studies in Thailand also indicated that floating ability is a recessive trait. Strong elongation ability is associated with tall plants but not with photoperiod sensitivity (Prechachart and Jackson 1975).

Tolerance to submergence under flood water is also a desirable trait in monsoonal Asia. Varieties tolerant to submergence do not necessarily have good elongation ability. The 2 traits appear to be independent of each other (IRRI 1976A).

Drought resistance in upland culture has been identified as associated with a deep and thick root system, greater ability to retain vigor under stress, higher cuticular resistance, greater stomatal resistance (Chang et al. 1972, 1974; IRRI 1973, 1974A, 1977A) and proline accumulation

(Chu and Li 1979; Li and Chu 1979). Although these attributes have not been studied genetically, the finding of drought resistant progenies largely in crosses involving at least 1 drought resistant parent of the traditional upland type indicates the heritable nature of drought resistance (IRRI 1976A). The polygenic nature of root length and number was indicated in one study (Bhaduri and Bairagi 1968). Low proline accumulation was controlled at least by two or three dominant genes. And high proline accumulation was positively associated with drought tolerance (Chu and Li 1979).

Disease Resistance

Among the rice diseases, resistance to the blast fungus (*Pyricularia oryzae*) has received the closest scrutiny. Mendelian analysis of many crosses where single isolates of the fungus were used indicated that resistance in a variety is controlled by any one of the following mechanisms: (1) from 1 to 4 dominant genes that could be independent or cumulative or complementary in action, (2) 1 major dominant gene and a few minor genes, (3) linked dominant genes, (4) a recessive gene for resistance, (5) a dominant gene inhibiting a resistant gene, or (6) from 5 to 9 effective factor pairs (see Chang 1964; Takahashi 1964; Ou 1972). Resistant progenies were found in crosses between susceptible parents (Woo 1965).

Systematic studies by Japanese workers have differentiated in Japanese and non-Japanese cultivars 14 major genes which confer the vertical type of resistance or "true resistance" to blast (Yamasaki and Kiyosawa 1966; Toriyama *et al.* 1968; Shinoda *et al.* 1971; Kiyosawa 1976). Some of the genes such as the *Pi-k*, *Pi-ta* and *Pi-z* are multiple allelic series. The resistance genes are distributed over 4 linkage groups, I, VII, VIII and X (Kiyosawa 1972, 1974).

For field resistance to blast, Japanese workers have found that either 1 major gene or a major gene and a few minor genes can reduce the number of lesions (Kiyosawa 1972; Toriyama 1972). Three major genes controlled the field resistance of Japanese upland varieties (see Ezuka 1977). Such field resistance genes are located on linkage groups IV and XI (Toriyama 1975).

Leaf reactions to the blast fungus generally agree with the reactions to neck blast. However, several exceptions to this have been found (Chang *et al.* 1965; Willis *et al.* 1968).

In another fungus disease, Hashioka (1951) reported that resistance to sheath blight (*Thanatephorus cucumeris*) is inherited as a dominant character in Japanese varieties and in some crosses a 3R:1S ratio was found in the F_2 populations.

Resistance to brown spot disease (*Helminthosporium oryzae*) in a Korean strain appeared to be a dominant characteristic (Nagai and Hara 1930). On the other hand, resistance was a recessive trait in a separate study by Adair (1941) and several genes could control the disease reactions.

For resistance to the narrow brown leaf spot pathogen (*Cercospora oryzae*), Adair (1941) postulated that 3 genes and other modifying genes could control the resistant reactions and that the resistant progenies were more predominant than the susceptible progenies in several crosses. Subsequent studies by Ryker and coworkers led to the finding that either 1, 2 or 3 dominant genes conferred resistance. Two of the genes were closely linked (Ryker and Jodon 1940; Ryker and Chilton 1942).

Studies in Japan indicate that resistance to the bacterial blight (*Xanthomonas oryzae*) pathogen is controlled either by 1 dominant gene (Xa_1) in the Kogyoku group, 2 linked genes (Xa_1 and Xa_2) in the Rantaj-Emas group or a third gene (Xa_3 or Xa_w) in the Wase Aikoku group for adult plant resistance (Sakaguchi 1967; Ezuka *et al.* 1975). Xa_4 is present in Sigadis, TKM-6, MTU-15, Syntha, Hom Thong and Semora Mangga (Heu *et al.* 1968; Petpisit *et al.* 1977; Librojo *et al.* 1977; Olufuwote *et al.* 1977). The recessive xa_5 gene is found in BJ-1, DZ 192, Kele, Chinsurah Boro II, Dular and Hashikalmi (Petpisit *et al.* 1977; Librojo *et al.* 1977). Xa_6 is in Zenith and Malagkit Sungsong (Sidhu and Khush 1968). The resistance of Lacrosse/Zenith-Nira selection to an Indian isolate is governed by a single recessive gene (AICRIP 1969). A more complex gene system involving 3 resistance genes and an inhibitor was described by Indian workers in the BJ-1/IR8 cross (Jayaraj *et al.* 1972). Complementary genes for resistance were reported by Moses *et al.* (1974). Resistance to enlargement of lesion is controlled by polygenes (Toriyama 1972). IRRI plant pathologists have found that resistance may be recessive or dominant in the F_1 hybrids, depending on the cross combination and the level of resistance in the resistant parent (IRRI 1967B). Indian workers and IRRI researchers also found differential reactions between tillering and flowering stages (Ratho *et al.* 1976; Olufuwote *et al.* 1977).

Although the genetic nature of varietal resistance to bacterial leaf streak (*Xanthomonas translucens* f. sp. *oryzicola*) could be under mono-, di- or tri-genic control (Shekhawat *et al.* 1972; Nayak *et al.* 1975), its heritable nature can be inferred from the rapid progress made at IRRI in incorporating resistance into the improved plant type (Khush and Beachell 1972).

Resistance to grassy stunt virus found in a strain of *O. nivara* (IRRI Acc. 101508) is governed by a single dominant gene (Khush and Ling 1974). However, this strain appears to be genetically different from several other strains of *O. nivara* which do not have the resistance, as indicated by the partial sterility in their hybrids with *O. sativa* cultivars (Ng Quat Nyat, 1977). On the other hand, this strain is cytogenetically closely affiliated with several tropical *O. sativa* cultivars (Dolores *et al.* 1978).

Resistance to the tungro virus appears more complicated. In crosses involving the highly resistant Pankhari-203 and several susceptible varieties, the F_1 hybrids were generally resistant and some of the F_2 populations segregated as 9R:7S (IRRI 1967A). In the cross of Peta/I-geo-tze, the F_2 population showed a continuous distribution in the level of resistance and relatively few resistant semidwarf progenies were recovered (Chang *et al.* 1975B). In the IR8/Latisail cross, Indian workers found resistant F_1 plants and 20 day old seedlings distributed as 9R:7S, but at 60 days following transplanting, the F_2 ratio became 15R:1S (Shastry *et al.* 1972). It appears likely that recovery was involved in the change in F_2 ratio and that different genes may control seedling reaction and adult plant resistance (see Chang *et al.* 1975B).

Resistance to the stripe virus in Japanese upland varieties is controlled by 2 complementary genes, St_1 and St_2. St_1 is located in linkage group I (see Toriyama 1972).

One dominant gene, *Bs*, controls the resistance to the black-streaked dwarf virus (see Toriyama 1972).

An incompletely dominant gene in Japanese varieties confers resistance to the yellow dwarf virus (see Toriyama 1972). The resistance in Firooz-1 and Kabara appeared to be controlled by 1 dominant gene (Lin *et al.* 1975).

Insect Resistance

Investigations on resistance to insect pests are relatively more recent when compared to those on disease resistance. One of the earliest reports on the heritable nature of resistance dealt with plant reactions to stem maggots (*Chlorops oryzae*) in which resistance appeared to be controlled by a single gene which lacked dominance (Fuke and Koyama 1955; Fukuda and Inoue 1962). Later studies indicated that 1 dominant gene controlled resistance (Koyama and Hirao 1971).

Resistance to the stem borers (striped, pink and white types) has been studied by several researchers. Resistance appeared to be generally dominant to susceptibility and could be ascribed to the role of 1 or few genes (see Chang *et al.* 1975B). However, since resistance mechanisms could include nonpreferrance, antibiosis and tolerance (Painter 1951), a multigenic interpretation would appear more plausible. Studies at IRRI showed that in resistant × susceptible crosses the F_2 distribution of plants according to dead heart incidence was continuous, indicating the role of multiple genes (IRRI 1966). By intercrossing 7 moderately resistant parents, a significant increase in striped borer resistance was obtained in subsequent generations (IRRI 1975A).

Studies made by Indian workers indicated that the resistance of TKM-6 to the borers at the flowering stage was controlled by a single recessive gene (Indian Counc. for Agric. Res. 1969).

Resistance to the green leafhopper (*Nephotettix virescens*) has been identified from at least 5 independent sources: Glh_1 in Pankhari 203 and other varieties, Glh_2 in ASD-7 and others, Glh_3 in IR8 and other varieties (Athwal *et al.* 1971; Siwi and Khush 1977), glh_4 in PTB-8, and Glh_5 in ASD-8 (Siwi and Khush 1977).

Resistance to the brown planthopper (*Nilaparvata lugens*) also falls under monogenic control: Bph_1 in Mudgo, MTU-15 and other varieties, bph_2 in ASD-7, PTB-18, H-105 and other varieties (Athwal *et al.* 1971; Chen and Chang 1971; Martinez and Khush 1974), Bph_3 in Rathu Heenati and bph_4 in Babawee (Lakshminarayana and Khush 1977). However, Bph_1 and bph_2 are tightly linked in the repulsion phase (Athwal *et al.* 1971). The allelic relationship between Bph_3 and bph_4 has not been determined. A dominant inhibitor $(I\text{-}Bph_1)$ was found in TKM-6. It is interesting that TKM-6 provided moderate resistance to IR26, IR28, IR30 and IR34, while the variety itself is susceptible to biotype 1 of the insect (Martinez and Khush 1974).

Resistance to the rice gall midge in Indian varieties appeared to be controlled by either a single dominant gene (Satyanarayaniah and Reddi 1972) or 3 complementary dominant genes, one of which is suppressed by a dominant inhibitor (Shastry *et al.* 1972; Sastry and Prakasa-Rao 1973).

Grain Quality

Quality characteristics of the rice grain are related to a complex of physicochemical properties: size and shape, milling recovery, endosperm properties and nutritional value. The inheritance of grain dimensions and shape has been discussed under the section on Quantitative Traits in this chapter. It is feasible to recombine any kind of grain length and shape with different levels of amylose content and gelatinization temperature.

Site and degree of chalkiness of the endosperm, expressed as white belly, white center or white back (see Chang and Bardenas 1965 for description), affect the percentage of whole milled kernels (head rice). In several studies made in India and in the United States, white belly or white center was controlled by a recessive gene each, *wb* or *wc*, respectively (see Anon. 1963). In other studies, white belly appeared to be a dominant trait (Nagai 1958; Chalam and Venkateswarlu 1965). Later studies suggest that a multigenic system interacting with environmental factors appeared more plausible (Seetharaman 1964; Nakata and Jackson 1973; Samoto and Hamamura 1973). Chalky kernels are more frequently associated with bold grain shape than with slender shape of comparable length (Nakata and Jackson 1973; Somrith 1974). Additive effects appeared to be a major component of F_2 variations (Somrith 1974).

High amylose content is incompletely dominant to low and controlled by 1 major gene and several modifiers (Seetharaman 1959; Kahlon 1965; Ghosh and Govindaswamy 1972; Bollich and Webb 1973; Somrith 1974; IRRI 1976A). The role of 2 complementary genes was indicated by another study (Stansel 1966). The amyloseless or glutinous condition has been mentioned earlier as under the control of a single recessive gene (*am* or *wx*) The *wx* gene has been shown to be a compound locus with ultrastructure (Li *et al.* 1965, 1968). Because the endosperm is a triploid tissue, differences in amylose content due to varying doses of the *Wx* alleles have been reported (Sugawara 1953; IRRI 1976A).

Gelatinization temperature of the rice starch presents an intriguing picture. High gelatinizing types were selected from crosses between intermediate and low gelatinizing varieties. Crosses between intermediate and high gelatinizing varieties gave only intermediate and high gelatinizing types. Some crosses between high and low gelatinizing varieties gave high, intermediate and low segregants (Beachell and Stansel 1963). Kahlon (1965) reported that the difference between a high parent and an intermediate parent could be attributed to a few genes and several modifiers. In a high × intermediate cross, Stansel (1966) recovered the parental types only. From a high × low cross, all 3 types were obtained but the intermediate type was deficient and it failed to breed true. Stansel used a system of 3 genes to interpret the concurrent changes in gelatinization temperature and amylose content. On the other hand, the dominance of high gelatinization temperature in such a cross combination was common to several F_2 populations (Heu and Choe 1973; Somrith 1974; IRRI 1976A). Additive gene effect was detected in 2 crosses but higher order interactions complicated the analysis of genetic variation (Somrith 1974).

The genetic control of brown rice protein content has been given some attention in recent years. Low protein content appears to be dominant to high (Erickson 1969; Hillerislambers *et al.* 1973; Chang and Lin 1974). Diallel analysis indicated dominance at some loci, epistasis expressed as overdominance or complementary gene action, and an unequal distribution of dominant and recessive loci. Maternal effect and metaxenia were observed. Both dominance and additive effects were noted in F_2 and F_3 data (Chang and Lin 1974). The heritability estimates for protein content were rather low, probably resulting from genotype × environment interactions. High protein content tends to be associated with early heading and light grains (Hillerislambers *et al.* 1973).

Linkage Groups

The symbols of oligogenic or polymeric genes enumerated in Table 3.1 have been standardized after the recommendations of the Food and Agricultural Organization of the U.N. (FAO) International Rice Commission (Int. Rice Comm. 1959). Some of the symbols may differ from those appearing in earlier publications. Descriptions of various mutant traits have been provided by Chang and Jodon (1963) and Chang and Bardenas (1965).

Among the 108 traits given in Table 3.1, only 30 have economic importance. For hybridization work, the *C-A-P*-gene complex, *Bf, Gh, gl, sd* and *wx* loci are useful as gene markers in distinguishing between true hybrids and selfed seeds.

There are 12 linkage groups as the rice plant has 12 pairs of chromosomes. Many workers have reported on instances of linkage. Through long years of painstaking studies, 12 linkage groups have been identified from Japanese varieties (Morinaga 1938; Morinaga and Nagamatsu 1942; Nagamatsu and Omura 1962; Nagao and Takahashi 1960, 1963; Iwata and Omura 1971A,B; Kinoshita 1976; Takahashi 1977), although 1 group (V) is represented by 2 genes only. An abbreviated diagram of

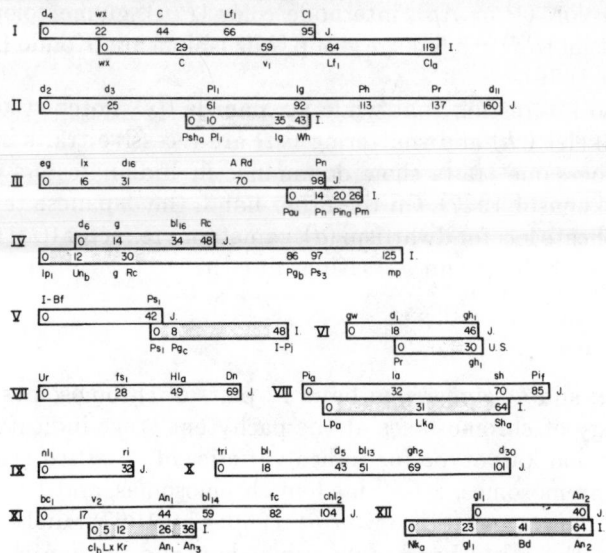

Partly adapted from Kinoshita (1976)

FIG. 3.1. THE 12 LINKAGE GROUPS OF RICE BASED ON MARKER GENES IDENTIFIED FROM 3 VARIETAL TYPES: INDIAN (*I.*) JAPANESE (*J.*) AND UNITED STATES (*U.S.*)

the 12 linkage groups is in Fig 3.1, while the whole set for Japanese varieties may be found in the recent publications of Kinoshita (1976) and Takahashi (1977). Recent studies with trisomics suggest that linkage groups VI, IX and XII may be combined into 1 group (Iwata and Omura 1976). The relationship between the linkage groups and individual chromosomes needs to be determined, however.

Different Indian workers have identified partial sets of linkage groups from different tropical varieties (Richharia *et al.* 1960; D'Cruz and Dhulappanavar 1963; Misro *et al.* 1966; Dhulappanavar 1973; Dhulappanavar *et al.* 1973; Thakur and Roy 1975). The information from various sources in India and the United States (Jodon 1957) has been combined in Fig. 3.1 to indicate their correspondence with the 12 linkage groups of Japanese testers. Additional instances of linkage have been summarized by Adair and Jodon (1966) and Hsieh (1976).

When the marker genes in Indian testers are compared with those in Japanese testers, the tropical varieties have more duplicate loci for several genes that control the distribution or localization of anthocyanin pigments. On the other hand, the Japanese testers have more duplicate loci of *gh* (gold hull) and *I-P1* (inhibitor for purple leaf) (Kinoshita 1976).

Figure 3.1 also shows that the Indian and Japanese testers differ in the relative position of several loci in the same linkage group, e.g. *C, Lg* and *Rc*. The 2 sets of testers also differ in the location of the genes for apiculus color, (*P* or *Ap*), internode color (*Pin*), glume color (*Pr*) and pericarp color (*Rd*) in a linkage group (Mizushima and Kondo 1959, 1962; Kinoshita 1976).

It is also interesting that while lax panicle (*lx*), notched kernel (*nk*), round spikelet (*rk*) and shattering (*sh*) are recessive traits in Japanese testers, the same traits show dominance in Indian testers (see Anon. 1963; Takahashi 1977). On the other hand, the Japanese testers carry more duplicate loci for dwarfism (*d*), gametophyte factor (*Ga*) and golden pigments on hull (*gf* and *gh*) (Kinoshita 1976).

Cytogenetics

O. sativa and *O. glaberrima* have 12 pairs of chromosomes each. The morphology of chromosomes at the pachytene stage indicates that the most common karyotype for Indica varieties of *O. sativa* is 7 to 8 submedian chromosomes, 2 to 3 median chromosomes, and 1 to 2 subtelocentric chromosomes (Shastry *et al.* 1960; Chu 1967; Sur 1975; Dolores 1976). On the other hand, only submedian and median chromosomes were observed in other studies (Das and Shastry 1963; Misra and Shastry 1967). The frequency of chromosome types classified according to arm ratios apparently varies with the strains being studied. The position of the nucleolus organizer also varies from variety to variety. From root tip smears, Khan (1974) observed 11 submedian and 1 subtelocentric chromosomes. The rice chromosomes are not favorable for detailed cytological studies because individual chromosomes do not have distinct and consistent morphological markers.

Since the question of secondary association among the 12 pairs of rice chromosome was brought up in the 1930s, rice researchers' opinions are still divided as to whether *O. sativa* is a primary diploid or a balanced secondary polyploid (see Chang 1964). While a number of duplicate loci has been identified (Table 3.1), critical evidence for reduplicated chromosomes from a presumably basic set of 5 is lacking.

Spontaneous as well as induced haploids, triploids and tetraploids have been reported by rice researchers. Summaries of cytological and morphological observation on the above plants may be found in the reviews by Chang (1964), Nayar (1973) and Khush (1974).

Aneuploids including monosomics and polysomics have been reported. Complete sets of primary trisomics have been obtained by a few workers but sustained efforts to make use of the aneuploids in cytogenetic studies have yet to be made.

Reciprocal translocations and trisomics have been used by Iwata and Omura (1971A,B,1975) to establish the independence of 10 linkage groups. However, the chromosomes of the translocation testers have not been characterized so as to lead to the development of a cytological map.

Cryptic differences in chromosome structure between geographical races of *O. sativa* were once widely believed to be the primary cause for the partial sterility found in interracial and intraracial hybrids (see IRRI 1964; Chang 1964; Hung and Chang 1976). Recent investigations made it clear that chromosome aberrations such as loose pairing at pachytene, univalent formation at metaphase, inversions and translocations together with other nonchromosomal aberrations could only account for a portion of the meiotic aberrations in interracial hybrids. The role of genic imbalance (Demeterio-Velasco *et al.* 1965; Engle *et al.* 1969; Dolores *et al.* 1975), "gametic development" genes and effect of certation (Oka 1953; Nakagahra *et al.* 1974) need to be investigated.

BREEDING

Recorded history may not mention man's earliest efforts to improve the rice plant. However, long before the advent of science, man undoubtedly had made good use of natural variousness in the crop and its wild relatives, spontaneous mutations, natural hybrids and introductions from foreign lands. Susrutha (c 1000 B.C.), in his Ayurvedic *Materia Medica*, recognized the differences among rices existing then in India and separated them into groups based upon their growth duration, water requirements and nutritional values (Ramiah and Rao 1953). Chinese classics show that Emperor Wen Ti of the Wei Dynasty (186–226 A.D.) discussed with his cabinet staff about a quality rice having strong and fragrant aroma. Another Emperor, K'ang Hsi (1662–1723) of the Ching Dynasty, selected an early maturing and aromatic mutant for a crop of rice grown in the imperial garden which later became the main staple of his household. The new strain was named Imperial Rice. The large-scale introduction, testing and extension of the early maturing Champa rices in central and east China during the 11th century marked the first massive government sponsored efforts to utilize efficient and productive genotypes.

Significant increases in rice yield through the adoption of improved varieties were invariably associated with improved irrigation facilities, use of fertilizers and other agricultural chemicals, better cultural practices and pest management, and favorable prices. Instances of such a combined package of technological improvements have been documented by Shen (1964), Hsieh and Ruttan (1967) and Akino and Hayami (1975).

Breeding in High Yielding Countries

According to recent FAO statistics, the high yielding countries of the world are in the order of: Japan, Spain, Italy, the Republic of Korea, Australia and the United States. The high yields attained in southern Europe, Australia and northern California may be partly attributed to the oasis-like environment of the crop season: intense solar radiation, long day length in summer months, moderately low night temperatures, abundance of irrigation water, low relative humidity and minimum incidence of diseases and insects. Another common feature is the long grain filling period of the predominant cultivars which leads to heavy grains.

Italy and Spain.—Breeding programs in Italy and Spain made use of introductions from China, Japan, Philippines and the United States. Parents of the principal commercial varieties are Originario, Chinese Originario and Lady Wright (Ciferri 1960; Tinarelli 1966; Campos et al. 1975). Recent breeding efforts in Spain are directed to developing nitrogen responsive and high yielding varieties with short culms, dark and erect leaves, dense panicles and grains of medium length and width. Emphasis is also placed upon improving milling recovery and adaptiveness to direct seeding (Campos et al. 1975).

Japan.—Perhaps Japan has had the longest history of scientific rice breeding among the major producing countries. The first variety developed by hybridization, Ominishiki, came out in 1906 (Matsuo 1957).

Since 1927 a network of breeding stations has been set up to meet the varying ecologic requirements of different areas. The network consisted of 7 national (regional) experiment stations, 9 prefectural experiment stations and 50 experiment farms. During 1962 the network was reorganized on a more refined ecoedaphic basis to include 3 breeding centers, 8 breeding stations and 40 experiment farms. Such a network provided an ample opportunity for breeders and their associates in related disciplines to tackle specifically district and regional problems such as cool tolerance, blast resistance, physiologic and insect problems associated with moderately high temperatures during summer, adverse soils and special quality for making rice wine. Moreover, experiment stations of the National Institute of Agricultural Sciences and of many universities conducted research of a more basic nature (Okabe 1972; Toriyama 1976). Annual conferences provided liaison and rapport among the rice workers.

High yield under heavy fertilization has been a major breeding objective during the last 60 years. Associated with high yield levels, emphasis was also given to blast, bacterial blight, stripe virus, short culms for lodging resistance, cold water tolerance, tolerance to the Akiochi physiological disorder and earliness for the warmer regions. More recently, adaptiveness to direct seeding, mechanized transplanting and combine harvesting as well as resistance to the brown planthopper and sheath blight are added breeding objectives. High panicle number was sought after until 1960, after which large spikelet number per panicle became an important objective. Almost imperceptibly the newer varieties are approaching the improved plant type advocated by Japanese crop physiologists: shorter culms, thick and erect leaves. Yield increases were concurrent with the intensive breeding efforts: more than 70% since 1925 (Okabe 1972).

In their experimental approach, Japanese breeders have made good use of the bulk method of selection, biometrical techniques for breeding research, induced mutations, disease and insect resistance from different sources and short day culture in the greenhouse for rapid generation advance (Kaneda 1976).

United States.—The historical aspects of rice breeding in the United States have been described by Jones (1936) and Adair *et al.* (1966). Since the breeding programs were initiated in each of the 4 major producing states in the south, the primary objectives were for high yield and stable production and a range of grain types (short, medium and long) for domestic and foreign markets. Disease resistance received more attention in the southern states than in California where cool tolerance, especially tolerance to cold irrigation water, is more critical. Milling, cooking and processing qualities are also prime criteria in the evaluation of hybrid progenies. During the period from 1951−61, the grain yields of 18 selected varieties grown at 6 locations have shown marked increases over the check varieties, ranging from 5–8% (Adair *et al.* 1966). The increases ranged from 2–31% for 5 locations during 1961–69 (Adair *et al.* 1973). A list of principal varieties developed between 1917 and 1971 was provided by Johnston *et al.* (1972).

Since the mid 1950s, U.S. breeders have devoted much attention to developing an improved plant type that would combine short plant stature, nitrogen responsiveness, lodging resistance, early maturity and high head rice yields with the desired grain quality (Johnston *et al.* 1972). In recent years, many crosses were made with the semidwarfs of tropical Asia and short culmed mutants. Recent success in California in developing short statured varieties made use of mutants induced from California varieties which provided a semidwarfing gene allelic to that of IR8, but which were easier to utilize (Rutger and Peterson 1976).

Strength of the breeding programs in the United States draws from cooperative federal and state efforts, acquisition and use of diverse germ plasm, and collaborative research with rice researchers in other disciplines. The increased and more efficient use of nitrogen fertilizers through growth stage specific timing, new chemical means of controlling weeds and ratoon cropping of early maturing varieties have contributed greatly to the recent increase in rice yields (Evatt and Beachell 1962; Johnston *et al.* 1972). The restriction on acreage has induced rice farmers to use the best fields for rice production.

Republic of Korea.—South Korea surpassed the 5 MT/ha mark in 1974. Undoubtedly the recent upsurge in rice yields in South Korea was largely due to the widespread adoption of the high yielding Tong-Il variety and other recent releases.

Tong-Il was selected from the cross Yukara/TN1//IR8 made at IRRI in 1965. The selection and evaluation processes were accelerated by growing winter nurseries at IRRI. Likewise, large plantings of seed increases were made in the Philippines and the seeds airlifted to South Korea in 1971. In 1975 the planted area of Tong-I1 was expanded to more than 450,000 ha since it was first introduced to Korean farmers in 1971 (Off. of Rural Dev., undated; IRRI 1976B).

The high yielding characteristics of Tong-I1 are its greater lodging resistance, higher number of spikelets per panicle and increased nitrogen responsiveness over the other commercial varieties. Tong-I1 is 10 to 15 days earlier than the other varieties and it also appears to be more efficient in the absorption of silica from the soil (Off. of Rural Dev., undated). Tong-I1 yielded 30% more than the other varieties. The yields averaged 6.6 MT/ha in 1974 and 7.0 MT/ha in 1975 (IRRI 1976B).

Rice research in South Korea was stimulated by the development of Tong-I1. Concerted efforts of researchers in several disciplines were

directed to the further improvement of rice varieties and cultural practices. Three short-statured varieties have recently been developed: Yushin (IR1317-392/Tong-I1), Milyang 22 (IR1317-316/IR24) and Milyang
23 (IR1317-316/IR24). From the cross of IR667, which gave Tong-I1,
an early maturing line was selected and named Early Tong-I1. The early
strain made possible the double cropping of rice with an early maturing
barley, Milyang 6 (IRRI 1976B).

The extension of Tong-I1 has also triggered Korean farmers' interests
in adopting improved cultural practices for rice and for other crops (Shin
and Shim 1975).

China.—Although the rice yields in mainland China and in Taiwan are
below the 5 MT/ha mark, double cropping of rice is of common practice.
In the Pearl River delta of Kwangtung Province, double cropping is the
general practice. Parts of Kwangsi Province, Fukien Province and Hainan Island also grow 2 crops of rice. Recently about 3% of the rice land in
Kwangtung Province is grown to 3 crops per year. In the Shanghai-Nanking area, double cropping now occupies 30% of the land. Parts of Kiangsi, Hunan, Chekiang and Yunnan provinces are traditional double cropping areas. In Taiwan more than 66% of the rice fields are double cropped every year (Shen 1951; Ru 1958; Ting 1961). In the above areas, the
rice yields per hectare per year certainly exceed the 5 MT/ha mark also.
Breeding efforts in China are described on pgs. 114—117 and 124.

Breeding in Tropical Asia up to the Early 1960s

For a period of nearly 30 years since 1934, national averages of rice
yield in south Asia and mainland Southeast Asia remained nearly constant at levels ranging from 1.09—1.41 MT/ha in 1934—38 to 0.75—1.94
MT/ha in 1965—66 (FAO 1949,1968). From the turn of the century to
1941, the bulk of the breeding programs in South and Southeast Asian
countries was devoted to mass or pure-line selection among those popular
farmers' varieties. One or a few experiment stations in a country were assigned this type of work. Grain quality was frequently a prime criterion
in evaluation. By 1941 several countries had identified a fairly large number of purified varieties which possessed a few advantages over the
unimproved ones, but none of them contributed substantially to yield
increases (Siregar 1951; Ramiah and Rao 1953; Umali et al. 1956; Parthasarathy 1972).

After World War II, much of the breeders' efforts were devoted to the
further evaluation of the pure-line selections made earlier in an attempt
to reduce the number of recommended varieties and to facilitate the
multiplication, distribution and certification of pure seed (Love 1955;
Dasananda 1960; Richharia et al. 1960). Some efforts were made to
identify varieties that would have wide adaptiveness within a country

(Parthasarathy 1959; Dasananda 1963). Meanwhile, the FAO's International Rice Commission sponsored the Indica × Japonica Hybridization Project in an attempt to transfer the high yielding traits of Japanese varieties into the tropical Indicas (Parthasarathy 1960). Because the unadapted Japanese varieties were largely used as parents, only a few varieties were developed by this cooperative venture. The more notable ones are Malinja (actually resulted from an Indica × Indica cross) and Mashuri of Malaysia and ADT-27 of India.

During the period from 1940 to the mid 1960s, several country programs made notable progress in developing varieties of hybrid origin. The widely planted varieties were CO-15, CO-16, TKM-6, MTU-15, ADT-8 and ADT-20 of India; Mas, Intan, Peta, Bengawan, Tjere Mas and Sigadis of Indonesia; Raminad Str. 3, BPI-76 and BPI-76 (N.S.) of the Philippines; and H-4 of Sri Lanka (see IRRI 1972).

The improved varieties mentioned above have contributed to modest increases in yield in their home or adopted countries. But the varieties were generally too tall and rather late maturing to allow heavy fertilization or double cropping. The yield increases had not kept pace with the growth in population during the period (see Parthasarathy 1972; Grist 1975). It remained for the semidwarfs to herald in a new era of significant yield increases.

Breeding for High Yield and Wide Adaptiveness

Efforts to combine a high yield with nonseasonal adaptiveness were initiated in subtropical Taiwan during the 1920s. Taichung 65 and other "Ponlai" varieties that followed represent a varietal group (kêng) that is insensitive to changes in photoperiod and temperature, early maturing, nitrogen responsive and high yielding—a product of continuous testing in 2 crop seasons. Ponlai varieties can be planted in both the first (spring) and second (summer) seasons of Taiwan and allow intensive multiple cropping between the rice crops (Iso 1954; Chang 1961; Huang et al. 1972). From a cross between a tall Indica variety and the semidwarf Deegeo-woo-gen made in 1949, the semidwarf Taichung Native 1 (TN1) was selected and named by the Taichung District Agricultural Improvement Station in 1956, tested in the Taichung district during 1957, widely tested in 3 districts in 1960 and officially recommended in 1960 (Chang 1961). Because of its short stature and early maturity, TN1 responded better to heavy nitrogen fertilization and outyielded the popular Ponlai varieties. By 1965 TN1 was planted on 10% of the total rice land of Taiwan. Because of its susceptibility to bacterial blight and the leafhoppers, TN1 has been largely replaced in subsequent years by other improved Indica and Ponlai varieties (Huang et al. 1972). TN1 was widely used in the hybridization programs of India (Hargrove 1976) and by IRRI.

Soon after IRRI began its research activities in 1962, the primary breeding objective was to develop high yielding, short and stiff culmed, photoperiod insensitive rice varieties that would resist lodging under heavy fertilization and would have wide adaptiveness for the tropical areas. The semidwarf IR8 derived from Peta/Dee-geo-woo-gen was named and released in 1966. Soon the slightly taller and later maturing IR5 followed. Between 1969 and 1977, 13 IR varieties (IR20, IR22, IR24, IR26, IR28, IR30, IR32, IR34, IR36, IR38, IR40 and IR42) were developed, representing further improvements in grain quality, disease and insect resistance while essentially maintaining the yielding ability of IR8 (Beachell et al. 1972A; Khush and Beachell 1972; IRRI 1975B).

From different country programs in tropical Asia, a large number of semidwarf varieties since 1968 have been developed, either using the semidwarfing gene in TN1 and IR8 or selecting among crosses made by IRRI. The more noted varieties are Jaya, Padma, Bala, Ratna, Sona, Cauvery and Annapurna of India; RD1 and RD2 of Thailand; Mehran 69 of Pakistan; CICA-4 and CICA-6 of Colombia; and Chandina and Mala of Bangladesh. Bahagia was selected from the IR5 cross and became one of the principal varieties in Malaysia; so was Pankaj of India. Indonesia bred Pelita I/1 and Pelita I/2 which have the IR5 plant type; sim-

Reciprocal translocations and trisomics have been used by Iwata and Omura (1971A,B,1975) to establish the independence of 10 linkage groups. However, the chromosomes of the translocation testers have not been characterized so as to lead to the development of a cytological map.

Plant type as a breeding objective for high yield was much discussed during the late 1950s and early 1960s (Tsunoda 1959; Beachell and Scott 1963; Jennings 1964; Tanaka 1965). The desired traits of short plant stature, erect leaves, moderate tillering, lodging resistance and nitrogen responsive were initially based on the high yielding potential of southern United States varieties, certain Japanese varieties and the Ponlai varieties. However, the vegetative growth vigor, high tillering capacity and short and stiff culms of the high yielding semidwarfs were largely contributed by Dee-geo-woo-gen (or its derivatives) and the tropical Indicas (Beachell et al. 1972A).

The consistently high yields attained by the semidwarfs and Ponlai varieties in the Asian and African tropics have been enumerated by Chang (1967A) and Chandler (1972). The genetic basis for wide adaptiveness has been discussed by Chang (1967B) and Chang and Vergara (1972) and later confirmed by the International Adaptation Trials of the International Biological Programme (Evelyn et al. 1971; Kikuchi et al. 1975). Extensive testing of locally bred semidwarfs in India not only confirmed the wide adaptiveness of the semidwarfs but also their superiority over the tall, traditional varieties in yield stability (Seshu et al. 1974).

Yoshida *et al.* (1972) and Yoshida (1972) have discussed the physiological basis for the high yielding potential of the semidwarfs.

Table 3.2 summarized the physiological and genetical interpretations on those plant characters that are associated with the high yielding potential of the semidwarfs and other HYVs of different countries (Chang 1977).

TABLE 3.2. PLANT CHARACTERS ASSOCIATED WITH HIGH YIELDING POTENTIAL IN WETLAND RICE

Plant part	Desirable characteristics	Physiological, ecological contribution	Genetical control, correlations
Leaf	Erect	Increases sun-lit leaf area, permitting even distribution of incident light; decreases the load of raindrops on leaf surface.	Erectness recessive to droopiness, but erectness largely associated with semidwarfism; erect leaves correlated with high yields.
	Short, small	Associated with erectness; even distribution of leaves in a canopy.	Polygenic control of leaf length, leaf area; broad leaf dominant to narrow.
	Thick	Associated with erectness, higher photosynthetic rate.	—
Culm, leaf-sheath	Short, thick culms; tight leaf-sheaths.	Provide lodging resistance, nitrogen responsiveness.	One recessive gene, modifiers control shortness; culm length correlated with panicle length.
Tiller	Upright, compact	Permits light penetration into canopy.	Multigenic control in most crosses; upright tillers associated with semidwarfism.
	High tillering	Adaptive to various spacings; compensating for missing hills; permits faster leaf expansion in transplanted crop.	High tillering is a dominant trait; largely additive effect; some dominance effect; associated with semidwarfism.
Panicle	High fertility at high nitrogen rates.	Allow high doses of nitrogen responsiveness.	Associated with semidwarfism, relatively short panicles.
	High harvest index	Associated with high grain yield.	Multigenic control; negatively correlated with growth duration; associated with early and sustained growth vigor.

Source: Adapted from Yoshida *et al.* (1972) and the publications and unpublished data of Chang and co-workers cited in this chapter.

Estimates on the impact of and probable benefits from growing the semidwarfs have been discussed by Chandler (1972) and Dalrymple (1977, 1978).

Efforts to develop high yielding semidwarfs on the China mainland began in the mid 1950s. Ai-chio-nan-te and Chen-chu-ai were 2 of the initial releases in 1959 (Shen *et al.* 1965; Kwangtung Prov. Acad. Agric. Sci. 1966). Ai-chio-nan-te, Chen-chu-ai, Kwang-chang-ai, Chiang-nan-ai and Nung-k'en 58 were the principal varieties grown during 1964—65 in central and south China and they gave rise to improved semidwarfs or "kêng" types such as Ai-nan-tsao 1, Hu-hsuan 17 and Yeh-li-ching (Stavis 1974). The development of early maturing semidwarfs (100 days or less) such as Ai-nan-tsao 1 and its derivatives made possible the triple cropping in Kwangtung Province and the double cropping in Shanghai-Nanking area. The above semidwarfs have the same recessive gene for short stature as in Dee-geo-woo-gen (IRRI 1978).

Breeding for Stable Yield and Improved Nutrition

After the marked jump in yield potential was attained in the late 1960s, the next problem was to stabilize the high yield level in the face of intensified disease and insect incidences. Rice fields in irrigated areas of the Asian tropics were ravaged by diseases and insects after increased nitrogen fertilization, denser planting, multiple cropping of rice and continuous planting of a few high yielding varieties. Examples of large-scale epidemics or zoototics are: (1) the tungro outbreaks of 1971 and 1975, the brown planthopper and grassy stunt epidemics of 1973—74 and the planthopper resurgence during 1976 in the Philippines, (2) the brown planthopper outbreak in Kerala state of India during 1973—74, (3) the tungro epidemic of southern Vietnam during 1973 and the brown planthopper outbreak in 1975, and (4) the brown planthopper outbreaks in Java and Sumatra of Indonesia during 1974—76.

Associated with the continuous planting of a few varieties of similar genetic background, a new or hitherto obscure biotype of a genetically plastic pest appears in quantity and the resistance in the major varieties breaks down. Instances of such cyclic events are the rapid appearance of the brown planthopper biotype 2 in the Philippines and Indonesia and the increasing susceptibility of previously resistant varieties such as IR26 and IR30.

Good progress has been made in transferring a major gene or a gene complex for resistance to a specific disease or pest into the high yielding background. The resistance gene or genes were initially identified by problem areas scientists in their systematic screening of rice germ plasm, a substantial portion of which has been carried out at IRRI since 1962 (see Chang *et al.* 1975B). The transfer of the desired gene or genes was made by way of a single, back or multiple crosses or a combination of them. The evaluation and selection of resistant progenies in the segregating generations were again made by multidisciplinary collaboration under the Genetic Evaluation and Utilization (GEU) program of IRRI

(IRRI 1974B). For bacterial blight, Xa_4 has been incorporated into IR20 and subsequent IRRI releases up to IR38; xa_5 into certain IR lines; xa_6 into several other IR lines. Blast resistance genes were transferred from Dawn, Gam Pai 105, Tetep and other varieties. The major gene for grassy stunt resistance from IRRI Accession 101508 of *O. nivara* has been incorporated into IR28 and the subsequent IRRI releases. Better success has been obtained by using the tungro resistance from Gam Pai and CR94-13 than from the highly resistant Pankhari 203. For green leafhopper resistance, most of the IRRI varieties have Glh_3; IR32 and IR38 have Glh_1. For resistance to the brown planthopper, Bph_1 is in IR26, IR28, IR30 and IR34; bph_2 in IR32, IR36, IR38, IR40 and IR42 (Khush and Beachell 1972; Khush *et al.* 1976; IRRI 1977C).

Most of the pest resistance genes incorporated into improved varieties of different sources belong to the major genes which confer the vertical type of resistance. The rapid breakdown of vertical resistance has been experienced with the blast disease in Japan and with the Bph_1 gene for brown planthopper resistance in the Philippines and Indonesia. Although there is some confusion on the definition of horizontal resistance and whether it occurs in rice or not, a more stable type of resistance can be realized by using parents that have a broad base of resistance (containing both major and minor genes), multiparent crosses and multi-site testing in the early segregating generations. Measures aimed to acquire such broad resistance to blast and to the brown planthopper have been recently discussed (IRRI 1977B,C).

National breeding programs have also developed a number of improved varieties that have 1 or more resistances to diseases and insects. Bred in Kerala state of India, Ptb-18 and -21, Vikram, Phalguna and Shakti are resistant to gall midge and stem borers; Ptb-33 and RP825-41 are resistant to brown planthoppers. Indian breeding lines like W-1263, CR94-13, R68-1 and RP260-288-1 were widely used in other breeding programs because of their multiple resistances. RD4 and RD9 of Thailand have resistance to the gall midge; RD9 is also resistant to the brown planthopper; RD5 and RD7 are resistant to bacterial blight. CICA-6, CICA-7 and CICA-9 of Colombia are resistant to the leafhopper *Sogatodes oryzicola*. Kaohsiung Sen 12 and Chianung Sen 11 of Taiwan have resistance to the brown planthoppers; Chianung-sen 11 and Taichung Sen 5 are resistant to bacterial blight; Kaohsiung 139 is resistant to blast.

Among environmental factors affecting yield stability, resistance to drought and tolerance to submergence of flood waters are also essential to stabilized crop yields during seasons of erratic precipitation. Discussion on these subjects will be made in the following section, Overcoming Stress Factors.

Milled rice has excellent digestibility and good protein quality, but the gross protein content (7 to 9%) is rather low for children. Efforts have been made at IRRI to increase the brown rice protein (BRP) content by 2% while retaining the other desired agronomic characteristics and pest resistances. Initial attempts in crossing the semidwarfs with unadapted and unproductive parents of high BRP contents turned out to be unrewarding (Beachell *et al.* 1972B). High protein contents appear to be correlated with low yields in fixed genotypes (IRRI 1974A). Subsequent attempts were made to identify high BRP genotypes in breeding lines already having other desired traits and to intercross the high BRP lines for recurrent selection. The progress has been slow (IRRI 1974A). The use of protein per seed appears to be a more effective selection criterion than the BRP content of a bulked seed sample of a plant or breeding line (IRRI 1976A).

Improved varieties have the aromatic quality such as Improved Sabarmati, Pusa 33, RP967-11 and RP107-8 have been developed from the traditional Basmati type (Seshu 1976).

Overcoming Stress Factors

A deficit or excess of precipitation, or both often appear during the crop season, resulting in drought or flood, or both in an alternate sequence. At high elevations in the tropics and subtropics, cool temperatures delimit rice yields. Cold irrigation water in some areas adversely affect seedling growth. Extremely high temperature at flowering may reduce seedset. In low lying deltas or valleys, deep water ranging from 1 to 5 m precludes the cultivation of moderately high yielding varieties. Edaphic factors, either as excesses or as deficiencies, adversely affect rice yields. In each case, resistant or tolerant cultivars have been identified from endemic areas, but such cultivars generally have very low yielding potentials.

Reaction to and recovery from drought constitute one of the most complex stress factors because different physiological mechanisms could operate and interact at specific growth stages (see O'Toole and Chang 1978). While the underlying mechanisms and related effects are yet to be fully elucidated, intensive studies on diverse germ plasm have led to the development of a mass screening test in upland soil (Chang *et al.* 1974) and the identification of promising resistant sources (Chang *et al.* 1974; IRRI 1974A, 1976A, 1977A). Under upland culture, an extensive system of thick and deep roots and the ability to maintain leaf turgor (expressed by a resilient leaf rolling and unrolling process in a diurnal cycle) are 2 principal traits associated with drought resistance (Loresto and Chang 1971; Chang *et al.* 1972, 1974; IRRI 1975B; Loresto *et al.* 1976). Under rainfed lowland culture, the ability to maintain leaf water potential is clearly associated with drought resistance although the root system may play a somewhat smaller role (IRRI 1978). Hybridization, selection and evaluation at IRRI have shown that drought resistance is a heritable trait when resistant sources are used as parents (IRRI 1974A). Slow but steady progress is being made in combining drought resistance with a higher yield potential while maintaining the stable base yield of the resistant parents. Incorporation of recovery ability from drought is also essential to the rainfed lowland culture (IRRI 1976A, 1977A). Other selection criteria of potential value to breeding have been discussed by O'Toole and Chang (1978).

Tolerance to submergence of flood water is different from tolerance to (or floating ability under) deep water. The 2 traits need to be evaluated separately but both qualities are needed for rices grown in deep water (IRRI 1976A).

For areas where the water depths range from 50 to 100 cm, high yielding types with limited internode elongation ability, such as T442-57 and IR442-2-58, have been identified. For areas with deeper water depths, progress by breeding has been slow. Since 1975 a regional project has been established with the participation of the Rice Division of Thailand, the Bangladesh Rice Research Institute and the IRRI. Progress in improving deep water rices for different areas may be found in the reports of 2 recent workshops (Bangladesh Rice Res. Inst. 1975; IRRI 1977D).

Tolerance to cool temperatures became necessary when population pressure forced farmers to cultivate rice at high elevations in tropical Asia. It is interesting to note that in Kashmir and other hilly tracts of north India, introductions from China have consistently outperformed the local rices and eventually replaced most of the local varieties (Anon. 1966), probably because of the efficacy of natural and human selection in the cooler areas of China during the past centuries. The bulu varieties of Indonesia probably have better cool tolerance than the lowland Indicas. An Indonesian line (RP Kn-2) has excelled the local rices in the Mountain Province of the Philippines (c 1000 in elevation) and made double cropping possible.

Good progress in breeding for cold water tolerance in northern Japan and in California has been attained. Success appears to be attributable to long years of empirical testing and selection, use of diverse germ plasm and development of efficient cultural practices (Okabe and Toriyama 1972; Carnahan et al. 1972; Toriyama 1976).

Under the GEU program of IRRI, many diverse sources of tolerance to different adverse soil factors— salinity, alkalinity, iron toxicity or deficiency, zinc deficiency, phosphorus deficiency, manganese and aluminum toxicity, and mineral stresses in acid sulfate soils, histosols and cold soils—have been identified. Some of the tolerant strains came from crosses that did not involve tolerant parents (Ponnamperuma and Castro 1972; IRRI 1975A, 1976A; Ponnamperuma 1977; Ponnamperuma and Ikehashi 1977). Much of the progeny testing is being conducted in areas with problem soils. Similar breeding efforts are being carried out in Colombia, India, Sri Lanka, Thailand, Liberia, Sierra Leone and Senegal (Katyal *et al.* 1975; Howeler and Cadavid 1976).

Conservation, Evaluation and Utilization of Genetic Resources

The extent of genetic diversity in *O. sativa* is remarkably great for a cultigen or any crop species. Its long history of cultivation, worldwide distribution and varied consumer's preferences have led to a tremendous proliferation of ecoedaphic and grain types that further differ in their reactions to diseases or insects or both (Chang 1976B,C). Based on estimates given by national research centers, as many as 100,000 cultivars or variants could have existed before the advent of the high yielding varieties.

Moreover, strains of the African cultivated rice (*O. glaberrima*), weed races of the 2 cultivated species and the 18 wild species further enrich the gene pools available to the rice researchers (Chang *et al.* 1977A). The genetic composition, relative productivity and potential value in breeding of different gene sources have been discussed by Chang (1976D) and Chang *et al.* (1977A).

It is fortunate for rice researchers that until recently the great majority of the rich germplasm was available to field collectors for conservation (Chang 1976A). Early efforts by national workers have led to sizeable collections maintained in India, Japan and the United States. FAO has assisted in setting up an Indica collection in India and a Japonica collection in Japan. The IRRI Germplasm Bank was initiated as soon as IRRI began its operations in late 1961. With the collaboration of many national and international organizations, the varietal collection of IRRI rapidly grew to 14,600 within the first decade of its operation (Chang 1972). A collaborative field collection was launched in 1972 (Chang *et al.* 1973; Chang and Perez 1975) and by 1978, about 38,000 samples were collected from farmers' fields in 13 countries in south and Southeast Asia. About 7970 samples were reputed to have 1 or more special features such as cool tolerance, drought resistance and salinity tolerance. As of late 1979, the IRRI Germplasm Bank has nearly 57,000 accessions of *O. sativa*, 2273 accessions of *O. glaberrima*, 908 strains of wild taxa, and 640 genetic testers and mutants (Chang *et al.* 1977B; IRRI 1980).

Collection for the African rices in West Africa has been underway since the early 1970s. Further collections in both tropical Africa and Asia are urgently needed in the face of the rapid spread of the HYVs.

Systematic evaluation of genetic materials in the IRRI Germplasm Bank has led to the identification of important sources of resistances or tolerances or quality features, as enumerated in the preceding sections of this chapter. Such multidisciplinary efforts are being expanded under IRRI's GEU program (IRRI 1974A,B).

Meanwhile, all the accessions in the bank are systematically characterized by their morphoagronomic characteristics. Seed stocks are kept in short, medium and long-term storerooms. Thousands of seed samples are being supplied to researchers in national programs every year. By preserving seed in cold storage, IRRI is truly serving the role of a genetic resources center (Frankel 1975; Chang et al. 1975A). The operations of the bank have been described (Chang 1972, 1976D).

IRRI's findings of many promising sources of resistances or tolerances have prompted several national research centers to systematically evaluate their indigenous germplasm. Since 1975 the implementation of the International Rice Testing Program has further expanded and systematized the interinstitutional collaboration in and exchange for genetic materials and information (IRRI 1974A, B). Thus, the rice researchers of different countries have access to both the original source of resistance (or tolerance) and advanced selections having the desired traits (IRRI, undated). In 1977, a series of collaborative research projects on resistances to diseases, insects and drought was initiated to further accelerate the region-wide research on problems of common interest and to provide a wide spectrum of early generation hybrid progenies for testing and selection at key sites in tropical Asia (IRRI 1978).

Although rice breeding started later than that of the other cereals, the speed at which rice researchers throughout the world have collaborated in the conservation, evaluation and utilization of rice germplasm exceeds the expectation of many researchers themselves. With the excellent collaboration attained to date, greater progress in using the diverse germplasm to develop improved varieties with many desired traits can be anticipated.

Breeding Methods

Being a self-pollinated crop, the methods of breeding rice share much of the features common to wheat, barley and oats. Discussions will refer to some of the past practices that could be improved and to a number of potential means that should be explored.

The pedigree method of selection at the experiment station site has been the predominant practice of rice breeders in handling segregating generations from the F_3 and beyond. Rice breeders may find it advantageous for certain situations to use the bulk method at sites where the environmental conditions would favor the identification and selection of the desired genotypes. The back-cross method may be expedient when a breeder wishes to make specific and limited improvement on a well adapted and preferred variety.

Many breeding programs in the past have suffered from a narrow base in choosing the parents, making a few crosses a year and in growing small F_2 populations. This trend is gradually being remedied, however.

Special efforts to breed for specific complex traits in the 1950s, such as wide adaptiveness, lodging resistance and high yielding potential from the Japonica varieties have failed to produce useful progenies partly because a lack of basic understanding about the complex traits and partly due to a limited choice of promising parents. Goal oriented research by a multidisciplinary team could greatly assist the plant breeders in carrying out effective hybridization and selection operations. Much of IRRI's progress in designing and developing high yielding, widely adapted, pest resistant or drought resistant strains derives from the findings of objective oriented research and the use of diverse germplasm.

Up to the late 1960s, many crosses made in tropical Asia were single crosses followed by selection until the next cross was made with one of the promising progenies selected from the cross. Experience at IRRI has shown that with the same set of parents, planning the single crosses in such a way so as to maximize the full range of recombination could produce more useful progenies than random pairing of parents in the double crosses (Chang 1976E). Repeated intercrossing of early generation progenies such as F_2 and F_3 plants is another means of obtaining a fuller range of recombinants. Recurrent selection following intercrossing could be helpful in improving traits that are under polygenic control and have low heritability.

To attain stable or general resistance to pathogens and insects, pyramiding different resistance genes, developing multilines, use of varietal mixtures or composite populations, and developing varieties with horizontal resistance have been proposed as alternate ways of coping with genetic changes in a pest or shifts within a population of biotypes. While little is known about the genetic mechanisms involved in conferring a given type of host resistance or in the genetic control of virulence (or avirulence) in the pest within the rice-pest syndrome, the new approaches of considering host resistance on a multigenic basis certainly deserve exploratory research. Intercrossing the different resistant sources, followed by multi-site or multibiotype evaluation and recurrent selection would be expected to confer more enduring resistance than the sequential release of varieties having the vertical resistance one at a time.

Some of the experiences gained from field resistance to blast in Japan indicate that a system of 1 or a few major genes plus several minor genes would provide adequate protection to changes in the population structure of the blast fungus (see IRRI 1977B). There is also the possibility that vertical (or biotype specific) resistance could be combined with general (or stable) resistance in a genotype to further enhance the life span of the resistant genotype. Viewing over time and space, adequate genetic diversity in the principal cultivars in a country or district could readily provide substantial protection against dramatic changes in the pathogen or insect.

The finding of heterosis in rice hybrids has been a frequent event, although some of the so-called hybrid vigor was not true F_1 superiority or heterobeltiosis (see Chang et al. 1973). The use of F_1 hybrids in commercial production hinges on the extent of heterobeltiosis, the ease and costs of producing F_1 seeds, and also the extent of inbreeding depression in the F_2. It is necessary to identify superior cross combinations, develop desirable cytosterile lines and produce high crossing pollen parents. While F_1 performance may be the primary target of such pursuits, the superior performance of certain parents themselves should not be overlooked. In U.S. wheats, some parents intended for hybrid seed production are as good as their F_1 hybrids (Hayward 1975).

However, a massive program of hybrid rice production began on mainland China in late 1974. The cytosterile source coming from a wild rice plant of Hainan Island was collected in late 1970 and used by researchers in Hunan province to produce the sterile male lines. Most of the useful restorers were of the Indica type. About 130,000 mou (8670 ha) of F_1 hybrids were grown in 1976. Hybrid seed production ranged between 1.5 and 5.25 MT/ha. The highest yields obtained from the F_1 hybrids were around 12 MT/ha. The yield increases ranged between 10 and 30%, averaging below 20%. The hybrid rices were described as having a vigorous root system, vegetative growth vigor, high tillering ability, large and dense panicles, heavy grains and wide adaptiveness in the middle Yangtze River basin and in south China (Hunan Prov. Rice Res. Inst. 1977; Yuan 1977). About 5 million ha were planted in 1978. However, the costs of producing the F_1 seeds by labor intensive practices were not discussed or considered (Chang 1979).

The use of induced mutations as a tool in rice breeding was very popular among rice workers of many Asian countries during the 1960s. The progress in using induced mutations as a direct product for varietal improvement has been rather disappointing in relation to manpower and research fund investments. Among the hundreds of promising mutants described in rice literature, only a few attained sizeable acreage: Reimei of Japan and Jagannath of India. Background information on the rather unrewarding practices of mutation breeding has been discussed in a survey by Gregory (1972). On the other hand, induced mutants could be more readily used in crosses than an exotic or incompatible source when a desired trait could not be found in the local variety group. Recent instances of such useful parents are the short statured mutants induced from Calrose in California and from the Ponlai varieties in Taiwan. Blast resistant mutants have been developed from adapted varieties of France, Italy and Hungary. While the induced mutation may belong to the same locus found in existing cultivars, the common genetic background facilitates the transfer of the desired gene (Hu 1973; Marie 1974; Woo *et al.* 1974; Rutger *et al.* 1976). On the other hand, mutagenic treatments could also serve as a powerful tool in breaking up tight linkages and in increasing recombinations in some wide crosses where sterility and differential gametic survival preclude the recovery of the desired recombinants (Chang 1976E; Brock 1976). Approaches to make more efficient use of induced mutations have been discussed by Gregory (1968, 1972) and Brock (1971).

During the last 20 years, rice breeders and their associates in plant physiology and agronomy have paid increasing attention to certain selection criteria other than the absolute grain yield. Experience to date has shown the promise of using one of the most stable yield components—grain weight, harvest index, leaf thickness or yield per hectare per day as additional selection criteria (IRRI 1972; Ito 1975). At the same time, workers must recognize that certain traits may concurrently vary either in parallel or opposite to changes in other traits. One of such negatively correlated responses is that while smaller culms are essential to stem borer resistance, lodging resistance of the semidwarf strain would be lowered when selection focused on borer resistance.

Rice breeders in Japan have recently adopted the use of greenhouses and darkrooms during the cooler months to advance the turnover of generations during the breeding process. While the breeders can certainly gain valuable time in advancing the generations, caution must be taken to provide such an environment and the appropriate selection criteria so that the process would not unconsciously select against certain desired genotypes.

Anther culture has been used by Japanese and Chinese researchers as a means to accelerate the identification of homozygous plants from hybrid progenies (Niizeki 1968; Woo and Su 1975; Kwangtung Inst. Bot. 1976; Mok and Woo 1976). Three improved varieties were recently developed by Chinese researchers: Hua Yu I and II and Tanfeng 1 (Tientsin Agric. Res. Inst. 1976; Yin *et al.* 1976).

Protoplasmic fusion is one technique that has yet to be tried in making extremely wide crosses in the genus *Oryza* or between 2 related genera.

Trends for the Future

Rice breeding for the irrigated areas has reached a temporary plateau in terms of grain yield. Further breakthrough in yield increase would depend on increased grain weight, better grain filling under heavy nitrogen application, higher spikelet number per unit area or more efficient translocation of photosynthates from shoot to the spikelets. Meanwhile, earlier maturing genotypes would allow a higher cropping index per year while retaining a high yield per hectare per day. Higher and more stable levels of resistance to the diseases and insects would sustain the high yields without incurring the boom and bust cycles that were prevalent in the 1970s.

For the rainfed areas, resistance to and good recovery from drought would markedly stabilize yield levels. Resistance to diseases and pests as well as competitiveness with weeds are equally important requisites. An elastic tillering ability may provide yield stability under variable soil moisture regimes. A plant type with culms taller than the semidwarfs and rather droopy leaves may be more efficient under rainfed culture. The disadvantages of droopy leaves may be partly compensated by an increase in photosynthetic efficiency.

For deep water or water logged areas, specific levels of photoperiod sensitivity are essential to synchronize crop growth with water regimes. Both elongation ability and submergence tolerance are essential for deep water areas. Tolerance to adverse soil factors will alleviate crop growth in water logged areas.

When population pressures necessitate rice cultivation in marginal areas, problems such as cool tolerance and mineral stresses of the soil would require rice breeders to develop environment specific adaptiveness. Chances are rather slim that the number of cultivars can be greatly reduced to suit the convenience of pure seed programs.

Concurrent with the breeding for specific adaptiveness in the less favored areas, genetic diversity needs to be reinstated in the favored areas. Directional selection for genetically diverse cultivars that are suited to a particular crop season would add genetic fallow between the 2 consecutive rice crops.

As population pressures increase, man in densely populated and adjoining areas will depend more on the calorie-rich rice as the staple food than on other starchy foods (Univ. of Calif. Food Task Force 1974). Increasing the nutritional quantity and quality of rice is an important task to be accomplished.

Increasing demands on rice will burden the rice researchers and especially the breeders with varied and complicated demands for rice genotypes having unprecedented combinations of gene complexes. Meanwhile, we can also look forward to greater and closer systems of collaborative research in tackling problems of the future. Much of the collaborative research will focus on the conservation, evaluation and utilization of the enormous diversity that exists in rice germplasm. Field testings are expected to be widely carried out in farmers' fields where a specific and endemic problem limits rice yield. Therefore, close collaboration between rice researchers and extension workers will be needed. Farmers in such areas would also constitute a member of the applied research team.

O. glaberrima and its wild relatives have not been intensively evaluated for their potential in meeting the diverse ecoedaphic niches and harsh conditions of West Africa. They could possibly be of use to other areas as well. Indigenous research as well as international and interinstitutional collaboration are needed to utilize the useful genes in the African rices.

The adoption of improved varieties and their replacement by newer releases could be greatly benefited by the establishment and management of a sound and efficient seed program in many developing countries. The lack of a coordinated system for the multiplication, testing and certification, storage, and distribution of pure seeds has hampered the rapid and large-scale use of improved varieties in the past. With the mounting interest in modern varieties in the developing countries, we can look forward to expansion of or improvement in pure seed programs, or in both aspects.

REFERENCES

ADAIR, C.R. 1941. Inheritance in rice of reaction to *Helminthosporium oryzae* and *Cercospora oryzae*. U.S. Dep. Agric. Tech. Bull. *772*, 1-18.

ADAIR, C.R., BEACHELL, H.M. and JODON, N.E. 1966. Rice breeding and testing methods in the United States. *In* Rice in the United States: Varieties and Production. U.S. Dep. Agric.—Agric. Handbk. *289*, 19-56.

ADAIR, C.R. and JODON, N.E. 1966. Distribution and origin of species, botany and genetics. *In* Rice in the United States: Varieties and Production. U.S. Dep. Agric.—Agric. Handbk. *289*, 5-18.

ADAIR, C.R., *et al.* 1973. Rice breeding and testing methods in the United States. *In* Rice in the United States. U. S. Dep. Agric.—Agric. Handbk. *289*, (revised), 22-75.

AKEMINE, M. 1925. On the inheritance of dwarf habits in rice. Proc. Jpn. Assoc. Adv. Sci. *1*, 308-314.

AKINO, M. and HAYAMI, Y. 1975. Efficiency and equity of public research: rice breeding in Japan's economic development. Amer. J. Agric. Econ. *57*, 1-10.

ALL-INDIA COORDINATED RICE IMPROVEMENT PROJECT (AICRIP). 1969. Progress Report, Kharif 1969, Vol. 3. Indian Counc. of Agric. Res., New Delhi.

ANON. 1963. Rice gene symbolization and linkage groups. U.S. Dep. Agric.—

Agric. Res. Ser. ARS *34*, 34-28, 1-56.

ANON. 1966. *Oryza. In* The Wealth of India. Raw Materials 7, 110-191. Publ. and Inf. Direct., New Delhi.

AQUINO, R.C. and JENNINGS, P.R. 1966. Inheritance and significance of dwarfism in an indica rice variety. Crop Sci. *6*, 551-554.

ATHWAL, D.S., PATHAK, M.D., BACALANGCO, E.H. and PURA, C.D. 1971. Genetics of resistance to brown planthoppers and green leafhoppers in *Oryzas sativa* L. Crop Sci. *11*, 747-750.

BALAKRISHNA RAO, M.J., SRINIVASULU, K. and CHOUDHURY, D. 1972. Inheritance of dwarfing in crosses involving rice variety Baok. Curr. Sci. *41*, 306.

BANGLADESH RICE RES. INST. (BRRI). 1975. Proc. of the Int. Sem. on Deep Water Rice, Aug., 1974. Joydebpur, Dacca.

BEACHELL, H.M., KHUSH, G.S. and AQUINO, R.C. 1972A. IRRI's international breeding program. *In* Rice Breeding. Int. Rice Res. Inst., Los Baños, Philippines.

BEACHELL, H.M., KHUSH, G.S. and JULIANO, B.O. 1972B. Breeding for high protein content in rice. *In* Rice Breeding. Int. Rice Res. Inst., Los Baños, Philippines.

BEACHELL, H.M. and SCOTT, J.E. 1963. Breeding rice for desired plant type. Proc. Rice Tech. Working Group, Houston, Texas, Feb. 21–23, 1962. Ark. Agric. Exp. Sta., Fayetteville, Arkansas.

BEACHELL, H.M. and STANSEL, J.W. 1963. Selecting rice for specific cooking characteristics in a breeding program. Int. Rice Comm. Newsl. Spec. Is., 25-40.

BHADURI, P.N. and BAIRAGI, P. 1968. Root characters in breeding higher yielding and better adapted varieties of rice. Sci. Cult. *38*, 357.

BOLLICH, C.N. 1957. Inheritance of several economic quantitative characters in rice. Diss. Abst. *17*, 1638.

BOLLICH, C.N. and WEBB, B.D. 1973. Inheritance of amylose in two hybrid populations of rice. Cereal Chem. *50*, 631-636.

BROCK, R.D. 1971. The role of induced mutations in plant improvement. Radiat. Bot. *11*, 181-196.

BROCK, R.D. 1976. Prospects and perspectives in mutation breeding. Ppr. presented at the Int. Symp. on Genet. Control of Divers. *In* Plants, March 1-7. Lahore, Pakistan.

CAMPOS, G.L., BALLESTEROS, R., CASTELLS, J. and BATALLA, J.A. 1975. Rice in Spain. Fed. Sindical Agric., Arroceros, Valencia, Spain. (Spanish)

CARNAHAN, H.L., ERICKSON, J.R. and MASTENBROEK, J.J. 1972. Tolerance of rice to cool temperature—United States. *In* Rice Breeding. Int. Rice Res. Inst., Los Baños, Philippines.

CHAKRAVARTI, A.K. 1948. A genetical study of the botanical characters of rice (*Oryza sativa* L.). Bull. Bot. Soc. Bengal *2*, 55-57.

CHALAM, G.V. and VENKATESWARLU, J. 1965. Introduction to Agricultural Botany in India, Vol. 1. Asia Publishing House, New Delhi.

CHANDLER, R.F. 1972. The impact of the improved tropical plant type on rice yields in south and southeast Asia. *In* Rice Breeding. Int. Rice Res. Inst., Los Baños, Philippines.

CHANDRARATNA, M.F. 1955. Genetics of photoperiod sensitivity in rice. J. Genet. *53*, 215-223.

CHANDRARATNA, M.F. 1964. Genetics and Breeding of Rice. Longman, Green and Co., London.

CHANDRARATNA, M.F. and SAKAI, K.I. 1960. A biometrical analysis of matroclinous inheritance of grain weight in rice. Heredity *14*, 365-373.

CHANG, T.M. 1971. Studies on the quantitative inheritance of *Oryza sativa* L. I. Diallel cross of semi-dwarf rice. J. Taiwan Agric. Res. Inst. *20* (2) 9-26.

CHANG, T.M. 1974. Studies on the inheritance of grain shape of rice. J. Taiwan Agric. Res. Inst. *23* (1) 9-15.

CHANG, T.T. 1961. Recent advances in rice breeding in Taiwan. *In* Crop and Seed Improvement in Taiwan, R. O. C., May 1959—Jan. 1961. J. Comm. Rural Reconstr. Pl. Ind. Ser. 22, Taipei.

CHANG, T.T. 1964. Present knowledge of rice genetics and cytogenetics. Int. Rice Res. Inst., Tech. Bull. *1*, Los Baños, Philippines.

CHANG, T.T. 1966. The need for genetical investigations to assist rice breeders in tropical Asia. Indian J. Genet. *26A*, 206-210.

CHANG, T.T. 1967A. The genetic basis of wide adaptability and yielding ability of rice varieties in the tropics. Int. Rice Comm. Newsl. *16* (4) 4-12.

CHANG, T.T. 1967B. Growth characteristics, lodging and grain development. *In* Symp. on Problems in Development and Ripening of Rice Grain. Int. Rice Comm. Newsl. Spec. Is., Food and Agric. Organ. U.N., Bangkok.

CHANG, T.T. 1967C. Analysis of short stature genes. Proc. 11th Rice Tech. Working Group, June 14—17, 1966, Little Rock, Ark. Div. of Agric. Sci., Univer. of Calif.

CHANG, T.T. 1972. International cooperation in conserving and evaluating rice germ plasm resources. *In* Rice Breeding. Int. Rice Res. Inst., Los Baños, Philippines.

CHANG, T.T. 1976A. Rice. *In* Evolution of Crop Plants. N. W. Simmonds (Editor). Longman, London and New York.

CHANG, T.T. 1976B. The rice cultures. *In* the Early History of Agriculture. Philos. Trans. R. Soc., London. B275,143-157.

CHANG, T.T. 1976C. The origin, evolution, cultivation, dissemination and diversification of Asian and African rices. Euphytica *25*, 425-441.

CHANG, T.T. 1976D. Manual on Genetic Conservation of Rice Germ Plasm for Evaluation and Utilization. Int. Rice Res. Inst., Los Baños, Philippines.

CHANG, T.T. 1976E. Exploitation of useful gene pools in rice through conservation and evaluation. SABRAO J. *8* (1) 11-16.

CHANG, T.T. 1977. Genetics and evolution of the Green Revolution. Paper presented at the UNESCO-CSIC sponsored Symp. on Genet. and Ethics, Oct. 10-14. Madrid.

CHANG, T.T. 1979. Hybrid rice. *In* Plant Breeding Perspectives. Sneep, J. and Hendriksen, J.T. (Editors). PUDOC, Wageningen.

CHANG, T.T. and BARDENAS, E.A. 1965. The morphology and varietal characteristics of the rice plants. Int. Rice Res. Inst. Tech. Bull *4*, Los Baños, Philippines.

CHANG, T.T. and JODON, N.E. 1963. Monitoring of gene symbols in rice. Int. Rice Comm. Newsl. *12* (4) 18-30.

CHANG, T.T., LI, C.C. and TAGUMPAY, O. 1973. Genotypic correlation, heterosis, inbreeding depression and transgressive segregation of agronomic traits in a diallel cross of rice (*Oryza sativa* L.) cultivars. Bot. Bull. Acad. Sin. *14*, 83-93.

CHANG, T.T. LI, C.C. and VERGARA, B.S. 1969. Component analysis of duration from seeding to heading in rice by the basic vegetative phase and the photoperiod-sensitive phase. Euphytica *18*, 79-91.

CHANG, T.T. and LIN, F.H. 1974. Diallel analysis of protein content in rice. *In* Agron. Abstr., Chicago.

CHANG, T.T. and LIU, P.T. 1967. Inheritance of lodging resistance. Proc. 11th Rice Tech. Working Group, June 14-17, 1966, Little Rock, Ark. Div., of Agric. Sci., Univ. of Calif.

CHANG, T.T., LORESTO, G.C. and TAGUMPAY, O. 1972. Agronomic and growth characteristics of upland and lowland rice varieties. *In* Rice Breeding. Int. Rice Res. Inst., Los Baños, Philippines.

CHANG, T.T., LORESTO, G.C. and TAGUMPAY, O. 1974. Screening rice germ plasm for drought resistance. SABRAO J. *6* (1) 9-16.

CHANG, T.T., MARCIANO, A.P. and LORESTO, G.C. 1977A. Morphoagronomic variousness and economic potentials of *Oryza glaberrima* and wild species in the genus *Oryza*. Ppr. presented at the IRAT-ORSTOM African Rice Species Meeting, Feb. 25-26. Paris.

CHANG, T.T. and OKA, H.I. 1976. Genetic variousness on the climatic adaptation of rice cultivars. *In* Climate and Rice. Int. Rice Res. Inst., Los Baños, Philippines.

CHANG, T.T. and PEREZ, A.T. 1975. Genetic conservation of rice germplasm in Southeast and South Asia, 1971-74. *In* Southeast Asian Plant Genetic Resources. L. T. Williams, C.H. Lamoureux and N. Wulijaini-Soetjipto (Editors). BIOTROP/LIPI/LBN, Bogor, Indonesia.

CHANG, T.T. and TAGUMPAY, O. 1970. Genotypic association between grain yield and six agronomic traits in a cross between rice varieties of contrasting plant type. Euphytica *19*, 356-363.

CHANG, T.T. and TAGUMPAY, O. 1973. Inheritance of grain dormancy in relation to growth duration in 10 rice crosses. SABRAO Newsl. *5* (2) 87-94.

CHANG, T.T. and VERGARA, B.S. 1971. Ecological and genetic aspects of photoperiod-sensitivity and thermo-sensitivity in relation to the regional adaptability of rice varieties. Int. Rice Comm. Newsl. *20* (2) 1-10.

CHANG, T.T. and VERGARA, B.S. 1972. Ecologic and genetic information on adaptability and yielding ability in tropical rice varieties. *In* Rice Breeding. Int. Rice Res. Inst., Los Baños, Philippines.

CHANG, T.T., VILLAREAL, R.L., LORESTO, G.C. and PEREZ, A.T. 1975A. IRRI's role as a genetic resources center. *In* Crop Genetic Resources for Today and Tomorrow. O.H. Frankel and J.G. Hawkes (Editors). Cambridge Univ. Press, Cambridge.

CHANG, T.T. and YEN, S.T. 1969. Inheritance of grain dormancy in four rice crosses. Bot. Bull. Acad. Sin. *10*, 1-9.

CHANG, T.T. *et al.* 1965. Genetic analysis of plant height, maturity and other quantitative traits in the cross of Peta X I-geo-tze. J. Agric. Assoc. China (N. S.) *51*, 1-8.

CHANG, T.T. *et al.* 1975B. The search for disease and insect resistance in rice germplasm. *In* Crop Genetic Resources for Today and Tomorrow. O.H. Frankel and J.G. Hawkes (Editors). Cambridge Univ. Press, Cambridge.

CHANG, T.T. *et al.* 1977B. IRRI's role as a genetic resources center II. *In* Proc. SE Asian Workshop on Plant Genet. Res. Int. Board for Plant Genet. Res. and Philipp. Counc. for Agric. and Res. Res., Los Baños, Philippines.

CHAO, L.F. 1928. Linkage studies in rice. Genetics *13*, 133-169.

CHEN, L.C. and CHANG, W.L. 1971. Inheritance of resistance to brown planthopper in rice variety Mudgo. J. Taiwan Agric. Res. *20* (1) 57-60.

CHU, Y.E. 1967. Pachytene analysis and observations of chromosome association in haploid rice. Cytologia *32*, 87-95.

CHU, T.M. and LI, C.C. 1979. Studies on drought resistance in rice. I. Proline accumulation and its inheritance. Natl. Sci. Counc. Mon., Taiwan. 7 (2) 167-179. (Chinese with English summary)

CIFERRI, R. 1960. Features of the rice history in Italy. Quaderno *8*, Milan. (Italian)

DALRYMPLE, D.G. 1977. Evaluating the impact of international research on wheat and rice production in the developing nations. *In* Resource Allocation and Productivity in National and International Agricultural Research. T.M. Arndt, D.G. Dalrymple and V.W. Ruttan (Editors). Univ. of Minn. Press, Minneapolis.

DALRYMPLE, D.G. 1978. Development and spread of high-yielding varieties of wheat and rice in the less developed nations. Sixth edition. Foreign Agric. Econ. Rep. *95*. U.S. Dep. of Agric. and U.S. Agency Int. Dev. Foreign Agric. Serv., Washington, D.C.

DAS, D.C. and SHASTRY, S.V.S. 1963. Pachytene analysis in *Oryza*. VI. Karyomorphology of *O. perennis* Moench Cytologia *28*, 36-43.

DASANANDA, S. 1960. Method of reducing the numbering of varieties in Thailand. Int. Rice Comm. Newsl. *9* (1) 7-11.

DASANANDA, S. 1963. Breeding for adaptability to greater range of ecological condition. Int. Rice Comm. Newsl. Spec. Is., 17-23.

DAVE, B.B. 1939. Annual Report Rice Research Station for 1939. Raipur,

Central Provinces, India.

D'CRUZ, R. and DHULAPPANAVAR, C.V. 1963. Inheritance of pigmentation in rice. Indian J. Genet. Pl. Breed. *23*, 3-6.

DEMETERIO-VELASCO, G.V., ANDO, S., RAMIREZ, D.A. and CHANG, T.T. 1965. Cytological and histological studies of sterility in F_1 hybrids of 12 indica-japonica crosses. Phil. Agric. *49*, 248-259.

DHULAPPANAVAR, C.V. 1973. The inheritance of pigmentation in certain plant parts of rice. Canad. J. Genet. Cytol. *15*, 867-870.

DHULAPPANAVAR, C.V., KOLHE, A.K. and CRUZ, R. 1973. Inheritance of pigmentation in rice. III. Auricle, junctura, pulvinus and leaf-axil. Indian J. Genet. Pl. Breed. *33* (3) 389-392.

DOLORES, R.C. 1976. The cytogenetics of *Oryza nivara* Sharma *et* Shastry × *O. sativa* L. F_1 hybrids. M. S. Thesis. University of the Philippines at Los Baños.

DOLORES, R.C., CHANG, T.T. and RAMIREZ, D.A. 1975. Cytogenetics of sterility in F_2 hybrids of Indica × Indica and Indica × Javanica varieties of rice (*Oryza sativa* L.). Cytologia *40*, 639-647.

DOLORES, R.C., CHANG, T.T. and RAMIREZ, D.A. 1978. The cytogenetics of F_1 hybrids from *Oryza nivara* Sharma *et* Shastry × *O. sativa* L. Cytologia. (in process)

ENGLE, L.M., RAMIREZ, D.A. and CHANG, T.T. 1969. The cytology of sterility in F_2, F_3 and F_4 hybrids of Indica × Japonica crosses of rice (*Oryza sativa* L.). Cytologia *34*, (4) 572-585.

ERICKSON, J.R. 1968. Rice genetic investigations. U.S. Dep. Agric. Ann. Rep., Rice Exp. Sta., Biggs, California.

ERICKSON, J.R. 1969. Rice genetic investigations. U.S. Dep. Agric. Ann. Rep., Rice Exp. Sta., Biggs, California.

EVATT, N.S. and BEACHELL, H.M. 1962. Second-crop rice production in Texas. Tex. Agric. Prog. *8* (6) 25-28.

EVELYN, S.H. *et al.* 1971. Preliminary report on the results of the International Rice Adaptation Experiments, 1969. Spec. Comm. for Rice Adap., Int. Biol. Prog./UM/Gene Pools, Int. Rice Adapt. Exp. 2.

EZUKA, A. 1977. Breeding for and genetics of blast resistance in Japan. Ppr. presented at IRRI Rice Blast Workshop, Feb. 21–23. Int. Rice Res. Inst., Los Baños, Laguna, Philippines.

EZUKA, A. *et al.* 1975. Inheritance of resistance of rice variety, Wase Aikoku 3 to *Xanthomonas oryzae*. Bull. Tokai-Kinki Natl. Agric. Exp. Sta. *28*, 124-130.(Japanese with English summary)

FAO (FOOD AND AGRIC. ORGAN. U.N.). 1949. 1948 Yearbook of Food and Agricultural Statistics, Vol. 1, Production. FAO, Washington, D.C.

FAO. 1968. 1967 Production Yearbook, Vol. 21. FAO, Rome.

FRANKEL, O.H. 1975. Genetic resources centers—a cooperative global network. *In* Crop Genetic Resources for Today and Tomorrow. O.H. Frankel and J.G. Hawkes (Editors). Cambridge Univ. Press, Cambridge.

FUKE, Y. 1955. On the genes controlling the heading time of leading rice varieties in Japan and their specific responses to day length and temperature. Bull. Natl. Inst. Agric. Sci. Ser. D5, 1-71. (Japanese with English summary)

FUKE, Y. and KOYAMA, T. 1955. On the resistance of rice to the rice stem maggot, *Chlorops oryzae*, and its inheritance. Rep. Tohoku Natl. Agric. Exp. Sta. *6*, 155-166. (Japanese)

FUKUDA, J. and INOUE, H. 1962. Varietal resistance of rice to the rice stem maggot. Int. Rice Comm. Newsl. *11* (1) 8-9.

FUTSUHARA, Y. and TORIYAMA, K. 1971. Genetic studies on cool tolerance in rice vs effectiveness of individual and line selection for cool tolerance. Jpn. J. Breed. *21*, 181-188.

GHOSE, R.L.M., GHATGE, M.B. and SUBRAHMANYAN, V. 1960. Rice in India. Indian Counc. of Agric. Res., New Delhi.

GHOSH, A.K. and GOVINDASWAMY, S. 1972. Inheritance of starch iodine blue value and alkali digestion value in rice and their genetic association. Riso *21*, 123-132.

GRANT, V. 1964. The Architecture of the Germplasm. John Wiley and Sons, New York.

GREGORY, W.C. 1968. A radiation breeding experiment with peanuts. Radiat. Bot. *8*, 84-147.

GREGORY, W.C. 1972. Mutation breeding in rice improvement. *In* Rice Breeding. Int. Rice Res. Inst., Los Baños, Philippines.

GRIST, D.H. 1975. Rice, 5th edition. Longman, London.

HARGROVE, T.R. 1976. The diffusion of genetic materials among rice breeders in India. Int. Rice Res. Newsl. *1*, 5.

HASHIOKA, Y. 1951. Inheritance of resistance to sheath blight in rice varieties. Ann. Phytopathol. Soc. Jpn. *15*, 98-99. (Japanese)

HAYWARD, C.F. 1975. The status of and prospects for hybrid winter wheat. 2nd Int. Winter Wheat Conf. Zagreb, Yugoslavia.

HECTOR, G.P. 1922. Ann. Rep. of Bengal Dep. of Agric. for 1921-23.

HEU, M.H., CHANG, T.T. and BEACHELL, H.M. 1968. The inheritance of culm length, panicle length, duration to heading and bacterial leaf blight reaction in a rice cross: Sigadis × Taichung (Native) 1. Jpn. J. Breed. *18*, 7-11.

HEU, M.H. and CHOE, Z.R. 1973. The inheritance of alkali digestibility of rice grain in Indica × Japonica crosses. Korean J. Breed. *5*, 32-36. (Korean with English summary)

HEU, M.H. and SUH, H.S. 1974. The segregation mode of plant height in cross of rice varieties. IV. The segregation in the F_2, F_3, BC_1F_1 and BC_1F_2 in some non-allelic combinations. Korean J. Breed. *6* (1) 34-41. (Korean with English summary)

HILLERISLAMBERS, D., RUTGER, J.N., QUALSET, C.O. and WISER, W.J. 1973. Genetic and environmental variation in protein content of rice (*Oryza sativa* L.). Euphytica *22*, 264-273.

HOWELER, R.H. and CADAVID, L.F. 1976. Screening of rice cultivars for tolerance to Al-toxicity in nutrient solutions as compared with a field screening method. Agron. J. *68*, 551-55.

HSIEH, S.C. 1962. Inheritance of mutations induced by irradiation in rice. Bot. Bull. Acad. Sin. *3*, 151-160.

HSIEH, S.C. 1976. Recent advances in rice breeding and genetical studies in Taiwan. Natl. Sci. Counc. Mon. *4* (12) 48-68. (Chinese)

HSIEH, S.C. and RUTTAN, V.W. 1967. Environmental, technological and institutional factors in the growth of rice production: Philippines, Thailand and Taiwan. Stanford Univ. Food Res. Inst. Stud. *7*, 307-341.

HU, C.H. 1973. Evaluation of breeding semidwarf rice by induced mutation and hybridization. Euphytica *22*, 562-574.

HU, C.H., WU, H.P. and LI, H.W. 1970. Present status of rice breeding by induced mutation. *In* Rice Breeding with Induced Mutation, Vol. 2. Tech. Rep. Int. Atom. Energy Agency, Vienna *102*, 13-19.

HUANG, C.H. 1961. Induction of mutations for rice improvement in Taiwan. Crop and Seed Improvement in Taiwan. R. O. C., May 1959—Jan. 1961. Pl. Ind Ser. *22*, JCRR.

HUANG, C.H., CHANG, W.L. and CHANG, T.T. 1972. Ponlai varieties and Taichung Native 1. *In* Rice Breeding. Int. Rice Res. Inst., Los Baños, Philippines.

HUNAN PROV. RICE RES. INST. 1977. Breeding success with hybrid rice is a song of triumph composed by Mao Tse Tung's idealistic score. Zhongguo Nongye Kexue (Chinese Agric. Sci.) *1*, 21-26. (Chinese)

HUNG, H.H. and CHANG, T.T. 1976. Aberrant segregation of three marker-genes in crosses between upland and semidwarf lowland varieties of rice. SABRAO, J. *8* (2) 127-134.

IKEHASHI, H. and ITO, R. 1971. Statistical property of the selection by the plant type index given by quotient of two traits. Jpn. J. Breed. *21*, 106-113.

INDIAN COUNC. FOR AGRIC. RES. (ICAR). 1969. Progress report of the All-India Coordinated Rice Improvement Project. Vol. 1, 1968, New Delhi. (mimeo)

INT. RICE COMM. (IRC). 1959. Gene symbols for rice recommended by the International Rice Commission. Int. Rice Comm. Newsl. *8* (4) 1-7.

IRRI (INT. RICE RES. INST.). Undated. Annual Report for 1963. IRRI, Los Baños, Philippines.

IRRI. 1964. Rice Genetics and Cytogenetics. IRRI, Elsevier, Amsterdam.

IRRI. 1966. Annual Report for 1965. IRRI, Los Baños, Philippines.

IRRI. 1967A. Annual Report for 1966. IRRI, Los Baños, Philippines.

IRRI. 1967B. Annual Report for 1967. IRRI, Los Baños, Philippines.

IRRI. 1972. Annual Report for 1971. IRRI, Los Baños, Philippines.

IRRI. 1973. Annual Report for 1972. IRRI, Los Baños, Philippines.

IRRI. 1974A. Annual Report for 1973. IRRI, Los Baños, Philippines.

IRRI. 1974B. IRRI's GEU program: tapping the genetic reservoir of rice. IRRI Reptr.

IRRI. 1975A. Annual Report for 1974. IRRI, Los Baños, Philippines.

IRRI. 1975B. Major Research in Upland Rice. IRRI, Los Baños, Philippines.

IRRI. 1976A. Annual Report for 1975. IRRI, Los Baños, Philippines.

IRRI. 1976B. How Tongil triggered a Korean rice revolution. IRRI Reptr.

IRRI. 1977A. Annual Report for 1976. IRRI, Los Baños, Philippines.

IRRI. 1977B. Rice blast workshop. IRRI, Los Baños, Philippines. (in process)

IRRI. 1977C. Brown planthopper symposium. IRRI, Los Baños, Philippines. (in process)

IRRI. 1977D. Deep-water rice workshop, Nov. 8−10, 1976. IRRI, Bangkok.

IRRI. 1977E. The International Rice Testing Program 1976. IRRI, Los Baños, Philippines.

IRRI. 1978. Annual Report for 1977. IRRI, Los Baños, Philippines.

IRRI. 1980. Annual Report for 1979. IRRI, Los Baños, Philippines. (in preparation)

ISO, E. 1954. Rice and Crops in its Rotation in Subtropical Zones. Jpn. Food and Agric. Organ., Tokyo.

ITO, R. 1975. Significance of grain/straw ratio in rice breeding. JARQ 9 (4) 1-190.

IWATA, N. and OMURA, T. 1971A. Linkage analysis by reciprocal translocation method in rice plants (*Oryza sativa* L.). I. Linkage group corresponding to the chromosome 1, 2, 3 and 4. Jpn. J. Breed. 21 (1) 19-28. (Japanese with English summary)

IWATA, N. and OMURA, T. 1971B. Linkage analysis by reciprocal translocation method in rice plants (*Oryza sativa* L.). II. Linkage groups corresponding to the chromosomes 5, 6, 8, 9, 10 and 11. Sci. Bull. Fac. of Agric. Kyushu Univ. 25 (3-4) 137-153. (Japanese with English summary)

IWATA, N. and OMURA, T. 1975. Studies on the trisomics in rice plants (*Oryza sativa* L.). III. Relation between trisomics and genetic linkage groups. Jpn. J. Breed. 25, 363-368.

IWATA, N. and OMURA, T. 1976. Studies on the trisomics in rice plants (*Oryza sativa* L.). IV. On the possibility of association of three linkage groups with one chromosome. Jpn. J. Genet. 51 (2) 135-137.

JAYARAJ, D., SESHU, D.V. and SHASTRY, S.V.S. 1972. Genetics of resistance to bacterial leaf blight in rice. Indian J. Genet. 32, 77-89.

JENNINGS, P.R. 1964. Plant type as a rice breeding objective. Crop Sci. 4, 13-15.

JODON, N.E. 1955. Present status of rice genetics. J. Agric. Assoc. China 10 (N.S.), 5-21.

JODON, N.E. 1957. Inheritance of some of the more striking characters in rice. J. Hered. 48, 181-192.

JOHNSTON, T.H., JODON, N.E., BOLLICH, C.N. and RUTGER, J.N. 1972. The development of early maturing and nitrogen-responsive rice varieties in the United States. *In* Rice Breeding. Int. Rice Res. Inst., Los Baños, Philippines.

JONES, J.W.   1928.   Inheritance of earliness and other agronomic characters in rice. J. Agric. Res. (U.S.) *36*, 581-601.

JONES, J.W.   1929.   Distribution of anthocyanin pigments in rice varieties. J. Amer. Soc. Agron. *21*, 867-875.

JONES, J.W.   1933.   Inheritance of characters in rice. J. Agric. Res. *47*, 771-782.

JONES, J.W.   1936.   Improvement in rice. U.S. Dep. Agric. Yearbook 1936, 415-454.

JONES, J.W. and ADAIR, C.R.   1938.   A "lazy" mutation in rice. J. Hered. *29*, 315-318.

JONES, J.W., ADAIR, C.R. and BEACHELL, H.M.   1935.   Inheritance of earliness and length of kernel in rice. J. Amer. Soc. Agron. *27*, 910-921.

KADAM, B.S.   1936.   Genic analysis of rice. I. Grain shedding. Proc. Indian Acad. Sci. Sect. B*4*, 224-229.

KADAM, B.S. and D'CRUZ, R.   1960.   Genic analysis in rice. III. Inheritance of some characters in two clustered varieties of rice. Indian J. Genet. Plant Breed. *20*, 79-84.

KAHLON, P.S.   1965.   Inheritance of alkali digestion index and iodine value in rice. Diss. Abst. B*25*, 5512-5513.

KANEDA, C.   1976.   Recent concepts and trends in rice breeding of Japan. Farming Jpn. *10* (2) 11-18.

KATYAL, J.C., SESHU, D.V., SHASTRY, S.V.S. and FREEMAN, W.H.   1975. Varietal tolerance to low phosphorus conditions. Curr. Sci. *44*, 238-240.

KAWANO, K. and TAKAHASHI, M.E.   1969.   Inheritance modes of quantitative traits in two rice crosses. J. Fac. Agric. Hokkaido Univ. *56* (2) 225-240.

KAWASE, T.   1961.   Studies on Effect of Inheritance of Heading Period and Genes Against Environmental Conditions. Plant Breed. Lab. Coll. of Agric., Kyoto University.

KHALEQUE, M.A. and EUNUS, A.M.   1975.   Inheritance of some quantitative characters in a diallel experiment of six rice strains. SABRAO J. *7* (21) 217-224.

KHAN, S.H.   1974.   The identification and characterization of a complete set of twelve primary trisomics in rice (*Oryza sativa* L.). Diss. Abst. B*35* (6) 2813-2814.

KHUSH, G.S.   1974.   Rice. *In* Handbook of Genetics. R.C. King (Editor). Plenum Press, New York and London.

KHUSH, G.S., AQUINO, R.C. and HERRERA, R.   1976.   Sources of disease and insect resistance in selections of improved plant type. Int. Rice Res. Newsl. *1*, 3.

KHUSH, G.S. and BEACHELL, H.M.   1972.   Breeding for disease and insect resistance at IRRI. *In* Rice Breeding. Int. Rice Res. Inst., Los Baños, Philippines.

KHUSH, G.S. and LING, K.C.   1974.   Inheritance of resistance to grassy stunt virus and its vector in rice. J. Hered. *65* (3) 135-136.

KIKUCHI, F., KUMAGAI, K. and SUZUKI, S.   1975.   Evaluation of adaptability by Finlay and Wilkinson's method. *In* Adaptability in Plants. T. Matsuo (Editor). JIBP Synthesis, Vol. 6. Japanese Comm. for the Int. Biol. Program, Tokyo.

KINOSHITA, T. 1976. Difference and similarity of linkage groups between Japonica and Indica types of rice. Recent Adv. Breed. *17*, 19-34. (Japanese)

KIYOSAWA, S. 1972. Genetics of blast resistance. *In* Rice Breeding. Int. Rice Res. Inst., Los Baños, Philippines.

KIYOSAWA, S. 1974. Studies on genetics and breeding of blast resistance in rice. Natl. Inst. Agric. Sci. Misc. Publ. Ser. D*1*, 1-58.

KIYOSAWA, S. 1976. Pathogenic variations of *Pyricularia oryzae* and their use in genetic and breeding studies. SABRAO J. *8* (1) 53-67.

KOYAMA, T. and HIRAO, J. 1971. Varietal resistance of rice to the stem maggot. Proc. Symp. on Rice Insects. Trop. Agric. Res. Ser. *5*, 251-266. Trop. Agric. Res. Cntr., Tokyo.

KRAMER, T. 1973. Inheritance of leaf characteristics in rice (*Oryza sativa* L.). Diss. Abst. B*35* (2) 886-B.

KUANG, H.H., TU, D.S. and CHANG, Y.H. 1946. Linkage studies of awn in cultivated rice (*Oryza sativa* L.). J. Genet. *47*, 249-259.

KUDO, M. 1968. Genetical and thermotological studies of characters, physiological or ecological in the hybrids between ecological rice groups. Bull. Natl. Inst., Agric. Sci. *19D*, 1-84. (Japanese with English summary)

KURIYAMA, H. 1965. Studies on the ear-emergence in rice. Bull. Natl. Inst., Agric. Sci. *13*D, 275-353. (Japanese with English summary)

KWANGTUNG INST. OF BOT. 1976. Studies on anther culture in *Oryza sativa* subsp. *shien*. II. The role of anther culture in purification and selection of rice cultivar. Acta Genet. Sin. *3*, 57-60. (Chinese with English summary)

KWANGTUNG PROV. ACAD. OF AGRIC. SCI. 1966. Preliminary report on breeding dwarf rice in Kwangtung. Zuowa Xuebao (Crop Sci. J.) *5*, 33-40.

LAKSHMINARAYANA, A. and KHUSH, G.S. 1977. Genes for resistance to the brown planthopper in rice. Crop Sci. *17*, 96-100.

LI, C.C. 1970. Inheritance of the optimum photoperid and critical photoperiod in tropical rices. Bot. Bull. Acad. Sin. *11* (1) 1-15.

LI, C.C. 1975A. Inheritance of cool-temperature seedling vigor in rice (*Oryza sativa* L.). and its relationship with other agronomic characters. Ph. D. Dissertation. University of California, Davis.

LI, C.C. 1975B. Diallel analysis of yield and its components traits in rice (*Oryza sativa* L.). J. Agric. Assoc. China (N.S.) *92*, 41-56.

LI, C.C. and CHANG, T.T. 1970. Diallel analysis of agronomic traits in rice (*Oryza sativa* L.). Bot. Bull. Acad. Sin. *11* (2) 61-78.

LI, C.C. and CHU, T.M. 1979. Studies on drought resistance in rice II. Utilization of proline accumulation in measuring drought resistance. Natl. Sci. Counc. Mon., Taiwan, 7 (4) 383-393. (Chinese with English summary)

LI, H.W., WANG, S. and YEH, P.Z. 1965. A preliminary note on the fine structure analysis of glutinous gene in rice. Bot. Bull. Acad. Sin. *6* (2) 101-105.

LI, H.W., WU, H.P., WU, L. and CHU, M.Y. 1968. Further studies of the interlocus recombination of the glutinous gene of rice. Bot. Bull. Acad. Sin. *9* (1) 22-26.

LIBROJO, V., KAUFFMAN, H.E. and KHUSH, G.S. 1977. Genetic analysis of bacterial blight resistance in four varieties of rice. SABRAO J. *8* (2) 105-110.

LIN, F.H. 1972. Inheritance of photoperiod sensitivity in the crosses between indica and japonica varieties of rice. Taiwan Agric. Q. *8,* 168-175. (Chinese with English summary)

LIN, M.H., CHU, C.L. and CHIEN, C.C. 1975. Inheritance of resistance to yellow dwarf disease in rice. J. Agric. Res. China *24* (3-4) 1-7.

LORESTO, G.C. and CHANG, T.T. 1971. Root development of rice varieties under different soil moisture conditions. *In* Proc. Second Annual Meeting, May 4–6. Crop Sci. Soc. of the Philippines, Los Baños, Philippines.

LORESTO, G.C. and CHANG, T.T. 1978. Half-diallel and F_2 analyses of culm length in dwarf, semidwarf and tall strains of rice (*Oryza sativa* L.). Bot. Bull Acad. Sin. *19,* 87-106.

LORESTO, G.C., CHANG, T.T. and TAGUMPAY, O. 1976. Field evaluation and breeding for drought resistance. Phil. J. Crop Sci. *1* (1) 36-39.

LOVE, H.H. 1955. Report of Rice Investigations, 1950–54. U.S. Operations Mission to Thailand, Agric. Div. and Minist. of Agric., Dep. of Rice, Bangkok.

MARIE, R. 1974. Rice mutants and resistance to blast disease. *In* Induced Mutations for Disease Resistance in Crop Plants. Int. Atom. Energy Agency, Vienna.

MARTINEZ, C.R. and KHUSH, G.S. 1974. Sources of inheritance of resistance to brown planthopper in some breeding lines of rice. *Oryza sativa* L. Crop Sci. *14,* 264-267.

MATSUO, T. 1957. Rice Culture in Japan. Minist. of Agric. and For., Japan.

MISRA, B., RICHHARIA, R.H. and THAKUR, R. 1966. Linkage studies in rice (*Oryza sativa* L.). VII. Indentification of linkage groups in indica rice. Oryza *3* (1) 96-105.

MISRA, R.N. and SHASTRY, S.V.S. 1967. Pachytene analysis in *Oryza*. VIII. Chromosome morphology and karyotypic variation in *O. sativa.* Indian J. Genet. Pl. Breed. *27,* 349-368.

MITRA, G.N. 1962. Inheritance of grain size in rice. Curr. Sci. *31,* 105-106.

MIZUSHIMA, U. and KONDO, A. 1959. Fundamental studies on rice breeding through hybridization between Japanese and foreign varieties. I. An anomalous mode of segregation of apiculus pigmentation observed in a hybrid between Japanese and an Indian variety. Jpn. J. Breed. *9,* 212-218. (Japanese with English summary)

MIZUSHIMA, U. and KONDO, A. 1962. Fundamental studies on rice breeding through hybridization between Japanese and foreign varieties. IV. On the mode of segregation of glutinous character in F_2 of crosses between varieties of remote origin. Jpn. J. Breed. *12,* 1-7. (Japanese with English summary)

MOHAMED, A.H. and HANNA, A.S. 1964. Inheritance of quantitative characters in rice. I. Estimation of the number of effective factor pairs controlling plant height. Genet. *49* (1) 81-93.

MOK, T. and WOO, S.C. 1976. Identification of pollen plants regenerated from anthers of an intersubspecific rice hybrid. Bot. Bull. Acad. Sin. *17* (2) 169-174.

MORINAGA, T. 1938. Inheritance in rice, *Oryza sativa* L. II. Linkage between the gene for purple plant color and the gene for ligulessness. Jpn. J. Bot. *9* (2) 121-129.

MORINAGA, T., FUKUSHIMA, E. and HARA, S. 1943. Inheritance on length of rice grain. Agric. Hort. *18,* 519-522. (Japanese)

MORINAGA, T. and KURIYAMA, H. 1948. Linkage between the genes for sparse caryopsis and depressed palea of *O. sativa* L. Jpn. J. Genet. *23,* 33-34.

MORINAGA, T. and NAGAMATSU, T. 1942. Linkage studies on rice (*Oryza sativa* L.) (A preliminary note). Jpn. J. Genet. *18,* 197-200.

MOSES, C.J., RAO, Y.P. and SIDDIQ, E.A. 1974. Inheritance of resistance to bacterial leaf blight in rice. Indian J. Genet. Pl. Breed. *34,* 271-279.

NAGAI, I. 1958. Japonica Rice: Its Breeding and Culture. Yokendo, Tokyo.

NAGAI, I. and HARA, S. 1930. On the inheritance of variation disease in a strain of rice plant. Jpn. J. Genet. *5,* 140-144.

NAGAMATSU, T. and OMURA, T. 1962. Linkage study of the genes belonging to the first chromosome in rice. Jpn. J. Breed. *12,* 231-236.

NAGAO, S. 1951. Genic analysis and linkage relationship of characters in rice. Adv. Genet. *4,* 181-212.

NAGAO, S. and TAKAHASHI, M. 1942. Type and inheritance of awnedness in rice. J. Soc. Agric. and For., Sapporo *34* (3) 36-43.

NAGAO, S. and TAKAHASHI, M. 1960. Genetical studies on rice plant. XXIV. Preliminary report of twelve linkage groups in Japanese rice. J. Fac. Agric. Hokkaido Univ. *51* (2) 289-298.

NAGAO, S. and TAKAHASHI, M. 1963. Trial construction of 12 linkage groups in Japanese rice. J. Fac. Agric. Hokkaido Univ. *53* (1) 72-130.

NAGAO, S., TAKAHASHI, M. and KINOSHITA, T. 1960. Genetical studies on rice plant. XXV. Inheritance of three morphological characters, pubescence of leaves and flora glumes, and deformation of empty glumes. J. Fac. of Agric. Hokkaido Univ. *51* (2) 299-314.

NAGAO, S., TAKAHASHI, M. and KINOSHITA, T. 1962. Genetical studies on rice plant. XXVI. Mode of inheritance and casual genes for one type of anthocyanin color character in foreign rice varieties. J. Fac. Agric. Hokkaido Univ. *52* (1) 20-49.

NAKAGAHRA, M., OMURA, T. and IWATA, N. 1974. New certation gene on the first linkage group found by inter-subspecific hybridization of cultivated rice. J. Fac. Agric. Kyushu Univ. *18,* 157-167.

NAKATA, S. and JACKSON, B.R. 1973. Inheritance of some physical grain quality characteristics in a cross between a Thai and Taiwanese rice. Thai J. Agric. Sci. 6, 223-235.

NANDI, H.K. and GANGULI, P.M. 1941. Inheritance of earliness in Surma Valley rices. Indian J. Agric. Sci. 11, 9-20.

NARAYANAN NAMBOODIRI, K.M. and PONNAIYA, B.W.X. 1963. Inheritance of seed dormancy in rice. Agric. Res. J. Kerala 2, 30-41.

NAYAK, P., REDDY, P.R. and MISRA, R:N. 1975. Pattern of inheritance of bacterial leaf streak resistance in rice. Curr. Sci. 44, 600-601.

NAYAR, N.M. 1973. Origin and cytogenetics of rice. Adv. Genet. 50, 373-382.

NEI, M. 1960. Studies on the Application of Biometrical Genetics to Plant Breeding. Mem. Coll. Agric. Kyoto Univ. 28.

NG, NYAT QUAT. 1977. A biosystematic study of Asian rices. Ph. D. thesis, University of Birmingham (U.K.).

NIIZEKI, H. 1968. Tissue culture of rice plants. Kagaku Seibutsu 6, 743-748.

OFF. OF RURAL DEV. (Undated). The Green Revolution in Korea is progressing. Suweon, Korea.

OKA, H.I. 1953. Calculation of recombination values from F_2 of semisterile hybrids and restriction of gene recombination due to gamete-development genes. J. Agric. Assoc. China 3/4 (N.S), 39-45.

OKABE, S. 1972. Breeding for high-yielding varieties in Japan. In Rice Breeding. Int. Rice Res. Inst., Los Baños, Philippines.

OKABE, S. and TORIYAMA, K. 1972. Tolerance to cool temperatures in Japanese rice varieties. In Rice Breeding. Int. Rice Res. Inst., Los Baños, Philippines.

OLUFUWOTE, J.O., KHUSH, G.S. and KAUFFMAN, H.E. 1977. Genetics of bacterial blight resistance in rice. Phytopathol. 67 (8) 772-775.

OSADA, A., SASIPRAPA, V., RAHONG, S. DHAMMANUVONG, S. and CHAKRABANDHU, H. 1973. Abnormal occurence of empty grains of indica rice plants in the dry, hot season in Thailand. Proc. Crop Sci. Soc. Jpn. 42, 103-109.

O'TOOLE, J.C. and CHANG, T.T. 1979. Drought resistance in cereals: Rice—a case study. Stress Physiology of Crop Plants. Wiley Interscience, New York.

OU, S.H. 1972. Rice Diseases. Mycolog Inst. Kew, Surrey, England.

OU, S.H. 1977. Breeding for resistance to rice blast—a critical review. Ppr. pres. at IRRI Blast Workshop, Feb. 21−23. Los Baños, Laguna.

PAINTER, R.H. 1951. Insect Resistance in Crop Plants. MacMillan, New York.

PARNELL, F.R., AYYANGAR, G.N.R., RAMIAH, K. and AYYANGAR, C.R.S. 1922. The inheritance of characters in rice. II. Mem. Dep. Agric. India. Bot. Ser. 11, 185-208.

PARTHASARATHY, N. 1959. The need for rice varieties with wider adaptability. Int. Rice Comm. Newsl. 8 (2) 20-25.

PARTHASARATHY, N. 1960. Final report on the international rice hybridization project. Int. Rice Comm. Newsl. 9 (1) 12-13.

PARTHASARATHY, N. 1972. Rice breeding in tropical Asia up to 1960. In Rice Breeding. Int. Rice Res. Inst., Los Baños, Philippines.

PETPISIT, V., KHUSH, G.S. and KAUFFMAN, H.E. 1977. Inheritance of resistance to bacterial blight in rice. Crop Sci. 17, 551-554.

PONNAMPERUMA, F.N. 1977. Screening rice for tolerance to mineral stresses. Int. Rice Res. Inst. Ppr. Ser. (6).

PONNAMPERUMA, F.N. and CASTRO, R. U. 1972. Varietal differences in resistance to adverse soil conditions. In Rice Breeding. Int. Rice Res. Inst., Los Baños, Philippines.

PONNAMPERUMA, F.N. and IKEHASHI, H. 1977. Breeding for adverse soils tolerance. Int. Rice Res. Newsl. 2 (4).

POONAI, P. 1966. Genetic behavior of spikelet length, spikelet breadth and plant height in cultivated rice (Oryza sativa L.) in the F_3 population. Diss. Abst. 27 (5) 1350B.

PRECHACHART, C. and JACKSON, B.R. 1975. Floating and deep-water rice varieties in Thailand. Proc. of the Int. Sem. on Deep-water Rice, Aug., 1974. Bangladesh Rice Res. Inst., Bangladesh.

RAM, J., JAIN, O.P. and MURTY, B.R. 1970. Stability of performance of some varieties and hybrid derivatives in rice under high yielding varieties program. Indian J. Genet. Pl. Breed. 30, 187-198.

RAMIAH, K. 1930. The inheritance of characters in rice. Part 3. Mem. Dep. Agric. Ind. Bot. Ser. B, 211-217.

RAMIAH, K. 1932. Madras Agricultural Station Reports, Paddy Breeding Section for 1931—32. Madras Dep. Agric., Coimbatore.

RAMIAH, K. 1933. Inheritance of flowering duration. Indian J. Agric. Sci. 3, 377-410.

RAMIAH, K. 1945. Anthocyanin genetics of cotton and rice. Indian J. Genet. Pl. Breed. 5, 1-14.

RAMIAH, K. and HANUMANTHARAO, K. 1936. Inheritance of grain shattering in rice. Madras Agric. J. 24, 240-244.

RAMIAH, K., JOBITHRAZ, S. and MUDALIAR, S.D. 1931. Inheritance of characters in rice. Part 4. Mem. Dep. Agric. Ind. Bot. Ser. 18, 229-259.

RAMIAH, K. and PARTHASARATHY, N. 1933. Inheritance of grain length in rice (Oryza sativa L.). Indian J. Agric. Sci. 3 (Part 5), 808-819.

RAMIAH, K. and RAMASWAMI, K. 1941. Floating habit in rice. Indian J. Agric. Sci. 11, 1-8.

RAMIAH, K. and RAO, M.B.V.N. 1953. Rice Breeding and Genetics. Sci. Monog. 19. Indian Counc. for Agric. Res., New Delhi.

RATHO, S.N., NAYAK, P., MISRA, R.N. and PADMANABHAN, S.Y. 1976. Diallel analysis of bacterial leaf blight resistance in rice. I. Nature of gene action. Riso 25, 65-74.

RICHHARIA, R.H., MISRO, B., BUTANY, W.T. and SEETHARAMAN,

R. 1960. Linkage studies in rice (*Oryza sativa* L.). Euphytica *9* (1) 122-126.

RU, S.K. 1958. Rice. Cheng Chung Book Co., Taipei. (Chinese)

RUTGER, J.N. and PETERSON, M.L. 1976. Improved short stature rice. Calif. Agric. *30* (6) 4-6.

RUTGER, J.N., PETERSON, M.L., HU, C.H. and LEHMAN, W.H. 1976. Induction of useful short stature and early maturing mutants in two Japonica rice cultivars. Crop Sci. *16*, 631-635.

RYKER, T.C. and CHILTON, S.J.P. 1942. Inheritance and linkage of factors for resistance to two physiological races of *Cercospora oryzae* in rice. J. Amer. Soc. Agron. *34*, 836-840.

RYKER, T.C. and JODON, N.E. 1940. Inheritance of resistance to *Cercospora oryzae* in rice. Phytopathol. *30*, 1041-1047.

SAKAGUCHI, S. 1967. Linkage studies on the resistance to bacterial blight, *Xanthomonas oryzae*, in rice. Bull Natl. Inst. Agric. Sci. Jpn. *D16*, 1-18. (Japanese with English summary)

SAKAI, K.I. and NILES, J.J. 1957. Heritability of grain shedding and other characters in rice. Trop. Agric. (Ceylon) *113*, 211-218.

SAKAI, K.I. and PINTO, E.R. 1959. Genetic studies on seed size in rice hybrids. Ann. Rep. Natl. Inst. Genet. (Japan) *9*, 72-73.

SAKAI, K.I. and SHIMAZAKI, Y. 1948. Genetical studies on abnormal hypertrophy of tapetal cells in *Oryza sativa* L. Kanchi-Nogyo *2* (2) 93-95. (Japanese)

SAMOTO, S. and HAMAMURA, K. 1973. Inheritance of the grain characters in rice plant. Hokuriku Crop Sci. *8*, 11-13. (Japanese)

SAMPATH, S. and SESHU, D.V. 1961. Genetics of photoperiod response in rice. Indian J. Genet. Pl. Breed. *21*, 38-42.

SASAKI, T., KINOSHITA, T. and TAKAHASHI, M.T. 1973. Estimation of the number of genes in germination ability at low temperature in rice—genetical studies in rice plant. LVII. J. Fac. Agric. Hokkaido Univ. *57* (3) 301-312.

SASTRY, M.V.S. and PRAKASA RAO, P.S. 1973. Inheritance of resistance to rice gall midge *Pachydiplosis oryzae* Wood Mason. Curr. Sci. *42*, 652-653.

SATYANARAYANAIAH, K. and REDDI, M.V. 1972. Inheritance of resistance to insect gall-midge (*Pachydiplosis oryzae* Wood Mason) in rice. Andhra Agric. J. 19 (1) 1-8.

SEETHARAMAN, R. 1959. The inheritance of iodine value in rice and its association with other characters. Diss. Abst. *20*, 856-857.

SEETHARAMAN, R. 1964. Certain considerations on genic analysis and linkage groups in rice. *In* Rice Genetics and Cytogenetics. Elsevier, Amsterdam.

SEN, P.K., MITRA, G.N. and BANERJEE, S. 1964. Inheritance of photoperiod reaction in rice. Indian J. Agric. Sci. *34*, 1-14.

SESHU, D.V. 1976. High yielding rice varieties from India. Int. Rice Comm. Newls. *25* (1-2) 40-41.

SESHU, D.V., SHASTRY, S.V.S. and FREEMAN, W.H. 1974. Phenotypic stability of performance in yield of dwarf rice varieties. Indian J. Genet. *34A*,

1130-1139.

SETHI, R.L., SETHI, B.L. and MEHTA, T.R. 1937. Inheritance of sheathed ear in rice. Indian J. Agric. Sci. 7, 134-148.

SHAH, H.M., BEACHELL, H.M. and ATKINS, I.M. 1961. Morphological and cytological changes in Century Patna 231 and Bluebonnet 50 rice resulting from X-ray and thermal neutron irradiation. Crop Sci. 1, 97-102.

SHANMUGASUNDARAM, A. 1953. Studies on dormancy in short-term rices. Madras Agric. J. 40, 477-487.

SHASTRY, S.V.S., JOHN, V.T. and SESHU, D.V. 1972. Breeding for resistance to rice tungro virus in India. In Rice Breeding. Int. Rice Res. Inst., Los Baños, Philippines.

SHASTRY, S.V.S., RANGA RAO, D.R. and MISRA, R.N. 1960. Pachytene analysis in Oryza. I. Chromosome morphology in Oryza sativa. Indian J. Genet. Pl. Breed. 20, 15-21.

SHEKHAWAT, G.S., SRIVASTAVA, P.N. and RAO, Y.P. 1972. Pattern of inheritance of bacterial leaf streak resistance in rice. Curr. Sci. 44, 600-601.

SHEN, T.H. 1951. Agricultural Resources of China. Cornell Univ. Press, Ithaca, New York.

SHEN, T.H. 1964. Agricultural Development on Taiwan Since World War II. Comstock Publishing Assoc., Ithaca, New York.

SHEN, T.T., LU, T.T. and LI, J.S. 1965. A genetical analysis of some characters in breeding early-maturing short-statured rice types. Zuowu Xuebao (Crop Sci. J.) 4, 391-402. (Chinese)

SHIN, D.W. and SHIM, Y.K. 1975. The effectiveness of the Tongil Rice Diffusion in Korea. Off. of Rural Dev., Suweon, Korea.

SHINODA, H. et al. 1971. Studies on the resistance of rice varieties to blast. VI. Linkage groups of genes for blast resistance. Bull. Chugoku Natl. Agric. Exp. Sta. 20A, 1-25.

SIDHU, G.S. and KHUSH, G.S. 1978. Dominance reversal of a bacterial blight resistance gene in some rice cultivars. Phytopathol. 68, 461-463.

SIREGAR, H. 1951. Variety-comparison and line selection in rice. Landb. Agric. J. Indonesia 23, 485-509. (Dutch with English summary)

SIREGAR, H. 1957. Some aspects of rice growing in Indonesia. C.R. 3rd Congr. Pan-Indian Ocean Sci. Assoc. 1975D, 115-136.

SIWI, B.H. and KHUSH, G.S. 1977. New genes for resistance to the green leafhopper in rice. Crop Sci. 17 (1) 17-20.

SOMRITH, B. 1974. Genetic analysis of traits related to grain yield and quality in two crosses of rice (Oryza sativa L.) Ph. D. Dissertation. Indian Agric. Res. Inst., New Delhi.

STANSEL, J.W. 1966. The influence of heredity and environment on endosperm characteristics of rice (Oryza sativa L.). Diss. Abst. B27, 48.

STAVIS, B. 1974. Making green revolution—the politics of agricural development in China. Cornell Univ., Rural Dev. Comm. Monog. Ser. 1.

SUGAWARA, T. 1953. Chemico-genetic basis for the formation of amylose and amylopectin in endosperm (Abst.). Jpn. J. Breed. *3* (1) 48. (Japanese)

SUGIMOTO, S. 1923. Examples of the genesis of abnormal forms in rice. Jpn. J. Genet. *2*, 71-75. (Japanese)

SUR, S. C. 1975. Identification and cytogenetic studies of primary trisomics in rice (*Oryza sativa* L.). Ph. D. Dissertation. University of Philippines at Los Baños.

SUZUKI, S. and KIKUCHI, F. 1975. Response of rice varieties to climatic factors and its relation to adaptability. *In* Adaptability in Plants. T. Matsuo (Editor). JIBP Synthesis, Vol. 6. Sci. Counc. of Jpn., Tokyo.

TAKAHASHI, M.E. 1957. Analysis of apiculus color genes essential to anthocyanin coloration in rice. J. Fac. Agric. Hokkaido Univ. *50*, 266-362.

TAKAHASHI, N. 1962. Physicogenetical studies on germination of rice seeds with special reference to the genetical factors governing germination. Bull. Inst. Agric. Tohoku Univ. *14* (1) 1-87. (Japanese)

TAKAHASHI, M. 1964. Linkage group and gene schemes of some striking morphological characters in Japanese rice. *In* Rice Genetics and Cytogenetics. Elsevier, Amsterdam.

TAKAHASHI, M. 1977. Linkage map of the rice plant. *In* Proc. 3rd Int. Congr. of Soc. for Adv. of Breed. Res. in Asia and Oceania (SABRAO), Canberra.

TAKEZAKI, Y. 1932. Genetics of the sterile swampy rice line. Pflanzenbau, Tiere *7*, 387-398, 2255-2267. (Japanese)

TANAKA, A. 1965. Plant characters related to nitrogen response in rice. *In* the Mineral Nutrition of the Rice Plant. Johns Hopkins Univ. Press, Baltimore.

THAKUR, R. and ROY, R.P. 1975. Linkage studies in Indica rice, *Oryza sativa* L. Euphytica *24* (2) 511-516.

TIENTSIN AGRIC. RES. INST. 1976. New varieties "Hua Yu I" and "Hua Yu II" developed from anther culture. Acta Genet. Sin. *3*, 19-24. (Chinese with English summary)

TINARELLI, A. 1966. Varietal catalogue of Italian Rices. *In* Contributo all Anno Internazionale del Riso. Food and Agric Organ. U.N., Rome. (Italian)

TING, Y. (Editor). 1961. Chinese Culture of Lowland Rice. Agricultural Publishing Soc., Peking. (Chinese)

TORIYAMA, K. 1972. Breeding for resistance to major rice diseases in Japan. *In* Rice Breeding. Int. Rice Res. Inst., Los Baños, Philippines.

TORIYAMA, K. 1975. Recent progress of studies on horizontal resistance in rice breeding for blast resistance in Japan. *In* Horizontal Resistance to the Blast Disease of Rice. Cntr. Int. Agric. Trop., Cali, Colombia.

TORIYAMA, K. 1976. National program of the rice breeding in Japan. Paper presented to the Assoc. for Sci. Co-op. in Asia Sem. on High-yielding Var. of Paddy Rice, Nov. 30—Dec. 2. Tokyo.

TORIYAMA, K. and FUTSUHARA, Y. 1960. I. Inheritance of cool tolerance. Jpn. J. Breed. *1*, 143-152.

TORIYAMA, K., YUNOKI, T. and SHINODA, H. 1968. Breeding rice varieties for resistance to blast, II. Inheritance of high field resistance of Chugoku

31 (Abst.). Jap. J. Breed. *18* (Suppl. 1), 145-146. (Japanese)

TSAI, K. 1976. Studies on earliness genes in rice, with special reference to analysis of isoalleles at the E locus. Jpn. J. Genet. *51,* 115-128.

TSAI, K.H. and OKA, H.I. 1966. Genetic studies of yielding capacity and adaptability in crop plants. II. Analysis of genes controlling heading time in Taichung 65 and other rice varieties. Bot. Bull. Acad. Sin. 7 (2) 54-70.

TSAI, K.H. and OKA, H.I., 1970. Genetic studies of yielding capacity and adaptability in crop plants. IV. Effect of an earliness gene, m_b, in the genetic background of a rice variety, Taichung 65. Bot. Bull. Acad. Sin. *11* (1) 16-26.

TSUNODA, S. 1959. A developmental analysis of yielding ability in varieties of field crops. II. The assimilation system of plants as affected by the form direction and arrangement of single leaves. Jpn. J. Breed. *9,* 237-244. (Japanese with English summary)

UMALI, D.L. *et al.* 1956. Cooperative rice improvement program in the Philippines. Int. Rice Comm. Newsl. *18,* 1-7.

UNIVERSITY OF CALIFORNIA. 1974. A Hungry World: the Challenge to Agriculture. Div. of Agric. Sci., Univ. of Calif.

VAN DER STOK, J.E. 1908. On the inheritance of grain colour in rice. Teysmannia *65,* 5.

VERGARA, B.S. and CHANG, T.T. 1976. The flowering response of the rice plant to photoperiod: a review of the literature, 3rd Edition. Int. Rice Res. Inst. Tech. Bull. *8.*

WILLIS, G.M., ALLOWITZ, R.D. and MENVIELLE, E.S. 1968. Differential susceptibility of rice leaves and panicles to *Piricularia oryzae.* Phytopathol. *58,* 1072 (Abst.).

WOO, S.C. 1965. Some experimental studies on the inheritance of resistance and susceptibility to rice leaf blast disease, *Pyricularia oryzae* Cav. Bot. Bull. Acad. Sin. *6* (2) 208-217.

WOO, S.C. and SU, H.Y. 1975. Doubled haploid rice from Indica and Japonica hybrids through anther culture. Bot. Bull Acad. Sin. *16* (2) 19-24.

WOO, S.C., WU, W.H. and TUNG, I.J. 1974. Two semidwarfness genes induced from Japonica rice, Chianung 242 and Tainan 5. Bot. Bull. Acad. Sin. *15* (1) 54-56.

WU, H.P. 1968. Studies on the quantitative inheritance of *Oryza sativa* L. I. Diallel analysis of heading time and plant height in F_1 progeny. Bot. Bull. Acad. Sin. *9* (2) 1-9.

YAMASAKI, Y. and KIYOSAWA, S. 1966. Studies on inheritance of resistance of rice varieties to blast. I. Inheritance of resistance of Japanese varieties to several strains of the fungus. Bull. Natl. Inst. Agric. Sci. Ser. D *14,* 39-69.

YANG, S.C. 1970. Variability of some agronomic traits in the F_2 population grown by generation acceleration method of rice. J. Taiwan Agric. Res. *19* (1) 1-5. (Chinese with English summary)

YIN, K.C. *et al.* 1976. A study of the new cultivars of rice raised by haploid breeding method. Sci. Sin. *19,* 227-242.

YOSHIDA, S. 1972. Physiological aspects of grain yield. Ann. Rev. Pl. Physiol. *23*, 437-464.

YOSHIDA, S., COCK, J.H. and PARAO, F.T. 1972. Physiological aspects of high yields. *In* Rice Breeding. Int. Rice Res. Inst., Los Baños, Philippines.

YU, C.J. and YAO, Y.T. 1957. Genetic germination time on rice. Jpn. J. Genet. *32*, 179-188, (Japanese)

YUAN, L.P. 1977. The execution and theory of developing hybrid rice. Zhong-guo Nongye Kexue (Chinese Agric. Sci.) *1*, 27-31. (Chinese)

4

Rice Culture

Duane S. Mikkelsen
Surajit K. DeDatta

Rice is a unique major food crop of the world by virtue of the extent and variety of uses and its adaptability to a broad range of climatic, edaphic and cultural conditions. It is usually grown under shallow flood or "wet paddy" conditions, but is cultured where flood waters may be several meters deep and to the opposite extreme as an upland cereal. Although rice appears to have a high water requirement, it is not much different from that of other field crops. Unlike most cereal crops, however, rice benefits from standing water. It has capability for anaerobic respiration and has aerenchyma tissue in the aerial organs through which oxygen diffuses to the roots.

Because of its unique ability to grow and produce high caloric food values per unit area on all types of land and water regimes, combined with its adaptation to a wide variety of climates and agricultural conditions, it is the world's most important cereal crop. Thousands of cultivars are grown throughout the world, representing a wide range of plant and grain characteristics. The crop occupies some 137,411,000 ha of land (1973–75) which produce 331,460,000 MT of rice with an average world yield of 2.4 MT/ha. The significance of rice is shown in its widespread use as a staple food by more than half of the world's population. Millions of people in Asia subsist almost entirely on rice. According to U.S. Department of Agriculture statistics (USDA 1976), 91.5% of the world's rice is produced and consumed in Asia. The estimated area and production of world rice, by selected countries and regions for the years 1973–75, is shown in Fig. 4.1 and 4.2. The average yields of paddy per hectare in each region of the world are shown as follows:

	MT/ha
East Asia	3.7
Southeast Asia	2.0

South Asia	1.8
West Asia, North Africa	3.8
Sub-Sahara	1.3
Europe	4.3
Latin America	1.8
North America	5.0
Oceania	5.9

According to Chang (1976), the 2 cultivated species of rice, *Oryza sativa* and *O. glaberrima,* developed from a common progenitor, progressively from a wild perennial to a wild annual and ultimately a cultivated annual. Through diversification accelerated by climatic changes, human dispersal and selection over a wide range of latitude and altitude, plus manipulation for cultural adaptation, many different cultivars have been developed. From its origin in ancient India, *O. sativa* was dispersed to north and central China, where puddling and transplanting techniques were developed. In Southeast Asia, upland culture occurred first. Planting progressed from shifting cultivation to direct sowing in prepared fields, with later transplanting into bunded fields. The dispersal of *O. sativa* has led to the development of 3 ecogeographical races, Indica, Japonica and Javanica, each grown with cultural practices ranging from upland, to lowland and to deep water cultures. *O. glaberrima* developed somewhat later than *O. sativa* and is grown largely in Africa. The tropical race Indica has spread through the humid tropics, the Middle East, Europe and Africa. Bold grain, tall Javanica races have spread through parts of Asia and many contiguous island areas, including Indonesia, the Philippines, Taiwan and Japan. The cool-season race, Japonica, was developed in the lower Yangtze River area of China, from where it was introduced into Korea, Japan and later to southern Europe, the USSR, the United States and South America.

With distribution of rice cultivars to the high latitudes, cultigenic changes occurred in response to low temperature and short photoperiod. This was accompanied by selection for short plant stature and plants with more determinate tillering and uniform heading. In the high latitudes of the world, rice can be grown only during the warm season of the year. Great genetic diversity exists as shown by the fact that more than 34,000 rice cultivars are currently identified at the International Rice Research Institute (IRRI) by differences in adaptability to various climates, soil and water requirements, agronomic characteristics of plant type, growth duration, photoperiodism, grain type and quality.

The main theater of rice production is found on the Asian continent and adjacent islands which lie in the tropical and subtropical regions of

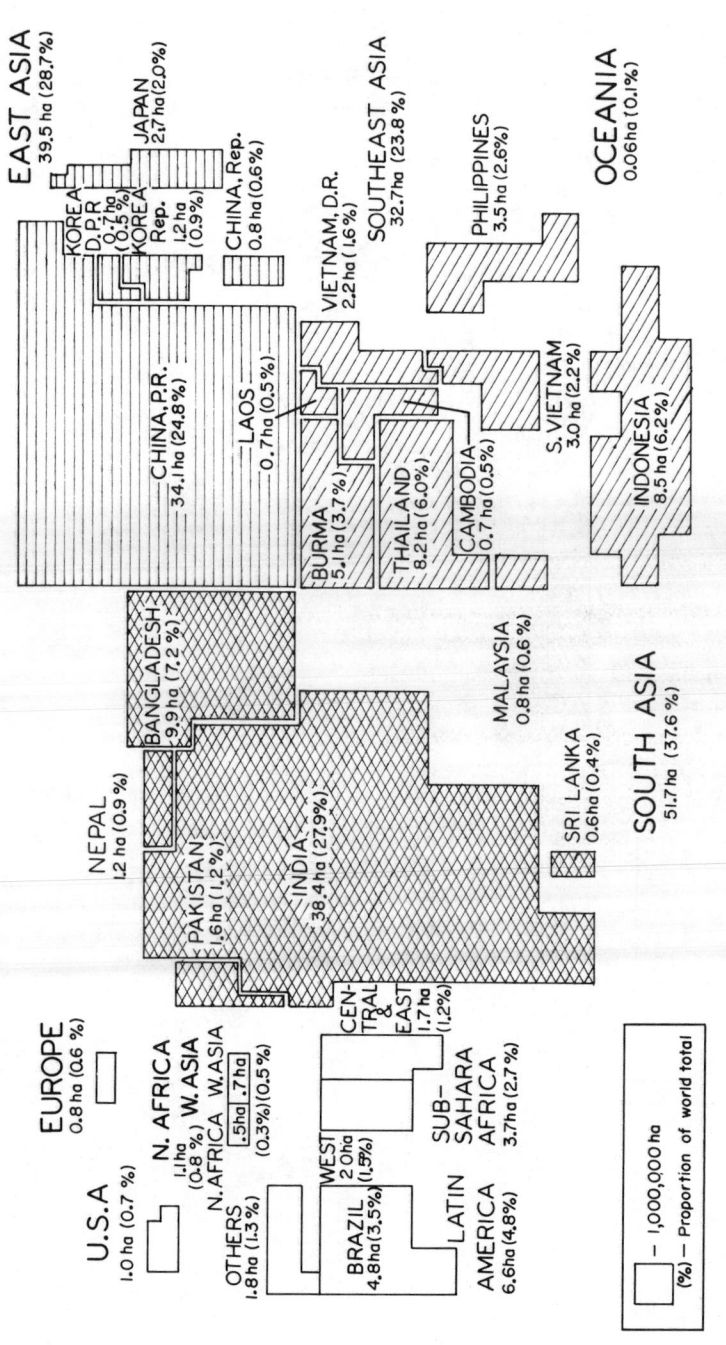

FIG. 4.1. THE WORLD'S RICE LAND

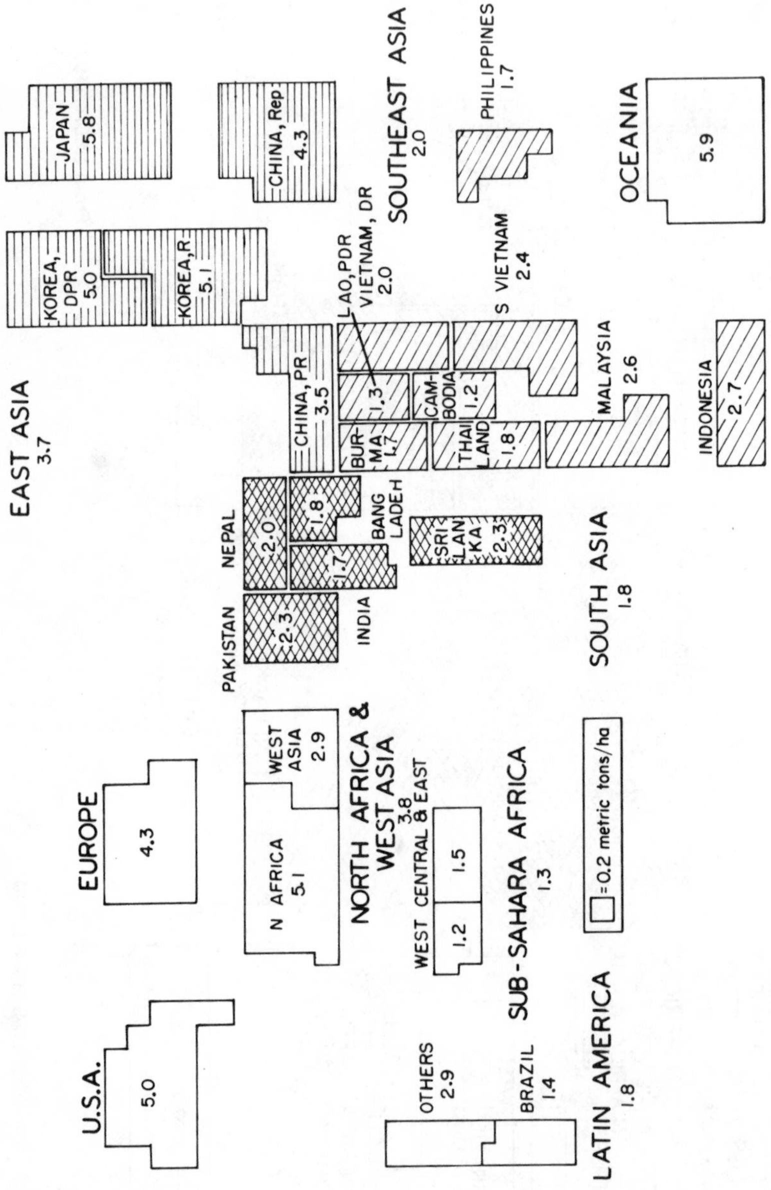

FIG. 4.2. THE WORLD'S RICE YIELDS

From USDA (1973–1975)

the world. Much of the Asian production lies in the monsoon area where the ease of securing water from rainfall plus some supplementary irrigation provides an edaphic advantage. Rice production is concentrated in areas where water management is convenient on flat lowlands, river basins and delta areas. The continental monsoon varies from year to year, however, creating an instability in rice production which will be overcome only when adequate supplementary irrigation facilities are constructed. Within the Asian area, regional monsoons, trade winds, typhoons and tropical depressions create a pattern of distinct wet and dry seasons and add variability to the climatic features from year to year. The monsoon season—from June to December in the tropics north of the equator and from November to April south of the equator— constitutes the area of the so-called "monsoon" rice crop.

While the bulk of the rice production is centered in wet tropical climates, the crop flourishes well in humid regions of the subtropics and temperate climates such as Africa, Japan, Spain, Portugal, Italy, South and Central America, Hungary, the USSR, the United States, France and Australia. The highest grain yields have been recorded between 30° and 45° latitude, while producing areas extend from about 45°N to 35°S latitude. Various explanations have been suggested for the higher rice yields obtained in the higher latitudes of the world. These include longer day length and higher solar energy, varietal differences (Japonica in higher latitudes), more favorable soil conditions, better water control, and fewer disease and insect problems. The rice plant is unique in that it has adaptability to a wide range of climatic and environmental conditions ranging from upland and lowland to deep water culture. Cultivars have been developed which meet the wide range of conditions for rice to be grown as a commercial crop.

Climatic Factors Affecting Rice Production

The distribution of rice production over the world is controlled primarily by climatic variables which ideally should provide adequate water during the entire growing season, relatively high air and soil temperatures, adequate solar radiation, the absence of destructive storms, a moderately long growing season and relatively rain free conditions during the ripening period. Important edaphic factors influenced by climate include the chemical and physical conditions of the soil and physiography of the land.

Precipitation and Water.—Water, often in the form of precipitation, is the most important factor influencing the distribution of rice in the world, and also its growth and yield potential. The growth of all rice varieties is favored by flooding, and yields of rice have been increased as much as 53% over adequately irrigated but unflooded culture (Senewiratne and Mikkelsen 1961). Although rice benefits from a flooded soil environment, it requires no more water than most upland field crops. The major benefits derived from a flooded soil are enhanced nutrient availability, especially of N, P, Fe and Mn, enhanced nitrogen fixation, less weed competition and favorable microclimatic regulation. Unflooded culture initially favors seedling growth but is unfavorable for tiller production, vegetative and reproductive growth, and ultimately yield. Consequently rice is preferentially grown only in the rainy season except where water storage and irrigation facilities are available. Total seasonal precipitation, its intensity and distribution, is extremely important.

A desirable rainfall pattern provides abundant precipitation at the beginning of the rice season to allow timely land preparation, transplanting, or broadcast seeding of the crop. Less frequent rainfall is needed during the vegetative growth period, but it should be sufficient to replace losses due to evaporation, transpiration, percolation and runoff. Ideally, during the reproductive and ripening growth stages, solar radiation should be high to favor photosynthesis and grain formation. Water stresses during the reproductive and ripening stages are particularly damaging to grain production. Rainy weather during the flowering period, which is often accompanied by a lowering of temperature, may also adversely affect normal pollination and increase the percentage of sterile spikelets.

Since rice is unable to make full use of the total annual rainfall, only the precipitation occurring during the cropping period is directly beneficial. As a rule, between 70 and 90% of the total rainfall during cropping effectively benefits the crop. The intensity and distribution of the rainfall, the water retention characteristics of the soil, various field losses and water conserving cultural practices all affect water availability.

With rainfed-flooded rice culture, the major determinants of crop water requirements are surface runoff, percolation losses and evapotranspiration. Where it is feasible, a portion of the potential surface runoff should be restricted by surrounding bunds, but when continuous precipitation exceeds 50—80 mm, runoff is likely to occur. Water is also lost by percolation depending upon soil type, land preparation practices and depth of the water table. Percolation rates may vary from 1 mm per day in well puddled clay soils to 3—10 mm per day in coarse textured soils with deep-water tables. Evapotranspiration values vary according to local climatic conditions, cropping practices and stage of plant development. Evapotranspiration in humid climates is fairly consistent between 3.5 and 6 mm per day where advective energy is relatively small to 8 mm per day with a high advective influence. Transpiration losses vary with temperature, humidity, wind velocity, water management practices and plant growth stages. Generally, transpiration increases with total crop leaf area. Transpiration is relatively slow after transplanting, increases with tillering, reaching a peak at heading, then decreases with grain ripening. Evaporation likewise varies with climatic and crop density factors but generally reaches its maximum rate during the seedling stage and decreases as the plant canopy shades the water.

The net water requirement for rice grown in tropical Asia varies from 750 to 2500 mm with the average about 1250 mm. Of this amount, 40 mm is allocated for seedling nursery purposes, 210 mm for land preparation and about 1000 for irrigation purposes. Typical water requirements in various areas include 2.5–4.5 acre-ft per season in Japan, 6.0 in Thailand, 4.0 in Indochina, and 4–6 in Sri Lanka, Australia and the United States.

Excessive rainfall can be a major source of crop loss where flooding cannot be controlled and crops are inundated. In parts of Asia and Africa, special deep water or "floating rice" varieties must be grown to tolerate deep flooding.

Temperature.—Rice is adaptable to areas with abundant sunshine and average temperatures above 20°–38°C (68°–100°F). Temperatures below 15°C (59°F) retard seedling development, delay transplanting, slow tiller formation, delay reproductive growth and consequently reduce grain yields.

The lower temperature limits for germination are difficult to estimate and vary with variety, but germination proceeds only slowly at 10°C (50°F). Optimum germination is in the range of 18°–33°C (64°– 91.4°F) with seeds of most varieties germinating more rapidly at the higher temperatures than the lower ones. At 42°C (108°F), germination is arrested; at 50°C (122°F) the seed is killed. Critical air temperature is 15°–15.5°C (59°–60°F) for transplanting seedlings from lowland nurseries, 14°–15°C (57.2°–59°F) from semi-irrigated beds, and 13°–13.5°C (35.4°–56.5°F) for seedlings from upland nurseries. Rice rooting occurs over a range of 19°–33°C (66.2°–91.4°F), 25°–28°C optimum (77°–82.4°F) with root inhibition below 16°C (60.8°F) and above 35°C (95°F).

Top growth is generally linear between 18° and 33°C (64.4° and 91.4°F), but above and below that range growth notably decreases. As a rule, water temperature is more critical than air temperature because the growing meristem is usually under water (Matsushima *et al.* 1966, 1968). Sasaki (1927) reported that leaf elongation increases with temperature in the range of 17° to 31°C (52.6° to 87.8°F), but decreases and practically ceases at 45°C (113°F). The lower limit for leaf elongation is 7° to 8°C (44.6°–46.4°F). Tillering is a complex relationship involving an interaction between carbohydrate metabolism, solar radiation and temperature.

Tillering is generally favorably affected in the range of 15°–33°C (59°–91.4°F).

Anthesis generally begins at the same time as panicle emergence from the sheath. Morning temperatures, rather than daily mean temperatures, determine the start of flowering. Sakai (1949) states that 29°–31°C (84°–88°F) appears to be the most favorable temperature for the opening of spikelets and for effective pollination. Flowering is usually stimulated by high temperatures, although variations occur among cultivars.

Patterns of temperature variation during each crop season and sequential changes from season to season are complex and affect plant growth and maturity in different ways. As a rule, the annual mean temperature decreases from lower to higher latitudes and altitudes. In the tropics, rice can usually be planted in any month of the year since temperature variations are only slight from season to season and day to night. Temperature differences do not usually cause yield fluctuations in the near sea level elevations in the tropics. Even so, low temperature at high altitudes may be a problem for tropical varieties, and near sea level, high temperature during flowering and grain formation may cause damage. In subtropical and temperate areas, temperatures usually increase from spring to summer and decrease again in autumn, thus limiting the growing season. Air and root-zone temperatures may change considerably from night to day and from season to season, affecting nutrient uptake, vegetative and reproductive development and grain ripening.

The effects of temperature on rice plant growth can be evaluated by relating phenological data to heat summation units or day-degrees. Heat summation data for all the growing days above the baseline of a mean daily temperature show that the lower range is probably between 1649° and 2204°C (898° and 1206°F) and that most varieties need an accumulated temperature of 2982° and 3593°C (1639° to 1978°F). Nuttonson (1961) computed day-degree summation with a 10°C (50°F) baseline from the sown to ripe period of nontropical cultivars showing a range of 704°C (373°F) in the USSR (Lat 43° 40'N) to over 2760°C (1516°F) in Texas (Lat 30° 04'N).

Photoperiod.—Photoperiod (day length) is a major factor influencing the development of the rice plant, especially its flowering characteristics (Chang and Vergara 1972). Although rice is considered to be a short day plant (short days decrease their growth duration), cultivars differ widely in sensitivity. Photoperiod is the duration of the light period between sunrise and sunset, including the twilight hours. The daylight hours fluctuate during the year, varying with latitude. The day length is shorter in winter, increasing gradually toward summer with the changes greatest in the high latitude areas. In the tropics the maximum photoperiod difference is less than 3 hr, but in the warm temperate zones, the difference in photoperiod may be as much as 5 hr.

Photoperiod sensitive cultivars flower when the decreasing day length reaches a critical length for induction of the flowering stage. Day length exerts a large effect on the growth duration of rice cultivars, depending upon their photoperiod sensitivity. The response occurs largely through its effect in altering the basic vegetative growth pattern and strongly affects the duration of reproductive growth. Photoperiod sensitivity (the critical day length required for flowering) varies greatly, some cultivars being referred to as nonseasonal or neutral, others weakly photoperiod sensitive and others strongly photoperiod sensitive. Weakly sensitive cultivars vary only slightly in growth duration from time of planting to harvest, while those termed strongly sensitive can be planted only during seasons with long day length. Photoperiods that are longer or shorter than the optimum delay the flowering of photoperiod sensitive cultivars. The range of diversity among cultivars is very large with virtually no cultivars showing a long day response.

The practical application is that farmers must select wisely among the cultivars they plant, since photoperiod sensitivity will determine the growth duration, maturity date, and potential yield of a crop as well as cultigenic adaptation to double cropping. In tropical areas, photoperiod-sensitive cultivars have traditionally been selected because they could be planted when the monsoon rains begin and are harvested at a fixed time after rains cease and flood waters recede. Such cultivars utilize the high solar radiation at the late growth stages which have beneficial effects on yield.

Solar Radiation.—Solar radiation and sunshine hours are important climatic determinants in rice production. Numerous studies have shown a close correlation between solar radiation, plant growth and yield (Moomaw et al. 1967; Stansel 1975). Young seedlings have a comparatively low solar radiation requirement (Tanaka et al. 1966), so shading at early stages exerts only small effects on ultimate yield. Light becomes progressively more important through the vegetative and reproductive stages, reaching maximum importance at the heading stage. The most critical period of solar energy requirement extends from panicle differentiation to about 10 days before maturity (Stansel 1975).

The amount of sunlight a crop receives depends on solar radiation intensity, day length, cloud cover and mutual shading by the plants in a population. Day length and solar radiation intensity are determined by geographical location and changing seasons as modified by cloud cover. Mutual shading is the shading of one plant part by others. Plant type and leaf arrangement play an important role on the energy absorption characteristics of the plant canopy. The effects of mutual shading in the tropics are most critical where days are shorter and midday intensity is high for relatively short periods. Cloudiness greatly reduces the period of energy absorption, especially during the monsoon. Many rice growing areas have distinct wet and dry seasons which greatly affect solar radiation. Results at IRRI (Moomaw *et al.* 1967) show that high grain yields were strongly correlated with total solar radiation between 30 and 45 days before harvest (De Datta and Zarate 1970).

Mean daily solar radiation values during the main rice growing season are lower in the tropics than in the temperate zone, which may account in part for yield differences in the 2 areas.

Tropical Storms.—Tropical storms of various types that are often violent and destructive to the rice crop include tropical cyclones with origins in the low-latitude oceanic areas, moving rapidly as violent hurricanes or typhoons, cyclonic depressions developing in the middle and high latitudes, and, to a lesser extent, tornadoes. If they occur after the heading of rice, they may cause severe lodging and shatter loss in some varieties. Strong winds just before heading may also cause a decrease in the number of spikelets per panicle. High winds during pollination will induce sterility, while continued strong winds have been shown to reduce photosynthesis and promote the spread of bacterial diseases both in temperate and tropical rice areas. Strong winds and rains which mechanically damage the leaves are harmful to rice growth and grain yields, and winds at the ripening stage cause grain losses through increased shatter. Cultivars that are susceptible to mechanical damage of leaves and the shattering of ripe grain have produced lower grain yields than less susceptible ones (Chang and Vergara 1972).

Soil Characteristics Affecting Rice Production

Over the centuries, a number of systems of rice cultivation have evolved to fit local conditions of climate, soils, water supply, economics and social conditions. Despite wide variations in culture, there are 2 main systems of soil management: (1) dry soil management, in which the land is prepared dry and the crop is seeded in the same manner as other cereal crops, referred to as upland rice culture; and (2) wet soil management, in which the land is flooded and all soil preparation is done in wet or submerged soil, referred to as paddy or wetland culture.

Upland Rice Soils.—Upland rice is grown under a wide range of soil conditions, on both flat and undulating fields that are not bunded to accumulate water. They are prepared for seeding in dry soil and are dependent on rainfall for crop moisture. The soil characteristics found in upland rice culture are nonspecific in respect to soil texture, pH, organic matter content, slope, and soil fertility variations encompass virtually all possible conditions (De Datta and Feuer 1975).

Moorman and Dudal (1965) suggest that soil texture and topography may be the most important soil characteristics for upland rice since they profoundly affect the moisture status of soil and its tilth for seeding. The textural profile of the surface and the subsoil layers are important to optimize water penetration and storage. Soil structure is important since it figures prominently in tillage, desirable seedbed characteristics, moisture interception and conservation, and crop management practices.

According to De Datta and Feuer (1975), vertisols and alfisols are the major soil groups used in upland rice culture in Southeast Asia and in tropical rainforest areas of Central and South America. Oxisols are of less importance in Southeast Asia except in Java, but are of major importance in Central and South America and parts of Africa. Hydromorphic soils with shallow ground water, seepage areas or surface waters during part of the growing season are often used for rice production. The gleyic cambisols, humic-gley soils and entric fluvisols represent large upland rice areas in West Africa. A surface soil with medium to fine texture overlying a subsoil of finer texture is considered most desirable overall for upland rice. Together with good soil structure which influences soil moisture storage potential and root penetration, these soils provide good conditions for upland rice production. Soil reaction for upland rice is most favorable over the pH range of 5.5–6.5, although most upland rice areas are more acidic. Ponnamperuma (1975) summarizes the growth limiting factors in upland rice soils as being related primarily to the variable soil moisture regimes, and a nutrient status with less soluble iron, phosphorus and silica than in flooded soils. Most plant nutrients occur in their oxidized forms, making nitrate-nitrogen and sulfate-sulfur susceptible to leaching. Potential problems on acid upland soils are manganese and aluminum toxicities and iron deficiency on alkaline soils.

Lowland Rice Soils.—A major portion of the rice in tropical, subtropical and warm temperate parts of the world is grown under conditions where the soil is flooded during the greater part of the growing season. Rice is physiologically, morphologically and anatomically adapted to grow in wet or flooded soil conditions. The flooded culture provides benefits of weed control, improved water and air microclimates and a root zone environment well suited for rice culture.

Soils used for lowland rice culture are usually on level terrain or on terraces where standing water can be maintained. By virtue of their physiographic position, often accentuated by inundation and soil management practices for rice, these soils usually have strong hydromorphic characteristics, poor internal drainage, but are capable of irrigation and drainage in rice culture management.

Lowland soils, particularly in Southeast Asia, are frequently prepared for planting while they are covered with several centimeters of water. The objectives of wet tillage (puddling) are to reduce the power requirements for land preparation, to control weeds and incorporate crop residues, to produce a "plow sole" to reduce percolation losses and to develop a soft soil suitable for transplanting. From a physical viewpoint, puddling destroys soil aggregates, reducing them to a slurry. Puddling decreases the volume of soil macrospores, increases bulk density and reduces internal drainage, thereby increasing the water holding capacity of the soil. Incorporating crop residues before transplanting rice lessens the likelihood of toxicities from anaerobic decomposition products, accelerates the onset of soil reductive processes and enhances the availability of some plant nutrients.

The diffusion of oxygen into water is only 10^{-4} of that in soil, so the exchange of oxygen from the atmosphere to the soil is extremely small. A few millimeters of soil beneath the soil water boundary does contain small amounts of oxygen and it remains in an oxidative condition. Beneath this thin oxygen sink the soil remains largely in a reduced state through the metabolic activity of aerobic and facultative anaerobic microorganisms which either act directly as electron acceptors in anaerobic dissimilation reactions or by forming various organic decomposition products.

The main chemical changes occurring in flooded soils are reviewed by Patrick and Mikkelsen (1971) and Ponnamperuma (1972). These include:

(1) Accumulation of gases, such as CO_2, CH_4, N_2 and H_2 from microbial dissimilation products;

(2) An increase in pH of acid soils and a decrease in pH of calcareous and sodic soils. This change in pH of acid soils occurs when Fe^{++} iron and Mn^{++} manganese begin to precipitate and equilibrate with CO_2 and HCO_3^- in the soil solution. In alkaline soils, the acidifying effect of CO_2 lowers the soil pH which is buffered by Ca and Mg carbonates;

(3) The change of NO_3^-, Mn^{++++}, Fe^{+++} and $SO^=$ to their reduced forms is controlled by an established thermodynamic redox sequence found to occur in flooded soils;

(4) Increased electrical conductivity as correlated with HCO_3^-, Fe^{++} and Mn^{++} ion concentrations in acid soils and with Ca and Mg bicarbonate concentrations in alkaline soils;

(5) Increased supply and production of NH_4^+-N both from the native soil nitrogen and heterotrophic and autotrophic nitrogen fixing organisms. Biological nitrogen fixation represents a direct addition of nitrogen to the soil, while organic matter is mineralized more slowly

and less immobilization occurs, leaving a net increase in available N;

(6) The increased availability of P, Fe, Mn, Si and Mo as a consequence of reduction, dissolution and desorption reactions;

(7) Accumulation of toxic substances and the associated anaerobic decomposition of soluble carbohydrate materials may cause the formation of gases such as CO_2, CH_4, N_2 and H_2, organic acids such as acetic, butyric, propionic and formic acids, and H_2S from reduced sulfates in flooded soils.

In the rice growing areas of the world, the full development of yield potential of rice production is often limited by soil problems of a diverse nature. Problems of salinity, strong acidity and alkalinity, various toxicities, and especially high levels of soluble iron and aluminum are most common. Some of these constraints on yield can be corrected by proper soil treatment and management, but the costs involved are often beyond the means of farmers in developing countries. Varieties that can tolerate adverse soils are sought in current research (IRRI 1974). Results from such research could bring vast areas of unused land into productive use.

Rice is grown over a wide range of soil characteristics often made possible by the ameliorating effects of flooding. In world rice cultivation about 80% of the production occurs on entisols and inceptisols where seasonal drainage problems occur. Vertisols, ultisols and alfisols are commonly used where available water is sufficient and the terrain is flat or gently sloping. Oxisols are sometimes used in rice production, in the tropical and subtropical areas, although the hilly terrain usually requires terracing. Organic soils are not used widely for rice and are quite variable in their production capability.

Biotic Factors Affecting Rice Production

Rice, like other plants, does not live alone in nature but in association with other organisms ranging from the soil microflora to man himself. This section briefly discusses the major biotic influences affecting the growth and yield of rice.

Weed Pests.—Weeds are universal competitors of rice, competing for moisture, light and plant nutrients that limit plant growth and yield. Weeds also create problems in harvesting, drying, cleaning and reduce the quality and marketability of the crop. Insects such as leafhoppers and stem borers affecting rice also live on weeds as alternate hosts and directly attack the crop and may spread virus diseases. Water management is often impeded when weeds block irrigation systems, slowing drainage. Mechanical harvest of rice is more expensive, and cleaning and drying costs are magnified when weeds grow in competition with the crop.

There are some 30,000 different weed species regarded as serious rice pests in the world, of which 30 species are very damaging and some 88 species are noxious. Matsunaka (1975), commenting on the world's worst rice pests, lists *Echinochloa colonum, E. crusgalli, Sphenoclea zeylanica, Ischaemum rugosum* and *Fimbristylis milacea vahl.* In Southeast Asia, the major rice weeds are reportedly *Echinochloa crusgalli, E. colonum, Cyperus rotundus, Fimbristylis miliaceae vahl* and *Monochoria vaginalis.* Species making up the total weed complex vary widely in economic importance in different rice growing regions and often differ between adjoining fields. Both the species, their density and the duration of their competition, affect rice yields.

Weeds are a major barrier to satisfactory rice production wherever they occur, but are especially troublesome in upland rice where their growth is less restricted and where conservation of available moisture is critical. Weeds in upland rice consist of both annual broadleafs, sedges and grasses and a wide variety of perennials including some shrub species. Since rice is grown under such a wide range of climatic conditions, soil types and crop rotations, it is not possible to identify the most damaging species. *Echinochloa crusalli, E. colonum,* and *Imperata cylindrica* are serious weed pests in most upland rice areas of the world.

The most common means of control is hand weeding, usually with a short handled hoe, requiring about 300 man-hr per ha. Cultivation with animal or machine power is also practiced, as well as weed burial and intercropping. Without satisfactory weed control, upland rice yields are severely restricted. Under moderate competition, weeds alone may reduce grain yields by 40 to 50%, and severe competition may cause complete crop failures (De Datta 1976; Smith *et al.* 1977).

Weed problems in upland conditions develop to serious proportions where intensive labor inputs are not available. In West Africa, Moody (1975) reports that present weed control methods (both cultivation and herbicides) are unsuitable for continuous or large-scale farming. Likewise, if cultivation areas are shifted once weeds become too prevalent, it is necessary to abandon the land to forest, a means used by early man to cope with upland weed problems.

Conditions in lowland rice culture favor the growth and reproduction of aquatic and semi-aquatic annual and perennial weeds. The most prevalent weeds are grasses, broadleaf weeds and sedges, but submerged aquatic weeds and algae likewise often have an adverse effect on rice growth.

Among the most damaging lowland and aquatic weeds in rice are: *Echinochloa* spp., *Eichornia crassipes, Fimbristylis* spp., *Cyperus* spp.,

Monochoria vaginalis, Sagittaria spp., *Salvinia* spp., *Polygonum* spp., *Marsilea quadrifolia, Althernanthera* spp., *Eclipta alba, Jussiaea* spp., *Sphenoclea zeylanica* and *Ischaemum rugosum.*

Control of weeds in flooded lowland is facilitated by thorough land preparation before planting. When properly done, most weed seeds fail to germinate and weeding becomes easier and less expensive. Straight row transplanting enables farmers to use push type rotary weeders along the rows, providing a distinct advantage over random transplanting. Hand weeding, involving pulling or trampling the weeds into the mud, is a satisfactory alternative where rotary weeders are not available. IRRI data show that under average field conditions, hand weeding requires 120 man-hr per ha, while similar rotary weeding takes 70 hr. Effective rotary weeding requires adequate flood water and a soft soil condition. Hand weeders cannot incorporate weeds if flood water is maintained too deep in the fields, since the plants float and remain alive.

Water management has long been known to be an effective means of weed control in both transplanted and direct seeded rice. The germination of weeds and the kinds of weeds that emerge are closely related to soil moisture content and depth of flooding. Research at IRRI has shown that continuous soil saturation (1–2.5 cm deep) up to the late dough stage of rice allows more sedges to grow than grasses or broadleaves. Flooding at depths of 15 cm from 4 days after transplanting to the late dough stage of rice suppresses grass and sedge germination. Broadleaf weeds are not controlled by flooding.

Many herbicides are effective against broadleaf weeds, sedges, grasses and submerged aquatic weeds. The concept of preemergence weed control with herbicides such as application of 2,4-D, 2-methyl-4-chlorophenoxyacetic acid (MCPA), and selected herbicides such as butachlor and thiobencarb, are rapidly changing weed control practices in tropical Asia.

Insect Pests.—Rice is grown under diverse conditions of climate and culture over a wide geographical range. Because of crop adaptability to warm humid conditions, the survival and proliferation of insects is very great in the tropics and semitropics, diminishing somewhat in the temperate rice areas. More than 70 species of rice pests are known and some 20 have major significance. They attack virtually all parts of the rice plant at all growth stages. Insects serve as vectors of virus diseases that attack rice, all contributing to the low rice yields obtained in the humid tropics. The continuous cropping and favorable environment of the humid tropics allow overlapping insect populations throughout the year, with no distinct diapause or dormancy period.

Among the insect pests of rice, the rice stem borers (*Chilo suppressalis* and *Tryporyza incertulas*) are generally considered the most serious in tropical rice production. Other important rice insects are green leafhoppers (*Nephotettix virescens*) and brown planthoppers (*Nilaparvata lugens*). The rice water weevil (*Lissorhoptrus oryzophilus*) is a common problem in the temperate zone in addition to the rice stink bug (*Oebalus bugnax*), planthopper (*Sogata orizicola*) and others. Other destructive groups include grasshoppers, locusts, gall midge, rice hispa, army worms, cutworms, whorl maggots and thrips.

Current control of rice insect pests depends largely on the use of pesticides, even though many traditional and new cultivars have some degree of resistance to insect pests. Chemicals will no doubt continue to play a major role in rice crop protection. With tropical rice where pesticides are costly and sometimes not available, it is increasingly important to combine various compatible management practices that provide crop protection at minimum costs. Entomologists recognize that the easiest integration of pest control methods is to combine cultivar resistance and the timely use of pesticides.

Diseases.—Diseases of rice are caused by a variety of organisms including fungi, bacteria, viruses and probably mycoplasma-like bodies. See Chapter 5 for a full discussion.

Other Rice Pests.—Among several pests often localized in their ecomomic damage are rodents, birds, snails, crabs and sometimes a variety of large animals.

Rodents, particularly rats, cause serious losses in rice and are considered as economically important as insects (Mochizuki 1975). Rodents damage rice by eating seeds and seedlings, gnawing off tillers, damaging plants and feeding on ripened grain. They attack rice in the field in all growth stages. They also cause large losses of grain in storage. Rats burrow in bunds and ditches causing loss of water, decreasing irrigation efficiency and increasing the costs of maintaining irrigation systems. Losses due to rats vary widely, but field losses vary from 3.5 to over 25% with complete crop failures occurring in some localized areas.

Birds cause greater rice crop losses than do most other vertebrate pests. They may attack ripening grain in all stages from the early "milk stage" through grain maturation. Most bird pests perch on the rice panicle and feed on the ripened grain. Large quantities of rice are often shattered, falling to the ground, and are lost. Some bird species feed on rice that falls to the ground after harvest, or rice that is direct seeded on dry prepared seedbeds, planted on mud or sown directly into water. Children, called "bird boys," are often employed to prevent the birds from attacking rice. A variety of devices to make a noise or otherwise scare the birds are likewise used to protect the crop.

The principal African bird pest is *Quelea quelea*, an abundant and widely spread sparrow-sized weaver bird. Large flocks of this nomadic bird migrate across Africa, descending on rice, millets and sorghum indiscriminately, often completely destroying entire crops. Control is very difficult because of the large populations involved, their migratory characteristics and their extremely broad feeding range. Frightening the birds by various devices provides some short-term relief, but the effort is very time consuming.

A wide variety of bird species eat seeds. This poses a problem in most areas of the world. Some are indigenous birds, omnipresent but often migratory, following the ripening of rice and various other grain crops. In tropical Asia the weaver birds of the genus *Ploeus,* parakeets, Munia and sparrows destroy rice crops. Blackbirds, cowbirds and the common grackle, permanently colonizing forest and swamp areas, cause great damage in Latin America.

Ducks, geese and various waterfowl damage rice in many countries, including Australia and the United States and countries of South America and West Africa.

Varieties of molluscs and crustacea, occurring in both fresh and brackish water, attack rice in many parts of the world. Snails such as *Lanistes ovum* and *Amulearia lineata* and *A. glauca* (usually growing in standing water) harm the rice crop by feeding on newly emerged seeds and seedlings. Snails are often difficult to control because of their soil-burrowing characteristics. Two types of crabs, land crabs (Gecarindal) and river crabs (Potamonidal), damage rice through their burrowing characteristics in bunds and irrigation systems. Crabs attack rice planted by transplanting as well as by direct seeding. Shrimp (*Triops granarius* and *T. longicaudatus*), often referred to as tadpole shrimp, cause damage especially in the semitropic and temperate zones of the Americas, West Indies, Hawaii and Japan. They do not attack drilled seed or transplanted seedlings, but cause damage to pregerminated seed sown directly in flooded fields. Chemicals applied directly to the flood water have been successful in combating this.

In some areas of rice production, disorders occur in rice that resemble the symptoms of fungus and virus diseases. Some of these disorders are called physiological diseases and are sometimes associated with physiological and nutritional abnormalities. These plant disorders have been called by names such as akagare, akiochi, aogare, bronzing and straight head.

SYSTEMS OF RICE CULTURE AND PRODUCTION METHODOLOGY

Despite the importance of rice as a world crop there are no complete quantitative data classifying rice lands by water regimes and predominant plant type. Four riceland classifications (upland regions, shallow water regions, regions of intermediate water depth and deep water including floating rice regions) are used for irrigation, irrigated double cropped, rainfed (bunded), upland (unbunded), deep water and floating rice culture. The world's rice land, classified by these water regimes and predominant rice types, is described schematically in Fig. 4.3. Further refinements of crop area and production statistics for South and Southeast Asia have been developed (Fig. 4.4) which divide the rainfed areas into 2 categories—shallow rainfed (less than 30 cm water depth) and medium-deep rainfed (30 cm—1 m water depth). Although there is inadequate documentation on which to base such a division, the best estimates suggest that about half of the world's rice is grown under rainfed lowland conditions in paddies but depending entirely on the monsoon rains for water (Fig. 4.4) (Barker *et al.* 1975). Although these definitions differ from many of those used throughout Asia, it seems important to divide categories on the basis of water control, the argument being that basic rice plant types suited to each area will differ. Furthermore, the

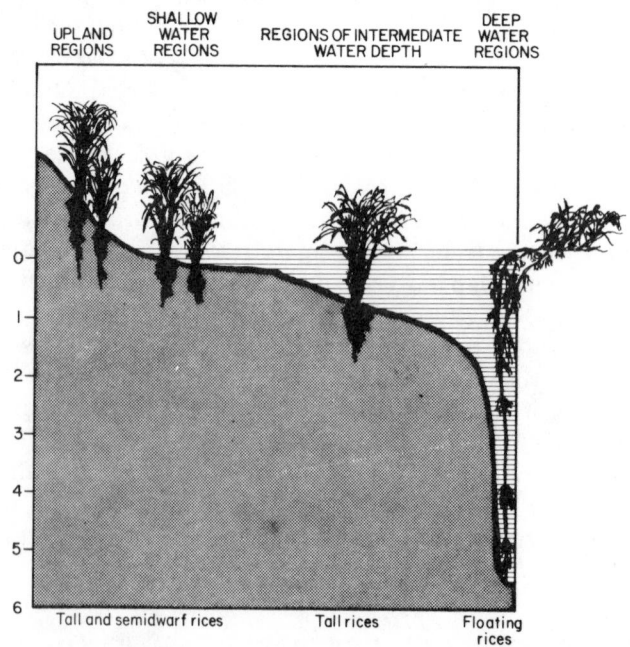

From IRRI (1974)

FIG. 4.3. THE WORLD'S RICE LAND CLASSIFIED BY WATER REGIMES AND PREDOMINANT RICE TYPES

irrigated rainfed environment can not be sharply divided but represents a continuum of conditions ranging from very good to poor water control (Barker *et al.* 1975). In the areas of poor water control, both drought and uncontrolled flooding can be a problem.

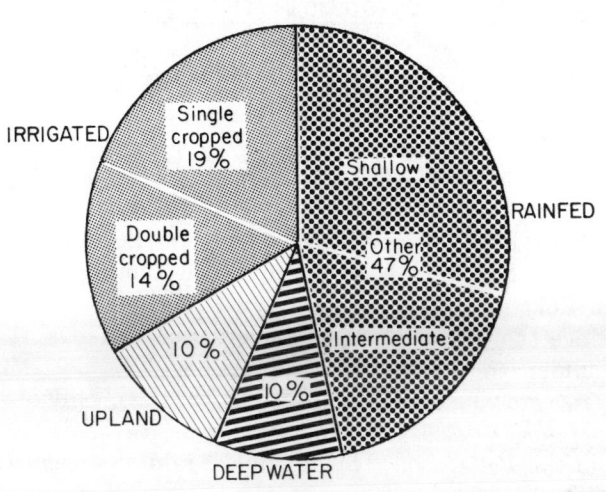

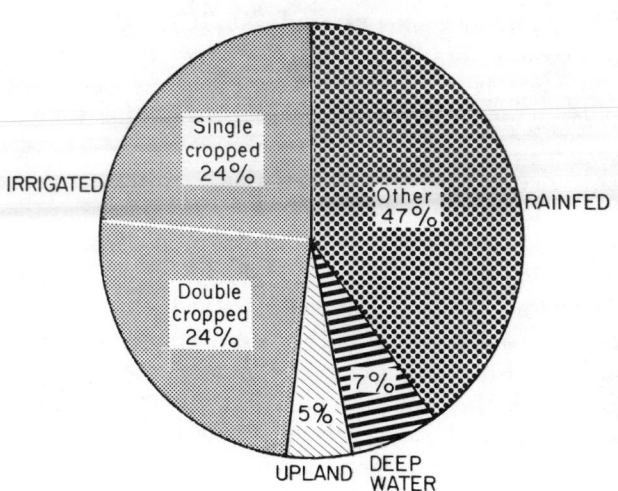

From Barker et al. (1975)

FIG. 4.4. ESTIMATE OF PERCENTAGE RICE CROP AREA AND PRODUCTION BY SPECIFIED LAND TYPE IN SOUTHEAST ASIA IN THE EARLY 1970s

In Africa and Latin America, upland rice is the major system of rice culture. On the basis of the type of rice culture, climate, soil conditions and technology level of agriculture, rice in Africa is divided into 14 major ecosystems (Table 4.1). All of West Africa's 1.77 million ha, however, comprise only 1.4% of the world's rice area.

TABLE 4.1. MAJOR RICE ECOSYSTEMS IN AFRICA

Rice culture type	Forest	Midaltitude		Sudan savannah	Guinea savannah			Mangrove
		Humid	Dry		Pluvial	Hydro.	Pluvial	
Upland	+	+			+	+		
Irrigated	+	+	+	+			+	
Shallow flooded	+			+			+	+
Deep flooded				+				

Source: Buddenhagen and ter Vrught (1977).

About 5% of the world's rice is grown on 6.5 million ha of land in Latin America. Brazil has 5.2 million ha, of which about 3.5 million ha are upland. In Colombia, South America, where rice yields have steadily gone up since the introduction of modern rice varieties, rice is grown primarily under irrigation.

In many rice growing areas of the tropics, the year is divided into fairly distinct wet and dry seasons. In most areas, the bulk of the rice is produced in the wet season. The amount of rainfall received during the dry season is usually not sufficient to grow a crop of rice, so irrigation is necessary. Because of the lack of irrigation facilities in most rice growing areas of the tropics, the hectarage planted to rice in the dry season is limited. Although many countries are trying to increase their irrigation facilities for rice (and for other crops), rice grown in the tropics will still have to depend largely on monsoon rains. The problem of variable rainfall is further compounded by the inflexible planting time in tropical Asia.

The principal rice growing seasons of various countries are shown in Fig. 4.5. Not shown in the figure is the area planted to rice in the Peoples' Republic of China, estimated at about 33.6 million ha, accounting for about 25% of China's total land area planted to cereal crops and tubers. Rice growing areas of China can be classified into 4 major regions: northern, central, southern and southwestern. Some examples of the rice growing season and rice based cropping systems are shown in Fig. 4.6.

The production methodology or cultural practices which have evolved or developed for rice are both complex and unusual. Because rice is grown under both irrigated and rainfed conditions, one set of cultural practices cannot be used effectively for all conditions. We will therefore discuss systems of rice culture separately and mention how cultural practices vary with different conditions.

FIG. 4.5. RICE SEASONS AND RICE BASED CROPPING SYSTEMS IN VARIOUS
COUNTRIES IN ASIA

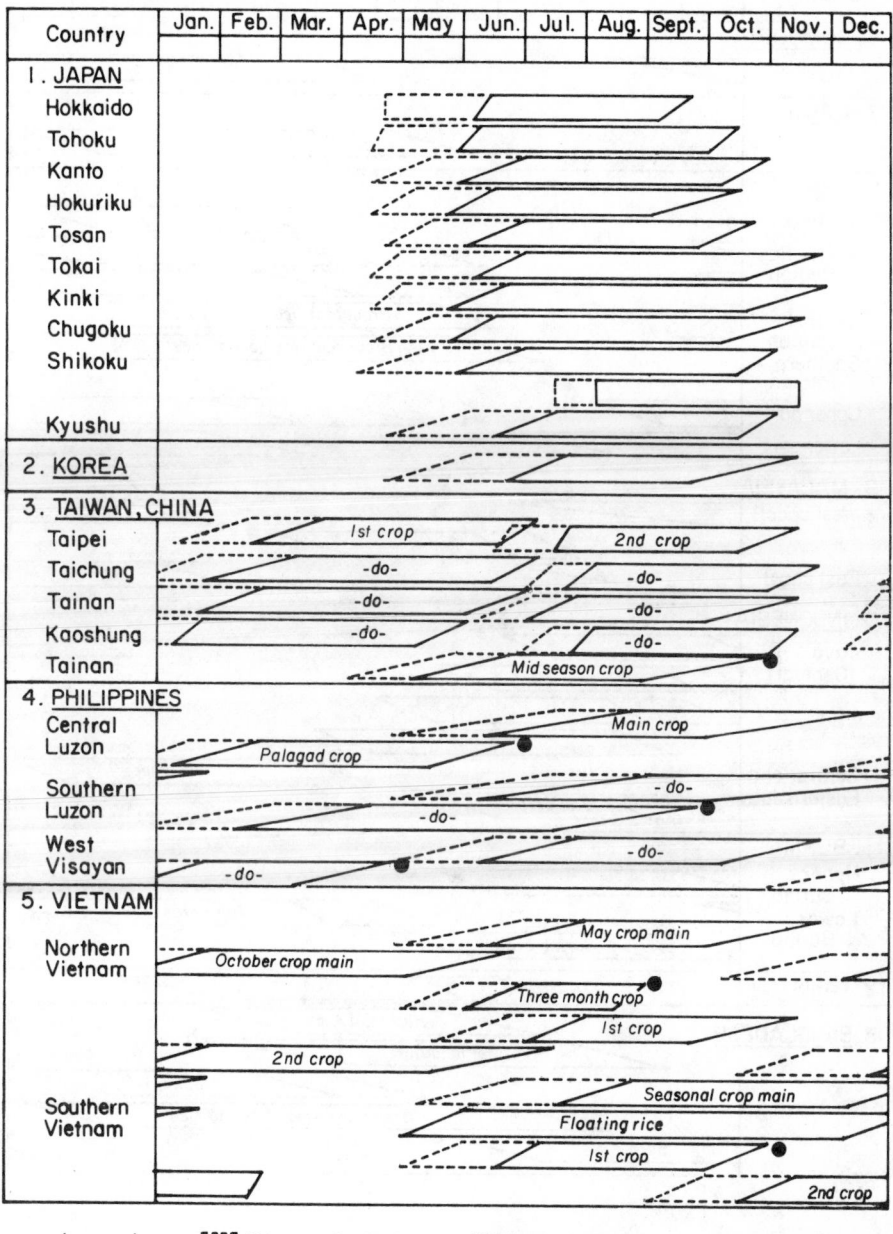

FIG. 4.5. (*CONTINUED*)

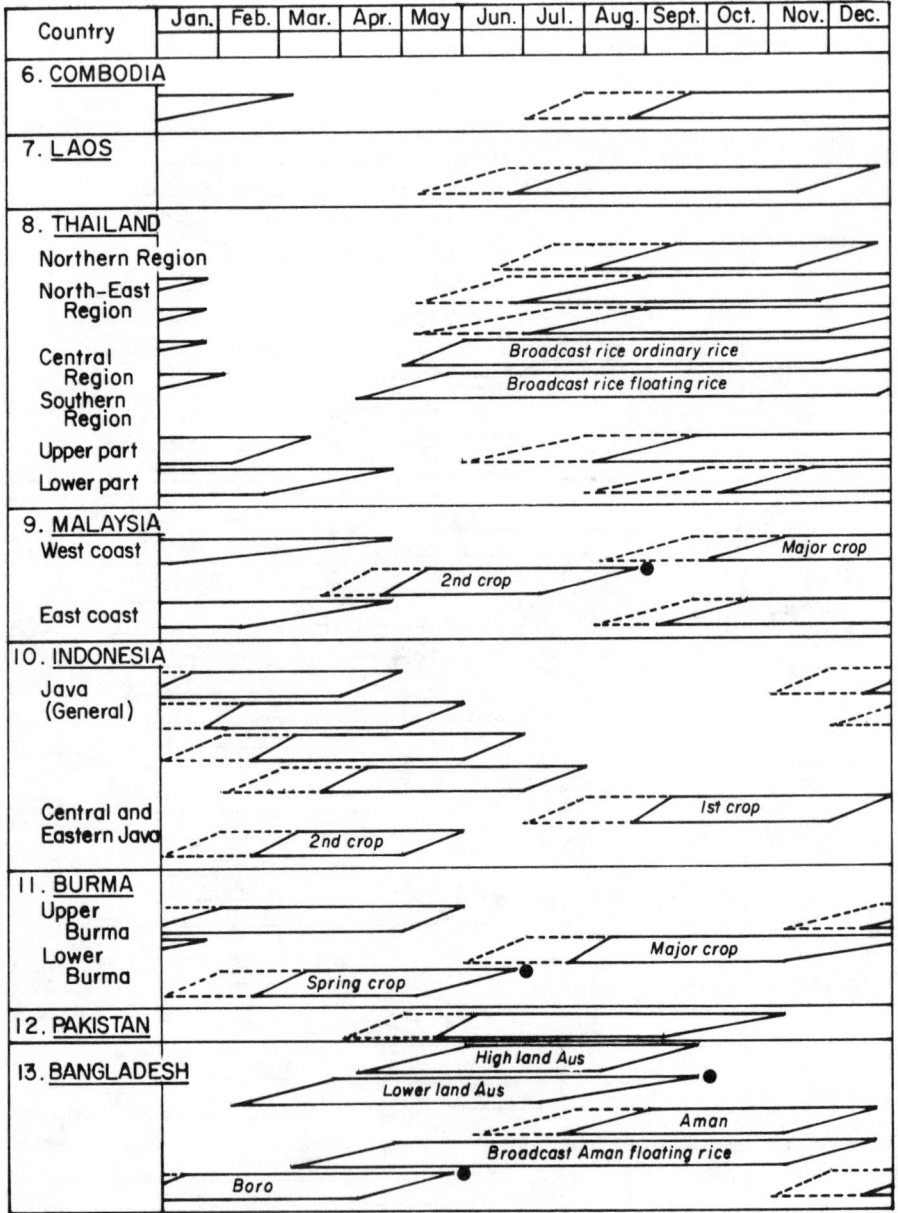

FIG. 4.5. (*CONTINUED*)

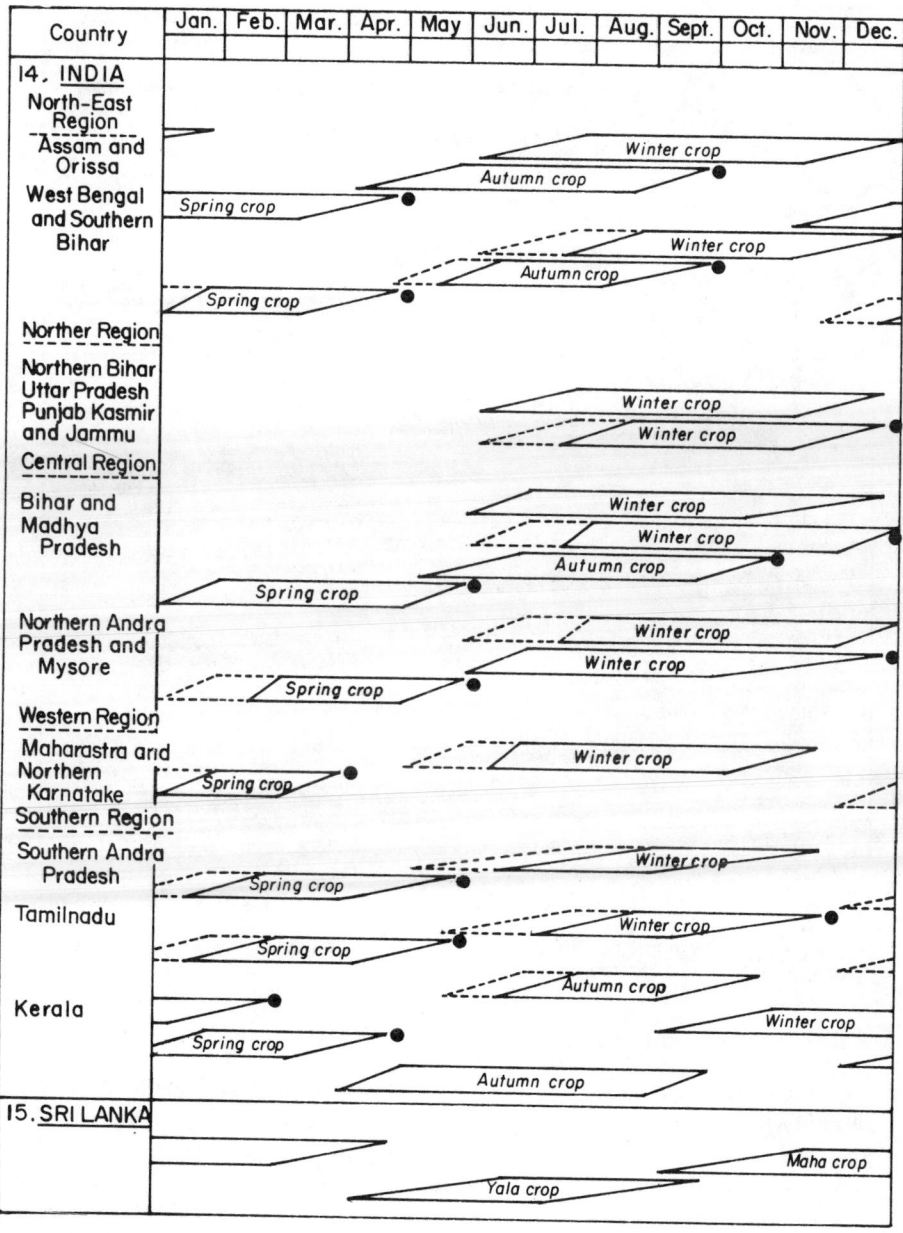

From Tanaka (1976)

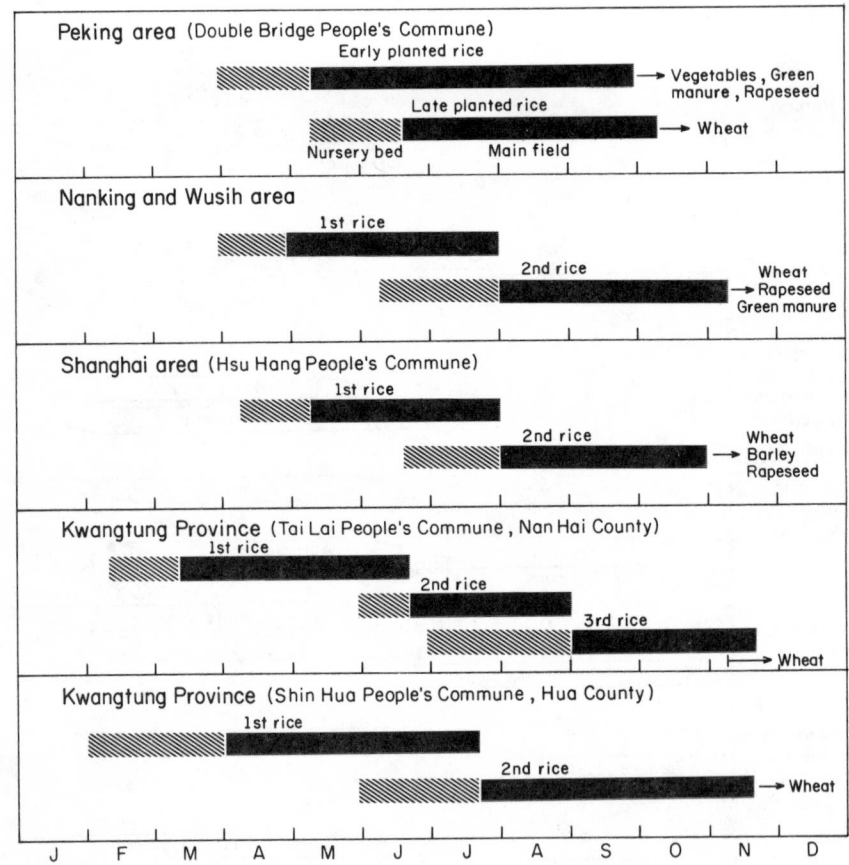

Adapted from report on IRRI Team's visit to PRC (1978)

FIG. 4.6. RICE BASED CROPPING SYSTEMS IN RICE GROWING AREAS OF THE PEOPLES' REPUBLIC OF CHINA (PRC)

Rainfed Lowland Rice

On most unirrigated farms, rice is grown during the wet season and the land lies fallow or is planted with low input crops through the dry season. Because of the farmers' dependence on both the fickle monsoon rains and conventional late maturing rice varieties, this has been the only feasible cropping system for centuries. The potential to intensify food production exists in most of these rainfed areas, although the technology and often the markets may be lacking. The practices followed are much more complex and variable in rainfed lowland areas than in irrigated areas.

Transplanted Rice.—The onset of the monsoon determines the planting time and thus the day length and solar radiation available during the growing period. In lowland rice culture, transplanting is the major system of rice culture.

Land Preparation: Effects of Puddling and Flooding.—The traditional method of preparing land for transplanted paddy rice consists of plowing and puddling the soil. The method has been widely adopted because it greatly reduces the weed population. In addition, puddling substantially increases the amount of water retained by the soil (Fukuda and Tsutsui 1968) and reduces the amount of water lost through percolation (Sanchez 1973B). In an experiment conducted (Maahas clay), water lost through percolation was considerably higher in unpuddled soil than in puddled soil (Fig. 4.7), so unpuddled soil received twice as much water (1180 mm) as puddled soil (588 mm). Rice in puddled soil had 2.5 times the efficiency of water use (7.9 kg rice/ha/mm of water) of unpuddled soil (2.9 kg rice/ha/mm of water) (De Datta and Kerim 1974).

Thus, in rainfed areas, puddling lowers the chance that yield-reducing moisture stress will occur between 2 successive rains. The rainfed puddled system should therefore result in better moisture conservation and

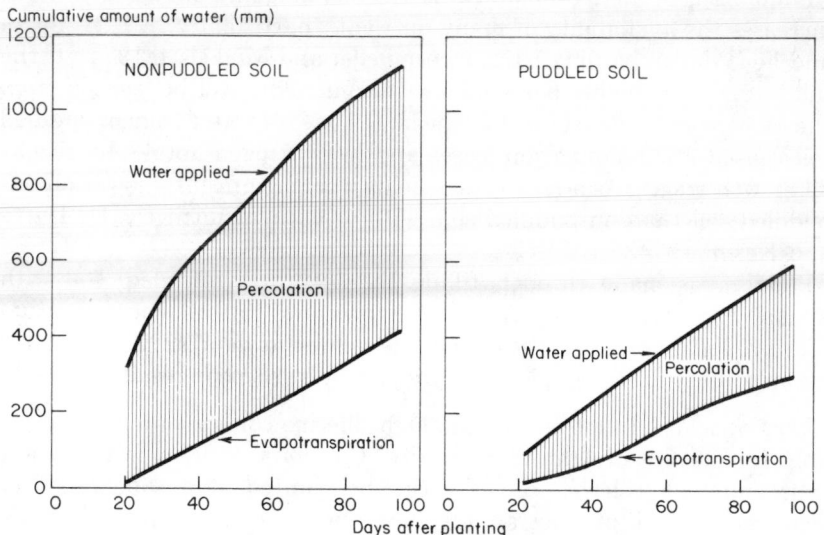

From De Datta and Kerim (1974)

FIG. 4.7. COMPARISON OF CUMULATIVE WATER APPLIED, EVAPOTRANSPIRATION AND PERCOLATION LOSS IN PUDDLED AND NON-PUDDLED SOILS CONTINUALLY FLOODED AT 5 CM (IRRI, 1972 WET SEASON)

possibly higher yield than the rainfed unpuddled system. Puddling, however, involved more labor.

The term puddling has several different meanings. Farmers in Asia consider it a method of soil preparation which facilitates transplanting and reduces moisture losses through percolation. Soil physicists describe puddling as the process of breaking down soil aggregates into a uniform mass when clay soils are plowed and harrowed at about soil saturation. Consequently, several changes take place in the soil structure (Koenigs 1961). Puddling changes the solid and liquid phases of the soil, decreasing the respective volumes occupied by macropores by 91 and 100%. Because most gaseous elements of the soil are in macropores, tillage at the moisture equivalent causes a corresponding reduction in the amounts present. According to Jamison (1953), puddling decreases the macroporosity of clay soils and increases their microporosity, thereby causing an increase in the so-called water holding capacity of the puddled soil.

For most crops, good soil structure is beneficial because it promotes infiltration of water and aeration of the soil. Although only a small amount of air is present in the micropores of well puddled soil (Buehrer and Rose 1943), no aeration problems occur with rice. A rice plant can tolerate low soil oxygen levels because its roots receive oxygen by transfer from the aerial parts.

Puddling is often accompanied by the submergence of rice soils. That increases the availability of many nutrients, particularly N, P, K, Ca, Si and Fe (Ponnamperuma 1972; Obermueller and Mikkelsen 1974). If the soil structure is highly porous, however, nutrients will be leached from the root zone. Studies show that the large losses of water from unpuddled soil caused greater nitrogen losses and less nitrogen uptake by rice at all growth stages. As a result, rice yields were significantly lower on unpuddled soil than on puddled soil, with or without nitrogen (De Datta and Kerim 1974).

Puddling is made through tilling the saturated (flooded) soil with animal- or power-driven plows and harrows until the soil macrostructure is destroyed. In practice, the land in the rainfed lowland areas is prepared by an animal drawn implement, frequently with poor results.

Plant Spacing.—In monsoon Asia, high tillering cultivars are very desirable for culture of transplanted rice. Cultivars with improved plant type and high tillering capacity can be planted at a wide range of spacings and still produce an adequate number of tillers per unit area. The tiller number per unit area in a rice population is largely a function of planting density. The tiller number is correlated with grain yield either positively or negatively depending on the rice cultivar and crop environment (Kawano and Tanaka 1968). A uniform stand containing an optimum plant population is essential for proper crop development and high grain yields. Management is found to be much easier if the field is planted in straight rows, since cultural practice such as weeding by hand or chemical methods, spraying insecticides, and topdressing with nitrogen fertilizers can then be done without damaging the rice plants significantly.

A great deal of work has been done in Japan on the effects of plant density and spacing on rice yield, and such work has lately assumed importance in the tropics. The conclusions from these findings vary. Some suggest that closer spacings produce higher yields, whereas others report opposite effects.

In Taiwan, Hsieh *et al.* (1966) reported increased yields with close planting. They attributed the increase to greater panicle numbers and total panicle weight. In Pakistan, Shafi and Ahmad (1967) tested IR8 and Jhona 349 at 4 spacings with 1 and 2 seedlings per hill in each case. They recorded highest grain yields from IR8 at the greatest plant population (spacing 22.5×10 cm) and the highest grain yields from Jhona at the lowest density (spacing 45×45 cm). Their results also showed that the yield of IR8 was not significantly greater with a 22.5×10 cm spacing than with 22.5×22.5 cm. They concluded that 22.5×22.5 cm is best for a short-saturated variety such as IR8 under heavy fertilization. Srinivasan *et al.* (1968), in India, obtained the highest yield with IR8 when it was spaced at 10×20 cm and received 200 kg N/ha.

Tanaka *et al.* (1966) reported that the optimum spacing for the lodging susceptible tall variety Peta was 50×50 cm, while optimum spacing for the nonlodging short variety Tainan 3 was 30×30 cm. Parao (1970) showed that optimum spacings were 10×10 cm for low tillering improved varieties, 20×20 cm for high tillering varieties, and 50×50 cm to 60×60 cm for the relatively high tillering tall Peta.

It is true that with varieties such as IR8, plant spacings between 15×15 cm and 30×30 cm would not give any significant yield difference. Recent evidence indicates that a relatively short-statured variety with moderate tillering capacity must have close spacing for maximum yield. For varieties such as IR28 and IR30, plant spacings of 15×15 cm give more significant yields than 20×20 and 25×25 cm. Further, under poor or no weed control, closely spaced rice competes better with weeds.

Water Management and Moisture-stress Effects.—Rice, like other crops, requires adequate water to grow and develop at its maximum potential rate. Unlike other crops, rice is usually grown in flooded soil. With an adequate water supply from rain, continuous flooding from 5 to 7 cm of water is desirable on most soils for the best moisture supply. It also gives the best weed and insect control with granular chemicals and high nutrient availability with minimum loss of nutrients from fertilizer and soil. Thus, wherever possible, flooding from 5 to 7 cm should be maintained even in rainfed rice. It is imperative that levees be kept in good repair to minimize surface runoff of excess water when it rains.

In rainfed areas, the paddies often become dry and the crop suffers from various degrees of moisture stress. Although rice can be grown under upland, lowland and deep water conditions, stable high yields occur only under a continuous flooded condition. Moisture stress is perhaps the chief factor that limits economical and stable yields of rainfed rice (De Datta et al. 1973B).

Senewiratne and Mikkelsen (1961) reported that grain yield was 53% lower under unirrigated unflooded conditions than a flooded condition. Moolani and Sood (1967) suggested that for rice, unlike other crops, the upper limit of the range of available moisture is not field capacity. Jana and De Datta (1971) reported that the optimum soil moisture condition for high grain yield in upland rice (should be true for rainfed lowland rice as well) is between the maximum water holding capacity and the field capacity. De Datta et al. (1973B) indicate that soil moisture tension as low as 15 centibars (cb) was enough to reduce grain yields of rainfed lowland rice. Part of the reduction in grain yield is due to a loss of nitrogen under the alternating dry and wet conditions which prevailed in the plots subjected to various stress levels. At the same time, there is enough evidence to demonstrate marked varietal differences between the modern rices in their tolerance to high level moisture stress (±30 bars).

Farmers in rainfed lowland areas have been growing varieties that fit in with the probable rainfall distribution. Early maturing rice varieties are often preferred to escape drought stress. Early maturing varieties such as IR36 should provide good stable yields under rainfed lowland culture.

Fertilizer Management.—The development and dissemination of fertilizer responsive cultivars of rice, wheat and other cereals have encouraged a steady increase in the use of fertilizer in the developing world. Fertilizer responsiveness is a key factor in differentiating among the traditional rices and the new high yielding cultivars. Only where the levels of fertility are at least modest do yield differences between the new and the old become significant; without fertilizer, the new cultivars yield little better than the old. With fertilizer, in contrast, the yield potential is often double or even triple that of the traditional ones (De Datta et al. 1974; Shastry et al. 1974).

The efficiency of applied nutrients in rice has long been studied by soil fertility scientists. Only recently, however, has concern been expressed over possible shortage and higher prices of fertilizer, particularly nitrogen and phosphorus. Serious concern about efficient use of these nutrients in rice production has stemmed from the realization that fossil fuel supplies needed for the manufacture of nitrogen fertilizer are not limitless and that phosphate deposits of commercial significance are also limited.

Fertilizer use efficiency can be described as the output of economic produce by any crop per unit of fertilizer nutrient applied under a specific set of soil and climatic conditions. The least amount of fertilizer needed to match plant use at maximum grain yield is the most efficient rate of fertilizer application. Bartholomew (1972) classified N needs according to plant use—good use efficiency, average use efficiency, poor use efficiency and polluting rates.

Despite agronomic efforts to improve plant nitrogen use efficiency, the recovery of fertilizer nitrogen applied to a rice crop is seldom more than 30 to 40% (De Datta and Magnaye 1969). Data indicate that among the processes which cause the loss of nitrogen in lowland soil, particularly rainfed lowland soil in the Asian tropics, the most important are nitrification and subsequent denitrification. One of the losses that has not been fully accounted for in most nitrogen balance studies is the ammonia that volatilizes when the NH_4 form of nitrogen is broadcast into water, especially at planting. In coarse textured soil, nitrogen losses by leaching may be significant.

These losses contribute heavily toward the poor and often variable results obtained from topdressed N application, particularly in rainfed lowland rice (Manguiat and Yoshida 1973; Yoshida and Padre 1974; De Datta et al. 1974). Understanding the mechanisms and pathways of nutrient losses has helped researchers develop practices to increase the efficiency of fertilizer use in rice.

Several factors determine fertilizer efficiency in rice. A few which influence the level of fertilizer efficiency achieved at the farm level are soil, cultivar, season, time of planting, water management, weed control, insect and disease control, cropping sequence, sources, and rates and time of fertilizer applications.

For India, Mahapatra et al. (1974) summarized the fertilizer efficiency in cereals to be as follows: dry season rice > wet season rice > wheat > corn > millet > sorghum.

A great deal of attention has been devoted to evaluation of the relative merits of various fertilizers for lowland rice. The yield response of rice to different sources of nitrogen is affected by soil condition and management practices, particularly water management and time and method of nitrogen application (Mikkelsen and Finfrock 1957; De Datta and Magnaye 1969).

In most situations, ammonium containing or ammonium producing (urea) fertilizers are similar in effectiveness as evaluated from grain yield (Engelstad 1967; De Datta and Magnaye 1969; De Datta et al. 1974).

Rice grown with broadcast planting and transplanting under both continuous flooding and rainfed conditions shows that grain yield differences are generally not significant among ammonium sulfate, urea, sulfur-coated urea (SCU) and other nitrogen sources. Nor are the differences between single and split applications significant.

In fertilizer tests conducted under continuous flooding or under rainfed conditions at 4 locations in the Philippines, all nitrogen sources tested— urea, ammonium sulfate, sulfur-coated urea (28% dissolution rate in 14 days)—gave comparable results for both irrigated and rainfed rice (Fig. 4.8).

Most of the rice in Peru is produced in dry coastal areas in the northern part of the country. The soils are high in pH and often saline. The typical water management system calls for intermittent flooding similar to rainfed paddies in tropical Asia. The flooding-drying sequences are rather short, permitting a number of cycles during the growing period of the rice. The paddies are not puddled and water infiltration rates are high. A number of experiments have been conducted in this area to evaluate SCU and urea as sources of N as a basal dressing and as a top dressing with rates of applied N ranging as high as 400 kg N/ha. The experiments show the greatest benefits from coating urea (Fig. 4.9). Results indicate that losses of urea-N from these soils are serious. Following nitrification, nitrate nitrogen is probably lost by leaching and/or denitrification. These and other results indicate that an increase in grain yield from ordinary urea and sulfur-coated urea would depend on water conditions in the field. The development of slow-release fertilizers such as SCU was to minimize nitrogen losses in rainfed areas where soils often get very dry.

Deep placed fertilizer is apparently less subject to volatilization and microbial oxidation. The loss of nitrogen is commonly more serious in rainfed lowland conditions than in irrigated fields. Experiments in several countries have shown that deep placement of fertilizer nitrogen gave higher yields than other methods of application. Results obtained in the Philippines are given in Fig. 4.10. The grain yield differences between deep placement and other methods were less at the IRRI farm, because rainfall distribution was more uniform there than in farmers' fields (De Datta *et al.* 1974).

Although research results indicate that nitrogen efficiency is higher when the fertilizer is deep placed, most farmers usually wait until the crop is established before applying any nitrogen fertilizer.

In rainfed areas, the best time to topdress nitrogen is at tillering (20–30 days after transplanting) and at panicle initiation.

Sources, method and time of application of phosphorus and potassium would not be any different for irrigated and rainfed areas. Still rainfed rice generally needs more phosphorus than irrigated rice does. This is because more phosphorus is brought into solution under good water conditions, such as continuous flooding, than under rainfed rice culture where moisture stress is a factor affecting nutrient availability.

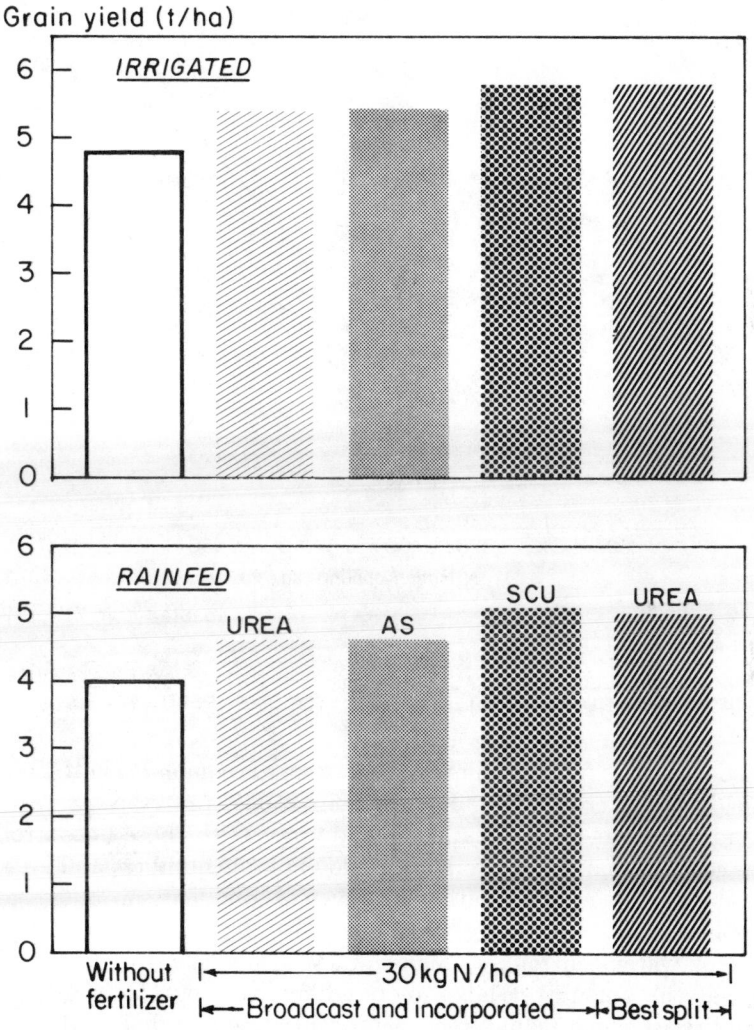

From De Datta et al. (1974)

FIG. 4.8. AVERAGE GRAIN YIELDS OF IR26 GROWN UNDER IRRIGATED AND RAINFED CONDITIONS

Grown without fertilizer nitrogen and with 30 kg nitrogen/ha as urea, ammonia sulfate (AS) and sulfur-coated urea (SCU) applied as basal treatment and urea applied in split doses. IRRI and farmers' fields in Malayantoc, Maligaya, Talavera and Nueva Ecija, Philippines.

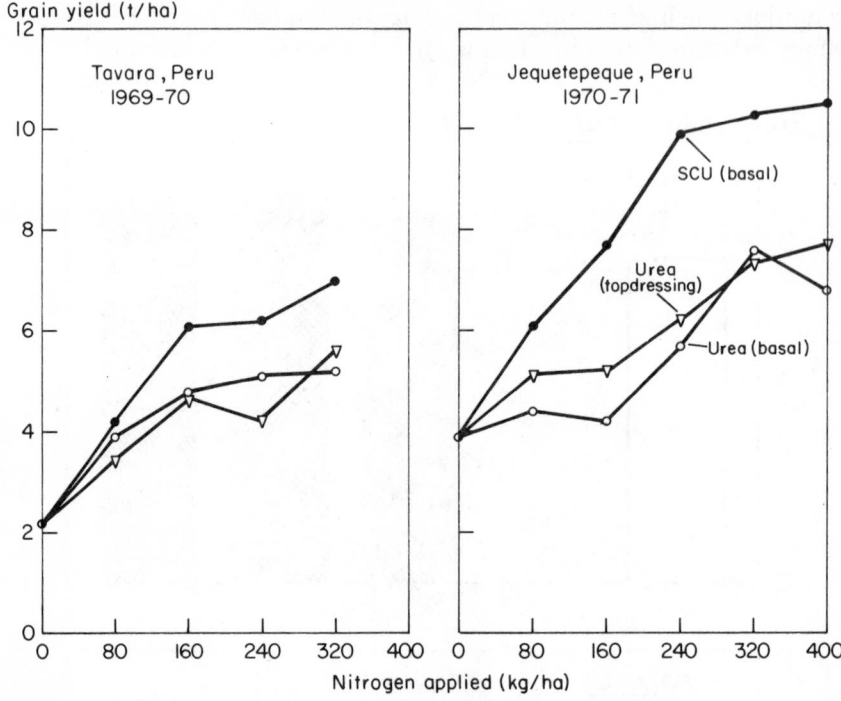

Adapted from TVA (1972)

FIG. 4.9. GRAIN YIELD OF RICE AS AFFECTED BY NITROGEN SOURCE, RATE AND TIME OF APPLICATION WITH INTERMITTENT FLOODING (PERU 1970–1971)

Weed Control.—In many rainfed areas, water accumulates in the bunded field as the crop grows. In other situations, the crop starts as a lowland crop and finishes as an upland crop. With those uncontrolled water conditions, weeds generally develop in large numbers and greater diversity of species than with rice grown in puddled soil under good irrigation.

Lack of water control is a major practical management factor that increases the amount of labor required for weeding (Moomaw *et al.* 1966). Precise water management with continuous flooding is ideal for a number of reasons, particularly to minimize weed growth. Since most fields in South and Southeast Asia could not be flooded continuously, indirect and direct complementary practices are essential for effective weed control at the farm level (De Datta 1976).

Like irrigated rice, many weed species in rainfed lowland rice are annuals, such as *Echinochloa crusgalli, E. crus-pavonis, Leptochloa chinensis, Fimbristylis littoralis, Cyperus difformis* and *C. iria.* There are also the perennials *Paspalum distichum* and *Scirpus maritimus.* The broadleaved weed *Monochoria vaginalis* also grows when water is standing in the field.

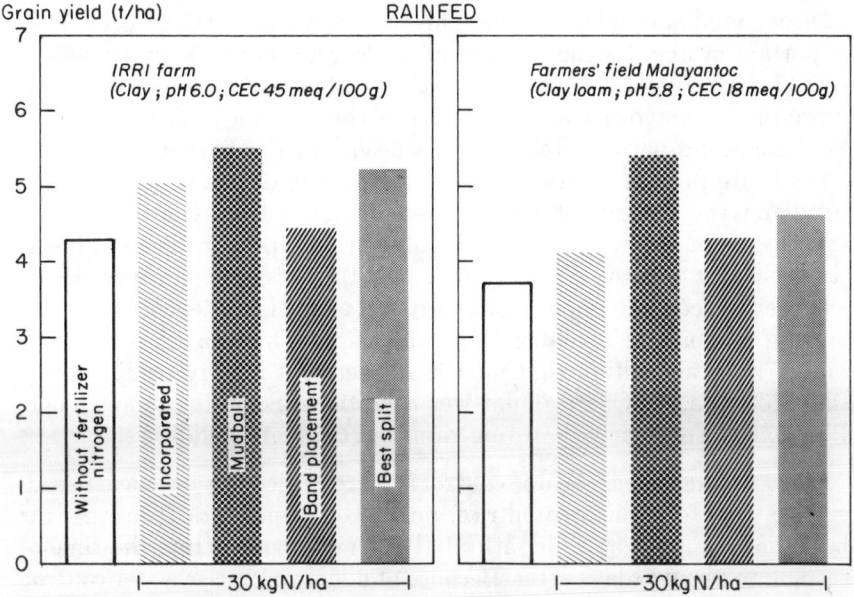

From De Datta et al. (1974)

FIG. 4.10. RELATIVE GRAIN YIELDS OF IR26 GROWN UNDER RAINFED CONDITIONS

Grown without fertilizer nitrogen and with 30 kg nitrogen /ha as urea under different methods of application. IRRI and farmer's fields in Malayantoc and Nueva Ecija, Philippines, 1974 wet season.

Under the existing conditions of high weed density, relatively low cost of labor and availability of some chemicals, the necessity of weed removal is more important than how it is achieved. In the Asian tropics, either hand weeding or use of a rotary weeder is effective in rainfed transplanted rice fields, although both are tedious, time consuming and somewhat expensive.

In tropical Asia, either 2,4-D or MCPA effectively control annual weeds (De Datta *et al.* 1968B). If the water control is not good, however, 2,4-D applied before emergence of the weeds (4 days after transplanting) often does not control weeds adequately. Thus, a supplemental hand weeding is often necessary to control surviving weeds. Selective herbicides such as butachlor or thiobencarb are somewhat more effective against a wide spectrum of weeds, particularly under the area of relatively poor water control in rainfed rice (De Datta 1977A).

Insect Control, Harvesting and Threshing.—Practices such as insect control, harvesting and threshing do not differ between rainfed and irrigated rice culture of a transplanted crop. These operations are discussed elsewhere in this chapter.

Direct Seeded Lowland Rice.—Direct seeded lowland rice culture is an important system for the Asian tropics. Pregerminated seeds are broadcast onto puddled fields without much standing water. Direct seeding is primarily by broadcast methods. Such is the case for most rice areas in Sri Lanka and parts of India, Bangladesh and the Philippines.

Fields are prepared under wet conditions. The degree of puddling depends on the amount of moisture accumulated from the rain. Stand establishment is often poor because of poor land preparation and insufficient water control. Farmers in Java, Indonesia, oftentimes decide whether to seed rice under wet or dry soil conditions, depending on the amount of moisture available during the planting month. If less than 200 mm of rain falls within that month, farmers opt for dry seeding; more than 200 mm, they seed under wet conditions. Some farmers in Iloilo, Philippines, seed early maturing rice varieties onto puddled soil.

Water Management.—Water control is more critical for broadcast seeded rice than for transplanted rice. For better stand establishment it is best that the field be kept saturated but not flooded from the time of seeding to about 6 days after. Because of a lack of precise water control, germinating rice seeds are often covered with muddy water thereby reducing stand. When that occurs, it is best to reseed with pregerminated seeds in the areas where germination is poor.

Weed Control.—In this system of rice culture, weeds grow vigorously. Selective herbicides such as butachlor and thiobencarb would control weeds under rainfed rice culture. Supplemental post-emergence weed control, either with 2,4-D (if the predominant weeds are broadleaves and sedges) or by hand weeding, is essential to control remaining weeds.

Other Cultural Practices.—In rainfed areas, fertilizer application, insect control, harvesting and threshing will not differ markedly between direct seeded and transplanted rice culture.

Dry Seeded Bunded Rice.—The greatest economic return to farmers in the rainfed areas may not come through an increase in the yield per crop, but rather by expansion through double cropping.

For many years, farmers in Bangladesh and Indonesia have grown rainfed lowland rice which is seeded directly on unpuddled soil at the beginning of the rainy season. This method of rice culture is known as "aus" cropping in Bangladesh and "gogo-rantja" in Indonesia. By doing this, the farmer takes advantage of early rainfall and may thus manage to grow an extra crop of rice.

Scientists have demonstrated that a similar concept is applicable in the Philippines. For example, despite rains that begin in May in Central Luzon, the farmers seldom have enough water to plow and puddle the soils until July or August. By the time crops are transplanted, often in August, the seedlings may have passed the optimum age. In an experiment involving 9 different sites in Central Luzon, the first crop of early maturing lines was sown in early May, before the heavy rains began. These crops were harvested in mid August—before the last 40% of the surrounding farmers had transplanted their first crops. Seedbeds for the second crops had been prepared about 3 weeks earlier and seedlings were transplanted immediately after harvest. Farmers were harvesting this second crop at the same time that local farmers using traditional crop cultures were harvesting their first (and only) crop (IRRI 1974).

Drought damage and weed infestation are the 2 most critical factors in good seedling establishment of dry sown rice on bunded fields. Good land preparation is difficult because of lack of power to cultivate the hardened soil. Drought damage is severe and weeds tend to outgrow rice under this system.

Recently, early-maturing modern cultivars such as IR36 with good drought tolerance minimized drought damage. Further research at IRRI and other organizations in Southeast Asian countries hopefully will develop improved technology for direct seeding of rice.

Despite the rate of unsatisfactory yields obtained in the recent direct seeding program of the Philippine government, the concept is important. It is important to develop the cultivars and practices that will reduce the farmers' risk and ensure a high probability of obtaining 2 crops where he had 1 before.

Irrigated Rice

Irrigated rice covers only 33% of the rice area but produces about 48% of the rice in South and Southeast Asia (Fig. 4.4). Of the 33% irrigated land, 14% of the land is double cropped and the rest is single cropped.

In temperate Asia, such as the People's Republic of China, Japan, Korea and Taiwan, most of the rice land is irrigated. In temperate America, Europe and Australia, rice is entirely an irrigated crop. In those temperate countries, rice is simply not grown if irrigation water is not available.

In irrigated areas of the world, rice is grown as follows: transplanted, direct seeded onto puddled soil, drill seeded into dry soil and direct seeded into water. Cultural practices in these systems of rice culture often vary a great deal.

Transplanted Rice.—In irrigated areas of Asia, most of the rice is transplanted.

Land Preparation.—Land preparation for irrigated transplanted rice is no different from that for rainfed transplanted rice discussed earlier. The water requirement for land preparation in the lowland field varies from 150 to 200 mm.

Water Management.—Rice, like any other crop, requires adequate water to grow and develop at its maximum potential rate. Unlike other crops, rice is usually grown in flooded soil.

Many researchers have installed metal tanks in the middle of the field to measure water use by flooded rice. An experiment during the 1968 dry season (De Datta and Williams 1968) showed that the most satisfactory regime in terms of grain yield was the intermediate continuous flooding (7.5 cm), although no significant difference was measured between it and 6 other treatments. Water use efficiency (liters of water required to produce 1 g of grain) was highest when soil was kept at continuous saturation. Experiments in the 1969 dry season confirmed our earlier findings, that continual flooding is not essential for high grain yield but that flooded rice can tolerate up to at least 15 cm of water depth if improved cultivars are grown. In this experiment, annual grain yields were similar in continually saturated plots and continually flooded plots. Submergence has other advantages, however, such as better weed control, higher efficiency of fertilizer and better insect and weed control with granular herbicide. Considering all factors, continuous submergence with 5 to 7 cm of water is probably best for irrigated rice (De Datta *et al.* 1973A).

Another method of applying water to fields is rotational irrigation as used in countries such as Taiwan. The field may often be without standing water between irrigations but ideally does not dry out enough for moisture stress to develop and reduce grain yield. Rotational irrigation is often recommended in locations where it is desirable to irrigate as large an area as possible with a limited water supply. In Taiwan, for example, rotational irrigation is as effective as the conventional method of irrigation, if not better (Chow 1959).

This method of irrigation has not been widely adopted because it requires competent irrigation personnel as well as good farmer cooperation, the existing conveyance system must be modified to include measuring devices. Weeds are also more difficult to control when the fields lack standing water for a time.

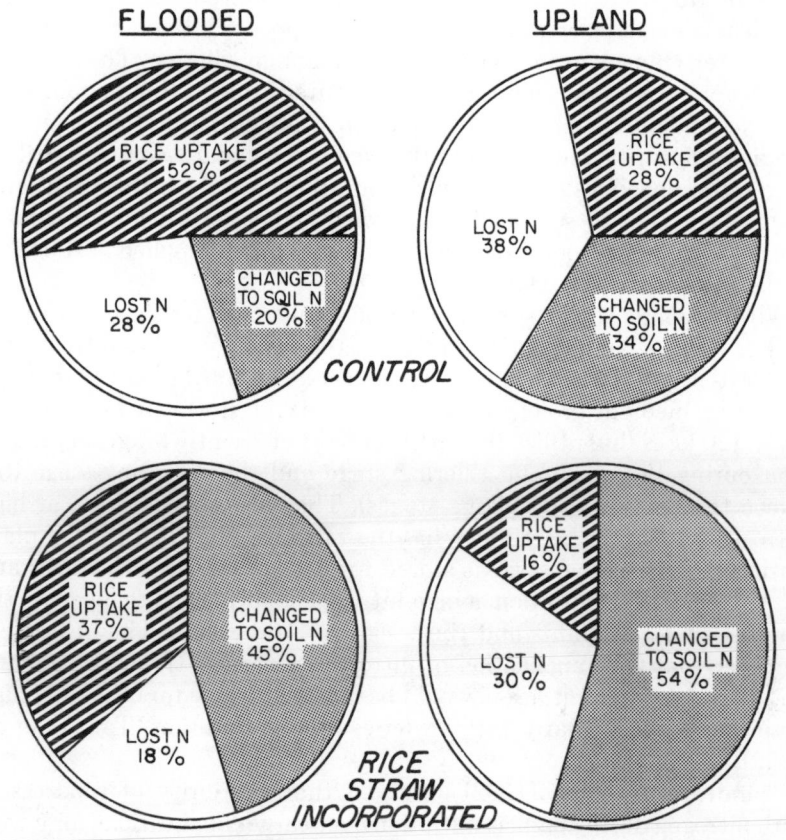

Adapted from IRRI (1974)

FIG. 4.11. BALANCE SHEET OF FERTILIZER NITROGEN APPLIED TO FLOODED AND UPLAND SOILS

Fertilizer Management.—Much of the fertilizer nitrogen applied for a rice crop is not taken up by the rice plants. Depending upon soil type and the method and time of fertilizer application, 20 to 60% of the fertilizer is utilized by the crop in a given growing season. Some of the remainder is combined with organic forms by soil microorganisms in the soil and it may be released too late for crop uptake. Some is lost to the atmosphere by the biological process of denitrification. In others, nitrogen is directly lost to the atmosphere by ammonia volatilization. Figure 4.11 shows the fate of fertilizer nitrogen under flooded and upland conditions in the Philippines. Obviously, the challenge is to reduce to a minimum the losses of fertilizer nitrogen and to increase the efficiency of the usage by the crop.

There are 2 stages in the growth of transplanted rice at which the efficiency of fertilizer nitrogen utilization appears to be highest. One is soon after transplanting to encourage maximum tillering. The second is just before or at panicle initiation to encourage the maximum number of panicles and grain numbers per panicle. It was shown that for 2 lines with medium maturity requirements (between 124 and 127 days), fertilizer nitrogen supplied just before panicle initiation gave the highest efficiency of use. The shorter seasoned line showed highest efficiency after basal application but responded well to the treatment applied just before panicle initiation (De Datta *et al.* 1974).

When properly timed, split applications of nitrogen fertilizer have given higher yields than has application of all the fertilizer as basal treatment (De Datta *et al.* 1974). In India, Tanaka *et al.* (1959A) reported that rice plants of medium growth duration (145 days) grown at low nitrogen levels (20 kg N/ha) utilize the fertilizer most efficiently for grain production during the maximum tillering stage and the flowering stage (between the booting and milking stages). They also showed that, at high nitrogen rates such as 120 kg/ha, the nitrogen absorbed by the plant during the vegetative stage is stored for use at later growth stages and that after panicle initiation, a high nitrogen supply tends to decrease the number of filled grains and the weight of 100 grains. They therefore consider that split application of nitrogen, 1 dose at transplanting and another at panicle initiation, would be most favorable for obtaining high grain yields, particularly with varieties of long duration (Tanaka *et al.* 1959B).

Chandraratna *et al.* (1962) observed the superiority of topdressing with nitrogen during panicle initiation over earlier applications for a wide range of Indica varieties indigenous to the tropics. They found that topdressing during panicle initiation increased the number of grains per panicle but not the number of panicles per plant.

Matsushima (1964) cited several studies on the nitrogen requirement of the rice plant at different growth stages. These included those concerned with the effect of time of nitrogen application on the trend and level of dry matter production, leaf color and leaf senescence, nutrient uptake and plant characteristics including date of maturity.

Water management may directly affect the efficiency of nitrogen utilization by lowland rice. When topdressing with nitrogen, it is generally believed that temporary drainage is necessary to bring the fertilizer material in contact with the soil particles. Data have shown that if the water depth of the field is maintained at 5 cm, temporary drainage at the time of nitrogen topdressing does not necessarily result in increased grain yields, so that it is not recommended in areas where there is no assured water supply (De Datta *et al.* 1969).

Coarse textured soils usually have high percolation rates and demand split applications of nitrogen (usually 2 or 3) to avoid serious losses of nitrogen that would prevent maximum crop growth.

Finer textured soils require a few nitrogen applications, often only 1 or 2, provided that water management practices are good.

Yield differences do not always correlate with differences in nitrogen uptake. Proper placement of nitrogen fertilizer within the flooded system is of extreme importance. Fertilizers containing ammonium nitrogen are stable provided they are placed 8 to 10 cm below the surface of the flooded rice field. This is because microorganisms capable of oxidizing ammonium nitrogen cannot function in an oxygen deficient reduced zone. Fortunately, rice readily uses ammonium as well as nitrate nitrogen.

Experiments as early as 1966 showed a recovery of fertilizer nitrogen of 68% when the fertilizer was placed at a 10 cm depth, compared with 28% recovery when the fertilizer was broadcast and incorporated by harrowing (De Datta et al. 1968A).

There appears to be a considerable advantage in keeping the nitrogen fertilizer concentrated rather than dispersing it throughout the soil. When nitrogen was concentrated in "mudballs," 60 kg N/ha gave a higher yield than 100 kg/ha applied by broadcasting and mixing with the soil. The greater efficiency of this mudball deep placement may be due to the fact that the nitrogen is better protected from microbial changes, including those which bind the nitrogen to soil organic matter (De Datta et al. 1974).

New equipment recently introduced into Japan, India and elsewhere in Asia may make deep placement of ammonium nitrogen feasible. In India, bullock drawn fertilizer drills have been developed to allow deep placement under puddled conditions. Special units have been designed and have successfully injected anhydrous ammonia into rice soils (Stangel 1970).

Recent work in India has shown that foliar application of a 2% urea solution is very effective for rice. Results obtained from these tests revealed that foliar applied N required less nitrogen to achieve equally superior yields. That was true for any combination of basal or split application of soil applied N (Stangel 1970). Foliar application of urea elsewhere did not give encouraging results (De Datta et al. 1974).

Sources of nitrogen are no different for irrigated transplanted rice than for rainfed transplanted rice. The topic is covered in the rainfed section.

Phosphorus deficiency is a widespread nutritional disorder of rice. It occurs on millions of hectares of ultisols, oxisols, vertisols and certain inceptisols. These soils are not only low in available phosphorus but also fix fertilizer phosphorus as highly insoluble minerals. Besides, soil submergence gives only a slight increase in availability of phosphorus in these soils.

Phosphorus deficiency is not only quite widespread but acute in certain areas among rice growing soils of India. In acute phosphorus deficient areas, response to P alone has been comparable to or even better than that to N at equal rates of application. Black, red, coastal alluvial and lateritic soils of southern Indian states and red and yellow soils of Orissa and Madhya Pradesh are more prone to phosphorus deficiency and likely to show high response to phosphate application (Goswami 1975). Results from India show a dramatic response to the first 40 kg P_2O_5/ha. In Rajendranagar, Hyderabad (pH 8.5), there was a 1.7 MT/ha increase in rice yield from applying 40 kg P_2O_5/ha. In West Bengal, India, response to added phosphorus was consistent up to 60 kg/ha. Phosphorus deficiency is quite widespread in parts of Southeast Asia.

In Indonesia, about 800,000 to 1 million ha of soils cultivated for rice are deficient in phosphorus, while many areas in Thailand, particularly in the northeast, are deficient in phosphorus. A report made by the Food and Agriculture Organization of the United Nations (FAO) shows that rice at Bankhen, Thailand, responded to phosphorus up to 25 kg P_2O_5/ha either as superphosphate or as basic slag. At Konkhaen, also in Thailand, phosphate response was dramatic (Moomaw and De Datta 1968).

In Sri Lanka, response to phosphorus in rice was recorded in all districts. The most dramatic responses were in districts such as Amparai, Hambantota, Mannar, Minepe and Polonnarea.

Few significant differences have been found among the effects of various phosphorus sources in flooded rice except in extremely acidic or alkaline soils. India, Pakistan, the Philippines, Korea and Burma use superphosphate as their primary sources of phosphorus on all soils, while Sri Lanka, Malaysia, Thailand, Khmer Republic and Vietnam use rock phosphate (Mukherjee 1962). On acid soils of South and Southeast Asia, phosphate rocks have been applied directly as sources of phosphorus for rice. These sources are lower in price than acidulated phosphates and can represent a means of reducing fertilizer cost where transport costs are not too great (Engelstad et al. 1974).

Experiments in the greenhouse at the Tennessee Valley Authority (TVA) and in fields in Thailand evaluated several phosphate rocks with varied citrate solubility. A procedure is developed which enables one to make the choice of P source on the basis of both agronomic effectiveness and prices of P as triple superphosphate (TSP) vs phosphate rocks (Engelstad et al. 1974).

In general, phosphorus is applied at planting. If needed, application can be postponed but not later than the time of active tillering. It has been reported that early application of phosphorus is essential for root elongation. Phosphorus applied during the tillering stage is most efficiently utilized for grain production.

In general, most phosphorus applications should be made as a basal treatment. Recent tests with radioisotopes have shown no advantage in deep placement of phosphorus. Split applications have not proved of value.

The phosphatic fertilizer needs of rice on deficient soils can be reduced if varieties that can extract phosphorus efficiently from the soil can be developed. That would be a real boon to the poor farmers of the tropics where most of the phosphorus deficient soils occur, especially since phosphate fertilizers have now tripled in price.

Most rice soils have not been shown to be particularly deficient in potash. That is because irrigation water usually contains significant amounts of potash. The majority of rice soils in Asia responds to more N and P, than to K. In Korea, although a large percentage of 566 experiments show positive response to K fertilizer, the actual increase in yield is extremely low.

In Japan, crops grown on degraded paddy soils respond well to potash, and yield increases of 20—30% can be expected. Further, response to K is obtained on light soils where leaching causes considerable loss of bases. Favorable responses to K have been demonstrated from sandy and coarse textured soils of Vietnam, Sri Lanka and Malaysia. In Sri Lanka, about 25% of the total rice fields suffer from K deficiency. Average increase in yield in Taiwan due to K is about 50%. Responses are most significant in red and yellow earths and the effects of K are generally greater in the second crop than in the first.

Common sources of potassium are potassium chloride and sulfate and complete fertilizers. Potassium should be applied during the final land preparation. Results from Taiwan occasionally indicate that split application of potassium is beneficial to rice grown on coarse textured soil Five year data on clay soils in the Philippines showed no beneficial effects from a split application of K over a single application (De Datta and Gomez 1974).

In view of recent fertilizer shortages, there is a move to reduce the doses of phosphorus and potassium and in some cases eliminate them altogether from recommendations for various crops in India. Results from the long-term fertility experiments indicate that in areas where 1 crop per year is grown with an adequate supply of N, the response of P and/or K may be marginal. Response may become marked, however, under intensive rice cultivation involving more than 1 crop per year (De Datta and Gomez 1974).

Because of the current fertilizer situation, inorganic fertilizers should be supplemented with green manure, compost and other organic manures, and rice straw. Quantitative experimental data are needed to assess the value of organic manures in realizing high yields from the modern rice varieties. Under Indian conditions, organic manures contain good amounts of micronutrients which are beneficial to rice production in certain rice growing areas (Joshi 1975). Dewan and Dongale (1974) estimated that 2200 MMT of compostable waste were generated annually in India which could contribute 22 MMT of plant nutrients.

FAO (1973) reported that there may not be enough compost or other organic manure available to treat large rice areas in Thailand. The available supplies of organic manures can easily be absorbed by the vegetable and fruit growers.

In some Asian countries, straw is available in large quantities as a source of organic matter and nutrients. In other countries, rice straw is used for cattle feed and hence not available for field use. In clay soils, straw incorporation has limited nutritional value, but in sandy soils of low CEC the effects may be of some significance (FAO 1973). Recent studies on management of straw and other organic sources of nutrients receive considerable attention. In long-term experiments conducted with 21 successive crops, compost applied at 10 MT/ha in each crop showed no yield advantage over inorganic N alone when compost was applied on an equal N level (IRRI Ann. Reports 1964−1975). These organic sources alone, however, would seldom serve as substitute for chemical fertilizers if modern high yielding rice cultivars are grown.

Next to nitrogen and phosphorus, zinc deficiency is perhaps the most important nutritional factor limiting the grain yield of wetland rice. It occurs in alkaline soils, calcareous soils, wet soils, peat soils and sandy soils. Millions of hectares of low-lying land in the humid tropics, well supplied with water, may be brought under rice if improved cultivars with tolerance to zinc deficiency are developed. Zinc deficiency in rice has been reported in India, Pakistan, the Philippines, Japan, the United States and Brazil.

In the Philippines, zinc deficiency is perhaps the second most important nutritional disorder limiting the yield of lowland rice. At least 13 provinces in the Philippines have large areas of zinc deficient soils (Orticio and Ponnamperuma 1977).

In these zinc deficient areas, NPK fertilizers alone will provide only a limited yield advantage unless zinc deficiencies are corrected first.

The incidence of zinc deficiency has increased in recent years because of replacement of old cultivars by modern ones with less tolerance to zinc deficiency, removal of large amounts of zinc by the new high yielding modern cultivars, replacement of ammonium sulfate by urea, increased use of phosphate fertilizers, increased use of concentrated fertilizers, and double and triple cropping of wetland rice.

In transplanted rice, zinc deficiency is easily corrected by dipping seedlings in 2% zinc oxide suspension (Yoshida et al. 1970).

Other methods of correcting zinc deficiency in transplanted rice include foliar sprays and soil application. Treatment of transplanted seedlings and soil treatment are the 2 most common methods of zinc application. Sprays are usually employed only in emergency situations when the crop is growing and shows deficiency symptoms (Mikkelsen and Kuo 1976). Cultivars with tolerance to zinc deficiency would reduce the problems of growing rice in zinc deficient soils. Results at IRRI showed that IR20, IR34, BG 90-2 and several IRRI elite breeding lines tolerated zinc deficiency. Cultivars tolerant to zinc deficiency should be grown on zinc deficient neutral soils. On organic and calcareous soils, it is advisable to use a combination of tolerant cultivars and a 2% zinc oxide dip (Orticio and Ponnamperuma 1977).

Weed Control.—Weed control does not differ greatly between irrigated transplanted rice and rainfed transplanted rice as discussed earlier. Good land preparation and good water control make weed control less difficult for irrigated transplanted rice than for rainfed transplanted rice. Hand weeding is commonly used in almost all areas in South and Southeast Asia. Rotary weeding seems common in some Southeast Asian countries although not in others. Herbicides are used most commonly in areas of inadequate labor and high wage particularly in East Asia—Taiwan, Korea and Japan.

In the Philippines and some tropical South and Southeast Asian countries, 2,4-D is the principal herbicide used because of its low cost. In the tropics, 2,4-D controls annual grasses if applied to transplanted rice 4 days after transplanting—before the emergence of weeds (De Datta et al. 1968B). It is apparent that at existing wage rates, herbicides are likely to have very limited acceptance unless the wage rate relative to the chemical cost rises fairly substantially as in Taiwan. In most of tropical Asia, developing integrated methods of weed control (using limited quantities of low cost chemicals in combination with direct and indirect methods) may be the most attractive alternative from agronomic, economic and ecological points of view (De Datta and Barker 1977).

Insect and Disease Control.—Insects do far more damage to rice in the tropics than in temperate regions. In fully irrigated areas in the tropics, where year round rice culture is practiced, pest problems became much more serious than in the cool temperate climates where only 1 crop can be grown each year.

The control of rice insect pests in the tropics currently depends largely on the use of pesticides, even though many traditional cultivars have some resistance to 1 or more insect pests or diseases. These have played and will continue to play a major role in successful rice production. In recent years, scientists have evaluated different insecticides and application methods to learn how farmers can maximize return on insecticidal investments. It was found that insecticides are more readily available to the plants (by systemic effect) when packaged in capsules and placed in the root zones. Inserting the insecticide capsules below the soil surface protects them from heat, sunshine, volatilization and draining out with the overflowing water (IRRI 1973B).

In biological control studies, it was found that a predator (Crytorhinus lividipennvis) can kill an average of 0.6 brown planthopper nymphs per day, or 50 green leafhopper nymphs per day, for at least 4 consecutive days. The predator prefers to prey on nymphs rather than adults, and prefers green leafhoppers to brown planthoppers (IRRI 1973B).

Entomologists have recognized that integration of pest control methods involves combining cultigenic resistance with pesticides when necessary. Plant resistance should reduce the pest population somewhat, if not considerably, and insecticides can be applied to reduce the number of these pests even further. Farmers in the tropics cannot afford the luxury of expensive plant protection measures. Crop resistance and minimal insecticide application must be the answer if increased rice production in the tropics is to be economic. Promising lines are being continually developed that combine broad genetic resistance to all major insects and diseases in tropical Asia.

For diseases of rice, resistant lines have been identified and incorporated in the elite breeding lines. Although chemical control of blast or of the vectors of virus diseases is possible, this control is neither practical nor economical. Cultigenic resistance to these diseases and to the vectors of virus diseases is the only economic solution to these problems.

Harvesting, Threshing, Drying and Storage.—In the tropics, rice matures about 30 days after heading. Harvest must occur on time otherwise grain may be lost through damage caused by rats, birds, insects, shatter and lodging. Timely harvest ensures optimum grain quality, a higher market value and greater consumer acceptance.

In the Asian tropics, harvesting is done by hand. The straw is usually cut about 15 to 25 cm above the ground, although in some areas, particularly Indonesia and some upland rice areas in the Philippines, the panicles are clipped off with a sharp knife.

Threshing is done by hand, by beating the rice heads on a perforated platform made of bamboo. Occasionally, threshing is done by having animals tread on the harvested rice crop. Threshers are rarely used in tropical Asia.

The grain is usually dried in the sun, although that is difficult during the wet season when clouds and rains often disrupt operations. Practical experience has shown that, in the Philippines, 4 days of sun drying are usually sufficient to reduce the moisture content of the grain to an acceptable level.

Most farmers save only a certain amount of harvested rice (rough rice) for family consumption and store the grain in sacks or in small barrels since proper storage facilities are rare in the Asian tropics. The problem of post harvest operations, including storage, has become more serious because of the higher yields obtained with modern cultivars. The IRRI Engineering Department has designed a number of small units of harvesters, threshers and dryers for the Asian tropics, and widespread use of these (like any mechanical equipment) will depend on availability, pricing and the profitability of production.

Direct Seeding onto Puddled Soils.—The practical difficulties of establishing good stands of direct seeded rice under monsoon conditions have been studied to determine whether transplanting is indeed superior to direct seeding under tropical conditions. Evidence suggests that grain yields are similar for direct seeded and transplanted rice. This finding, along with the increasing cost of transplanting, may lead to serious consideration of direct seeding as a desirable alternative at least in areas with controlled irrigation. Although machines are available to drill pregerminated rice onto puddled soil, the broadcast method is most common.

Direct seeding onto puddled soil is practiced in some parts of India, Bangladesh and the Philippines. In Sri Lanka, 80 to 90% of the rice acreage is broadcast. Jayasekera (1966) concluded from a number of experiments that high grain yields are possible with properly managed broadcast seeded rice. With broadcast seeding, weed control is a serious problem at low plant density.

From the Philippines, de Jesus (1959) noted the prevalence of lodging with broadcast rice, while Castillo (1962) observed that lodging occurred earlier in direct seeded rice. Ramiah and Hanomontha (1936) made similar observations in India and attributed this to poor root development.

In Fiji, Bharat (1953) noted that acreage planted to rice increased over the years because more farmers planted rice by broadcasting.

Cultural practices for direct seeding onto puddled soils are similar to those for transplanted rice in some operations, and distinctly different in other operations.

Land Preparation.—Preparation of land is closely related to the method of planting rice. The conventional method of tillage for transplanting aims at making a soft puddled soil. This degree of land preparation is often considered a requirement for direct seeded rice, although this is not usually required except to aid air water conservation.

The practice of puddling in broadcast seeded rice culture, similar to the transplanted system, reduces water percolation. Puddling is also reported to aid in creating a level soil surface, incorporate uniformly added fertilizer, incorporate weeds and hasten mineralization of organic nitrogen.

Water Management.—Without leveling of fields, water management is difficult. Depth of water determines successful seedling establishment. Systems of water management for direct seeding onto puddled soil vary widely. In 1 system, fields are flooded just before seeding and kept flooded until harvest. They may be drained soon after seeding and flooded again. In another system, pregerminated seeds are sown onto a moist but drained field, and the field is reflooded with 3 to 5 cm of water when rice is at the 4 leaf stage and weeds at the 3 leaf stage.

Fields are often drained during the growing season. The reasons for drainage include control of aquatic weeds, algae and water-weevil, and application of plant nutrients and agricultural chemicals.

From the literature, continuous submergence appears suitable under most situations for direct seeded rice.

Fertilizer Management.—Management of fertilizer does not greatly differ between direct-seeded irrigated rice and direct-seeded rainfed rice. However, nitrogen fertilizer efficiency is greater in direct-seeded irrigated rice than in rainfed rice under a similar method of planting.

Weed Control.—Weed control is more critical and difficult in rice grown from pregerminated seeds broadcast directly into the field than in transplanted rice. Laborers cannot move through broadcast rice to weed by hand without destroying some rice plants. Furthermore, laborers cannot distinguish between young grassy weeds and young rice.

For direct-seeded flooded rice, the herbicides, butachlor and thiobencarb, which are commercially available in several Asian countries, give excellent weed control without injuring the rice crop (De Datta and Bernasor 1971). Propanil and molinate are also effective in direct seeded rice production (Smith *et al.* 1977). Among the new herbicides, a mixture of piperophos and dimethametryne has shown excellent selectivity and broad spectrum activity in both the tropics (De Datta and Bernasor 1973) and temperate regions (Green and Ebner 1972). Another new selective herbicide for direct-seeded flooded rice is butralin (De Datta and Bernasor 1973). In trials in India, Pillai (1973) reported direct-seeded to that used for drilling on dry soil, except that the seedbeds are usually left very coarse and dry before flooding. Soil clods 2 to 6 cm in diameter are found in a typical good seedbed. After flooding, these and nitrogen effectively control weeds in direct-seeded flooded rice. In the

Philippines, direct-seeded flooded rice culture will become popular as new herbicides are identified and marketed. The direct-seeded flooded rice culture will become an acceptable alternative as the cost of labor rises, as less expensive selective herbicides become available and as water control becomes better.

Insect and disease control and other cultural practices are the same for direct seeded as for transplanted rice in most tropical areas. Harvesting, threshing and storage are also similar.

Direct Seeding into Water.—In areas where a highly mechanized agriculture is justified or where the labor requirement for transplanting is excessive, direct seeding into flooded fields is practiced. The practice is ancient in parts of Asia, including India, Sri Lanka, Malaysia and Thailand, and is now done on a wide scale in the Americas, southern Europe, the USSR and Australia. Direct seeding is economical in time and saves labor in both land preparation and in actual sowing.

More care must be exercised in management of water in drill seeded and water sown rice than with most other methods of cultivation. Water depth, excessively high or low water temperatures, and muddy water are occasionally sources of problems in stand establishment. Direct seeding requires 40–50% more seed than transplanting; the seed must have good viability and the cultivars must possess good seedling vigor. Oxygen deficiency does not appear to be a limiting factor in stand establishment of rice in flooded soils (Chapman and Mikkelsen 1963). Adequate chemical weed control measures must be available since mechanical cultivation is not possible. While transplanting is advisable in areas where multiple cropping is practiced for protection against certain pests, for better insurance against crop failure and for saving seed, in a number of instances no significant differences in rice yields are reported between direct seeding and transplanting (Adair *et al.* 1942; IRRI 1967B).

Land Preparation.—Land preparation for rice seeded directly into water is similar to that used for drilling on dry soil except that the seedbeds are usually left very coarse and dry before flooding. Soil clods 2 to 6 cm in diameter are found in a typical good seedbed. After flooding, these clods slake down to smaller aggregates leaving an irregular surface that reduces seedling drift and helps establish root penetration. A suitable seedbed should be relatively level, devoid of low and high spots. Water may be too deep for good emergence in the low spots, and weeds (especially barnyardgrass) may be favored in soil not covered with water.

Fertilizer.—Fertilizer placement by band drilling nitrogen 5 to 10 cm deep in dry soil or broadcast applications incorporated to the same depth have given significant yield increases and higher nitrogen use efficiency than surface applied nitrogen (Mikkelsen and Finfrock 1957). Surface applied nitrogen is nitrified and subsequently lost by denitrification after the flood water is applied. Ammonium nitrogen drilled 5 to 10 cm deep where reducing conditions develop within a few days of flooding, remains effective in the soil during the entire crop season. Broadcast applications to typical rice seedbeds show that ammonium nitrogen is nitrified if it is applied too far in advance of flooding. With soil temperatures that average 22°C (71.6°F), as much as 60% of the ammonical nitrogen can be converted to nitrate in 7 to 9 days (Mikkelsen and Miller 1963). The average effect of nitrogen placement at 4 locations was to increase rice yields by 34% over broadcast in water applications and 19% over broadcast applications on a dry seedbed (Mikkelsen and Miller 1963).

Split applications of nitrogen, with part applied as a basal application and the balance top dressed at panicle initiation, has proved effective in some parts of the world. Top dressing experiments in California have shown no superiority over adequate preplant soil applications (Mikkelsen and Miller 1963). When the basal nitrogen application was not sufficient to meet crop needs, top dressings have been effective in increasing rice yields.

Phosphorus, potassium and zinc are best applied to the soil before flooding. Phosphorus and potassium can be incorporated during seedbed preparation with good effectiveness. Zinc, however, should be applied to the soil surface without incorporation. Since water sown rice has a great need for zinc during the first 3 weeks after planting, it is essential that it be applied in the soil surface for maximum efficiency of recovery at the lowest levels of application. Surface applied zinc adequately meets the needs of rice during the growing season (Mikkelsen and Kuo 1976).

Weed Control.—Weed control in water sown rice is accomplished by proper management of water and water depth and the judicious use of herbicides. The same types of herbicides and methods of application described for drill seeded rice apply with relatively few exceptions (Smith *et al.* 1977). Barnyardgrass and other annual grasses do not survive if water is properly maintained between 5 and 10 cm deep. Surviving grass seedlings are controlled by propanil applied in actively growing plants but has no residual value. Treatments at recommended rates do not normally cause permanent injury, although propanil applied wih certain seed treatments or carbamate insecticides may cause severe injury. Propanil may cause damage to sensitive crops, so spray drift must be avoided. Molinate in a granular or emulsifiable form, also used as a pre-emergence and post-emergence herbicide, is especially effective against *Echinochloa* spp. For maximum effectiveness water in rice fields treated with molinate should be held for at least 7 days. Phenoxy herbicides, particularly 2,4-D, MCPA and silvex, control many broadleaf weeds, sedges and aquatic weed species. As reported for drilled rice, consideration is necessary for the proper rate and time of application.

Harvesting and Drying.—Harvesting and drying requirements are the same as described for drill seeded rice. Conventional harvesters are widely used where artificial systems of rice drying are available in many parts of the world. In Japan and other Asian countries, small combines are available as regular types with single row and head threshing units (Stout 1966).

Drill Seeding into Dry Soil.—Drill seeding into dry soil is widely used in the Americas, southern Europe, the USSR and Australia where land holdings are large and labor scarce. Typically the preplanting climate is relatively dry, land is level and irrigation can be controlled with some precision. Drill seeding is used also in rice producing areas of South and Southeast Asia, Africa and South America where seed remains in the soil until rainfall is sufficient to initiate germination and sustain seedling growth. In most drill seeding, fields are irrigated to ensure adequate moisture for stand establishment, and after an appropriate time the fields are placed in permanent flood.

Land Preparation and Planting.—Drill seeding of rice is usually accompanied by other highly mechanized farming practices. The land is frequently plowed in the off-season with a large tractor drawn moldboard or disk plow, then disked again and harrowed before planting. When fields remain wet during the off season, the entire land preparation may be done just before planting. Harrowing is usually done with a minimum of tillage unless the soil contains large amounts of clay. A roller packer is often used to break up clods and firm the soil, helping retain moisture. Soil preparation usually involves machine travel over the gently sloping levees which are usually rebuilt after seeding and subsequently seeded.

Conventional grain drills are commonly used for drilling seed in rows spaced 15–20 cm apart. Drills are capable of planting seed and when supplied with a fertilizer hopper, place seed at a uniform depth and simultaneously place fertilizer below and to the side of the seed. In Australia, direct seeding is sometimes done into pasture sod without previous land preparation. A planting depth of 3 to 5 cm is common. Uniformity of seeding is important to uniform stand establishment and crop maturity.

Water Management.—When soil moisture is not adequate for seed germination, fields are flushed with water. The irrigation water is drained soon after the fields are completely submerged, since seed cannot germinate and emerge if covered with both soil and water. If soil moisture becomes limiting or if crusting occurs, additional irrigation and drainage is used. Irrigation practices vary, but fields are usually completely submerged with 5 to 10 cm of water when seedlings are 8−12 cm tall. If the fields are weedy, water is applied to inhibit weed growth as soon as the rice has emerged. After the permanent flood is applied, the water depth is held nearly constant until several weeks before harvest. Water is usually drained only when special problems develop as with algae, water weevil or weed control.

Fertilizer Management.—Fertilizer use on drilled rice takes on many variations, depending on variables of climate, cultivars and crop and water management practices. The required plant nutrients, except nitrogen are commonly either applied during land preparation or are drilled with the seed along with a part of the nitrogen requirement. Up to half the nitrogen requirement is usually applied by drilling or broadcast just before permanent flood is applied to the rice crop. Midseason nitrogen applications, applied broadcast into the water, are made sometimes using internode elongation as a guide to proper timing as reviewed by Mikkelsen and Patrick (1968). Proper timing reduces plant height and grain yields.

Zinc deficiency is a common problem on alkaline soils, causing chlorosis which affects plant survival and crop maturity. Soil applications of zinc are usually applied in combination with other fertilizer materials or applied independently in a broadcast application. Applying sulfuric acid to either the soil or flood water has provided some relief from zinc deficiency. Where rice is grown in rotation with soybeans, timing must be carefully controlled to avoid inducing zinc deficiency in rice.

In areas of southern United States with long growing seasons, the growth of a ratoon or "stubble" crop is possible. After harvest, the ratoon crop is stimulated by applying nitrogen.

Weed Control.—Weed control is accomplished with judicious management of both water and herbicides. Cultivation between drill row is difficult except with rotary weeders, and the practice is not widely used. Herbicides are an important production resource to reduce weed populations and to increase the effectiveness of other practices. The herbicides propanil and molinate, used most widely, are applied early after seeding to control grass seedlings, sedges and some aquatic weeds. Broadleaf weeds and sedges are usually controlled with phenoxy herbicides, especially 2,4-D and MCPA.

Propanil is applied by both ground and aerial equipment as medium-fine droplets to obtain good plant coverage. *Echinochloa* spp. are controlled most effectively when they are in the 3−5 leaf stage and when growing actively.

Molinate is applied after the emergence of rice, usually in a granular form applied to the water. It controls *Echinochloa* spp. up to about 7 cm tall, but is not effective against broadleaf, sedges or aquatic weeds. For effective control, water should be held on the field for at least 7 days after treatment. During cool weather or where irrigation water is to be drained, molinate is more effective against barnyard grass than is propanil.

Phenoxy herbicides, including 2,4-D and MCPA, are effective against many broadleaf weeds, sedges and aquatic plants, but are ineffective against grasses. These herbicides are usually applied about midseason when rice is in the early jointing stage. At that time, weed control is effective with minimum damage to the rice plants. A variety of new chemicals have potential for weed control in drilled rice, including thiobencarb, oxadiazon, butachlor, bifenox and bentazon.

Harvesting and Drying.—Self-propelled combine harvesters are used extensively in the United States, southern Europe, the USSR and Australia where mechanized drill seeding is practiced. Combines mounted on tracks or equipped with oversized pneumatic tires with large lugs for traction and flotation move effectively over drained fields. Harvesters are provided with the component parts necessary for cutting the crop, threshing and separating the grain, and holding it in field storage. The equipment must sometimes be specially adapted for certain harvesting conditions. Mechanical failures from working in wet soils in rainy weather sometimes make combine harvesting difficult. Since rice is usually harvested at 20—24% moisture content to preserve high milling yields, artificial drying must be used to reduce the grain moisture below 14% for safe storage.

Rice in Medium-deep Water Areas

Vast areas of rice in Asia are grown in uncontrolled water regimes ranging from 30 cm to 1 m deep. The margins of the deep water rice areas, the low-lying areas, the so-called "stagnant water" areas, and the tidal swamp areas are of this nature. India, Bangladesh, Thailand and Indonesia have large flooded areas grown to rice. Presently available cultivars and technology are not suited for such variable water regimes. The cultivars used are usually the tall Indica rice type, known as "flooded rice." In some areas, floating rice types are planted. Intermediate statured cultivars, such as Pankaj, IR5 and IR442-2-58, are having a limited success in some areas. Mashuri is another cultivar which is popular in such areas in India, Burma, Malaysia, Vietnam and Indonesia.

It is anticipated that by breeding and research efforts it will be possible to expand the area where improved cultivars can be grown.

Deep Water and Floating Rice Areas

About 10 million ha of rice land is a conservative estimate of the area subjected to annual floods (Table 4.2). The depth of water, duration of flooding, the rate of increase in water level, temperature, turbidity, time of occurrence, etc., vary at different locations, so that the term deep water may have different meanings (Vergara 1977).

TABLE 4.2. SUMMARY OF DEEP WATER AREAS PLANTED TO RICE[1]

Country	Area in ha
Bangladesh	
Deep water	2,100,000
Medium deep	3,300,000
Vietnam	
Deep water	500,000
Medium deep	
Single transplanted	1,200,000
Double transplanted	220,000
Indonesia	
Kalimantan	163,000
Tidal swamp	
Burma	
Deep, medium	
deep water	486,000
West Africa	
Mali	132,000
Niger	5,000
Nigeria	77,000
Dahomey	—
Gambia	8,000
Sierra Leone	
Deep water	14,350
Medium deep	14,350
Cambodia	800,000
Thailand	5,500,000
India	
West Bengal	
Deep water	20,000
Assam	
Deep Water	100,000
Bihar	
Deep water	500,000
Medium deep	1,500,000
Manipur	—
Orissa	—
Kerala	—
Andhra Pradesh	100,000
Madras	—
Uttar Pradesh	600,000
Tamilnadu	—
Mysore	—

Source: Vergara (1977).
[1]The "submerged areas" or areas where deep water remains for only a short period of time are not included in this table.

Deep water rice is a general term used to refer to a rice culture or cultivar which is planted where the standing water for a certain period is more than 50 cm, resulting sometimes in complete but usually partial submergence of the plant. New high yielding cultivars cannot be used in areas with uncontrolled water regimes where flood waters annually rise from 50 cm to 4 or 5 m deep.

For our discussions, these areas with uncontrolled water supply are divided into deep water rice and floating rice areas. Nearly half of the rice in Asia is grown in uncontrolled water regimes with maximum water depths varying from 50 to 150 cm. We would like to call part of this area (depth of 100 to 150 cm) the deep water rice area.

When the water level exceeds 150 cm and the water usually remains in the field for 3 to 4 months, special rice types known as "floating rice" are planted. Such conditions of uncontrolled water regimes, whether a deep water or a floating rice area, prevail in densely populated deltas, estuaries and river valleys of Asia in India, Bangladesh, Burma, Thailand, Cambodia and Indonesia. In Africa, deep water rice is grown in Mali, Niger, Nigeria, Dahomey, Gambia and Sierra Leone. Unlike the local cultivars, the semidwarf rices cannot elongate with rising flood water, nor can they withstand submergence; most simply drown out if grown in such deep water. The few plants that may survive will mature when the water is still high. Harvest is difficult or impossible, and panicles may rot on the stalk. It is easy to see why farmers in deep water rice areas prefer their traditional cultivars, mostly photoperiod sensitive types, even if they yield only about 1 MT/ha.

In deep water rice areas, dry seeding is fairly common—as well as drought stress (De Datta and O'Toole 1976). For example, in Bangladesh, sowing starts in mid March, but with the advent and amount of rainfall as factors determining the actual date of sowing (Hasanuzzaman 1974). Further, farmers are fully dependent on uncertain rainfall during March–April that causes frequent drought damage and poor stand. In West Bengal, seed is broadcast in April–May on dry soil (Mukherjii 1974). In Thailand, deep water rice is grown under dry conditions until the monsoon achieves full force, any time from June to August. During this period, rice frequently suffers from severe drought (Prechachart and Jackson 1974). In Vietnam, farmers often compare deep water rice culture to gambling. Farmers in deep water rice areas often have to reseed once or twice if it does not rain for several days after seeding (Xuan and Kanter 1974). Early seeding is desirable to ensure that the plants are of sufficient size to elongate when water levels rise. However, the erratic onset of the monsoon rains and the possibility of subsequent drought increase risks in early planting. If drought tolerance is bred into rices meant for deep water areas, early seeding will ensure rapid establishment, thereby reducing risks that a sudden rise of flood water will destroy young seedlings.

Another problem of deep water rice cultivars is that they need 6 weeks of growth before their internodes elongate when submerged. Floods that occur before the plant has grown sufficiently are major factors causing low plant density obtained in the deep water rice crop. Although the rise in water level is generally 3 to 10 cm per day, in certain years and in certain places the increase may be as much as 60 cm per day during the first 2 days. The deep water rices must flower after the water level has reached its peak to minimize the danger of the panicles being submerged. The traditional deep water rices for a particular area generally flower at a time selected by farmers to ensure less crop damage and easier harvesting (Vergara 1977).

Weed control is a major factor limiting the grain yields of deep water rices. The problem of weeds usually occurs during the early stages of growth when the deep water rice is under upland conditions before the rise in flood water. Although early weed competition may be decreased by heavy seeding, tall weeds that grow faster than the rice will still offer competition. Farmers generally do not weed the deep water rice areas (Vergara 1977). The deep water rices with vigorous seedling growth and spreading tillers provide better competition with weeds. In Bangladesh, weeding is done to a small extent under either preflooded or flooded conditions (Hasanuzzaman 1974). Except for *Ipomoea aquatica, Eichornia crassipes* Salons. (water hyacinth) and wild rices, weeds are not a serious problem at later stages of growth. The rise in water level eliminates most of the weeds.

Most of the pests and diseases in shallow water rice culture have also been reported in deep water rice culture. There are very few reports on outbreaks of plant pests and diseases in deep water rice.

Salinity is another problem which affects some deep water rice areas.

Because most of rice cultivars grown in deep water rice areas are photoperiod sensitive, harvesting is done after flood waters have receded.

It is important that breeding, agronomic efforts and relevant physiological studies should be directed to improve the yields of deep water rice. This effort should help a large number of subsistence farmers. It is anticipated that these breeding and research efforts will expand the area where improved cultivars can be grown.

Upland Rice

Upland rice refers to rice grown on both flat and sloping fields that are not bunded, that are prepared and seeded under dry conditions and that depend on rainfall for moisture. Upland rice is grown on 3 continents, mostly by small subsistence farmers in the poorest regions of the world. Grain yields are generally low: 0.5 to 1.5 MT/ha in Asia, about 0.5 MT/ha in Africa and 1 to 4 MT/ha in Latin America.

The area planted to upland rice, however, is so large (nearly ⅙ of the world's total rice land) that even a small increase in yield would substantially influence total rice production (De Datta 1975).

A recent publication discussed most aspects of upland rice production problems and summarized current research findings that are available to solve those problems (IRRI 1975B).

The following sections summarize updated information on factors that limit the grain yield of upland rice and discuss the prospects of overcoming them.

Irregular Rainfall.—Upland rice is grown under a wide range of rainfall, from as low as 400 mm (in some years) in Faizabad, Uttar Pradesh, India to 4000 mm in Amazon Basin, Peru. Naturally, the factors that limit the grain yield of upland rice are as diverse as the amount of rainfall under which it is grown. The distribution of rainfall is as diverse as the amount of rainfall. For example, in the upland areas of Burma, rainfall from May to November can be as low as 500 mm or as high as 2000 mm; in Sri Lanka it varies from 875 to 1000 mm (De Datta and Vergara 1975). In the southern portion of West Africa, 1200 mm in 2 rainy seasons is separated by a short dry season (FAO Inventory Mission 1970; Cocheme 1971). Some dry northern parts of West Africa have 600 mm of rain during the 4 month rainy season (Cocheme 1971). Latin America has 200 mm of monthly rainfall during the growing season (Brown 1969; Sanchez 1972). These total amounts are usually adequate for growing a crop of upland rice, although they are not indicative of upland rice yields. The daily rainfall is actually more critical than the monthly or annual rainfall. Plants in an area that receives as much as 200 mm of precipitation in 1 day can be damaged or even killed by moisture stress if no rain falls for the next 20 days. A precipitation of 100 mm/month, distributed evenly, is preferable to 200 mm/month which all falls in 2 or 3 days (De Datta and Vergara 1975).

More important than variations in the seasonal distribution of rainfall is the variability in efficiency of a given volume of precipitation which substantially depends on temperature, day length, atmosphere humidity, wind movement and other factors that determine the rate of evapotranspiration (Jana and De Datta 1971). Tropical cyclones (typhoons and hurricanes) and torrential rains usually occur during the monsoon season when upland rice is planted. The strong winds may cause lodging, sterility and grain shatter, thereby adding to the problem of too little or too much rain.

The prospects for overcoming irregular rainfall with varietal development will depend on 2 major factors: agronomic characteristics, such as plant type and maturity group, and drought tolerance. Many upland cultivars currently grown are early maturing types. Quite likely they were bred or selected primarily for drought tolerance. Nevertheless, they provide valuable material to obtain low but stable yields under upland rice culture.

Researchers differ about the ideal plant type for upland rice. Most of them, however, believe that rice strains with intermediate stature (125–130 cm under well-watered conditions) will withstand the adverse conditions better than semidwarfs (about 100 cm). However, many of high yielding rices were semidwarfs. For example, in 19 trials conducted under diverse conditions in the Philippines, India, Nepal and Liberia in the 1975 wet season in the International Testing Program (IRTP), the semidwarf line IR1529-430-3 produced an average grain yield of 2.9 MT/ha.

Major factors that often determine the partial success or failure of the crop are plant type, maturity group and yielding ability, and tolerance to drought. For screening rices for drought tolerance, several techniques are available that provide an opportunity to mass screen a large number of cultivars and breeding lines. Information is meager on practices that minimize drought damage in upland rice. Since upland rice depends on rainfall, some cultural practices have been suggested to decrease damage from water shortage (De Datta et al. 1976).

It is clear that poor to imperfect drainage conditions are preferred for the growth and production of rice. All soils that are well or excessively drained are risky to a varying degree, but the risks may be diminished by conserving rain water through leveling and bunding of the land as defined earlier. In low rainfall areas (400–1200 mm of annual rainfall), much of the rice area is planted to dry-seeded bunded rice which is loosely called *upland rice*. Similar dry seeding in bunded fields is practiced in the *aus* crop areas in West Bengal, India and Bangladesh. It is also practiced in Java where a choice between a system of upland/lowland culture combination (gogo-rantja) and purely lowland rice culture will depend on the amount of rain that falls at the time of seeding. If less than 200 mm rain is received during the planting month, the land is prepared dry and seeded dry. Fields are bunded to hold rainwater in later months. Such agronomic practices are being tried in the "sabog-tanim" method in the Philippines to grow a direct seeded crop in bunded dry fields. Similar water management was suggested on land with level topography in West Africa (Moorman et al. 1975). Bunding and leveling of land was suggested to increase the hectarage of rainwater fed riceland that exists outside the optimum climatic zone.

Very few studies have been made on the effects of mulching on moisture conservation in upland rice. Studies in Nigeria showed that moisture retention was improved by straw mulching (Int. Inst. Trop. Agric. 1972). A surface straw mulch of 4 MT/ha was found to be more effective in increasing soil moisture retention than the same organic matter of 5.0 MT/ha buried 10 cm below the surface (Int. Inst. Trop. Agric. 1973).

A well designed windbreak often influences crop productivity and makes water use efficiency more favorable (Van Eimern 1964).

IRRI physiologists have theorized that upland rice grown under coconut trees is less subject to drought damage primarily because the crop is shaded. It is believed that, besides providing shade, coconut trees may serve as windbreaks, reducing wind damage to upland rice crop. Such planned tree plantings around upland rice fields may be worth considering in areas with high frequency of drought and where strong wind aggravates drought damage.

This is the practice in some areas of Java, Indonesia, where a tall legume species is planted around upland rice fields, providing shade and serving as a windbreak for upland rice. Flowers from the legume plants are eaten by the farmers. The leaves of the legumes are used as cattle feed and a green manure to supplement commercial fertilizers.

No matter what the annual rainfall or its distribution may be, progressive farmers sow upland rice at the onset of the rainy season and harvest it about 100 days after seeding. With the residual moisture, many farmers in Batangas province in the Philippines sow another crop that may have a high value in cash, such as onion, garlic, corn or sorghum (De Datta *et al.* 1976). In parts of north India (in Uttar Pradesh), upland rice is followed by wheat. All these examples convince us that to make upland rice economically viable, it should be considered as part of the cropping system and not a monoculture subsistence farming, as it is considered today.

Diseases and Insects.—It is mandatory that all upland rice cultivars have a high level of blast resistance. In fact, most upland types currently grown resist blast.

Sheath blight is a disease becoming serious in upland rice in the Philippines.

Other diseases that may affect the growth of upland rice are *Helminthosporium* and *Cercospora* leaf spots, bacterial leaf streak and bacterial leaf blight (Ou *et al.* 1975).

The picture of insect damage is less clear in upland rice than in lowland rice. Some examples are grasshoppers, stem borers, leaf folders, white backed planthoppers, green leafhoppers, zigzag leafhoppers and brown planthoppers. Considerable progress has been made in incorporating insect resistance into lowland rices. Nevertheless, many of the crosses with intermediate stature and high insect resistance should generate progenies with desirable plant type and resistance to pests and diseases.

Soils on Which Upland Rice is Grown.—The frequency and duration of moisture stress is affected not only by rainfall distribution but also by the capacity of the soils to retain water (De Datta and Feuer 1975). The texture of a soil affects its moisture status more than any other factor except topography.

Texture is particularly important in upland rice fields which have no bunds to retain surface moisture. Rooting depth depends not only on texture of the surface soils but also on the textural profile, including subsoil texture.

Many upland rice soils have a sandy or loam texture and have poor moisture retention capacity. In such soils, drought stress commonly occurs unless the rainfall distribution is perfect, which is very rare.

In West Africa, soils on which upland rice is grown are either freely drained or hydromorphic (with high water tables and receiving supplemental water through surface runoff). In these freely drained soils, upland rice frequently suffers from drought stress. In addition, they are highly weathered and generally have low nutrient content because they are derived mainly from acid rocks and are more strongly leached under the high rainfall regime of the areas in which they occur (Moorman *et al.* 1975). Raising a good upland rice crop on these freely drained soils requires cultivars and technology that would tolerate these extremely harsh conditions. On the other hand, suitable management technology should be developed to take advantage of hydromorphic soils on which upland rice would suffer fewer moisture and nutrient deficiency problems.

Deficiency or Toxicity of Elements.—In general, alternate wetting and drying of soils leads to losses of both native and applied nitrogen (De Datta and Magnaye 1969). It was reported that rice takes up less nitrogen under upland conditions than under lowland conditions (Shapiro 1958).

An upland soil has less water than has a submerged soil, iron phosphate and silica are less soluble, available nitrogen is present as a highly mobile nitrate, and soil acidity and alkalinity may be problems (Ponnamperuma 1975). Upland soils, unlike submerged soils, are not able to adjust their pH levels to the favorable range of 6.5 to 7.0. This means that manganese and aluminum toxicities can occur in strongly acid soils and iron deficiency in alkaline soils (Ponnamperuma 1975). It seems that upland rice will do best on lower numbers of the topo sequence of slightly acid soils. Sodic, calcareous and saline soils, acid sulfate soils and soils low in organic matter are not suitable for upland rice.

It is entirely possible that interacting factors, such as lack of adequate moisture and soil problems, affect the growth and yield of rice. A need to study these factors has been demonstrated together (De Datta *et al.* 1975). It was reported that the reduction in grain yield caused by increased soil moisture tension may be due to either the direct effects of moisture stress or moisture stress-induced soil problems, such as iron deficiency, or to a combination of both. Therefore, moisture stress and soil problems should be considered together in evaluating the suitability of rices for upland culture.

The content of water soluble iron in most aerobic soils is too low for ready detection. The apparent iron requirement of rice is higher than that of other plants, and rice suffers from iron deficiency in well drained upland soil. Several cultivars and breeding lines that are tolerant to iron deficiency were identified by IRRI soil chemists. These rices would perform well in most iron deficient soils on which rice is grown.

Competition from Weeds and Control Measures.—More weeds grow under upland than under lowland conditions. Sometimes, the infestation in untreated controls in upland rice is so heavy that no grains can be harvested (De Datta 1973). The traditional puddling and subsequent flooding which minimizes weed infestations in lowland rice is irrelevant to upland rice. Flooding has been recognized as an effective weed control method for many years (Jenkins and Jones 1944). Under heavy weed competition, weeds alone may reduce grain yields by as much as 83% or even 100% (Vega *et al.* 1967). If perennial weeds, such as purple nutsedge *(Cyperus rotundus* L.), are present, competition for moisture, light and nutrients is severe, reducing grain yields of upland rice (Okafor and De Datta 1976A).

Much future expansion in rice areas may have to be in upland areas, since most potential lowland areas are already under cultivation. Suitable weed control practices, including use of herbicides, would play an important role in increasing rice production in such new areas.

For example, Indonesia has 1.28 million ha under upland rice, and the upland area is gradually expanding in southern Sumatra and Borneo. If weeds such as *Imperata cylindrica* and *C. rotundus* are controlled, further increases in areas under upland rice are possible.

For West Africa, reports indicate that present weed control methods are unsuitable for continuous or large-scale farming. In shifting cultivation areas, once weeds become too great a problem, the land is abandoned to forest, a cheap and effective means of control. To make upland rice culture meaningful, highest priority should be given to research aimed at finding chemicals that are economical and effective (Moody 1975).

The most feasible and economical schedule for weed control in upland rice is one which uses both chemical and mechanical methods as well as other cultural practices. One hand weeding often takes 350 to 600 man-hr per ha. However, several hand weedings are necessary to remove weeds completely so that chemical weed control may be essential for successful upland rice cultivation anywhere in the world. The following herbicides look promising for upland rice culture in the Philippines: butachlor, AC 92553 (a coded compound), dinitramine, oxadiazon, thiobencarb and the combination herbicide piperophos plus dimethametryne. They were effective when used at 2 kg active ingredient per hectare and sprayed before the weeds emerged. Although propanil, alone or in combination, was found effective in controlling weeds in upland rice in India and in West Africa (Aryeetey 1973), the results in the Philippines have been inconsistent. With all chemical treatments, a subsequent hand weeding is often essential to remove weeds either not controlled by the chemicals or a fresh regrowth (De Datta 1977B).

The perennial nutsedge, *Cyperus rotundus* L., is a problem in upland rice because it germinates just before or simultaneously with upland rice; it grows with, and thereby competes with, the rice crop. When broadcast-sprayed with a herbicide K-223 at 8 to 10 kg active ingredient per hectare and immediately mixed into the soil just before seeding rice, control of perennial nutsedge was excellent (Okafor and De Datta 1976B).

Other cultural practices, such as good land preparation, timely hand or mechanical weedings, choice of proper rice cultivars, seeding rates and fertilizer management would greatly facilitate and complement at times completely substitute for chemical weed control in upland rice (De Datta and Ross 1975).

Other Cultural Practices.—Practices such as land preparation, sowing method, seed rate, row spacing, date of planting and fertilization affect the grain yields achieved by the upland rice farmers.

Land Preparation.—In most of Asia, little mechanization is used to prepare land for planting. As soon as enough rain has fallen to permit initial land preparation, the field is plowed with an animal drawn implement, and then harrowed with a comb harrow to prepare a good seedbed and to firm up the soil. Sometimes the weed seeds are allowed to germinate for a week, and then the field is harrowed for the final time, destroying all germinated seeds.

In Thailand, slightly elevated areas are plowed by water buffalo and cattle, and then hoed. On hills, the soil is hardly cultivated.

Indonesian farmers generally prepare the land with animal drawn plows during June through August. Indian farmers simply turn the soil over with country plows and pulverize it no more than 10 cm deep.

About 98% of the rice land in Africa is prepared manually because draft cattle are scarce, most being susceptible to the trypanosome disease (FAO Inventory Mission 1970).

Land-preparation methods vary greatly from country to country in Latin America. In the shifting cultivation areas of Peru, for example, mature secondary forests are cut and burned during the drier months from July to September. Upland rice is then seeded by dibbling without further land preparation (Sanchez 1972). Shifting cultivation in Peru is quite similar to the slash-and-burn methods that precede planting in Malaysia, Burma or Thailand.

Sowing Time, Methods, Seed Rate and Row Spacing.—Upland rice farmers have learned through experience to plant early where rainfall distribution is unimodal and the rainy season lasts about 4 months. Upland rice gives the highest grain yield and best nitrogen response if planted shortly after the first monsoon shower.

In most of the Philippines, an animal-drawn wooden plow called a "lithao" is used to open furrows. Dry seeds are then broadcast. An implement called a "Kalmut" is then used to divert the seeds into rows and to cover them. In most of India, seeds are broadcast on either dry or moist soil in roughly prepared fields. Indonesia farmers seed rice broadcast on dry soils soon after the rainy season begins. In the shifting cultivation areas of Asia, the land is cleared by the "slash and burn" method and seeds are then dibbled into the soil. In West Africa, rice is sown by broadcasting or dibbling. On the 40% of the upland area with annual rainfall of less than 1500 mm, seeds are dibbled into rows made with a pointed stick or a narrow bladed hoe. On the 60% that has more than 1500 mm annual rainfall, seeds are broadcast in dry soil (FAO Inventory Mission 1970).

Seeding methods vary greatly among Latin American countries. In Peru, 8 to 10 rice seeds are normally planted in holes dug with a pointed stick called a "tacarpo" at irregularly wide spacings, about 50 × 50 cm. The seeds are not covered with soil. Sanchez (1972) considers this system of seeding inefficient because the rice competes poorly with weeds and ripens unevenly.

In parts of Brazil, seeds are drilled with a tractor driven seed drill, at spacing as wide as 60 cm.

Upland rice in Latin America, whether grown under shifting or under semimechanized cultivation, is generally spaced widely to discourage the spread of blast disease and to help the crops tolerate drought.

Optimum row spacing for upland rice should be developed locally depending on 2 major factors: severity of drought in the area and varietal resistance to blast disease.

To get high yields of upland rice, it is essential to plant a lodging resistant and high yielding cultivar. On the other hand, if the area suffers from annual drought, or the cultivar grown is susceptible to blast, it is essential to plant at wide spacings (45−60 cm) as in Brazil.

Variety and Fertilizer Application.—Most upland rices do not respond well to nitrogen; nitrogen fertilization increases susceptibility to blast and lodging. However, considerable progress has been made through plant breeding in developing rices with higher level of blast resistance. There seems to be trade-off for plant type requirement for drought tolerance and shading of weeds with semidroopy leaves versus lodging resistance. Many intermediate statured cultivars with high tillering ability will lodge under heavy showers. On the other hand, these rices fare better than short strawed varieties under less favorable moisture status and in weedy areas. Fortunately, damage from lodging is less severe under upland than under lowland conditions. For this reason, the taller types are favored.

Regardless of varietal type, it is better to apply nitrogen in split doses to minimize lodging and to get maximum fertilizer nitrogen efficiency. Results further indicate that banding the fertilizer close to the seed (10 cm deep) greatly helps increase nitrogen efficiency in upland rice. In phosphorus deficient areas, banding of both nitrogen and phosphorus complements and increases the efficiency of both elements under upland rice culture.

Harvesting.—Many different systems of harvest have been devised for rice, depending upon environmental, cultural, religious and economic factors. Over the world the major portion of rice is harvested by hand sickle, though various knives are also used. In highly mechanized rice producing areas, rice is harvested entirely with high capacity self-propelled combine harvesters.

Hand harvest is widely used because it can be done under a wide range of weather and field conditions, is adapted to small plots where rice maturity varies from plot to plot, can harvest as panicles ripen, and can eliminate weeds from the harvested material. Mechanical equipment is often inoperable under these conditions.

The type of harvest equipment depends on such factors as method of planting, plant height, lodging problems, shatter losses, timeliness of harvest, method of threshing and weather conditions during the harvest season. Rice is very susceptible to lodging, especially with the tall native types, where the soil is unusually fertile, and where strong winds and rain may cause the crop to lodge. Harvesting is slow and difficult when the crop is lodged. Transplanted rice, properly spaced and managed, is less susceptible to lodging than direct seeded rice.

Shattering of ripened grain is a problem with certain cultivars, especially when the grain is too dry (below 15—20% moisture). The harvest techniques employed must minimize shatter losses, including timely harvests. Rice subjected to excessive drying may shatter badly and may undergo "sunchecking," which causes minor cracks to develop in the kernel, favoring breakage during hulling and milling. A grain moisture value between 20 and 25% is considered satisfactory for an acceptable harvest.

The various possible uses of rice straw for animal feed, fuel, bedding, compost and rope or mat making also affect the possible selection of harvest methods.

The sickle is used widely for harvest, although the size and design varies from country to country. Harvesting requires workers to hold the straw with one hand and cut with the other. The length of cut varies with local conditions. After cutting, the harvested portion is either dried further in the field or collected for threshing.

In parts of Southeast Asia, only the panicles (plus 5—6 cm of straw) are harvested. A special harvest knife, consisting of a soft metal blade set in a crescent-shaped wooden handle, is used. Individual panicles are sometimes harvested as they mature, lessening shatter losses and facilitating drying.

Animals and engine-powered dropper type harvesters and reaper-binders have been developed for use in Japan and Taiwan. The machines travel down the rows, cut the plants and with the reaper-binder tie the bundles for transport.

Tractor-powered and self-propelled reapers, windrowers and threshers have been developed and are used for small plots in Europe, Japan, the Republic of China and elsewhere. Modern combine harvesters, suitable for large-scale mechanized rice production, are used in many areas, especially in the United States, Europe and Australia. The modern combine cuts the straw and threshes out the grain which is separated from the straw, cleaned and stored temporarily in self-contained bins. A high capacity combine can harvest 6—12 ha per day. A part of combine harvesting essential for high milling quality is artificial drying.

Threshing of hand-harvested sheaves usually requires adequate pre-threshing or post-threshing drying. The sheaves may be laid on the ground, stacked or hung on racks for drying prior to threshing or storage. Threshing involves separating the grain from the panicle by a variety of methods involving manual beating or treading, animal treading or engine powered threshing devices. Cleaning foreign materials from the grain and drying to 13—16% moisture is accomplished before rice is placed in storage.

FUTURE OUTLOOK IN RICE PRODUCTION TECHNOLOGY

Rice truly means life itself to the most densely populated regions of the world. A third of mankind—1.3 billion people—depends on rice for more than half of its food. Rice is the primary or secondary staple food of $\frac{9}{10}$ of the low-income people of the most densely populated regions of the world. The average annual income of those who depend primarily on rice is only $80.

Population increases in most rice consuming countries are among the highest in the world. Consequently, demands for rice are expected to increase by 30% by 1987.

In spite of the impressive technological advances of the past decade, national production figures show increases that are barely high enough to meet population growth in the developing countries. The experiences of the past few years remind us that the production revolution needed to feed rice consumers has only begun.

Rice output can be increased through the expansion of cultivated area or through an increase in the productivity of existing land. In South and Southeast Asia, prior to 1960, the expansion of land area provided the principal source for output growth. New lands were opened up at a pace roughly in keeping with the growth in the agricultural labor force.

The gradual closing of the frontier land after 1960—more pronounced in rice than in upland crops—necessitated a shift toward the use of modern yield increasing inputs (Herdt and Barker 1977).

Modern Cultivars

The most important single factor for increasing rice production was the development of modern cultivars that respond to yield increasing inputs. The introduction of the IR8 plant type in the late 1960s generated widespread hope for coping with the chronic rice production gap that plagued many developing countries. The introduction of modern cultivars also generated improved technology to exploit the yield potential of these cultivars. More fertilizers were used in these modern cultivars than in traditional ones and the practice was termed the *green revolution.* It is still not uncommon to refer to so-called green revolution technology as synonymous with seed-and-fertilizer-technology. Often, little emphasis is given to complementary management practices such as control of insects and weeds. Nevertheless, the greatest breakthrough took place in the development of modern cultivars which take from 120 to 125 days to mature from seed to harvest, unlike the old types, which took at least 150 days. Shortening the growth period while obtaining a marked increase in grain yield potential was perhaps the most significant breakthrough in agriculture as a whole, and rice research in particular. These groups of modern cultivars increased rice production considerably in irrigated areas and to some extent in rainfed areas.

In the following 8 to 10 years (1967−1976), varietal development focused on 2 issues: (1) building up insect and disease resistance in modern cultivars, and (2) shortening the growth duration further, to around 100 days. As a result, a series of cultivars have been introduced in South and Southeast Asia which met both objectives. The examples are IR28, IR29, IR30 and IR36. These cultivars provided a unique opportunity to intensify rice cropping. It is true that future developments might generate cultivars with a still shorter growth period, such as those that would mature between 85 and 90 days. In fact, there are cultivars grown in India with extremely short growth duration, such as "Tinpakhia" which matures in about 75 days. There is, however, enough evidence to indicate that shortening the growth period from 85 to 90 days sacrifices a certain amount of yield potential. Despite this shortcoming, rice with an extremely short growing period would perform a valuable role in certain environments. In many instances it means an extra crop instead of nothing. With the present level of technology, 100-day growth duration is perhaps the lowest limit without greatly sacrificing the grain yield potential.

Rice Based Cropping Systems

Several terms have been used by various researchers and research administrators to explain the basic production systems of intensive cropping. For example, terminology such as double cropping, triple cropping, multiple cropping, cropping intensity, cropping patterns, continuous cropping and cropping systems all have one thing in common: they deal with more than 1 crop culture.

No matter what the terminology is, increased cropping intensity will continue to play a significant role in future rice production in South and Southeast Asia. Results from agronomic research focused on crop intensification show that under most conditions in irrigated areas, and partly in rainfed areas, at least 1 additional crop can be grown profitably.

The Peoples' Republic of China has in fact partially met this objective by growing 1 additional crop in some rice growing regions. For example, in northern China, rice has been added to the traditional wheat area as a second crop. In the Yangtze River Valley, 2 crops of rice are grown where 1 was grown before. In the Canton area, 3 crops of rice are being grown in some areas, and 3 crops of rice and a wheat crop are being considered in limited areas.

In many areas of successful intensive cropping, the essential features are the availability of improved crop land with good irrigation and drainage, improved cultivars and management practices, the availability of farm machinery, and a high demand for farm produce. Further, a favorable ratio of labor to land must be present in the Peoples' Republic of China.

In some parts of tropical Asia, cropping systems can be intensified by extending the growing season and by utilizing the available growing season more efficiently (Zandstra and Price 1977). Extension of the growing season can be accomplished by supplemental irrigation, early crop establishment, conservation of residual soil moisture and growing drought tolerant crops into the dry season.

Better utilization of the growing season can be achieved by growing shorter maturing varieties, reducing turnaround time, harvesting at physiological maturity and using relay cropping and intercropping.

Yield increases can be partitioned between increases from irrigating a higher proportion of the area and from using more yield increasing inputs, such as new seed and more fertilizer.

In the Philippines, the major contribution of increased production in the earlier period was irrigation (double cropping and improved average land quality). Land area expansion, seed and fertilizer were not important factors for increased production. In the latter period, the contribution of land area became negative principally because of a decline in upland rice. The contribution of irrigation expanded with the area that is double cropped, more than off-setting the decline in net area (Herdt and Barker 1977).

There are vast areas where rice is grown under rainfed conditions. In these rainfed areas, the greatest returns may not come through an increase in yield per crop but through expansion of the area that is double cropped (Barker *et al.* 1975).

Rainfed farms in Indonesia, Bangladesh and northeast India have been taking advantage of premonsoon rain to establish a direct seeded rice crop. IRRI scientists have demonstrated that rainfed lowland rice which is seeded directly on unpuddled soil at the beginning of the rainy season takes advantage of early rainfall and may allow an extra crop of rice to be grown (Fig. 4.12).

Continuous cropping of rice in irrigated areas and increasing rice cropping intensity in the rainfed areas are not free of problems. Evidence indicates that continuous cropping of rice causes a buildup of some pests in greater numbers. At times, under intensive cropping of rice, insecticide use must be increased. More and better insecticides must be used to cope with more and diverse pests or different biotypes or strains of the same pest. At times, insects develop resistance to the insecticide used or change to another biotype. Further, insect vectors of virus diseases take a heavy toll by spreading the virus disease. The same is true for blast and

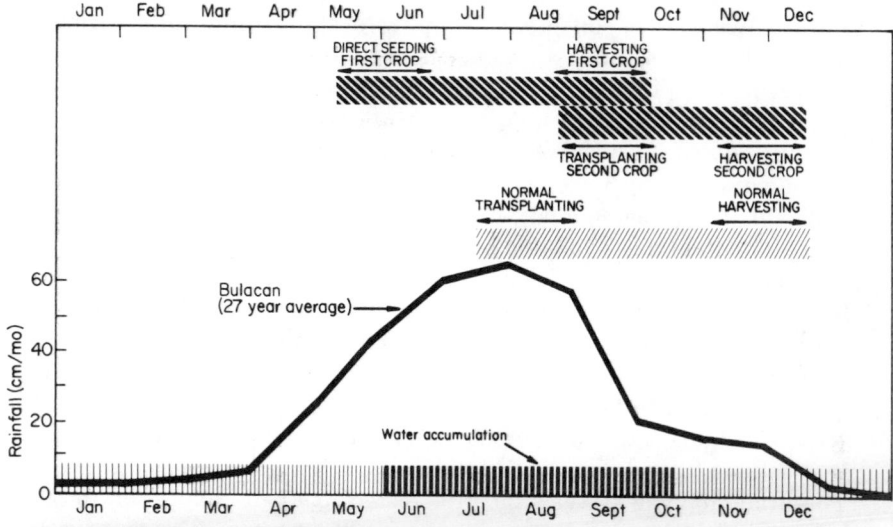

Adapted from IRRI (1974)

FIG. 4.12. EARLY MATURING RICE LINES DIRECT SEEDED IN FARMERS' FIELDS IN EARLY MAY BEFORE HEAVY RAINS BEGAN

Some of these crops were harvested before surrounding farmers had transplanted using regular cropping systems. Second crops were harvested when local farmers were harvesting their first and only crops. In Bulcan and Central Luzon, Philippines.

bacterial diseases. These diseases spread because sources of inoculum are always present. Rodent damage at times becomes overriding in determining an extra crop versus total crop failure.

Problems of continuous croppings are more serious and more complex, but farmers in rice growing areas would grow rice wherever it can be grown with reasonable success. The greatest challenge of the future would be to develop technology to minimize these hazards.

This buildup of complex and diverse insect, disease and weed problems could force us to consider cropping intensity that would include crop diversity.

Modern rice cultivars mature a month or more earlier than traditional rices, leaving enough time and soil moisture to grow other crops. Residual fertilizer left in the ground after rice harvest can be converted into more food rather than be wasted.

Alternate crops will use to the fullest extent the long and warm growing season of the tropics.

Multiple cropping and rice based cropping systems are not new. Farmers have grown other crops with or following rice for a thousand years. But the need to intensify cropping is far greater today than ever before. Studies are under way at the IRRI and elsewhere to find out the biological potentials of different cropping systems. Intensive farming schemes that are slanted to the needs of small-scale farmers rank high in development priority.

Since cropping intensity and pattern are controlled by the physical, economic and social resources, and by environment. A knowledge of the environments in which we work and how they limit cropping potential is important. The researcher who understands the environment and the existing and potentially available resources is equipped to design improved patterns to make more efficient use of the available physical resources. He must, at this point, have a source of improved component technology (cultivars, pest management, new crops, etc.). He puts potentially feasible cropping patterns together with socioeconomic resource requirements that can be used in his target area. These cropping patterns, together with their management and input requirements, become potential cropping systems ((Zandstra and Price 1977).

Potential cropping systems must be tested for their adaptation to the physical and socioeconomic environments. The testing of cropping systems thus requires farmer participation at an early stage. Improved cropping systems aim at more efficient and more complete use of farmer resources. To be successful, they must both produce more agricultural products and improve farmer welfare.

The traditional practice of intercropping (growing 2 compatible crops, such as corn and upland rice, simultaneously in alternate rows) is highly suited to situations of limited land and surplus labor. Intercropping corn and upland rice produces as much as 1⅓ ha of corn and upland rice planted separately. Intercropping uses inputs efficiently.

Besides increased food production, intercropping and multiple cropping provide less risk of total crop failure, higher income and more balanced nutrition to farm families than does monoculture of rice.

A future goal, then, is to find the sequences and combinations of crops that will increase productivity and improve farmer welfare while using resources more effectively. The first task is to provide a choice of alternative cropping patterns which fit the soil type, drainage and water availability pattern f the farmer. More than one alternative cropping pattern is needed, because as intensity increases, labor and power usually become limiting. Efficient use of both of these resources depends on diversification of farm enterprises.

Diversification also compensates for the loss of stability in any given cropping pattern as intensity is increased.

In the future, it will be necessary to develop sufficient component technology for crop management, such as tillage method, and control of insects, diseases, weeds and nematodes. Further, it is necessary to develop varieties with suitable plant type, maturity and management characteristics, and develop the know-how to manage the crops under a range of conditions of fertility, tillage, water, and cash and labor availability (Harwood 1975).

Turnaround Time.—In most parts of monsoon Asia, rice is transplanted during the early part of July and will be harvested toward the middle of November, a period when a crop may not suffer from major stress problems. However, in many situations the rainy season continues through December. In these long rainy season areas, the major upland crop production potential exists from the middle of October until the middle of February. After harvesting rice in November, the fields are planted to a fairly secure short duration upland crop or a risky medium duration upland crop with a 45 day turnaround time between crops. In this cropping pattern, the rice crop may not suffer from severe drought stress and the upland crop is exposed to drought stress only toward the end of its growing cycle. Farmers often establish a low input legume crop, such as mung beans, with very little land preparation. This then reduces turnaround time and places this crop in a production environment that minimizes damage due to drought stress. The low yields obtained with second crops such as mung beans, however, often make them less attractive than 2 crops of shorter maturing rice varieties.

Crop intensification in this typical environmental complex can be achieved by using: (1) early crop establishment techniques, (2) using a premonsoon upland crop during the short period just before the main rice growing season, (3) introducing a second rice crop necessarily in combination with an early seeded first rice crop, (4) reducing the turnaround time between the first and second rice crops, (5) ratoon cropping of the first rice crop, (6) planting upland crops (after a single rice crop, a double crop, or a premonsoon upland crop followed by a rice crop), (7) reducing the turnaround time between rice and upland crops, and (8) relay cropping or intercropping.

All of the above systems (Zandstra and Price 1977) are possibilities. Research data and confirmation at the farm level would be required under farmers' socioeconomic and physical environments before they are adopted.

Recent data at IRRI indicate that land preparation at the end of the previous wet season saved up to 3 weeks of potential growing time for the direct seeded crop of the following wet season. By comparison, up to 6 weeks of potential growing time was lost through transplanting the crop. It was concluded that in regions with a distinct dry season, generation and maintenance of a dry soil mulch would be a valuable cultural practice for securing early establishment of the dry seeded crop and reduce turnaround time.

Whether irrigated or rainfed, turnaround times between rice crops can occupy up to 50 days of prime growing season.

With these views in mind, minimum and zero tillage techniques of land preparation have been evaluated for reducing turnaround time between rice crops. These techniques have been shown to be capable of replacing conventional tillage operations under widely different conditions of climate, weeds and soils. By using these techniques, it has also been demonstrated that considerable saving in time, labor, capital and power can be achieved without loss in yield.

Recent studies at IRRI indicate that minimum tillage is a dependable alternative to conventional tillage. This would require a few modifications in present management practices, at least where difficult to control weeds are absent or have been previously controlled. The use of minimum tillage in rice in tropical Asia would become more relevant if turnaround time between crops has to be reduced. At the present level of technology, zero tillage needs to be treated with more caution.

Further research is required to make a judgment on the extent and causes of lengthy turnaround time in farmers' cropping systems.

Some questions that need answering are:

(1) What is the relation between turnaround time and rainfall?
(2) What are the benefits of reduced turnaround time or the production of the second crop or the production of upland crop following it?
(3) Are delayed turnaround times caused by a lack of suitable technology?
(4) What effects do staggered harvests have on turnaround time?
(5) Do farmers consider that decomposition of straw or weed or insect or disease incidence contributes to a long turnaround time?

It seems that the greatest bottleneck in increasing cropping intensity and reduced turnaround time in South and Southeast Asia is time taken for land preparation. In most of those areas, the turnaround time is normally 2 months. Even on mechanized farms, land preparation activities require 3 weeks. Nevertheless, mechanization or part mechanization did help minimize turnaround time. For example, in some parts of the Philippines and elsewhere, land is partly or fully prepared by tractors— small and large. Two surveys made by the IRRI Agricultural Economics Department indicate that farmers in Laguna Province in the Philippines use power tillers for harrowing. Often, the plowing is done with water buffalo. In Central Luzon, on the other hand, 7-hp 4-wheel tractors are very common and rented for primary tillage, plowing or rotovating (Cordova and Barker 1977). Either for the primary or the secondary tillage operations, mechanization of any kind will have tremendous impact on shortening the turnaround time.

This, in fact, is being practiced in the Peoples' Republic of China. For example, during the visit of the IRRI team in 1976, harvesting of the rice crop and sowing of the winter crop were taking place—a peak period of demand for labor. Land preparation was being done almost entirely by tractors, in some cases by large tractors but primarily by small 12-hp tillers. By mechanization, land preparation between the harvest of 1 crop and planting of another could be reduced to a few days for an individual crop and 2 weeks for most communes (IRRI 1978).

Why is it that in the Peoples' Republic of China, turnaround time could be shortened to a few days whereas in most of tropical Asia that cannot be accomplished? There are perhaps 3 factors for these differences: (1) water control, (2) power and labor availability, and (3) organization of labor. The ability not only to deliver water but to irrigate and drain the field when needed enhances the efficiency in time and use of machinery and labor.

Despite the sharp contrast in turnaround time between tropical rice growing Asia and the Peoples' Republic of China, the new rice technology that relates to increasing cropping intensity and crop diversity would make important issues of turnaround time and mechanization to reduce it.

Development of New Cultivars and Technology

We know that the areas under irrigation are increasing, so efforts are intensified to develop cultivars and technology that would produce still higher yields than those obtained with existing cultivars and technology. We also know that even under irrigated conditions, the grain yield potential of existing cultivars and technology is not achieved under farmers' conditions. We must know why we are not able to get as high a yield in farmers' fields as at experiment stations. Are the conditions in the experiment stations so different from those in farmers' fields that variety and technology would not give similar performance in farmers' fields? If so, we must learn what are the constraints to high yields under farm conditions. If the variety and technology are not entirely suitable under farm environments—biophysical and socioeconomic—then we must develop appropriate variety and technology to suit the farmers' environments.

The problem of increasing grain yield is far greater for rainfed farming than for irrigated farming. Constraints on high yields in rainfed farms are innumerable. Even today, for about 3 rice farms out of 4, there is no improved cultivar and technology that can significantly increase production beyond current levels. For these millions of subsistence farmers, the best available technology consists of hardy but low yielding local rices

and ancient farming methods. These farmers depend solely on the unpredictable monsoon rains to water their crops. Some grow upland rice and
manage it like wheat. Others bund their fields to hold water on the land
in paddies. But the monsoon rains often fail and drought sets in.

The new short statured rices and modern technology are not suited for
hundreds of millions of farmers—those on whose fields flood waters
annually rise to depths of 50 cm to 4 or 5 m.

Then there are farmers with problem soils such as saline and alkaline
soils or toxic soil with excess iron content and farmers in mountains and
hilly regions where rice suffers from cold damage. This shows that in
drought- and flood-prone areas in mountainous areas and with problem
soil areas—there is no green revolution.

Therefore, efforts of individual and national programs in South and
Southeast Asia are focused on 2 issues: increasing further yield level of
the irrigated areas and emphasizing the problem of the by-passed
farmers—farmers in the rainfed areas. The job is extremely difficult and
complex. Therefore, scientists at international centers should join forces
to work with their counterparts in national programs to develop cultivars
and technology that would tolerate each of these stresses, providing
many new rices for farmers in these harsh conditions.

The issues of varietal development and technology development for the
future are discussed in the following sections.

Varietal Improvements of Cultivars.—At many locations efforts are
continuing to develop rice cultivars with different plant types and with
different lengths of growing seasons. Efforts are increasing to develop
improved rice with early maturity. Using early maturing cultivars, farmers can often grow an additional crop before or after that rice crop. The
early rices are also generally more efficient producers.

Rice of intermediate height may be best for medium-deep water, where
semidwarfs cannot grow. They may compete with tall weeds better than
the semidwarfs. But development of somewhat taller rice cultivars that
do not lodge is a challenge.

The following are a set of characteristics that should be incorporated
into future rice cultivars to meet the needs of various environmental and
water regimes:

Irrigated Rice with Good Water Control.—The traits for the irrigated
conditions should emphasize:

(1) early maturity,
(2) higher yield potential, and
(3) introduction of diverse genes for resistance to disease and insects.

The basic plant type of IR8, representing short, sturdy, high tillering plants with dark green erect leaves and highly responsive to nitrogen, is suitable for irrigated areas with good water control.

Physiologists, agronomists and plant breeders would, however, like to see the yield potential of rice cultivars for irrigated areas be increased beyond the present level establishment with IR8. How that could come about is still not certain. Efforts should be continued to raise the yield potential of modern cultivars, not perhaps a great increase but matching the grain yield level that has been recorded with a single rice crop in temperate countries such as southern Australia, Italy and Spain.

Rainfed Lowland Rice.—There are vast areas where rice is grown under rainfed conditions. In these areas, cultivars with intermediate height and with various degrees of photoperiod sensitivity would be more suitable than semidwarfs.

Many rice growing areas have maximum water depths of 30 to 50 cm, and stagnant water lasts for 3 to 4 months. Experience with cultivars with intermediate height and essentially insensitivity to photoperiod, such as IR5, Bahagia, Pelita 1, C4-63, IR442-2-58 and Mashuri, is worth considering in these areas. In many, various degrees of photoperiod sensitivity are essential to cope with harvesting problems in standing water.

The fertility of the soil and production inputs used are generally low. Therefore, a high level of lodging resistance would be desirable but may not be essential.

Deep Water Rice.—There are large areas in the river deltas of Thailand, Burma, Bangladesh and India where maximum water depth varies from 50 cm to about 150 cm. In those areas, the monsoon season with its heavy rains starts around the first week of June, and there are continuous rains and standing water in the field until the end of October. Varieties which mature before the first week of November are difficult to harvest because of standing water. Therefore those areas require varieties with photoperiod sensitivity and having intermediate stature, ability to elongate with the increase in water level in the field, tolerance to submergence and resistance to major diseases and insects. These varieties should have good seedling height because water depth in the field is sometimes more than optimum during transplanting time.

Floating Rice.—Essential in floating rice areas, where maximum water depths vary from 150 cm to 5 m, are rapid elongating ability, tolerance to submergence and photoperiod sensitivity. However, very little can be done to improve rice in these areas.

Saline, Alkaline and High-temperature.—Suitable for inland areas where saline and alkaline soils occur and high temperatures prevail, are cultivars with traits similar to those for irrigated areas but tolerance to salinity and alkalinity and, for some areas, tolerance to high temperature.

Upland Rice.—A productive upland type is characterized by good seedling vigor, moderate tillering ability, semidwarf to intermediate plant height, semidroopy leaves, tolerance to and recovery from drought, resistance to blast and other diseases, tolerance to problem soils, and varying growth duration to fit a particular situation.

Cold-affected Areas.—Plants with intermediate height with good panicle exsertion and good seedling vigor and resistance to panicle sterility are most suitable. In addition to cold tolerance at appropriate stages of growth, resistance to blast is essential (Khush *et al.* 1977).

Multiple Resistance to Diseases and Insects.—It is known now that if only a few resistant varieties are planted intensively over wide areas, new biotypes of insects that could destroy varieties previously considered resistant may undergo high natural selection. Therefore, it is essential to study closely the races, strains and biotypes of rice plants and identify new genes for resistance. By incorporating these genes into the elite breeding lines, it would be possible to diversify rice of genetic differences.

New Technology.—If production in irrigated areas needs to be increased further and new cultivars are introduced into rainfed areas, it is important to address our attention to the development of new technology that would encourage more efficient use of agricultural inputs.

Fertilization.—One area of recent concern is increasing fertilizer efficiency. Energy based commodities such as fertilizers and pesticides were badly hit by the unrealistic rise in prices in 1974. Urea, for example, was sold at U. S. $350 per MT—but now the price has fallen to U. S. $125 per MT. While the former price is unrealistically high because of the sudden drop in input demand, the current price may be unrealistically low. The era of cheap fertilizers and pesticides is probably at an end. Studies of farm yield constraints in 4 major rice growing areas in the Philippines indicate that farmers in the study area would be able to increase their grain yield further if higher rates of fertilizers and insecticides were to be used. This and other studies indicate that the millions of small-scale rice farmers who have reaped the benefits of the new rice technology often do not have the capital to use high rates of agricultural chemicals.

Agronomists have demonstrated that deep placement of nitrogen fertilizer increased efficiency, sometimes saving 40% of fertilizer nitrogen with no reduction in grain yield. Similarly, entomologists have demonstrated that systemic insecticides applied once to the root zone at 2 kg/ha as capsules can control insects as well as 8 kg/ha conventionally applied in 4 doses.

The greatest challenge for the future is to develop a machine which will place fertilizer and insecticide in the root zone where losses are minimum. This increase in fertilizer and insecticide efficiency would reduce the dependence of the rice crop on petroleum-based agricultural chemicals. That would greatly help the densely populated and land scarce nations of Asia where more than ½ of the earth's people live and most depend on rice as their staple food.

Mechanization.—The modern technology is referred to as seed fertilizer technology. However, throughout Asia, the concept of modernization is closely associated not only with biological and chemical technology but also with mechanical technology. The biological and chemical technology is thought of primarily as land saving and the mechanical technology as labor saving.

To raise production at the farm level in rainfed and irrigated areas, some degrees of mechanization are essential. Some of these mechanization processes are already taking place. For example, in the area of land preparation under the puddled system, mechanization of some form is spreading rapidly in the Philippines and some other South and Southeast Asian countries. What is needed badly is to mechanize land preparation operations to establish a dry seeded crop in rainfed areas.

Weed control is another operation that needs to be partially mechanized, more so in the rainfed areas under dry seeding. Threshing is another operation where mechanization would be helpful. At present, losses in threshing are high in most farms in tropical Asia.

Despite some mechanization in land preparation and threshing, there is a substantial gain in the use of hired labor, partially substituting family labor. That is because mechanization is complementary to existing methods of land preparation, weeding and threshing. Further relief of labor use in farm operation is substituted for another farm operation. As a result, there is a net gain of hired labor utilization in South and Southeast Asia (Cordova and Barker 1977; IRRI 1975C).

It appears, then, that the emphasis on mechanization in tropical Asia would help increase rice production. In the future, if the availability of labor decreases, then mechanization would help minimize labor costs. In other words, the experiences in East Asia would be repeated in tropical Asia if alternative employment of labor is generated.

With the shorter-duration rice varieties, a distinct possibility is growing 2 crops where 1 grew before or 3 crops where 2 grew before. We not only have to reexamine the issue of land preparation but also relate it to seeding method, pest and disease control and harvesting practices in order to maximize the potential of the new rice technology. In short, the changes witnessed to date in rice growing in tropical Asia are only a beginning. The future portends great prospects for increased rice production through improved technology and national efforts to minimize production constraints.

In the developed countries of the world, a relatively advanced degree of rice production technology already exists. Annual rice yields are reasonably stable each year, primarily because the production resources are available and used at near optimum levels. Extreme climatic variables do not usually occur, complete irrigation control is possible and disease and insect pests cause fewer problems. Annual rice production in these developed countries tends to be influenced more by general national agricultural policies than production constraints. Rice production and trade policies are closely linked together, usually with production support programs which guarantee to domestic producers prices somewhat above normal world price levels. These policies have multiple objectives in order to achieve certain broad social and economic gains to insure a desired level of income and stability to farms and equitable pricing policies for consumers. Production policy is sometimes further complicated by the use of rice surpluses in world markets on concessional terms, creating irregularities in normal commercial trade channels.

Rice yields per hectare in developed countries are not likely to make large short-term gains in the foreseeable future. With a high level of technology already in existence, future yield increases will be small and largely geared to improved rice varieties, more careful farm management and the skillful combination of the various production resources.

Among the major objectives for improved rice production in the developed nations would be:

(1) development of high yielding cultivars with short stature and lodging resistance, earlier maturity, resistance to major insect and disease pests, better adaptability to climatic conditions, and grain qualities preferred by various consumer groups; and

(2) development of improved cultural practices, including seeding techniques, insect, disease and weed control, improved water management and desirable crop rotations which effectively deal with residues from the rice crop.

In a situation similar to that in existence in developing countries, in the future, we can look to greater prospects of rice production and increased grain yields per hectare through improved cultivars, technology and greater national efforts to control the environment for rice production.

PROJECTION OF RICE PRODUCTION IN THE COMING DECADE

The worldwide rate of food production has recently been increased along with population. The green revolution of the late 1960s raised the hope against massive hunger. In Asia, where rice is mostly produced and consumed, the green revolution is no less dramatic. Several developing countries now feed their own people.

Still what appears overwhelming is the population increase. The world population is about 4 billion now and heading for 6 billion by the end of the century. The basic question still exists: can the world population feed itself? As early as 1966, based on some extrapolation, Lester R. Brown, formerly with the USDA, published a paper projecting grain production and trade in the year 2000. Brown's projections were that even with tripled production of grains in the developing countries by the year 2000, export from the developed countries to less developed ones would need to be quadrupled to meet the rising demand for food grains (Wortman 1976).

Cereals dominate the consumption patterns in most of the countries but noncereals are particularly important in Sri Lanka, Malaysia and Taiwan. Rice contributes an appreciable fraction of total calories in all countries, and makes up an especially large portion of total calories in the mainland Southeast Asian countries of Bangladesh, Burma, Thailand, Laos, Cambodia and Vietnam (Herdt et al. 1977).

There have been a number of attempts to determine whether future food availability in Asia will be *adequate* to meet future demands.

The projections of the International Food Policy Research Institute (IFPRI 1976), the Asian Development Bank (1977) and the World Bank (Hadler 1976) differ somewhat in absolute levels because they differ in the basis of calculation (Herdt et al. 1977).

Despite these differences, all 3 studies indicated that a gap between 25 (World Bank) and 40 (IFPRI) MMT of food grains will exist between production and demand in the Asian region by 1985 (Fig. 4.13). Although the projection for production and consumption is for all cereals in Asia, rice being the principal cereal, the trend should hold true for rice.

What is the prospect of narrowing the gap between the projected consumption at suitable level of sufficiency and nutrition and the production? Wortman (1976) summarized the problem by bringing our attention to the following changes that are occurring in most developing countries:

(1) The nature and magnitude of the problem has become known in recent years;
(2) Tailoring of technology for each locality and developing it make the probability of success greater;

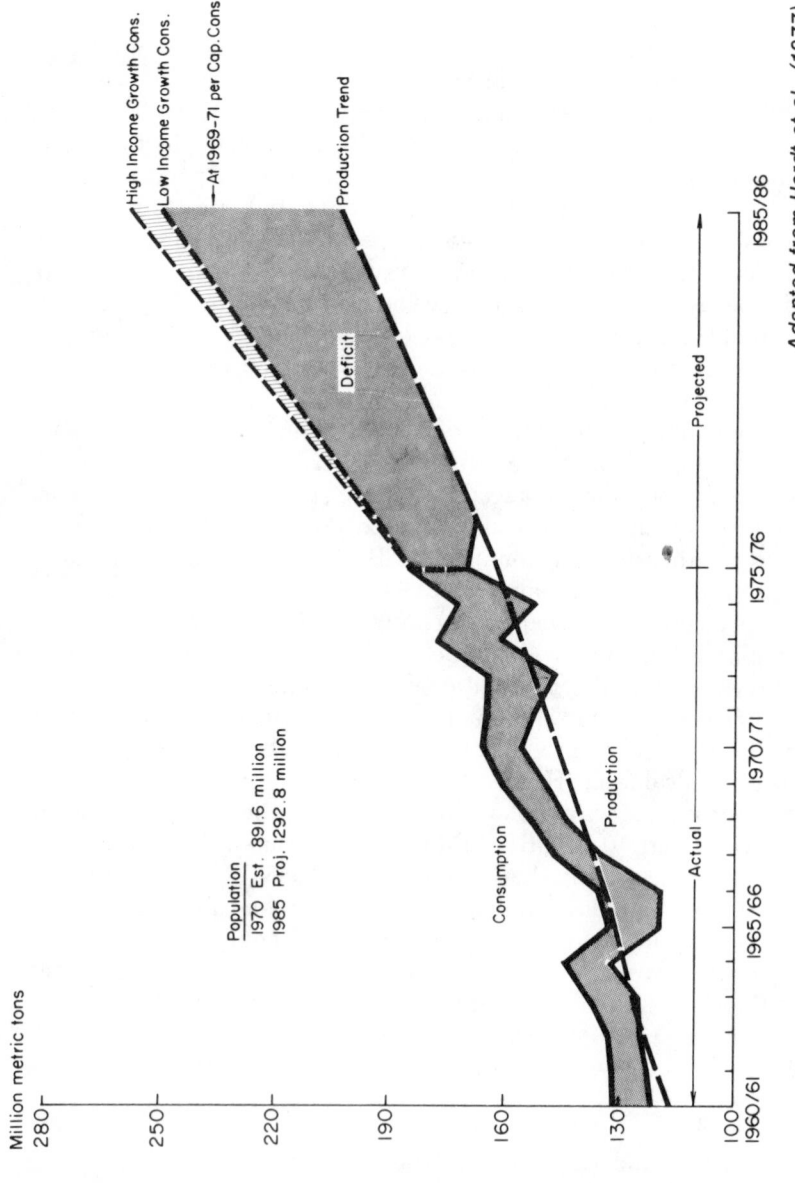

FIG. 4.13. ASIAN CEREALS: PRODUCTION AND CONSUMPTION 1960—1975 AND PROJECTED 1985 (IFPRI)

Adapted from Herdt et al. (1977)

(3) The potential for raising yields is great;

(4) For the first time chemical fertilizers in sufficient quantities for most food crops are now available;

(5) Most governments are taking effective steps for the farmers to adopt new technology;

(6) Many financial institutions to help finance national food programs are now present;

(7) Several institutes and organizations have emerged capable of and willing to provide technical and managerial development of national programs;

(8) Some governments of developing countries are paying greater attention to rural development; and

(9) There still remains considerable arable but uncultivated land that can be brought into production except perhaps in Europe and parts of Asia (Wortman 1976). In a recent report it was also pointed out that in densely populated South and Southeast Asia, millions of hectares of land physiographically and climatologically suited for rice production lie uncultivated largely because of soil problems such as salinity, alkalinity, strong acidity, or excess organic matter (IRRI 1977). In these areas, suitable technology may bring about some significant change in rice production.

One has to recognize the tremendous costs needed to bring these large areas under cultivation or improve the yield level of the existing rice area. Developing cultivars capable of producing some yields under these harsh conditions may help millions of subsistence farmers. Technology development should sharply focus on the problems which have reasonable chance of success at a cost which developing countries can afford.

Herdt and Barker (1977) point out that in the decade ahead, more research will be needed to provide answers to critical questions regarding the further expansion of modern technology and development of water control. Without improvement in efficiency with which irrigation and modern inputs are developed and used, output growth will certainly slow down, thereby the cost of food grains will rise.

REFERENCES

ADAIR, C.R., BEACHELL, H.M. and JODON, N.E. 1942. Comparative yields of transplanted and direct sown rice. Agron. J. *34*, 129-137.

ARYEETEY, A.N. 1973. Chemical weed control in rice in Ghana. Ghana J. Agric. Sci. *6*, 199-204.

ASIAN DEV. BANK. 1977. Second Asian Agricultural Survey. Asian Dev. Bank, Manila, Philippines. (preliminary)

BARKER, R., KAUFFMAN, H.E. and HERDT, R.W. 1975. Production constraints and priorities for research. Int. Rice Res. Inst. Agric. Econ. Dep. Ppr. 75-8. (unpublished mimeo)

BARTHOLOMEW, W.V. 1972. Soil nitrogen supply and space processes and crop requirements. N. Car. State Univ. Tech. Bull. 6..

BHARAT, S. 1953. Padi planting in Vannua Sevu. Colony of Fiji Agric. J. 24, 3-4.

BROWN, F.B. 1969. Upland rice in Latin America. Int. Rice Comm. Newsl. 18, 1-5.

BUDDENHAGEN, I.W. and TER VRUGT, J. 1977. Rice ecosystem research and evaluation in tropical agriculture—IITA Approach. In Proc. "Rice in Africa" Conf. Int. Inst. Trop. Agric. March 7—11. Ibadan, Nigeria, Academic Press, London.

BUEHRER, T.F. and ROSE, M.S. 1943. Studies on soil structure. V. Bound water in normal and puddled soils. Univ. Ariz. Agric. Exp. Sta. Bull. 110.

CASTILLO, P.S. 1962. A comparative study of directly seeded and transplanted crops of rice. Philipp. Agric. 45 (10) 560-566.

CHANDRARATNA, M.F., FERNANDO, L.H. and WEERARATNA, H. 1962. Fertilizer responses of rice in Ceylon. I. Effect of method and time of nitrogen application. Empire J. Exp. Agric. 30 (117) 16-26.

CHANG, T.T. 1976. The origin, evolution, cultivation, dissemination and diversification of Asian and African rices. Euphytica 25, 425-441.

CHANG, T.T. and VERGARA, B.S. 1972. Ecological and genetic information on adaptability and yielding ability in tropical rice varieties. In Rice Breed. Int. Rice Res. Inst., Los Baños, Philippines.

CHAPMAN, A.L. and MIKKELSEN, D.S. 1963. Effect of dissolved oxygen supply on seedling establishment of water-sown rice. Crop Sci. 3, 392-397.

CHOW, L. 1959. Rotational irrigation for rice—a revolution in Taiwan. In Rice Improvement in Taiwan. Joint Comm. on Rural Reconstr., Taipei, Taiwan.

COCHEME, J. 1971. Notes on the ecology of rice in West Africa. In Notes on the Ecology of Rice and Soil Suitability for Rice Cultivation in West Africa. U.N. Dev. Prog., Food Agric. Organ. U.N., Rome. (French)

CORDOVA, V.G. and BARKER, R. 1977. The effect of modern technology on labor utilization in rice production. Int. Rice Res. Inst. Sat. Sem. May 28. (unpublished mimeo)

DE DATTA, S.K. 1972. Chemical weed control in tropical rice in Asia. Pest. Arctic. News Summ. (PANS) 18 (4) 433-440.

DE DATTA, S.K. 1975. Upland rice around the world. In Major Research in Upland Rice. Int. Rice Res. Inst., Los Baños, Laguna, Philippines.

DE DATTA, S.K. 1977A. Weed control in rice in Southeast Asia: methods and trends. Crop Weed Sci. Bull. 4, 39-65.

DE DATTA, S.K. 1977. Some relevant research on weed control and soil and crop management in rainfed rice at IRRI and other locations in tropical Asia. *In* Proc. "Rice in Africa" Conf. March 7–11. Int. Inst. Trop. Agri., Ibadan, Nigeria. Academic Press, London.

DE DATTA, S.K., ABILAY, W.P. and KALWAR, G.N. 1973B. Water stress effects in flooded tropical rice. *In* Water Management in Philippine Irrigation Systems: Research and Operations. Int. Rice Res. Inst., Los Baños, Philippines.

DE DATTA, S.K. and BARKER, R. 1977. Economic evaluation of modern weed control techniques in rice. *In* Integrated Control of Weeds. J.D. Fryer and S. Matsunaka (Editors). Univ. of Tokyo Press, Tokyo.

DE DATTA, S.K. and BERNASOR, P.C. 1971. Selectivity of some new herbicides for direct-seeded flooded rice in the tropics. Weed Res. *11* (1) 41-46.

DE DATTA, S.K. and BERNASOR, P.C. 1973. Chemical weed control in broadcast-seeded flooded tropical rice. Weed Res. *13,* 351-354.

DE DATTA, S.K., CHANG, T.T. and VICENCIO, P.G. 1976. Prospects for overcoming yield-limiting factors of upland rice. *In* Proc. from 2nd Sem. on Variety Improv. of the West Africa Rice Dev. Assoc. (WARDA), Sept. 13–17. Monrovia, Liberia.

DE DATTA, S.K., FAYE, F.G. and MALLICK, R.N. 1975. Soil-water relations in upland rice. *In* Soil Management in Tropical America. *In* Proc. of Sem., Feb. 10–14. CIAT, Cali, Colombia.

DE DATTA, S.K. and FEUER, R. 1975. Soils on which upland rice is grown. *In* Major Research in Upland Rice. IRRI, Los Baños, Laguna, Philippines.

DE DATTA, S.K. and GOMEZ, K.A. 1974. Changes in soil fertility under intensive rice cropping with improved varieties. Soil Sci. *120* (5) 361-366.

DE DATTA, S.K. and KERIM, M.S.A.A. 1974. Water and nitrogen economy of rainfed rice as affected by soil puddling. Soil Sci. Soc. Amer. Proc. *38* (3) 515-518.

DE DATTA, S.K., KRUPP, H.K., ALVAREZ, E.I., and MODGAL, S.C. 1973A. Water management practices in flooded tropical rice. *In* Water Management in Philippine Irrigation Systems: Research and Operations. IRRI, Los Baños, Philippines.

DE DATTA, S.K. and MAGNAYE, C.P. 1969. A survey of forms and sources of fertilizer nitrogen for flooded rice. Soils Fert. Abstr. *32* (2) 103-109.

DE DATTA, S.K., MAGNAYE, C.P. and MAGBANUA, J.T. 1969. Response of rice varieties to time of nitrogen application in the tropics. Proc. from Symp. on Trop. Agric. Res., Sept. Tokyo

DE DATTA, S.K., MAGNAYE, C.P. and MOOMAW, J.C. 1968A. Efficiency of fertilizer nitrogen (^{15}N-labelled) for flooded rice. 9th Int. Congr. Soil Sci., August. Adelaide, Australia.

DE DATTA, S.K. and O'TOOLE, J.C. 1976. Screening deep water rices for drought tolerance. *In* Proc. of Deep Water Rice Workshop. Int. Rice Res. Inst., Rice Division, Nov. 8—10. Bangkok, Thailand.

DE DATTA, S.K., PARK, J.K. and HAWES, J.E. 1968B. Granular herbicides for controlling grasses and other weeds in transplanted rice. Int. Rice Comm. News. *17*, (4) 21—29.

DE DATTA, S.K. and ROSS, V.E. 1975. Cultural practices for upland rice. *In* Major Research in Upland Rice. IRRI, Los Baños, Laguna, Philippines.

DE DATTA, S.K., SALADAGA, F.A., OBCEMEA, W.N. and YOSHIDA, T. 1974. Increasing efficiency of fertilizer nitrogen in flooded tropical rice. *In* Proc. Fertilizer Assoc. of India—FAO Sem. on Optimizing Agric. Prod. Under Limited Avail. of Fert. New Delhi.

DE DATTA, S.K. and VERGARA, B.S. 1975. Climates of upland rice regions. *In* Major Research in Upland Rice. IRRI, Los Baños, Laguna, Philippines.

DE DATTA, S.K. and WILLIAMS, A. 1968. Rice cultural practices. B. Effects of water management practices on the growth characteristics and grain yield of rice. *In* Proc. and ppr. from 4th Sem. on Econ. and Soc. Stud. (Rice Production), Comm. for Coordination of Investigations of the Lower Mekong Basin, Oct. 7—11. Los Baños, Laguna, Philippines.

DE DATTA, S.K. and ZARATE, P.M. 1970. Environmental conditions affecting the growth characteristics, nitrogen response and grain yield of tropical rice. Bimeterol. *4*, 71-89.

DE JESUS, J.C. 1959. Comparative performance of transplanted seedlings, drilled seeds, transplanted ratoon tillers and ratooned crops in Milbuen 5(3). B.S.A. thesis. University of the Philippines College of Agriculture, Laguna, Philippines.

DEWAN, M.L. and DONGALE, J.H. 1974. Composting of Urban and Rural Wastes. *In* Proc. from FAI-FAO Sem. on Optimizing Agric. Prod. Under Limited Avail. of Fert. Fert. Assoc. India, New Delhi.

ENGLESTAD, O.P. 1967. Nitrogen Sources for Rice in Flooded Soils. Tenn. Val. Auth., Muscle Shoals, Alabama.

ENGELSTAD, O.P., GETSINGER, J.G. and STANGEL, P.J. 1972. Tailoring of fertilizers for rice. Tenn. Valley Authority Final Rept. for Agency Int. Dev.

ENGELSTAD, O.P., JUGSUJINDA, A. and DE DATTA, S.K. 1974. Response of flooded rice to phosphate rocks varying in citrate solubility. Soil Sci. Soc. Amer. *38* (3) 524-529.

FAO INVENTORY MISSION (FOOD AND AGRIC. ORGAN. U.N.). 1970. Development of rice cultivation in West Africa. Preliminary report of the Food and Agric. Organ. U.N. Inventory Mission, July. *In* Conf. of Plenipotentiaries for the establishment of a West Africa Rice Dev. Assoc., Sept. 1—4 Dakar, Senegal. (mimeo)

FAO. 1973. Research on rice. Food and Agric. Organ. U.N. Proj. Working Ppr. *1*, Bangkok, Thailand.

FUKUDA, H. and TSUTSUI, H. 1968. Rice Irrigation in Japan. Food and Agric. Organ. U.N., Rome.

GOSWAMI, N.N. 1975. Understanding and utilizing problem soils—phosphorus-deficient rice soils of India. Ppr. presented at Int. Rice Res. Conf., April. Los Baños, Philippines. (unpublished mimeo)

GREEN, D.H. and EBNER, L. 1972. A new selective herbicide for rice, S(2-methyl-1-piperidylcarbonylmethyl)-0,0-di-N-propyl dithio-phosphate, for use alone or in mixtures. *In* Proc. 11th Br. Weed Control Conf., Brighton, U.K.

HADLER, S. 1976. Developing county foodgrain projections for 1985. Int. Bank for Reconstr. and Dev., Bank Staff Working Ppr. *247,* Washington, D.C.

HARWOOD, R.R. 1975. Farmer-oriented research aimed at crop intensification. *In* Proc. Cropping Systems Workshop, March 18–20. Int. Rice Res. Inst., Los Baños, Philippines.

HASANUZZAMAN, S.M. 1974. Cultivation of deep-water rice in Bangladesh. *In* Proc. Int. Sem. on Deep-Water Rice. Bangladesh Rice Res. Inst., Joydevpur, Dacca, Bangladesh.

HERDT, R.W. and BARKER, R. 1977. The status of potential of food production in Asia. *In* Proc. Int. Rice Res. Inst.–U.N. Univ. Workshop on "Interfaces Between Agriculture, Nutrition and Food Sciences," Feb. 28–May 3. IRRI, Los Baños, Philippines.

HERDT, R.W., TE, A. and BARKER, R. 1977. The Prospects for Asian Rice Production. Ppr. presented at Int. Rice Res. Conf. April 18, Int. Rice Res. Inst., Los Baños, Philippines. (unpublished mimeo)

HSIEH, C.F., KAO, S. and CHIANG, C. 1966. Effects of plant densities, amounts of fertilizer application, and planting forms on the yield of rice. J. Taiwan Agric. Res. *15* (4).

INT. INST. TROP. AGRIC. (IITA). 1972. Farming systems program. Annual report for 1972. Ibadan, Nigeria.

IITA. 1973. Farming systems program. Annual report for 1973. Ibadan, Nigeria.

IRRI (INT. RICE RES. INST.). (Undated). Annual report for 1964. IRRI Los Baños, Philippines.

IRRI. 1966. Annual report for 1965. IRRI, Los Baños, Philippines.

IRRI 1967A. Annual report for 1966. IRRI, Los Baños, Philippines.

IRRI. 1967B. Annual report for 1967. IRRI, Los Baños, Philippines.

IRRI. 1969. Annual report for 1968. IRRI, Los Baños, Philippines.

IRRI. 1970. Annual report for 1969. IRRI, Los Baños, Philippines.

IRRI. 1971. Annual report for 1970. IRRI, Los Baños, Philippines.

IRRI. 1972. Annual report for 1971. IRRI, Los Baños, Philippines.

IRRI. 1973A. Annual report for 1972. IRRI, Los Baños, Philippines.

IRRI. 1973B. Research highlights for 1973. IRRI, Los Baños, Philippines.

IRRI. 1974. Annual report for 1973. IRRI, Los Baños, Philippines.

IRRI. 1975A. Annual report for 1974. IRRI, Los Baños, Philippines.

IRRI. 1975B. Major Research in Upland Rice. Los Baños, Philippines.

IRRI. 1975C. Changes in Rice Farming in Selected Areas of Asia. IRRI, Los Baños, Philippines.

IRRI. 1977. International Rice Research Newsl. 2 (2). IRRI, Los Baños, Philippines.

IRRI. 1978. Rice research and Production in China: An IRRI team's view. IRRI, Los Baños, Philippines.

INT. FOOD POLICY RES. INST. (IFPRI). 1976. Meeting food needs in the developing world: The location and magnitude of the task in the next decade. IFPRI Rep. 1, Washington, D.C.

JAMISON, V.C. 1953. Changes in air-water relationships due to structural improvement of soils. Soil Sci. 76, 143-151.

JANA, R.K. and DE DATTA, S.K. 1971. Effects of solar energy and soil moisture tension on the nitrogen response of upland rice. In Proc. Int. Symp. on Soil Fert. Eval. J. S. Kanwar, N. P. Datta, S. S. Baius, D. R. Bhumbla, T.D. Biswas (Editors). Indian Soil Sci. Soc., Ind. Agr. Res. Inst., New Delhi.

JAYASEKERA, E.H.W. 1966. Varieties and cultural practices. In Proc. Symp. on Res. and Prod. of Rice in Ceylon. Colombo Apoth., Ceylon.

JENKINS, J.M. and JONES, J.W. 1944. Results of experiments with rice in Louisiana. Louisiana Exp. Sta. Bull. 384.

JOSHI, A.B. 1975. Utilization of organic materials. In Proc. FAI-FAO Seminar on Optimizing Agricultural Production Under Limited Availability of Fertilizers. Fert. Assoc. India, New Delhi.

KAWANO, K. and TANAKA, A. 1968. Studies on the interrelationship among plant characters. I. Effects of varietal difference and environmental conditions on the correlations between characters. Jp. J. Breed. 18 (2) 75-79.

KHUSH, G.S. et al. 1977. GEU 102, Agronomic characteristics. Int. Rice Res. Inst. Ann. Prog. Rev. Handout. (unpublished mimeo)

KOENIGS, F.F.R. 1961. The mechanical stability of clay soils as influenced by the moisture conditions and some factors. Wageningen. Verst. Lanb. Orinderz., Holland.

MAHAPATRA, I.C., BAPAT, S.R. and SINGH, M. 1974. Economics of fertilizer use. Fert. Market. News 5 (12) 1-17.

MANGUIAT, K.J. and YOSHIDA, T. 1973. Nitrogen transformation of ammonium sulfate and alanine in submerged Maahas clay. Soil Sci. Pl. Nutr. 19 (2) 95-102.

MATSUNAKA, S. 1975. Weed control and herbicides in rice culture. In Rice in Asia. Univ. of Tokyo Press, Tokyo.

MATSUSHIMA, S. 1964. Nitrogen requirements at different stages of growth. In The Mineral Nutrition of the Rice Plants. Johns Hopkins Univ. Press, Baltimore, Maryland.

MATSUSHIMA, S., TANAKA, I. and HOSHINO, T. 1966. Analysis of yield determining process and its application to yield prediction and cultural improvement of lowland rice. LXXV. Temperature effects on tillering in case of leaves and culm, culm bases and roots independently treated. Proc. Crop Sci. Soc. Jpn. *34*, 478-483.

MATSUSHIMA, S. *et al.* 1968. Analysis of yield determining process and its application to yield prediction and cultural improvement of lowland rice. LXXXI. Combined effects of air temperature, water temperature, shading and the amount of fertilizers in seedling period on the characteristics of seedlings of rice plants. Proc. Crop Sci. Soc. Jpn. *37*, 169-174.

MIKKELSEN, D.S. and FINFROCK, D.C. 1957. Availability of ammoniacal nitrogen to lowland rice as influenced by fertilizer placement. Agron. J. *49*, 296-300.

MIKKELSEN, D.S. and KUO, S. 1976. Zinc fertilization and behavior in flooded soils. *In* The Fertility of Paddy Soils and Fertilizer Applications for Rice. Asian-Pacific Food and Fertilizer Tech. Cntr., Taipei, Taiwan.

MIKKELSEN, D.S. and MILLER, M.D. 1963. Nitrogen fertilization of rice. calif. Agric. *17*, 9-11.

MIKKELSEN, D.S. and PATRICK, W.H., JR. 1968. Fertilizer Use on Rice. *In* Changing Patterns in Fertilizer Use. Soil Sci. Soc. Amer., Madison, Wisc.

MOCHIZUKI, M. 1975. Field rat problems in Southeast Asia. *In* Rice in Asia. Univ. of Tokyo Press, Tokyo.

MOODY, K. 1975. Weed control systems for upland rice production. *In* Rept. Expert Consultation Meet. on the Mechanization of Rice Production, Int. Inst. Trop. Agric. June 10–14, 1974. Ibadan, Nigeria.

MOOLANI, M.K. and SOOD, A.P.R. 1967. Response of rice to various moisture regimes. Il Riso *16*, 85-90.

MOOMAW, J.C., BALDAZO, P.G. and LUCAS, L. 1967. Effects of ripening period environment on yields of tropical rice. Int. Rice Comm. Newsl. Spec. Is., 18-25.

MOOMAW, J.C. and DE DATTA, S.K. 1968. The response of high yielding varieties to improved practices. Ppr. *8* from 12th Ses. of Int. Rice Comm. Working Party on Rice Production and Protection. Kandy, Ceylon. (unpublished mimeo)

MOOMAW, J.C., NOVERO, V.P. and TAURO, A.C. 1966. Rice weed control in tropical monsoon climates: problems and prospects. Int. Rice Comm. News. *15* (11) 1-18.

MOORMAN, F.R., CURFS, H.P.F. and BALLAUX, J.C. 1975. Classification of conditions under which rice is grown in the tropics with special reference to mechanization in West Africa. *In* Rept. Expert Consultative Meet. on the Mechanization of Rice Production, June 10–14, 1976. Int. Inst. Trop. Agric., Ibadan Nigeria.

MOORMAN, F.R. and DUDAL, R. 1965. Characteristics of soils on which paddy is grown in relation to their capability classification. Soil Surv. Rep. *32.* Land Dev. Dep., Bangkok.

MUKHERJEE, D.K. 1974. Problems of deep-water rice cultivation in West Bengal and possibilities for evolving better varieties. *In* Proc. Int. Sem. on Deep-Water Rice, August 1974. Bangladesh Rice Res. Inst., Joydevput, Dacca, Bangladesh.

MUKHERJEE, H.N. 1962. Problems of soils of the paddy fields. Int. Soc. Soil Sci. Trans. Comm. Ch. 4–5. New Zealand.

NUTTONSON, M.Y. 1951. Ecological crop geography and field practices of Japan. Japan's natural vegetation and agro-climate analogues in North America. Amer. Inst. of Crop Ecol., Washington, D.C.

OBERMUELLER, A.J. and MIKKELSEN, D.S. 1974. Effects of water management and soil aggregation on the growth and nutrient uptake of rice. Agron. J. *65*, 627-632.

OKAFOR, L.I. and DE DATTA, S.K. 1976A. Competition between upland rice and purple nutsedge for nitrogen, moisture, and light. Weed Sci. *24* (1) 43-46.

OKAFOR, L.I. and DE DATTA, S.K. 1976B. Chemical control of perennial nutsedge *(Cyperus rotundus* L.) in tropical upland rice. Weed Res. *16*, 1-5.

ORTICIO, M.R. and PONNAMPERUMA, F.N. 1977. Zinc deficiency: a widespread nutritional disorder of rice in the Philippines. Ppr. presented at 8th Ann. Sci. Meeting of the Crop Sci. Soc. of the Philipp., May. La Trinidad, Benguet. (unpublished mimeo)

OU, S.H., LING, K.C., KAUFFMAN, H.E. and KHUSH, G.S. 1975. Diseases of upland rice and their control through varietal resistance. *In* Major Research in Upland Rice. IRRI, Los Baños, Philippines.

PALACPAC, A.C.. 1977. World Rice Statistics. Int. Rice Res. Inst., Dep. of Agric. Econ., Los Baños, Philippines.

PARAO, F. 1970. Performances of low and high tillering improved varieties. Int. Rice Res. Inst. Sat. Sem. Ppr. (unpublished mimeo).

PATRICK, W.H. and MIKKELSEN, D.S. 1971. Plant nutrient behavior in flooded soils. *In* Fertilizer Technology and Use. Soil Sci. Soc. Amer. Inc., Madison, Wisc.

PILLAI, K.G. 1973. Recent results of herbicide trials on rice in India. Ppr. presented at Int. Rice Res. Conf., April 1978. Int. Rice Res. Inst., Los Baños, Philippines. (unpublished mimeo)

PONNAMPERUMA, F.N. 1972. The chemistry of submerged soils. Advan. Agron. *24*, 29-96. Academic Press, New York

PONNAMPERUMA, F.N. 1975. Growth limiting factors of aerobic soils. *In* Major Research in Upland Rice. IRRI, Los Baños, Laguna, Philippines.

PRECHACHART, C. and JACKSON, B.R. 1974. Floating and deep-water rice varieties in Thailand. *In* Proc. Int. Sem. on Deep-Water Rice, Aug. Bangladesh Rice Res. Inst., Joydevpur, Dacca, Bangladesh.

RAMIAH, K. and HANOMONTHA, K. 1936. Broadcasting versus transplanting. Trop. Agric. *26* (5) 310-313.

SAKAI, K. 1949. Cytohistological and thermmatological studies on sterility of rice in northern part of Japan, with special reference to abnormal hypertrophy of tapetal cells to low temperature. Rep. Hokkaido Agric. Exp. Sta. *42*, 1-46.

SANCHEZ, P.A. 1972. Agricultural techniques to optimize the potential production of new varieties of rice in Latin America. *In* ppr. presented at Seminar on Rice Policies in Latin America. Oct. 10—14, 1971. Int. Cntr. Trop. Agric. (CIAT), Cali, Colombia.

SANCHEZ, P.A. 1973.A Puddling topical rice soils: 1. Grwoth and nutritional aspects. Soil Sci. *115, (2)*, 149—158.

SANCHEZ, P.A. 1973B. Puddling tropical soils: Effects of water losses. Soil Sci. *115* (4) 303-308.

SASAKI, T. 1977. On the relation of temperature to the longitudinal growth of leaves of rice plant. Proc. Crop Sci. Soc. Jpn. *1,* 23-42.

SENEWIRATNE, S.T. and MIKKELSEN, D.S. 1961. Physiological factors limiting growth and yield of *Oryza sativa* under unflooded conditions. Plant Soil *14* (2) 127-146.

SHAFI, M. and AHMAD, B. 1967. Effect of spacing and number of seedlings per hill on yield of paddy. Rice Res. Sta., Kala Shah Kaku, Pakistan. (unpublished mimeo)

SHAPIRO, R.E. 1958. Effect of flooding on availability of phosphorus and nitrogen. Soil Sci. *85,* 190-197.

SHASTRY, S.V.S., FREEMAN, W.H. and PILLAI, K.G. 1974. New rice varieties need rational management for optimum yields. Yojana *18* (2) 33-36.

SMITH, R.J., FLINCHUM, W.T. and SEAMAN, D.E. 1977. Weed control in U.S. rice production. Agric. Handbk. *497,* Washington, D.C.

SRINIVASAN, V., KALAYANIKUTTY, T. and NARAYANAN, T.R. 1968. Optimum spacing and nitrogen dose for high yielding rice varieties in Madras State. Indian Farming *18* (6) 12-17.

STANGEL, P.J. 1970. Modern chemical fertilizers, their potential and method of application. Asia Food and Fert. Tech. Cntr. Extens. Bull. *2.*

STANSEL, J.W. 1975. Effective utilization of sunlight. *In* Six decades of rice research in Texas. Res. Mono. No. *4.* Texas Agric. Exp. Sta.

STOUT, B.A. 1966. Equipment for Rice Production. FAO Agric. Dev. Ppr. *84.* Food and Agric. Organ. U.N., Rome.

TANAKA, A. 1976. Comparison of rice growth in different environments. *In* Climate and Rice. Int. Rice Res. Inst., Los Baños, Philippines.

TANAKA, A., KAWANO, K. and YAMAGUCHI, J. 1966. Photosynthesis respiration and plant types of the tropical rice plant. Int. Rice Res. Inst. Tech. Bull. *7.*

TANAKA, A., PATNAIK, S. and ABICHANDANI, C.T. 1959A. Studies on the nutrition of rice plant. III. Partial efficiency of nitrogen absorbed by rice plant at different stages of growth in relation to yield of rice *(Oryza sativa* var. indica). Proc. Indian Acad. Sci. Sec. B *49* (4) 207-216.

TANAKA, A., PATNAIK, S. and ABICHANDANI, C.T. 1959B. Studies on the nutrition of rice plant. IV. Growth and nitrogen uptake of rice varieties *(O. sativa* var. indica) of different durations. Proc. Indian Acad. Sci. Sec. B. *49* (4) 217-226.

USDA (U.S. DEP. AGRIC.). 1976. Economic Research Service. Twenty six years of world cereal statistics by County and Region. USDA, Washington, D.C.

VAN EIMERN, J. 1964. Windbreak and Shelterbelts. Tech. Note *59* WMO (145) Tech. Ppr *70*, Geneva, Switzerland.

VEGA, M.R., ONA, J.D. and PALLER, JR., E.C. 1967. Evaluation of herbicides for weed control in upland rice. *In* Proc. 1st Asian-Pacific Weed Control Interchange. Honolulu.

VERGARA, B.S. 1977. Deep water rice. II. Problems of deep-water rice culture. *In* Genetic Evaluation and Utilization (GEU) Training Program, March. Int. Rice Res. Inst., Los Baños, Philippines.

WORTMAN, S. 1976. Food and agriculture. Sci. Amer. *235* (3) 31-39.

XUAN, V. and KANTER, D.G. 1974. Deep-water rice in Vietnam: Current practices and prospects for improvement. *In* Proc. Int. Sem. on Deep-water Rice. Bangladesh Rice Res. Inst., Joydevpur, Dacca, Bangladesh.

YOSHIDA, S., McLEAN, G.W., SHAFI, M. and MUELLER, K.E. 1970. Effects of different methods of zinc application on growth and yields of rice in a calcareous soil, West Pakistan. Soil Sci. and Pl. Nutr. *16* (4) 147-149.

YOSHIDA, T. and PADRE, JR., B.C. 1974. Nitrification and dentrification in submerged Maahas clay soil. Soil Sci. Pl. Nutr. *20* (3) 241-247.

ZANDSTRA, H.G. and PRICE, E.C. 1977. Research topics critical for the intensification of rice-based cropping systems *In* Background paper for research programming meeting on cropping systems research at Int. Rice Res. Inst. May 3. Los Baños, Philippines. (unpublished mimeo)

5

Rice Plant Diseases

Shu-Huang Ou

Rice diseases are major biological constraints in rice production. More than 60 diseases are well described and numerous species of fungi are reported to be associated with rice. They are caused by various pathogens such as viruses, bacteria, fungi, nematodes and others (Ou 1972). New diseases continue to appear. In 1977, a new and potentially destructive rice virus disease was identified in the Philippines and possibly in neighboring countries. New maladies of unknown origin were reported in Indonesia, Korea and Taiwan. Bacterial blight was discovered recently in Latin America where it was not known before (Ou 1978; Lozano 1977).

Some diseases are cosmopolitan while others are confined to certain regions. A summary of the general distribution and importance of the more common rice diseases of the rice growing world (Table 5.1) shows that more diseases occur in Asia where rice has been grown for centuries. Several major diseases may occur together in the same fields.

Crop losses from diseases can be severe. The brown spot disease was considered a major factor contributing to the Great Bengal Famine in India in 1942 (Padmanabhan 1973). In the 1950s, hoja blanca threatened rice production in Latin America. The tungro virus spread to about 33% of the rice land in Thailand in the 1960s and infected tens of thousands of hectares in the Philippines in 1971 and in Indonesia, India and other countries in recent years. Grassy stunt devastated rice production in several areas in the Philippines, Indonesia and India during the early 1970s. Bacterial blight caused major crop losses in India and other tropic countries of Asia in the late 1960s. Blast is the most widely distributed disease in rice growing areas. It caused severe losses in upland rice areas in Latin America, Africa and many parts of Asia. In lowland (flooded) rice in areas of Asia, a 5–10% incidence of neck-blast is common. It usually escapes detection and aggregated losses are considerable because of its wide occurrence.

235

TABLE 5.1. THE RELATIVE PREVALENCE AND IMPORTANCE OF RICE DISEASES IN MAJOR GROWING REGIONS

	Asia		America		Africa	
	Tem-perate	Trop-ical	North	Latin	West	East
Virus Diseases						
Dwarf	++3[1]	−0	−0	−0	−0	−0
Stripe	++3	−0	−0	−0	−0	−0
Black-streak dwarf	1	−0	−0	−0	−0	−0
Yellow dwarf	+++2	+++1	−0	−0	−0	−0
Tungro	−0	+++3	−0	−0	−0	−0
Grassy stunt	−0	+++2	−0	−0	−0	−0
Transitory yellowing	−0	+2	−0	−0	−0	−0
Hoja blanca	−0	−0	±0	+++3	−0	−0
Yellow mottle	−0	−0	−0	−0	+1	+1
Bacterial diseases						
Leaf blight	+++3	+++3	−0	+1	+1	+2
Leaf streak	−0	+++2	−0	−0	−0	−0
Fungus diseases						
Blast	+++3	+++3	++2	+++3	+++3	++2
Sheath blight	+++3	+++3	++2	++1	+1	
Brown spot	+++2	+++2	++2	++2	+1	+1
Cercospora leaf spot	+++2	+++2	+++2	++1	+1	
Stem rot	+++2	+++2	++2	+1	++1	
Bakanae disease	+++2	++1	+1	+1	+1	
Sheath rot	++2	++2	+1	+1		
Leaf scald	+1	++1	−0	++1	++1	+1
Seedling dumping off	++1	+1	+2			
False smut	+1	+++1	+1	+++1	++1	++1
Kernel smut	+1	++1	++1	++1		
Nematode diseases						
White tip	++2	+++2	++2	+++2	++2	++2
Stem nematode	−0	++2	−0	−0	−0	−0
Root nematode	++1	++1	++1	+1	++1	+1

[1]Not present
+Occasionally present
++Commonly present
+++Abundantly present

0: Absent or rarely found
1: Of minor importance
2: Of moderate importance
3: Of major importance

Two major strategies for controlling rice diseases currently employed are: (1) chemical protection and (2) breeding for resistant rice cultivars. Chemicals have been used extensively in Japan for the last 20 years and are commonly used in other countries. The use of chemicals in Japan is encouraged by high yields per hectare and government policy of a high subsidy for rice prices. The subsidized price is usually more than double that of the world market prices. Other countries use chemicals because there are no practical alternative methods.

Chemicals are costly, particularly in those countries where pesticides must be imported. Pollution to soil and water from continuous use of pesticides is a serious environmental problem. Japan has adopted strict regulations on the use of agricultural chemicals. Accumulation of arsenical compounds in soil from the use of organic arsine compounds for sheath blight control has created problems in some countries.

Many pathogens have developed resistant strains to chemicals, particularly the blast fungus, for which most known chemicals have been used (Sakuri and Naito 1976).

In developing countries, the lack of a forecasting system and strong extension services, unavailability of chemicals and application equipment in rural areas, and a lack of ready cash by farmers have made the use of chemicals difficult. High input technology, such as chemical control of rice diseases, is difficult for small farmers in developing countries to adopt.

A more suitable strategy of controlling plant diseases is the use of resistant cultivars. This is particularly appropriate for the small rice farmers in developing nations. Since about 1962, the development of disease resistant rice cultivars has received greater attention in both national and international food production research programs. For some diseases, the development of resistant cultivars is relatively simple, for example some of the virus diseases. For other diseases, it is a more complex situation. In general, when the pathogen is variable and has many strains or physiological races, the breeding procedures are complicated; such is also the case with wheat rusts and potato late blight.

In tropical Asia where several diseases occur together, rice cultivars must have resistance to major diseases and to insect pests. The effort required is geometrically increased each time a disease or insect resistance is to be incorporated into new cultivars. Nevertheless, many new rice cultivars and breeding lines have multiple resistance to diseases (Ou 1977A). Genes for disease resistance appear to be independent; no close linkage of genes resistant to one disease but susceptible to another have been detected as yet.

In regions where disease problems are simple, breeding for resistance may focus on only 1 or 2 major diseases. In Latin America or Africa for instance, the major problem at present is blast. In all circumstances, developing resistant cultivar remains the basic approach for controlling rice diseases. Epidemiological studies also suggest means for developing cultural methods to help control certain diseases.

Asia produces more than 90% of the world's rice. Thus, rice diseases receive much attention in tropical Asia. Some of the major diseases are discussed in detail in the following.

VIRAL DISEASES

About 15 diseases of rice caused by virus or mycoplasm have been identified (IRRI 1969; Ling 1972; Ou 1972; Chekiang Acad. Agric. Sci. 1975). A new virus disease appeared in the Philippines in 1977 and possible others have been reported in Africa. Some have been known since the early 1900s. The dwarf virus was the first plant virus demonstrated to be transmitted by an insect. It was the first case of transovarial transmission of virus—the virus passed from one generation of the vector to the next through eggs. Tungro, grassy stunt, orange leaf and transitory yellowing viruses were identified in the early 1960s. Mentek disease, now believed to be similar to tungro, has been known in Indonesia since 1859. The yellow mottle of Kenya and Waika of Japan have been reported only in recent years.

Among this group of diseases, the yellow dwarf and grassy stunt are believed to be caused by myoplasm-like organisms and the others by viruses. A giallume disease reported from Italy is also thought to be caused by mycoplasma.

Distribution and Vectors

The distribution of rice viruses appears to be regional, probably because of the regional distribution of the respective insect vectors. All known viruses causing rice disease are transmitted by leafhoppers and planthoppers, except yellow mottle (by beetles), mosaic (mechanical) and necrosis mosaic (soil). The genus *Nephotettix,* the rice green leafhopper, is the most important insect group in transmitting rice viruses. Insect vectors of known rice virus diseases and the worldwide distribution of viruses are summarized in Table 5.2.

Transmission, Vector-virus Relationship

Usually only a portion of a field population of vectors is able to transmit the virus. The active transmitters vary from 5–20% in hoja blanca, grassy stunt, orange leaf and others to 50–80% in black-streaked dwarf, tungro, yellow dwarf, etc. Different vector species carrying the same virus have a different percentage of active transmitters; for instance, in the case of tungro virus, *Recilia dorsalis* has 4–8% active transmitters, while *N. virescens* has 80–85%.

While a virus can be acquired in a few minutes to half an hour by a small percentage of a vector population, generally seven hours to a day are needed for most of the active transmitters to acquire the virus.

After acquiring the virus, the incubation periods in the vectors vary. In a few cases, the vectors can transmit the virus to new plants immediately (tungro). Many require 1 or 2 weeks, while others require 1 month or longer to be able to transmit the virus (dwarf, yellow dwarf).

Most viruses persist in the vectors for a long period and some for life (persistent) but a few vectors lose the ability to transmit the virus in a few days (nonpersistent), such as tungro. Dwarf, stripe and hoja blanca

TABLE 5.2. RICE VIRUS DISEASES: THEIR DISTRIBUTIONS AND INSECT VECTORS

Disease	Distribution	Insect vector-transmission
Dwarf	China mainland, Japan, Korea	*Nephotettix apicalis, N. cincticeps, Recilia dorsalis*
Stripe	China mainland, Japan, Korea	*Laodelphax striatellus, Riboutodelphax albifascia, Unkanodes saporonus*
Black-streak dwarf	China mainland, Japan, Korea	*L. striatellus, R. albifascia, U. saporonus*
Necrosis mosaic	Japan	soil
Waika	Japan	*N. cincticeps, N. virescens*
Transitory yellowing	Taiwan	*N. apicalis, N. cincticeps, N. virescens*
Yellow dwarf	China mainland, Japan, Korea, Taiwan, Philippines, Thailand, Malaysia, Indonesia, India, Pakistan, Sri Lanka	*N. apicalis, N. cincticeps, N. virescens*
Tungro	Philippines, Thailand, Malaysia, Indonesia, India, Pakistan	*N. virescens, N. apicalis, R. dorsalis*
Grassy stunt	Philippines, Thailand, Malaysia, Indonesia, India, Sri Lanka	*Nilaparvata lugens*
Orange leaf	Philippines, Thailand, Malaysia, Indonesia, India, Sri Lanka	*R. dorsalis*
Mosaic	Philippines	mechanical
Hoja blanca	Latin American countries	*Sogatodes cubanus, S. orizicola*
Yellow mottle	Kenya and West Africa	*Sesselia pusilla*
Giallume	Italy	

Source: Ou (1972).

viruses are the transovarial type. Once the virus has been acquired, the insects remain infective for many generations through the females. In the persistent type, the virus has to be acquired once in the life of the insect. In the nonpersistent type, the insect has to acquire the virus repeatedly to remain infective continuously. In the transovarial type of disease, virus continuity may be accomplished by the insect alone, whereas in the persistent and nonpersistent types, diseased plants are required to complete the cycle under natural conditions. These vector-virus relationships have important bearings on the development of virus diseases in the field. Disease epidemics are greatly dependent upon climatic conditions, cultural practices and other factors. However, the nonpersistent type of virus diseases are more prone to fluctuations from year to year.

The incubation period in plants, from infection to the appearance of symptoms, varies from a few days to about 2 weeks for most viruses. The incubation time of the yellow dwarf virus is 1 month or more, depending upon the ambient temperature.

Symptoms

Generally virus infections are systemic throughout the plants. The most common symptom of rice viruses is stunting of the rice plant. Also plant leaves are often discolored and sometimes spotted. When plants are infected early in their growth cycle, no panicles are formed. However, when the plant is infected late in the growth period, panicles may be formed but maturity is delayed and the grain quality is poor.

Ling (1972) devised a key by which to identify different rice virus diseases primarily by their symptoms. The following key describes the major disease symptoms:

Key for Identifying Rice Virus Diseases by Symptoms

A_1 Plants showing inconspicuous stunting, but reduced tillering
  B_1 Upright growth habit, premature death, orange colored and rolled leaves *Orange leaf*
  B_2 Spreading growth habit, oval to oblong faint chlorotic patches or fine faint mottling on leaves, brown necrotic lesions on basal parts of culms at later stages *Necrosis mosaic*

A_2 Plants showing stunting and reduced tillering
  C_1 Leaves with chlorotic spots and white stripes
    D_1 New leaves not unfolding properly but twisted and droopy *Stripe*
    D_2 New leaves unfolding normally *Hoja blanca*
  C_2 Leaves with mottling and yellowish streaks, crinkling of the first newly formed leaves when infected at an early stage of growth *Yellow mottle*
  C_3 Leaves with yellow or yellow-orange discoloration
    E_1 Virus particles are bullet-shaped and persist in the vector *Transitory yellowing*
    E_2 Virus particles are spherical or polyhedral and do not persist in the vector *Tungro*

A_3 Plants showing severe stunting and excessive tillering
  F_1 Galls on leaves and culms *Black-streaked dwarf*
  F_2 No galls

G$_1$ Leaves with chlorotic to whitish specks forming interrupted streaks *Dwarf*

G$_2$ Narrow, stiff, light-green leaves often with rusty spots *Grassy stunt*

G$_3$ Leaves showing general chlorosis *Yellow dwarf*

Host Range

Among the better known viruses, dwarf, stripe, black-streaked dwarf and yellow dwarf have a wide host range (Ling 1972; Ou 1972) in the Graminaeae family including barley, oats, wheat and maize. Several species of the grass family showed symptoms when inoculated with tungro, but the virus was difficult to recover from inoculated plants. For other viruses, the host range is limited or unknown. Rice is also a host of several viruses of some cereals and grasses when artificially inoculated.

Damage

Stripe and dwarf viruses continue to cause considerable damage in Japan and Korea, particularly in the southern parts of these nations.

Damage from hoja blanca in Latin America at present is limited because host resistance to the virus has been found and used. However, the disease remains a potential threat. Yellow dwarf is widespread and the vectors are commonly present. Because of its long incubation period both in the vector and in the rice plant, damage is generally moderate, although in some areas and in some years, disease damage is appreciable.

In tropical Asia, damage caused by tungro and grassy stunt virus is marked. In Thailand, tungro was first noticed in 1964. By 1966, 660,000 ha were infected and 50% of these were severely diseased. It disappeared gradually and by 1970, there was little tungro in Thailand. In the Philippines, there was a severe outbreak in Central Luzon in 1971. Tungro has infected tens and hundreds of thousands of hectares in Indonesia, India, Malaysia and other countries during the last decade. Earlier, a disease known as "mentek" (believed to be tungro) in Indonesia destroyed 30,000 to 50,000 ha of rice from 1934 to 1936. The "accep na pula" or "stunt" disease (tungro) in the Philippines caused a 30% crop loss equivalent to 1.4 million MT of rice annually in the early 1940s.

Grassy stunt was identified in 1964 when there were only scattered infected plants in the Philippines, Indonesia and nearby countries. With an increase in the brown planthopper (the vector) population in recent years, the disease has become more prevalent. In 1973, severe outbreaks occurred in several provinces in South Luzon and Mindanao of the Philippines, in Java of Indonesia and in Kerela in South India.

Control

Efforts to control virus disease by chemical control of the insect vectors in Japan and in other rice growing countries show that generally when the vector population is low, reasonably good disease control is achieved. However, when vector population is high, the efficiency of chemicals is poor. Usually, the vectors transmit the virus before they are killed by the pesticide. In areas where an insecticide has been repeatedly applied, the insects (green leafhopper and brown planthopper) develop resistance to the chemicals (Asakawa and Kazano 1976). Instead of suppressing the insect, it fostered insect development because the pesticides reduce or remove the natural enemies of the vector.

Rice cultivars resistant to viruses have been developed for several virus diseases. In Japan, Chugoku 31 and St 1 have been developed for resistance to stripe virus. Tongil and other rice cultivars are highly resistant to stripe in Korea. Many old cultivars from the Orient are resistant to hoja blanca. New cultivars developed in recent years in Latin America are resistant, for example, the CICA series of cultivars developed in Colombia.

Tens of thousands of rice cultivars have been tested for resistance to tungro and grassy stunt at IRRI and elsewhere. Many new cultivars and breeding lines developed at IRRI and elsewhere are resistant to tungro. Many resistant sources exist so that breeders may use them for developing tungro resistant cultivars. Only a strain of a wild rice, *Oryza nivara*, is resistant to the grassy stunt virus among many thousands of rice cultivars tested at IRRI. The resistance gene was quickly transferred to many new cultivars. Most of the newly developed rice cultivars are resistant to grassy stunt. Resistance to the viruses seems to be stable; no sign of breakdown of resistant cultivars has yet been recognized.

Yellow dwarf resistance has been tested extensively in Taiwan. Many rice cultivars are known to be resistant to most of the viruses. However, breeding for resistance to other viruses is needed.

Resistance to insect vectors may also be used in plant breeding. However, resistance to vectors alone is often not sufficient to prevent the spread of viruses in epidemic areas. Many rices are resistant to both the virus and the vector.

BACTERIAL DISEASES

Several bacterial diseases of the rice plant are known: bacterial blight *(Xanthomonas oryzae)*, bacterial leaf streak *(X. translucens* f. sp. *oryzicola)*, bacterial stripe *(Pseudomonas setariae)*, bacterial sheath rot *(P. oryzicola)* and several other bacteria that attack rice grains *(X. itoana, P. glume, X. cinnamona, X. atroviridigenum)* (Ou 1972). Bacterial blight and bacterial leaf streak damage rice crops severely; other bacterial diseases do not appear to be economically important.

Distribution and Damage

Bacterial blight, first reported in Japan, is widely distributed in Asia, from Japan to Sri Lanka and from the Philippines to Pakistan. It was reported in Malagasy of East Africa and was observed in the Ivory Coast, Liberia and Niger of West Africa by C. C. Chien. Recently it was observed and confirmed to be present in several Latin American countries (Ou 1977B; Lozano 1977).

Bacterial leaf streak is confined to tropical Asia, including south China. Fang *et al.* (1957) in China were the first to distinguish this disease from bacterial blight. It was first reported from the Philippines by Reinking in 1918. Earlier workers had confused this disease with bacterial blight (Ou 1972.)

Bacterial blight is one of the most important rice diseases of Asia. Yield losses in severely infected fields range from 20 to 30%. In the tropics, when the "kresek" type of symptoms occur (for symptoms, see following section), losses to individual fields are heavy. In the late 1960s when some high yielding but susceptible rice cultivars (IR8, Taichung Native 1) were extensively cultivated, the yield losses ranged from 6 to 60% in some states of India.

Bacterial leaf streak can occasionally cause as much damage as bacterial blight when climatic conditions such as rain storms or typhoons in the tropics favor the spread of the disease. Infected plants often outgrow the disease when the weather becomes less favorable to the disease, because bacterial leaf streak is not a vascular disease as is bacterial blight.

Symptoms

Bacterial blight usually becomes conspicuous in the field during the later stages of rice growth, usually at booting or flowering. Typically, part or an entire leaf becomes straw yellow due to the death of the leaf tissue. The elongated lesions with wavy margins start from one or both edges of the leaf and gradually extend and kill the entire leaf (Fig. 5.1). When the lesions advance quickly, the leaf tissues wilt. Bacterial exudates may be observed in the morning as milky dew drops on young lesions. They spread by rainstorm to cause new infections. In severe cases, almost all leaves in a rice field are blighted.

In 1950, Reitsma and Schure described kresek disease in Indonesia, caused by a bacterium which was later called *X. kresek* Schure in 1953.

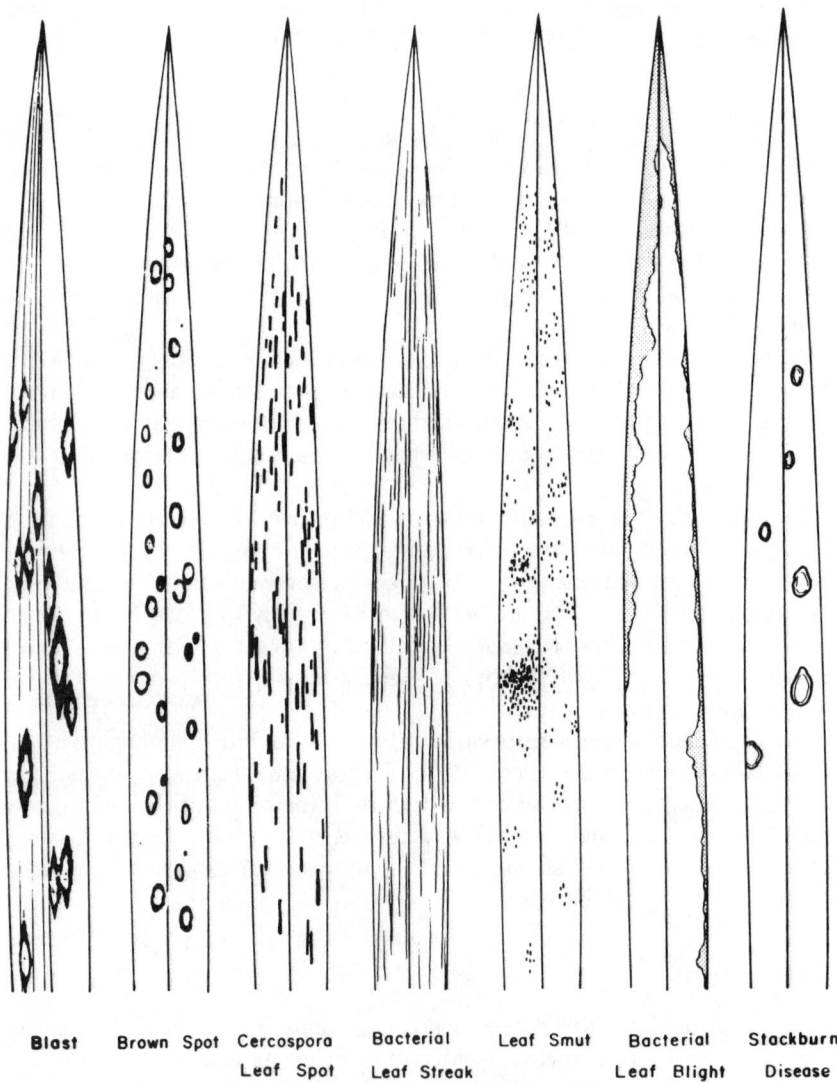

| Blast | Brown Spot | Cercospora Leaf Spot | Bacterial Leaf Streak | Leaf Smut | Bacterial Leaf Blight | Stackburn Disease |

From Ou (1972)

FIG. 5.1. SKETCHES SHOWING TYPICAL LESIONS OF SOME OF THE COMMON LEAF SPOT DISEASES OF RICE

Rice seedlings died 2 to 3 weeks after transplanting, and the dead leaves floating on the water surface caused the disease to be called kresek. Goto (1964) found it to be a severe form of bacterial blight and in fact it is common in tropical Asia. Goto also described a pale yellow system—a mature plant with 2 or 3 pale yellow young leaves.

Bacterial leaf streak begins with fine translucent streaks between the veins of leaves. The lesion enlarges lengthwise and may also advance laterally across the veins. Bacterial exudates appear on the surface of lesions under moist conditions and dry up to form numerous minute yellow beads. These exudates are spread by water and wind to initiate new infections on the same leaf or leaves of other plants. When numerous lesions occur, the leaves dry in a manner similar to that of bacterial blight. The dead leaves become bleached and appear gray white, because many saprophytic microorganisms have colonized. At this stage the 2 diseases cannot be distinguished.

Infection and Spread

The bacterial blight organism enters the plant through water pores or hydathodes on leaf edges of leaves, cracks from the growth of young roots and through wounds. It is not known to enter through stomata. From the hydathodes, the bacterium reaches terminals of vascular tissues. It progresses and invades the vascular system of the plant, up and down and from one leaf to another. Thus it causes partial wilting of leaves.

The kresek symptoms start with the infection either from the leaves or from roots broken when seedlings are pulled from the seedbed for transplanting. When seedlings are too tall, the leaf tips are also cut (trimmed) for transplanting in the tropics. When field water contains a sufficiently high bacterial population, kresek occurs.

In bacterial leaf streak, the pathogen enters the leaves through stomata and wounds but does not invade the vascular system.

In the tropics, rainstorms or typhoons are major climatic factors that cause the spread of both diseases in fields. After a storm, the diseases may spread to the entire field from a small cluster of infected plants. Irrigation water also carries the bacteria from field to field.

Although diseased seed and straw are known to be sources of primary infection for bacterial blight in temperature regions, no evidence of seed transmission of bacterial blight has been found in the tropics. The pathogen does not seem to live long at high temperatures in infected tissues nor in soil or water.

Bacterial leaf streak has been shown to be transmitted via seeds in the tropics. In Japan, the bacterial blight organism was found to colonize on nonhost weeds. Several grass weeds also serve as hosts of the bacterial pathogen. The source of the primary inoculum after a long dry season in extensive areas in some tropical countries has not yet been identified.

The isolates (strains) of both the bacterial blight and leaf streak organisms produced lesions of varying sizes (length on leaves) when inoculated on healthy rice plants. Many strains of the bacterial blight organism from the tropics produce longer lesions than Japanese strains on tropical rice cultivars. Until quite recently the variation in virulence was known to be unidimensional in that some bacterial strains are more virulent and some are less virulent to any rice cultivar. Recent findings at IRRI, however, show that the reactions of many rice cultivars are differential. A cultivar may be resistant to one strain of the pathogen but susceptible to another and vice-versa. It is similar to the physiologic races of many fungi. This information is important in understanding host-plant resistance and in breeding for host resistance. Detailed studies on the bacterial leaf streak organism have yet to be made.

Bacterial phages of both the bacterial blight and leaf streak pathogens are commonly found in diseased leaves, irrigation water and soil. Many lysotypes can be identified when many bacterial isolates and phages are used. The morphology and host range of the phages vary markedly. Mizukami and Wakimoto (1969) studied the phages of bacterial blight organism extensively. They used the phages to estimate the bacterial population to forecast development of the disease. Goto and Okabe and others, however, found that the phages change after passing a host strain. Phage-resistant strains have been found.

Control

The Bordeaux mixture, other copper compounds and mercury compounds were the first control methods used in early years. Many chemicals have been tried and continue to be tested in Japan for controlling bacterial blight. Among the numerous antibiotics tested are: penicillin, chloromphenicol, collocidin, aureomycin, blasticidin and others. Some of the synthetic compounds, such as dethianon, dimethyl-nickel, carbamate, fertiazon and phenazine that have also been tested are moderately effective. However, most are short lived and none have been recommended for general use.

Integrated measures of control have been suggested in Japan, including: (1) the use of more resistant cultivars, (2) avoidance of flooding in the nursery, (3) removal of primary inoculum sources, (4) spraying with chemicals, (5) avoidance of applying excess nitrogen fertilizer, and others. In the People's Republic of China cultural methods are stressed—disinfect contaminated seeds, control irrigation and drainage water to preclude passage of the water from one field to another, remove infected plant tissues, etc. Many of these methods are not practical under the farming conditions practiced in most tropical areas.

During the last decade, the development of rice cultivars resistant to bacterial blight has been emphasized. Tens of thousands of the rice germ plasm collection at IRRI have been screened for sources of resistance.

Several new resistant cultivars were developed in breeding programs. Recently, however, some resistant cultivars appear to be breaking down; virulent strains of the bacterium have developed or their populations have increased in fields. Cultivars resistant to different strains of the bacterium have been identified and breeding to incorporate different resistance genes is under way. Because as yet there are only a few strains of the bacterium known, breeding for a stable type of resistance to the disease is feasible.

Studies in recent years have provided a better understanding of the genetics of resistance. Some resistant genes are dominant, others are recessive; some control the resistance at seedling stage while others effect control at the adult stage of the plant. Some plant genes cause resistance for both stages. Three genes for resistance are known in Japan, and at least 4 or 5 are known in the tropics. Genetic studies will help greatly in programs of plant breeding for resistance.

Many rice cultivars are resistant to bacterial leaf streak. Special efforts for breeding for resistance have not yet been initiated.

FUNGAL DISEASES

Rice diseases caused by fungus are numerous (Ou 1972). A few are discussed in the following—brown spot, *Cercospora* leaf spot, stem rot, sheath blight and blast. These diseases are cosmopolitan. Some are more common in Asia where rice has been grown for centuries, and less common in Latin America and Africa. Blast and sheath blight are the more serious fungus rice diseases. The symptoms of common leaf spot diseases of rice are shown in Fig. 5.1.

Brown Spot

Brown spot, a disease well known for many years, appears as small oval brown spots distributed evenly over the leaf surface (Fig. 5.1). Fully developed spots have a gray or white center. Dark brown spots appear on the glumes and sometimes cover a large portion of the glumes. Under favorable conditions, the growth of the fungus on the glumes gives a velvety appearance and can occur also on panicle branches and other plant parts.

The disease is caused by the infection of *Helminthosporium oryzae*; the perfect stage, *Cochliobulus miyabeanus,* is seldom found. The disease causes: (1) blight of seedlings grown from heavily infected seeds, (2) weakening of the growth of seedlings and older plants, and (3) lowering of the grain quality and weight.

Brown spot was considered a major causal factor in the Bengal Famine in 1942. Recent studies in Japan on the physiological disease "akiochi" and the fact that the disease can rarely be found on rice plants growing in normal soils suggest that brown spot is associated with abnormal or poor soils. It is used as an index for abnormal and poor soils. Japanese studies found that it is almost impossible to distinguish between the crop losses caused by brown spot and those caused by abnormal soil factors. It appears that brown spot is basically a physiological disorder. Under the proper conditions, fungus growth causes additional damage.

The brown spot fungus penetrates directly into the host tissues from appressoria formed after the germination of spores. The conidial spores produced on the fully developed lesions are spread by air currents. The fungus also infects several other hosts of the grass family.

Chemicals have been used to treat rice seeds to prevent seedling blight and to protect rice plant from secondary infections. Many rice cultivars were found to be resistant to the disease. There have been no extensive studies on control. Correction of soil abnormalities should be the basic approach to the solution of the problem of brown spot.

Cercospora Leaf Spot

Cercospora leaf spot or narrow brown leaf spot is caused by *Cercospora oryzae;* the perfect stage, *Sphaerulina oryzina,* seldom occurs in nature. The disease generally produces short linear brown lesions on leaves; the lesions are about 2−10 mm long and 1 mm wide (Fig. 5.1). The lesions tend to be narrower, shorter and a darker brown on the more resistant cultivars and wider and a lighter brown on susceptible rices. The lesions may appear in large numbers during the later stage of growth. Recently the fungus was found to infect both young and old plants. The late appearance was attributed to a long incubation period in the plant (20−30 days) and a favorable microclimate in the fields.

In the 1930s and 1940s, considerable attention was paid to the disease in the United States because some commercial rice cultivars such as Blue Rose, commonly grown at that time, were very susceptible. When many lesions occur on the leaves, the effectiveness of the leaf surface for photosynthesis is reduced and subsequently, the yield is lowered. However, no precise data are available on this.

The fungus enters the leaves through stomata and becomes established and grows in the paranchyma cells. The conidiophores emerge from the stomata to produce conidia which causes a secondary infection. Several physiologic races of the fungus were found in the United States with specific resistance or susceptibility to certain rice cultivars.

Many rice cultivars are resistant to *Cercospora* leaf spot. Several resistant cultivars were developed in the United States in the 1950s—Sunbonnet, Bluebonnet, Toro and others. Some genetic information on the inheritance of resistance is also known in the United States. The disease has not been studied in any detail elsewhere in recent years, although it appears to be common.

Stem Rot

Two similar fungi cause stem rot; they are generally referred to as *Helminthosporium sigmoideum* (or *Sclerotium oryzae)* and *H. sigmoideum* var. *irregulare.* The perfect stage of *H. sigmoideum,* formerly called *Leptosphaeria salvinii,* is seldom found. The perfect stage was found recently in California and the name was changed to *Magnaporthe salvinii.*

The disease begins to appear in the field during later growth stages of rice. First, a small, blackish, irregular lesion develops on the outer leaf sheath near the water line. The lesion enlarges as the disease progresses and the fungus penetrates into the inner leaf sheath and the stem. One or 2 internodes of the stem rot collapse and only the epidermis remains intact. The next lower internode may, however, be free of disease. The plants lodge as the stems become weak.

Usually numerous small (⅓ mm) black sclerotia are formed inside the infected node and leaf sheath. The presence of the characteristic sclerotia is usually a positive way of diagnosing the presence of the disease. The sclerotia of *H. sigmoideum* are spherical and those of *H. sigmoideum* var. *irregulare* are somewhat irregular under the microscope. The 2 fungi often occur together. The fungi also grow saprophytically on rice stubbles and the sclerotia remain in soil to infect a new crop of rice.

Controversy exists about the extent of the damage caused by stem rot of rice. On one hand, heavy damage is reported from the United States and Asian countries; it varies from 10 to 80% yield losses on susceptible cultivars in some fields. On the other hand, reports from Italy indicate that the fungus is endemic and virtually ubiquitous but seldom causes appreciable crop damage. Recently, a similar observation was made in the Philippines, although earlier reports showed severe losses.

Work at IRRI indicates the importance of the presence of wounds on the rice plants in the development of the disease. Reports from Japan also showed that lodging and stem borer injuries accelerated crop damage due to stem rot. When artificially inoculating rice plants by scattering sclerotia on the water surface, some reported severe infections while others reported that the inoculations were not successful. These facts indicate that the severity of the disease may depend largely upon related factors, such as lodging, insect or other types of wounds on the plant rather than upon the presence of the fungi only.

Rice cultivars resistant to the disease have been found by many workers. Cultivar Raminad was bred for stem rot resistance in the Philippines in the 1930s. Not much work has been done on breeding for host resistance in recent years. Burning of rice stubbles after harvest was recommended in several countries many years ago. Fungicides have been tried in Japan but are not recommended for general use. The use of resistant, nonlodging rice cultivars is perhaps the most satisfactory solution to the problem at the present time.

Sheath Blight

The disease causes large, oval or oblong spots on leaf sheaths near the water line during the later stages of rice growth. It spreads to higher leaf sheaths and to leaf blades and reaches the flag leaves under favorable conditions on susceptible varieties. The spots have a gray-white center and a brown margin. Sclerotia are formed at or near the spots and are easily detached. The presence of several spots on a leaf sheath usually causes the death of the whole leaf and in severe cases, all the leaves may be blighted.

The development of the disease is rapid under conditions of high humidity, high temperature and heavy nitrogen application. All leaves of a plant may be blighted in 1 week after artificial inoculation. When the fungus reaches the upper leaves, it spreads to other leaves by contact; it is similar to the spread of web blight.

When the disease reaches the flag leaves, a loss of 20−25% in yield can be expected. A recent experiment at IRRI confirmed that on a susceptible rice cultivar heavily inoculated in high nitrogen plots, the yield loss was about 23%. On a moderately resistant cultivar in the same experiment, the highest loss was about 14%. The disease has become of great concern in Asia in recent years because of its wide occurrence. The use of high yielding rice cultivars which have large numbers of tillers, or increased plant stands per unit area, and the use of more fertilizers which promote plant growth have increased the humidity of the plant layer and consequently promoted a greater incidence of sheath blight.

The causal organism belongs to the *Rhizoctonia solani* group. The perfect stage is often produced on rice growing in upland conditions. It agrees with *Thanatephorus cucumeris* (Frank) Donk. Several other names were previously used: *Hypochnus sasakii, Corticium sasakii, Pellicularia filamentosa,* etc. The straight runner mycelium grows rapidly under favorable conditions, and produces much branched, short and clustered lobated mycelium which infects the rice plant. When cultured, the fungus from different localities has different color and type of mycelium and size and number of sclerotium. Young sclerotia sink in water but the older sclerotia float on paddy water and infect rice plants. Sclerotia sink into the soil or are mud decayed because of the action of other microorganisms.

The fungus has a wide host range, including cereal crops, such as maize, sorghum, sugarcane, etc., and grass weeds.

Varietal resistance tested in Japan, Taiwan and other countries revealed no resistant cultivars in Japan, but some moderately resistant cultivars were identified in Taiwan. A renewed effort was made recently at IRRI where many thousands of rice cultivars were screened. While many rices were found to be susceptible, many were classified as moderately resistant. A high level of resistance has not yet been found. An international testing program was started a few years ago. Plants seem to be more susceptible to sheath blight at the late stage of growth. Thus, early maturing cultivars appear more susceptible in a field at one time.

Fungicides have been extensively tested in Japan (Kozaka 1961), including 3 inorganic copper fungicides, 13 organic mercury compounds, 10 organic arsine compounds and several organic sulfur compounds. The organic arsines were most effective in inhibiting mycelial growth, infection and lesion enlargement, particularly methylarsine sulfide and urbacid (methylarsine bisdimethyl dithiocarbamate). The organic arsines are often phytotoxic but it may be reduced by adding iron to yield such compounds as ferric-methyl arsenate which provides good control without plant damage. Antibiotic preparations such as polyoxin and validimycin are also commercially used. Several new test compounds were also found effective at IRRI.

The possibility of farm practices which would hasten the decay of sclerotia in soil is being explored.

Blast

Blast is an ancient disease of rice. Similar maladies were recorded in ancient literature in China. Symptoms of blast on the rice plant include spots or lesions on leaves, nodes, the panicle base and branches, and on grains, but seldom on the leaf sheath. The leaf spots are typically elliptical with pointed ends (Fig. 5.1). The center of the spots is usually gray and the margin, brown or red brown. The shape, size and color of the spots vary, however, depending upon the age of the spot, the degree of susceptibility of the rice cultivar and the environmental conditions.

The spots usually begin as small, water-soaked, white, gray or blue dots. They enlarge quickly under moist conditions on susceptible cultivars and remain gray in color for some time. Fully developed spots can be 1−1.5 cm long, 0.3−0.5 cm wide, and have a brown margin. Under shaded conditions, the spots tend to have a yellow halo rather than the brown border. Generally, only minute brown specks occur on resistant cultivars.

The spots are round or short elliptic, a few millimeters long, and with a brown margin on cultivars moderately resistant to blast. The types of spots are usually an indication of the degree of the cultivar's resistance to blast. Leaves with many spots are killed; frequently seedlings and young plants are killed in fields because of blast infection.

When the nodes are infected, the sheath pulvillus rots, turns black, dries, often breaking apart and remaining connected by the nodal septum only. All parts above the infected node die.

Any part of the panicle may be infected, manifesting brown lesions. Areas near the panicle base often affected, causing "rotten-neck" or "neck-rot." When infection occurs late or only a portion of the panicle base is infected, the grains are partly filled.

The dominant symptoms of blast in a given area depend upon the climatic conditions at a specific stage of rice growth. When climatic conditions are favorable for the pathogen, leaf blast is conspicuous at an early stage of plant growth and shows rotten-neck at later stages. The disease may occur at both stages. For some areas, only node blast near the water surface may occur, such as "shara" disease in Iraq.

Blast is generally considered the principal disease of rice because of its wide distribution and destructiveness under favorable climatic conditions. Plants at the tillering stage are often completely killed. Heavy neck-rot kills all plants in many fields. Often, fields with an incidence of only 5—10% neck-rot go undetected. Many millions of dollars are spent annually on chemicals to control rice blast.

Blast development depends heavily on conditions in which rice grows. It is universally severe in upland rice. Climatic factors greatly influence blast development in lowland or flooded rice, and the dew period is probably the most important controlling factor in the tropics. It affects 3 vital processes of the disease cycle: conidial germination, infection and conidial release (El Refaei 1977). Heavy applications of nitrogen fertilizer favor disease development.

Blast is caused by *Pyricularia oryzae* of the fungi imperfecti group. The conidia germinate, form appressoria and penetrate the host cells directly or enter through stomata. The conidia germinate in 3—4 hr in free water and infection may take place in 6 hr. Prolonging the time the leaf is wet greatly increases the number of lesions produced. Only 4 days are needed between the time conidia are deposited on the leaves to the appearance of leaf lesions. A new crop of conidia can be produced in 5 to 6 days.

A typical lesion produces 4000 to 6000 conidia each night for about 2 weeks or more. The release of conidia from the conidiophore occurs at night and requires free water. The longer the dew period is, the greater the number of conidia released. Conidia are disseminated by air currents but most conidia travel short distances only—within 1 or 2 or a few meters. The short disease cycle, the great reproductive capacity and airborne spores make the disease potentially very destructive. In the tropics, the airborne conidia are present year round, but are present only 3 months in temperate regions (Fig. 5.2).

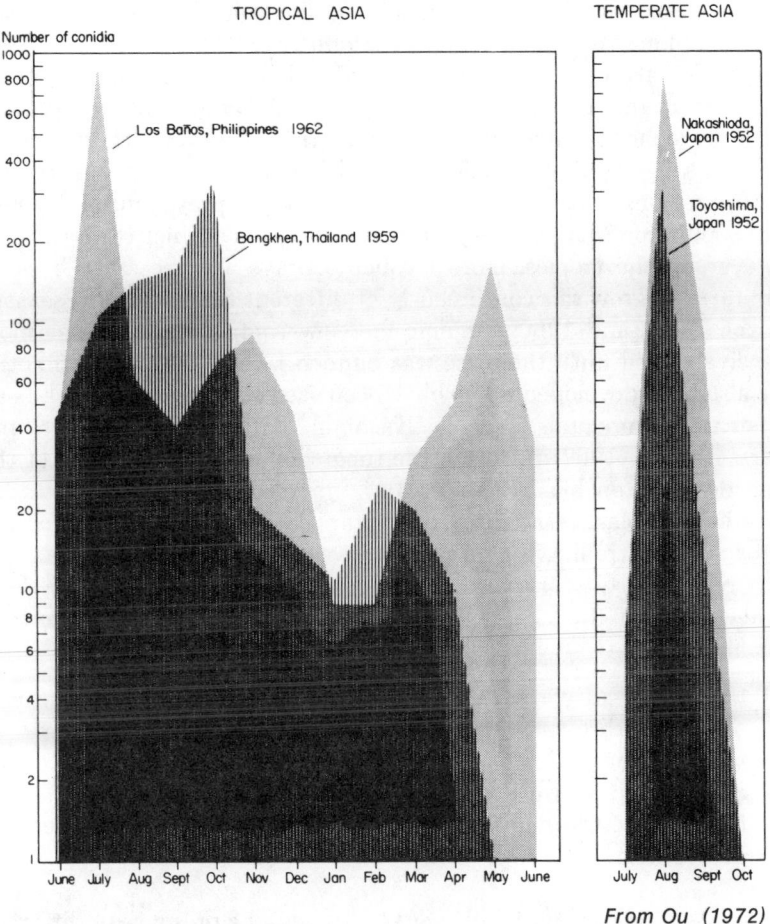

From Ou (1972)

FIG. 5.2. AIRBORNE CONIDIA POPULATION DURING THE YEAR IN TROPICAL AND TEMPERATE REGIONS OF ASIA

The blast fungus has marked pathogenic variability. It has many physiologic races and the single conidium can produce many new races. The races present in fields differ not only from country to country but also between locations within a few kilometers. The race compositions also differ from month to month in the same location (Ou and Ayad 1968;

Quamaruzzaman and Ou 1970; Ou *et al.* 1970; Ou 1972). This is one of the difficulties in screening for blast resistant progenies in rice breeding programs.

Many species of the *gramineae* family are hosts of the blast fungus, including some cereals. Similar fungi are also found on many members of *Gramineae, Zingiberaceae, Cannoceae, Musaceae* and *Cuperaceae* families. The taxonomy of these fungi is not completely understood particularly among the fungi found on *Gramineae* (Ou 1972).

Fungicides are used extensively to control blast in Japan and other temperate countries and recently in Latin America. Many tests and studies were made in Japan. The seasonal fluctuation of airborne spores (Fig. 5.2) suggests that chemical control is more advantageous in temperature regions than in the tropics. Two or three sprays during the peak spore population in Japan may protect the plants against the pathogen. However, in the tropics, more frequent sprays are necessary because small farmers grow rice continuously at different times during a season.

In the 1950s and 1960s, organic mercury compounds were the most extensively used until their use was banned recently. Many antibiotics have also been developed; some have been used extensively after the ban of mercury compounds, such as Kasumin, Kitazin, Blasticidin-S and others. Kozaka (1969) reviewed the fungicidal activities of some of the fungicides used on blast (Table 5.3).

Breeding for blast resistant rice cultivars seems to be the fundamental approach to controlling the disease and was initiated some 50 years ago. Many resistant cultivars were developed but they became susceptible to the new races of the fungus which subsequently developed. Recent efforts have sought:

(1) to identify cultivars with a broad spectrum of resistance to be used as donors through international cooperative tests,
(2) to study the variability of the pathogen to understand better the relationship between host and the pathogen, and
(3) to elucidate the genetics of resistance to develop feasible breeding procedures.

Because of the high variability of the fungus, the presence of different races in different locations, resistance donors cannot be identified in single localities by single institutions. A recently developed cultivar, Tongil, was highly resistant in Korea for several years, but was susceptible in the Philippines. In 1976, Tongil, was highly resistant in Korea for several years, but was susceptible in the Philippines. In 1976, Tongil began to breakdown in Korea.

An international blast nurseries program was started in 1963, and through the tests in 30 countries for the last 12 years, several rice

TABLE 5.3.   RELATIVE FUNGICIDAL ACTIVITY OF THE FUNGICIDES IN CURRENT USE IN JAPAN FOR THE CONTROL OF RICE BLAST

Commercial name	Protective activity against		Eradicative activity against			Systematic activity through	
	Penetration	Residual action	Development of lesion	Sporulation	Residual action	Foliage application	Root application
Blasticidin-S	++	++	++++	+++	++	—	—
Kasumin	+	±	+++++	++++	+++	+++	+++
Blastin	+++++	+++++	±	+?	+++	+?	—
Oryzon	++++	+++++	+	+?	+++	+?	—
Rabcon	++++	+++	+++	+++	+++	±	
Kitazin P	+++	+++	+++	+++	+++	±±	++++
Hinosan	+++	+++	+++	+++	++	±±	++
Inezin	+++	+				±	

Source: Kozaka (1969).

cultivars have been identified as having a broad spectrum of resistance. Cultivars such as Tetep, Carreon and others are generally resistant in all tests, only occasionally were a few spots observed. These cultivars are much better donors than those used earlier. Many studies have indicated that resistance in these cultivars are controlled by several or many genes. Kiyosawa (1972) found 13 genes in rice cultivars governing resistance to 7 isolates of the fungus. To transfer all or most of the resistance genes in such cultivars as Tetep, repeated or multiple crosses are necessary and several donors should be used in order to pyramid the resistance.

The modified multiline varieties concept is used to breed disease-resistant wheat cultivars at the International Center for Maize and Wheat Improvement in Mexico. It involves putting as many resistance genes in the field by using several or many genotypically different but phenotypically similar cultivars. This approach may also be tried with rice.

Most of the past work in breeding for blast resistance in rice employed only simple crosses. The whole complement of resistance present in the donors was not transferred to new cultivars. Another important procedure in breeding is multilocational tests. Because the pathogen has numerous races and not all the races are present at a single location (a site such as Taluca Valley in Mexico for potato late blight is not known for blast), the progenies of breeding materials are tested in many localities to select new lines with a broad spectrum of resistance, in a similar way the resistant donors are identified. International cooperative testing is indispensable.

The level of resistance to blast required may vary according to ecosystems. Upland rice requires a higher level of resistance to blast than does lowland rice. Rice in epidemic areas also requires high resistance.

NEMATODE DISEASES

Several nematodes cause diseases on rice. Two are rice plant nematodes, the white tip (*Aphelenchoides besseyi*) and stem nematode (*Ditylenchus augustus*). Others are soil inhabitants: (1) root nematode (*Hirschmanniella oryzae*), (2) root-knot nematode (*Meloidogyne* spp), (3) cyst nematode (*Heterodera oryzae*), (4) stunt nematode (*Tylenchorhynchus martini*), (5) other migratory parasitic nematodes. The white tip disease, first found in Japan, occurs in all major rice growing regions. The stem nematode is restricted to certain areas of the tropics. Other nematodes do not appear as major problems on rice (Ou 1972).

White Tip

As the name implies, the characteristic symptoms are chlorotic or white leaf tips up to several centimeters long; these become brown and tattered in appearance. Diseased plants are stunted, lack vigor and produce small panicles. The affected panicles show high sterility, distorted glumes and small distorted kernels. Leaf symptoms usually become evident after the maximum tillering stage. The upper leaves, particularly the flag leaves, of severely diseased plants are markedly twisted and are darker green in color than normal plants. Diseased plants sometimes give rice to tillers at the upper nodes. Not all infected plants show symptoms, but the yields are reduced.

The nematode is seed-borne and survives on seeds for 1 or 2 years. It is an ectoparasite, occurring within the young folded leaves in the early stages of plant growth. The nematode enters the spikelets after ear formation and most occur within the glumes when the grains are mature. Infections from the soil are rare. However, when healthy and diseased seeds were sown in adjacent plots, many seedlings from healthy seeds became infected indicating movement of nematodes in the soil.

The nematode is known to occur on a wide range of hosts among grasses and nongrasses.

Hot water treatment and chemicals have been tried to disinfect the seed in Japan and the United States. Several chemicals, including nicotine sulfate, organo-phosphorus insecticides, etc. and fumigation with methyl bromide are effective. However, most of the chemicals do not completely eliminate the nematode and sometimes damage the rice seed.

Tests for host resistance to nematodes have been conducted in Japan, Taiwan and the United States. Some rices are very susceptible to nematodes while others are resistant.

CONCLUSIONS

The world food problem is a great concern today. Rice feeds more than ½ the world population, and many rice eating people are hungry. Rice diseases are major constraints to rice production. Rice disease control is still in the developmental stages in the tropics. More efforts in research and extension are required to reduce the losses caused by rice diseases. Presently, there are only limited numbers of personnel engaged in solving disease problems. Much work still needs to be done.

REFERENCES[1]

ASAKAWA, M. and KAZANO, H. 1976. Resistance to carbamate insecticides in the green rice leafhopper, *Nephotettix cincticeps* Uhler. Rev. Pl. Prot. Res. *9*, 101-123.

---

[1]Note: Significant information has become available in several areas of rice diseases since the completion of this manuscript. Readers are requested to consult also current literature.

CHEKIANG ACAD. AGRIC. SCI., INST. PLANT PROT. 1975. Virus diseases of rice. Hsinhwa Book Co., Peking. (Chinese)

EL REFAEI, I. 1977. Epidemiology of rice blast disease in the tropics with special reference to the leaf wetness in relation to disease development. Ph.D. Dissertation. University of the Philippines at Los Baños.

FANG, C.T., REN, H.C., CHEN, T.Y., CHU, Y.K., FAAN, H.C. and WU, S.C. 1957. A comparison of the leaf blight organism with the bacterial leaf streak organism of rice and *Leersia hexandra* Swartz. Acta Phytopathol. Sia. *3*, 99-124. (Chinese with English summary)

GOTO, M. 1964. "Kresek" and pale yellow systemic symptoms of bacterial leaf blight of rice caused by *Xanthomonas oryzae* (Uyeda et Ishiyama) Dowson. Pl. Dis. Reptr. *48*, 858-861.

IRRI. (INT. RICE RES. INST.). 1965. The rice blast disease. Proc. Symp. at Int. Rice Res. Inst., July 1963. Johns Hopkins Univ. Press, Baltimore.

IRRI. 1969. The virus diseases of rice plant. Proc. Symp. at Int. Rice Res. Inst. April 1967. Johns Hopkins Univ. Press, Baltimore.

KIYOSAWA, S. 1972. Genetics of blast resistance. *In* Rice Breeding. Int. Rice Res. Inst., Los Baños, Philippines.

KOZAKA, T. 1961. Ecological studies on sheath blight of rice plant caused by *Pellicularia sasakii* (Shirai) S. Ito, and its chemical control. Chugoku Agr. Res. *20*, 1-133. (Japanese with English summary)

KOZAKA, T. 1969. Chemical control of rice in Japan. Rev. Pl. Prot. Res. *2*, 53-63.

LING, K.C. 1972. Rice virus diseases. Int. Rice Res. Inst., Los Baños, Laguna Philippines.

LOZANO, J.C. 1977. Identification of bacterial leaf blight in rice, caused by *Xanthomonas oryzae,* in America. Int. Rice Res. Inst. Newsl. *4* (77) 4.

MIZUKAMI, T. and WAKIMOTO, S. 1969. Epidemiology and control of bacterial leaf blight of rice. Ann. Rev. Phytopathol. *7*, 51-72.

OU, S.H. 1972. Rice Diseases. Commonw. Myc. Inst. Kew, Surrey, England.

OU, S.H. 1977A. Genetic defense of rice against disease. Ann. N.Y. Acad. Sci. *287*, 275-286.

OU, S.H. 1977B. Possible presence of bacterial blight in Latin America. Int. Rice Res. Inst. Newsl. *2*, 5-6.

OU, S.H. and AYAD, M.R. 1968. Pathogenic races of *Pyricularia oryzae* originating from single lesions and monoconidal cultures. Phytopathol *58*, 179-182.

OU, S.H., NUQUE, F.L. and EBRON, T.T. 1970. The international uniform blast nurseries, 1968−69 results. Int. Rice Comm. Newsl. *19* (4) 1-13.

QUAMARUZZAMAN, M. and OU, S.H. 1970. Monthly changes of the pathogenic races of *Pyricularia oryzae* (Cav.) in a blast nursery. Phythopathol. *60*, 1266-1269.

PADMANABHAN, S.Y. 1973. The great Bengal famine. Ann. Rev. Phytopathol. *11*, 11-26.

REITSMA, J. and SCHURE, P.S.J. 1950. "Kresek," a bacterial disease of rice. Contr. Gen. Agric. Stn. Bogor *117.*

SAKURI, H. and NAITO, H. 1976. A cross-resistance of *Pyricularia oryzae* Cavara to Kasugamycin and Blasticidin S. J. Antibiot. *29,* 1341-1342.

SCHURE, P.S.J. 1953. Attempts to control the Kresek disease of rice by chemical treatment of the seedlings. Contr. Gen. Agric. Res. Stn. Bogor *136,*

6

Insect Pests of the Rice Plant

Clarence C. Bowling

Insects are an important part of the ecosystem that man seeks to maintain for the production of rice, a major source of human food. The rice plant is vulnerable to attack by various kinds of insects from the time the seed is planted until the grain is harvested. The germinating seed, roots, seedlings, leaves, stems and developing grains may serve as a temporary or permanent host for some species of insect pests.

Activity of insect pests on the rice plant host produces various symptoms. Destroyed seeds or seedlings, damaged root system, reduced tillering, desiccation, loss of chlorophyll, reduced vigor, uneven or delayed maturity, physical removal of leaf surface or stem all can be indications of insect activity. These symptoms in turn may result in reduced quantity and/or quality of grain produced by the rice plants. In addition to direct damage, insects damage rice plants indirectly by transmitting plant diseases.

Detailed taxonomic and biological aspects of insect pests of the rice plant have been reported in the books "The Major Insect Pests of the Rice Plant" (Johns Hopkins Univ. Press), "Pest Control in Rice" (PANS Manual *3)* and in an article by M. D. Pathak (1968) in the Annual Review of Entomology (Anon. 1967, 1976).

Numerous recent articles consist of reports of research directed toward the control of insect pests of rice. Much of this valuable information is not included due to the limitation of space.

In this chapter, the author proposes to list the more important species of insects that are rice pests, to describe the symptoms of insect activity on the plant and to discuss the effects on yield and quality of rice produced. For convenience of discussion, insects have been grouped according to the part of the rice plant on which they feed. Some insects may damage more than one part of the plant. In these instances the insects are grouped according to the importance of the damage to the host plant.

SEED AND SEEDLING FEEDERS

Description of Damage

A variety of insects inhabit or enter the soil at planting time and feed on germinating rice seeds or seedlings. Aquatic larvae of Chironomids are pests on rice that are directly seeded in fields or nurseries (Borema 1967–68). The larvae of *Cricotopus sylvestris* (Fabr.) burrow into the germinating seed and consume all or part of the contents (Darby 1962). The larvae of Chironomids also feed on leaves and primary roots and sometimes uproot young seedlings (Abul-Nasr *et al.* 1970A,B; Clement *et al.* 1977).

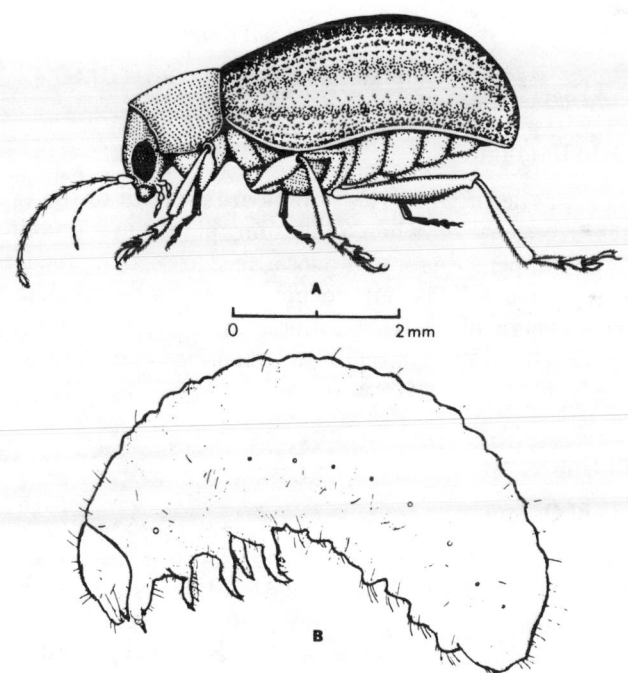

Larva Illustration Courtesy of L.H. Rolston; Adult Illustration by Mrs. H.R. Broad, Courtesy of D.S. Hill

FIG. 6.1.   GRAPE COLASPIS, *COLASPIS BRUNNEA* (FAB.)

(A) Adult;   (B) larva.

Overwintering larvae of grape colaspis, *Colaspis brunnea* (Fab.), feed on germinating seed and seedling. Damage occurs primarily when rice follows lespedeza. Damage is restricted to dry seeded rice and rice growing on levees in fields of water seeded rice (Rolston and Rouse 1960, 1965). White grubs, mole crickets and scarab beetles feed on germinating seeds or seedlings in upland or rainfed rice fields or in direct seeded fields before irrigation (Amaral 1949; Rincon 1967; Chatterjee 1973; Kennard 1965).

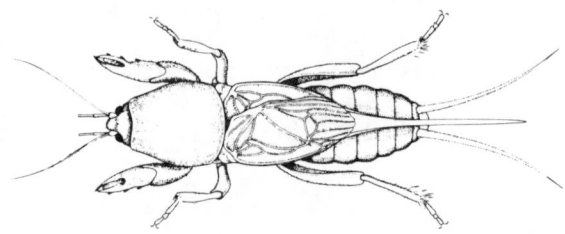

By Mrs. H.R. Broad, Courtesy of D.S. Hill

FIG. 6.2. AFRICAN MOLE CRICKET, *GRYLLOTALPA AFRICANA* BEAUVOIS

Losses Due to Damage

Economic losses occur when rice stands are reduced to the extent that reseeding is necessary or when remaining plants are unable to tiller sufficiently to compensate for stand loss. Seed or seedling feeders may be a problem in nursery beds but do not have the permanent effect of reducing yields as in direct seeded fields. A list of the more important insect pests that feed on rice seeds and seedlings and their distribution are shown in Table 6.1.

ROOT FEEDERS

Description of Damage

Several species of aphids and mealybugs feed on the roots of the rice plant (Hsieh 1970; Rai 1975; Mannen 1976). Infestations on roots are limited to rice plants growing in nonflooded fields. Light infestations cause only a yellowing of leaves but heavy feeding may result in wilted or dead plants. *Rhopalosiphum padi* has recently been identified as a vector of "Giallume," a virus disease of rice (Belli *et al.* 1975). Root aphids and mealybugs normally cause little economic loss.

The water weevils, typified by the rice water weevil *Lissorhoptrus oryzophilus* Kuschel, are probably the most widely distributed and economically important root feeders on rice. The rice water weevil attacks the rice plant in both its adult and larval stages. The adult feeds on the leaves of young rice plants. This activity leaves a longitudinal scar or stripe where the leaf surface has been removed. This damage is generally believed to be of no economic importance.

TABLE 6.1. INSECT PESTS THAT FEED ON THE SEED, SEEDLING OR ROOTS OF THE RICE PLANT

Scientific Name	Common Name	Distribution
Phyllophaga sp.	White grubs	South America
Gryllotalpa hexadactyla	Mole cricket	South America
Gryllotalpa africana (P. deB)	Mole cricket	Africa, Tropical Asia, Europe, Japan
Heteronychus oryzae Britten	Scarab beetle	Nigeria, Sierra Leone
Colaspis brunnea (Fab.)	Grape colaspis	Southern United States
Lissorhoptrus oryzophilus Kuschel	Rice water weevil	North, South, and Central America, Japan
Chironomus spp.	Crane fly	Egypt, Europe, USSR, Australia, Calif.
Cricotopus spp.	Crane fly	Egypt, Europe, USSR, Australia, Calif.
Rhopalosiphum rufiabdominalis Sasaki	Rice root aphid	Most rice growing areas
Syntermes molestus Burmeister	Termite	Brazil
Echinocnemus oryzae Mshl.	Rice root weevil	India

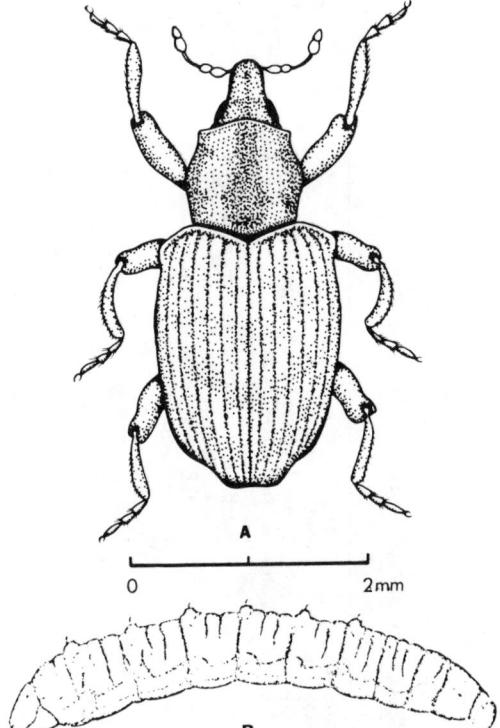

Larvae illustration by D. Isely and H. Schwardt;
Adult illustration by Mrs. H.R. Broad, Courtesy of
D.S. Hill.

FIG. 6.3. RICE WATER WEEVIL, *LISSORHOPTRUS*
ORYZOPHILUS, KUSCHEL

(A) Adult; (B) larva.

The larvae of the rice water weevil feed within and upon rice roots. Where larval populations are dense, the root system is reduced to the extent that wind may cause the plant to lean or in extreme cases, to become dislodged and float. It also has been noted in some instances that the reduced root system appeared blackened and decayed. Rice water weevil damage to the root system of the rice plant causes a reduction in tillering, and in plant height, a slight delay in maturity and a reduction in grain yield (Bowling 1957; Gifford *et al.* 1975).

Rice water weevils, *L. oryzophilus* and *L. isthmicus* Kuschel, have recently been discovered in Japan and Haiti, respectively (Sommeijer 1975; Tsuzuki *et al.* 1976).

The paddy root weevil, *Echinocnemus oryzae* Mshll., destroys the roots of rice plants causing them to wither and dry up (Srivastava *et al.* 1975). Insects that feed on the roots of the rice plants and their distribution are included in Table 6.1.

Losses Due to Damage

Root aphids and other sucking type insects that feed on roots normally cause little economic loss. Yield increases resulting from control of the rice water weevil in Texas ranged from 0 to 846 kg/ha in studies conducted from 1956 to 1962. The average for all tests was 380 kg/ha. The rice water weevil population in untreated plots in all tests was approximately 40 larvae per 30.48 cm of drill row (Bowling 1967). The average yield increase in tests conducted by Rolston and Rouse (1960) in Arkansas to control the rice water weevil and grape colaspis was approximately 230 kg/ha. Yield decreases cannot always be accurately predicted from larval populations. In some tests, populations as high as 30 larvae/30.48 cm of drill row did not significantly reduce yield (Bowling 1961). Unknown soil conditions or biological factors, in conjunction with rice water weevil infestations, are believed to influence yield losses.

LEAF FEEDERS

Description of Damage

Insects that feed on the leaves of rice plants are numerous and varied. The types of damage to rice plants by leaf feeding insects include mechanical removal of a portion of the photosynthetic surface, interference with photosynthesis by rolling or tying leaves, extraction of juices and excretion of wastes, blockage of the vascular system, mechanical damage to protective cells allowing entry of disease organisms and direct transmission of plant pathogens. (See Table 6.2.)

Chewing Insects.—The mechanical removal of portions of the leaves and rolling or tying of the leaves are the most easily recognized damages. Leaf feeding adults and larvae may remove only minute portions of the leaves or in extreme instances entire leaf surfaces. Feeding activity may occur during any stage of growth and may persist for various periods of time. The fall armyworm, *Spodoptera frugiperdia,* (J. E. Smith) slug caterpillar, *Parasa bicolor,* and rice green caterpillar, *Naranga aenescens* Moore, are typical of a number of Lepidopterous larvae that feed on the leaves of the rice plant (Kishino and Sato 1975; Raha 1975; Navas 1976). The rice leaf roller, *Cnaphalocrocis medinalis* (GN) and rice case worm, *Nymphula depunctalis* (GN), not only feed on leaves but utilize leaves

TABLE 6.2. INSECTS THAT FEED ON LEAVES OF THE RICE PLANT

Scientific Name	Common Name	Distribution
Spodoptera exigua (Hbst.)	Beet armyworm	Africa, Asia, North America, Australia, Hawaiian Islands, West Irian, USSR, West Indies
Spodoptera mauritia (Boisd.)	Rice armyworm	Africa, Australia, India, Indonesia, Malaysia, Pakistan, Philippines, Sri Lanka, Thailand
Spodoptera frugiperda (J.E. Smith)	Fall armyworm	North, Central and South America, West Indies
Mythimna separata (Wlk.)	Ear cutting caterpillar	Asia, Pacific Islands, East Australia, Fiji, New Zealand
Naranga aenescens Moore	Rice green caterpillar	Japan
Parasa bicolor Wlk.	Slug caterpillar	India
Baliothrips biformis (Bagn.)	Rice thrips	Bangladesh, India, Indonesia, Japan, Malaysia, Sri Lanka, Thailand, Vietnam
Latoia bicolor (Wlk.)	Slug caterpillar	India, Indonesia
Chaphalocrocis medinalis (Gn.)	Rice leaf roller	India, Indonesia, Korea, Malaysia, Pakistan, Philippines
Nymphula depunctalis (Gn.)	Rice case worm	Argentina, Australia, Brazil, Gambia, Ghana, India, Indonesia, Malagasy Republic, Malaysia, Malawi, Mauritius, Mozambique, Nigeria, Pakistan, Philippines, Sri Lanka, Uruguay, Venezuela, Zaire
Parnara guttata (Bremer and Grey)	Rice skipper	Celebes, China, Himalayas, Indonesia, Japan
Agromyza oryzae (Munukata)	Japanese rice leaf miner	Japan, Java, East Siberia
Hydrellia griseola (Fall.)	Rice leaf miner	North Africa, North and South America, Europe, Japan, Malaysia
Hydrellia philippina Ferino	Rice whorl maggot	Philippines
Hydrellia sasakii Yuasa & Isitani	Paddy stem maggot	Japan
Dicladispa armigera (Ol.)	Rice hispa	Bangladesh, Burma, South China, India, West Malaysia, Nepal, Pakistan, Sumatra, Thailand, West Irian

TABLE 6.2. *(CONTINUED)*

Scientific Name	Common Name	Distribution
Trichispa sericea (Guer.)	Africa rice hispa	Africa
Oulema oryzae (Kuway.)	Rice leaf beetle	China, Japan, Korea, Manchuria, Ryuku Islands, East Siberia, Taiwan
Oligonychus oryzae (Hirst)	Paddy mite	India
Laodelphax striatellus (Fall.)	Small brown planthopper	Europe, Japan, Korea, Philippines, Siberia, Taiwan
Sogatella furcifera (Horv.)	White backed planthopper	China, Fiji, India, Indonesia, Japan, Korea, Malaysia, Philippines, Sri Lanka, Taiwan, USSR, Vietnam
Nilaparvata lugens (Stål.)	Brown planthopper	China, Fiji, India, Indonesia, Korea, Malaysia, New Guinea, Philippines, Sri Lanka, Taiwan, Thailand
Sogatodes orizicola (Muir)	Rice delphacid	Mexico, Central America, Argentina, Brazil, Colombia, Surinam, Caribbean
Nephotettix cincticeps (Uhl.)	Green rice leafhopper	Japan, Korea, Manchuria, Ryuku Islands, Australia, Bangladesh, Burma, Cambodia, Caroline Islands, China, India, Indonesia, Laos, Malaysia, Mariana Islands, Nepal, Nicobar Islands, Pakistan, Papua, New Guinea, Philippines, Singapore, Sri Lanka, Taiwan, Vietnam, West Irian
Nephotettix virescens (Dist.)	Green rice leafhopper	Bangladesh, Burma, Cambodia, China, India Indonesia, Laos, Malaysia, Pakistan, Philippines, Taiwan, Thailand, Vietnam
Nephotettix nigropictus (Stål.)	Green rice leafhopper	Bangladesh, Burma, Cambodia, China, India, Indonesia, Laos, Malaysia, Pakistan, Philippines, Taiwan, Thailand, Vietnam
Recilia dorsalis (Motsch.)	Zigzagged leafhopper	India, Japan, Malaysia, Philippines, Sri Lanka, Taiwan, Thailand
Draeculacephala sp.		North, South and Central America

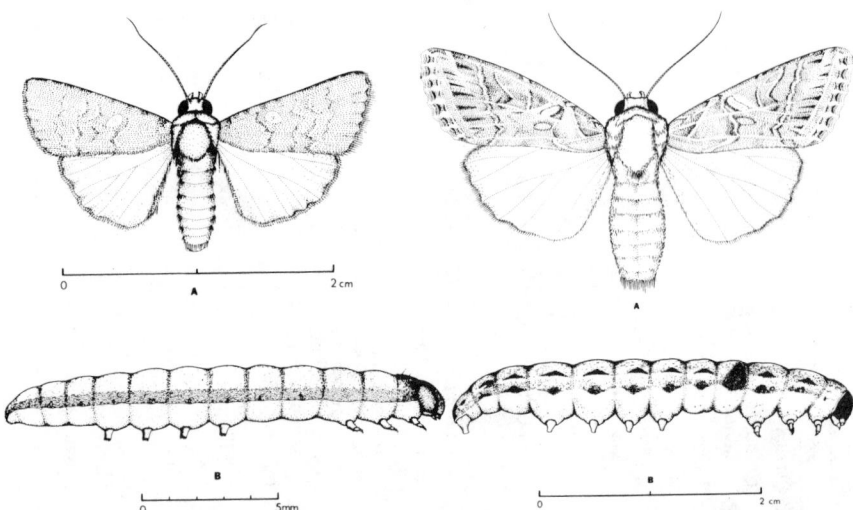

By Mrs. H.R. Broad, Courtesy of D.S. Hill

FIG. 6.4. BEET ARMYWORM,
SPODOPTERA EXUIGUA (HBST.)

(A) Adult; (B) larva.

By Mrs. H.R. Broad, Courtesy of D.S. Hill

FIG. 6.5. RICE CUTWORM, *SPODOPTERA LITURA* (F.)

(A) Adult; (B) larva.

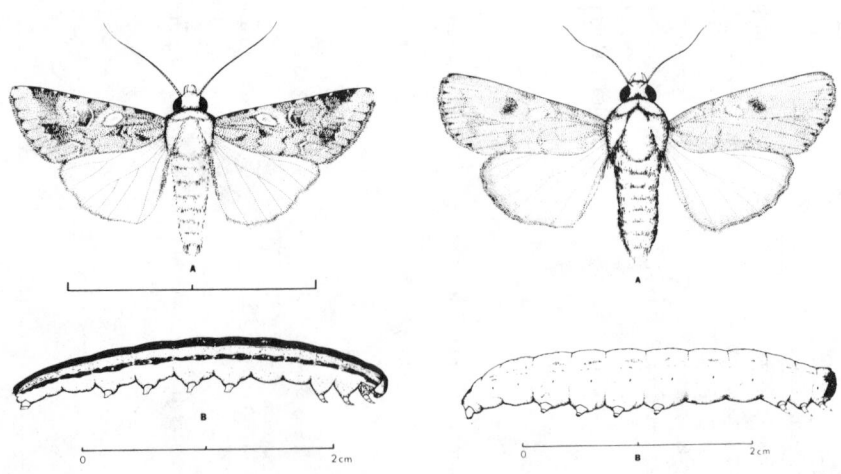

Mrs. H.R. Broad, Courtesy of D.S. Hill

FIG. 6.6. NUTGRASS ARMYWORM,
SPODOPTERA EXEMPTA (WLK.)

(A) Adult; (B) larva.

By Mrs. H.R. Broad, Courtesy of D.S. Hill

FIG. 6.7. PADDY ARMYWORM,
SPODOPTERA MAURITIA (BOISD.)

(A) Adult; (B) larva.

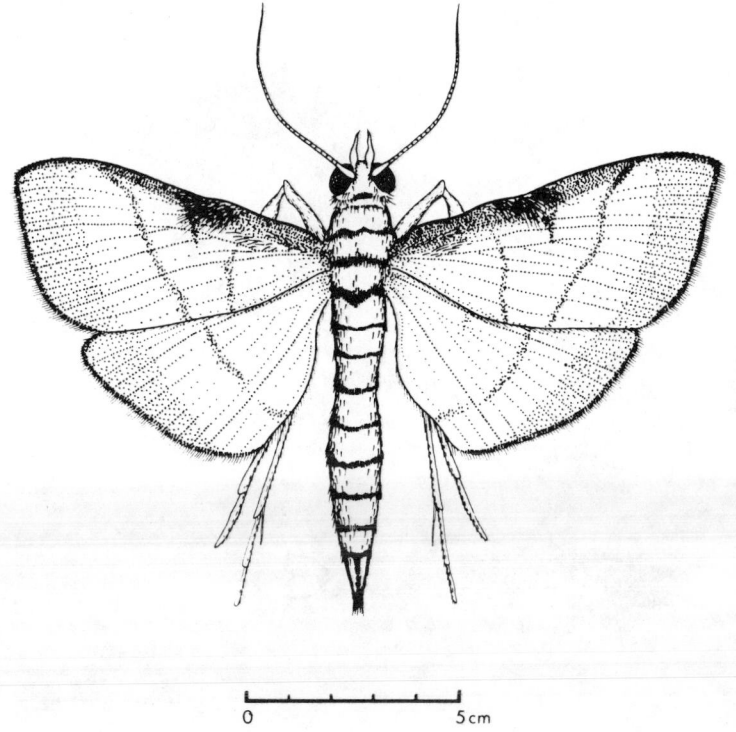

By Mrs. H.R. Broad, Courtesy of D.S. Hill

FIG. 6.8. RICE LEAF ROLLER, *CNAPHALOCROCIS MEDINALIS* (GN.)

to construct protective coverings for themselves (Das and Nair 1974). Both the adult and larval stages of rice hispa *Dicladispa armigera* (Oliv.) feed on leaves. The adults consume the tissue between the veins of the leaves leaving parallel white streaks. The larvae mine into the mesophyll and cause white streaks or patches (Sajjan *et al.* 1975).

In Kenya, 3 species of beetles belonging to the family Chrysomelidae feed on rice leaves and mechanically transmit rice yellow mottle virus (Bakker 1971). The smaller leaf miner, *Hydrellia griseola* Fallen, feeds on the mesophyll of rice leaves. The leaf blades become transparent, dry out and lie prostrate on the water surface. Larvae also occasionally mine the leaf sheath and stem (Kuwayama 1958; Grigarick 1959). The rice whorl maggot, *Hydrellia spp.*, feeds in a similar fashion but primarily on the inner margins before the leaves emerge and expand (Azam and Tejkumar 1971).

Sucking Insects.—Leafhoppers and planthoppers feed by extracting fluids from the leaves of rice plants and are probably the most destruc-

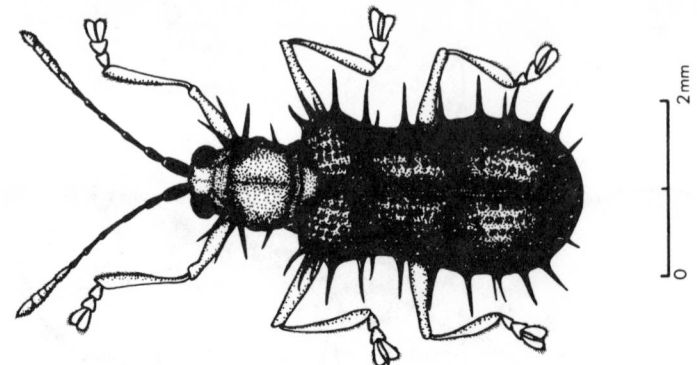

By Mrs. H.R. Broad, Courtesy of D.S. Hill

FIG. 6.9. PADDY HISPA, *DICLADISPA ARMIGERA* (OL.)

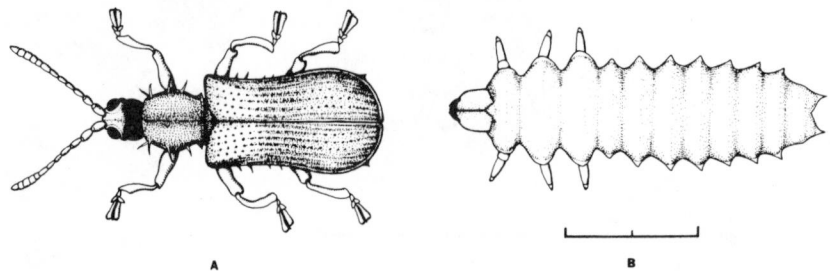

By Mrs. H.R. Broad, Courtesy of D.S. Hill

FIG. 6.10. RICE BEETLE, *TRICHISPA SERICEA* (GUER.)

(A) Adult; (B) larva.

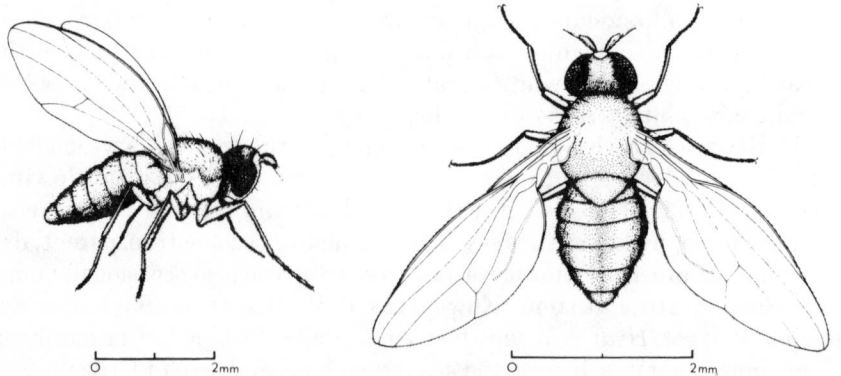

By Mrs. H.R. Broad; Courtesy D.S. Hill

FIG. 6.11. RICE LEAF MINER, *HYDRELLIA GRISEOLA* (FALL.)

tive group of rice insect pests. The brown planthopper, *Nilaparvata lugens* Stal., damages rice directly by plugging the phloem tissue with their feeding sheaths, injuring epidermal cells and depositing honeydew. Black sooty moulds develop on the honeydew deposited on plant surfaces which in turn reduces the amount of sunlight which reaches the leaves. Oviposition activity damages the epidermal cells allowing entry of plant pathogens and by blocking the plant nutrient transport system. Direct damage from the brown planthopper results in reduced tillering, plant height, number of productive tillers per plant and increased number of unfilled grains (Pathak *et al.* 1969; Sogawa 1970, 1972; Kulshreshtha *et al.* 1974). Symptoms of direct damage by the rice delphacid, *Sogatodes oryzicola* (Muir), have not been clearly described separately from the virus disease hoja blanca. However, yield increases from insecticide applications to control the rice delphacid have been attributed to reduction of direct damage as well as reduction of damage from the virus.

Several species of leafhoppers damage rice by feeding and oviposition activities similar to planthoppers. A comparison of damage symptoms by various species of leafhoppers and planthoppers has been reported by Misra and Israel (1968).

Loss Due to Damage

Chewing Insects.—The economic importance of leaf surface removal or rolling and tying of leaves is difficult to assess. Alam (1967) estimated that annual yield loss from rice hispa is 20% of the yield on 60–80,000 ha in Pakistan. Lange *et al.* (1953) estimated damage from leaf miner at 10 to 20% on c 80,000 ha in California in 1953. In the past it has been generally assumed that there was a yield loss proportional to the amount of leaf surface removed. Prakasa Rao *et al.* (1971) were able to correlate yield losses due to feeding of hispa with percentage green leaf area loss during the flowering stage.

Navas (1976) concluded that rice plants in early stages of growth could compensate for severe damage by fall armyworm. Artificial removal of 25% and 50% of leaf surface in both the seeding tillering stage resulted in yield reductions of 168, 442, 282 and 725 kg/ha, respectively (Bowling 1978). Yield losses due to leaf removal are greatest when the rice plant is in the reproductive stage (Anon. 1977A).

Sucking Insects.—Yield losses due to direct damage by leafhoppers and planthoppers are difficult to separate from losses attributable to the diseases transmitted, however, losses appear to be directly related to population levels. Yield increases from insecticide applications may result from reduction of direct damage by the insect as by reducing virus diseases.

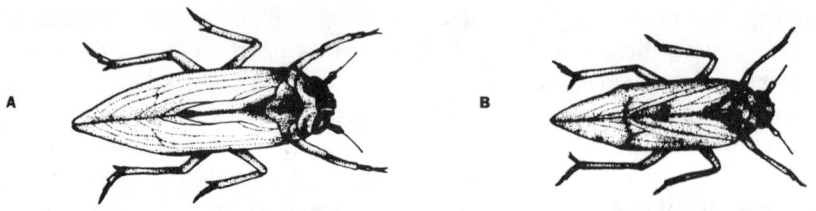

By Mrs. H.R. Broad, Courtesy of D.S. Hill

FIG. 6.12. BROWN PLANTHOPPER, *NILAPARVATA LUGENS* (STAL.)

(A) Female; (B) male.

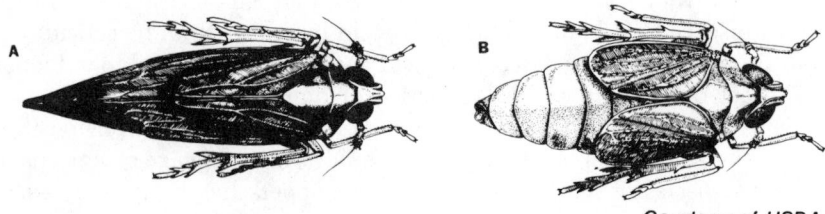

Courtesy of USDA

FIG. 6.13. RICE DELPHACID, *SOGATODES ORIZICOLA* (MUIR)

(A) Male; (B) female.

Courtesy of IRRI

FIG. 6.14. ZIG-ZAGGED LEAFHOPPER, *RECILIA DORSALIS* (MOTSCH)

(A) Male; (B) female.

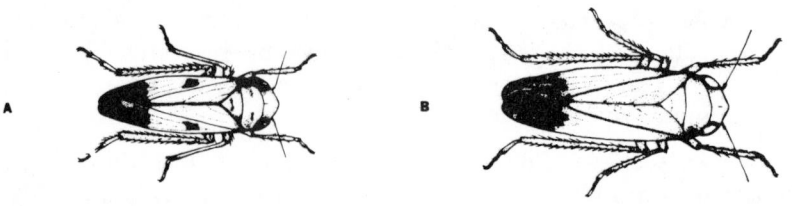

Courtesy of IRRI

FIG. 6.15. GREEN RICE LEAFHOPPER, *NEPHOTETTIX VIRESCENS*

(A) Male; (B) female.

Transmission of plant pathogens by sucking type insects is probably responsible for the largest economic loss caused by leaf feeding insects on rice plants. The symptoms of the various diseases transmitted by insects are discussed in another chapter. Insects that feed on leaves are listed in Table 6.2.

STEM FEEDERS

Description of Damage

Stem Borers.—Several species of Lepidoptera of the families Pyralidae and Noctuidae constitute one of the major groups of insect pests of rice referred to collectively as stem borers. All moths of this group place their eggs on plant surfaces except *Sesamia inferens* Wlk., which oviposite between the leaf sheath and the stem. Newly hatched larvae enter the space between the leaf sheath and the stem and begin feeding on the tissue inside the leaf sheath. Usually, only the inside surface of the leaf sheath is damaged. The exterior surface of the damaged areas remains intact and may appear transparent. Later, the damaged tissue dries and appears as a brown discolored area. Small larvae have also been found feeding in the basal portion of the midrib of the leaf. Larvae feed for a few days and then penetrate the stem and feed upward or downward within the stem. Some species may leave the plants infested initially and move to nearby healthy plants. When larval growth is completed they prepare an exit hole and pupate within the stem, straw or stubble. Detailed taxonomy and habits of each species vary and have been summarized by Kapur (1967), Kiritani and Iwao (1967) and by Banerjee and Pramanik (1967).

Stem borers may feed on rice plants from seedling stage to maturity. The effect varies according to size and stage of growth. When borers infest rice plants in the vegetative stage their feeding activity severs the plant nutrient transport system. Consequently, the central leaf whorl does not unfold but turns brown and dies while the lower leaves remain green. Plants infested in the vegetative stage show this type of damage and are called "dead hearts."

When rice plants in the reproductive stage are infested by borers the feeding activity may prevent the development of a head or the head may emerge, die and dry out. These heads are rapidly bleached of color and are referred to as "white heads." Panicle growth and grain filling may be stopped at any stage by stem borer feeding, resulting in partially filled heads and grains (Koyama 1973). Newly hatched larvae feed on the rice plant at the uppermost node after the head has emerged and partially

TABLE 6.3. INSECTS THAT FEED ON THE STEMS OF THE RICE PLANT

Scientific Name	Common Name	Distribution
Chilo loftini (Dyar)		Mexico, California, Arizona
Chilo plejadellus Zinken	Rice stalk borer	Southern United States, Mexico
Chilo suppressalis (Wlk.)	Striped rice borer	Indian subcontinent, Southeast and East Asia, China, Manchuria, Egypt, Spain
Chilo polychrysus (Meyrick)	Dark headed rice borer	Malaya, India, China, Philippines
Scirpophaga albinella (Cramer)	White stem borer	Neotropical region and Neartic southern United States, South America
Diatraea saccharalis F.	Sugarcane borer	Southern United States, South America
Tryporyza incertulas (Wlk.)	Yellow rice borer	Southeast and East Asia, China, Indian sub-continent, Afghanistan
Zeadiatraea lineolata (Wlk.)	Neotropical corn stalk borer	Neotropical region. Recorded infesting rice in Venezuela
Sesamia calamistis (Hmps.)	African pink borer	Ethiopian region
Diopsis thoracica Westw.	Stalk-eyed borer	Tropical West and East Africa, Somalia, Zanzibar
Antherigona spp.	Rice stem fly	India, Indonesia, Japan, Malaysia, New Guinea, Philippines, Sri Lanka
Orseolia oryzae (Wood-Mason)	Rice gall midge	Tropical Asia except Philippines and Malaysia, Nigeria, northern Cameroon, Sudan
Scotinophara coarctata (Thnb.)	Black paddy bug	Indonesia, Malaysia
Scotinophara lurida Burm.	Japanese black rice bug	India, Japan, all Asia
Nezera viridula L.	Southern green stink bug	All rice growing areas, but does not always attack rice

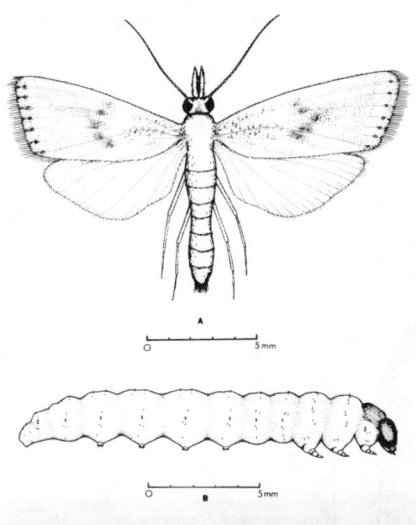

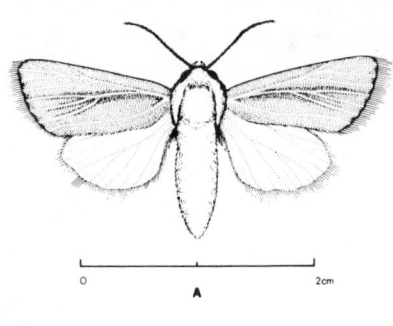

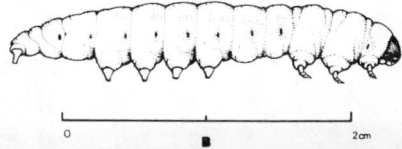

By Mrs. H.R. Broad, Courtesy of
D.S. Hill

By Mrs. H.R. Broad, Courtesy of
D.S. Hill

FIG. 6.16. DARK-HEADED STRIPED
BORER, *CHILO POLYCHRYSUS* (MEYR.)

(A) Adult; (B) larva.

FIG. 6.17. PINK BORER, *SESAMIA
INFERENS* (WLK.)

(A) Adult; (B) larva.

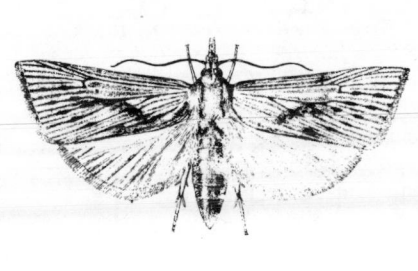

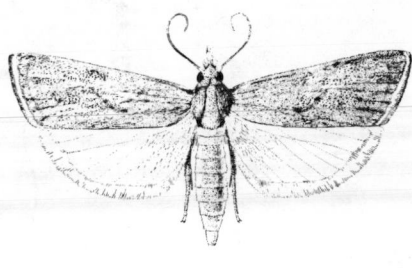

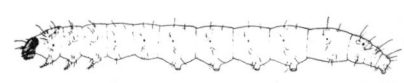

Courtesy of USDA

Courtesy of USDA

FIG. 6.18. SUGARCANE BORER,
DIATRAEA SACCHARALIS (FABRICIUS)

(A) Adult; (B) larva.

FIG. 6.19. RICE STALK BORER, *CHILO
PLEJADELLUS* (ZINCKEN)

(A) Adult; (B) larva.

filled. These heads break at the injured point and are lost during mechanical harvesting (Bowling 1967).

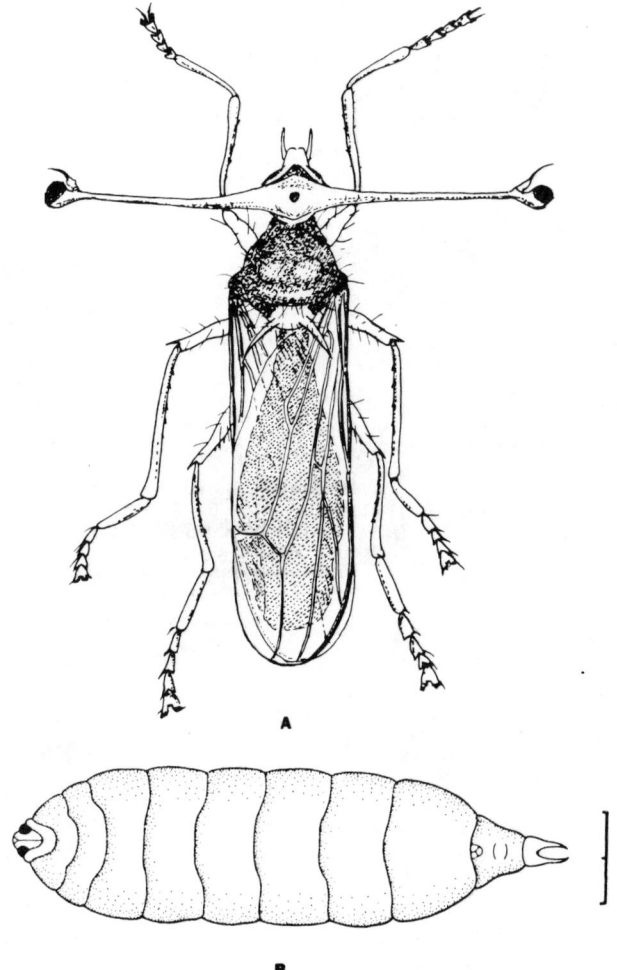

By Mrs. H.R. Broad, Courtesy of D.S. Hill

FIG. 6.20. STALK-EYED BORER, *DIOPSIS THORACICA* (WESTW.)

(A) Adult; (B) pupa.

Two species of Diptera, the stalk eyed borer *Diopsis thoracica* Westw. (Fig. 6.20) and the rice stem fly, *Aterigona exigua* Stien, infest rice and inflict damage similar to the Lepidoptera stem borers (Morgan and Abu 1973; Anon. 1976).

Rice Gall Midge.—The rice gall midge, *Orseolia oryzae* (Wood-Mason), is a pest causing an unusual effect on the leaf sheaths of rice in the tillering stage. The female midge oviposits primarily on undersides of leaves of rice and grass plants. The newly hatched larvae find their way to the growing points of tillers, establish themselves in the growing bud and feed until pupation. The larval feeding activity stimulates the leaf sheath of the infested tiller to elongate into a tubular gall commonly called silver shoot which resembles an onion leaf. Damaged tillers dry out and do not produce panicles. Rice plants infested early may produce numerous new tillers which are reduced in height and vigor. Plants can be infested from the seedling to panicle initiation stage (Perera and Fernando 1970). Every infested tiller results in the induction of one more tiller from the node just below the silver shoot. Severe infestations result in prolonged tillering phase, stunted growth, delayed flowering and uneven maturity. Ultimately the number and length of panicles and grain weight are drastically reduced (Israel and Prakasa Rao 1968; Prakasa Rao 1972).

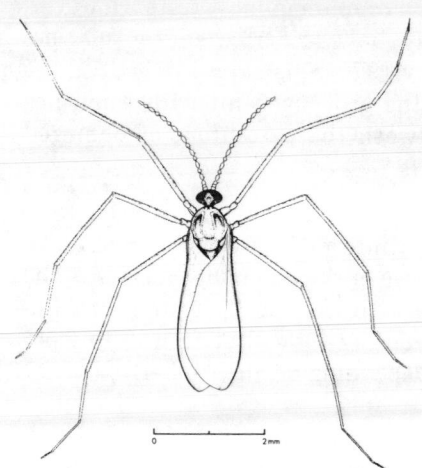

FIG. 6.21. RICE GALL MIDGE, *ORSEOLIA ORYZAE* (WOOD-MASON)

By Mrs. H.R. Broad, Courtesy of D.S. Hill

FIG. 6.22. RICE SHOOT FLY, *ATHERIGONA EXIGUA* STIEN.

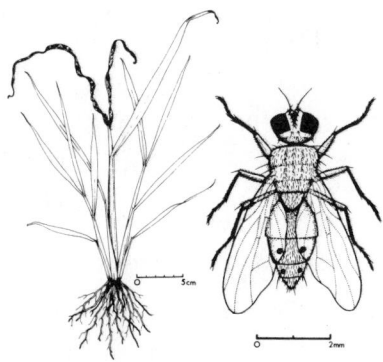

By Mrs. H.R. Broad, Courtesy of D.S. Hill

Other Stem Feeders.—The black paddy bug, *Scotinophara coarctata* (Thnb.), and the Japanese black rice bug, *Scotinophara lurida* Burm., feed on rice stems at the base of the plant. Their feeding activity causes the leaves and panicle to turn brown (Anon. 1976). The southern green stinkbug, *Nezara viridula* L., has also been reported to cause similar damage (Gifford *et al.* 1968).

Losses Due to Damage

Stem Borers.—Yield loss due to stem borer injury is difficult to estimate. Ishikura (1967) reviewed the problem and concluded that no precise method of assessing losses had been established. Gomez and Bernardo (1967) found that percentage of yield loss varied with factors that affect yield potential. They found that 2% white heads caused *c* 4.4% yield loss in fields with a yield potential of 3 MT/ha and 6.4% yield loss in fields with a yield potential of 4 MT/ha. Without controls, yields may be reduced 80% (Anon. 1977B). Studies with insecticide application on rice show yield increases of 1000 kg/ha or more in most tests. However, insects other than stem borers may have contributed to this loss (Prakasa Rao *et al.* 1970A,B; Satpathy 1970; Singh *et al.* 1974). Studies with simulated damage by piercing the growing point with a pin has helped establish the relationship between the percentage of damaged stems and yield losses (Anon. 1977A).

Gall Midge.—Visual estimates of losses from gall midge have been based on average yield in years when infestations occur. Israel *et al.* (1959) found that for every unit increase in rice gall midge incidence the yield loss was 0.502% or 26.55 kg/ha with the variety tested. Prakasa Rao (1974) demonstrated that different varieties lost from 0 to 2.5% yield per unit increase in incidence, depending on their inherited tolerance.

GRAIN FEEDERS

Several species of Heteroptera damage the rice crop by removing portions of the fluid material from the grain during the filling period. Feeding activity of Heteroptera may reduce yield, milling yield and grade of rice.

Description of Damage

Rice Bugs.—Adult rice bugs develop on weedy grasses or adjacent rice

FIG. 6.23. BLACK PADDY BUG, *SCOTINOPHARA COARCTATA* (THNB.)

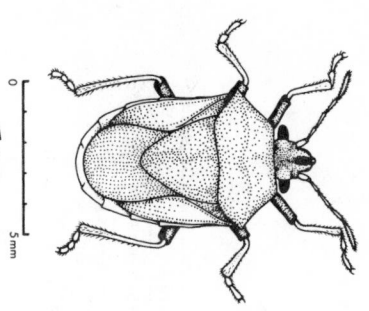

By Mrs. H.R. Broad, Courtesy of D.S. Hill

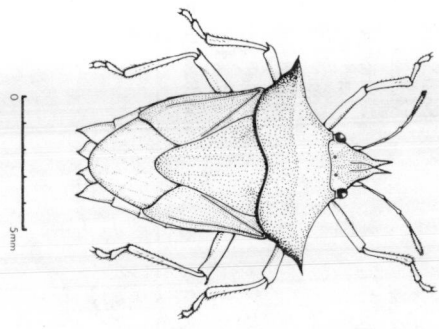

FIG. 6.24. RICE SHIELD BUG, *DIPLOXS FALLAX* (STAL.)

By Mrs. H.R. Broad, Courtesy of D.S. Hill

FIG. 6.25. RICE BUG, *STENOCORIS APICALIS*

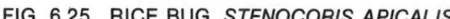

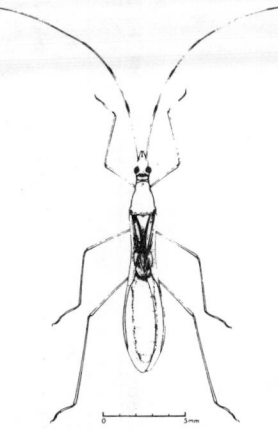

By Mrs. H.R. Broad, Courtesy of D.S. Hill

TABLE 6.4. INSECTS THAT FEED ON THE DEVELOPING GRAINS OF THE RICE PLANT

Scientific Name	Common Name	Distribution
Leptocorisa sp.	Rice bugs	Widely distributed in world where rice is grown
Cletus trigonus (Thnb.)	Slender rice bug	North Borneo, India, Japan, Malaysia, Philippines, Sri Lanka, Taiwan
Oebalus pugnax (F.)	Rice stink bug	Southern United States, Cuba, Dominican Republic
Oebalus ornata (Sailer)	Rice stink bug	Brazil, Haiti, Dominican Republic, Puerto Rico
Oebalus poecilus (Dall.)	Padi bug	South America
Diploxys fallax Stal.	Rice shield bug	Malagasy, Swaziland
Nezara viridula L.	Southern green stink bug	All rice growing countries but does not always feed on rice.
Eusarcoris inconspicuus (H.S.)	La chinche del arroz	Spain
Trigonotylus colestialium Kirkaldy	Rice leaf bug	Japan
Stenodema rubrinerve Horvath	Red veined leaf bug	Japan
Haplothrips ganglbaueri Schmutz	Thrips	India, Pakistan, Ceylon
Haplothrips aculeatus (F.)	Thrips	Palearctic region

FIG. 6.26. RICE BUG, *LEPTOCORISA ACUTA*
(THUB.)

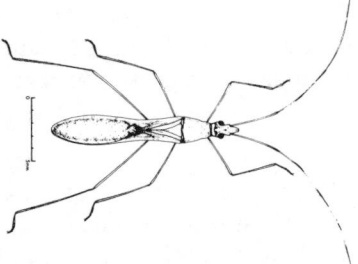

By Mrs. H.R. Broad, Courtesy of D.S. Hill

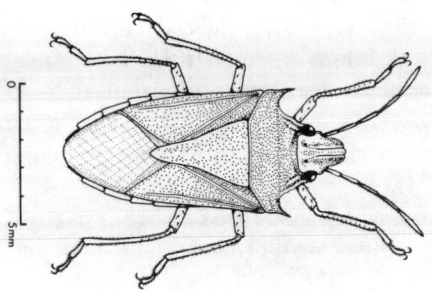

FIG. 6.27. RICE STINK BUG, *OEBALUS PUGNAX* (F.)

By Mrs. H.R. Broad, Courtesy of D.S. Hill

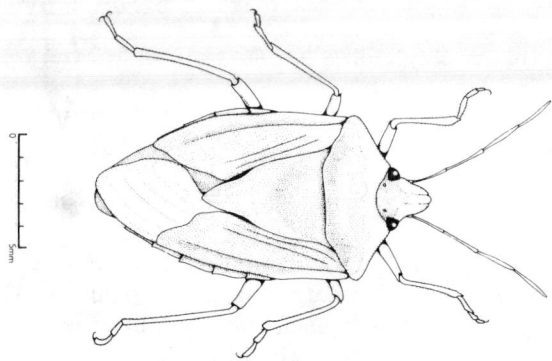

By Mrs. H.R. Broad, Courtesy of D.S. Hill

FIG. 6.28. SOUTHERN GREEN STINK BUG, *NEZARA VIRIDULA* L.

fields and migrate to plants as the grains begin to fill. Both adults and nymphs of rice bugs feed on developing grains by inserting their piercing, sucking mouth parts between the lemma and palea. A visible brown spot remains at the point of insertion (Corbett 1930). The damaged grain and straw from heavily infested fields has an objectionable odor and flavor (Ghosh 1971).

Stink Bugs.—Stink bugs insert their stylets directly through the glume of the rice grain. Grains fed upon in the early milk stage may be depleted of all or most of the fluid material and result in an empty floret or an atrophied grain (Swanson and Newsom 1962). Grains fed upon in the late dough stage are weakened structurally and may break during milling, resulting in lower head rice yield. Fungi may be mechanically introduced on the mouth parts as the rice stink bug *Oebalus pugnax* (F.) feed (Douglas and Tullis 1950). When these fungi grow they form dark round spots that remain on the milled grain. This damage is referred to as "pecky" rice. The southern green stink bug *Nezara viridula* L. is a pest on rice and imparts an objectional odor to rice that is apparent at cooking (Kiritani 1971). The feeding activity of several other species of Pentatomids cause discolored spots on rice grains in Japan (Okuyama and Inouye 1974; Tokuo *et al.* 1974).

Recently, *Leptocorisa acuta* (Thnb.), which also feeds on other parts of the rice plant, has been implicated in the spread of bacterial leaf blight, *Xanthomonas oryzae* (Mohiuddin *et al.* 1976).

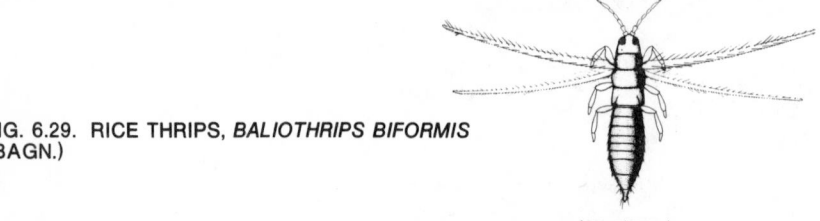

FIG. 6.29. RICE THRIPS, *BALIOTHRIPS BIFORMIS* (BAGN.)

By Mrs. H.R. Broad, Courtesy of D.S. Hill

In addition to Heteroptera, adult and larval stages of the rice thrips, *Haplothrips ganglbaueri* (Schmutz), infest rice panicles. Their feeding activity may result in severe damage to lemma, palea and ovarian tissue (Abraham *et al.* 1972; Ananthakrishnan 1973).

Losses Due to Damage

Rice Bugs.—Estimates of economic losses from infestations of grain feeding Heteroptera have been reported by various workers. Srivastava and Saxena (1967) reviewed the available information pertaining to losses from rice bug infestations. Estimates of 10 to 100% yield losses are mostly visual estimates without correlation with specific infestation levels. Rothschild (1970) estimated than an infestation level of 100,000 *L. oratorius* (F) per acre would result in a 25% yield loss. One rice bug per 35 panicles reportedly reduced yield 14% or 2.2 g/hill (Anon. 1977A).

Stink Bugs.—Liang (1964) reported that 50% of the rice crop is sometimes lost to feeding by *Oebalus ornata* (Sailer) in the Dominican Republic. Quantitative loss from *Oebalus pugnax* (F.) infestations in cage tests were found to be greater than qualitative losses. Total monetary loss per acre was $6.53, $12.75 and $19.93 for population levels of 1, 2 and 4 *O. pugnax* per 30.48 cm^2, respectively (Bowling 1963). These data vary considerably from similar studies by Swanson and Newsom (1962) and Odglen and Warren (1962). Qualitative losses from *O. poecilus* (Dall.) and "red rice" in Guyana was estimated as 6,000,000 Guyanan dollars in 1970 (Rai 1974).

Estimates of qualitative losses are complicated by grading systems that include pecky grains with other discolored grains and with red rice. Qualitative losses based on percentage of damaged grains by a given population level will change as the yield levels change.

SUMMARY

The major insect pests of the rice plant have been identified and are fairly easy to separate into groups according to the part of the plant on which they feed. The visual damage that different groups such as stem feeders, leaf feeders, etc., inflict upon the rice plant may be readily associated with a specific insect or insects of that particular group. In some instances the identification of the disease transmitted may be more indicative of the presence of an infestation than the damage the insect inflicts.

Many references to yield losses of rice due to insect damage are estimates based on the general level of yield from an insect infested area compared with yields from other areas. Yield or quality losses resulting from insect damage are difficult to predict. Numerous factors other than the number of insects present may influence the yield or quality loss due to insect damage.

Recent studies with insecticides to control insects have been beneficial in providing information on yield and quality losses. However, in many studies multiple species are involved, making evaluations of damage from individual species more difficult. Additional information on yield and quality losses are needed for individual species to assist in establishing economic threshold levels for use in integrated control programs on rice.

REFERENCES

ABRAHAM, C.C., THOMAS, B., KARUNAKARAN, K. and GOPALAKRISH-NAN, R. 1972. Occurrence of *Haplothrips ganglbaueri* Schmutz (Phlaeo-thripidae: Thysanoptera) as a serious pest of rice earheads in Kerala. Curr. Sci. *41* (721).

ABUL-NASR, S., ISA, A.L. and EL-TANTAWY, A.M. 1970A. Control measures of the bloodworms, *Chironomus* Sp. in rice nurseries. Bull. Entomol. Soc. of Egypt, Econ. Ser. *4,* 127-133.

ABUL-NASR, S., ISA, A.L., KIRA, M.T. and EL-TANTAWY, A.M. 1970B. Biological studies on the bloodworms, *Chironomus* sp., injurious to rice seedlings in United Arab Rep. Bull. Entomol. Soc. Egypt *54,* 381-388.

ALAM, M.Z. 1967. Insect pests of rice in East Pakistan. *In* Major Insect Pests of the Rice Plant. Proc. of Symp. Int. Rice Res. Inst., Los Baños, Philippines. 1964. Johns Hopkins Univ. Press, Baltimore.

AMARAL, S.F. 1949. Biology and economic importance of "precejo" on rice in the State of São Paulo. O Biologico *15* (3) 47-58.

ANANTHAKRISHNAN, T.N. 1973. Thrips: Biology and Control. Macmillan, India.

ANON. 1967. Major insect pests of the rice plant. Proc. Symp. Int. Rice Res. Inst. 1964. Los Baños, Philippines. John Hopkins Univ. Press, Baltimore.

ANON. 1976. Pest Control in Rice. Cntr. for Overseas Pest Res., Pans Man. *3.*

ANON. 1977A. Control and Management of Rice Pests. Int. Rice Res. Inst. Ann. Rep. for 1975. IRRI, Los Baños, Philippines.

ANON. 1977B. Rice Insects. Insect Control in the People's Republic of China. Committee on Scholarly Communication with the People's Republic of China Rep. *2*

AZAM, K.M. and TEJKUMAR, S. 1971. Whorl Maggot, *Hydrellia* spp. *(Diptera: Ephydridae)* as a serious pest of rice in Andhra Pradesh. Andhra Agric. J. *18* (4) 176.

BAKKER, W. 1971. Three new beetle vectors of rice yellow mottle virus in Kenya. Neth. J. Pl. Pathol. *77,* 201-206.

BANERJEE, S.N. and PRAMANIK, L.M. 1967. The lepidopterous stalk borers of rice and their life cycles in the tropics. *In* Major Insect Pests of the Rice Plant. Proc. Symp. Int. Rice Res. Inst., 1964. Los Baños Philippines. Johns Hopkins Univ. Press. Baltimore.

BELLI, G., CORBETTA, G. and OSLER, R. 1975. Research on epidemiology and control of rice giallume. Riso *24* (4) 359-363.

BOREMA, E.B. 1967 to 1968. Annual report of the rice section. Yanco Agric. Coll. and Res. Sta. New South Wales Dept. of Agric.

BOWLING, C.C. 1957. Seed treatment for control of the rice water weevil.

J. Econ. Entomol. *50* (4) 450-52.

BOWLING, C.C. 1961. Chemical control of the rice water weevil. J. Econ. Entomol. *54* (4) 710-12.

BOWLING, C.C. 1963. Cage tests to evaluate stink bug damage to rice. J. Econ. Entomol. *56* (2) 197-200.

BOWLING, C.C. 1967. Insect pests of rice in the United States. *In* Major Insects of the rice plant. Proc. Symp. Int. Rice Res. Inst., 1964. Los Baños Philippines. John Hopkins Univ. Press, Baltimore.

BOWLING, C.C. 1978. Simulated insect damage to rice: Effect of leaf removal. J. Econ. Entomol. *71*, 378-379.

CHATTERJEE, P.B. 1973. Two harmful insects of rice in North Bengal. Indian Farming *22* (12) 35.

CLEMENT, S.L., GRIGARICK, A.A. and WAY, M.O. 1977. Conditions associated with rice plant injury by Chironomid midges in California. Environ. Entomol. *6* (1) 91-96.

CORBETT, G.H. 1930. The bionomics and control of *Leptocorisa acuta* Thunberg with notes on *Leptocorisa* spp. Malaya Dep. Agric. Straits Settle and Fed. Malaya States Sci. Ser. *4,* 1-40.

DARBY, R.O. 1962. Midges associated with California rice fields, with special reference to their ecology. Hilgardia *32* (1).

DAS, N.M. and NAIR, M.R.G.K. 1974. Studies on the chemical control of the rice leaf roller, *Cnaphalocrocis medinalis* Guenee. Agric. Res. J. Kerala *12* (1) 44-48.

DOUGLAS, W.A. and TULLIS, E.C. 1950. Insects and fungi as causes of pecky rice. U.S. Dep. Agric. Tech. Bull. *1015.*

GHOSH, C.C. 1917. *Leptocorisa varicornis* Fabricius. Proc. 2nd Entomol. Meet. Pusa, India.

GIFFORD, J.R., OLIVER, B.F. and TRAHAN, G.B. 1968. New insect pests of rice in Louisiana. *In* Louisiana State Univ. Rice Exp. Stn. 60th Ann. Prog. Rep., Crowley, La.

GIFFORD, J.R., OLIVER, B.F. and TRAHAN, G.B. 1975. Rice water weevil control with pirimiphos ethyl seed treatment. J. Econ. Entomol. *68* (1) 79-81.

GOMEZ, K.A. and BERNARDO, R.C. 1974. Estimation of stemborer damage in rice fields. J. Econ. Entomol. *67,* 509-513.

GRIGARICK, A.A. 1959. Bionomics of the rice leaf miner, *Hydrellia griseola* Fallen, in California. Hilgardia *29* (1).

HSIEH, CHAO-YEN. 1970. The aphids attacking rice plants in Taiwan. (II) Studies on the biology of the red rice root aphid, *Rhopalosiphum rufiabdominalis* (Sasaki). Plant Prot. Bull. (Taiwan) *12* (2) 68-77.

ISHIKURA, H. 1967. Assessment of rice loss caused by the rice stem borer. *In* Major insect pests of the rice plant. Proc. Symp. Int. Rice Res. Inst., 1964. Los Baños, Philippines. Johns Hopkins Univ. Press, Baltimore.

ISRAEL, P. and PRAKASA RAO, P.S. 1968. Influence of gall midge incidence in rice on tillering and yield. Int. Rice Comm. Newsl. *17* (3) 24-32.

ISRAEL, P., VEDAMOORTHY, G. and RAO. Y.S. 1959. Assessment of field losses caused by pests of rice. 8th Meet. Working Party on Rice Prod. and Protect. Int. Rice Comm., Peradeniya, Ceylon.

KAPUR, A.P. 1967. Taxonomy of stem borers. *In* The Major Insect Pests of the Rice Plant. Proc. Symp. Int. Rice Res. Inst., 1964. Los Baños, Philippines. Johns Hopkins Univ. Press. Baltimore.

KENNARD, C.P. 1965. Pests and disease of rice in British Guiana and their control. Food and Agric. Organ. U.N. Pl. Prot. Bull. *13* (4) 73-78.

KIRITANI, K. 1971. Distribution and abundance of the southern green stink bug, *Nezara viridula. In* Proc. Symp. on Rice Insects. Trop. Agric. Res. Cntr., Nishigahara, Japan.

KIRITANI, K. and IWAO, S. 1967. The biology and life cycle of *Chilo suppressalis* (Walker) and *Tryporyza (Schoenobius) incertulas* (Walker) in temperate-climate areas. *In* Major Insect Pests of the Rice Plant. Proc. Symp. Int. Rice Res. Inst., 1964. Los Baños, Philippines.

KISHINO, K. and SATO, T. 1975. Ecological studies on the rice green caterpillar, *Naranga aenescens* Moore. Tohuku Natl. Agric. Exp. Stn. Bull. *50,* 27-62.

KOYAMA, JURO. 1975. Studies on the diminution of insecticide application to rice stem borer, *Chilo suppressalis* Walker. Jpn. J. Appl. Entomol. Zool. *17,* 147-153.

KULSHRESHTHA, J.P., ANJANEYULU, A. and PADMANABHAN, S.Y. 1974. The disastrous brown plant hopper attack in Kerala. Indian Farming *24* (9) 5-7.

KUWAYAMA, S. 1958. The smaller rice leaf-miner, Hydrellia griseola Fallen in Japan. *In* Proc. of 10th Int. Congr. of Entomol., Aug. 17−25, 1956. Montreal.

LANGE, W.H., INGEBRETSEN, K.H. and DAVIS, L.L. 1953. Rice leaf miner. Calif. Agric. *7* (8) 8-9.

LIANG, C.J. 1964. Ecological studies on the rice stink bug *(Solubea ornata* Sailer) during the period of 1964 in Republic Dominican. Pl. Prot. Bull. (Taiwan) *8,* 9-26.

MAEDA, H., TAKIHIRO, T., NAKAYABU M. and KIMURA, H. 1974. Studies on the causes and the control of the spotted rice. II. On the bionomics and the control of the red veined leaf bug. Bull. Hir. Pref. Agric. Exp. Sta. *33,* 23−32.

MANNEN, K.V. 1976. Occurrence of rice mealybug in Kerala. Int. Rice Res. Inst. Newsl. *2* (14).

MISRA, B.C. and ISRAEL, P. 1968. Leaf and planthoppers of rice. Int. Rice Res. Comm. Newsl. *17* (2) 7-12.

MOHIUDDIN, M.S., RAO, Y.P., MOHAN, S.K. and VERMA, J.P. 1976. Role of *Leptocorisa acuta* in the spread of bacterial blight of rice. Curr. Sci. *45*, 426-427.

MORGAN, H.G. and ABU, J.F. 1973. Seasonal abundance of *Diopsis* (Diptera, Diopsidae) on irrigated rice in the Accra Plains. Ghana J. Agric. Sci. *6*, 185-191.

NAVAS, D. 1976. Fall armyworm in rice. *In* Proc. Tall Timber Conf. Ecol. on Animal Control and Habitat Manag., Aug. 17−25. Tallahasee.

ODGLEN, G.E. and WARREN, L.O. 1962. The rice stink bug, *Oebalus Pugnax* F. in Arkansas. Ark. Exp. Sta. Rep. Ser. *107.*

OKUYAMA, S. and INOUYE, H. 1974. On the relation between the occurrence of the black rot of rice grains and insects. Hokkaido Prefectural Agric. Exp. Sta. Bull. *30*, 85−94.

PATHAK, M.D. 1968. Ecology of common insect pests of rice. Ann. Rev. Entomol. *13*, 257-294.

PATHAK, M.D., CHEN, C.H. and FORTUNO, M.E. 1969. Resistance to *Nephotettix impicticeps* and *Nilaparvata lugens* in varieties of rice. Nature *223*, 502-504.

PERERA, N. and FERNANDO, H.E. 1968. Infestation of young rice plants by the rice gall midge, *Pachydiplosis oryzae* (Wood-Mason) (Dipt. Cecidomyiidae), with special reference to shoot morphogenesis. Bull. Entom. Res. *59*, 605-613.

PRAKASA RAO, P.S. 1972. Ecology and Control of *Tryporyza incertulas* Walker and *Pachydiplosis oryzae* Wood-Mason in rice. Ph.D. dissertation. Utkal University, Vani Vihar, Bhubaneswar, India.

PRAKASA RAO, P.S. 1974. Recent ecological studies of rice insects: stem-borers, gallmidges and rice hispa. *In* Proc. of Int. Rice Res. Conf., April 1974. Manila, Philippines. (mimeo)

PRAKASA RAO, P.S., ISRAEL, P. and RAO, Y.S. 1971. Epidemiology and control of the rice hispa, *Dicladispa armigera* Olivier. Oryza *8* (2) 345-359.

PRAKASA RAO, P.S., KALODE, M.B., DAS, P.K., VERMA, A. and ISRAEL, P. 1970B. Field evaluation of new insecticides applied in standing water in the control of rice stem borers. Oryza *7*, 113-120.

PRAKASA RAO, P.S., RAO, Y.S. and ISRAEL, P. 1970A. Problems and prospects in the chemical control of rice stem borers. Oryza *7*, 89-102.

RAHA, S.K. 1975. A note on the effectiveness of some modern insecticides against *Parasa bicolor* on paddy rice. Sci. Cult. *41* (2) 80-82.

RAI, B.K. 1974. Losses caused by the paddy bug and "red rice" in Guyana. Food and Agric. Organ. U.N. Pl. Prot. Bull. *22* (4) 82-86.

RAI, P.S. 1975. *Tetraneura radicicola* Strand, a new aphid on rice seedlings. Curr. Sci. *44* (13) 486-487.

RINCON, A.B. 1967. Principal pests of rice in Colombia. *In* Studies of the Federation of National Rice Growers. Cooperative Program of ICA−Fedearroz, Bogota, Colombia.

ROLSTON, L.H. and ROUSE, P.   1960.   Control of grape colaspis and rice water weevil by seed or soil treatment.   Ark. Exp. Sta. Bull. *624.*

ROLSTON, L.H. and ROUSE, P.   1965.   The biology and ecology of grape *colaspis flavida* Say, in relation to rice production in the Arkansas Grand Prairie. Ark. Agric. Exp. Sta. Bull. *694.*

ROTHSCHILD, G.H.L.   1970.   Some notes on the effects of rice ear bugs on grain yield.   Trop. Agric. Trin. *47* (2) 145-149.

SAJJAN, S.S., SINGH, J. and KHANGURA, J.S.   1975.   Chemical control of rice hispa, *Hispa armigera* Olivier.   J. Res. Punjab Agric. Univ. *12* (3) 272-275.

SATPATHY, J.M.   1970.   Further studies on the insecticidal control of the rice stem borer and gall midge.   Oryza 7, 85-88.

SINGH, S.R., SUTANDAR, A. and BENJAMIN, B.   1974.   Performance of different galecron formulations applied to paddy water vs *Tryporyza incertulus* in Indonesia.   J. Econ. Entomol. *67*, 131-132.

SOGAWA, K.   1970.   Studies on the feeding habits of the brown plant hopper. II.   Honeydew excretion.   Jpn. J. Appl. Entomol. and Zool. *14*, 134-139.

SOGAWA, K.   1972.   Studies on the feeding habits of the brown plant hopper. III.   Effects of amino acids and other compounds of sucking response.   Jpn. J. Appl. Ent. Zool. *16*, 1-7.

SOMMEIJER, M.J.   1975.   Rice water weevils on rice.   Food and Agric. Organ. U.N. Pl. Prot. Bull. *23*, 161-162.

SRIVASTAVA, A.S., NIGAM, P.M., LAL KATIYAR, S.S. and AWASTHI, B.K. 1975.   Chemical control of paddy root weevil, *Echinocnemus pryzae* Mshll. Labdev. J. Sci. and Tech. Life Sci. *13-B,* 79-81.

SRIVASTAVA, A.S and SAXENA, H.P.   1967.   The rice bug, *Leptocorisa varicornis* Fabricius and allied species.   *In* Major Insect Pest of the Rice Plant. Proc. Symp. Int. Rice Res. Inst., 1964. Los Baños, Philippines. Johns Hopkins Univ. Press, Baltimore.

SWANSON, M.C. and NEWSOM, L.D.   1962.   Effect of infestation by rice stink bug, *Oebalus pugnax,* on yield and quality in rice.   J. Econ. Entomol. *55* (6) 877-879.

TSUZUKI, H., ISOGAWA, K. and WATANABE, N.   1976.   Occurring of the new insect "Rice water weevil" in Aiti Prefecture. Ecology of the new insect "Rice water weevil."   Food and Agric. Organ. U.N. Pl. Prot. Bull. *30* (9) 341-346.

7

Insect Pests of Stored Rice[1]

Robert R. Cogburn

Losses incurred by agriculture after crops are harvested are particularly significant because all costs of production—labor, fuel, fertilizer, etc.—are already invested. Thus, the value of the crop is at its highest point and loss of relatively small percentages can be financially significant.

Stored-product insects damage agricultural produce during storage, processing or transportation. Many insect species are found associated with grain and grain products but only about 50 are injurious, either occasionally or frequently (Cotton 1956). The more common species are listed in Table 7.1.

Most stored-product insects are cosmopolitan in distribution. The same species do the same types of damage everywhere in the world and many species are omnivorous with respect to stored commodities. For example, the cigarette beetle, *Lasioderma serricorne* (F.), breeds in tobacco, red pepper, cinnamon, other spices, flour, cornmeal, grains such as rice, wheat and sorghum, and processed commodities such as breakfast cereals or animal feeds. The saw-toothed grain beetle, *Oryzaephilus surinamensis* (L.), attacks all grains, processed farinaceous commodities (flour, cornmeal, etc.), dried fruits (i.e., raisins) and nuts of various kinds. The relative abundance of particular species may vary from country to country. Storage insects are most abundant in areas that have warm, humid climates, but they can and do cause severe problems anywhere that grain is stored. Unfortunately, most rice is produced and stored in tropical or subtropical areas where conditions are ideal for the growth of insects.

Rice is thought to be somewhat less susceptible to infestation than other grains because of protection afforded by the hull. Breese (1960) examined samples of rice infested with rice weevils, *Sitophilus oryzae* (L.), and found no insects developing in any grain that had an intact hull.

[1]Cooperative investigations with the U.S. Department of Agriculture, Texas Agricultural Experiment Station and Texas Rice Improvement Association.

TABLE 7.1.   STORED-PRODUCT INSECTS ASSOCIATED WITH RICE AFTER HARVEST

Common name[1]	Scientific name	Types of storage infested[2]
Insects often encountered in rice in the United States		
Angoumois grain moth (L)	Sitotroga cerealella (Olivier)	RR
Lesser grain borer (C)	Rhyzopertha dominica (F.)	RR,RC
Rice weevil (C)	Sitophilus oryzae (L.)	RR,RC
Red flour beetle (C)	Tribolium castaneum (Herbst)	*
Almond moth (L)	Ephestia cautella (Walker)	*
Sawtoothed grain beetle (C)	Oryzaephilus surinamensis (L.)	RR,W,M,RC
Cigarette beetle (C)	Lasioderma serricorne (F.)	W,M,RC
Flat grain beetle (C)	Cryptolestes pusillus (Schönherr)	RR,RC
Rusty grain beetle (C)	C. ferrugineus (Stephens)	
"Flour mill beetle" (C)	C. turcicus (Grouvelle)	
Indian meal moth (L)	Plodia interpunctella (Hübner)	
Hairy fungus beetle (C)	Typhaea stercorea (L.)	RR,RC
"Sap beetle" (C)	Carpophilus pilosellus (Motschulsky)	RC,W
Psocids (book louse) (P)	Liposcelis spp.[3]	*
Dermestids (C)	Trogoderma spp.	RC,M
Insects occasionally or rarely encountered in rice in the United States		
Confused flour beetle (C)	Tribolium confusum Jacquelin duVal	RC,W
American black flour beetle (C)	Tribolium audax Halstead	RC
Granary weevil (C)	Sitophilus granarius (L.)	RR,RC,MR
Maize weevil (C)	Sitophilus zeamais Motschulsky	RR,RC
Larger black flour beetle (C)	Cynaeus angustus (Le Conte)	RC
Black carpet beetle (C)	Attagenus megatoma (F.)	RC,M

TABLE 7.1. (CONTINUED)

Common name[1]	Scientific name	Types of storage infested[2]
Cadelle (C)	Tenebroides mauritanicus (L.)	RC
Foreign grain beetle (C)	Ahasverus advena (Waltl)	RC
Squarenecked grain beetle (C)	Cathartus quadricollis (Guérin-Méneville)	RC
Merchant grain beetle (C)	Oryzaephilus mercator (Fauvel)	*
Dried fruit beetle (C)	Carpophilus hemipterus (L.)	RC
Longheaded flour beetle (C)	Latheticus oryzae Waterhouse	RC
Smalleyed flour beetle (C)	Palorus ratzeburgi (Wissman)	RC
Yellow mealworm (C)	Tenebrio molitor L.	RC
Dark mealworm (C)	Tenebrio obscurus F.	RC
Pink scavenger caterpillar (L)	Pyroderces rileyi (Walsingham)	RC
Meal moth (L)	Pyralis farinalis L.	W
Spider beetles (C)	Ptinidae	RC,W,M
	Other insects associated with stored rice	
Khapra beetle (C)	Trogoderma granarium Everts	
"Rice moth" (L)	Corcyra cephalonica (Stainton)	
Lesser mealworm (C)	Alphitobius diaperinus (Panzer)	
"Siamese grain beetle" (C)	Lophocateres pusillus (Klug)	
Moths (L)	Ephestia spp.	
Grain mite (Acarina)	Acarus siro L.	

[1]The order of the species is indicated in parentheses: (C) Coleoptera, (L) Lepidoptera, (P) Pscocoptera

[2]The author has observed these species infesting the type of storage indicated: RR—rough rice (paddy) in bins; MR—bagged milled rice, RC—rail cars, W—warehouses, M—mills, *indicates all storage situations

[3]Probably the cereal psocid, L. divanatorius (Mueller)

Cogburn (1974) found that adult rice weevils did not reproduce on rice with intact hulls and starved to death within 2 weeks. However, lesser grain borers, *Rhyzopertha dominica* (F.), and Angoumois grain moths. *Sitotroga cerealella* (Olivier), both effected some reproduction on rice grains that had been microscopically examined and selected for perfect hulls. Larvae of these 2 species were either able to penetrate through the hull or to force entry through gaps between the palea and lemma that were too small to be detected even under high magnification.

Thus, while the rice hull is a partially effective barrier to insect penetration of the grain, it cannot be depended upon to preclude infestation in rice that has been harvested and stored as paddy (rough rice). The proportion of grains with hull defects will depend on many things, i.e., method of harvesting, subsequent handling, environmental conditions during the growing season, fertilization rate, combine settings and variety of rice. In addition, rice is subject to infestation after the hulls have been removed in the milling process.

Milled rice (white rice, polished rice) is a rather poor growth medium for storage insects because essential nutrients in the bran coat and germ are missing (Le Cato 1975). However, infestability of milled rice relates directly to the degree of milling (McGaughey 1970) and rice is sometimes deliberately lightly milled in order to retain more nutrients and vitamins. Such rice will support much more infestation than will rice from which all of the bran has been removed (McGaughey 1974). In any milled rice, however, the major problems associated with infestation stem from contamination rather than from destruction of the commodity. Contaminants of milled rice include live and dead insects, insect fragments, excreta, webbing, cast larval skins, eggs, etc. The significance of such contaminants varies with local laws and traditions, but in most countries allowable foreign matter and especially insect contamination in processed food commodities is strictly limited.

While infestation is somewhat limited in rough rice because the hull affords partial protection and in milled rice because it often lacks essential nutrients, brown rice is an excellent medium for the growth of storage insects. Brown rice has had the hulls removed, but the germ and bran coat remain in place. Fortunately, rice is not normally stored as brown rice, but the United States exports a significant amount of brown rice to countries that have milling facilities. This appears to be a growing industry. Rice remaining in the brown form for extended periods will eventually be subject to severe problems of infestation.

In the United States, infestation of milled rice is a greater problem in rice that is exported than in rice consumed locally. Rice for local consumption is usually "well milled," i.e., removal of the bran coat is total and the rice is milled, packaged and distributed in a continuous operation. Thus, even if a few eggs or young larvae survive the milling process, they usually do not develop to the adult stage and breed because the rice normally is consumed within a relatively short time after milling. Rice for export is often "lightly" or "reasonably well" milled, depending upon the desires of the recipient country, and there is a longer time between processing and consumption. This rice will support insect development because some of the bran coat remains on the kernels. Also, this rice is frequently exposed to sources of infestation and is almost certain to contain a few insects when it leaves the port (Cogburn 1973 A,B). Thus, the time spent in transit and the temperature of the rice are the factors that determine whether these few insects develop into large, damaging populations. A damaging insect population cannot develop if the area within which the rice is transported has a mild climate and transit time is short. If the transit time is protracted and transportation is largely within tropical waters, extremely large, multispecies insect populations can develop. When they do, huge financial losses result and international incidents may be precipitated. These often result in angry words between diplomats, suspicion of the quality of United States rice and general dissatisfaction among buyers in the recipient countries. Such financial losses are real and measurable but losses in terms of international relations and good will are no less important, even though they defy absolute measurement (Howe 1965; Cogburn 1977).

Of course, the problem is not peculiar to the United States (Freeman 1965). Elimination of insect contamination in exported foodstuffs, whether rice or another commodity, is one of the major problems faced by any country that exports agricultural produce (Freeman 1974).

STORED RICE INSECTS

The insect species discussed in this chapter include the most common pests of stored rice in North America. Literature available suggests that these same species are also the major pests of rice throughout the world. References to these species from various countries include Trinidad (Breese 1960), Ceylon (Easter 1954), Taiwan (Li 1953), Sierra Leone (Prevett 1971), Egypt (Bishara et al. 1973), all of western Europe (Freeman 1973) and many points in Asia (Freeman 1974).

To many people, any insect that is found in grain or grain products is a "weevil." There are indeed several species of weevils that damage stored grain but the term is frequently a misnomer. In addition, such vague and inconsistent terms as "bran bugs" or "millers" are used commonly. Use of such terms often leads to misidentification of insects and/or misapplication of control measures. All people involved in the storage, handling or processing of grain should become adept in the proper identification of storage insects and should learn something of the life history and habits of each important species. This knowledge would facilitate better insect control at reduced cost. In other words, "know thine enemy."

All of the really significant storage insects are either beetles (Coleoptera) or moths (Lepidoptera). Thus, all undergo complete metamorphosis. They have 4 distinct life stages: egg, larva, pupa and adulthood. The type and extent of damage that the species can inflict on rice depends on where they develop in relation to the grain.

Rice infesting insects are divided into 2 groups. One group develops from larva to adult inside kernels of rice so the endosperm is consumed by the developing insect. The other group, which includes most species, develops outside the kernels. These insects feed on the bran coat, germ, dust or other debris and do relatively little damage to the endosperm. Members of the first group are quite destructive because of the rice they consume. Damage caused by the second group usually relates more to contamination rather than to destruction of the grain, although some of the insects can cause rather large losses of grain when infestations cause heating and spoilage. Some moth species form clumps of kernels with their webs. In any case, insect carcasses must be removed from the rice. All these factors can result in loss of rice during the cleaning process and infestations in milled rice sometimes necessitate remilling or recleaning.

Insects that Develop Inside Kernels of Rice

Weevils.—The weevils that infest rice belong to the family Curculionidae and are characterized by the front part of the head being elongated into a snout or proboscis. The mouthparts are located on the tip of the snout. This structure is utilized in both the feeding and reproductive activities.

There are 3 weevils that are found in stored rice: the rice weevil, the maize or corn weevil, and the granary weevil. Of the 3, the rice weevil (Fig. 7.1) is by far the most common in rice in the southern United States. The maize and granary weevils are rare in this area, but both are widely distributed and are significant in other parts of the world.

Rice and maize weevils are strong fliers but granary weevils do not have functional flight wings. All are dark brown to black and about 3 mm long, although the size can vary with the type of grain on which the insects developed and possibly among different strains of the same species. Rice weevils and maize weevils have 4 lighter colored spots on the back, 2 on each wing cover. The granary weevil has a uniform color. The species can be separated by the shape and arrangement of the pits on the pronotum, the shape of the genitalia and other characteristics. For an exact account of the taxonomy of the rice and maize weevils, the reader is referred to Boudreaux (1969).

FIG. 7.1. ADULT RICE WEEVIL, *SITOPHILUS ORYZAE*, WITH
KERNELS OF LONG GRAIN RICE SHOWING TYPICAL
DAMAGE

All 3 species of weevils have similar habits and cause the same type of
damage. The female chews a hole into a kernel of rice, inserts her
ovipositer and deposits an egg at the bottom of the hole. As the ovipositer is withdrawn, a mucous-like substance is secreted and fills the hole
to the surface of the grain. This substance dries into a hard plug that
protects the egg and later the larva from desiccation. When the egg
hatches, the larva feeds on the endosperm and hollows out most of the
kernel as it grows. The pupal stage and early adult stage are also spent
inside the kernel. When the exoskeleton becomes hard and resistant to
desiccation, the adult weevil chews its way out of the kernel and seeks a
mate and the cycle begins anew.

Lesser Grain Borer.—The lesser grain borer (Fig. 7.2) is perhaps the
most potentially destructive insect that infests stored rice. It is a member of the family Bostrichidae (powder post beetles) and has the characteristic shape of this family—elongated and cylindrical with the head
deflexed beneath the hoodlike prothorax. Lesser grain borers are about
3 mm long and 1 mm wide and are dark brown. They can live deep within a grain mass, develop large populations and destroy many kilograms
of rice before their presence is known.

The eggs of the borers are not implanted in kernels, but are deposited
loose in the grain mass. The larvae, as they hatch, bore into individual
kernels and from then on develop much as weevils do. The adults are
relatively long lived, capable of flight, and feed and reproduce for several months. Each female lays from 300 to 500 eggs (Cotton 1956).

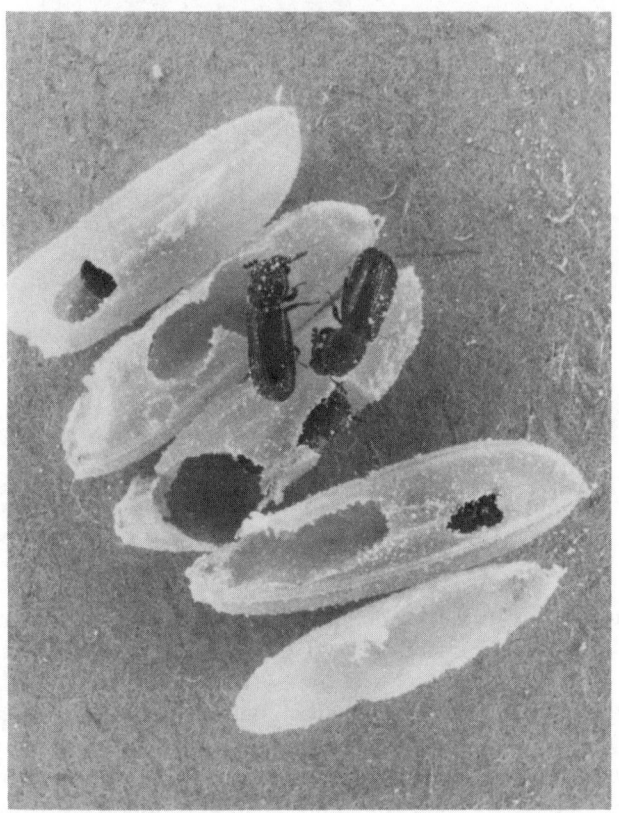

FIG. 7.2. ADULT LESSER GRAIN BORERS, *RHYZOPERTHA DOMINICA*, WITH KERNELS OF LONG GRAIN BROWN RICE SHOWING TYPICAL DAMAGE

Angoumois Grain Moth.—The Angoumois grain moth (Fig. 7.3) is probably the most abundant of the insects that infest stored rough rice in the southern United States. However, its capacity for damage is limited by its inability to infest rice deep within a grain mass. Infestations are almost invariably limited to the top 30 or 40 cm of the grain surface.

The moth is small (5–6 mm) and tan, and the posterior edges of the wings are fringed with fine hairs. The females deposit eggs, usually in clusters, in protected places in the grain or storage structure (beneath the husk of grains with broken hulls, small crevices in bin walls, etc.). When the eggs hatch, each larva bores into a rice kernel and there develops to the adult stage. Just before pupation, the larva cuts a circular "window" to the outside of the grain and webs it over lightly. When the adult moth emerges, it pushes the window aside and thus escapes from the grain. The damaged grains in Fig. 7.3 show typical damage by the Angoumois grain moth. This type of damage is unique among storage insects.

FIG. 7.3. ADULT ANGOUMOIS GRAIN MOTH, *SITOTROGA CEREALELLA*, WITH KERNELS OF MEDIUM GRAIN ROUGH RICE SHOWING TYPICAL DAMAGE

Insects that Develop Outside Kernels of Rice

The other insect species that will be discussed here do not penetrate the kernels. They may be found in stored paddy, processed rice, rice bran, or processed products such as breakfast cereals. They usually do not consume significant amounts of grain, but they can cause grain to be lost through heating, mold growth and spoilage or, in the case of some moths, webbing.

Over half the rice produced in the United States is exported as processed (milled) rice. The following species are particularly troublesome to the export industry.

FIG. 7.4. ADULT RED FLOUR BEETLES, *TRIBOLIUM CASTANEUM*, ON KERNELS OF LONG GRAIN BROWN RICE

Flour Beetles.—The flour beetles, members of the genus *Tribolium,* are perhaps the most common of all the stored product insects. In the southern United States rice belt, the red flour beetle, *Tribolium castaneum* (Herbst) (Fig. 7.4), is by far the most abundant. The confused flour beetle, *T. confusum* Jacquelin duVal, is generally more abundant in the colder regions. The 2 species closely resemble each other; they are about 3−4 mm long, 1 mm wide, dark reddish brown and somewhat flattened. The red flour beetle is a strong flier, but the confused flour beetle, although it has apparently functional wings, is flightless. Magnification is required to distinguish between these 2 species. The antennal segments of the confused flour beetle enlarge gradually from base to tip. The terminal 3 segments of the antenna of the red flour beetle are abruptly larger than the preceeding segments.

Neither species consume the endosperm of rice, but both feed on the bran coat, germ and loose flakes of bran or on grain dust. In the southern United States, the red flour beetle is abundant in rough rice stored on farms or in commercial silos, in mills and milling machinery, and in rail cars, ships and warehouses.

Two other species of *Tribolium* infest grain in the United States (Cotton 1956) but are rare in rice. These are the black flour beetle, *T. Audax* (Halstead), and *T. destructor* Uyttenboogaart. Presumably, either species will infest rice if they find it available.

FIG. 7.5. ADULT CIGARETTE BEETLES, *LASIODERMA SERRICORNE*, ON GRAINS OF LONG GRAIN RICE

Cigarette Beetle.—As the name implies, the cigarette beetle (Fig. 7.5) is a major pest of tobacco but is an omnivorous feeder and extremely abundant in rice mills and port warehouses on the United States Gulf Coast (Cogburn 1973A). Strangely, it was relatively uncommon in box-cars that delivered rice and wheat flour to Gulf Coast ports (Cogburn 1973B). Between 1953 and 1963, it was the third most common insect found infesting cargoes of rice and rice bran shipped from the United States to Great Britain (Freeman 1965).

The insect is about 2.5 mm long, more or less rounded and reddish-brown. It is a strong flier and in badly infested tobacco warehouses large numbers can be caught in light traps (Childs 1958). In rice mills and warehouses it maintains populations in trash bran or other farinaceous substances and infests clean rice as it is being processed or shipped.

Sawtoothed Grain Beetle.—The sawtoothed grain beetle is occasionally found in rice mills and warehouses (Cogburn 1973A) and was reported by Freeman (1965) to occur frequently in milled rice from the United States that was imported into England. The adults are about 2.5 mm long and flattened. The lateral edges of the thorax bear 6 tooth-like spines on each side which gives the insect its name. A closely related species, the merchant grain beetle, looks almost like the sawtoothed grain beetle, occurs in similar habitats and does similar damage. The feeding habits of both are similar to those of the flour beetles.

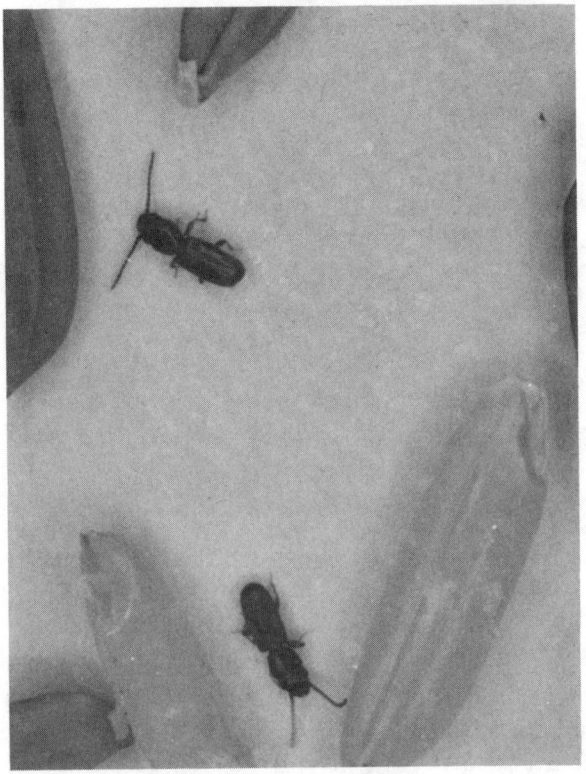

FIG. 7.6. ADULTS OF *CRYPTOLESTES* SPP. WITH LONG GRAIN RICE

Cryptolestes Complex.—The smallest of the grain infesting beetles are several members of the genus *Cryptolestes* (Fig. 7.6). Three species were recently identified from stored rough rice in Texas: the flat grain beetle,

C. pusillus; the rusty grain beetle, *C. ferrugineus;* and the "flour mill beetle," *C. turcicus.* The adult insects are 1.5 to 2 mm long and less than 1 mm wide. All 3 species do similar damage. For specific identification the reader is referred to Biege and Partida (1976). Although these insects are quite small, they can build up to huge populations in stored grain and thus cause heating, spoilage and contamination of the commodity.

Cryptolestes larvae normally crawl about among the interstices between kernels of grain and feed on flakes of bran, grain dust, etc. Occasionally, however, a feeding larva will tunnel into the germ of a rice kernel and thus be hidden from inspectors or others examining the grain for infestation. They do not normally pupate inside the germ or damage the endosperm. These insects infest rough or milled rice, rice bran, rice flour or other processed rice products.

Moths.—Except for the Angoumois grain moth, which has been discussed previously, the species of Lepidoptera that infest stored rice do not penetrate the kernels and usually do not damage the endosperm. Rather, they feed on the bran and germ, flakes of bran, grain dust or other debris. However, moth infestations can and do cause some of the most severe financial losses experienced by those who store, process or transport rice.

In bulk storage of rough rice, infestations are restricted to the grain surface. In milled rice, each package is subject to infestation, usually on the surface area, but sometimes throughout, depending on the size of the package and the type of storage.

In the southern United States, almost all moth infestations in rice are caused by the almond moth, *Ephestia cautella,* (Fig. 7.7). The Indian meal moth, *Plodia interpunctella,* (Fig. 7.8) is significant in this area, but the species is not nearly as abundant as is the almond moth. The "rice moth," *Corcyra cephalonica,* is often thought to be a pest of rice in the southern United States, but in 10 years of experience, this author has never encountered this species. It apparently is a major pest of rice in other areas of the world because it is listed by Freeman (1974) in his tabulation of insects in imported foods in Britain. In addition to those species named, other members of the *Ephestia* genus [*E. kuhniella* (Zeller), *E. elutella* (Hübner), and *E. figuliella* Gregson] probably can infest rice but rarely do so in the United States.

Regardless of the species, the habits and the damage caused by the stored-product moths are similar. An adult female produces about 200 eggs (rarely, as many as 400). The eggs are laid singularly, scattered at random over the surface of the commodity and usually hatch in about 4 or 5 days. The larvae (Fig. 7.9) are equipped with spinnerets located near the mouthparts, so as they feed, they weave together particles of dust or

FIG. 7.7. ADULT ALMOND MOTH, *EPHESTIA CAUTELLA*, ON KERNELS OF LONG GRAIN BROWN RICE

bran or kernels of rice with fine, silken threads. This habit accounts for most of the damage other than contamination. Heavy infestations can completely mat over the surface of grain in a bin (Cotton 1956), which restricts air movement in the grain mass and promotes an increase in moisture content, heating and spoilage. Even if infestations are not that severe, grains that are webbed together are removed by grain cleaning equipment prior to milling and are discarded along with straw and other trash. Thus, moth infestations cause the loss of far more grain than is consumed by the insects.

The larval stage lasts about 4 weeks. The pupa transforms into the adult stage in about 7 days.

FIG. 7.8.   ADULT INDIAN MEAL MOTH, *PLODIA INTERPUNCTELLA*, ON KERNELS ON LONG GRAIN BROWN RICE

Insects that Feed on Fungi

The hairy fungus beetle, *Typhaea stercorea,* and a sap beetle, *Carpophilus pilosellus,* are extremely abundant in rail cars, warehouses and sometimes farm storages. Both species feed on fungi or other microorganisms and do not damage sound rice, but their presence indicates that the grain is going out of condition or that unsanitary conditions exist in the storage facility. However, the species occasionally do cause financial losses, not from any damage that they do, but because they are misidentified as damaging pests. For example, when rice lodges in the field, some of the fallen heads become moistened by rain or irrigation water, and fungi or molds grow on these heads, thus providing a source of food for fungus feeding insects. When the grain is harvested, hairy

FIG. 7.9. LARVAE OF THE ALMOND MOTH, *EPHESTIA CAUTELLA*, ON KERNELS OF LONG GRAIN BROWN RICE

fungus beetles are collected along with the rice by the combine. When the grain is being dried, heat and desiccation drives the insects from the grain. Then, farmers or dryer operators often mistake them for damaging species and sometimes apply unneeded fumigations. As noted, both species are common in rail cars, probably because trash grain is seldom removed from boxcars and often spoils (Cogburn 1973B). The adults are active insects. Movement of the car disturbs them and causes them to migrate throughout the car. Then, they are seen on the commodity by inspectors who often mistake them for damaging species. As a result the rice is fumigated unnecessarily.

ORIGINS OF INFESTATIONS

Residual Insect Populations

Most infestations of storage insects in any commodity initiate when the clean commodity is stored or processed near trash grain or grain products that remain in storage areas after previously stored commodities have been moved out. For example, rough rice in the United States is frequently stored on farms in cylindrical metal bins with perforated metal floors that are elevated about 1 m above a concrete slab. The bins are equipped with aeration fans that blow air into the void between the concrete and the floor through the perforations and the grain mass. Dust, flakes of bran, particles of grains, etc., fall through the perforations into the void area where they remain indefinitely. Such trash maintains populations of insects between one storage period and the next. When fresh rice is binned, the insects immediately invade the new grain.

Residual populations will remain in almost any type of storage area or transportation carrier unless the unit is properly cleaned between uses. For example, boxcars delivering commodities to Gulf Coast ports were found to contain farinaceous debris 77% of the time (Cogburn 1973B). Grain dust that collects in elevators or rice mills is nutritious and supports insect populations. Spillage and trash of any kind is a potential source of insects.

Migration.—"Cross-contamination" occurs when clean commodities are stored adjacent to infested ones. Many storage insects are strong fliers and can migrate to areas remote from their origin. Rice weevils, lesser grain borers, red flour beetles and all moths are capable of sustained flight and undoubtedly migrate from one food source to another. However, little information is available concerning the migratory powers of storage insects and the conditions that stimulate the insects to migrate.

Field Infestation.—All rice in the United States is machine-harvested. This practice has largely eliminated field infestation as a source of many storage insects though years ago when rice was cut, tied into bundles and left in the field to dry, it frequently became infested before threshing. Today, rice is harvested by combine when the moisture content of the grain is about 22% and is then dried with heated air to about 12.5% moisture. Thus, the grain does not remain long enough in the field in a state suitable for insect development. However, Angoumois grain moths can be trapped in fields of standing rice and undoubtedly, some eggs or larvae are in the harvested grain. Whether these survive harvesting and drying is not known. Some countries do not use mechanized harvesting equipment, so the significance of field infestation depends upon local cultural practices.

CONTROL OF STORAGE INSECTS

Cultural Control.—Sanitation is the cheapest and among the most effective of the control measures that can be used against storage insects. The rationale behind this practice is quite simple—if the insects are deprived of food, they cannot survive, breed and disseminate to stored commodities. A sanitation program must be designed to remove all trash grain, grain dust or other farinaceous debris from the storage premises. All bins, floors, window sills, milling machinery, harvesting machinery, conveying equipment, etc., must be cleaned thoroughly after each use and in some cases, each day. Trash materials must not be allowed to collect or be left in or near the storage facility, they must be taken to a place remote from the facility and buried, burned or otherwise destroyed.

The importance of sanitation cannot be overemphasized. The amount of food and money that could be saved worldwide by the application of strict, conscientious sanitation programs is virtually inestimable.

Chemical Control.—Chemical insecticides are used to supplement sanitation programs, but they are not cure-alls that can overcome dirty conditions. Space does not permit a completely comprehensive discussion of the various insecticides and fumigants that can be used or of the various methods of application. For a complete discussion of useful chemicals, the reader is referred to Monro (1961) and Harein and de las Casas (1974).

Insecticidal Sprays.—Insecticidal sprays may be applied to exposed surfaces of empty grain bins, warehouses, mills or other areas as residual treatments after the facilities have been cleaned and before new commodities are stored or processed. These sprays are usually formulated as water emulsifiable liquids or powders. Care must be exercised to treat all cracks and crevices where farinaceous debris may lodge or where insects can hide.

Many chemicals, including malathion, methoxychlor, synergized pyrethrins, lindane, fenitrothion, pirimiphos methyl, chlorpyriphos methyl and dichlorvos, have been used as residual sprays for grain bins and other facilities. Specific materials are not recommended here because permissable chemicals, methods of use, dosages and residue tolerances are strictly governed by local laws, and any user of an insecticide must first determine what is approved for use in his country or area.

Some insecticides can be applied directly to bulk rough rice as it is being put into final storage. These materials are often called "protectants." They are applied only to rough rice (paddy) and not to brown or milled rice. This type of treatment is meant to protect the rice from infestation, not to eliminate an established infestation. Malathion has been used as a protectant for many years with remarkable success (Freeman 1974), but recently, many species of storage insects around the world have developed resistance to malathion (Bengston et al. 1975; Zettler 1975). Thus, the utility of this chemical is now questionable and an alternative material is badly needed (Cogburn 1976).

Some insecticides are applied in mills or warehouses as aerosol space treatments for control of flying insects, particularly moths, and as a supplement to sprays and protective treatments. These aerosols are mists, fogs or vapors that are blown into the building with specialized equipment and special formulations of the insecticide. Dichlorvos is especially effective when used in this manner (Cogburn 1975). Contamination of the commodities stored in the area must be avoided and the treatments must be applied in the absence of personnel other than the applicator. Automatic systems that dispense aerosols at predetermined times are available in some areas.

Fumigants.—People often confuse aerosol smokes and fogs with fumigants. Aerosols consist of very small particles of solutions suspended in air. Fumigants are gasses and act as individual molecules (Monro 1961). Aerosol particles impinge on the setae of the insects and puddle on the membranes at the base. The insecticide enters the insect by penetration through the integument (Anon. 1971). Fumigants enter the insect almost entirely through the respiratory system.

Fumigation is the preferred method of eliminating an established insect population from stored rice or rice products. Mills and warehouses, if properly sealed, can be fumigated to clean up indigenous populations of insects that were not entirely eliminated by sanitation procedures. A successful fumigation, regardless of the fumigant, always depends on holding a lethal concentration of the fumigant in air long enough to kill all metamorphic stages of the insects. When the insects are in the egg or the pupal stage, respiration is slight and these 2 stages are the most difficult to kill by fumigation. Although some fumigants will penetrate the chorion of the egg or the integument of the pupa, lethal concentrations must usually be maintained until eggs become larvae or pupae become adults—stages that are highly susceptible to the lethal action of fumigants.

Fumigants can be formulated as liquids or solids but must reach the gaseous state in order to become effective. Liquid fumigants are mixtures of highly volatile substances, commonly, carbon tetrachloride, carbon bisulfide and ethylene dichloride. All these materials, after volatilization, weigh more than air. Liquid fumigants are usually applied to the surface of a grain mass and penetrate downward by gravity. They are particularly useful in farm bins and very tall concrete silos when other methods of fumigation are often impractical.

Methyl bromide, one of the most common fumigants in the world, has a wide variety of uses. These include space fumigation of mills, warehouses, etc.; fumigation of grain in large, flat storages equipped with recirculation systems; fumigation of bagged rice under plastic tarpaulins; and fumigation of various carriers such as rail cars, ships, trucks and barges. The chemical is available in pressurized canisters that contain from 454 g to 170 kg or more. While in the canister under pressure, methyl bromide is a liquid. However, the boiling point is 3.6°C (38.5°F), so it volatilizes to the gaseous state almost as soon as the cylinder is opened and the pressure released.

Methyl bromide is colorless and odorless and thus cannot be detected by man's senses. Therefore, it is usually formulated with about 2% chloropicrin (tear gas) so the applicator or other personnel can detect its presence should they accidently enter an area under fumigation. This material must be used with extreme caution. Not only is the acute toxicity to man and other mammals extremely high; sublethal exposure can also result in permanent, chronic damage to the liver. In addition, the liquid or high concentrations of the gas may cause severe skin blisters in man (Monro 1961).

Phosphine gas (PH_3), a fumigant introduced in the mid 1950s, is generated from tablets, pellets or powders of aluminum phosphide. It has become one of the most widely used fumigants worldwide. Upon contact with atmospheric moisture, aluminum phosphide undergoes a chemical reaction, the product of which is hydrogen phosphide (phosphine gas). The gas is flammable and under certain conditions, it will ignite spontaneously; thus, the label instructions must be followed implicitly to avoid fires. Although the gas has a distinctive odor, it is not imparted to the product being fumigated.

Phosphine can be used to fumigate rice, both in bags and in bulk (Cogburn and Tilton 1963; Tilton and Cogburn 1965). Successful fumigation with this material is closely linked to temperature. Thus, applicators must monitor ambient conditions and take appropriate steps to insure that lethal concentrations of gas are maintained long enough to kill all stages of the target insects.

REFERENCES

ANON. 1971. Untitled. National Pest Control Operators Newsl. *31*, (10) 3-7.

BENGSTON, M., COOPER, L.M. and GRANT-TAYLOR, F.J. 1975. A comparison of bioresmethrin, chlorpyrifosmethyl and pirimiphos-methyl as grain protectants against malathion-resistant insects in wheat. Queensland J. Agric. and Animal Sci. *32*, (1) 51-78.

BIEGE, C.R. and PARTIDA, G.H. 1976. Taxonomic characters to identify three species of *Cryptolestes* (Coleoptera: Circujidae). J. Kan. Entomol. Soc. *49*, 161-164.

BISHARA, E.I., KOURA, A. and EL-HALFAWY, M.A. 1973. Oviposition preference of the grainary and rice weevils on Egyptian rice varieties and recommendations for grain protection. Bull. Soc. Entomol. Egypte *56*, 145-150.

BOUDREAUX, H.B. 1969. The identity of *Sitophilus oryzae*. Ann. Entomol. Soc. Amer. *62*, 169-172.

BREESE, M.H. 1960. The infestibility of stored paddy by *Sitophilous sasakii* (Tak.) and *Rhysopertha dominica* (F.). Bull. Entomol. Res. *51*, 599-630.

CHILDS, D.P. 1958. Warehouse fumigation of flue-cured tobacco with HCN to control the cigarette beetle. J. Econ. Entomol. *51*, 417-421.

COGBURN, R.R., and TILTON, E.W. 1963. Studies of phosphine as a fumigant for sacked rice under gas-tight tarpaulins. J. Econ. Entomol. *56*, 706-708.

COGBURN, R.R. 1973A. Stored-product insect infestations in port warehouses of the Gulf Coast. Environ. Entomol. *2*, 401-407.

COGBURN, R.R. 1973B. Stored-product insect infestations in box-cars delivering flour and rice to Gulf Coast ports. Environ. Entomol. *2*, 427-431.

COGBURN, R.R. 1974. Domestic rice varieties: Apparent resistance to rice weevils, lesser grain borers and Angoumois grain moths. Environ. Entomol. *3*, 681-685.

COGBURN, R.R. 1975. Dichlorvos for control of stored-product insects in port warehouses: Low-volume aerosols and commodity residues. J. Econ. Entomol. *69*, 361-365.

COGBURN, R.R. 1976. Pirimiphos-methyl as a protectant for stored rough rice: Small bin tests. J. Econ. Entomol. *69*, 369-373.

COGBURN, R.R. 1977. Susceptibility of varieties of stored rough rice to losses caused by storage insects. J. Stored Prod. Res. *13*, 29-34.

COTTON, R.T. 1956. Pests of stored grain and grain products. Revised Edition. Burgess Publishing Co., Minneapolis, Minn.

EASTER, S.S. 1954. Infestation control in stored rice. Trop. Agric. Peradeniya *110*, 217-219.

FREEMAN, J.A. 1965. On the infestation of rice and rice products imported into Britain. Proc. 12th Int. Congr. Entomol., 1964. London.

FREEMAN, J.A. 1973. Problems of infestation by insects and mites of cereals stored in Western Europe. Ann. Technol. Agric. *22* (3) 509-530.

FREEMAN, J.A. 1974. A review of changes in the pattern of infestation in international trade. Eur. Mediterr. Plant Prot. Organ. Bull. *4* (3) 251-273.

HAREIN, P.K. and DE LAS CASAS, E. 1974. Chemical control of stored grain insects and associated micro- and macro-organisms. *In* Storage of Cereal Grains and Their Products, *V*, Revised, Am. Assn. Cereal Chem., St. Paul, Minn.

HOWE, R.W. 1965. Losses caused by insects and mites in stored foods and feeding stuffs. Nutr. Abst. Ev. *35*, 285-293.

LE CATO, G.L. 1975. Red flour beetle: Population growth on diets of corn, wheat, rice or shelled peanuts supplemented with eggs or adults of the Indian meal moth. J. Econ. Entomol. *68*, 763-765.

LI, C.S. 1953. A preliminary study with stored rice insect pests and their control in Taiwan. Memoirs Natl. Taiwan Univ., College of Agric. *2* (5) 99-103.

MCGAUGHEY, W.H. 1970. Effect of variety and degree of milling on insect development in milled rice. J. Econ. Entomol. *63*, 1375-1376.

McGAUGHEY, W.H. 1974. Insect development in milled rice: Effects of variety, degree of milling, parboiling and broken kernels. J. Stored Prod. Res. *10*, 81-86.

MONRO, H.A.U. 1961. Manual of fumigation for insect control. Food and Agric. Organ. U.N. *57.*

PREVETT, P.F. 1971. Storage of paddy and rice (with particular reference to pest infestation). Trop. Stored Prod. Inf., *22*, 35-49.

TILTON, E.W. and COGBURN, R.R. 1965. Phosphine fumigation of rough rice in upright bins. Rice J. *69* (11) 8-9.

ZETTLER, J.L. 1975. Malathion resistance in strains of *Tribolium castaneum* collected from rice in the U.S.A. J. Stored Prod. Res. *11*, 115-117.

8

Harvest, Drying and Storage of Rough Rice

James F. Steffe
R. Paul Singh
George E. Miller, Jr.

The importance of post harvest management of food grains is underscored by the fact that large losses of food occur in the world due to poor grain handling techniques. Therefore, harvesting, drying and storage of rice should receive serious attention. This chapter provides some guidelines to achieve this goal.

Harvesting of rice is carried out by manual labor in most rice growing countries. On the other hand, combines are used to harvest and thresh rice in some developed countries with mechanized agriculture. The section on harvesting in this chapter addresses the mechanized rice harvesting conditions prevalent in California.

The theory of rice drying is the major content of this chapter. It is important to know the physical and thermal properties of rice to understand the drying mechanism in rice. Such information is reviewed and basic theoretical considerations of rice drying phenomenon are presented. Some engineering parameters necessary for design and operation of rice drying and storage facilities are also discussed.

Rice processing consists of several unit operations such as drying, husking, milling and polishing. During these operations various parts of the kernel are removed. Over the years a terminology has developed to describe the grain at different stages in the process. The grain, after it comes from the field is called rough rice or paddy. Amounts of rough rice may be expressed as bags or cwt (100 lb), bushels (45 lb), or barrels (162 lb). After the hull or husk is removed the product is called brown rice. Milled or white rice is what remains after the bran and part of the germ have been removed during milling. White rice is composed of whole and broken kernels. Whole kernels are referred to as head rice and head yield refers to the quantity of unbroken kernels milled from a specified amount of rough rice. The various parts of rice during processing are shown in Fig. 8.1.

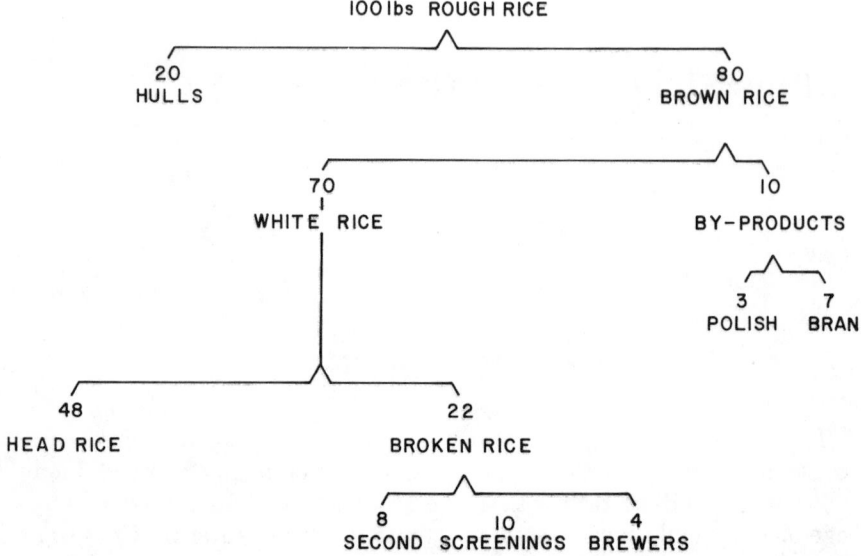

From Henderson (1976)

FIG. 8.1. PRODUCT FRACTIONS FROM THE STANDARD MILLING OF RICE

(Consult Appendix for appropriate table on metric conversions.)

An important quantitative factor of rice, in drying, is moisture content. The measure of moisture content can be expressed as a wet basis or a dry basis number. Wet basis measure specifies the quantity of water in a solid as a percentage of the total wet weight. Thus the percentage of moisture content wet basis (% MC wb) is

$$\% \text{ MC wb} = \frac{\text{Total wet weight}-\text{Dry weight}}{\text{Total wet weight}} \times 100$$

and the percentage of moisture content dry basis (% MC db) is

$$\% \text{ MC db} = \frac{\text{Total wet weight}-\text{Dry weight}}{\text{Dry weight}} \times 100$$

The relationship between wet and dry basis moisture contents is shown in Fig. 8.2. Moisture content expressed on a wet basis has traditionally

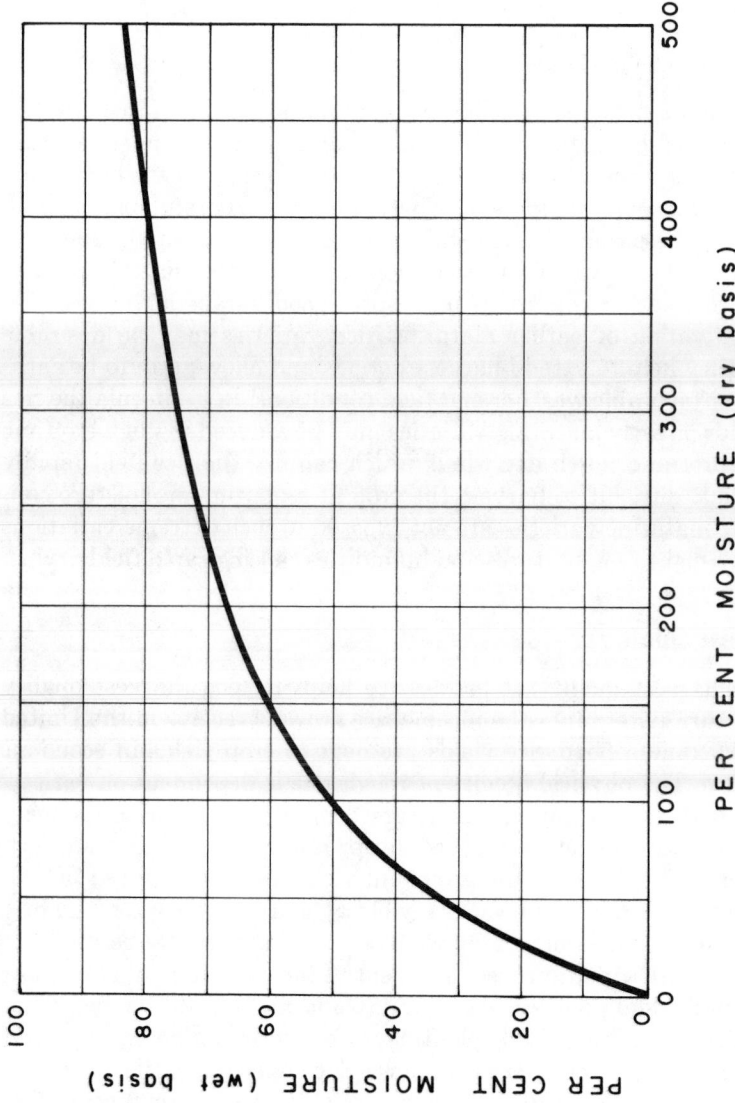

FIG. 8.2. RELATIONSHIP BETWEEN WET AND DRY BASIS MOISTURE CONTENTS

been used by rice growers, drier operators and processors. Dry basis moisture content is generally used in research on drying rice because the denominator in the working equation is independent of the quantity of water in the grain.

RICE HARVESTING

Many factors must be considered to obtain an optimum rice harvest. The grain must be mature, high in quality and have the proper moisture content. The soil should also be sufficiently dry to support the harvesting and field transport equipment. There should be careful planning to ensure that these conditions prevail simultaneously. It cannot, however, be denied that in spite of all precautions and best planning, unusual or unexpected weather conditions may upset good harvesting.

The cultivation of earlier maturing rice varieties may be desirable because they help to extend harvest periods and allow grain to be gathered under favorable weather and field conditions. In California the rice head yields of early maturing varieties may be reduced by high daytime temperatures and north dry winds which can dry the rice field rapidly and make it difficult to harvest the crop at an optimum moisture content. Late maturing varieties are not exposed to such extreme variations in the day-night temperatures and humidities causing rapid field drying.

Preharvest Quality

The preharvest quality of paddy rice helps its post harvest quality which in turn determines the income rice growers receive in the United States. Maximum head rice yields are next to crop yields in economic importance. The physical quality of rough rice is dependent on variety, stage of maturity at harvest, moisture content and physical damages (impact, abrasion and stress) caused during harvest, handling, transport, processing and storage of the grain. All varieties of rice tested in California reach a maximum head rice yield at a moisture content slightly higher than when maximum crop yield occurs. This is partly because rice does not mature simultaneously in an entire field, an individual plant or even on individual panicles. Variety, day and night grain temperatures during the maturing period, planting rates, germination rates, water management, plant population and weed competition affect the uniformity of maturation and consequent grain quality. Differences of 5 to 10% MC wb have been observed between grains on the same plant and even on the same panicle. If the harvest were delayed for all kernels to be mature, the early maturing kernels would become overripe and overall quality would decrease. Overripe kernels, generally identified by moisture contents below 22% MC wb for early maturing varieties in California, are subject to field cracking if not harvested promptly when they are mature. Most early maturing rice grown in California should be harvested at an average moisture content of 24–26% wb when the highest daytime temperatures are over 23.9°C (75°F) as shown in Fig. 8.3 and 8.4. Later in the season (after Oct. 10 in California) when maximum

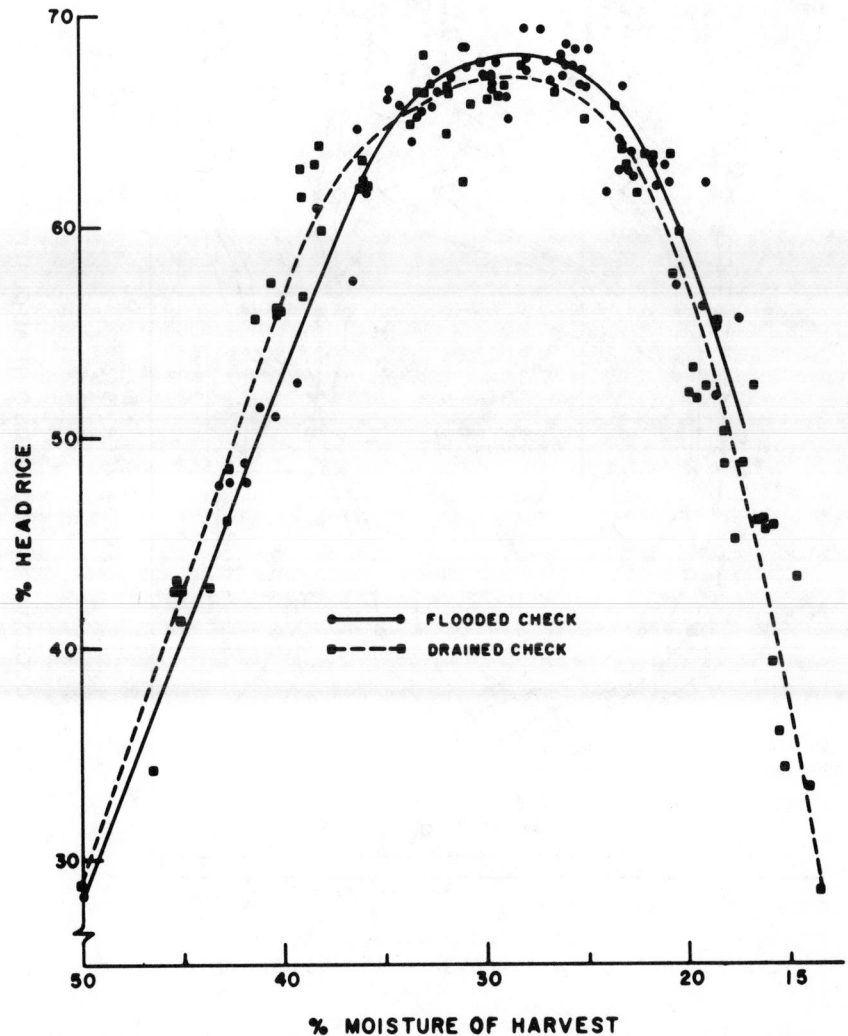

FIG. 8.3. QUALITY EVALUATION OF RICE SAMPLES FROM 1975 FIELD TESTS OF S-6 VARIETY

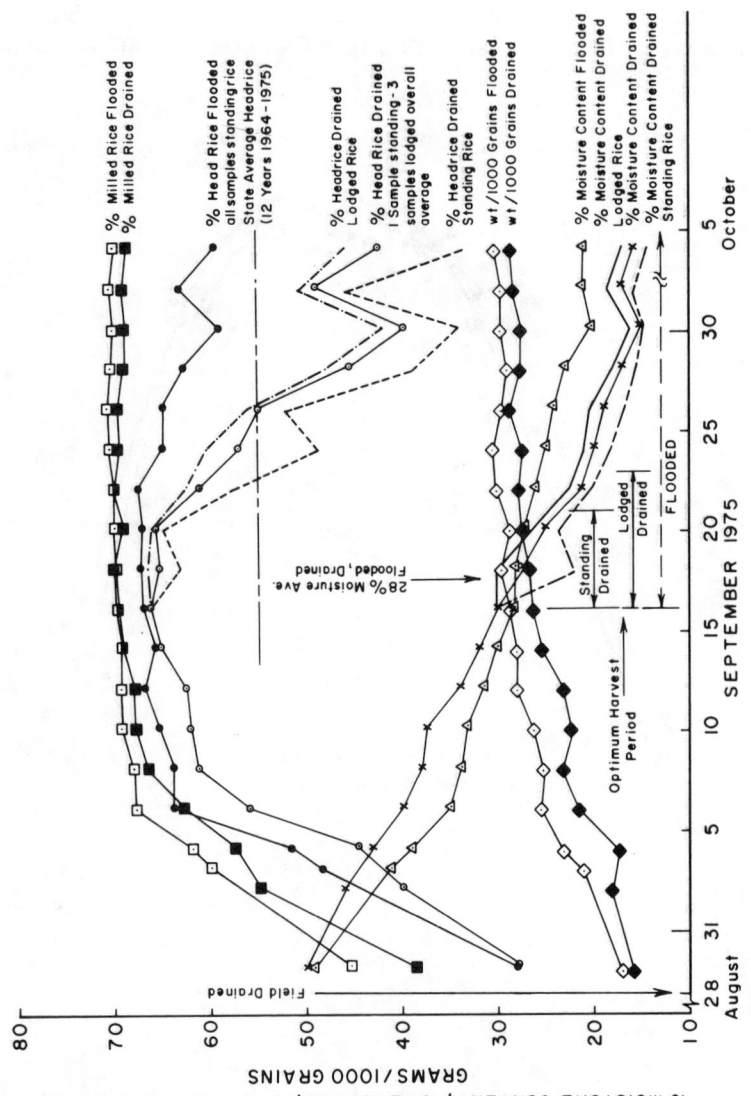

FIG. 8.4.   INFLUENCE OF FLOODED VS DRAINED FIELDS ON HEAD QUALITY IN FIELD TESTS OF S-6 VARIETY

daily air temperatures drop below 23.9°C (75°F), the later maturing varieties of rice can be harvested at low cylinder speeds (609.6 m/min) even when the grain moisture content is as low as 13–14% wb while maintaining high head rice yields. Most varieties of California rice, if carefully harvested with proper controls and under proper conditions, will produce high head rice yields as shown in Table 8.1.

Rice grown in tropical and semitropical climates is not subjected to wide variation in day-night temperature and humidity. Thus, the rapid drying experienced in early season harvests in California is not present in the tropics and the potential for damage in these areas (preharvest and during harvest) is not as great.

Research on rice grown in southern United States, India, Surinam and the Philippines show that most varieties have reduced head rice yields after reaching a certain maturity. These studies indicate that even in subtropical climates, the exposure to varying temperatures and humidities causes greater kernel stresses as lower moisture contents are reached. It appears that harvesting procedures should be determined for each variety grown in each climatic region. Appropriate harvest timing, harvester cylinder speed and clearance adjustments should be prescribed to minimize impact damage in the harvester at various stages of maturity.

Harvest Preparation

Various harvest related factors such as variety, planting time, local weather patterns, soil type, field water temperature and management and date of heading affect maturity. After the planting date is fixed, the next critical observation regarding harvest is the date of the "first heading" when approximately 10% of rice heads have emerged. In California, crops of average yield should be ready for harvest 45–55 days after the first heading.

After the date of maturity is estimated, field draining is given primary consideration. This draining must be related to the maturation characteristics of the variety, the expected weather, the soil type, the field characteristics for draining surface water and the effects of plant evapotranspiration on drying the soil surface. Proper selection of draining date is important because the soil must be sufficiently dry to support the harvesting and transport machinery when the grain is ready to be harvested. The final decision for draining is made with the knowledge and experience of individual growers and farm advisors. If the field is drained too early, the rice plants will run short of moisture before maturing.

In California, field draining usually begins in the latter part of August for early maturing varieties. The precise timing for draining each field is determined by the drooping of the heads (panicles) which indicates the

TABLE 8.1. PERCENTAGE DISTRIBUTION OF TESTED SAMPLES FOR CALIFORNIA HEAD RICE MILLING YIELD

Class/Crop yr		Head Rice Yield						
		Under 40	40–44.5	45–49.5	50–54.5	55–59.5	60–Over	Ann. Avg.
		(%)	(%)	(%)	(%)	(%)	(%)	(%)
Short Grain								
Oct. 1975–Sept.	1976	2.2	1.8	8.1	28.6	38.3	21.0	55.1
" 1974– "	1975	3.7	4.4	10.4	26.0	37.0	18.5	54.1
" 1973– "	1974	4.7	3.9	6.4	16.5	33.7	34.8	55.8
" 1972– "	1973	2.0	3.1	4.4	11.9	36.4	42.2	57.3
" 1971– "	1972	2.9	4.5	13.7	23.2	39.6	16.1	54.3
" 1970– "	1971	2.3	3.9	6.6	22.8	38.6	25.8	55.8
" 1969– "	1970	1.1	2.4	4.2	16.3	43.0	33.0	57.2
" 1968– "	1969	4.3	3.6	7.9	18.1	34.5	31.6	56.2
" 1967– "	1968	3.7	3.3	11.6	26.3	39.1	16.0	54.2
" 1966– "	1967	5.3	6.0	14.1	26.8	34.0	13.8	53.1
" 1965– "	1966	.5	1.5	3.6	20.2	54.2	20.0	56.5
" 1964– "	1965	3.0	5.9	24.7	42.6	19.8	4.0	51.4
Average		3.0	3.7	9.6	23.3	37.4	23.1	55.1
Medium Grain:								
Oct. 1975–Sept.	1976	5.0	4.0	9.6	26.5	31.4	23.5	54.1
" 1974– "	1975	4.3	5.9	13.7	26.1	33.5	16.5	53.0
" 1973– "	1974	2.1	2.1	5.4	14.7	34.6	41.1	57.1
" 1972– "	1973	1.9	2.2	4.9	13.4	37.2	40.4	57.2
" 1971– "	1972	2.4	1.9	5.2	14.4	35.8	40.3	57.4
" 1970– "	1971	1.5	1.0	3.4	10.5	32.1	51.5	58.8
" 1969– "	1970	4.3	1.8	4.4	12.7	38.6	37.8	56.9
" 1968– "	1969	2.4	2.5	5.6	12.8	28.4	48.3	58.1
" 1967– "	1968	7.2	7.0	12.0	21.3	32.1	20.4	53.4
" 1966– "	1967	1.1	2.7	8.2	30.7	40.5	16.8	55.1
" 1965– "	1966	.9	3.2	9.7	29.9	43.5	12.8	54.8
" 1964– "	1965	5.1	5.8	16.7	38.5	29.3	4.6	51.9
Average		3.2	3.3	8.3	21.0	34.8	29.5	55.7

size and number of rice kernels. This is usually about 2 to 3 wks before harvest. Some rice growers drain the fields when 10–20% of the heads are drooping and others wait until all of the heads droop. The supply of water to the field is cut off 3 or 4 days before the field is drained by removing the water level control boards in the check boxes between the paddy fields. Slow draining helps to reduce lodging and the stem is stronger if it hardens slowly as it dries out. In some cases checks (levees) are blown with dynamite in strategic locations to expedite the final drainage of the surface water from the fields. Since the rice plant is still growing, transpiration aids in removal of surface water and moisture in the upper layers of soil. This drying is required to develop sufficient soil strength for efficient movement of harvesting machinery.

Muddy conditions in the fields, due to incomplete drainage or from rains during the harvest period, create difficult harvesting conditions which may leave the field rutted and make its preparation for the next crop unfavorable. After the field has been drained and the surface sufficiently dry, a bulldozer is used to prepare roadways or paths through the levees for the harvesters and bankout wagons to pass from check to check and carry the harvested grain to the waiting bulk grain trucks. Usually a harvester will make a pass around the check so that the bulldozer can get to strategic locations to level areas in the check for access of harvesting equipment. Most harvesters currently in use are self-propelled rubber tire units; others use either half tracks or have full tracks with independent hydrostatic drives to improve floatation and maneuverability under wet soil conditions.

After the field has been drained, the moisture content of rice is monitored as it reaches maturity. When draining, the moisture content may be 40–50% wb. The grain moisture content will drop approximately 1% wb per day when the grain has a high moisture content and the weather is hot and dry. When the weather is cool and overcast or the rice moisture content is 20 to 25% wb, the moisture loss may be as low as 0.5% wb per day. Rain or dew may cause an increase in grain moisture content.

Harvesting should commence at 24–26% MC wb, particularly with early maturing varieties harvested during hot dry periods in California. Minimum cylinder speeds should be used except in an emergency, especially when moisture contents drop below 22% wb and daytime temperatures exceed 23.9°C (75°F). This precaution is not that important when the moisture content is 24–26% wb. Research carried out at the International Rice Research Institute (IRRI) on cylinder types and speeds indicate that about 25% of the kernels will be ruptured if the rice is at 22% MC wb and the cylinder speed is 1280.2 m/min compared to 5% at 609.6 m/min. Related work being conducted at the Agricultural Engineering Department at the University of California, Davis, indicates similar trends. There appears to be a relationship between rate of drying resulting in internal kernel stresses, and the potential for damage from impact in the cylinder of the rice harvester.

Harvester Management

Rice harvesters have been developed over the years from binders and stationary type threshers used in early days of grain threshing to self-propelled combines that directly cut and thresh rice. Conventional harvesters consist of a pickup reel, a 4.3 to 6 m sickle bar, and auger conveyors or draper belts that carry the harvested grain into the threshing cylinder. Some new models have double sickle bars for faster cutting in heavy crops. The rotating cylinder contains rows of spike-toothed elements that carry the straw and grain panicles between stationary rows (usually 2 or 3) of spike teeth so that the grain is separated from the stem by impact and abrasion between the rotating and stationary spike teeth. Most of the grain, along with some leaves and pieces of broken straw, falls through a grate beneath the cylinder and is conveyed to an oscillating cleaning shoe. A special sieve for rice and a fan aid in separating foreign material from the grain. The leaves and chaff are blown out the back of the harvester. Some unharvested panicles are conveyed to the rear of the shoe and diverted into a conveyor that carries them back into the cylinder for additional threshing. The bulk of the straw from the cylinder is transported to the oscillating straw walkers which lift and drop the straw as it moves up the stair-stepped slope. The straw walkers shake out the loose grain and partially harvested panicles. This material is conveyed by gravity or auger conveyors to the cleaning shoe. The straw is discharged at the rear of the machine in a windrow or onto a straw spreader. Straw placed on the ground by a spreader will dry rapidly and is eventually burned. The grain from the cleaning shoe is conveyed to a bulk bin, usually located on top of the harvester. When the bin is full, the grain is loaded into a bankout wagon which transports the rice to a waiting truck for delivery to the rice drier.

The operator of a combine has many important responsibilities. The operator controls the forward speed of the harvester, speed of the pickup reel, height of the sickle, cylinder speed, cylinder clearance and speed of the fan on the cleaning shoe. In addition, the operator examines the straw and soil surface behind the harvester to ensure that all the grain is being removed from the straw and that no unnecessary quantity of grain is lost through the straw discharge of the harvester. The operator occasionally examines the harvested grain to see if any hulling or cracking of the grain is taking place. Kernel damage may indicate an improper clearance or cylinder speed or a malfunction of an auger conveyor. In a field containing grain with varying maturity, the operator will examine the color of the grain to determine if it is mature enough to be harvested.

Harvesting begins each day when the dew is off the plants. In California, this is usually at about 10 to 11 a.m. (Pacific Standard Time). Harvesting is usually stopped when dew begins to form at 8 to 9 p.m. If an overcast or dry north wind has been blowing all night, these schedules may change. Much of the knowledge of harvester settings comes from the experience of those who grow and harvest rice. Good operators will adjust the height of cut, forward speed and reel speed as these may change with a change in the height of the plant. Other factors affecting the harvesting operation include stem thickness, lodging, yield and rice variety.

The predominant cylinder type utilized in current harvesters in California is the spike tooth cylinder with the shaft crossways to the harvester. Some rasp bar or rub bar cylinders are also used and until 1950 rasp bar harvesters were common in California. The spike tooth cylinder provides gentler action, plugs up less frequently and is easier to clean when plugged.

Post Harvest Management

When widely varying maturity occurs in a rice field at harvest, the harvested grain should be quickly transported to the drier and given priority in the drying schedule. The importance of this has been demonstrated by Kunze and Parsad (1976). In their tests, grain ruptured in storage when high moisture kernels were stored in contact with low moisture kernels. The high humidity produced in the vicinity of the high moisture grain was sufficient to rupture low moisture kernels. This may occur within a few hours depending on the moisture differentials.

Field transport equipment has been developed for bulk handling of rice. These units allow high speed transport from the harvester to waiting trucks while keeping the harvester in continuous operation. Road transport trucks move the crop to the driers, since the rice must be reduced from approximately 24−26% MC wb down to 13−14% MC wb for safe storage. The risk of heat damage is always present in high moisture rice and early transport to the drying facility is therefore important.

Rice Harvesters

Most of the major farm machinery manufacturers make grain harvesters for small grain harvesting. With certain modifications and special traction units, headers, conveyors, straw walkers and shoe designs, these machines are used for rice harvesting. In California there are also 2 shortline manufacturers of combines designed and specifically built for rice harvesting. They are large, high capacity, self-propelled combines with full track support that can generally operate under wet field conditions (Fig. 8.5 and 8.6).

FIG. 8.5. COMBINE IN A GRAIN FIELD WITH A 4.87 M HEADER

Two types of cylinders are available for rice harvesting in the United States. These are the rasp bar and the spike tooth cylinder. Practically all rice in California is harvested with spike tooth cylinders. This is partially due to the large volumes of straw produced in the high yielding California rice fields and the easier cleaning allowed with a spike tooth cylinder. In addition, the harvesting process may be a little more efficient with the spike toothed cylinder in removing the grain from the straw. Small Japanese harvesters often have wire loop cylinders which, for slow speed cylinder operation with hold on systems (the straw does not pass through the cylinder), result in an efficient harvest with a high quality product. It is not certain whether this principle can be applied to the large high capacity commercial harvesters used in the United States.

IRRI has been conducting research on axial flow harvesters and stripper harvesters for a number of years. Axial flow type harvesters feed the grain and straw through inclined cylinder(s) mounted longitudinally in the harvester compared to conventional cylinders which are mounted crossways to the movement of the harvester. This method eliminates the conventional straw walker. Several of these types of harvesters are

FIG. 8.6. A RICE COMBINE WITH AN 5.48 M HEADER

becoming commercially available for wheat, barley and corn harvesting. It is expected that such harvesters will become available for rice as well.

Research on stripper harvesters, done at the University of California, Davis, indicates that they have an advantage of removing the grain from the plant without cutting the stem. This allows the grain to be removed at lower cylinder speeds with less grain damage. The total machinery and overall weight are considerably less than the conventional harvesters. These harvesters appear to operate well but there are problems with lifting lodged rice and feeding it into the stripper cylinder. New short stature varieties which have more standing rice at harvest may improve the prospects for this type of machine, although the shorter stem may create feeding problems.

For the present, it appears that the conventional spike tooth cylinder harvester with suitable modifications made by manufacturers and growers will be used for harvesting most of the rice in the highly mechanized agricultural areas of the world. Manual methods and small handfed machine harvesters continue to be best suited for areas where rice is grown on small individual plots of ground and where labor is available. For more information on rice harvesting methods in developing countries the reader may refer to Araullo *et al.* (1976).

RICE DRYING

Equilibrium Moisture Content

The equilibrium moisture content of grain, commonly abbreviated as EMC, refers to the quantity of moisture in the product when it is at equilibrium with the surrounding environment, usually air. The EMC of rice will depend on the air temperature and humidity, grain variety, maturity and previous history. In addition, EMC will depend on whether or not the rice had to adsorb or desorb moisture to achieve equilibrium. The EMC achieved by desorption will be higher than that achieved by adsorption. This phenomenon is referred to as a *hysteresis effect*.

Pfost *et al.* (1976) have summarized data on equilibrium moisture content for rice and other grains. These researchers computed constants for the following 2 EMC prediction equations for use with reference to rice:

(1) The first equation was presented by Henderson (1952) and modified by Thompson (1972):

$$M_e = \left(\frac{\ln(l - RH)}{-1.918 \times 10^{-5}(T + 51.161)} \right)^{0.409} \tag{1}$$

where: M_e = Equilibrium moisture content (% db)
RH = Relative humidity (decimal)
T = Temperature (°C)

(2) The second equation was originally given by Chung and Pfost (1967) and modified to include an extra constant by Pfost *et al.* (1976):

$$M_e = [0.325535 - 0.046015 \, ln\,[-1.987(T + 35.703 \, ln\,(RH)]] \times 100 \tag{2}$$

where: T   = Temperature (°C)
RH = Relative humidity (decimal)

Pfost *et al.* (1976) computed the constants for the above equations using data given by Brooker *et al.* (1974) and Kososki (1976). The collection of data is from many sources and includes adsorption and desorption information. Figure 8.7 shows EMC curves generated from the modified Chung-Pfost Equation (2).

The physical properties of rough rice with reference to individual kernels and in bulk are given in Table 8.2. The thermal properties for a medium grain variety of rice are presented in Table 8.3. Properties for

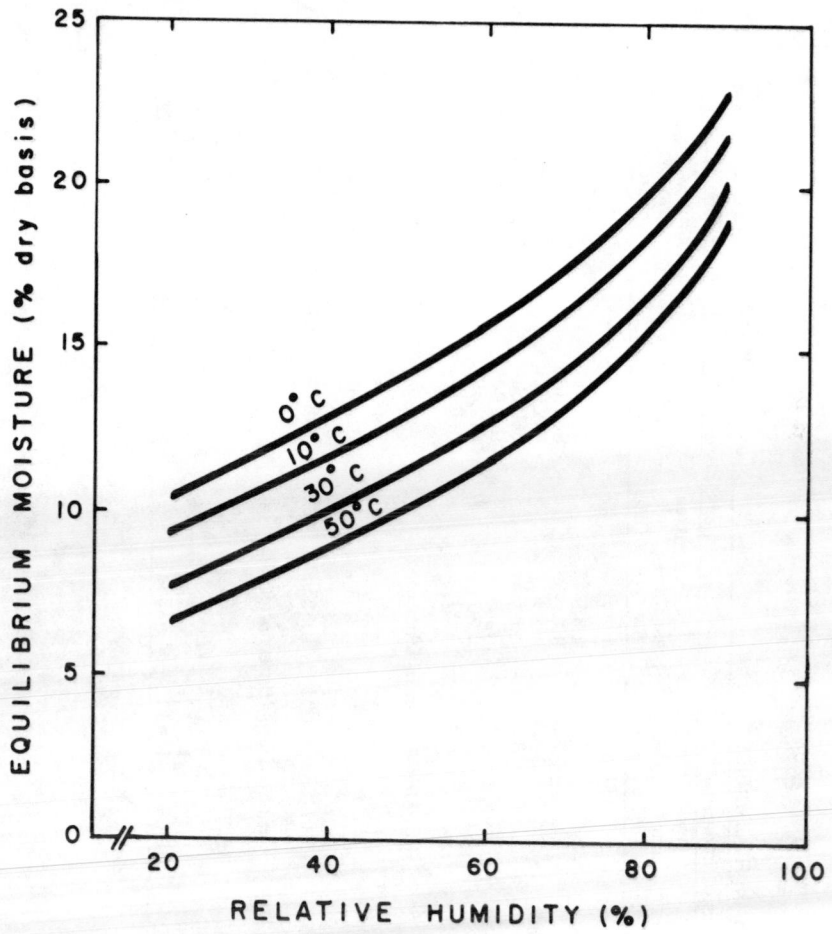

From Pfost et al. (1976)

FIG. 8.7. EQUILIBRIUM MOISTURE CONTENT OF ROUGH RICE

specific varieties will vary, but the values for the same grain types (short, medium or long grain) will usually be similar. Smooth hulled and rough hulled varieties of the same type of rice will have a different bulk density. Smooth hulled varieties are 5 to 10% denser than rough hulled varieties (Henderson and Parsons 1974).

Grain Drying Methods

In this section different types of grain driers are discussed and some recommendations for operating them are presented. It should be noted

TABLE 8.2. PHYSICAL PROPERTIES OF ROUGH RICE

Moisture content % wb	Individual grain properties							Bulk properties		
	Length m ×10³	Width m ×10³	Thickness m ×10³	Volume cu m ×10⁶	Density kg/m³	Specific gravity	Area m² ×10⁶	Density kg/m³	Porosity %	Specific gravity (apparent)
Short Grain (Caloro)[1]										
12	7.464	3.465	2.294	—	—	1.35		634.71	—	—
14	7.489	3.483	2.312	—	—	1.34		643.36	—	—
16	7.513	3.500	2.329	—	—	1.33		652.01	—	—
18	7.538	3.518	2.347	—	—	1.32		660.50	—	—
20	7.562	3.536	2.365	—	—	1.31		669.00	—	—
22	7.568	3.554	2.383	—	—	1.30		677.49	—	—
24	7.611	3.572	2.401	—	—	1.29		686.14	—	—
Medium Grain (Saturn)[2]										
12	7.899	3.124	1.956	16.06	1324.37	1.374	40.2	598.35	58.5	0.599
14	7.925	3.124	1.956	16.71	1337.19	1.355		618.05	56.5	0.618
16	7.950	3.124	1.981	17.53	1354.17	1.350		633.75	55.0	0.630
18	7.976	3.175	2.007	19.17	1371.95	1.325	42.5	648.65	53.1	0.653
Long Grain (Bluebonnet-50)[3]										
12	9.677	2.591	1.905	18.35	1362.50	1.384		585.69	59.6	0.586
14	9.754	2.616	1.930	18.52	1371.15	1.373		588.25	59.3	0.589
16	9.855	2.642	1.930	19.17	1377.72	1.372		605.40	57.9	0.606
18	10.033	2.692	1.981	19.66	1383.17	1.358		615.17	56.9	0.615

[1]Source: Morita (1977).
[2]Source: Wratten et al. (1969).
[3]Source: Wratten et al. (1969).

that no single set of operating instructions apply to any particular type of drier. Establishing a drying procedure is a situational proposition; many factors such as ambient air conditions, quantity of grain which has to be dried and expected use of the grain must all be considered. Drier operating methods established for a certain location, rice variety and time of year may be inappropriate for another situation.

TABLE 8.3. THERMAL PROPERTIES OF ROUGH RICE

Moisture content, (% wb)	Specific heat, kJ/kg°C	Bulk conductivity, W/m°C × 10^2	Bulk diffusivity, m²/s × 10^7
Short Grain (Caloro)[1]			
12	1.69	.113	1.06
14	1.76	.116	1.03
16	1.83	.118	.993
18	1.90	.120	.960
20	1.97	.122	.928
22	2.04	.124	.895
24		.127	.863
Medium Grain (Saturn)[2]			
12	1.60	.102	1.05
14	1.70	.105	.999
16	1.80	.108	.950
18	1.89	.111	.903
20	1.99	.113	.857

[1]Source: Monita (1977).
[2]Source: Wratten *et al.* (1969).

Fixed-bed Driers.—Fixed-bed driers are used for complete on-farm drying, or for finish-drying after the major step of moisture removal has been completed in a continuous flow drier. After drying, these driers may also be used for grain storage. Two common types of fixed-bed driers used in California, circular and rectangular, are shown in Fig. 8.8 and 8.9. The depth of grain in fixed-bed driers may vary, but 4.3 to 6 m is the maximum practical depth (Henderson and Parsons 1974). Grain in a fixed-bed system is dried with forced air, which may or may not be heated. Grain quality and drying time are affected by the temperature and relative humidity of the drying air. Figure 8.10 indicates that the number of whole kernels in milled rice increases with the humidity of the drying air and decreases with the drying air temperature. Figure 8.11 shows the effect of air temperature and humidity on drying time.

Circular bins (Fig. 8.8) are constructed with perforated floors and a high pressure area is created under the grain by a fan attached to the structure. Stirring devices may be added to the bin to aid drying by grain mixing. Dry air is pulled in from outside and forced up through the rice; the moistened air exits through the top of the bin. Supplemental heaters are usually added to systems of the type shown in Fig. 8.8. These heaters

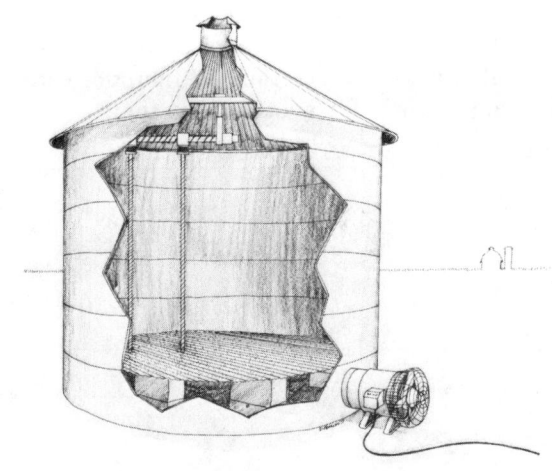

FIG. 8.8.   A CIRCULAR BIN WITH A STIRRING DEVICE

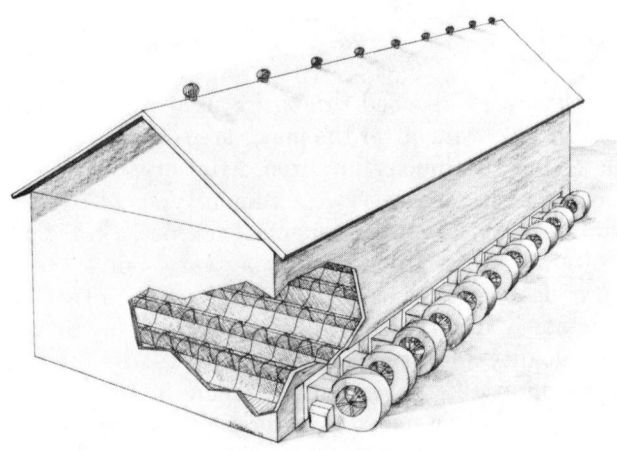

FIG. 8.9.   A RECTANGULAR BIN FOR DRYING RICE

are used when the relative humidity of ambient air is too high to make it suitable for drying. When high moisture air is used in a fixed-bed drier, it can rewet the rice and cause serious quality deterioration. Wasserman and Calderwood (1972) have made the following recommendations for California and Texas rice drying: "Use supplemental heat only when the relative humidity is above 75% for a prolonged period, provide enough heat to raise the air temperature 6.7°C (12°F), and limit drier inlet air temperatures to 29.4°C (85°F) in California and 35°C (95°F) in Texas." Information presented in Table 8.4 can be used to estimate fuel requirements when supplemental heating is necessary.

Large rectangular bins (sometimes called flats) of the type shown in Fig. 8.9 are usually not designed with supplemental heating equipment. These systems are used for finish-drying (removal of moisture after column drying) and storage of rice. The walls and floor of these structures are normally made of concrete. Large fans are located outside the bin and air distribution tunnels are placed from wall to wall on the floor of the bin.

It is vital to maintain the proper airflow when drying rice. If the air flow is insufficient, the rice may be spoiled before the storage moisture

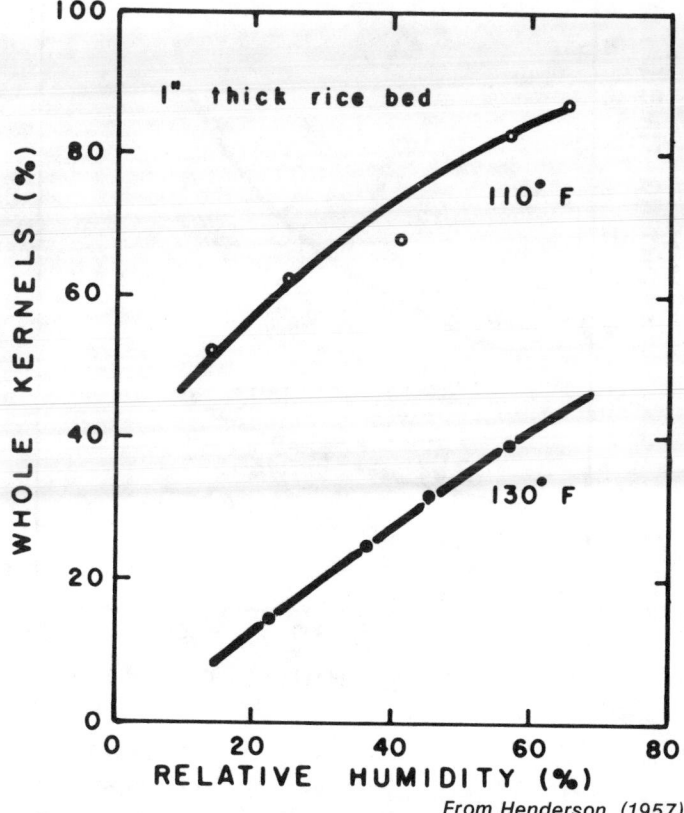

From Henderson (1957)

FIG. 8.10. EFFECT OF DRYING–AIR TEMPERATURE AND RELATIVE HUMIDITY ON PERCENTAGE OF WHOLE KERNELS IN MILLED RICE

(Consult Appendix for appropriate tables on metric and °F to °C conversions.)

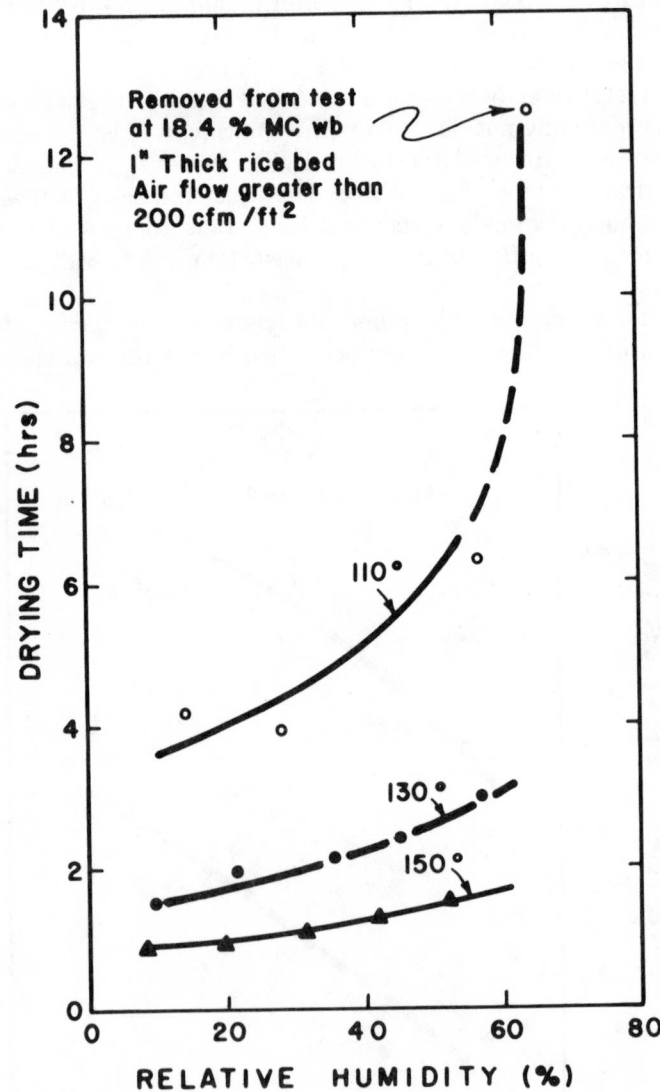

From Henderson (1957)

FIG. 8.11. TIME REQUIRED TO DRY ROUGH RICE FROM 24%
TO APPROXIMATELY 13.5% MOISTURE CONTENT AT AIR
TEMPERATURE AND RELATIVE HUMIDITY NOTED

(Consult Appendix for appropriate table on metric conversions.)

content is reached. Rice provides a resistance to airflow which must be overcome by applying positive pressure at the air inlet to the grain or negative pressure at the air outlet. Resistance to airflow is evidenced as a drop in pressure as the air travels through the bed.

TABLE 8.4. FUEL REQUIRED PER 45,359 KG FOR 5.6°C INCREASE IN AIR TEMPERATURE

Air Rate				
$m^3/s \cdot kg$ $\times 10^5$	m^3/s	Watt	Natural gas m^3/s $\times 10^4$	L.P. fuel m^3/s $\times 10^7$
2.1	0.94	7,033	2.12	2.92
3.1	1.42	10,550	3.15	4.38
4.2	1.89	14,067	4.25	5.94
5.2	2.36	17,583	5.27	7.40

Source: Henderson and Parsons (1974).

Empirical data presented in Fig. 8.12 can be used to determine the pressure drop necessary to achieve a particular airflow rate. The depth of rice in a deep-bed drier can be observed and the system static pressure can be measured between the fan and the rice. Knowing the necessary airflow rate, the required pressure drop per meter of depth can be read from Fig. 8.12. With this information and knowledge of the bed depth, the static pressure which the fan must produce can be calculated. For instance, suppose there is a bin of long grain rice filled to a depth of 1.8 m and we wish to maintain an airflow rate of 0.025 $m^3/s \cdot m^2$ of bin floor area. From curve H (Fig. 8.12), it can be seen that a pressure drop of 63.68 Pa/m of rice will result if an airflow rate of 0.025 $m^3/s \cdot m^2$ is maintained. Therefore, the total pressure drop through the rice will be 121.9 Pa (63.68 × 1.8). The system fan must maintain a pressure of at least 1.3 cm of water on the air inlet side of the grain. Airflow and static pressure requirements can also be estimated for drying deep beds of rice (under California conditions) with the information presented in Tables 8.5 and 8.6.

A number of factors regarding airflow in rice need emphasis. Resistance to airflow is greatly affected by the presence of foreign matter in the grain. In general, if the foreign material is larger in size than the grain, then less static pressure will be required to achieve a fixed airflow. If the foreign matter is smaller in size, the static pressure requirement will have to be increased. In addition, resistance to airflow depends on the kind of "fill" that is obtained in the rice bed. Data represented by the curves in Fig. 8.12 are for a loose fill, i.e., when rice was simply poured into the bins at a reasonably fast rate. There will be a packed fill if rice is poured in slowly, the bin is vibrated after filling or people walk on top of the rice

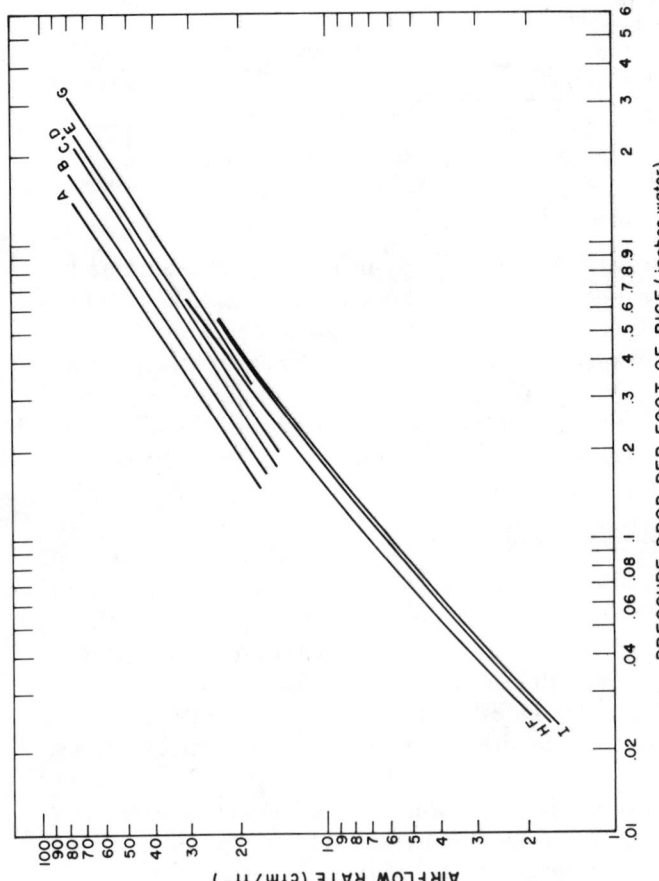

FIG. 8.12. RESISTANCE OF ROUGH RICE TO AIRFLOW

(A) Calrose, medium grain, rough hulled (with trichomes), 24.4% MC wb 0.66% foreign matter. Henderson and Parsons (1974); (B) Same as A but 12.7% MC wb; (C) CSM3, medium grain, smooth hulled (glabrous), 27.6% MC wb, 0.88% foreign matter. Henderson and Parsons (1974); (D) Caloro, short grain, dry (exact MC unknown) clean. Cervinka (1971); (E) Same as C but 12.7% MC wb; (F) Bluebonnet (Hukill (1974)), long grain, 13.4% MC wb, clean. Shedd (1953); (G) CSM3, dry (exact MC unknown), clean. Cervinka (1971); (H) Belle Patna, long grain, 15.2% MC wb, clean. Calderwood (1973); (I) Nato, medium grain, 16.5% MC wb, clean. Calderwood (1973).

(Consult Appendix for appropriate table on metric conversions.)

after filling. A packed fill means that significantly higher pressures will be required to obtain an adequate amount of air flowing through the bed.

TABLE 8.5. QUANTITY OF AIR (M³ /S PER 1 M² OF FLOOR) REQUIRED AT VARIOUS AIRFLOW RATES

Grain depth (m)	Air rate—m³/s · kg × 10⁵ Rough Rice[1]			
	2.1	3.1	4.2	5.2
6.1	.0752	.1128	.1504	.1880
5.5	.0676	.1016	.1351	.1692
4.9	.0600	.0904	.1204	.1504
4.3	.0529	.0787	.1052	.1316
3.7	.0452	.0676	.0904	.1128
3.0	.0376	.0564	.0752	.0940
2.4	.0300	.0452	.0600	.0752
1.8	.0224	.0340	.0452	.0564

Source: Henderson and Parsons (1974).
[1]Air rates can be increased 5 to 10% for the denser, smooth hulled varieties.

When drying with unheated air, the optimum airflow rate will depend on the psychrometic conditions of the air and the rice moisture content. Wet rice will require faster drying to prevent microbial deterioration. Airflow rates will be lower for geographical areas which are usually dry. The recommended airflow rate for Texas and Louisiana is 5.83×10^{-5} m³/s·kg (Wasserman and Calderwood 1972). Henderson and Parsons (1974) have mentioned the following for conditions found in the California Central Valley:

(1) The optimum airflow rate is 3.12×10^{-5} to 4.16×10^{-5} m³/s·kg if the rice moisture content must be reduced to below 13.5% MC wb before December 1.
(2) Rice above 26% MC wb should not be dried in a deep-bed drier.
(3) A deep-bed drier airflow rate of 2.08×10^{-5} m³/s·kg satisfactory if the initial moisture content of rice is 18% MC wb or lower.

Continuous Flow Driers.—Most of rice produced in the United States is dried commercially in continuous flow driers. These driers use forced

TABLE 8.6. STATIC PRESSURE (PASCALS) REQUIRED TO ACHIEVE VARIOUS AIRFLOW RATES IN CLEAN RICE

Grain depth (m)	Air rate—m³/s · kg × 10⁵							
	Rice - rough hulled				Rice - smooth hulled			
	2.1	3.1	4.2	5.2	2.1	3.1	4.2	5.2
6.1	796	1468	2289	3135	1244	2040	2986	4106
5.5	572	1120	1717	2414	970	1692	2339	3135
4.9	448	921	1269	1792	697	1294	1742	2464
4.3	274	597	921	1269	547	921	1369	1817
3.7	224	398	622	871	398	672	946	1244
3.0	149	249	398	547	299	398	622	896
2.4	75	149	224	323	149	274	348	498
1.8	50	75	100	149	100	149	199	249

Source: Henderson and Parsons (1975).

heated air as drying medium. Two common continuous flow driers are the mixing and nonmixing types. A nonmixing columnar type drier is shown in Fig. 8.13. The rice flows by gravity in a straight path between 2 screens. The grain flow rate is controlled by a variable speed discharge

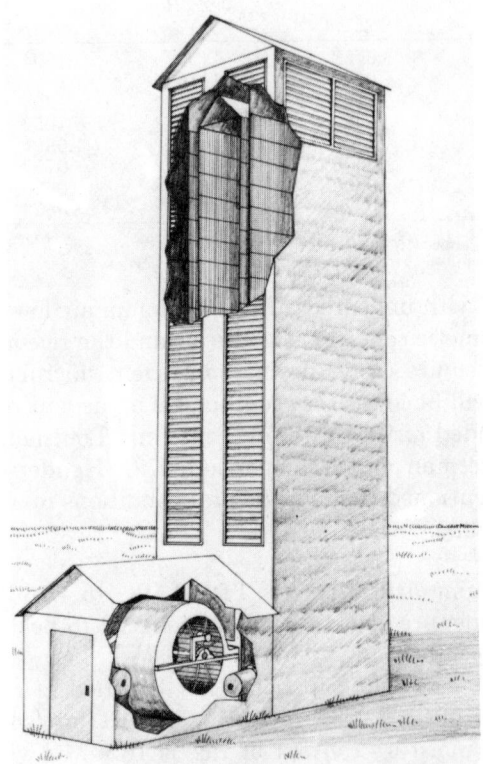

FIG. 8.13.   A NONMIXING TYPE RICE DRIER

roll. The grain is taken away from the drier with a screw conveyor. This drier is sometimes called a "cross-flow" drier because air is forced to flow across a moving bed of rice. The screens are generally 15.2 to 22.9 cm apart and the drier may be 12.2 m high and 3 to 3.7 m wide. The nonmixing column type drier is probably the most common commercial rice drier in use today.

A mixing type columnar drier which uses baffles is presented in Fig. 8.14. Another mixing type drier designed at Louisiana State University is shown in Fig. 8.15. In this drier, rice flows downward over inverted V-shaped air channels. Air flows in and out alternate rows of channels and mixing is accomplished because the inlet and outlet air ducts are offset from one another.

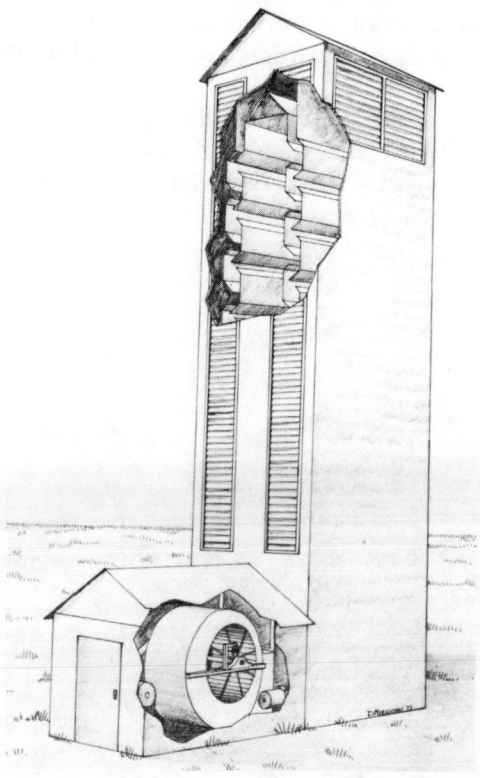

FIG. 8.14. A MIXING TYPE (BAFFLE) RICE DRIER

In terms of grain quality the mixing type driers have an advantage over the nonmixing type. In the nonmixing columnar drier (Fig. 8.13) the grain flows straight downward continuously exposing the grain on the air inlet side of the screen to the hottest air. Thus the grain on the air-exit side of the drier will be cooler and wetter than the grain on the air inlet side. This may result in some of the rice being overdried while the remaining portion would be underdried. The mixing type driers do not have this limitation. In the baffle type drier (Fig. 8.14) rice takes a zigzag path downward. With this type of movement individual kernels are not continuously exposed to the hottest drying air. The same effect occurs in the Louisiana State University drier (Fig. 8.15) because the air channels divert the path of any particular kernel. Grain mixing promotes more uniform drying of rice.

No single set of operating instructions can be formulated for controlling a continuous flow drier. Optimum conditions must be determined at each site location by test drying the particular variety of rice being processed.

The procedure basically involves determining the head yield obtained when rice is dried under various drier conditions. Detailed instructions on testing methods for crossflow driers are presented by Wasserman *et al.* (1958A,B, 1965).

Continuous flow rice driers are usually operated on a multipass basis. The moisture content of rice may be reduced 2 to 4% (dry basis) each time it passes through the drier. Between passes, rice is held for a short

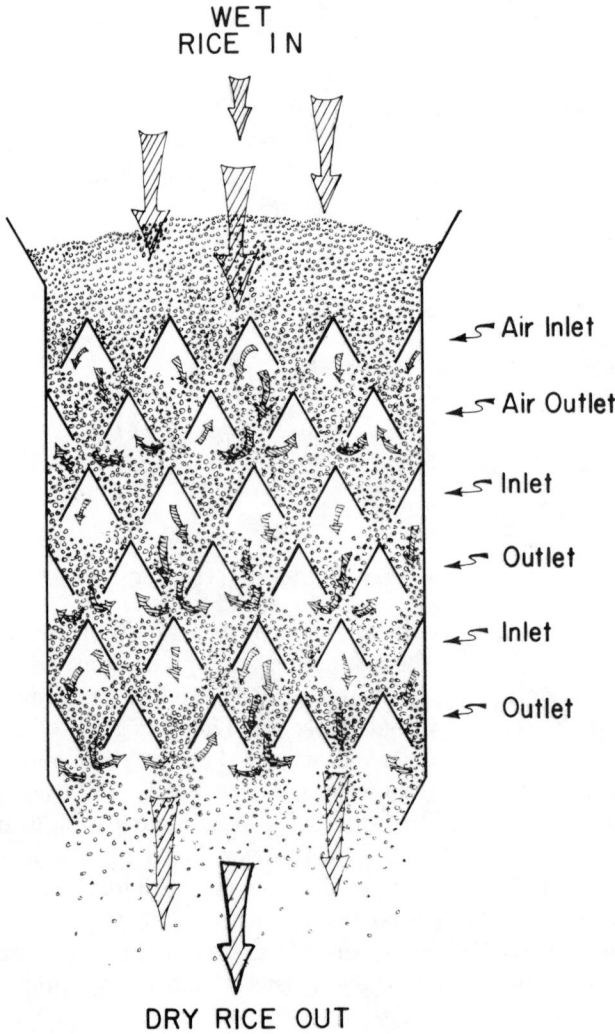

FIG. 8.15. A MIXING TYPE DRIER DESIGN AT LOUISIANA STATE UNIVERSITY

period to allow the kernel moisture gradients developed during drying time to be reduced. This holding period, which may be as long as 24 hr, is referred to as tempering. To increase drier output it may be desirable to temper the rice for as short a period as possible. Figure 8.16 shows the effect of time and temperature on the head yield of short grain rice. The data indicate that 4-hr tempering periods may be adequate if the rice is tempered at 40.6°C (105°F). Further work is needed to fully evaluate the effect of tempering conditions on head yield.

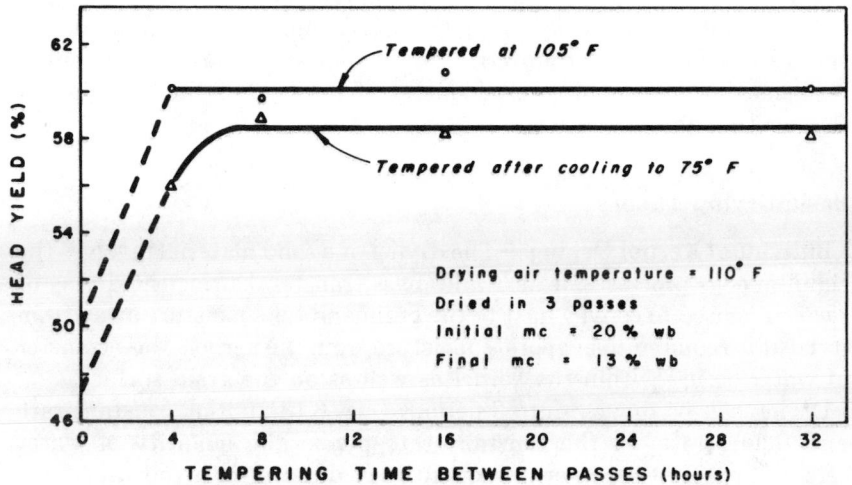

From Wasserman et al. (1964)

FIG. 8.16. HEAD YIELD WHEN TEMPERING SHORT GRAIN RICE AT VARIOUS TIMES AND TEMPERATURES

(Consult Appendix for appropriate table on °F to °C conversions.)

Fast drying induces rice checking which ultimately reduces head yield. In multipass drying, the number of drier passes and the quantity of moisture to be removed during each pass is usually determined by the individual drier operator. There are many factors such as drier capacity, quantity of rice to be dried and moisture to be removed which should be considered in making this decision. When deep-bed driers are used for finish drying, only one pass may be necessary. According to Wasserman and Calderwood (1972), typical airflow rates are 187×10^{-5} to $416 \times 10^{-5} m^3/s \cdot kg$ in nonmixing type driers and 73×10^{-5} to $161 \times 10^{-5} m^3/s \cdot kg$ in the mixing type; air inlet temperatures up to 65.6°C (150°F) in mixing type driers and 54.4°C (130°F) in nonmixing type driers are common.

A discussion of continuous flow rice driers would not be complete without mention of alternate drier configurations which may gain prominence in the near future. Drier configurations other than crossflow, such as concurrent flow and countercurrent flow, have been investigated for drying corn. Thompson *et al.* (1969) made a comparison of crossflow, concurrent flow and counterflow grain driers. Counterflow and concurrent flow driers are shown in Fig. 8.17. In a counterflow drier the grain and the air flow are in opposite directions; in a concurrent flow drier the grain and the air flow are in the same direction. The concurrent flow drier has the advantage of causing the hottest air to come in contact with the wettest grain. The counterflow drier appears to have the potential for removing more moisture per meter of drying bed than concurrent flow or the crossflow driers.

Basic Drying Theory

Individual Kernel Drying.—The drying of a solid material involves the simultaneous processes of heat and mass transfer. During the drying of rice, air is used to convey heat to the grain and take moisture away from it. Heat is required to evaporate moisture from the kernel. Mass transfer of water occurs within the kernel as well as on the grain surface.

Drying can be divided into 3 periods (Fig. 8.18): initial, constant rate and falling rate. In the constant rate period the quantity of water removed per unit time per unit quantity of dry matter is constant. This pattern may occur in high moisture foods where the product surface will remain saturated with water. Rice may exhibit a constant rate period if it is very wet at the onset of drying. Such conditions could prevail if rice is harvested after rain or heavy dew. In the falling rate drying period the quantity of water removed per unit time per unit quantity of dry matter is not constant. In this case the product surface is no longer water saturated and rate of drying is governed by movement of internal moisture. The initial period, which is short and occurs at the start of drying, should be considered as a kind of warmup time. In the constant rate period the mathematical derivative of moisture content in relation to time is constant. This derivative steadily decreases during the falling rate period. The moisture content which marks the change from the constant rate to the falling rate period is generally referred to as the "critical point."

The drying of a rice kernel principally occurs in the falling rate period. The inner movement of moisture within a grain kernel during the falling rate period controls the drying process. This internal moisture movement is a complex phenomenon which is not clearly understood. Hall (1957) listed the following physical mechanisms as possibilities in controlling the transfer of moisture within agricultural products:

(1) Diffusion (liquid or vapor),
(2) Capillary action,
(3) Shrinkage and vapor pressure gradients,
(4) Gravity, and
(5) Vaporization of moisture.

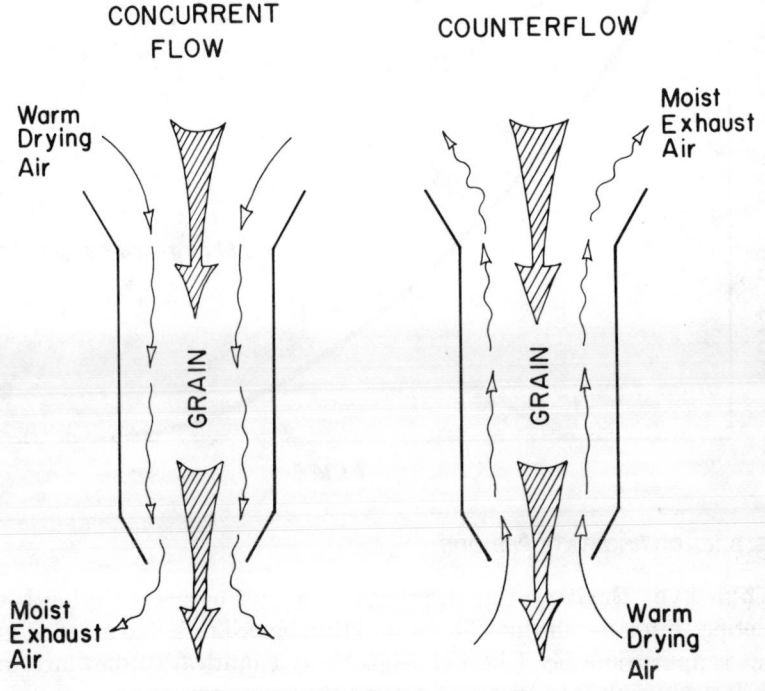

FIG. 8.17. CONCURRENT FLOW AND COUNTERFLOW GRAIN DRYING SYSTEM

In their studies with rice, Yamazawa *et al.* (1971) found capillarity to be the principal mechanism of moisture transfer between the hull and kernel during the initial stage of drying. They also discovered that moisture was transferred by vapor diffusion through the gap between the hull and the kernel. It is generally accepted that moisture movement in a grain kernel is a diffusion (liquid or vapor) type process.

Brooker *et al.* (1974) and Henderson and Perry (1976) have discussed the shortcomings of the theoretical diffusion type approach in grain drying. The problems are centered around the simplifying assumptions that have been made in development of diffusion models. These simplifications have introduced significant error into the theoretical diffusion analysis. Thus, most of the drying equations which have appeared in the literature over the years are of an empirical nature.

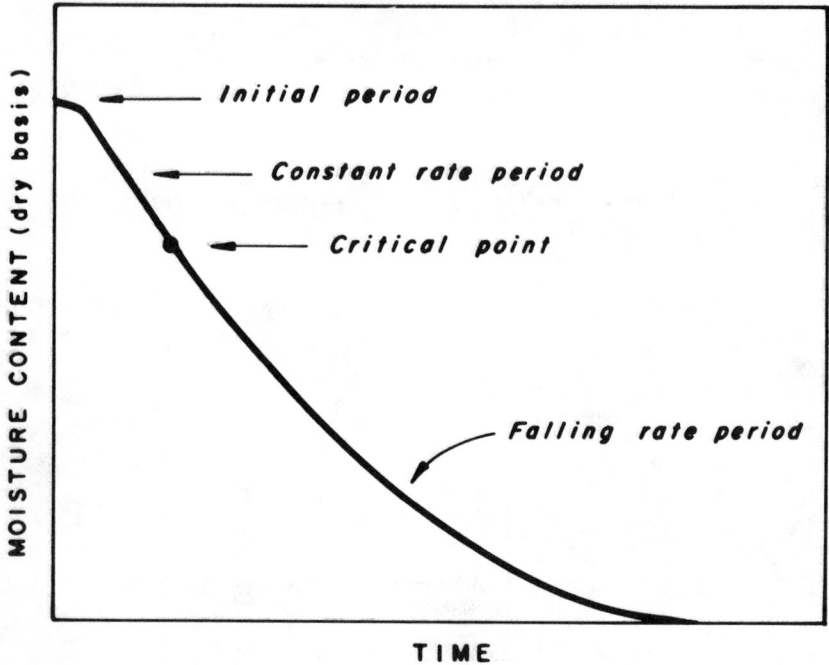

FIG. 8.18. DRYING RATE PERIODS

Thin-layer Drying.—One approach in analyzing grain drying is considering thin layer drying. The word "thin layer" refers to a bed of grain that is approximately 1 kernel deep. Many empirical studies have been conducted using thin layers of grain to examine the drying behavior of the particular product in question. Data are collected specifying the quantity of moisture removed from the grain as a function of time when dried with air having certain psychrometric properties such as temperature and relative humidity. When thin beds of grain are dried with air at different conditions, the resulting data can be treated to yield an equation which specifies moisture removal as a function of time and air properties. The equation may take various forms and is accurate within a specific range of drying air temperatures, relative humidities and grain moisture contents.

Many authors have discussed Newton's law of cooling in reference to the thin layer drying behavior of small grains. Newton's law states that the rate of change of temperature of a body is proportional to the temperature difference between the body and the surrounding medium. The analogy of drying to heat transfer is made by substituting a moisture content difference for a temperature difference. Thus, moisture instead of temperature would be the driving force. The drying equation obtained from the Newton analogy would be:

$$\frac{dM}{dt} = -k(M-M_e) \qquad (3)$$

Integration yields

$$\frac{M-M_e}{M_o-M_e} = K \exp(-kt) \qquad (4)$$

where: $\dfrac{dM}{dt}$ = Drying rate

M_o = Initial moisture content at time 0 (% dry basis)
M = Moisture content at time t (% dry basis)
M_e = Equilibrium moisture content (% dry basis)
t = Time (hr)
k = Drying rate constant (hr^{-1})
K = Constant of integration

The drying equation (3) was used by Kachru *et al.* (1970) to analyze the thin layer drying of 4 rice varieties. They found a close relation between this equation and the drying rate of paddy. The authors also concluded that the drying rate constant (k) was a characteristic of the paddy and not a function of the drying rate or grain moisture content. The drying rate constants determined for the varieties of paddy considered were: IR8, k= 0.6504 hr^{-1}; Dular, k = 0.8040 hr^{-1}; Patnai-23, k = 0.8424 hr^{-1}; and Taichung Native 1, k = 0.6480 hr^{-1}. Henderson and Pabis (1961) also applied a Newton-type equation and studied the effect of temperature on the drying rate constant (k). They concluded that k, the drying rate constant, was related to temperature by an Arrhenius-type equation stating that *ln* k is proportional to the negative inverse of the absolute temperature.

The Newton-type drying equation has also been discussed by Allen (1960) with regard to shallow bed drying (1.9 cm thin layer) of rice. He found from experimental work that modifications and restrictions would be necessary before the equation could be applied to the drying of rice. He concluded that moisture content and the drying rate constant are functions of the drying air temperature, drying air humidity, the quantity of drying air supplied and the initial and instantaneous values of the grain moisture content. Due to this variability in M_e and k, he suggests that the first drying period (up to ½ hr) be given a treatment different from the latter drying period. The Newton-type equation was applied to each drying period but with different values of k and M_e.

Hukill and Schmidt (1960) also investigated the application of Newton's Law of Cooling to drying. The authors found that Newton's concept does not adequately model grain drying and stated that the assumption that moisture flow is proportional to the moisture gradient must be abandoned. They felt (taking support from Babbitt 1940) that moisture flow in grain appears to be proportional to the vapor pressure gradient. The contention that vapor pressure is the main driving force for moisture movement during drying has also been supported by Wang (1958).

A number of equations which describe the thin layer drying of rice can be found in existing technical literature. Henderson and Henderson (1968) presented the following equation to describe the drying history of rice (Colusa 1600) when the drying was carried out with air at 41.7°C (107°F):

$$\frac{M - M_e}{M_o - M_e} = 0.650 \left[\exp\left(-0.220\ t\right) + \frac{1}{9}\left(-0.220\ (9)\ t\right)\right] \qquad (5)$$

Chancellor (1968), using rice drying data from various references, developed the following equation on thin layer drying which is similar to the one mentioned above:

$$\frac{M - M_e}{M_o - M_e} = 0.75 \left[\exp\left(-Gt\right) + \frac{1}{9}\exp\left(-Gt\right)\right] \qquad (6)$$

where: $G = 8860 \exp\left(-6147/T_{abs}\right)$
              T_{abs} = Absolute grain temperature (degrees R)

There is no universally accepted equation to describe thin layer drying of rice. The reason for this is that the drying characteristics of rice may vary with types of rice (short, medium, or long grain), its variety, previous grain history and drying environment.

Deep-bed Drying.—A deep-bed drier may be defined as one in which there is a finite moisture gradient through the drying layer at any time except time 0 (Henderson and Perry 1976). Deep-bed drying occurs in all situations where the drying bed is more than thin-layer deep. All fixed-bed and continuous flow driers are of the deep-bed type.

Hukill (1947, 1954) developed a procedure to relate drying time to moisture content at any location within a grain bed. This is a generalized method and will apply to the fixed-bed drying of any grain. Hukill developed the following simplified form of the general drying equation:

$$m = \frac{2^D}{2^D + 2^Y - 1} \tag{7}$$

where: m = Moisture ratio $\dfrac{M - M_e}{M_o - M_e}$

Y = Time unit (dimensionless)
D = Equivalent depth

This equation is pictorially shown in Fig. 8.19. The D value is calculated from:

$$D = \frac{X\,W\,h_{fg}\,(M_o - M_e)}{6000\,(Q)\,C_a\,P\,(\Delta T)} \tag{8}*$$

where: X = Distance from the bottom of the bin (ft)
        W = Density of dry matter in the grain (lb/ft^3)
        h_{fg} = Heat of vaporization of moisture in the grain (Btu/lb)
        Q = Mass rate of dry airflow (lb/min ft^2)
        C_a = Specific heat of the air (Btu/lb°F)
        P = Period of half response (hr)
        ΔT = Maximum temperature drop which may occur in the air (°F)

The period of half response, P, is the time required for fully exposed grain to be reduced from a moisture ratio of 1 to a moisture ratio of 0.5. The numerical value of P can be estimated with experimental data or from a thin drying equation such as any of those presented in the last section of this chapter. ΔT is the maximum temperature change which may occur in the air as it passes through the bed of rice. Hukill suggests that the value of ΔT used in the analysis should be the dry-bulb temperature of the inlet air minus that of the exhaust air. The exhaust air temperature is found (from a psychometric chart) by reading the dry-bult tempera-ture at the point where the initial wet-bulb temperature of the air inter-sects the relative humidity corresponding to the initial moisture content of the grain. The value of relative humidity used above is taken from experimental data. If this value is not available, a value of 85% RH will be a reasonable estimate. The Y value given in equation (7) is defined as:

*See Appendix for tables on metric and °F to °C conversions.

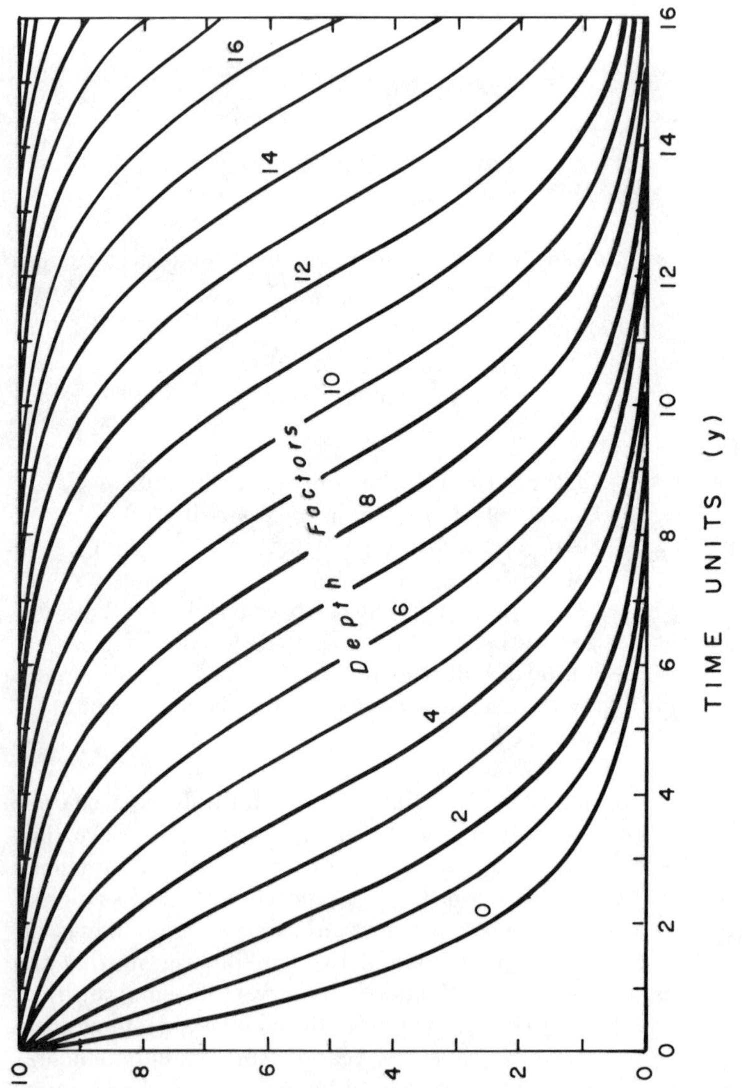

From Hukill (1954)

FIG. 8.19.   BULK DRYING CURVES

$$Y = t/P \tag{9}$$

where: t = time from start of drying (hr)

The usefulness of Fig. 8.19 and equations (7), (8) and (9) can be seen in the following examples:

Rice is to be dried in a circular bin. The level of rice in the bin is 0.6 m and the initial moisture content is 21.9% db. The airflow rate is 0.1 m³/s·m² and the inlet air is 41.7°C (dry-bulb) and 20% RH. Assume: h_{fg} = 2722 kJ/kg, C_a = 1.0 kJ/kg·°C, and W = 540.6 kg/m³. The equilibrium moisture content can be estimated from equation (2).

$$100 \, (0.325 - 0.046 \ln [-1.987 \, (41.7 + 35.703) \ln (0.2)]) = 7.15 \text{ MC \% db.}$$

From the psychrometric chart, the specific volume of the inlet air is found to be 0.905 m³/kg dry air and the exhaust air temperature is estimated to be 25.0°C. With this information, Q is calculated as:

$$Q = \frac{0.1}{0.905} = 0.11 \text{ kg dry air/m}^2\text{·s}$$

P can be estimated from equation (5) as 1.3 hr. With the prerequisite information available, D may be obtained from equation (8). Assume a column of grain with bottom area of 1 m².

$$D = \frac{0.6 \times 540.6 \times 2722 \times 14.75}{3.6 \times 10^5 \times 0.11 \times 1.0 \times 1.3 \times (41.7 - 25)} = 15.15$$

This means the 0.6 m rice bed is represented by 15.15 depth factors. Each depth factor spans 0.6/15.15 = 0.0396 m of the rice bed. The drying behavior, moisture ratio vs dimensionless time, at the top of the bed is described by the line corresponding to a depth factor of 15.15. Knowledge of how the depth factors correspond to bed position allows one to relate (Equation (7) or Fig. 8.19) moisture ratio to dimensionless time anywhere in the bed. To find the time required for the middle of the bed to reach 18% MC db, we would first compute:

$$m = \frac{18 - 7.15}{21.9 - 7.15} = 0.74$$

For a depth factor of 7.6 and m = 0.74, we find from Fig. 8.19, Y = 6.1. Therefore,

$$t = YP = 6.1(1.3) = 7.9 \text{ hr}$$

The middle of the rice bed will dry to 18% MC db in approximately 7.9 hr.

Hukill's method can be used to estimate the moisture content and drying time for rice if the necessary data are available. Mention must be made that Allen (1960) developed a method to approximate the mean grain temperature of a fixed-bed of rice during drying.

Since the advent of computer technology, a number of researchers have attempted to simulate the drying of grain, specifically corn. Analog computers have been used in simulation of fixed-bed grain driers by Baughman *et al.* (1971) and Hamdy and Barre (1970). There are also a number of simulation schemes which utilize digital computers to model the fixed-bed drying of grain. Spencer (1969) developed a drying model which appears to give a good prediction of drying rates from both high and low initial moisture contents. Sutherland (1975) has prepared a digital computer program to relate airflow rate, grain moisture content and time in a fixed-bed drier. Henderson and Henderson (1968) proposed a computational procedure for the analysis of fixed-bed drying which uses thin-layer drying theory. This study is of particular interest because simulation results and data are presented for the fixed-bed drying of rice.

Due to the complexity of the drying phenomenon, it appears that computer simulation offers the greatest potential for predicting grain drier performance. In the past, very little work has been done in applying computer drying models to rice. Bakker-Arkema *et al.* (1974) developed the fixed-bed grain drier simulation model for corn based on the fundamental laws of heat and mass transfer. This model was adapted by Chan (1975) to simulate rice drying. Drying simulation models are valuable as an aid for new drier design and for increasing our basic understanding of the drying process. Insight into the drying behavior of a bed of rice can be gained by looking at the results of a fixed-bed drier simulation (Fig. 8.20). The simulation results presented in Fig. 8.21 show the effect of drying air temperature on the average moisture content of corn. Simulation models can also be used to predict drier performance. General performance information determined by Thompson *et al.* (1969) for a crossflow drier is presented in Fig. 8.22.

In current technical literature one can find simulation models for drier configurations other than the fixed-bed. Cross-flow, concurrent flow and counterflow drier simulation models for corn can be found in publications by Thompson *et al.* (1969) and Bakker-Arkema *et al.* (1974). The mathematical models presented by them and others should be very useful to people interested in rice drying. These models may be modified for use with rice and they also serve as guidelines for the development of new simulation techniques.

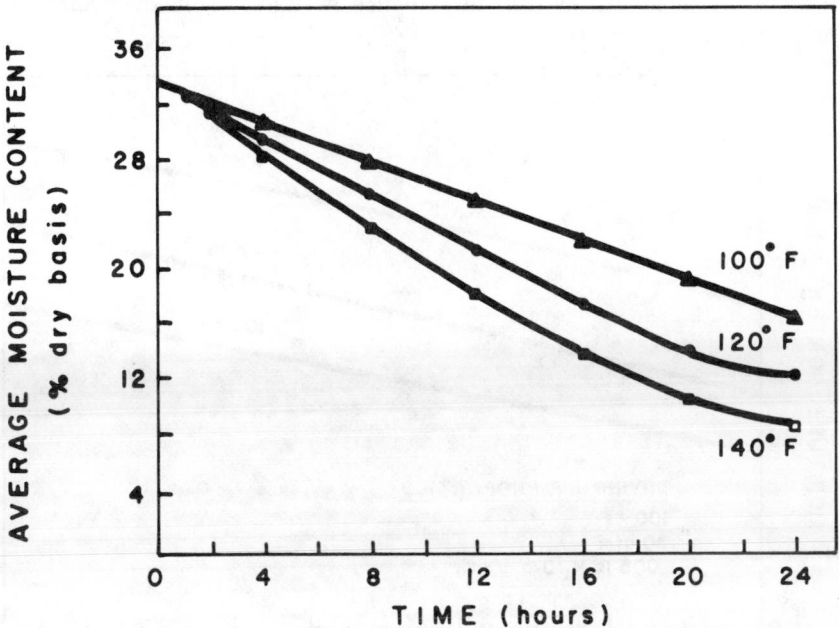

FIG. 8.20. THE RELATIONSHIP BETWEEN RICE MOISTURE CONTENT AND BED POSITION

(Consult Appendix for appropriate tables on metric conversions.)

Rice Cracking

The purpose of this review of literature on rice cracking is to investigate the causes, mechanisms and characteristics of cracking which are important in rice drying. Suncracks, fissures and checks are some words commonly used to refer to cracks which are sometimes found in rice. The word suncrack is a misnomer because it implies that cracking is simply the result of sundrying. Although sundrying is an important factor that causes field cracking, there are many other factors which cannot be ignored. This is particularly true when rice is commercially dried because it is removed from the field when it still has a high moisture content. In this case, it is mechanical drying and not natural sundrying which is the contributing factor.

Rice cracking structurally weakens the kernel making it more susceptible to breakage during milling and handling operations. The economic consequences of this are significant because the value of broken rice is much less (⅓–½) than that of whole rice. Cracked rice is also more susceptible to insect infestation. Cracking has the added limitation that it may reduce the vitality of the seed rice. Cracks which occur across the kernel, as most cracks do, may reduce seedling vigor by decreasing endosperm availability.

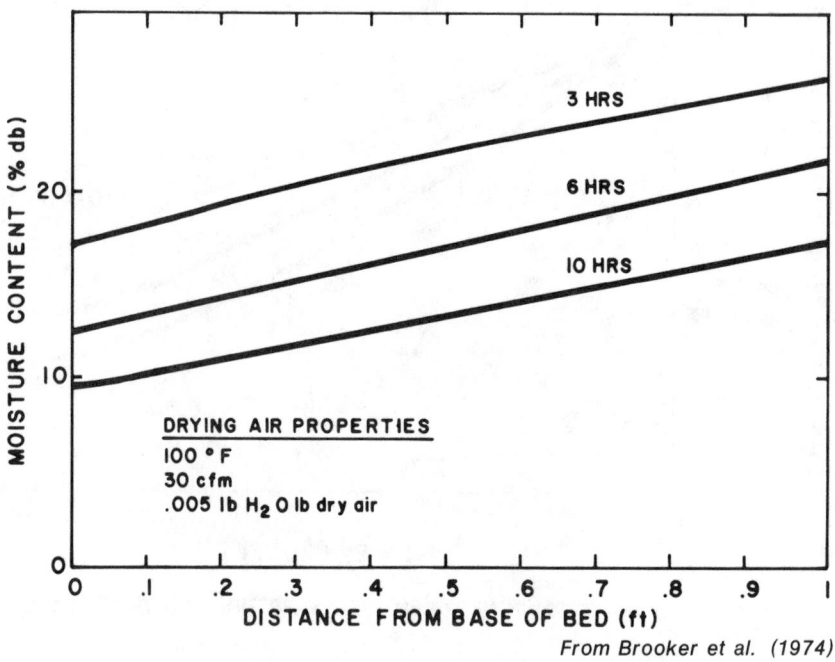

From Brooker et al. (1974)

FIG. 8.21. EFFECT OF DRYING-AIR TEMPERATURE ON THE AVERAGE MOISTURE CONTENT OF A FIXED BED OF CORN AFTER VARIOUS DRYING TIMES

(Consult Appendix for appropriate table on °F to °C conversions.)

The mechanism of rice cracking has been studied by a number of researchers. Henderson (1954), in his studies with short grain rice concluded that cracking during fast drying was due to an increase in temperature rather than a decrease in moisture in portions near the surface of the kernel. He found that cracking could also be caused by a rapid increase in moisture which could occur in the field if dew accumulated on the kernels. Kunze and Hall (1965) found that cracking occurred when brown rice, equilibrated at a particular humidity, was subjected to a high humidity environment. The degree of cracking was dependent on the magnitude of change in relative humidity. The researchers hypothesized that adsorptive fissures were caused when external cells expanded by adsorbing moisture and produced compressive stresses in surface layers.

The internal parts of the kernel would resist these stresses and fail if the material tensile strength was exceeded.

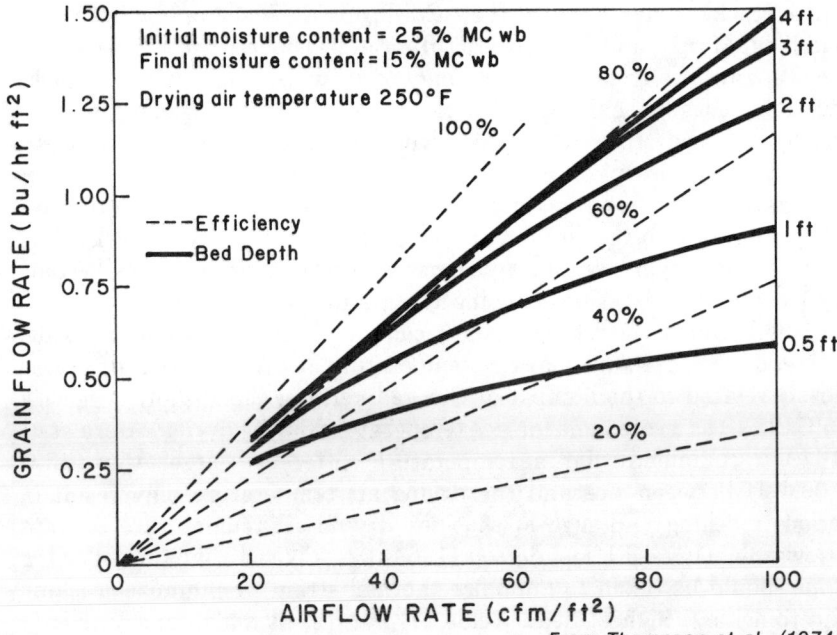

FIG. 8.22. GENERAL PERFORMANCE OF A CROSSFLOW DRIER FOR CORN

(Consult Appendix for appropriate tables on metric and °F to °C conversions.)

Rhind (1962) maintains that moisture changes in rice result in unequal volumetric changes that induce internal stresses. He reports that the highest tensile stresses can be expected in the surface layers during drying and in the inner grain parts during adsorption. A similar conclusion has also been reached by Nayato *et al.* (1964) and Kunze and Choudhury (1972). They found that cracking from hydration and drying was due to unequal swelling or shrinking within the rice kernel. The force distribution within a rice kernel was more fully explored in a hypothetical 2-dimensional stress analysis presented by Kunze and Choudhury (1972).

Rhind (1962) suggests that a change in the physical behavior of rice starch occurs at 15% MC wb and this change of state is a factor in the cracking phenomenon. MacDonald (1967) has suggested that a physical change within the rice kernel may also be a factor which contributes to rice cracking. This author observed that cracks usually occur between the cell walls of the endosperm and rarely pass through them. He hypothesized that it is not the cell walls which fail when overstressed but the gum holding the walls together. MacDonald (1967) also mentions that the intercellular gums are somewhat plastic above 21% MC wb and become increasingly inelastic with lower moisture contents.

Rice cracking may occur in the field during drying, storage, milling or handling. Kunze and Parsad (1976) have considered situations where moisture adsorbing environments may occur which can cause rice cracking. They suggest that harvested rice may have grains which are sufficiently dry that they will fissure when subjected to an environment conducive to adsorption of moisture. The researchers specified 2 situations where the low moisture rice in harvested grain may experience a moisture-adsorbing environment: (1) in the combine hopper or holding tanks before drying, and (2) in a drier where the drying air has become highly humidified before reaching the grain.

A number of authors have made recommendations to reduce cracking. Itoh *et al.* (1974) found that rice temperature at the exit of the drier was closely related to the number of checked kernels; the authors suggest a 35°C (95°F) as a safe limit for rice temperature during drying. Arora *et al.* (1973) recommend that a temperature difference larger than 43°C (109.4°F) between rice and the drying air temperature may result in serious cracking and suggest that the drying air temperature be kept below 53°C (127.4°F). Henderson (1954) has advised that the following steps should be taken to minimize thermal strain to minimize cracking and to achieve highest head yields: (1) harvest at a high moisture content, (2) dry at as low an air temperature as possible, and (3) use as many passes as convenient in multipass continuous flow drying.

RICE STORAGE

A commercial bulk storage system designed for long-term safe storage of rough rice requires that: (1) it must provide proper aeration to prevent spontaneous heating, and (2) it must maintain the rice grain at low moisture content (around 13.5% wet basis) to protect it from fungi and insects. The operation of aeration systems for bulk storages in high humidity environments calls for the operator's constant care and attention. The following discussion is aimed at developing engineering parameters for the design and selection of an aeration system for storing rough rice.

Storage structures, usually classified as "flat" or "vertical," are used to store rice for long durations. A flat storage refers to a structure with height less than the width or diameter whereas a vertical storage has height greater than width or diameter. Vertical storages have the advantage of easy unloading by gravity since they have hopper bottoms. Mechanical means, such as power shovels or sweep augers, are needed to unload flat bins.

Respiration of rice grain must be taken into account when designing storage systems. Since rice is a biological product, there will be some deterioration during storage even if it is stored at the 13.5% "safe" moisture content.

Aerobic respiration, requiring atmospheric oxygen can be expressed as:

$$C_6H_{12}O_6 \rightarrow 6CO_2 + 6H_2O + 673 \text{ KCal.}$$

The respiration process as expressed by the above equation involved utilization of oxygen, evolution of carbon dioxide, liberation of thermal energy and a decrease in weight. The spontaneous generation of heat is of serious consequence in bulk storage since rice has a porous, granular structure and a low thermal conductivity. If there are no convective currents present, the grain mass practically insulates itself. Thus, respiration could lead to heat damaged grain if proper care is not taken.

An aeration system for storage, if provided, has several advantages including: (1) equal storage temperatures thus avoiding the occurrence of hot spots, (2) cooling of rice (depending on the weather) thus minimizing insect activity and mold growth, (3) removal of odors from stored rice, and (4) distribution of fumigants inside the stored product.

Schroeder and Calderwood (1972) recommend that aeration for cooling of dry rice during winter months or for equalizing temperatures during other months can be accomplished with airflow rates as low as 1.3×10^{-6} m^3/s·kg. Although typical airflow rates for aeration range between 1.0×10^{-6} to 2.1×10^{-6} m^3/s·kg a higher airflow rate is needed to prevent spontaneous heating.

An excellent discussion on aeration-system design is given by Holman (1966). In the following discussion, a summary of Holman's recommendations, applicable to designing aeration systems for rice, is presented.

The grain must be adequately cooled during storage to prevent mold growth and insect activity. Typically, during late summer and fall harvest, rice dried with heated air may go in storage with grain temperatures up to 37.8°C (100°F). Although there is no one optimum storage temperature, mold growth is inhibited below 21.1°C (70°F) (for rice at 13 moisture content, wet basis). Insect activity is considerably reduced at temperatures below 15.6°C (60°F). It should be emphasized that aeration will not entirely replace fumigation as a means of insect control.

Selection of aeration systems for rice requires consideration of: (1) type of storage structure in which the aeration system is to be installed, (2) air-flow rate (e.g., m^3/s·kg), and (3) number of storage structures and quantity of grain to be aerated. Based strictly on the above information, the following should be determined: (1) total air volume to be supplied, (2) static pressure against which the fan must operate, (3) size and type of fan and motor, and (4) size of ducts and supply pipes.

Types of Aeration Systems

Satisfactory aeration of grain in upright and flat storages can be accomplished through following arrangements of fans and ducts.

A fixed fan for each storage or for each duct in the storage is simple and efficient as shown in Fig. 8.23. The disadvantage of the system is its high cost since several fans and motors are required. In addition, fumigants must be introduced into each individual fan.

A manifold system consists of 1 fixed fan connected to several storages or ducts with 1 manifold as shown in Fig. 8.24. This type of system is less expensive. The use of dampers allows aerating 1 or several storages at the same time. In designing such a system, friction losses in the manifold and pressure losses in the bends and dampers must be considered. It needs mention that the manifold systems are usually poor air distributors.

A relatively cheaper system is to use a portable fan to aerate several storages. In such a system, the fan and motor are mounted on wheels and moved to adjoining storages when aeration is complete.

FIG. 8.23. A FIXED FAN FOR EACH STORAGE STRUCTURE

Airflow Requirements

An airflow rate of 1.3×10^{-6} m³/s·kg is suitable for purposes of cooling dry rice during winter months or for equalizing temperatures in the storage during other months. A higher airflow rate is needed to prevent spontaneous heating for undried rice. In addition, higher airflow rates are required when the aeration system is used for supplementary drying by cooling rice after a drier pass. In selecting an airflow rate, proper consideration must be given to the lowest rate than can be permitted in any portion of the stored grain.

FIG. 8.24. A MANIFOLD SYSTEM FOR SEVERAL STORAGE STRUCTURES

The lowest airflow rate in a well designed aeration system for an upright storage is usually not less than 90% of the selected rate. However, in flat storages the lowest airflow rate may be only ½ or less of the selected rate. This emphasizes the importance of considering the lowest airflow rate when designing aeration systems.

Fan Selection

The following parameters are important in selection of fans for an aeration system: (1) the volume of air to be moved, (2) the static pressure at which the fan must operate, and (3) the noise level of the fan. The factors influencing static pressure include airflow rate, grain depth, variety (such as rough hulled or smooth hulled) and size of the grain. The fan speed influences the noise.

Both propeller (axial flow) fans and centrifugal (radial flow) fans are used for aeration where the static pressure of the grain and system is less than 10 cm of water. Centrifugal fans are used to obtain higher pressures.

A "forward-curve" centrifugal fan, consisting of a large number of blades, operates at a relatively low speed. The motor in this fan may overload if the static pressure is decreased. A centrifugal fan with straight blades also overloads when static pressure is reduced. A "backward-curve" centrifugal fan is a high speed fan more efficient, but more expensive than the previously described centrifugal fans. Operating "backward-curve" centrifugal fans near the point of greatest efficiency will avoid the danger of overloading the motor.

The performance of the aeration system can be seriously affected if the fan is not properly connected. Short elbows at fan inlet and exit restrict the fan output. It is important to keep the pipe size uniform.

Duct Design and Selection

Semicircular and inverted V-shaped perforated ducts are most common. Perforated false floors are relatively expensive. In perforated ducts the area of the perforations should equal at least 10% of the total duct surface area.

The following 3 dimensions are important in designing a duct system: (1) both the cross-sectional area and length of the duct influence the uniformity of airflow and air velocity within the duct, (2) the circumference of the duct influences static pressure losses in grain surrounding the duct, and (3) the distance between the ducts influences the uniformity of airflow through the grain between ducts.

The duct cross-sectional area can be determined by the following:

$$\text{cross-sectional area, m}^2 = \frac{\text{Total air volume (m}^3/\text{s)}}{\text{Air velocity (m/s)}}$$

An air velocity of 10.16 m/s per min is permissible for upright storages. The maximum recommended air velocities in ducts for flat storage are given in Table 8.7.

Excessive pressure losses in grain surrounding the duct can be avoided by limiting the velocity of air entering or leaving the grain. It is obvious that this velocity will be influenced by both the surface area of the duct and the airflow rate.

For upright storages, the duct surface area should be selected to limit the air velocity through the grain near the duct to 0.15 m/s or less. For flat storages, the air velocity through the grain near the duct should be limited to 0.1 m/s by proper selection of duct surface area.

A maximum velocity of 12.7 m/s is permissible in supply pipes that provide passage of air between the duct and the fan.

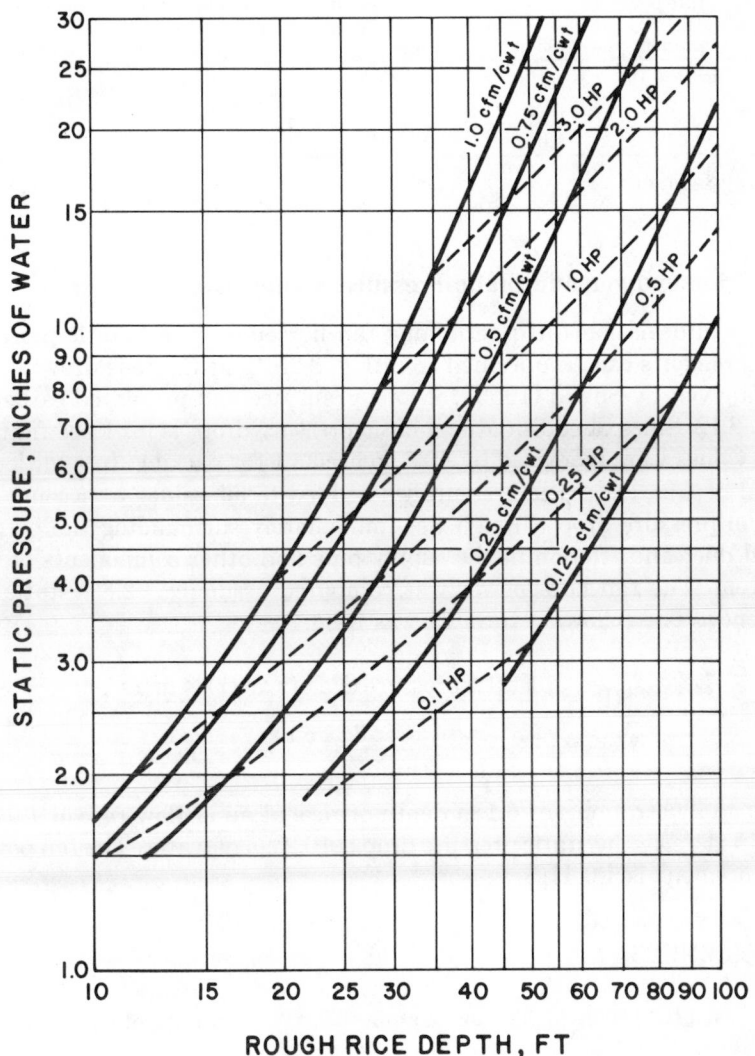

FIG. 8.25. STATIC PRESSURE AND FAN HORSEPOWER REQUIRED FOR AERATING ROUGH RICE (LONG GRAIN, BLUE BONNET) AT DIFFERENT RATES OF AIRFLOW AND AT DEPTHS FROM 10 TO 100 FEET

(Consult Appendix for appropriate table on metric conversions.)

TABLE 8.7. RECOMMENDED MAXIMUM ALLOWABLE AIR VELOCITIES WITHIN DUCTS INSTALLED IN FLAT STORAGES

Airflow	Corn, Soybeans, and Other Large Grains				
rate per m³ of product	Air velocity within ducts (m/s) for grain depths of				
	3.05 m	6.1 m	9.14 m	12.19 m	15.24 m
m³/s × 10⁴	m/s	m/s	m/s	m/s	m/s
6.70	—	5.08	7.62	8.89	10.16
13.40	3.81	7.62	10.16	—	—
26.79	5.08	10.16	—	—	—

Source: Holman (1960).

Fan Horsepower and Static Pressure Requirement

The information for determining fan horsepower and static pressure requirements can be obtained from Fig. 8.25. This figure is drawn from data given by Shedd (1953) for long grain rice and presented earlier in Fig. 8.12. Since Shedd's data on basic pressure drops is for loose-fill rice, the values were increased by 40% to account for packing. In addition, a total of 2.54 cm static pressure was added to all values to account for radial pressure drop through rice immediately surrounding the perforated duct and friction in the supply pipe and other components of the system. The fan horsepower was computed assuming 62.8% static efficiency. The following formula was used:

$$HP = \frac{\text{Static pressure (Pa of water)} \times m^3/s}{34{,}065}$$

As an illustration, for 6.1 m depth of rice, at an airflow rate of 1.04×10^{-5} m³/s·kg the static pressure drop is 10.2 cm of water. The fan power requirement is 1.0 HP.

REFERENCES

ALLEN, J.R. 1960. Application of grain theory to the drying of maize and rice. J. Agric. Eng. Res. 5 (4) 363-386.

ARAULLO, E.V., DEPADUA, D.B. and GRAHAM, M. 1976. Rice Postharvest Technology. Int. Dev. Res. Cntr., Ottawa, Canada.

ARORA, V.K., HENDERSON, S.M. and BURKHARDT, T.H. 1973. Rice drying cracking versus thermal and mechanical properties. Trans. Amer. Soc. Agric. Eng. (ASAE) 16 (2) 320-323.

BABBITT, J.D. 1940. Observations on the permeability of hygroscopic materials to water vapor. Amer. J. Res. *18* A, 105-121.

BAKKER-ARKEMA, F.W., LEREW, L.E., DE BOER, S.F. and ROTH, M.G. 1974. Grain Dryer Simulation. Michigan State Univ. Agric. Exp. Sta., Res. Rep. *224.*

BAUGHMAN, G.R., HAMDY, M.Y. and BARRE, H.J. 1971. Analog computer simulation of deep-bed drying of grain. Trans. ASAE *14* (6) 1058-1060.

BROOKER, D.B., BAKKER-ARKEMA, F.W. and HALL, C.W. 1974. Drying Cereal Grains. AVI Publishing Co., Westport, Conn.

CALDERWOOD, D.L. 1973. Resistance to airflow of rough, brown and milled rice. Trans. ASAE *16* (3) 525-527, 532.

CERVINKA, V. 1971. Resistance of seeds to airflow. Unpublished research report. Dep. of Agric. Eng., Univ. of Calif., Davis.

CHAN, N.K. 1975. Simulation of the batch drying of rice. M.S. thesis. Michigan State University, East Lansing, Michigan.

CHANCELLOR, W.J. 1968. Characteristics of conducted heat drying and their comparison with those of other drying methods. Trans. ASAE *11* (6) 863-867.

CHUNG, D.S. and PFOST, H.B. 1967. Adsorption and desorption of water vapor by cereal grains and their products. Trans. ASAE *10* (4) 545-551, 555.

HALL, C.W. 1957. Drying Farm Crops. Agric. Consulting Assoc., Reynoldsberg, Ohio.

HAMDY, M.Y. and BARRE, H.J. 1970. Analysis and hybrid simulation of deep-bed drying of grain. Trans. ASAE *13* (16) 752-575.

HENDERSON, S.M. 1952. A basic concept of equilibrium moisture content. Agric. Eng. *33* (1).

HENDERSON, S.M. 1954. The causes and characteristics of rice cracking. Rice J. *57* (5) 16, 18.

HENDERSON, S.M. 1957. Milled rice yields. Calif. Agric. (7), 6.

HENDERSON, S.M. 1976. Unpublished data. Agric. Eng. Dep., Univ. of Calif., Davis.

HENDERSON, J.M. and HENDERSON, S.M. 1968. A computational procedure for deep-bed drying analysis. J. Agric. Eng. Res. *13* (2) 87-95.

HENDERSON, S.M. and PABIS, S. 1961. Grain drying theory I. J. Agric. Eng. Res.

HENDERSON, S.M. and PARSONS, R.A. 1974. Rice drying and storage. Unpublished research report. Dep. of Agric. Eng., Univ. of Calif., Davis.

HENDERSON, S.M. and PARSONS, R.A. 1975. Deep-bed grain drying. Univ. of Calif., Davis Coop. Ext. Ser. Lflt. *2103.*

HENDERSON, S.M. and PERRY, R.L. 1976. Agricultural Process Engineering. Third edition. AVI Publishing Co., Westport, Conn.

HOLMAN, L.E. 1966. Aeration of grain in commercial storages. U.S. Dep. of Agric. Mktg. Res. Rep. *178.*

HUKILL, W.V. 1947. Basic principles in drying corn and grain sorghum. Agric. Eng. *28* (8) 335-338.

HUKILL, W.V. 1954. Drying grain. *In* Storage of Cereal Grains and their Products. Amer. Assoc. Cereal Chem., St. Paul, Minnesota.

HUKILL, W.V. 1974. Personal communication. Iowa Agric. Exp. Sta., received by S.M. Henderson, Univ. of Calif., Davis.

HUKILL, W.V. and SCHMIDT, J.L. 1960. Drying rate of fully exposed grain kernels. Trans. ASAE *2* (3) 71-77, 80.

ITOH, K., TERAO, H., IKEUCHI, Y. and YOSHIDA, T. 1974. Drying of rough rice II. J. Soc. Agric. Mach. Jpn. *35* (4) 393-398.

KACHRU, R.P., OJHA, T.P. and KURUP, G.T. 1970. Drying characteristics of Indian paddy varieties. Bull. Grain Tech. *813*, 98-102.

KOSOSKI, A.R. 1976. Two methods of comparing equilibrium moisture of grains. M.S. thesis. Kansas State University, Manhattan, Kansas.

KUNZE, O.R. and CHOUDHURY, M.S.U. 1972. Moisture adsorption related to the tensile strength of rice. Cereal Chem. *49* (6) 648-696.

KUNZE, O.R. and HALL, C.W. 1965. Relative changes that cause brown rice to crack. Trans. ASAE *8* (3) 396-399, 405.

KUNZE, O.R. and PARSAD, S. 1976. Grain processing potentials in harvesting and drying of rice. Trans. ASAE 76-3560.

MACDONALD, D.J. 1967. Suncracking in rice. M.S. thesis. University of Sydney, Sydney, Australia.

MORITA, T. 1977. Unpublished data. Dep. of Agric. Eng., Univ. of Calif., Davis.

NAYATO, K., EBATA, M. and ISHIKAWA, M. 1964. On the formation of cracks in rice kernels during wetting and drying of paddies. J. Soc. Agric. Machinery *33*, 82-89. (Japanese)

PFOST, H.B., MAURER, S.G., CHUNG, D.S. and MILLIKEN, G.A. 1976. Summarizing and reporting equilibrium moisture for grains. Trans. ASAE 76-3520.

RHIND, D. 1962. The breakage of rice in milling. Rev. Trop. Agric. *39* (1) 19-38.

SCHROEDER, H.W. and CALDERWOOD, D.L. 1972. Rough rice storage. *In* Rice Chemistry and Technology. D. F. Houston (Editors). Amer. Assoc. Cereal Chem., St. Paul, Minnesota.

SHEDD, C.K. 1953. Resistance of seed grains to airflow. Agric. Eng. (9) 616-619.

SPENCER, H.B. 1969. A mathematical simulation of grain drying. J. Agric. Eng. Res. *14* (3) 226-235.

SUTHERLAND, J.W. 1975. Batch grain drier design and performance prediction. J. Agric. Eng. Res. *20* (4) 423-432.

THOMPSON, T.F. 1972. Temporary storage of high-moisture shelled corn using continuous aeration. Trans. ASAE *15* (2) 333-337.

THOMPSON, T.L., FOSTER, G.H. and PEART, R.M. 1969. Comparison of concurrent-flow, crossflow, and counterflow grain drying methods. U.S. Dep. Agric. Mrktg. Res. Rep. *841*.

WANG, J. 1958. Theory of drying. Ph.D. dissertation. Michigan State University, East Lansing, Michigan.

WASSERMAN, T. and CALDERWOOD, D.L. 1972. Rough rice drying. *In* Rice Chemistry and Technology. Amer. Assoc. Cereal Chem., St. Paul, Minnesota.

WASSERMAN, T., MILLER, M.D. and GOLDEN, W.G. 1965. Heated air drying of California rice in column driers. Calif. Agric. Exp. Sta., Univ. of Calif., Davis.

WASSERMAN, T., FERREL, R.E., KAUFMAN, V.F., SMITH, G.S., KESTER, E.B. and LEATHERS, J.G. 1958A. Improvements in commercial drying of western rice. I. Mixing type drier. Rice J. *4* (36) 30-32, 34-36.

WASSERMAN, T., FERREL, R.E., KAUFMAN, V.F., SMITH, G.S., KESTER, E.B. and PIERCE, S. 1958B. Improvements in commercial drying of western rice. II. Non-mixing columnar-type drier. Rice J. *5* (40) 42-46.

WASSERMAN. T., FERREL, R.E., HOUSTON, D.F., BREITWIESER, E. and SMITH, G. 1964. Tempering western rice. Rice J. (Feb. 16) 17, 20-22.

WRATTEN, F.T., POOLE, W.D., CHESNESS, J.L., BAL, S. and RAMARAO, V. 1969. Physical and thermal properties of rough rice. Trans. ASAE *12* (6) 801-803.

YAMAZAWA, S., YSHIZAKI, S. and MAEKAWA, T. 1971. Studies on the drying of agricultural products (I). Moisture transfer between hulled and unhulled rice during drying pause. J. Soc. Agric. Mach. Jpn. *32* (2) 192-199.

9

Milling

James J. Spadaro
Jack Matthews
James I. Wadsworth

The purpose of rice milling is to remove the hulls and bran from harvested, dried rough rice and to produce a milled, polished or white rice with a minimum of breakage and with a minimum of impurities such as weed seeds in the final product.

The meaning of the term *milling* varies appreciably, not only in the many different industries in which the term is used, but also within the grain industry. For example, milling of wheat refers to a grinding operation to produce a flour whereas in the rice industry milling can refer to either the overall operations in a rice mill—cleaning, shelling, bran removal, size separation—or it can refer simply to the one operation of removal of the bran or outer layers from the brown rice to produce a whole grain white rice product.

The term *bran* as used by the rice industry includes the soft germ and several histologically identifiable soft layers surrounding the hard endosperm.

The following terminology is also somewhat specific to the rice industry and clarification at this point may help to understand better the discussions on rice milling in this chapter. *Rough* rice, which is also known as *paddy* rice, is the harvested, unshelled rice as it comes from the field or the dryer. *Shelling* refers to the removal of the outer shell. The operation is conducted in machines which have been known by many different names such as shellers, hullers, huskers, dehuskers and decorticators. Similarly, the shells are also known as hulls, husks and chaff.

The term *hulling* in most parts of the United States has the same meaning as shelling, however, in some areas of the United States and other countries hulling also refers to the removal of both hulls and bran. This probably stems from the fact that earlier machines such as the Engelberg Huller[1] removed either the bran or both hulls and bran in one

---

[1]Names of companies or commercial products are given solely for the purpose of providing specific information; their mention does not imply recommendation or endorsement by the U.S. Department of Agriculture over others not mentioned.

operation. Consequently, one occasionally hears that "bran is removed in a hulling machine." After removal of the hulls, the rice is called *brown rice*. This terminology does not refer necessarily to the actual color of the rice but simply infers that the bran has not been removed. The brown rice is also known as shelled rice, husked rice, cargo rice and loonzian (Houston and Kohler 1970). The brown rice is *milled* in a *milling machine* whereby all or most of the bran is removed to produce the white *milled rice*. This *milling* operation to remove the bran is sometimes called *scouring* or *whitening*. In *polishing* of the milled rice, traces of bran that may remain on the rice are removed and the rice surface is given a smoother finish. *Total* milled rice includes both the head rice and the broken rice. *Head rice* or *head yield* refers to the milled whole rice grains. The broken rice is generally subdivided into 3 sizes—*second heads* which are the largest of the broken kernels, *screenings* which have an intermediate size, and *brewers* rice which are the small broken kernels.

It should also be understood that, in the rice industry, when one refers to a *rice mill* he is including a series of processing operations which, in general, consists of cleaning, shelling, milling, polishing, separation of whole rice and broken rice, sizing of broken rice, packaging, and several auxilliary operations such as husk aspiration, paddy separation and glazing of rice.

The objective of these rice mill operations is to produce a white, whole grain or head rice product that is essentially free of bran and foreign matter and which contains a minimum of broken kernels.

Rice milling is of great economic importance to the rice industry particularly since rice breakage is attributed primarily to the milling operation and since broken rice is worth about ½ that of whole rice. Although the actual milling operation (bran removal) has been thought to be responsible for most of the breakage in the past, Matthews and Spadaro (1975) have shown that about ½ of the breakage occurs during harvesting and the remainder occurs in subsequent handling and milling. The overall economic effect of rice breakage can best be shown by citing the following example. For the rice mills in the United States the decreased returns due to breakage amounted to over $115,000,000 for the 1974–75 processing season. This figure is based on the following data: the milling of 4767 million kg of rough rice, an average of 11 kg of broken milled rice produced per 100 kg of rough rice and a price difference of $22 per 100 kg between the whole grain and brokens. Based on 40 rice mills in the United States, the decreased revenue amounts to almost $2,900,000 for the average rice mill. If it is assumed that ½ of the total rice breakage occurs during milling, this amounts to a decreased revenue of $57,000,000 per year or about $1,425,000 per mill. It is realized that all breakage cannot be eliminated; however, even a 1% decrease in brokens would result in an increased return of about $10,500,000 or $262,500 for the average mill.

In this chapter emphasis will be placed in the following related areas of rice milling. A brief review of the early milling equipment and a description of current milling equipment will be presented. Factors that effect rice breakage will be discussed such as combine harvesting, thickness of paddy rice, drying, handling and milling machine variables which include rotor speed, feed gate opening, back pressure, single versus multiple passes, air flow, relative humidity, rice moisture content and rice temperature. The advantages and disadvantages of lightly milled rice as related to legal requirements, nutrition, stability, color and cooking characteristics will be presented. Well-milled rice and its demand by the United States and European consumers and its U.S. Department of Agriculture (USDA) rice grades and pricing structure will be discussed. Deep milled rice and the production of high protein flour, techniques for measuring degree of milling and the process for solvent extractive milling—its products and advantages and disadvantages—will also be included in this discussion.

RICE PROCESSING MACHINERY

In some rice growing areas rice milling is accomplished by very primitive methods such as pounding the rough rice in a wooden mortar and pestle followed by winnowing. At the other extreme are very modern methods where milling is accomplished in large, highly automated plants with the manifold operations being monitored from a central console (Tolson and Robe 1977). Thus, there is no typical rice mill. In addition, rice processing equipment is manufactured in a number of countries throughout the world with minor to major differences in the design of individual functional pieces of machinery. However, the modern processing of rice consists of essentially the same steps: (1) cleaning the incoming rough rice, (2) shelling the rough rice, (3) milling to remove the bran from the brown rice, (4) grading the milled rice by length into whole grain and brokens, (5) mixing milled whole grain and brokens to meet specifications of buyers, and (6) packaging. Other descriptions of rice milling equipment include Witte (1970, 1972) and Van Ruiten (1976).

Cleaning

Rough rice is first cleaned to remove foreign material such as straw, soil particles and weed seeds. Cleaning is an important step in producing a high quality milled rice. According to USDA standards (USDA 1976), the highest grade (U.S. No. 1) of milled rice has a maximum limit of 1 weed seed in 500 g. If the milled rice contains 2 weed seeds in 500 g the rice is down graded to a U.S. No. 2 rice, although all other specifications of No. 1 rice are met. Separation of foreign matter is based on differences in gross size, weight or density, and shape (principally length) of the impurities compared to the rough rice. Figure 9.1 shows in cross section a scalping machine (scalperator) which removes straw and other large objects by screening and dust and light weight material by aspiration. Rough rice enters through hopper 1. Feed rate is regulated by gate 2. The rice falls on the rotating cylindrical screen or scalping reel 4. The mesh size of the screen is such that straw, sticks and other large foreign matter are carried over the screen to discharge 13 while the rice falls through the screen. The rice is then subjected to a stream of air which removes low density matter such as dust, sterile florets, and stemmes and discharges these in settling chamber 8. The rough rice is discharged at 12. Flat inclined screens (screen separators) having a horizontal reciprocating motion may be used to further clean the rice. An example may be used to illustrate the separating action of screens based on length of the objects to be separated. A roughly spherically-shaped weed seed having a diameter equivalent to the width of a rough rice kernel may fall through the round perforations of a screen while the rough rice by laying flat on the screen will not because its length is longer than the diameter of the perforations. The rough rice thus bridges across the perforations in its movement down the inclined screen. In the same fashion sticks or straw longer than the rough rice may be separated by passage over a slotted screen. Indented disc separators may also be used to separate rough rice from longer or shorter impurities. The operation of indented discs are discussed in more detail in connection with the separation of whole grain milled rice from brokens. Permanent or electromagnets are generally installed in the stream of rice to remove ferrous metals. Stoners may also be included to remove dense materials. These machines essentially operate by aspirating the rough rice away from the denser objects.

Shelling

The first step in the actual milling of rice is to remove the hulls (shelling). In Fig. 9.2, a cross-sectional side view of a modern rubber roll sheller is illustrated. The feeder 2 meters a falling stream of rough rice between 2 closely spaced rubber rollers which are turning in opposite directions and at different speeds. As the rice passes between the rollers it is subjected to a shearing force which separates the 2 hulls (glumes) from the brown rice as shown in the exploded view on the right of Fig. 9.2. The distance between the rollers may be regulated by a handwheel 7 as illustrated or in the most modern machines by a pneumatic mechan-

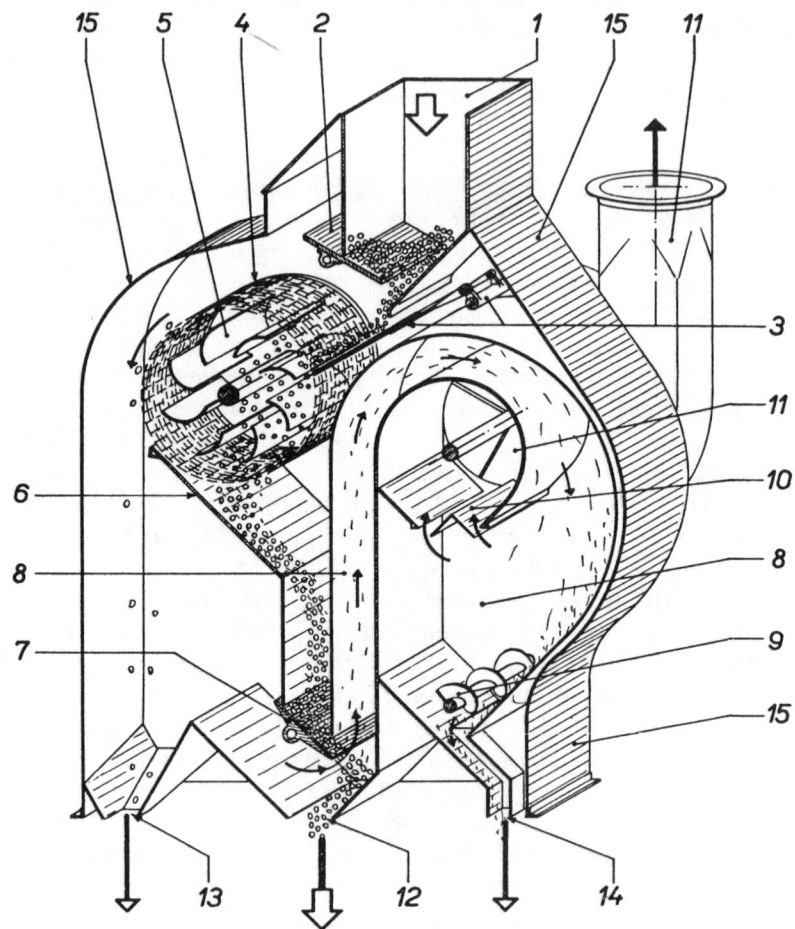

Courtesy of FAO (Borasio and Garibaldi 1957)

FIG. 9.1. SCALPING MACHINE (SCALPERATOR)

(1) Feed hopper, (2) feed regulation gate, (3) adjustable rack, (4) wire mesh scalping reel, (5) baffle plates, (6) grain receiving rack, (7) seal gate, (8) aspirating leg and settling chamber, (9) screening screw conveyer, (10) air outlet baffle plates, (11) suction fan, (12) grain outlet, (13) scalpings outlet, (14) screenings outlet, (15) machine frame and housing.

ism which automatically separates the rolls and turns off the driving motor if there is an interruption of rice flow through the machine. The sheller is normally set to shell about 95% of the incoming rough rice. Excessive pressure and increased percentages of shelling cause excessive breakage of the grain. The main disadvantage of this sheller is that the rollers

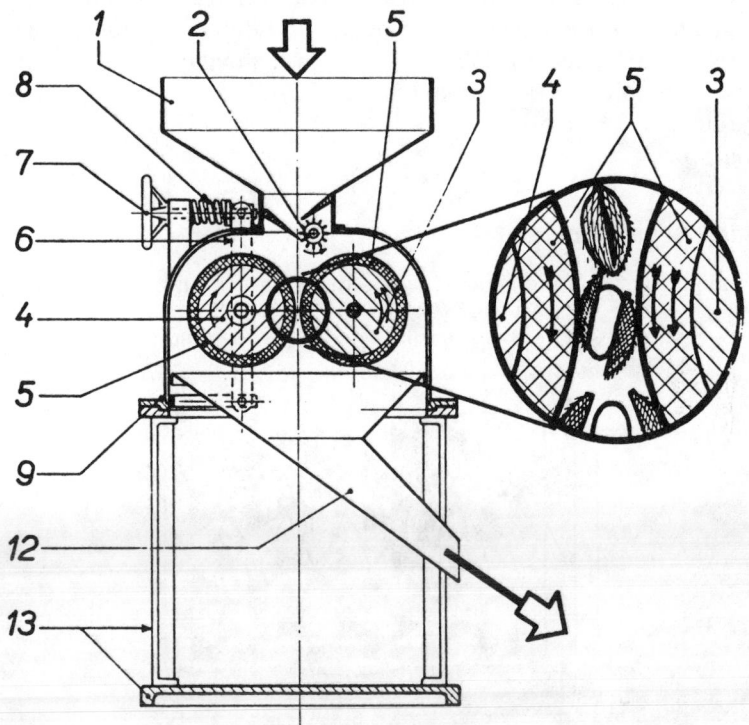

Courtesy of FAO (Borasio and Garibaldi 1957)

FIG. 9.2. RUBBER ROLL SHELLER

(1) Feed hopper, (2) feed roller, (3) fast roll, (4) slow roll, (5) rubber surface, (6) roll shifting arm, (7) roll clearance adjustment, (8) roll tension
spring, (9) husker housing, (12) husked rice and husk outlet, (13) machine
base and frame.

wear rapidly and have to be replaced or the surfaces renewed quite frequently, particularly in hot weather. However, the efficiency of the rubber roll sheller in shelling grain of widely varying thickness more than
offsets the expense of the rollers. Prior to the introduction of the rubber
roll sheller an earlier design called a stone sheller or under-runner disc
sheller was used. This sheller is discussed in detail by Gariboldi (1974).
The stone sheller has the disadvantages of being dependent on a rather
uniform length of the grain to be shelled, of breaking more grain than the
rubber roll sheller and abrading the bran on the brown rice. However, the
disc sheller is still very commonly used in developing countries and has
the advantage that the abrasive surfaces of the machine can be repaired
by the miller himself using a mixture of emery or silicon carbide and an
appropriate cementing compound.

The product of the sheller is a mixture of whole grain brown rice, broken brown rice, unshelled rough rice and hulls. The hulls, which have a large surface area for their weight, are removed from the mixture by aspiration. The further processing and uses of hulls are considered in Chapter 22. The remaining rice is conveyed to a paddy machine (paddy separator).

Paddy Separators

The paddy machine separates the brown and rough rice received from

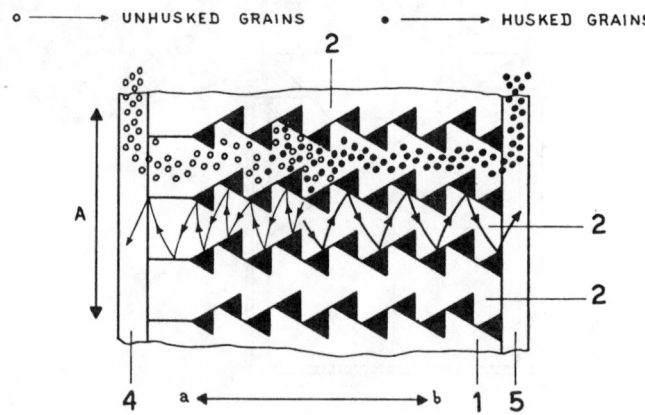

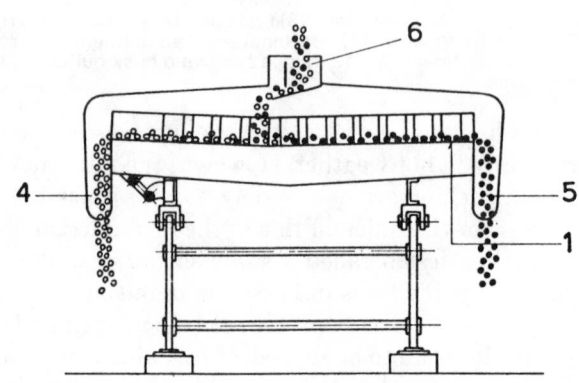

Courtesy of FAO (Garibaldi 1975)

FIG. 9.3. COMPARTMENT SEPARATOR TYPE PADDY MACHINE

(1) Table (arrow A shows direction of motion of table), (2) compartments, (3) triangular raised portions of table (in solid black), (4) unhusked grain outlet, (5) husked grain outset, (6) intake hopper.

the aspirator. There are 2 types of paddy machines in general use. The compartment type is illustrated in Fig. 9.3. The upper diagram shows a top view of the separator which has a horizontal reciprocating motion in the plane of the paper in the directions of arrow A. The table (1) is divided into compartments (2) extending crosswise to the direction of motion. Each compartment has zig-zag sides and a smooth steel bottom slightly inclined with the higher side being on the left. The grains slide from side to side of the compartment. The lower density rough rice stratifies as a layer above the brown rice. The oblique sides of the compartment tend to impart an upward thrust to all the grains. However, the greater density of the brown rice causes it to slide down the incline and is discharged into the collecting trough 5 while the rough rice is discharged at the top of the incline into collecting trough 4. The lower diagram of Fig. 9.3 is a cross-sectional view with the direction of motion being perpendicular to the plane of the paper. The stroke length, frequency of strokes, and the inclination of the table may be adjusted to achieve the separation. The feed to each compartment is adjustable by feed gates 4 and 5 as shown in Fig. 9.4 so that each compartment receives the same

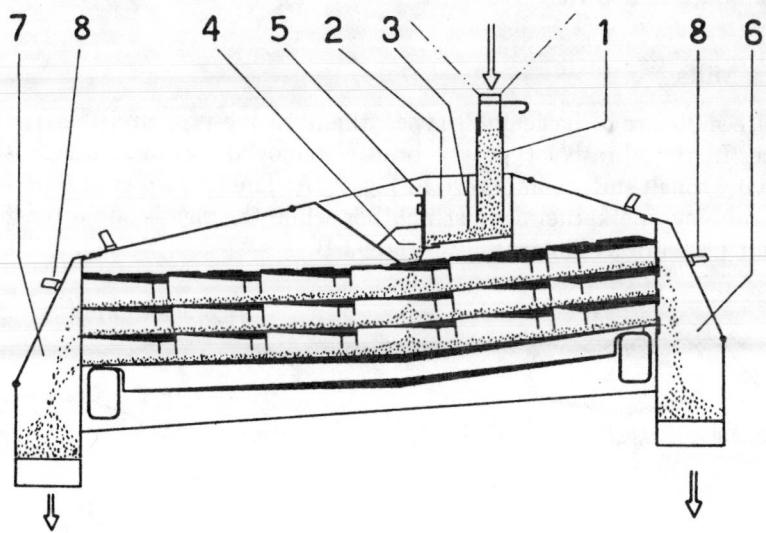

Courtesy of FAO (Garibaldi 1974)

FIG. 9.4. COMPARTMENT SEPARATOR TYPE PADDY MACHINE

(1) Feed inlet, (2) feeding box, (3) baffle with adjustable gates, (4) opening into compartment, (5) adjustable sliding gate, (6) collecting trough for unhusked rice, (7) collecting trough for husked rice, (8) inspection doors.

amount of rice. The table may have a number of tiers (usually 3 to 5) as illustrated in the figure. The bottom of the collecting troughs 6 and 7 may be inclined in various ways so that the discharge may be at either end or at the middle of the machine to suit space requirements in the mill. The inspection doors 8 provide for examination of the performance of compartments in the various tiers.

A more recent Japanese design of paddy machine is called a *tray separator*. Briefly, this machine consists of several tiers of indented steel trays inclined at an angle. The mixture of brown and rough rice is introduced at the top of the tray. The tray has an upward and forward motion towards the top of the tray. The kernels of rice bounce up and down on the trays which impart an upward motion to the brown rice while the rough rice tends to slide down the incline. This machine discharges 3 streams of material: (1) pure brown rice, (2) pure rough rice, and (3) a mixture of brown and rough rice which is recycled through the machine to achieve a complete separation.

The rough rice from the paddy machine is sent back to the shelling machines or in many instances to what is called a return sheller set aside for the specific purpose of shelling rough rice returned from the paddy machines. The brown rice is conveyed to rice mills which remove the bran to produce milled rice.

Rice Mills

Rice mills are of basically 2 types: the abrasive type and the friction type. In the abrasive type the bran is removed by contact against a moving rough surface as shown in Fig. 9.5A. The friction type functions by rubbing one kernel against another while the rice is subjected to a slight pressure as illustrated in Fig. 9.5B.

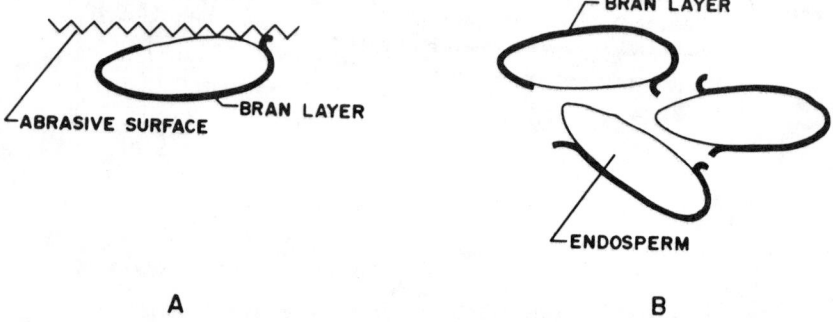

A   B

FIG. 9.5. MILLING PRINCIPLE OF ABRASIVE (A) AND FRICTION (B) TYPE RICE MILLS

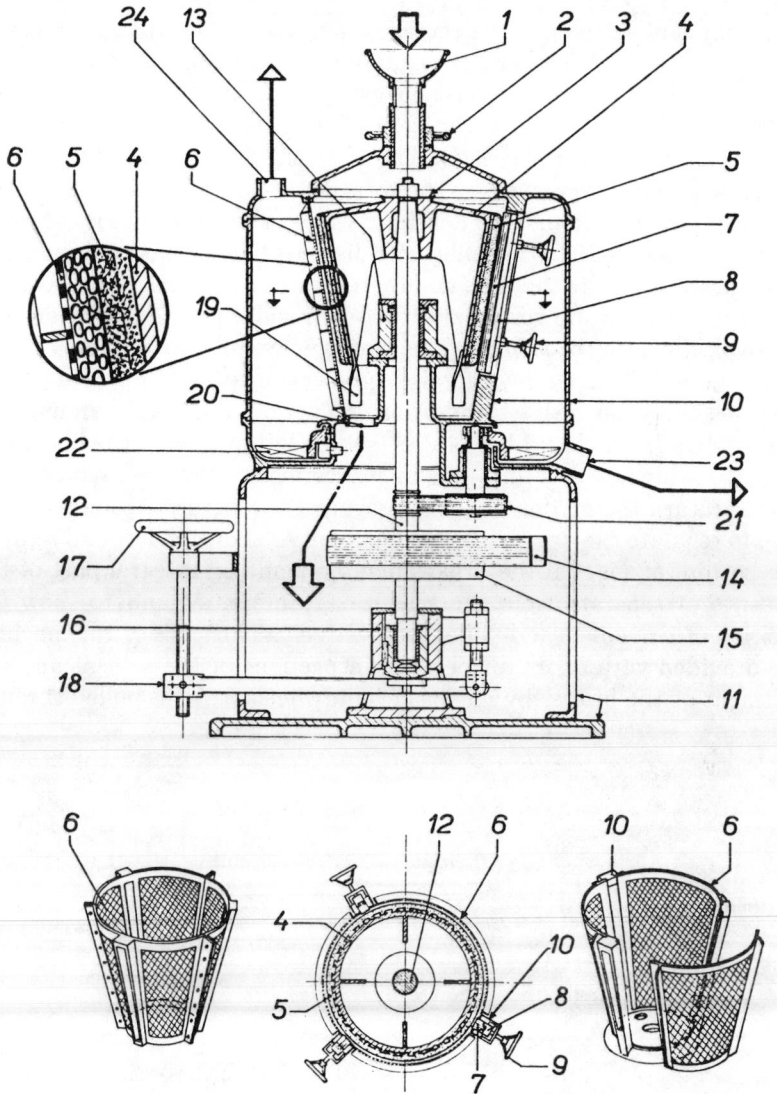

Courtesy of FAO (Borasio and Garibaldi 1957)

FIG. 9.6. ABRASIVE CONE MILL

(1) Feed hopper, (2) feed control handwheel, (3) rotating cone clamping level, (4) iron cone, (5) cone abrasive surface, (6) screen case (independent segment type) (bottom right), (6) screen case (rubber brake frame segment type) (bottom left), (7) rubber brakes, (8) rubber brake frame, (9) rubber brake regulation handwheel, (10) rotating cone housing, (11) frame, (12) cone shaft, (13) upper bearing (central bearing type cone), (14) drive belt, (15) drive pulley, (16) lower bearing, (17) screen and cone clearance adjustment handwheel, (18) shaft supporting arm, (19) pearled rice conveyor, (20) pearled rice outlet, (21) bran conveyer drive pulley, (22) bran conveying flight, (23) bran outlet, (24) air suction outlet.

Abrasive Types.—An example of the abrasive type machine is a vertical cone mill shown in cross-sectional side view in Fig. 9.6. The feed of brown rice through hopper *1* is regulated by raising or lowering sleeve *2*. The rice falls on the top of iron cone *4* and then on the abrasive surface of the cone *5*. The rice passes down between the abrasive cone and screen *6* (see exploded view on the left). The abraded bran passes through the screen where it is conveyed by the rotating blades *22* to the bran outlet *23*. After passing through the polishing section the milled rice is conveyed by moving blades *19* to the milled rice discharge *20*. Provision is made for elevating or lowering the cone and thereby altering the distance between the abrasive cone and screen and thereby regulating the degree of bran removal. Elevating or lowering the cone is accomplished by means of the handwheel *17*. The handwheel raises or lowers bearing *16* upon which the shaft of the cone rests. In addition, at periodic intervals around the screen are adjustable rubber brakes (shown in top view at the bottom of Fig. 9.6) which project into the space between cone and screen. The brakes retard the motion of the rice against the abrasive cone and thus help to regulate the degree of bran removal. As shown to left and right at the bottom of Fig. 9.6 the brakes may be made an integral part of the screen *6* or separate from the screen *6*. Numbers *14* and *15* show the position of drive belt and pulley, respectively. The sides of the machine are provided with doors for periodic inspections of the screens and adjustment of the brakes by means of handwheels *9*. Air is pulled through

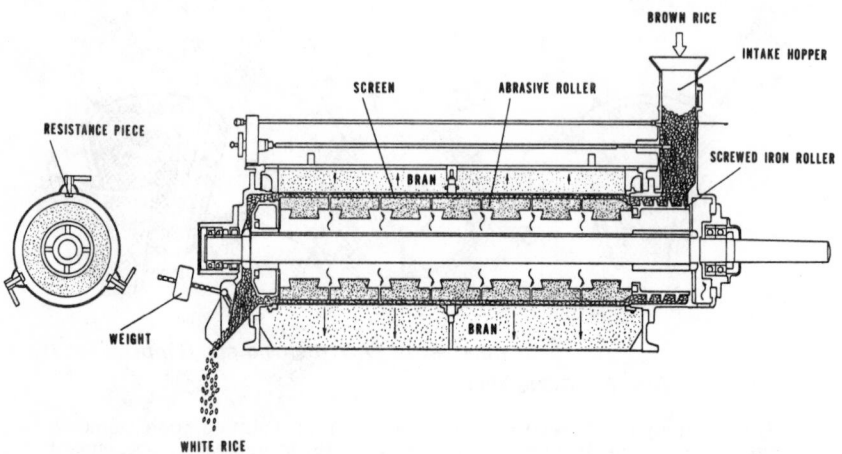

Courtesy of Satake Engineering Co.

FIG. 9.7. HORIZONTAL ABRASIVE TYPE MILL

exit *24* to remove the warm moist air generated by the heating of the rice in the mill. Heat and moisture cause the bran to cake and thus to clog the holes in the screen.

In a horizontal abrasive mill (Fig. 9.7) brown rice is conveyed under a slight pressure into the space between a cylindrical abrasive surface and screen. No provision is made for altering the distance between abrasive surface and screen. Suitable brakes (resistance pieces) are provided. After passing through the mill the rice emerges against an outlet plate which provides an adjustable back pressure by means of the weight to retain the rice for a longer or shorter time in the mill. The pressure and retention time in the mill control the degree of milling of the rice.

Generally, rice is not milled in 1 pass through an abrasive mill but rather by passing the rice through 2, 3, or 4 mills in series. With several mills in series the pressure may be decreased and clearances in each mill increased with less formation of broken rice. With increased clearances, mills in series have a larger throughput than a single mill producing the same degree of milling.

Friction Type.—A friction type mill is illustrated in Fig. 9.8. Brown rice enters the mill through an adjustable feed gate. A short screw conveyer propels the rice under slight pressure into the milling chamber. A rotor

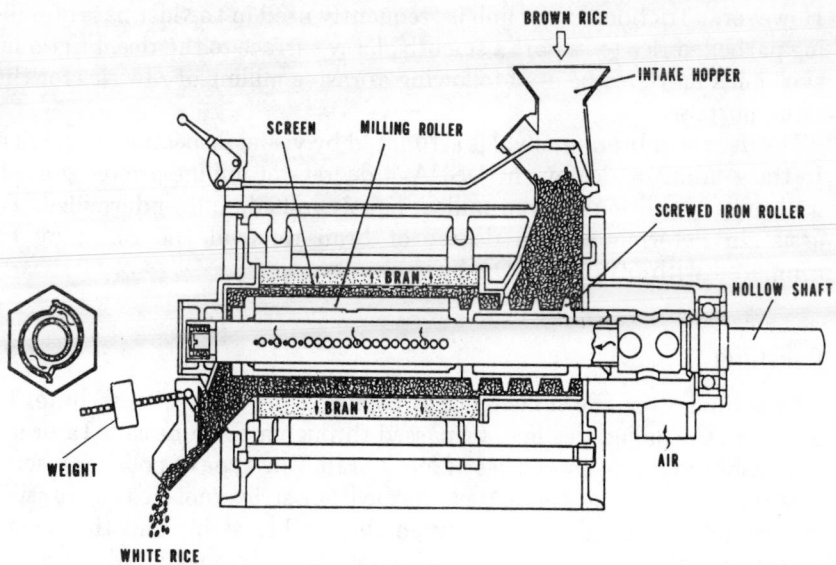

Courtesy of Satake Engineering Co.

FIG. 9.8. FRICTION TYPE RICE MILL

(roller) with projecting ridges imparts motion to the rice within a hexagonal slotted screen as shown in the cross section perpendicular to the shaft (on the left of Fig. 9.8). The rotor in this case rotates clockwise. The rice emerges from the mill against a plate which is adjustable by weights to provide more or less back pressure. The amount of back pressure affects the pressure and retention time in the mill and in conjunction with the pressure applied by the screw conveyer regulates the degree of bran removal. The rotor is hollow and is mounted on a hollow shaft so that air can be passed through the milling chamber. The air serves several functions: (1) it prevents the parts of the mill from heating up, (2) it cools the rice during passage through the mill, and (3) most importantly it assists in separating the bran from the milled rice by blowing the bran through the slotted screen. This machine is capable of producing a well milled rice in 1 pass and is frequently used in this manner. However, 2 machines in series reduce breakage because pressures in the 2 machines can be decreased. Since rice bran is relatively soft compared to the hard endosperm, the frictional type machine has an advantage in that there is a sort of natural stopping point in producing a well milled rice. Removal of peripheral endosperm layers for various purposes, as discussed in the section on degree of milling, must be accomplished with abrasive type machines. The bran of parboiled rice is very sticky and must be removed in small increments using a series of abrasive type mills. However, a frictional type mill is frequently used in the last pass of milling parboiled rice to impart a smooth glossy surface to the rice. Frictional type mills may also be used following abrasive milling of raw rice for the same purpose.

The degree of bran removal is estimated by visual inspection of the rice. In the grading of rice by the USDA, 4 degrees of milling are recognized: well milled, reasonably well milled, lightly milled, and undermilled. To assist in determining the degree of bran removal the rice may be compared with official USDA line samples.

Polishing

As a final step, the milled rice is aspirated to remove loose bran. In addition, the milled rice may be passed through a machine called a brush or polisher to remove particles of loose bran. One type of polisher resembles the vertical cone mill except that leather strips replace the abrasive surface. By rolling the rice between the leather strips and the screen under very mild pressure remaining loose bran is removed and the rice is given a more polished surface. Passing the milled rice through trumbles, which by causing the mass of grains to slide over one another, also helps to impart a smooth surface to the rice.

Sizing

The milled rice is sized by length to separate the whole kernels (head rice) from the brokens. Based upon length, the brokens in the United States are separated into second heads, screenings and brewers rice. United States standards for these classes are discussed further in Chapter 15. A preliminary separation of whole grain from brokens may be made by use of screens. However, final separation is made by means of indented cylinders or indented discs shown in Fig. 9.9 and 9.10, respectively. Milled rice is introduced into the indented cylinder through hopper 1 to form a bed of rice along the bottom of the cylinder as shown in the bottom illustration of Fig. 9.9. The inside surface of the cylinder consists of small indented pockets whose size is determined by the separation to be made. As the cylinder turns clockwise the indents become filled with both the shorter kernels to be separated and the longer kernels. As the side of the cylinder approaches a vertical position the longer kernels fall out because the weight of the kernel extending past the lip of the indent is heavier than the portion within the indent. The shorter kernels fit within the indent, are carried further up the side of the cylinder, and fall into the collecting tray when the indent becomes inverted and unable to further hold the shorter kernel. The position of the tray 6 is adjustable by handwheel 10. A screw conveyer 7 inside the tray discharges the shorter kernels through outlet 13. A grain spreader 12 helps to insure contact of all the grains with the indented surface. The longer kernels are discharged through outlet 14.

An indented disc separator consists of a number of indented discs rotating (counterclockwise in Fig. 9.10) through a bed of rice in the bottom of a U-shaped trough. Depending on the size of the indent pockets the short kernels are picked up as the disc turns through the bed of rice. The short kernels remain in the indents until the turning of the disc results in an inverted (upside down) position of the indent. At this point the grain falls out into collecting trough 4. The longer kernels are conveyed by conveying flights 3 to grain outlet 9.

After separation the whole grain and the various classes of brokens are weighed on automatic dump scales and then conveyed to separate bins. Bins are arranged along a conveyer belt so that rice from 1 or more bins can be combined and mixed to meet a buyer's specifications. The final step is to package the milled rice.

In the case of parboiled rice the percentage of breakage with modern processing is very small. However, parboiling enhances the discoloration of grains which have been damaged (discolored) by insects and/or microflora in the field. Thus, to produce a high quality parboiled rice the discolored grain must be removed by photoelectric color sorting machines.

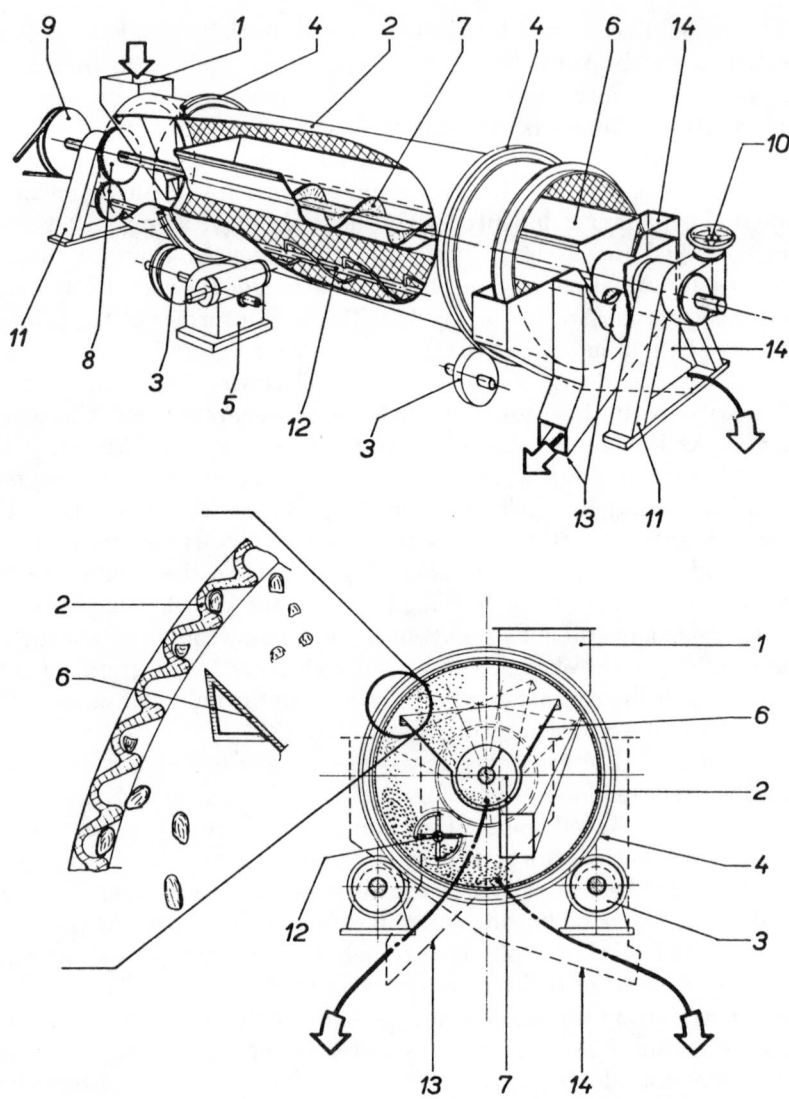

Courtesy of FAO (Borasio and Garibaldi 1957)

FIG 9.9. INDENTED CYLINDER SEPARATOR

(1) Feed hopper, (2) indented cylinder, (3) cylinder supporting roll, (4) outer cylinder ring, (5) speed reduction unit, (6) collecting tray, (7) screw conveyer, (8) screw conveyer and spreader driging gears, (9) screw conveyer and spreader drive pulley, (10) tray position adjustment handwheel, (11) machine frame, (12) grain spreader, (13) liftings outlet, (14) grain outlet.

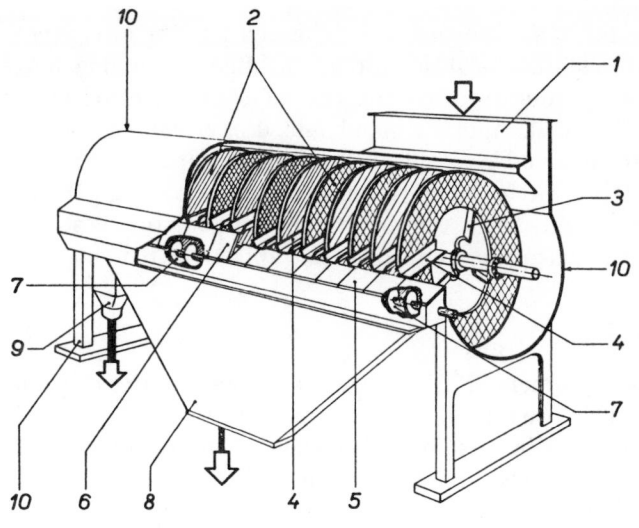

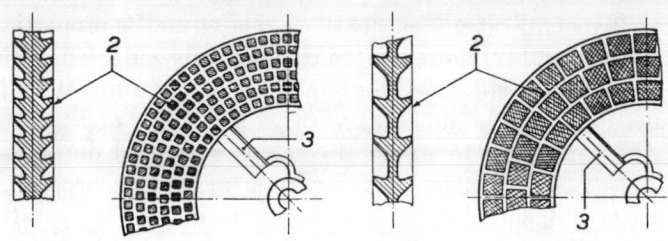

Courtesy of FAO (Borasio and Garibaldi 1957)

FIG. 9.10. INDENTED DISC SEPARATOR (1) Feed hopper, (2) undercut pocket disks, (3) grain conveying flights, (4) liftings collecting trough, (5) liftings trap door, (closed), (6) liftings trap door (open), (7) liftings return screw conveyer, (8) liftings discharge hopper, (9) grain outlet, (10) machine housing and frame.

Small Capacity Mills

In developing countries small capacity mills operated by local entrepreneurs may omit several steps of the milling process discussed above. Particularly to be mentioned is the so-called huller mill. This is a friction type mill which simultaneously removes both hulls and bran from rough rice. This machine has the disadvantages of producing high percentages of broken rice and discharging a by-product mixture of bran and hulls. The bran is thus not as useful as a feed supplement or as a source of vegetable oil recovery. These mills frequently only act as service mills to the farmer for home consumption. The mill has the advantage in that investment in storage facilities and in machinery is minimal and being in proximity to producing areas the cost of transportation of the bulky rough rice is small. Another small mill uses a rubber roll sheller, an aspirator to remove loose hulls and a friction type machine to mill the approximately 95% brown−5% rough rice mixture from the sheller. In making a well milled rice, the rough rice in the mixture is shelled during the bran removal process and does not appear in the final product.

Coated Rice

Some rice consumers prefer a very glossy or shiny rice called coated or glazed rice. This rice is produced by adding dry talc and a glucose solution to well milled rice in a trumble. Rotation of the trumble evenly distributes the glucose-talc mixture over the grain.

RICE BREAKAGE DURING MILLING

The factors contributing to rice breakage during milling may be classified into 2 general categories: (1) those related to the properties of the rice grain itself, and (2) those related to conditions under which the grain is milled. The properties of the grain itself at the time of milling are determined by both the inherent varietal properties of the rice kernel and by the conditions to which the rice is subjected during growing, harvesting, drying, storage, transportation and other processing. The conditions of milling affecting breakage are the rice temperature and moisture content during milling, the temperature and humidity of the milling environment, the degree of milling, and the operating conditions and mechanical settings of the milling machinery.

Premilling Factors

Harvesting.—To obtain high quality as well as high grain yields, rice must be harvested at the proper stage of maturity. If the crop is harvested too early (immature), breakage during milling is excessive due to the large quantity of thin, light and chalky kernels which are very fragile. If it is harvested too late, breakage during milling is excessive due to a phenomena known as "sun checking" which is the development of cracks in the individual kernels. Smith *et al.* (1938) reported that delayed harvesting, which led to increased breakage of individual grains, was accompanied by decreased moisture content of the paddy rice.

Apparently rice varieties differ in the range of harvest time moisture content at which they yield the best quality milled rice. Davis (1944) found that the optimal range for Caloro was 20 to 25%. Kester (1959), Kester and Pence (1962) and Kester *et al.* (1963) reported the optimal range for Calrose was 22 to 27%. Working with various short and medium grain varieties they found that milling head yields decreased between 0.9 and 1.7 percentage points for each 1 percentage point decrease in harvest moisture content within the harvest moisture range of 26 to 14%. When they hand harvested rice over a period of time while the field moisture content was decreasing from 47 to 15%, the milling head yields were best between field moisture contents of 32 and 25%. McNeal (1950), working with medium grain (Zenith) and long grain (Rexark, Nira and Prelude) varieties reported maximum milling head yields were obtained when the harvest moisture content was between 16 and 24%. Morse *et al.* (1967) reported maximum milling head yield for Caloro at between 26 and 30% harvest moisture. But, he also reported that total milled rice yield per acre continued to increase as the harvest moisture decreased to 12%. Below 26% harvest moisture the increase in total yield was not very rapid. Johnston and Miller (1973) concluded that the optimal harvest moisture content, considering both milling head yield and total milled rice yield per acre is between 18 and 24%.

Another harvesting factor which affects milling head yield is the mechanical action of the combine on the rice kernels. Matthews and Spadaro (1975) measured the effects of combining on rice breakage and head yield. In each of 5 fields 1 rice sample was collected from a combine and another was collected by hand harvesting a 1.2 m by 9.1 m plot in the field adjacent to the combined area. X-ray photographs were used to determine the percentage of cracked and broken rough rice kernels. The combine harvested samples averaged over 5 percentage points more cracked and broken rough rice kernels than the hand harvested samples (6.1 and 0.4% respectively). This difference in breakage carried through the subsequent milling operations with the combine harvested samples averaging 11.2% breakage while the hand harvested samples averaged only 6% breakage. However, current practice is to allow a small percentage of the kernels to crack during combining in order to obtain a maximum total yield per acre.

Drying.—The optimal moisture content for harvesting rice is too high for safe storage, thus the rough rice must be dried. The drying conditions to which the rice is subjected have a pronounced effect on breakage during milling. As the rice kernel dries, the outer portion shrinks, setting up stresses within the kernel. The magnitude of these stresses depends upon the moisture and temperature gradients within the rice kernel while the tensile strength of the rice kernel depends upon its moisture content and temperature. If the rice is dried too rapidly, the internal stresses will exceed the tensile strength resulting in checking or cracking of the rice kernel (Henderson 1954, 1957; Kunze 1964; Kunze and Hall 1965; Choudhary 1970). These checked kernels are more susceptible to breaking during milling resulting in reduced head yield (Rhind 1962).

Two rice drying methods commonly used are: (1) bin drying with unheated or slightly heated air (on-the-farm drying), and (2) multi-pass drying in continuous-flow heated air driers (commercial drying). A considerable amount of research has been done with these methods to determine economically feasible drying conditions which will minimize the breakage of rice during subsequent milling operations. Wasserman *et al.* (1958A) reported on drying procedures which increased drier capacity and simultaneously improved milling head yield for western United States rice varieties. Pominski *et al.* (1961) and Spadaro (1961) extended this work to varieties grown in the southern United States. McNeal (1961) investigated the effects of drying techniques and temperatures on milling head yields for 4 rice varieties—Arkrose, Bluebonnet, Nato and Zenith. Sorenson and Crane (1960) reported on the practicality of drying rice in storage bins and presented equipment requirements and operating conditions to prevent loss in grade and milling yields. Calderwood *et al.* (1975) summarizes the methods and adjustments for increasing capacity of continuous-flow heated air dryers without loss of milling quality.

Wasserman (1960) reported that tempering rice after drying without cooling (40.5°C, 105°F) gave higher head yields than when the rice was cooled immediately after drying (23.8°C, 75°F) and then tempered.

Stipe *et al.* (1976) reported on a radically different method of drying rough rice which actually increases the milling head yields. Before drying, the naturally moist rough rice is exposed to live steam (2.07×10^5 Pa, 15 psig) for a short time interval (0.05 to 5.0 min). The effect is to lightly parboil the undried rice. The steam treated rough rice is then dried in a single pass with high temperature air (80°C, 176°F) to a moisture level of less than 13%. The key step in the process, which prevents breakage during milling, is to temper the rice immediately after drying for 1 hr at 70°C (158°F) in a sealed vessel. Whole grain milling yields can actually be increased in a low quality rice, such as one subjected to adverse harvesting conditions. The only negative effect of the procedure is that the color of the milled rice is affected by the steam treatment and is similar to lightly parboiled rice.

Storage and Handling.—The conditions to which rice is subjected during storage and transportation also have an influence on the milling head yield. Several investigators have studied the effects of storage time after drying on milling quality with varied results. Wasserman *et al.* (1958A) reported that rice grown in the western United States could be milled immediately after drying without any loss in milling yields. Five lots of rice were each milled after drying and after 3, 6, and 9 weeks of storage. Milling yields did not change during these periods. Pominski *et al.* (1965), studying long grain and medium grain rice grown in the southern United States, also concluded that rice can be milled immediately after drying with no loss in head yield. He found milling yields did not significantly change over a 4 week storage period. In addition he reported that there was some indication in his data that higher head yields could be obtained with medium grain rice by milling immediately following drying. The data of Wasserman (1961) on short grain rice also shows higher head yields immediately after drying. Sorenson (1973) reported substantial increases in milling head yields (4−6 percentage points) for rice stored over a 10 month period. This is in agreement with the work of Choudhary (1970) who found that the tensile strength of rice increased during storage. He concluded that during storage in a constant environment, the residual stresses which were induced by moisture gradients during drying may be gradually relieved resulting in a stress-free kernel which has a higher tensile strength and is therefore less likely to break during milling.

McNeal (1960) studied the milling of rice which was harvested and stored without drying for various periods of time at various moisture levels. He reported that the undried rice declined in grade before there was any decline in milling yields. Calderwood and Schroeder (1975) investigated the use of chemical preservatives for maintaining undried rough rice in storage. They found that propionic acid, at levels sufficient to prevent moist rough rice from deteriorating, did not affect milling yields.

The environmental conditions to which the rice is exposed during storage and transportation are very important factors affecting milling yield. Recent research has indicated that a significant amount of rice breakage during milling is due to rice kernels which have been weakened by stress cracks resulting from moisture adsorption or desorption. Kunze (1964) observed the development of fissures when rice kernels, equilibrated at storage moisture, were exposed to higher relative humidities. Kunze and Hall (1965, 1967) investigated the moisture adsorption characteristics of brown rice and relative humidity changes that cause brown rice to crack. They found that approximately 50% of the rice kernels, initially at storage moisture content, fissured within a 0.8 to 1.6 hr period when exposed to 100% relative humidity at 33.3°C (92°F). The exposure time required for fissuring depended on the rice variety. Stermer (1968B) developed a relationship between stress crack damage and the magnitude of the change in equilibrium moisture content. Kunze and Choudhary (1971) found that the tensile strength of the rice kernels gradually decreased as the time of exposure to moisture-adsorbing conditions increased even before physical fissures appeared. Therefore, even the kernels which have not developed cracks are more susceptible to breakage after exposure to adverse environmental conditions.

Since most rice is stored and transferred in bulk or in 45.4 kg burlap bags, it is important to minimize exposure to environments where moisture transfer may occur. Storage bin design and size are important for controlling the environment and preventing moisture migration. Bin height must be limited to prevent excessive breakage due to the weight of the stored rice. Transfer by belt, screw conveyors and bucket elevators can also lead to added kernel breakage by mechanical action. Louvier and Calderwood (1972) studied the breakage of rice due to falling impact such as would occur in the bulk loading of a ship. They reported that: (1) the amount of breakage increased with dropping height up to 18.3 m, (2) long grain rice was more susceptible to impact breakage than medium grain, (3) breakage increased as the moisture content of the rice was reduced, and (4) the relative humidity and temperature of the air through which the rice fell had little effect on impact breakage.

Kernel damage due to improper storage and/or insect infestation can lead to increased breakage. Schroeder (1967) reported reduced head and total milling yields resulting from invasion by storage fungi in experimental rice storage studies.

Parboiling.—It is common knowledge in the rice processing industry that parboiling will greatly improve the milling quality of rice such that head yields will approach total yields (i.e., 0 breakage). Bhattacharya and Subba Rao (1966) and Bhattacharya (1969) have conducted comprehensive quantitative studies on the effects of parboiling on the breakage of rice. They reported that kernel defects such as cracks, chalkiness and incomplete grain filling are completely healed during the parboiling process. When properly dried, the rice kernels are very resistant to mechanical breakage. Thus, the milling quality of parboiled rough rice is determined solely by its drying conditions following parboiling and is completely independent of the previous history or condition of the rice. Consequently, for rough rice which is to be parboiled, the optimum harvesting, drying and storage conditions should be selected on some basis other than that of preserving the milling quality. For the same reasons, parboiling is an excellent means for salvaging any rough rice, raw or previously parboiled, whose milling quality has been inadvertently damaged by improper handling or processing, as well as rough rice containing a high proportion of immature kernels which are extremely fragile and would break into minute fragments when milled.

Milling Factors

During the shelling and milling (bran removal) operations, the rice kernels are subjected to mechanical forces in order to remove the husk and bran and produce white milled rice. Milling will cause breakage of some kernels even in rice lots which have been harvested, dried and stored with extreme care. For lots which have been subjected to poor treatment, milling may break every grain. The ideal milling operation is one in which the mill environment, rice properties and machine settings are controlled to minimize breakage (maximize head yield) while producing a rice with the desired degree of milling.

Defective Kernels.—Several investigators have studied the relationship between the amount of checking or cracks in the rice kernels and the amount of breakage occurring in the milling operations. Reports on the degree of correlation between defective kernels and breakage vary. Autrey *et al.* (1955) did not find any correlation for medium grain or long grain rice between percentage broken during shelling with that broken during milling. Nor was there any correlation between the defective kernels as received and breakage during milling. Hogan *et al.* (1954), using X-ray technique (radiographs) to evaluate the percentage of cracked rough rice, also did not find a correlation between the percentage of cracked before milling and broken milled rice. He concluded that, either there are cracks in the grains which could not be detected radiographically or milling head yields were affected by other factors. McDonald (1967), who studied factors influencing rice checking in Australia and the effects of checking on milling quality, concluded that milling quality (head yield) is not related to the percentage of cracked rough rice kernels and there is no way of assessing it except by test milling. In contrast to the above findings, Ten Have (1958) found for 5 varieties of long grain rice that the percentage of cracked rough rice kernels, determined by hand shelling and visual inspection of the brown rice, had a good correlation with breakage during milling. In rice lots with low percentage cracked and low breakage, the percentage brokens generally exceeded the percentage cracked. In rice lots with 20% or more breakage during milling the percentage of brokens was less than the percentage of cracked rough rice kernels. Henderson (1954), who determined cracks in rough rice by X-ray examination, reported that many of the internally or incipiently cracked rough rice kernels of short grain Caloro did not break when milled. The amount which may be cracked but unbroken after milling may range from 0 to 86%. He examined a sample of short grain rice obtained from a local retail grocer and found 65% of the whole kernels had cracks. Stermer (1968B), who determined stress cracks with transmitted polarized light, reported that in limited mechanical breakage tests on long grain Belle Patna and medium grain Nato varieties, rice kernels showing stress-crack damage were easily shattered. Bhatta-charya (1969), using transmitted light to visually determine cracked kernels, reported that breakage during milling was related quantitatively to the amount of cracked and immature kernels which were very fragile and almost disappeared on milling. He concluded that it is principally the defective grains that ultimately fail in rice milling. Matthews *et al.* (1970) made a comprehensive investigation of the relationships between breakage during milling and defective rough rice kernels as determined by X-ray examination. He reported that grain type (long, medium or short) is extremely important in the relationship between defective kernels and milling breakage. For long grain rice he demonstrated a proportionality between the percentage of cracked rough rice kernels and the percentage of broken brown and milled rice. Only ⅓ of the breakage after milling could be attributed to rough rice kernels that had an observable crack in a position to give rise to a broken milled rice kernel. The low proportion of cracked rough rice to broken milled rice could be due to the inability of the X-ray technique to show all cracks. An alternative explanation is that the percentage of cracks is indicative of a residual stressed condition in the apparently sound kernels. They recommend additional research to determine the relationship between cracks observable by X-ray examination and cracks observable by visual methods such as described by Stermer (1968B) and to determine the nature of the fragility of long grain kernels which break but do not have an observable crack. For medium grain rice, Matthews *et al.* (1970) report that the percentage of cracked rough rice kernels was approximately equal to the percentage of broken after milling. About ½ of the cracked kernels broke during shelling. The greater resistance to breaking of cracked medium grain kernels is probably related to the more rounded thicker kernel as compared to long grain rice. Henderson (1954) has shown that cracked short grain rice kernels can be very resistant to breakage with a high percentage of them surviving milling intact. Soren-son (1973) confirmed the importance of grain type in the relationship between cracked rough rice and broken milled rice. He reported that long grain stress-cracked rice kernels are very fragile and break easily during milling. The medium grain stress-cracked kernel very often withstands milling without breaking. They conclude that breakage is only partially related to the quantity of cracked kernels in the rough rice.

Kernel Thickness.—Matthews and Spadaro (1976) investigated the relationship between kernel thickness and breakage for long grain rice. Six lots of 3 varieties of long grain rough rice were each separated by thickness into 4 fractions. X-ray photographs were used to determine the percentage of cracked kernels in each rough rice thickness fraction. They demonstrated that, for a given lot of rice, the thinner kernels are more susceptible to breakage during milling. They also demonstrated relationships between cracked rough rice and broken milled rice which were dependent upon the thickness of the kernels.

Rice Moisture Content.—The effect of rice moisture content on milling head yield has been the subject of several investigations. Pominski *et al.* (1961) reported that moisture content had a very significant effect on milling yields of Bluebonnet 50 long grain rice. Over the range of 10 to 14% moisture, head yields and total yields increased approximately 3 and 0.7 percentage points, respectively, for each 1 percentage point decrease in rice moisture. Wratten (1960) also reported that lower moisture content increased the milling yields of Bluebonnet 50. Wasserman (1960, 1961) demonstrated that milling yields of Caloro short grain rice varied inversely with moisture content. Over the range of 10 to 14% moisture, the head and total yields increased 1.8 and 1.2 percentage points, respectively, for each 1 percentage point decrease in moisture. He obtained similar results with California Pearl short grain rice. He also showed that the total yield of the milled rice could be further increased by remoistening it to normal moisture content. The remoistening was done by exposing the milled rice to humidified air under carefully controlled conditions such that there was little increase in breaking or cracking. Bhatia (1969), using Saturn medium grain rice, found that with each 1 percentage point decrease in moisture from 12 to 10%, the breakage decreased by 1.6 percentage points.

Webb and Calderwood (1977) investigated the relationship of moisture content to the degree of milling in rice. They reported that rice with low moisture levels (6–10%) was markedly more resistant to milling at a specified mill setting than samples at higher moisture levels. The low moisture samples required considerably more milling pressure in order to obtain an equivalent degree of milling. As expected, head yield of the low moisture samples at a conventional mill setting was greater than that of the higher moisture samples, but the difference in head yield between low moisture and higher moisture samples was greatly reduced or eliminated when mill settings were adjusted to obtain an equivalent degree of milling.

Stipe *et al.* (1972) investigated the effects on milling yields of shelling rough rice at higher than normal moisture content, drying the brown rice to normal milling moisture content and milling. They found that with Saturn medium grain rice there was a considerable decrease in both total and head yield when the moisture content was above 18% during shelling. Shelling tests with 2 long grain varieties, Dawn and Starbonnet, showed significant reductions in head yields when shelled at moisture contents of 16 to 18% as compared with the 12 to 14% levels.

Temperature.—Autrey *et al.* (1955) investigated the effect of temperature on breakage during milling. He reported that, while the actual temperature of the rice or the mill room was not related to breakage, the difference between the temperature of the rice entering the mill and the temperature of the mill room itself could be correlated with head yield. When the temperature of the rice, Zenith medium grain, was 5.5°C (10°F) higher than that of the mill room, head yields decreased an average of 1.6 percentage points. When the rice temperature was 5.5°C (10°F) below the room temperature, the decrease in head yield averaged 0.9 percentage points. Rexark variety rice was less sensitive to the temperature than the Zenith. With a 5.5°C (10°F) differential, higher or lower, Rexark showed 0.8 or 0.7 percentage points decrease in head yield. Rhind and Tin (1933) reported that for each degree centigrade higher rice temperature at time of milling there was a decrease in whole kernel yield of between 0.5 and 1.0 percentage points. Henderson (1954) also observed that higher temperatures enhanced breakage. Sorenson (1973) milled rice with liquid nitrogen being passed through the mill rotor and through the rice while it was being milled. They found that higher head yields were realized with the liquid nitrogen cooling; however, because of mechanical problems encountered during milling, they believe that the study is inconclusive and the use of liquid nitrogen is impractical for cooling rice during milling. Bhatia (1969) varied the temperature of rice before milling from 21.1° to 32.2°C (70° to 90°F) with various rice moisture contents. His data showed that breakage during milling actually decreased slightly as the initial rice temperature increased.

Later work has indicated that the effects correlated with temperature changes may actually be related to moisture gain or loss in the rice kernel which accompanies the temperature changes. Kunze and Hall (1967) reported that subjecting rice kernels to a large temperature change—up to 34.4°C degrees (62°F degrees)—while controlling the relative humidity such that no moisture changes occurred, did not produce any cracks in the kernels. They concluded that the effect of temperature gradient on breakage is small. Matthews *et al.* (1971) subjected long grain rough rice to a combination of time-temperature treatments in hermetically sealed containers to minimize moisture transfer. Duration of heat treatments varied from 2 to 19 hr and temperature from 60° to 120°C (140° to 248°F). Results indicated that heat by itself was not detrimental in causing breakage of the rice kernel.

Relative Humidity.—It is recognized in the rice industry that moisture changes in the rice kernel are closely related to stress cracking. The relative humidity of the environment in which the rice is milled can be an important factor in the moisture changes and several investigators have studied the effect of this aspect of milling on head yield. Autrey *et al.*

(1955) investigated this in a pilot plant with closely controlled humidity conditions using 3 varieties of raw rough rice and 1 sample of parboiled. The mill room temperature was maintained at $29.4° \pm 1.1°C$ ($85° \pm 2°F$). The optimum relative humidity for processing the rice was found to be approximately 70%. The rice used in these tests had a moisture content of around 13% and, thus, was in moisture equilibrium with $29.4°C$ ($85°F$) air at about 70% relative humidity. These conditions would minimize moisture transfer to or from the rice. When relative humidities above or below 70% were used the resulting head yields were decreased. The effect of humidity on head yield was greatest for Bluebonnet long grain rice and lowest for Zenith medium grain. Rexark long grain reacted similar to the Bluebonnet. For Bluebonnet the range in head yield was 3.2 percentage points over a range in humidity from 50 to 70%; for Zenith, 1.8 percentage points over the same humidity range. There was little effect of the relative humidity on the head yield of the parboiled sample. The results obtained in the pilot plant were verified in commercial mills where it was found that there was a pronounced decrease in head yields when milling was done on days with low relative humidities. In another series of pilot-plant experiments Autrey regulated the humidity in various pieces of equipment at 70% while the mill room was at 30%. He concluded that near optimum head yields could be obtained by regulating the humidity just in the pieces of equipment where the rice is exposed to large volumes of air.

Aeration During Bran Removal.—A milling modification which has received considerable attention recently is that of blowing air through the rice while the bran is being removed. Several investigators have reported increased milling head yields due to aeration. Bhatia (1969), who studied milling in a ventilated Engelberg huller, reported that breakage could be reduced by approximately 3 percentage points by providing 4.34 m³/sec of ventilating air per m³ of annular space in the huller. The aeration lowered the temperature of the rice during milling. He concluded that breakage is a function of the increase in temperature of the grain during the milling process. Bhatia also controlled the humidity of the ventilating air. The controlled humidity air was in contact with the rice only for the short time the rice was in the mill. While Bhatia concluded that the effect of humidity on head yields was negligible his data show that the trend is similar to that found by Autrey et al. (1955). With ventilating air at 50% relative humidity the average breakage during milling was 13.7%, with 70% humidity the breakage was 12.8%, and with 90% humidity the breakage was 14.3%. Wasserman et al. (1974) reported that aerating the rice during milling with a CeCoCo continuous abrasion type mill, decreased milling time by 6% but had little effect on head yield. Matthews and Spadaro (1974) reported that for the Satake Jet Pearler, the head yield increased as the ventilating air flow rate through the mill was increased. Childers *et al.* (1975) studied the effects of aerating during milling in a McGill No. 3 batch type mill. In tests with long grain rice, head yields increased from 2.0 to 4.5 percentage points with aeration. For medium grain rice the increase in head yield was about 3 percentage points. Childers also controlled the relative humidity of the aerating air but did not find any relationship between humidity and head yield. However, the rice was aerated for only 1 min at a relatively low airflow rate 0.0033 m³/sec and they believe additional tests at higher flow rates are needed to determine the full effect of relative humidity on head yield with a batch type mill.

Both Childers *et al.* (1975) and Bhatia (1969) concluded that because the unaerated rice reached higher temperatures during milling, it was subjected to greater thermal stresses which probably caused the rice kernels to be broken more easily. Another possible explanation for the decrease in breakage with aeration is that the lower breakage is the result of decreased moisture gradient stresses rather than thermal stresses. Matthews *et al.* (1971) and Kunze and Hall (1967) both report that subjecting rice to large temperature gradients but under conditions where no moisture transfer occurred did not result in the development of cracks due to thermal stresses. Mannapperuma (1974) analysed the stresses inside a rice grain due to temperature and moisture gradients. An axisymmetric model with an elliptical longitudinal section was developed using finite difference methods to represent the rice kernel. He calculated that exposure to a 41.7°C temperature gradient results in a thermal tensile stress of 1.72×10^6 Pa (250 psi) at the center of the kernel which is well below the rupture stresses in tension and compression. It has been shown by several investigators that moisture-gradient stresses alone can cause a rice kernel to crack. When the rice heats up during milling, the vapor pressure of the water in the rice increases which results in a higher rate of water transfer from the rice kernel to the surrounding air. Aerating the rice during milling reduces the rice temperature and removes heat which would otherwise be available to evaporate the rice moisture thereby reducing the rate of moisture transfer and the resulting stresses due to the moisture gradient. To clarify this point additional precise experimental data need to be collected on the changes occurring during milling in rice moisture content and temperature as well as changes in the temperature and humidity of the air surrounding and flowing through the rice.

Milling Time.—One area of milling which has considerable conflicting information in the literature is the effect of the length of time of milling on breakage. Autrey *et al.* (1955), working with both medium grain and long grain rice, reported that only about ⅕ of the breakage occurs during the time period required to remove 75% of the bran. When the milling time is extended to remove the remaining 25% of the bran, about ⅘ of the breakage occurs. Rhind (1962) stated that his experience has been that stress-cracked kernels are present in every sample and these defective grains break as soon as milling starts. Thereafter breakage proceeds steadily as the rice is subjected to more and more force as milling continues. He believes that kernels which are badly cracked are destined to break and do so with a minimum of milling. Smith and McCrea (1951) milled numerous large samples of many different varieties of rice for times ranging from 15 to 90 sec. They reported that the highest percentage of breakage occurred during the first 15 sec of milling in all of the tests and from that time on the breakage was small. Bhattacharya and Subba Rao (1966) reported that most of the breakage in a lot of rice occurred at the earliest stage and increased little with continued milling. Bhatia (1969) concludes that on the average 70 to 80% of the total breakage occurs during the first 5 sec of mill operation and from that time on the breakage is comparatively small. Matthews *et al.* (1970), working with both long grain and medium grain rice, reported that almost all of the breakage occurred in the first 10 sec of milling. About 70% of the bran was removed in the first 10 sec. Bran removal continued although at a decreasing rate for the entire milling period. The polishing period could be extended for about 150 sec without increasing breakage significantly over that taking place in the first 10 sec of milling. They conclude that the breakage is due to mechanical stresses and not to moisture or thermal stresses which develop continuously as milling proceeds. They hypothesize that the major reason that breakage is not completed in the first 1 sec of milling is probably that not all of the kernels have been mechanically stressed by this time. For example, the rice kernels at the end of 5 sec of milling consisted of about 40% well milled grains and 60% which had hardly been milled at all, rather than consisting of a uniform collection of grains each having about 40% of the bran removed (such as might be inferred from the 40% total bran removal achieved with 5 sec milling time). Hogan (1969), investigating the effects of deep milling rice by abrasion to produce a high protein rice flour, reported that kernel breakage progressively increased as over milling proceeded. The breakage for long grain types was greater than that for medium grain and short grain varieties.

Additives.—Matthews *et al.* (1971A) stated that in present commercial practice the amount of long grain rice broken at the end of the shelling process represents only 25 to 45% of the total broken rice after milling. Therefore, a chemical or physical treatment of the brown rice to facilitate bran removal could potentially reduce a major portion of the total breakage occurring in conventional milling. Autrey *et al.* (1955) studied the effects of steaming brown rice before milling. While steaming they found that droplets of water should not be allowed to come into contact with the rice and that the temperature of the rice should not be allowed to rise appreciably. When either of these conditions occurred, the result was a decrease in the head yield. Optimum results were obtained when the brown rice came out of the steam bath slightly damp and at a temperature near its entering temperature. With properly dried rice, steaming had no significant effect upon head yield, however, the mill capacity was increased. With rice which was improperly dried or stored, the head yield was increased significantly by steaming. They conclude that steaming results in lesser adherence of the bran to the endosperm thus requiring less mechanical force for bran removal. Obtaining higher head yields by exposing the brown rice to moisture before milling appears to be in direct conflict with the work of Kunze and Hall (1965) who showed that moisture adsorption caused brown rice to develop stress cracks. However, Choudhary (1970) reported that the initial short-term effect of moisture adsorption was an increase in the tensile strength of the rice kernel. He stated that this behavior was not anticipated and additional research is required for an explanation of this phenomenon. Autrey *et al.* (1955) also studied the effects of adding abrasives (calcium carbonate, fine bauxite, coarse bauxite and Attaclay) to the rice during milling. The most efficient milling additive was calcium carbonate. However, like steaming of properly dried and stored rice, abrasives had no significant effect upon head yield but did increase mill capacity. The head yield was increased by using abrasives on rice that had been improperly dried or stored. When both steam and calcium carbonate were used together, there was greater improvement in mill capacity than when either was used alone.

Matthews *et al.* (1971B) reported on the treatment of rough rice with ammonia or sulfur dioxide gases in an effort to effect a loosening or softening of the bran. Treatments ranged from 3 to 15 hr in hermetically sealed containers. Milling yields were not improved over the untreated control. Ammonia treated samples acquired a yellow color in the endosperm and it was difficult to expel the gas from the rice under vacuum. Sulfur dioxide bleached the endosperm and the resulting milled kernels were an attractive white color, but it was also difficult to rid the rice of the residual gas odor.

Roberts and Wasserman (1977), Wasserman and Roberts (1972) and Wasserman *et al.* (1974) investigated wet milling and wet milling with various additives including calcium carbonate, whole or ground rice hulls, and rough rice. The effects of most of the additives showed reductions in milling time and energy consumption and increases in total and head rice yields when compared to runs with no additives. The addition of water gave favorable results in every case. The addition of 5% rough rice was not as effective as whole or ground hulls. Because mill settings were changed for the tests using calcium carbonate, the effect could not be directly compared with those of other additives, but in combination with water the results improved markedly. In the case of water addition, breakage was reduced as holding time between the water addition and the milling operations was shortened. When abrasive solids were used, with or without water, the Engleberg mill settings had to be adjusted to remove less bran and polish per pass in order to prevent stalling of the equipment.

Morgan *et al.* (1966) reported on the chemical milling of rice. The bran is chemically loosened by strong warm lye and then physically removed by vigorous water washing. The product retains all the endosperm, including the aleurone layer, but loses most or all of the colored bran layers. Warm dilute acid restores the surface whiteness and the grain is dried in warm air.

Wayne (1966) developed the solvent extractive milling (SEM) process which uses rice oil to chemically soften the bran before milling. Then the rice is wet milled in the presence of a rice oil-hexane solution. Lynn (1966) reported that increased head yields of up to 10 percentage points are obtained. Details of the SEM process are presented later in this chapter.

Machinery Settings.—Other factors which affect breakage during milling are the mechanical mill settings. In commercial mills the mill settings are adjusted by the operator based on his visual inspection of the milled rice, a technique which can be very inconsistent. Ideally, it is desired to adjust the mill such that a given degree of milling (percentage of bran removal) is obtained while minimizing breakage and maximizing milling capacity. The settings which are adjustable vary with the type of mill being used.

Matthews and Spadaro (1974) investigated the factors contributing to breakage in a Satake Jet Pearler Model BA-3 which is a small model of the Satake BA-15 used in rice mills in the United States. They reported on the effects of rotor speed, feed-gate opening, back pressure and number of passes on breakage, degree of milling, energy consumption and production rate. Higher rotor speeds resulted in less breakage. The wider feed-gate openings (and production rate) increased breakage. Higher back pressures resulted in a greater degree of bran removal and breakage. Two passes were superior to one pass in effecting the same degree of bran removal with less breakage.

Wasserman *et al.* (1974) and Roberts and Wasserman (1977) investigated the effects of mill settings for a CeCoCo mill which removes bran by abrasion against an emery cone, and a small Engelberg mill which rubs grains against each other under pressure. They found that mill settings had little effect in the CeCoCo. Mill settings in the Engelberg were more critical. As the feed gate was adjusted from half open to fully open, milling time and energy use were reduced 23% and 21%, respectively, but head yield decreased about 1.5 percentage points. As the discharge gate was opened gradually over its operating range, milling time was shortened up to 60% and energy requirements were reduced up to 55%. Yields of total rice changed very little but head rice yields increased about 5 percentage points. As the milling pressure was increased by moving the pressure bar closer to the rotor, milling time was reduced up to 59% and energy consumption was reduced up to 34%. However, total yield was about 1 percentage point lower and head yield decreased about 6 percentage points.

DEGREE OF MILLING

The term "degree of milling" refers to the percentage of bran removed on a rough rice basis. In USDA grading practice the degree of milling is one of the factors that determines the grade assigned to the milled rice. For each grade there are maximum limits of objectionable factors such as weed seeds, damaged (discolored) kernels, chalky kernels, etc. The maximum allowable limits of these objectionable factors increase with decrease in degree of milling so that the less well milled rices are of generally lower quality and bring a lower price in the market.

Effect on Nutritional Properties

Generally, a well milled rice is preferred by the majority of rice consumers. Brown rice is preferred by a relatively small percentage of consumers even though it contains more protein, vitamins, minerals and lipid than milled rice. Part of the reluctance of consumers to eat brown rice is its more chewy texture. Brown rice also requires a cooking time of about 45 min compared to about 20 min for milled rice. Between the extremes of brown rice and well milled rice is rice that has been lightly milled or undermilled, thus retaining part of the bran. In some developing countries, such as India (Desikachar 1967), the government has enforced the production and consumption of undermilled rice in order to make the maximum possible use of rice resources and an improvement in the nutritional value of the rice. The Indian government has fixed as a maximum a 3 to 4% degree of milling for rice to be consumed in India. These figures may be compared to the 8 to 9% degree of milling of a well milled rice. The 3 to 4% degree of milling allows for rapid water penetration and a cooking time comparable to well milled rice. The main drawback of lightly or undermilled rice is its tendency to become rancid particularly after storage of about two months. Since in developing countries the rice is generally consumed within this period there are no practical difficulties encountered.

Measurement of Degree of Milling

In conventional rice milling and grading practice the degree of milling is estimated subjectively by visual observation. However, this subjective means of measurement is not considered completely satisfactory. A number of investigators have been concerned with the development of objective methods for evaluating degree of bran removal. These methods may generally be classified as chemical or physical in nature.

Chemical methods that have been investigated consist of differential dye staining procedures to enhance the visual differences between endosperm and bran, and of chemical analysis for constituents which occur predominately in the bran. Desikachar (1955A) and Borasio (1955) developed differential dye staining techniques. Although the presence of bran is enhanced there is still an element of subjectivity in the assignment of a degree of milling. Differential dye staining has been developed into a quantitative procedure by Barber and Benedito de Barber (1976). Milled rice kernels are stained with a methylene blue and eosine solution in methyl alcohol. Areas covered with bran stain blue while the endosperm stains pink. The bran and total kernel areas of magnified plane images are measured by planimetry. Colored Bran Balance (CBB) values range from 100 for brown rice to 0 for complete milled rice. This method is of particular interest in that it permits evaluation of the homogeneity of milling. Measurements of the degree of milling through analysis of constituents which occur predominantly in the bran include assays for thiamin (Kik 1951A; Desikachar 1955B), phosphorus (Desikachar 1955B), surface lipid (Autrey et al. 1955; Hogan and Deobald 1961), surface lipid, ash and protein (Watson et al. 1975).

Physical methods for determining the degree of milling have generally utilized the optical characteristics of the kernel. These methods depend upon the reflection of light from and the transmission of light through the milled rice at selected wavelengths. Kik (1951B) and Angladette (1957) were early developers of optical devices for measuring the degree of milling. More recently Stermer and coworkers (1962, 1968A) developed improved optical instruments. The main difficulty encountered in the use of optical devices is the presence of chalky kernels in the rice to be measured. The chalky kernels are not optically translucent and thus interfere with accurate measurements. This drawback can be remedied by removing the chalky kernels by photoelectric sorting machines before making the measurement of degree of milling.

A new physical method that shows promise is based on the total lipid content of the rice measured by nuclear magnetic resonance (Pomeranz *et al.* 1975).

Deep Milling

The removal of peripheral layers of well milled rice (endosperm) is termed *deep milling* or *overmilling*. Because of the hardness of the rice endosperm, deep milling must be accomplished with abrasive type mills. Research (Hogan 1969) on deep milling has shown that: (1) there is a heterogeneous distribution of protein and other major noncarbohydrate nutritional constituents in the milled rice kernel; (2) below the bran coat and aleurone cells, which are removed during conventional milling, there exists a peripheral layer unusually high in protein, minerals, lipids and vitamins; (3) the high protein and naturally enriched layer may be removed by presently available commercial machinery in the form of a finely divided rice flour, which may be considered an extension of the regular milling operation; and (4) the intact residual kernels, although reduced slightly in the nonstarch nutrients, are essentially an attractive white and well-milled rice suitable for regular table and food uses.

Normand *et al.* (1966) determined the amount of certain nutrients and the amino acids pattern in high protein rice flour produced by deep milling. They reported that the prepared material, which was removed in 12 fractions from successive layers of commercially milled rice consisted of approximately 18% of the original kernel weight as finely divided flour with a 22 to 12% protein content. Fat, thiamin, riboflavin, niacin, phosphorus and calcium similarly decreased progressively from outside layers towards the center of the kernel. Starch and amylose, by contrast, progressively increased towards the center of the rice kernel. The increase (approximately 19%) in amylose content of the rice starch from outer to inner layers reflects a significant difference in rice starch composition. The data indicated no essential differences in amino acids content of the rice protein except for a decrease in tryptophan toward the center of the kernel.

Organoleptic evaluation of the residual kernels remaining after different degrees of deep milling showed an increase in cooking quality preference with increasing degrees of milling up to a maximum after which further milling gave a loss of quality (Hogan 1969). Cooked rice samples were evaluated for cohesiveness (i.e., lack of tackiness), integrity of the kernels (i.e., degree of fragmentation), color, flavor and texture by a taste panel. An optimum preference was expressed by the panelists for those residual kernels resulting from between approximately 4 and 9% weight removal by overmilling.

The nutritional quality of rice endosperm is reported in detail in Chapter 11.

SOLVENT EXTRACTIVE MILLING

A relatively new development in rice milling technology is the radically different solvent extractive milling (SEM) process (Wayne 1965A, B, 1966, 1968). The SEM process differs from conventional rice milling in the method of bran removal and in the processing of by-products. Conventional rice milling uses abrasion and/or frictional pressure to remove the outer layers of bran from the rice kernel. With the SEM process, the bran layers are first softened, then wet milled in the presence of a rice oil-hexane miscella. In addition to milled rice, by-products produced by the SEM process include crude rice oil and defatted rice bran. There is presently one rice mill, Riviana Foods, Inc., Abbeville, La., which is using the process on a commercial basis.

Process Details

Lynn (1966) reported details of the SEM process and Lynn and Lawler (1966) presented a detailed flow sheet for the 1.5 million bbl/yr plant at Abbeville, La. A simplified schematic diagram of the process is shown in Fig. 9.11.

Rough rice with 10–14% moisture is cleaned and shelled with conventional rice mill machinery. The shelling must be very efficient and produce hull free brown rice for the SEM process because the gentle milling which it receives may not remove any adhering hull particles.

The variation from conventional milling begins with the bran removal operation. First the bran layers are chemically softened so that they are more easily removed. This chemical softening is accomplished by spraying warm rice oil onto the brown rice in closely controlled amounts (approximately 0.5%) and with uniform distribution. The treated brown rice is held for a period of time (2–4 hr) to allow the rice oil to permeate the bran membranes and soften them.

After the holding period, the treated rice with the softened bran is conveyed to Satake Rice Pearling machines which have been sealed and provided with solvent ports to permit the flow of large volumes of temperature controlled solution of rice oil in hexane (miscella) throughout the rice during milling. The miscella acts as a washing and rinsing medium to aid in flushing the bran from the rice endosperm as well as a conveying medium for continuously transporting the removed bran away

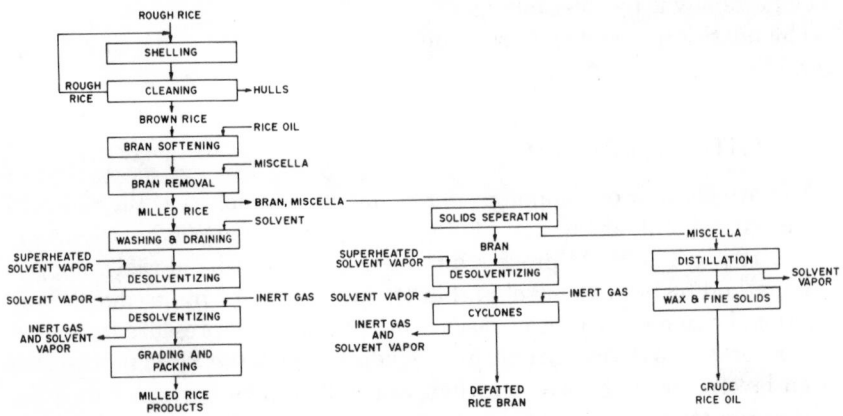

FIG. 9.11. SCHEMATIC DIAGRAM OF THE SOLVENT-EXTRACTIVE-MILLING PROCESS

from the rice. In addition, the miscella lends additional lubricity to the rice and prevents the increase of rice temperature which tends to make rice more susceptible to breakage. These factors reduce the amount of pressure required during the bran removal operation which significantly reduces the amount of breakage.

Following the bran removal step, the rice is rinsed with additional hexane on vibratory screens. It is then drained of the excess solvent and conveyed to desolventizing equipment. Superheated hexane vapor is used to flash evaporate the bulk of the hexane remaining on the rice. Then the rice is treated with flowing inert gas which removes the last traces of solvent. At this point the SEM process for the milled rice is completed and the rice is sized, graded and packaged with conventional rice mill equipment.

Bran Processing

The slurry of bran, rice oil and hexane (15−20% bran) is pumped from the milling equipment to vessels where it is allowed to settle and the extraction of the oil from the bran is completed. The remaining bran processing steps are used to separate the solid bran from the liquid stream. A settled bran slurry (30−35% bran) is drawn off the bottom of the first settling vessel and pumped to a second settling unit. This settling tank is more quiescent and allows a fairly heavy bran concentration slurry to be drawn off and pumped to the first of 2 centrifuges which remove the bran from the miscella. The centrifuged miscella-wet bran is mixed in another tank with hexane. The slurry from this tank is recentrifuged and the hexane-wet bran is conveyed to a desolventizer which operates on the same principle as the rice desolventizer, using superheated hexane vapor to vaporize the liquid solvent. The bran is then purged with warm inert gas and pneumatically conveyed through cooling cyclones and then into conventional bran handling equipment.

Oil Processing

The oil rich miscella obtained from the tops of the bran settling vessels and centrifugate from the bran centrifuging operations is collected and passed through additional solids removal equipment. This stream passes through a series of liquid cyclones to remove the coarse bran particles. The miscella is then cooled to precipitate part of the wax which is removed along with the fine bran particles by centrifugation. Next the miscella (approximately 10% oil) is heated and sent to a single stage, one pass, continuous flow evaporator where the oil concentration is raised to approximately 92%. It now goes to a typical solvent stripper that removes the remaining hexane. The rice oil is clarified further by chilling and centrifuging to remove the remainder of the nonfat solids.

Excess hexane vapors from the 2 superheated vapor desolventizers for bran and milled rice, and vapors from the oil evaporator and stripper, are passed through water-cooled condensers. These are vented through mineral oil to trap any residual hexane and allow the escape of the inert, noncondensible gases through a vent to the atmosphere. The condensate is collected in a hexane separating unit which contains an internal overflow barrier. This permits water to separate and collect on one side of the barrier and the lighter hexane to overflow to the dry hexane side of the vessel. The dry hexane is returned to a central hexane tank for redistribution.

Comparison of the SEM Process with Conventional Milling

The SEM process has some distinct advantages over conventional abrasion and frictional pressure milling (Lynn and Anderson 1967). The rice is whiter and more attractive in appearance than conventionally milled rice and has a significantly lower fat content. The white color is believed to result from more complete removal of the surface fats and color bodies by exposure to solvents.

Consumer evaluation in direct comparison of the SEM rice with conventionally milled rice has shown no major dissimilarities between the 2 products. Color of the SEM rice is generally cited as a preference factor. Taste and smell are about the same as regularly milled rice. There is less "stickiness" in the SEM product. The cooked flavor, texture and appearance have not been altered materially.

The lower fat content also contributes to excellent storage stability for the SEM rice. This is particularly important in packaging of rice since problems of fat "bleed through" have frequently occurred resulting in obliteration or smearing of the printed labels on transparent packaging materials. Consequently, less packaging problems and fewer plys are required for the packaging of SEM rice.

Improved brewing performance is another indirect product advantage obtained in the "brewers" fraction because of reduced fat content. Excessive fat level interferes with the fermentation process in brewing beer.

In view of the importance of the role played by microorganisms in the deterioration of rice, it is of interest to note that conventionally milled white rice shows a plate count of 5000 or more per g; the SEM rice plate count averages 1000 to 3000 per g. An indirect result of the lower bacteria levels is that higher quality rice flour can be produced. Low bacteria counts are important when rice flour is used as a dusting flour in refrigerated baked goods to prolong shelf-life.

A major advantage of the SEM process is that the yield of head rice (whole grains) can be increased by as much as 10 percentage points over conventional milling, as well as an increase in total milled rice yield. Tests were run with 4 combinations: long grain and medium grain rices which were each lightly milled and well milled. The average increase in total yield was 1 to 2 percentage points and the average increase in head yield was 4 to 5 percentage points for all 4 combinations of milling. The advantage in head yield for the medium grain rice tended to decrease as the yields from both SEM and conventional milling approached the maximum theoretical yield. Thus, for those lots showing poor milling yields (because of environment, cultural practices, handling, etc.) the advantage obtained by use of the SEM process is greater.

A significant difference in the SEM process and conventional milling is that 2 essentially new products are directly produced from SEM—defatted rice bran and crude rice oil. The defatted rice bran produced by SEM contains 17 to 21% protein and approximately 1.5% fat. The SEM bran is creamy white in color, more stable and clean and therefore is much more suitable for human food use than conventional rice bran, which is typically darker and bitter. Extensive research and development efforts have gone into developing products and applications for the SEM bran, however, the major use is still in animal feeds. The SEM rice bran has been shown to contribute to a higher rate of weight gain than does conventional rice bran in animal feeding studies. The protein efficiency ratio was found to be 1.9 based on casein at 2.5.

The traditional means of producing rice oil involves the solvent extraction of the oil from the bran after the bran has been removed from the rice grain. During conventional milling, immediately upon rupture of the bran layers, lipolytic enzymes and the lipid substrates are brought together. Hydrolysis of the fats starts immediately and a significant increase in the free fatty acid content is observed within a few hours. Traditional supplies of crude rice bran oil have suffered from problems of high refining losses due to the high levels of free fatty acids. The SEM rice oil is extracted from the rice during milling rather than from the bran itself. It is removed before the active lipolytic enzymes have time to split the triglyceride esters into component fatty acids. This results in SEM crude rice oil having free fatty acid levels (2–5%) comparable to other vegetable oils produced commercially.

The SEM crude rice oil produces a light colored edible oil when refined by conventional alkali methods followed by customary bleaching and deodorization. The refined rice oil has excellent stability as a frying fat. It can be hydrogenated for use in specialty products or winterized to yield high quality salad oils.

The disadvantages of the SEM process are its higher capital costs and operating costs. The solvent milling of brown rice is a sophisticated process and requires careful operation and more technical knowledge on the part of the operator who must understand not only the means available for controlling and adjusting the operation of equipment but also the hazards inherent in the use of flammable solvents and solvent vapors. Laboratory control is essential to maintain quality of the finished products and to guide the plant operation by analysis of in-process products. Close liason between operating and technical personnel is a necessity for safety as well as for optimal production.

Economic advantages accrue due to greater efficiency in the separation of the products of milling and the increased yield of higher priced products. However, the market for human food use of the defatted rice bran has not developed as anticipated and the cost of custom refining of the crude rice oil has made this practice economically questionable for the quantity of oil currently being produced.

REFERENCES

ANGLADETTE, A. 1957. Note on the subject of photoelectric control of the milling of rice. Riz et Rizic *3* (4) 139-143. (French)

AUTREY, H.S., GRIGORIEF, W.W., ALTSCHUL, A.M. and HOGAN, J.T. 1955. Effects of milling conditions on breakage of rice grains. J. Agric. Food Chem. *3*, 593-599.

BARBER, S. and BENEDITO DE BARBER, C. 1976. An approach to the objective measurement of the degree of milling. Rice Process Eng. Cntr. 2 (2) 1-8.

BHATIA, K. 1969. Effect of environmental conditions during milling on breakage of rice grains. M.A. thesis. Louisiana State University, Baton Rouge, Louisiana.

BHATTACHARYA, K.R. 1969. Breakage of rice during milling, and effect of parboiling. Cereal Chem. 46, 478-485.

BHATTACHARYA, K.R. and SUBBA RAO, P.V. 1966. Processing conditions and milling yield in parboiling of rice. J. Agric. Food Chem. 14, 473-475.

BORASIO, L. 1955. Macrocolorimetric method for evaluation of the final product of rice processing. Annali, Stazione Spermentale di Risicolture e della Colture Irrigue, Vercelli, Italia 3, 51-56. (Italian)

BORASIO, L. and GARIBOLDI, F. 1957. Illustrated glossary of rice processing machines. Food and Agric. Organ. U.N., Rome.

CALDERWOOD, D.L. and SCHROEDER, H.W. 1975. Chemical preservatives for maintaining moist rice in storage. In Tex. Agric. Exp. Sta., PR-3322.

CALDERWOOD, D.L., SORENSON, J.W., Jr. and SCHROEDER, H.W. 1975. Drying, storing and handling rice. In Six Decades of Rice Research in Texas, Tex. Agric. Exp. Sta. Res. Monograph 4.

CHILDERS, R.E., SORENSON, J.W., Jr. and STERMER, R.A. 1975. Effects of aerating rice during laboratory milling on milled rice yields. Tex. Agric. Exp. Sta. PR-3323C.

CHOUDHARY, M.S. 1970. The effects of moisture adsorption on the tensile strength of rice. Ph.D. dissertation, Texas A & M Univ., College Station, Texas.

DAVIS, L.L. 1944. Harvesting rice for maximum milling quality in California, Rice J. 47 (3) 3-4, 17-18.

DESIKACHAR, H.S.R. 1955A. Determination of the degree of polishing in rice. I. Some methods for comparison of the degree of milling. Cereal Chem. 32, 71-77.

DESIKACHAR, H.S.R. 1955B. Determination of the degree of polishing in rice. II. Determination of thiamin and phosphorus for processing control. Cereal Chem. 32, 78-80.

DESIKACHAR, H.S.R. 1967. Some aspects of the processing and storage of rice in India. In Symp. on Problems in Development and Ripening of Rice Grain. Int. Rice Comm. Newsl., Spec. Is. 1967.

GARIBOLDI, F. 1974. Rice milling equipment operation and maintenance. Food and Agric. Organ. U.N. Agric. Serv. Bull. 22.

HENDERSON, S.M. 1954. The causes and characteristics of rice checking. Rice J. 57 (5) 16, 18.

HENDERSON, S.M. 1957. Milled rice yields. Calif. Agric. 11 (7) 6, 15.

HOGAN, J.T. 1969. Rice processing and products research, Southern Division. Rice J. 72 (7) 54, 56, 58-62.

HOGAN, J.T. and DEOBALD, H.J. 1961. Note on a method of determining the degree of milling of whole milled rice. Cereal Chem. *38,* 291-293.

HOGAN, J.T., LARKIN, R.A. and MACMASTERS, M.M. 1954. X-Ray and photomicrographic examination of rice. J. Agric. Food Chem. *2,* 1235-1239.

HOUSTON, D.F. and KOHLER, G.O. 1970. Nutritional Properties of Rice. Natl. Acad. Sci., Washington, D.C.

JOHNSTON, T.H. and MILLER, M.D. 1973. Rice in the United States: Varieties and production. U.S. Dep. Agric. Handbk. *289,* Rev. 1973.

KESTER, E.B. 1959. Effects of certain preprocessing and cultural variables upon milling and other processing qualities of rice. Proc. of Calif. Rice Res. Symp. Albany, California.

KESTER, E.B., LUKENS, H.C., FERREL, R.E., MOHAMMED, A. and FINFROCK, D.C. 1963. Influence of maturity on properties of western rices. Cereal Chem. *40,* 323.

KESTER, E.B. and PENCE, J.W. 1962. Rice investigations at Western Regional Research Laboratory. Rice J. *67* (7) 45-47.

KIK, M.C. 1951A. Nutritive studies of rice. Arkansas Agric. Exp. Sta. Bull. *508.*

KIK, M.C. 1951B. Determining the degree of milling by photoelectric means. Rice J. *54* (12) 18-22.

KUNZE, O.R. 1964. Environmental conditions and physical properties which produce fissures in rice. Ph.D. dissertation. Michigan State University, East Lansing, Michigan.

KUNZE, O.R. and CHOUDHARY, M.S.U. 1971. Moisture adsorption and strength of rice. ASAE Ppr. *71-374.* Ppr. presented at 1971 SW Reg. Meeting of the Amer. Soc. of Agric. Eng., April 1–2. Wagoner, Oklahoma.

KUNZE, O.R. and HALL, C.W. 1965. Relative humidity changes that cause brown rice to crack. Trans. Amer. Soc. Agric. Eng. (ASAE) *8,* 396-399.

KUNZE, O.R. and HALL, C.W. 1967. Moisture adsorption characteristics of brown rice. Trans. ASAE *10,* 448-450, 453.

LANGFIELD, E.C.B. 1959. Time of harvest in relation to grain breakage on milling of rice. Rice J. *62* (8) 34.

LOUVIER, F.J., Jr. and CALDERWOOD, D.L. 1972. Breakage of processed rice due to falling impact. Cereal Sci. Today *17* (4) 98-101.

LYNN, L. 1966. The SEM rice milling process. Rice J. *69* (10) 19-20.

LYNN, L. and ANDERSON, R.M. 1967. X-M milling expands use of rice. Food Eng. *39* (1) 77-79.

LYNN, L. and LAWLER, F.K. 1966. Revolutionizes rice milling. Food Eng. *38* (11) 68-73.

MANNAPPERUMA, J.D. 1974. Analysis of thermal and moisture stresses inside a rice grain. Proc. 15th Rice Tech. Working Group, March 12–14. Fayetteville, Arkansas.

MATTHEWS, J., ABADIE, T.J., DEOBALD, H.J. and FREEMAN, C.C. 1970. Relation between head rice yields and defective kernels in rough rice. Rice J. 73 (10) 6-12.

MATTHEWS, J., HOGAN, J.T., MOTTERN, H.H. and DEOBALD, H.J. 1971B. Southern laboratory is studying rice milling. Rice J. 74 (6) 28-30, 34.

MATTHEWS, J. and SPADARO, J.J. 1974. Rice breakage during milling. Ppr. presented at 15th Rice Tech. Working Group Meetings, March 12—14. Fayetteville, Arkansas.

MATTHEWS, J. and SPADARO, J.J. 1975. Rice breakage during combine harvesting. Rice J. 78 (7) 59-63.

MATTHEWS, J. and SPADARO, J.J. 1976. Breakage of long-grain rice in relation to kernel thickness. Cereal Chem. 53, 13-19.

MATTHEWS, J., VEAL, D.M. and DEOBALD, H.J. 1971A. Comparative head rice yields from commercial and laboratory milling equipment. Rice J. 74 (4) 5-9.

MCDONALD, D.J. 1967. Suncracking in rice—some factors influencing its development and the effects of cracking on milling quality of the grain. M.A. thesis. University of Sydney, Sydney, N.S.W., Australia.

MCNEAL, X. 1950. When to harvest rice for best milling quality and germination, Ark. Agric. Exp. Sta. Bull. 504.

MCNEAL, X. 1960. Rice storage—effect of moisture content, temperature and time on grade, germination and head rice yield. Ark. Agric. Exp. Sta. Bull. 621.

MCNEAL, X. 1961. Effects of drying techniques and temperatures on head rice yields. Ark. Agric. Exp. Sta. Bull. 640.

MORGAN, A.I., Jr., BARTA, E.J. and GRAHAM, R.P. 1966. Chemical peeling of grain. Chem. Eng. Progr. Sym. Series S-69, 138-141.

MORSE, M.D., LINDT, J.H., OELKE, E.A., BRANDON, M.D. and CURLEY, R.G. 1967. The effect of grain moisture at time of harvest on yield and milling quality of rice. Rice J. 70 (11) 16-20.

NORMAND, F.L., SOIGNET, D.M., HOGAN, J.T. and DEOBALD, H.J. 1966. Content of certain nutrients and amino acids pattern in hi-protein rice flour. Rice J. 69 (9) 13-18.

POMERANZ, Y., STERMER, R.A. and DIKEMAN, E. 1975. NMR-oil content as an index of rice milling. Cereal Chem. 52, 849-853.

POMINSKI, J., SCHULTZ, E.F. and SPADARO, J.J. 1965. Effects of storage time after drying on laboratory milling yields of southern rice. Rice J. 68 (5) 20-23, 43.

POMINSKI, J., WASSERMAN, T., SCHULTZ, E.F., Jr. and SPADARO, J.J. 1961. Increasing laboratory head and total yields of rough rice by milling at low moisture levels. Rice J. 64 (10) 11-15.

POMINSKI, J., WASSERMAN, T., SPADARO, J.J., DECOSSAS, K.M. and DORE, A.B., Jr. 1961. Improvements in commercial drying of southern grown rice. I. Zenith—a medium-grain variety. Rice J. 64 (9) 10.

RHIND, D. 1962. The breakage of rice in milling: A review. Trop. Agric. *39* (1) 19-28.

RHIND, D. and TIN, U. 1933. The effect of temperature on the breakage of rice in milling. Indian J. Agric. Sci. *3,* 658.

ROBERTS, R.L. and WASSERMAN, T. 1977. Effect of milling conditions on yields, milling time and energy requirements in a pilot scale Engleberg rice mill. J. Food Sci. *42,* 802-3, 806.

SCHROEDER, H.W. 1967. Milling quality of Belle Patna rice in experimental storage: A study of the effects of field fungi on subsequent invasions by storage fungi. J. Stored Prod. Res. *3,* 29-33.

SMITH, W.D., DEFFES, J.J. and BENNETT, C.H. 1938. Effect of date of harvest on yield and milling quality of rice. U.S. Dep. Agric. Cir. *484.*

SMITH, W.D. and McCREA, W., Jr. 1951. Where breakage occurs in the milling of rice. Rice J. *54* (2) 14-15.

SORENSON, J.W., Jr. 1973. Design principles for the development of laboratory rice milling equipment. Texas A & M Univ., Dep. Agric. Eng. Final Rep. Coop. Agr. No. *12-14-100-10.073 (51),* College Station, Texas.

SORENSON, J.W., Jr. and CRANE, L.E. 1960. Drying rough rice in storage. Tex. Agric. Exp. Sta. Bull. *B-952.*

SPADARO, J.J. 1961. Evaluation of WRRL drying methods on southern rice. Ppr. Presented at 2nd conf. on Rice Utilization Proc., May 18−19. U.S. Dep. Agric.—Agric. Res. Serv. ARS*74-24.* Western Regional Research Center, Albany, Calif.

STERMER, R.A. 1968A. An instrument for objective measurement of degree of milling and color of milled rice. Cereal Chem. *45,* 358−364.

STERMER, R.A. 1968B. Environmental conditions and stress cracks in milled rice. Cereal Chem. *45,* 365-73.

STERMER, R.A., SCHROEDER, H.W., HARTSTACK, A.W. and KINGSOLVER, C.H. 1962. A rice photometer for measuring the degree of milling of rice. Rice J. *65* (5) 22-29.

STIPE, D.R., WRATTEN, F.T. and MILLER, M.F. 1972. Effects of various methods of handling brown rice on milling and other quality parameters. Proc. 14th Rice Tech. Working Group, June 20−22, Davis, California.

STIPE, D.R., WRATTEN, F.T. and MILLER, M.F. 1976. Drying steam-treated naturally moist rough rice. *In* 68th Annual Progress Rep. Louisiana State Univ. Rice Exp. Sta., Crowley, La.

TEN HAVE, H. 1958. Investigation on the percentage of crack and breakage in rice. Surinaamse Landb. *6,* 77-83. (Dutch)

TOLSON, W.J. and ROBE, K. 1977. World's first push-button rice mill. Food Process. *38* (5) 132-133.

U.S. DEP. AGRIC. (USDA)—AGRIC. MRKTG. SERV. 1976. United States Standards for rough rice, brown rice for processing, milled rice. Washington, D.C.

VAN RUITEN, H. 1976. Milling. *In* Rice Post Harvest Technology. E.V. Araullo, D.B. Padua and M. Graham, (Editors). Int. Dev. Res. Cntr, Ottawa, Ontario, Canada.

WASSERMAN, T. 1960. Heated air drying of western rice, Proc. 8th Rice Tech. Working Group, June 29—July 1. Lafayette, Louisiana.

WASSERMAN, T. 1961. Low moisture milling. ARS-*74-24,* Proc. 2nd Conf. on Rice Utilization, May 18—19. Albany, California.

WASSERMAN, T., FERREL, R.E., KAUFMAN, V.F., SMITH, G.S., KESTER, E.B. and LEATHERS, J.G. 1958A. Improvements in commercial drying of western rice. I. Mixing type dryer—Louisiana State University design. Rice J. *61* (4) 30; *61* (5) 40.

WASSERMAN, T., FERREL, R.E., KAUFMAN, V.F., SMITH, G.S., KESTER, E.B. and PIERCE, S. 1958B. Improvements in commercial drying of western rice. II. Non-mixing columnar-type dryer. Rice J. *61* (7) 9.

WASSERMAN, T. and ROBERTS, R.L. 1972. Effects of surface wetting of brown rice immediately prior to milling. Proc. 14th Rice Tech. Working Group, June 20—22. Davis, California.

WASSERMAN, T., ROBERTS, R.L., McCREADY, R.M. and BAILEY, G.F. 1974. Changes in yields, milling time and conditions, and energy requirements in commercial-type rice mills. Proc. 15th Rice Tech. Working Group, March 12—14. Fayetteville, Arkansas.

WATSON, C.A., DIKEMAN, E. and STERMER, R.A. 1975. A note on surface lipid content and scanning electron microscopy of milled rice as related to degree of milling. Cereal Chem. *52,* 742-747.

WAYNE, T.B. 1965A. Apparatus for milling rice and the like. U.S. Pat. *3,165, 134.* January 12.

WAYNE, T.B. 1965B. Apparatus for milling cereal grains. U.S. Pat. *3,217,769.* November 16.

WAYNE, T.B. 1966. Extractive milling of rice in the presence of an organic solvent. U.S. Pat. *3,261,690.* July 19.

WAYNE, T.B. 1968. Horizontal rice mill. U.S. Pat. *3,401,731.* July 17.

WEBB, B.D. and CALDERWOOD, D.L. 1977. Relationship of moisture content to degree of milling in rice. Cereal Food World. *22,* 484.

WITTE, G.C. 1970. Rice milling in the United States. Assoc. Oper. Millers Bull. *2,* 3147-3159.

WITTE, G.C. 1972. Conventional rice milling in the United States. *In* Rice: Chemistry and Technology, D.F. Houston (Editor). Amer. Assoc. Cereal Chem., St. Paul, Minnesota.

WRATTEN, F.T. 1960. Effects of milling rice at low moisture levels. Proc. Rice Tech. Working Group Meeting, June 29—July 1. Lafayette, Louisiana.

10

Properties of the
Rice Caryopsis

Beinvenido O. Juliano

Because rice is a cereal that is consumed mainly as a whole grain, the knowledge of the structure and composition of the caryopsis is as important as physicochemical and biochemical properties of its constituents in the study of the processing properties of the grain. Recent ultrastructure studies of the rice caryopsis used scanning (Evers and Juliano 1976; Watson and Dikeman 1977) and transmission electron microscopy (Hoshikawa 1973, 1974; Bechtel and Pomeranz 1977, 1978A, B). There are also reviews on the grain composition (Houston and Kohler 1970; Juliano 1966, 1967, 1972B, 1977A) and on correlation between physicochemical, and functional and nutritional properties of the grain (Juliano 1972A,C, 1973A,B, 1976, 1977B, 1978; Juliano *et al.* 1964). Reference to these reviews includes the earlier publications cited in them.

PHYSICAL PROPERTIES

Structure of the Rice Caryopsis

The rice grain consists of the outer covering structures—the hull or husk (Chapter 22)—and the enclosed edible portion, the rice caryopsis or brown rice. Brown rice constitutes 72 to 82% of rough rice weight. A brown rice recovery of 86% was reported for an Italian rice (Baldi *et al.* 1974). The surface of the caryopsis has ridges that correspond to those of the hull. Brown rice varies widely among cultivars in size, shape and weight.

Outer Layers.—Detailed studies have been made on the gross structure of brown rice (see Juliano 1972B; Bechtel and Pomeranz 1977). The caryopsis is enveloped by the caryopsis coat composed of the pericarp, seed coat and nucellar layers (Bechtel and Pomeranz 1977) (Fig. 10.1).

403

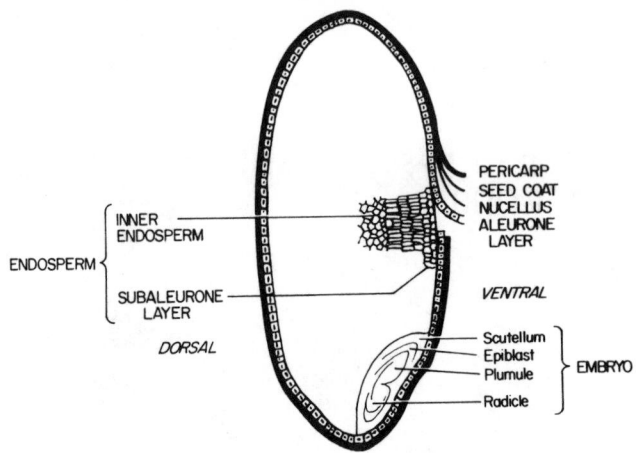

Adapted from Juliano (1972B)

FIG. 10.1. DIAGRAMATIC REPRESENTATION OF THE RICE CARYOPSIS

The fibrous pericarp consists of crushed cells and is about 10 μm thick. Next to the pericarp is a layer of cells representing the remains of the inner integuments and termed the seed coat. The seed coat has a thick cuticle. Interior to the seed coat are the remains of the nucellus, which is about 2.5 μm thick and has a thick cuticle adjacent to the seed coat cuticle.

The next enclosing layer is the aleurone, which completely encloses the endosperm and the embryo (Cho 1956). The aleurone layer is composed of quadrangular or rectangular parenchyma cells with 2 μm cell walls. The aleurone may have from 1 to 7 cell layers and is thicker on the dorsal (back) than along the lateral and ventral (embryo) surfaces. The aleurone of coarse or bold, short grain cultivars has more cell layers than that of slender long grain cultivars. The aleurone cells that surround the endo-sperm are filled with protein-rich aleurone grains with a single membrane containing globoids or phytate bodies in a protein-carbohydrate "matrix" (Ogawa *et al.* 1975), and lipid droplets or spherosomes around a centrally located nucleus (Bechtel and Pomeranz 1977). The modified aleurone cells that surround the embryo lack aleurone grains, have fewer spherosomes than the other aleurone cells, and contain filament bundles. Plastids in the aleurone cells exhibit a unique structure in which the 2 membranes invaginate to form tubules and vesicles within the plastids. Transfer aleurone cells are not observed, but transfer cells are present in the node of the rice spikelet (Zee 1972).

The embryo or germ is extremely small and is located on the ventral side of the caryopsis (Barber *et al.* 1976). In a longitudinal section the outlines of the embryonic leaves (plumule) and the embryonic primary root (radicle) are seen joined together by a very short stem (hypocotyl) (Fig. 10.1). The plumule is enclosed by a protective covering, the coleoptile, and the root is enclosed by a root cap and the coleorhiza. The embryo is bound in the inner side by the scutellum (cotyledon) (Tanaka *et al.* 1976A), which lies next to the endosperm. Immature cells of scutellum and mesocotyl are found between mature differentiated parenchyma cells. Three general categories of parenchymatous cells are identified based on protein body characteristics and on lipid body distribution: (1) cells with inclusions of protein bodies and numerous lipid bodies scattered throughout the cytoplasm and which probably function in a storage capacity; (2) cells with protein bodies with or without electron-dense inclusions and with peripheral lipid bodies with or without electron-dense deposits on the sectioned surface. These cells are primarily epidermal except for the root apex and root cap; and (3) cells lacking protein bodies and having peripheral lipid—the provascular system, plumule and radicle (Bechtel and Pomeranz 1978A). Plastids are similar in all cells; many have osmiophilic globules and phytoferritin.

Endosperm.—The starchy endosperm consists of thin-walled parenchyma cells, usually radially elongated and heavily loaded with compound starch granules and some protein bodies (Fig. 10.2) (del Rosario *et al.* 1968; Mitsuda *et al.* 1967). Protein bodies are more abundant in the outer cell layers particularly in the subaleurone or 2 outermost cell layers where the compound starch granules are small. Protein bodies are single membraned, whereas compound starch granules are in double-membraned amyloplasts. They have deeper staining near the periphery, at the center, or both. Isolated protein bodies showed a lamellar structure in some of them (Mitsuda *et al.* 1969). Endosperm protein bodies show 2 distinct components in a medium of refractive index 1.66, but only 1 component at 1.58 (Wolf and Khoo 1970). Crystalline protein bodies with crystal-lattice structure are located exclusively in the subaleurone layer together with the usual protein bodies (Harris and Juliano 1977; Bechtel and Pomeranz 1978B).

The starch granules have a pentagonal dodecahedron surface (Watabe and Okamoto 1960). On HC1 hydrolysis, the residual starch shows a lamellar structure the same as other cereal starch granules on transmission electron microscopy (Buttrose 1969), but no change is noted by scanning electron microscopy (Evers and Juliano 1976).

Opacity in both immature and mature nonwaxy or nonglutinous grains has been shown to be due to loose cell contents in the endosperm portions (del Rosario *et al.* 1968). Simple spherical starch granules are observed in the cells by scanning electron microscopy (Utsunomiya *et al.* 1975A;

Tashiro and Ebata 1975). In "crumbly" rice, which has a soft and opaque endosperm, most granules are spherical (del Rosario *et al.* 1968; Evers and Juliano 1976). In glutinous or waxy rice, opacity of endosperm is due mainly to pores in the granules as reflected by micropores on the surface of the compound granules by replica transmission electron microscopy (Watabe and Okamoto 1960). Pores are seen by scanning electron microscopy at the center of the compound granules of waxy rice (Utsunomiya *et al.* 1975B; Tashiro and Ebata 1975).

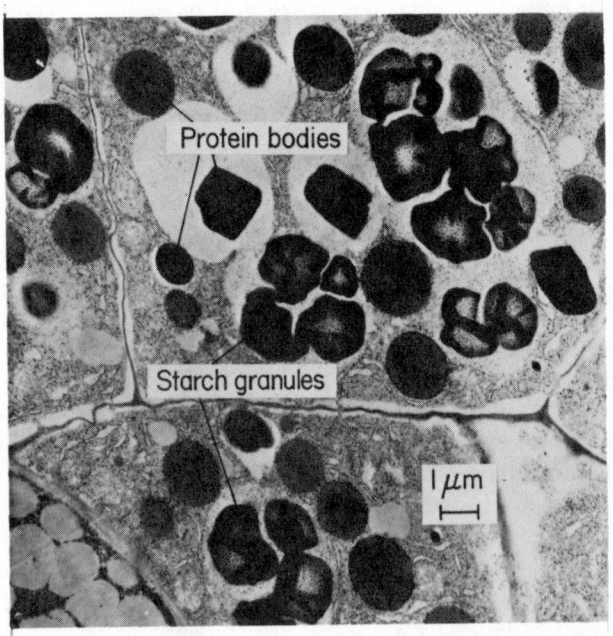

Courtesy of N. Harris (IRRI 1976)

FIG. 10.2. PROTEIN BODIES AND COMPOUND STARCH GRANULES IN SUBALEURONE LAYER OF DEVELOPING IR26 RICE GRAIN

Sample was fixed in glutaraldehyde, postfixed in osmium tetroxide and stained with uranyl acetate and alkaline lead citrate.

Changes During Grain Development

After fertilization, the caryopsis becomes structurally mature in 14 days. It attains full length first, then full width and then full thickness (del Rosario *et al.* 1968; Hoshikawa 1973, 1974). A longer ripening period results from uneven flowering of photoperiod insensitive rices. Tropical rices are usually harvested 30 days after flowering for optimum moisture content for drying with minimum cracking. The pericarp is initially well developed but degenerates later. The aleurone layer and embryo are fully developed earlier than the endosperm. Translocation of nutrients occur through the vascular bundle in the nucellar tissue at the dorsal line of the endosperm (Hoshikawa 1973,1974).

Starch granules are first seen in the endosperm 4 days after flowering, while protein bodies are seen by 7–8 days after flowering (Harris and Juliano 1977). Increase in starch content is mainly due to an increase in size of starch granules since only 1 set of granules are formed (del Rosario *et al.* 1968; Hoshikawa 1973,1974; Sato and Ehara 1974). Protein bodies increase in number but not in mean size during grain development. Thus most of the compositional changes ocurring 7 or more days after flowering are mainly due to endosperm accumulation.

Processing Effects

The weight distribution of the various parts of brown rice is pericarp, 1–2%; aleurone, nucellus and seed coat, 4–6%; embryo 2–3%; and endosperm 89–94% (Juliano 1972B). In the embryo fraction of brown rice, the weight distribution is 0.26% epiblast, 0.18% coleorhiza, 0.34% plumule, 0.18% radicle, and 1.18–1.4% scutellum (Hinton 1948). A related study reported 6.5% pericarp, testa, nucellus and aleurone; 2–2.1% scutellum; 0.8–1.1% embryo, and 90.4–90.6% endosperm (Wang *et al.* 1950). Abrasive milling removes the outer layers of the rice caryopsis, producing milled or polished rice and the byproducts bran and polish. Usually 6 to 10% by weight of brown rice is removed on milling. With a McGill mill, bran is about 8%, polish 1–2% and milled rice 89–90%. Well milled rice should have a smooth surface because the ridges of its outer layer corresponding to those of the hull have been removed during milling.

The process of steaming wet grain or parboiling (Chapter 14) gelatinizes the starch granules and destroys the spherosome structure. This may contribute to the gummy nature of bran obtained from parboiled rice during milling. Protein bodies remain intact on cooking. Degree of parboiling determines the puffed volume of parboiled rice (Antonio and Juliano 1973).

Cooking essentially results in starch gelatinization and water absorption and volume expansion of the grain with splitting of cell walls (Little and Dawson 1960). In most varieties, the expansion is girthwise but in certain low-to-intermediate amylose varieties of low-to-intermediate gelatinization temperature, expansion is lengthwise from cooking as exemplified by Basmati rices of Pakistan and India, D25-4 variety from Burma, and Sadri varieties from Iran (Juliano 1972A). The exact cause for the extreme elongation is not fully understood. The grains of varieties such as D25-4 and Basmati tend to have chalky endosperm.

PHYSICOCHEMICAL PROPERTIES

Proximate Analysis

Proximate analysis of brown rice and its milling fractions show uneven distribution with the outer layers richest in nonstarch constituents and the endosperm richest in starch (Table 10.1). In 41 brown rice samples, protein content (6.6–10.7%) was not correlated with oil content (1.9–2.9%; $r = 0.05^{ns}$) (IRRI 1975). The first or second outer 5% fraction had the highest protein content in rice with average protein (Barber 1972) but only the third outer 5% fraction had the highest protein content in high protein rice (Fig. 10.3) (Mitsuda and Murakami 1969; IRRI 1977B).

TABLE 10.1.   PROXIMATE ANALYSIS OF BROWN RICE AND ITS FRACTION (% DRY BASIS)

Constituent	Brown rice	Milled rice	Rice bran	Rice embryo	Rice polish
Protein (N × 5.95)	7.1–13.1	5.6–13.3	12.1–17.2	17.7–24.1	12.8–16.4
Crude fat	1.8– 4.0	0.2– 1.1	14.6–21.7	15.2–23.8	8.8–15.3
Dietary fiber	0.2– 2.6	0.1– 0.6	8.7–13.1	2.0– 4.8	2.1– 5.3
Ash	1.0– 2.4	0.3– 0.7	9.0–12.2	6.1–10.1	5.0– 9.3
Nitrogen-free extract	74.5–90.2	84.0–93.5[1]	40.9–49.1	36.2–57.3	53.7–71.3

Source: Juliano 1972B; Baldi et al. 1974; Fossati et al. 1976.
[1]Mainly starch.

Carbohydrates

Starch Granules.—Starch is the major constituent of the endosperm and makes up more than 90% of dry matter. It exists in the form of polyhedral granules 3–10 μm in size, several in a compound granule inside an amyloplast (Hoshikawa 1973,1974; Sato and Ehara 1974). In waxy rice the polyhedral nature of the granule remains, but pores are noted on the surface. Likewise, waxy starch granules have lower specific gravity than nonwaxy starch (Juliano 1967). The properties of rice starch has been reviewed (Juliano 1972B; Schoch 1967) and a general reference on starch was recently published (Banks and Greenwood 1975).

Starch granules exhibit birefringence under polarized light and an A-type pattern on X-ray diffraction, which are characteristics of cereal starches. Differences are noted in final gelatinization or birefringence end-point temperature (BEPT) when at least 95% of granules have swollen irreversibly in hot water with loss of birefringence and crystal-

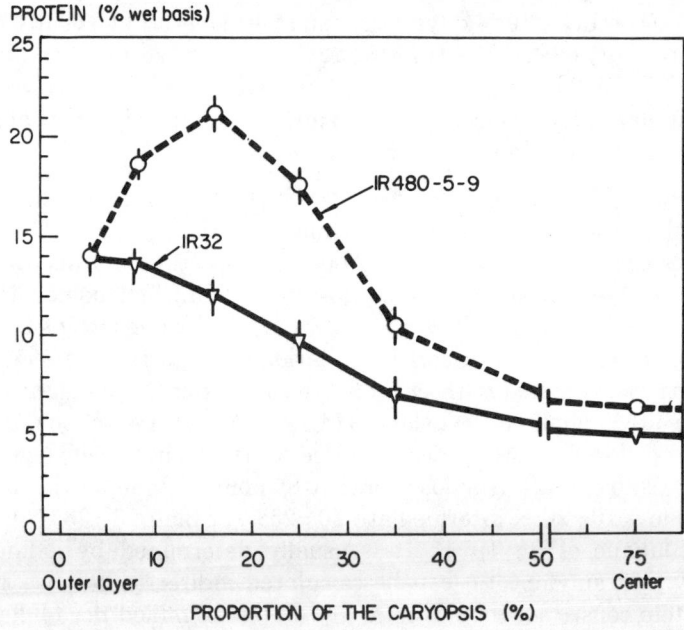

PROTEIN (% wet basis)

PROPORTION OF THE CARYOPSIS (%)

From IRRI (1977B)

FIG. 10.3. PROTEIN DISTRIBUTION IN MILLING FRACTIONS FROM BROWN RICE OF LOW PROTEIN (7.4%) RICE IR32 AND A HIGH PROTEIN (10.7%) RICE IR480-5-9

linity. Final BEPT ranges from 58°–79°C (136.4°–174.2°F) and is classified as low 58°–69.5°C (136.4°–157.1°F) intermediate 70.5°–74°C (158°–165.2°F); and high > 74°C (> 165.2°F) (Juliano *et al.* 1974). BEPT is a varietal characteristic but is affected directly by ambient temperature during grain development (Nikuni *et al.* 1969; Resurreccion *et al.* 1977).

BEPT is estimated in the breeding program by the alkali test in which the chemical gelatinization of head milled rice in 1.7% potassium hydroxide is read after a 23-hr soaking and is discussed in Chapter 15. The degree of disintegration correlates negatively with BEPT. It is possible that BEPT may reflect general grain hardness because Century Patna 231 with high BEPT showed the best Vickers hardness of varieties tested (Nagato and Kono 1963) and was more resistant to fissuring when exposed to high humidity than other United States long grain varieties of intermediate BEPT (Kunze and Hall 1965).

BEPT correlates with cooking time during boiling of rice. Extent of 2.2 N-hydrochloric acid corrosion correlated negatively with BEPT regardless of starch composition (Juliano 1973B; Evers and Juliano 1976). Although maximum varietal spread of acid corrosion in 2.2 N acid at

35°C (95°F) occurs after 4 days corrosion rate levels off at about 15 days and residual starch ranges from 6−22%. Residual starch was directly affected by BEPT and amylose content (IRRI 1977B). α-Amylolysis gave a similar but less consistent trend among samples of similar amylose content (Evers and Juliano 1976).

Physicochemical Properties of Starch.—Starch is composed of a branched fraction, amylopectin and a linear fraction, amylose. The major linkage is α-1,4-D-glucopyranosidic but amylopectin contains, in addition, the α-1,6-D-glucopyranosidic linkage at branched points. Degree of branching of amylopectin is 4−5% or a mean chain length of 20 to 28 anhydroglucose units by periodate oxidation (Juliano 1967). Its multiple branching was verified with enzymic method utilizing pullulanase and β-amylase (Marshall and Whelan 1974). Amylopectin is the major fraction of rice starch. In waxy rice, amylose content ranged between 1 and 2% only (Table 10.2). Amylose content of nonwaxy milled rice is classified as low 10−20%, intermediate 20−25%, or high 25−33% (Juliano 1972A; Juliano *et al.* 1974). It is usually determined by iodine colorimetry and amylopectin can be calculated indirectly because starch and protein constitute 98% of total dry matter of milled rice (Juliano *et al.* 1964). Amylose:amylopectin ratio is the single most important factor determining the eating quality or texture of cooked milled rice (Juliano 1972A, 1973A; Juliano *et al.* 1965, 1972A, 1974) and rice cakes (Perdon and Juliano 1975B). Amylose content varies directly with ambient temperature during grain development particularly in low amylose rices (Nikuni *et al.* 1969; Resurreccion *et al.* 1977).

TABLE 10.2. PHYSICOCHEMICAL PROPERTIES OF WAXY AND NONWAXY RICE STARCH AND THEIR FRACTIONS

Property	Waxy		Nonwaxy	
BEPT (°C)	58 −	78.5	58 −	79
Granule size (μm)	1.9 −	8.1	1.6 −	8.7
Density (xylene displacement)	1.48−	1.50	1.49−	1.51
Iodine binding capacity (%)	0.15−	0.86	2.36−	6.96
Residual protein (% N × 5.95)	0.1 −		0.1 −	0.7
Residual phosphorus (mg/g)	0.02−	0.03	0.12−	0.45
Gel (6%) viscosity (cps)	64	−1890	140	−2200
$[\eta]_{0.15N\ KOH}$ (ml/g)	46	− 164	160	− 194
Starch fractions	Amylopectin		Amylopectin	Amylose
$S°_{20,\ DMSO}$(Svedbergs)	28	−500	30 − 1400	3.5− 5.8
$S_{20,\ DMSO}$(Svedbergs)	27	−242	73 − 237	2.0− 6.6
$[\eta]_{0.15N\ KOH}$ (ml/g)	46	−164	83 − 221	55 −242
Mean chain length (anhydroglucose units)	20	− 27	20 − 28	No data
β-amylolysis limit (%)	49	− 56	49 − 58	73 − 76
Iodine binding capacity (%)	0.07−	0.09	0.37− 2.74	17.4− 20.2
Gel (6%) viscosity (cps)	19	−330	290 −740	13 −160

Source: Juliano 1967; Reyes *et al.* 1965; Juliano and Perdon 1975; Perdon and Juliano 1975A; Juliano *et al.* 1969; Kongseree and Juliano 1972; Tabata *et al.* 1975; IRRI 1977B.

Amylopectin interferes with amylose-iodine colorimetry at the pH 4.5 used in the simplified assay (Juliano 1971). By the use of carbonate-bicarbonate buffer pH 10.2 a deep blue color is obtained similar to the original Williams method, instead of greenish blue at pH 4.5 (IRRI 1977B). In addition, similar absorbance of starch-iodine color is obtained at 590 and 630 nm. Residual fat or lipid in the endosperm, however, reduces the iodine blue color probably by helical complex to amylose, but the decrease was much higher at pH 10.2 than at pH 4.5. "Absolute" amylose values are obtainable by defatting rice flour with refluxing 95% ethanol, or by ambient temperature defatting with water-saturated butanol (Bolling and El Baya 1975; Sowbhagya and Bhattacharya 1971).

Starch is normally prepared from milled rice by protein removal with 0.1 N sodium hydroxide or 1.2% sodium dodecylbenzene sulfonate-0.12% sodium sulfite (Reyes et al. 1965). Sodium hydroxide is a more drastic reagent and results in less crystalline starch granules (Lugay and Juliano 1965). However, purified starch granules still contain 0.1 to 0.7% residual protein and phosphorus, and both cannot be further extracted. Residual protein and phosphorus are lowest in waxy starch and highest in high amylose rices (Juliano 1973B). Residual ash is about 0.1% (Chikubu 1975). Acid-corroded starch still contained residual protein. Enzyme proteins are known to be bound to starch granules such as starch synthetase (Baun et al. 1970). Sodium dodecylbenzene sulfonate is claimed, however, to partially replace bound lipid in complexing with starch granules (Fujii 1972).

In addition, bound lipids or fat-by-hydrolysis occur in rice starch at about 0.6%–1.0% after ether extraction of 0.6% lipids and are only extracted with 85% methanol or 95% ethanol, or with water-saturated butanol (Acker and Schmitz 1967; Yasumatsu and Moritaka 1964). Bound lipid is reported to be only 16% lysolecithin in nonwaxy rice starch (Acker and Schmitz 1967), but lysolecithin is absent in bound lipids of waxy rice and corn starch (Nakamura et al. 1958A; Acker and Schmitz 1967). Amylose can complex with as much as 6% lysolecithin and amylopectin with 2% lysolecithin (Nakamura et al. 1958B). Bolling and El Baya (1975) found lysocephalin in addition to lysolecithin in the butanol-water extract of milled rice, together with variable amount of lecithin. Hirayama and Matsuda (1973) reported 61% of phospholipids of rice starch to be lysolecithin.

Although amylose content and BEPT relationship is presumably poor, rices with the following combination of starch properties have not been generally obtained among cultivated rices: intermediate BEPT, waxy and low amylose starches; and high BEPT, intermediate and high amylose starches.

Starch fractionation can be achieved by gelatinizing and dispersal of starch, in water or dimethylsulfoxide-water, and precipitation of amylose with slow cooling after addition of 1-butanol, isoamyl alcohols or thymol (Juliano and Perdon 1975). Amylose can be further purified by recrystallization from 1-butanol saturated water. The first crystallization is preferably done with isoamyl alcohols or by thymol to obtain pure amylopectin. Amylopectin is precipitated as amorphous powder from the mother liquor by pouring, with stirring, into 3–4 volumes of ethanol.

The X-ray crystallinity of rice amylose show varietal differences in the ratio of the V-anhydrous and V-hydrate forms, which are characteristic of helical complex of amylose (Lugay and Juliano 1965).

Tabata *et al.* (1975) reported ester-bound phosphate in waxy rice starch in the order of 14–15 ppm, or about 85% of total phosphorus of native starch. This is mainly 6-phosphoglucose residue. A sample of nonwaxy Japonica starch had only 11 ppm 6-phosphoglucose residue out of 350 ppm, or only 3%, indicating that phosphate esters are probably in the amylopectin fraction, and the presence mainly of phospholipid (e.g., lysolecithin) bound to starch, in nonwaxy rice starch.

Differences in mol wt of the starch fractions particularly amylopectin have been reported (Table 10.2). Among waxy starches, mol wt based on intrinsic viscosity and sedimentation constant correlates positively with BEPT (Perdon and Juliano 1975A). Gel consistency which measures the relative rate of retrogradation of waxy rice gels also correlates with BEPT. For rice cake, the rices with BEPT 64°–66°C (147.2°–150.8°F) had better cake quality than those with BEPT 57°–60°C (134.6°–140°F) (Palmiano and Juliano 1972A). Among Philippine rices, those preferred for flattened, parboiled rice (Antonio and Juliano 1974) and rice cake (Antonio *et al.* 1975) had lower BEPT (<70°C, <158°F) than those with high BEPT (>74°C) (>165.2°F). High BEPT rices produced processed (gelatinized) products that harden faster during storage particularly under refrigeration. In waxy rice, gel consistency is best determined by dispersing 200 mg milled rice flour in 2 ml 0.15 N potassium acetate and measuring the length of the cooled gel after 30 min on a horizontal position (Perdon and Juliano 1975A). BEPT estimation by alkali test in 1.7% and 1.3% potassium hydroxide readily identifies waxy rices of good eating quality (IRRI 1977B).

Among nonwaxy starches, gel consistency correlated with mol wt of amylopectin but was not simply correlated with BEPT (Juliano and Perdon 1975). Amylopectin contributes more than amylose to gel consistency and viscosity. Gel consistency is determined in nonwaxy rices by dispersing 100–120 mg flour in 2 ml 0.2 N potassium hydroxide and measuring the length of the cooled gel after 1 hr on a horizontal position and scored as hard 27–40 mm; medium 41–60 mm; and soft 61–100 mm (Cagampang *et al.* 1973; IRRI 1977A). Hard gel consistency has been observed exclusively in high amylose rices with high Brabender amylograph consistency (>400 Brabender Units) and setback (>300 B.U.) and low BEPT. In the alkaliviscogram, those rices with hard gel consistency have exceptionally high peak viscosity (Suzuki and Juliano 1975). These rices also have lower content of both amylopectin and amylose soluble on boiling rice in water (Maningat and Juliano 1978) and are generally less preferred than those of the same amylose content with soft gel consistency (Juliano et al. 1974). Intermediate and low amylose rices require more rice (120 mg instead of 100 mg) to produce differences in gel consistency (IRRI 1977A).

Although some correlation has been found between BEPT and intrinsic viscosity of amylose and amylopectin in nonwaxy rices (Juliano 1972B) the relationship was not consistent when studied on sister lines grown in 2 seasons differing both in amylose content and BEPT (Kongseree and Juliano 1972). The importance of BEPT on physicochemical properties of cooked nonwaxy rice is still not well appreciated and understood.

Hemicellulose, Cellulose and Proteoglycans.—Very little work has been done on the hemicellulose of rice grain, particularly that of milled rice (Cartaño and Juliano 1970; Gremli and Juliano 1970). As in other cereals, the hemicelluloses, which constitute dietary fiber with cellulose, are highest in the rice bran and polish. Leonzio (1967) reported 1.4–2.1% pentosans in brown rice, 0.6–1.1% in milled rice, 8.6–10.9% in bran, and 3.2–6% in polish. The embryo has 4.8–7.4% pentosans. The corresponding distribution of pentosans is 43% in bran, 8% in embryo, 7% in polish and 42% in milled rice.

A bran polish contained only 0.1% water soluble hemicellulose and 1% hemicellulose extractable with 0.5 N sodium hydroxide (Gremli and Juliano 1970). Milled rice contains 0.02% water-soluble and 0.1% 0.5 N-sodium-hydroxide-soluble hemicelluloses (Cartaño and Juliano 1970). Enzymic studies indicate that rice hemicelluloses are of the arabinoxylan type containing galactose in the bran polish fraction and galactose and glucose in the milled rice fraction. The water soluble hemicelluloses are probably more branched arabinoxylans, as is evident from their higher arabinose:xylose ratio.

The distribution of cellulose in brown rice is 62% in bran, 4% in embryo, 7% in polish, and 27% in milled rice (Leonzio 1967). The high cellulose content in bran is consistent with the thick cell walls of the pericarp, the seed coat and the aleurone layer. In a study of the carbohydrates of brown and milled rices (Fraser and Holmes 1958), brown rice had 84% starch, 1.2% pentosans, 0.9% dietary fiber, and 0.7% sugars. The mean carbohydrate content for 5 milled rice samples was 89% starch, 0.4% pentosans, 0.3% sugars, and 0.4% dietary fiber.

Proteoglycans have been isolated and characterized from rice bran (Yamagishi *et al.* 1975, 1976). They are rich in hydroxyproline and arabinose. Pronase and hemicellulase hydrolysis of proteoglycan yielded a sugar-amino acid compound, *O*-α-L-arabinofuranosyl hydroxyproline (Fig. 10.4).

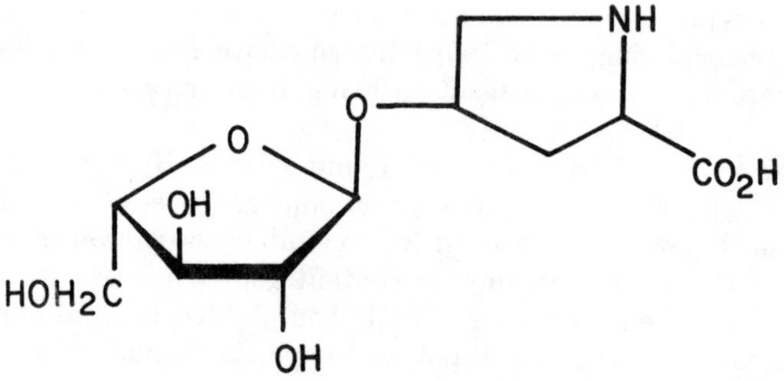

From Yamagishi et al. (1976)

FIG. 10.4. STRUCTURE OF *O*-α-L-ARABINOFURANOSYL HYDROXYPROLINE FROM THE ENZYMIC AND ALKALINE DEGRADATION OF RICE BRAN PROTEOGLYCANS

Free Sugars.—The main sugars of rice embryo and endosperm are sucrose, glucose and fructose, with small amounts of raffinose. Free sugars are concentrated in the aleurone layer and have higher levels in waxy (0.52%) than in nonwaxy (0.25%) rice (Singh and Juliano 1977; Pascual *et al.* 1978). Bran polish contains 80% of nonreducing sugars in IR28, a nonwaxy rice, and 74% in IR28, a waxy rice, but accounts for only 40% of reducing sugars in IR28 and 41% in IR29; the rest is in milled rice. Bran has 6.4% free sugars in IR28 and 6.5% in IR29. Sucrose is the major nonreducing sugar followed by raffinose. The principal reducing sugars are glucose and fructose together with melibiose and maltooligosaccharides. Maltose, maltotriose and maltotetraose were only noted in developing rice grain (Pascual *et al.* 1978).

Phytin or myo-inositol hexaphosphate salt is an important constituent of the aleurone layer and embryo. Reported phytin phosphorus contents are brown rice 0.21−0.28%; milled rice 0.04−0.06%; bran polish 2.0−2.6%, and embryo 0.8−1.9% (Juliano 1972B; Hayakawa 1977). At least 80% of the phosphorus content of brown rice, 90% of bran phosphorus, and 40% of the phosphorus in milled rice is phytate phosphorus. Hayakawa (1977) reported that phytate phosphorus in a nonwaxy Japanese rice accounted for 78% of total phosphorus of brown rice, 38% of milled rice phosphorus, 94% of bran phosphorus, and 88% of embryo phosphorus.

In a sample of rice embryo, 98% of the phytin phosphorus is hexa-phosphate and only 1% each is pentaphosphate and tetraphosphate; no mono, di and triphosphates of phytin were detected (Hayakawa 1977). Free myo-inositol of 12 brown rice ranged from $24-45$ $\mu g/g$. Phytic acid content of milled rice ranged from $1.23-1.45$ mg/g (Kennedy and Schel-straete 1975), 3.8 mg/g in bran, and 3.7 mg/g in polish (Fossati *et al.* 1976), mainly as a mixed magnesium and potassium salt (Tanaka *et al.* 1974,1976B).

Nitrogenous Compounds

Whole Protein.—Protein content of the rice caryopsis in brown rice is about 8% and that of milled rice is 7%. It is the second major constituent of milled rice and is the major protein source in the diets of tropical Asians, for whom it contributes from 30 to 80% of calories and protein. The factor 5.95 converts Kjeldahl nitrogen to rice protein based on the 16.8% nitrogen content of the major rice protein, glutelin.

Milled rice protein consists of 5% albumin (water soluble protein), 10% globulin (salt soluble protein), > 80% glutelin (alkali soluble protein), and < 5% prolamin (alcohol and soluble protein) (Juliano 1972B). Increase in protein content, whether genetic or environmental, results mainly from an increase in glutelin content. Bran and embryo proteins have more albumin and globulin than milled rice protein (Baldi *et al.* 1976; Cag-ampang *et al.* 1966).

Amino acid composition of rice protein includes a lysine content of about 4 g/16.8 g N, one of the highest among cereal proteins. Lysine is the first limiting essential amino acid of rice protein as in other cereals. The amino acid score of milled rice protein ranges from 58 to 74% of the Food and Agriculture Organization of the United Nations (FAO) pattern based on lysine at 5.5 g/16 g N as 100% (FAO 1973). Amino acid score amply estimates protein quality of rice, indication that protein content mainly determines the utilizable protein of milled rice (Bressani *et al.* 1971; Eggum and Juliano 1973,1975; Hegsted and Juliano 1974; Murata *et al.* 1978; Roxas *et al.* 1975, 1976; Nishizawa *et al.* 1977). Other *Oryza* species have an amino acid composition similar to rice (Ignacio and Juliano 1968). In a survey of more than 10,000 entries of the world rice collection, lysine content did not vary more than 0.5% from the mean at any protein level (Juliano *et al.* 1973). Lysine content of protein dropped slightly with increasing protein content up to a protein level of about 10% in both brown and milled rices, but leveled off above 10% protein (Fig. 10.5).

The *waxy* gene has no effect on amino acid composition of rice caryopsis (Vidal and Juliano 1967).

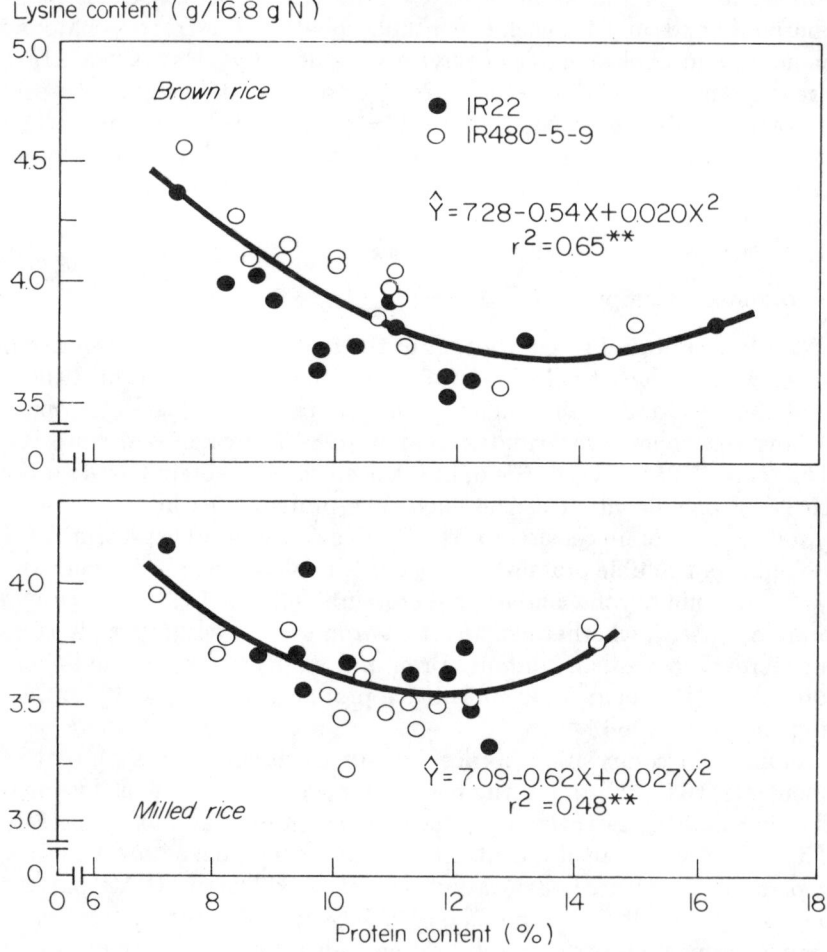

Lysine content (g/16.8 g N)

$$\hat{Y} = 7.28 - 0.54X + 0.020X^2$$
$$r^2 = 0.65^{**}$$

$$\hat{Y} = 7.09 - 0.62X + 0.027X^2$$
$$r^2 = 0.48^{**}$$

Protein content (%)

From Juliano et al. (1973)

FIG. 10.5. RELATIONSHIP BETWEEN PROTEIN CONTENT AND LYSINE CONTENT OF PROTEIN IN IR26 AND IR480-5-9 BROWN AND MILLED RICE

Samples above 13% protein have low grain weight.

Protein Bodies.—Protein bodies, or aleurins, in the aleurone layer have been characterized by Dr. Kasai's group at Kyoto University (Tanaka *et al.* 1973). The aleurins are 1–3 μm in size and contain in addition to 11.7% protein, 7.9% carbohydrate and globoids or phytate bodies (9.4% myo-inositol and 11.3% acid-soluble phosphorus). They differ distinctly from the protein bodies of the endosperm in the absence of strata structures and the presence of globoids. About 70% of its protein is water soluble (albumin).

Endosperm protein bodies have been characterized as containing from 10 to 28% of lipids and from 12 to 29% carbohydrate (Mitsuda *et al.* 1967). However, all preparations have been from the outer layer of milled rice, the fraction of the endosperm that has the highest lipid content (Barber 1972). Increase in protein content results mainly from more protein bodies in the endosperm (Harris and Juliano 1977; del Rosario *et al.* 1968).

Milled rice protein has a lower lysine content than brown rice protein because the protein in the pericarp, embryo and aleurone layers are richer in lysine due to their higher content of albumin (Baldi *et al.* 1976; Cagampang *et al.* 1976) (Table 10.3).

TABLE 10.3. PROPERTIES AND AMINO ACID COMPOSITION OF IR8 BROWN RICE AND ITS FRACTIONS (G / 16.8 G N)

Property	Brown rice	Milled rice	Rice pericarp	Rice embryo	Rice aleurone layer
Alanine	5.8	5.8	6.7	6.6	6.2
Arginine	9.0	8.5	7.8	9.6	8.7
Aspartic acid	9.0	9.0	11.2	9.1	9.4
Cystine	2.4	2.6	2.9	2.0	2.7
Glutamic acid	16.8	18.3	12.4	15.1	13.9
Glycine	4.8	4.7	6.1	6.0	5.5
Histidine	2.6	2.7	2.9	3.8	3.1
Isoleucine	4.5	4.7	4.5	3.8	4.3
Leucine	8.3	8.5	8.2	6.8	7.8
Lysine	4.4	4.0	5.7	6.8	5.1
Methionine	2.6	2.6	1.7	1.9	1.8
Phenylalanine	5.3	5.5	5.3	4.2	5.0
Proline	4.9	5.3	5.8	4.5	4.9
Serine	5.8	5.9	5.8	5.3	5.7
Threonine	3.9	3.9	4.6	4.5	4.0
Tryptophan	1.2	1.3	1.0	1.4	1.3
Tyrosine	4.0	3.8	3.1	3.3	3.6
Valine	6.6	6.8	6.9	6.3	6.3
Ammonia	2.8	3.3	2.3	3.4	2.2
Protein (% N × 5.95)	7.8	7.2	16.0	25.3	15.8
Wt distribution	100	92.1	2.2	2.3	3.4
Protein distribution	100	82.0	4.3	7.2	6.5

Source: Cagampang *et al.* 1976.

Albumin.—Very little study has been made on the albumin fraction. Albumin may be prepared by direct water extraction of milled rice flour, dialysis against water, centrifugation and precipitation with ammonium sulfate to 80% saturation. This method is preferred to 5% sodium chloride extraction, dialysis against water, centrifuging the globulin precipitate and using the supernate as albumin. It has the highest lysine content among the rice protein fractions (Table 10.4). Unlike wheat flour albumin, no α-amylase inhibitor is detected in milled rice albumin (Silano 1976; IRRI 1977B).

TABLE 10.4. PURITY AND AMINO ACID COMPOSITION (G/16 OR 16.8 G N) OF PROTEIN FRACTIONS OF MILLED RICE (IR480-5-9)

Property	Whole (g/16.8 g N)	Crude Albumin (g/16 g N)	Globulin Sol. at pH 4.5 (g/16 g N)	Globulin Insol. at pH 4.5 (g/16 g N)	Purified Prolamin (g/16.8 g N)	Purified Glutelin (g/16.8 g N)
Protein (% N × 5.95 or 6.25)	10.4	29.2	72.2	87.1	99.6	100
Alanine	5.8	7.8	7.7	4.2	7.1	5.4
Arginine	8.5	9.5	12.8	15.2	6.1	9.5
Aspartic acid	10.2	9.7	9.5	4.9	8.2	10.7
Cystine	1.8	1.9	3.1	3.0	0.2	1.8
Glutamic acid	20.8	12.1	12.5	27.8	29.0	20.8
Glycine	4.6	6.6	7.9	5.1	3.1	4.3
Histidine	2.5	3.2	2.7	0.4	1.1	2.6
Isoleucine	4.2	3.3	2.4	1.8	4.9	4.7
Leucine	6.8	7.6	7.3	6.6	13.7	8.1
Lysine	3.5	4.9	1.9	1.1	0.1	4.5
Methionine	1.6	1.8	3.0	4.0	0.2	1.6
Phenylalanine	5.7	3.2	3.0	3.4	6.6	5.4
Proline	4.6	6.8	7.0	5.0	5.6	6.2
Serine	6.1	4.7	3.8	8.8	5.4	5.9
Threonine	3.8	4.0	2.8	2.2	2.3	4.1
Tryptophan	1.2	1.4	2.0	0.6	0.5	1.2
Tyrosine	4.9	4.9	6.1	6.7	10.2	5.2
Valine	6.1	7.4	5.8	4.0	7.0	6.9
Ammonia	2.5	1.5	1.1	2.3	3.7	2.2

Source: Perdon and Juliano 1978; Mandac and Juliano 1978; Villareal and Juliano 1978.

Electrophoretically, albumin of brown rice has a very heterogeneous pattern composed of 4 major bands of intermediate mobility and 6 minor ones (Cagampang et al. 1976). Sodium dodecyl sulfate (SDS)-polyacrylamide gel electrophoresis showed 12 bands ranging in apparent mol wt from 8500 to 95,000. The 3 major subunits have mol wt 8500, 11,000 and 16,000. Most of the albumin of milled rice is eluted in a sodium chloride gradient on diethylaminoethyl cellulose chromatography into 3 broad peaks which are still heterogeneous (IRRI 1977B). Albumin preparations readily change in solubility during storage, even as ammonium sulfate precipitate at −20°C (−4°F). Incubation of albumin overnight at 40°C (104°F) results in the disappearance on SDS-polyacrylamide electrophoresis of 1 major and 1 minor band, from 13 to 11 bands.

Globulin.—Comparatively more is known about globulin than about albumin of rice grain, particularly the γ-globulins present mainly in the bran layers which have been studied in detail by Dr. Morita's group at Kyoto University (Horikoshi and Morita 1975). In milled rice the lysine content of globulin is similar to that of whole protein but globulin is richer in sulfur amino acids (Perdon and Juliano 1978) (Table 10.4). Globulin may be precipitated from 5% sodium chloride extract either by addition of ammonium sulfate to 30% saturation, or by dialysis against distilled water. The former preparation is relatively free of nucleic acid. The major fraction obtained by isoelectric precipitation at pH 4.5 is characterized as an α-globulin, with sedimentation coefficient $S^\circ_{20,w}$ of 1.6. It has subunit mol wt 18,000 based on SDS-polyacrylamide disc gel electrophoresis and mol wt 20,000 based on Sephadex gel filtration. It elutes as a single peak on diethylaminoethyl cellulose chromatography at 0.22 M sodium chloride. The α-globulin dissociates into 2 components between pH 6 and 9 as indicated by polyacrylamide gel electrophoresis, Sephadex gel filtration and ultracentrifugation. The higher mol wt component was estimated to be 98,000 by gel filtration indicating a pentamer. The globulin fraction soluble at pH 4.5 is also rich in sulfur amino acids (Table 10.4).

Prolamin.—Prolamin, the lowest lysine-content protein, can also be extracted directly from milled rice with 70% ethanol (Mandac and Juliano 1978). When extracted directly and precipitated by stripping the ethanol, the crude prolamin (50% protein) is contaminated with lipid (30%) and phenols (6%) (Tecson *et al.* 1971; Mandac and Juliano 1978). Preliminary 95% ethanol extraction removes phenolic contamination, and acetone precipitation of prolamin from the subsequent 70% ethanol extract leaves the lipid and phenol in solution. Prolamin purified by repeated precipitation from 70% ethanol by acetone addition has almost no lysine and sulfur amino acids (Table 10.4), and gives only a single peak eluted at 0.42 M sodium chloride in diethylaminoethyl cellulose chromatography.

Prolamin shows 1 major band and 1 minor band on analytical and SDS polyacrylamide disc gel electrophoresis. Mol wt are 17,000 for the major subunit and 23,000 for the minor subunit. Bran polish prolamin also shows 1 major and 1 minor protein band on electrophoresis, mol wt 17,000 and 23,000 (Mandac and Juliano 1978).

Glutelin.—Glutelin, the major protein fraction of milled rice has not been amenable to characterization due to its extreme insolubility in neutral solvents. By use of 0.5% SDS-0.6% β-mercaptoethanol, glutelin is readily dissolved and then alkylated with acrylonitrile. SDS-polyacrylamide disc gel electrophoresis showed the major subunits of glutelin have mol wt 38,000, 25,000 and 16,000 (Juliano and Boulter 1976). The ratio of the subunits is similar in 12 milled rice samples tested at 1:1:1, but 3 samples of *O. nivara* tested gave a ratio of 1:1:2 (Villareal and Juliano 1978). Of the 3 subunits, mol wt 38,000 subunit had the highest lysine content and the mol wt 16,000 had the lowest. However, the subunits with mol wt > 38,000 had higher lysine content than the major subunits. Several hours exposure of rice flour to 70% ethanol reduces extraction efficiency of glutelin with SDS-β-mercaptoethanol (Juliano and Boulter 1976).

Glutelin of aleurone layer is close in subunit composition to milled rice glutelin (Villareal and Juliano 1978). Those from pericarp and embryo showed subunit composition different from milled rice glutelin. Of the 4 solubility fractions of endosperm glutelin, the major (74%) fraction extracted by 0.5 M sodium chloride-0.6% β-mercaptoethanol-0.5% SDS in 0.1 M sodium borate buffer pH 10 was closest in properties to glutelin. Prolamin- and globulin-like fractions are first extracted with the usual solvents in the presence of 0.6% β-mercaptoethanol.

Free Amino Acids and Nucleic Acids.—The major free amino acids in developing rice grain include alanine, aspartic acid-asparagine, valine, glutamic acid, histidine and ornithine whereas the principal amino acids of rice protein are alanine, arginine, aspartic acid, glutamic acid, leucine and valine (Cagampang *et al.* 1971). Free amino acids constitute 0.7% of brown rice protein, 0.2% of milled rice protein, 1.35% of bran polish protein and 4.6% in embryo protein, equivalent to a distribution of 53% in embryo, 30% in bran polish and 17% in milled rice (Tamura and Kenmochi 1963).

The nucleic acid of brown rice is mainly ribonucleic acid: 0.1−0.3% of brown rice, 0.9% in embryo and 0.01% in milled rice (Juliano 1972B). The deoxyribonucleic acid content is about 0.01% in brown rice. Various nucleotides have also been identified in the developing rice grain.

Lipids

The lipids of all brown rice fractions have similar physical and chemical properties and are presented in Chapter 23. Lipids of brown rice have also been reviewed recently (Juliano 1977A). Recently Hartman and Lago (1976) reported that lipids on the surface of the rice caryopsis had a composition of fatty acids and unsaponifiable matter closer to that of rice hull than to bran and brown rice lipids. It is present in spherosomes (lipid droplets) about 0.1−1 μm in size in the embryo and the aleurone layer (Bechtel and Pomeranz 1977,1978A). In the endosperm, the lipids are associated with protein bodies and starch granules probably in the membrane fraction as lipoprotein (Mitsuda *et al.* 1967; Hirayama and Matsuda 1973). In addition to membrane lipids, the starch granules contain bound lipids (fat-by-hydrolysis) mainly phospholipids particularly lysolecithin (Nakamura *et al.* 1958A; Hirayama and Matsuda 1973).

Vitamins and Minerals

Vitamins.—Vitamins are generally present in higher levels in brown rice than in milled rice (Juliano 1972B) (Table 10.5). The contributing factor is the concentration of vitamins in the embryo and aleurone layer. Thus, milling results in the loss of more than 50% of the vitamin B complexes of brown rice to human consumption. The level of thiamin in high protein brown rice, IR480-5-9, was similar to that of the low protein rice, IR8, but a greater proportion of it is retained in milled rice (IRRI 1977B). Protein distribution showed the same trend (Fig. 10.2). Rice contains little or no vitamin C and vitamin D.

Parboiled milled rice tends to have higher vitamin content than milled rice despite the partial thermal decomposition of vitamin during parboiling (Padua and Juliano 1974). Parboiling results in inward diffusion of water-soluble vitamins to the endosperm. By contrast, no redistribution of aleurone protein and oil is observed.

The values for inositol and choline are total values, since inositol is mainly in the form of phytin and choline is mainly in the phospholipids, lecithin and lysolecithin.

Minerals.—The mineral composition of rice grain depends considerably on nutrient availability of the soil in which the crop is grown and on the different analytical methods used by various investigators. The ash distribution in brown rice is calculated to be 51% in bran, 10% in embryo, 11% in polish, and 28% in milled rice (Leonzio 1967). Iron, phosphorus and potassium show a similar distribution as total ash. However, some minerals, such as sodium and calcium show a relatively more even distribution in the grain (Table 10.6). Phosphorus and potassium are the major mineral elements of brown rice followed by silicon and magnesium.

Distribution of mineral elements in the bran layers follows closely the phytin distribution in the aleurone layer and scutellum as shown by electron microprobe X-ray analysis (Tanaka *et al.* 1974, 1976A).

Changes During Grain Development

During grain development the starch granules increase in amylose content in nonwaxy rice, but a slight decrease in amylose content is noted in waxy rice (Singh and Juliano 1977). BEPT and X-ray diffraction pattern of starch granules remain unchanged during grain development (Briones *et al.* 1968). Amylopectin shows a slight increase in intrinsic viscosity during grain development, but no change in gel consistency is noted.

TABLE 10.5. VITAMIN CONTENT OF BROWN RICE AND ITS FRACTIONS (μG /G)

Vitamin	Brown rice	Milled rice	Rice bran	Rice embryo	Rice polish
Vitamin A	0.1	trace	4.2	1.3	0.95
Thiamin	2.1 – 4.5	trace – 1.8	10 – 28	45 – 76	16 – 30
Riboflavin	0.4 – 0.9	0.1 – 0.4	1.7 – 3.4	2.7 – 5.0	1.4 – 3.4
Niacin (nicotinic acid)	44 – 62	8 – 26	241 – 590	15 – 99	228 –385
Pyridoxine	1.6 – 11.2	0.4 – 6.2	10 – 32	15 – 16	10 – 31
Pantothenic acid	6.6 – 18.6	3.4 – 7.7	28 – 71	3 – 13	26 – 92
Biotin	0.06– 0.13	0.005– 0.07	0.16– 0.47	0.26– 0.58	0.14– 0.66
Inositol Total	1194 –1220	100 –125	4600 –9270	3725 –6400	4280;5436
Free	24 – 45	11	197	342	
Choline	1120;1220	450–713	1279;1700	2030;3000	1020;1130
p-Amino benzoic acid	0.30	0.14;0.16	0.75	1	0.73
Folic acid	0.20;0.60	0.06 – 0.16	0.50– 1.46	0.9 – 4.3	0.4 – 1.9
Vitamin B$_{12}$	0.0005	0.0016	0.005	0.0105	0.003
Vitamin E (tocopherols)	13	trace	149	87	63

Source: Juliano 1972B; Hayakawa 1977; Kennedy *et al.* 1975.

TABLE 10.6. INORGANIC CONSTITUENTS OF BROWN RICE AND ITS FRACTION (µG/G)

	Brown rice	Milled rice	Rice bran	Rice embryo	Rice polish
Aluminum	—	0.73— 7.23	54— 369	—	—
Calcium	65— 400	46 — 385	250— 1310	510— 2750	90— 910
Chlorine	203— 275	163— 372	510— 970	1520	—
Iron	7— 54	2 — 27	130— 530	110— 490	100— 280
Magnesium	380—1400	170 — 700	8,600—12,300	6,000—15,300	5,700— 7,600
Manganese	13— 42	10 — 33	110— 880	120;140	50;80
Phosphorus	2500—4400	860 —1920	14,800—28,700	17,100—27,300	15,300—25,100
Potassium	1200—3400	140 —1200	13,200—22,700	3,800—21,500	9,300—18,000
Silicon	190—1900	50 — 370	1,700— 7,600	460— 1,900	560— 2,400
Sodium	31— 176	22 — 85	180— 290	160— 240	65— 210
Zinc	15— 22	3 — 21	50— 160	100;300	40;60

Source: Juliano 1972B; Baldi et al. 1974; Fossati et al. 1976; Kennedy and Schelstraete 1975.

Intrinsic viscosity of amylose also increases progressively during grain ripening while maintaining high blue value (IRRI 1975).

Protein accumulation has been studied in detail in the developing rice grain (Palmiano *et al.* 1968; Perez *et al.* 1973; Cagampang *et al.* 1976). Albumin and globulin are the major proteins accumulated at early stages together with free amino acids. Appearance of protein bodies by 7 days after flowering coincides with the increase in glutelin and prolamin per rice endosperm. However, although the percentage of glutelin in brown rice increased (Villareal and Juliano 1978), the percentage of prolamin in the rice dropped during grain development (Mandac and Juliano 1978). Biochemical factors affecting protein accumulation in the developing rice grain have been studied (Cruz *et al.* 1970; Perez *et al.* 1973). High protein rice differs from low protein rice in more efficient nitrogen translocation from leaves to developing grain, not in nitrogen absorption (Perez *et al.* 1973; Cagampang *et al.* 1971).

Protein content of rice grain is influenced by various environmental factors such as time and rate of nitrogen fertilizer application, solar radiation during grain development and spacing (Juliano *et al.* 1972B) and application of herbicides at subherbicidal level (Esmama and Juliano 1976). Such variability in protein content in a variety makes breeding efforts to improve protein content of the rice grain from 7 to 9% very difficult (Beachell *et al.* 1972; Juliano and Beachell 1975; Coffman and Juliano 1976).

The metabolic and structural relationships of lipids of rice were recently reviewed (Fujino *et al.* 1976).

Changes Due to Storage and Parboiling

Storage of the grain for as long as 3 months results in aging and the loss of dormancy. The changes include increase in volume expansion and water absorption during cooking, reduction in extracted solids during cooking, more flaky (less sticky) cooked rice, and more resistance to breakage during milling (Barber 1972; Juliano 1970; Villareal *et al.* 1976). Tensile strength of grain increases during storage (Kunze and Choudhury 1972).

The exact mechanism of aging is not fully understood but molecular oxygen is probably not required. In addition, waxy rice and petroleum ether surface-defatted nonwaxy milled rice do not increase in hardness index during storage (Villareal *et al.* 1976). The increase in water absorption and decrease in extractable solids during cooking were similar regardless of the form in which the rice was stored (rough, milled and defatted milled) or of the amylose content of rice (1–26%). The stickiness of nonwaxy rice decreased during storage but that of waxy rice did not. The amylograph peak viscosity of all samples including starch increased. Corresponding increases in gel viscosity were noted for the milled rice but not for starch. Salt-soluble protein of all samples decreased during storage both at 2°C (35.6°F) and 29°C (84.2°F). The highest levels of free fatty acids and carbonyl compounds were found in the stored milled rice and the lowest occurred in the defatted milled rice. Protein and starch changes contributed to the aging of rice to a marked extent. Lysine availability remains high during storage (IRRI 1976).

Parboiling is a modified form of aging because starch granules are gelatinized without much volume expansion. Changes are mainly physical resulting from the retrogradation of gelatinized starch (Raghavendra Rao and Juliano 1970; Watson and Dikeman 1977). The treatment hardens the endosperm making milled parboiled rice slower to cook and more resistant to disintegration than raw milled rice (Raghavendra Rao and Juliano 1970). The grain translucency and hardness improve. Cooked parboiled rice is shorter but thicker girthwise than cooked raw rice. Only high amylose rices show considerable drop in amylograph peak viscosity on parboiling in contrast to waxy and low amylose rices.

Parboiling destroys the characteristic A-type X-ray diffraction pattern of rice starch (Raghavendra Rao and Juliano 1970), and leaves only the weak V-pattern of amylose helix complex similar to cooked rice (Priestley 1974,1976). Priestley (1976) contends that this amylose helix complex is the major factor instead of retrogradation, which makes parboiled rice more water-insoluble than raw rice, and that this V-pattern is determined by both amylose content and oil content of rice (Priestley 1974). Using a waxy rice and a low and high amylose nonwaxy rice, no V-pattern was found in cooked waxy rice, but the V-pattern of cooked nonwaxy rice was higher in Khao Dawk Mali (16% amylose, 0.7% oil) than in IR28 (28% amylose, 0.9% oil) (IRRI 1977B). It is also known that bound lipids of starch do not increase on storage (Yasumatsu and Moritaka 1964). Thus, retrogradation of amylose without any appreciable V-pattern must be an important mechanism of parboiling. Degree of parboiling and percentage of parboiled grains are measured by a modified alkali test in 1% potassium hydroxide for 1 hr, in which only parboiled grains disintegrate (Ali and Bhattacharya 1972).

Parboiling and cooking reduce the digestibility of milled rice protein in rats and probably in man, but result in an increase in the quality of digested protein (Eggum et al. 1977). Hence, net protein utilization remains high, because the proteins rendered indigestible are the poorer quality ones. Indigestible protein of cooked rice in rat and man may be recovered as fecal protein particles representing the central portion of rice protein bodies (Tanaka et al. 1975; IRRI 1977B). This residual protein has less than 2% lysine and is essentially of mol wt <35,000.

Similar drops in lysine content are obtained by incubating raw milled rice in α-amylase- or glucoamylase-acid protease mixtures (IRRI 1977B).

The flavor of cooked rice was shown by Obata and Tanaka (1965) to be derived from the photolysis of L-cysteine and L-cystine to produce hydrogen sulfide, ammonia, carbon dioxide and acetaldehyde. The vapor of cooked rice contains hydrogen sulfide, ammonia and carbon dioxide. The stale flavor of stored milled rice is due primarily from the auto-oxidation of unsaturated fatty acids of rice oil to form carbonyl compounds such as C_3, C_5 and C_6 aldehydes (Chikubu 1970).

BIOCHEMICAL PROPERTIES

Developing and Mature Grain

Rice enzymes have been reviewed previously by Akazawa (1972). Of the enzymes of starch metabolism, phosphorylase, branching or Q-enzyme, and debranching or R-enzyme, have peak activities 10 days after flowering in developing IR8 rice grains (Baun et al. 1970). The α- and β-amylases have maximum activity 14 days after flowering, whereas starch synthetase (ADP glucose-α-glucosyl transferase) bound to starch granule increased in activity up to 21 days after flowering. In lines differing in amylose content, the samples differed only in the level of starch-granule-bound synthetase, which was proportional to amylose content. Granule-bound starch synthetase may be solubilized from rice starch granules by sonication of amorphous granules in 75% dimethyl sulfoxide (Perdon et al. 1975). The activities of the enzymes related to conversion of sucrose to ADP-glucose were high throughout starch deposition in developing rice grain and were highest 8 to 18 days after flowering (Perez et al. 1975). Soluble primed phosphorylase and starch synthetase were both present during starch accumulation.

The content of 3-phosphoglycerate, an effector of ADP-glucose pyrophosphorylase, and chlorophyll was highest in the grain 7 to 8 days after flowering when starch synthesis was at a maximum (Villareal and Juliano 1977). The enzyme 3-phosphoglycerate phosphatase was found to be soluble and concentrated in the pericarp aleurone layer and had maximum activity 12 to 14 days after flowering.

Myo-inositol 1-phosphate synthetase has also been characterized in mature and developing rice grain (Hayakawa et al. 1977) and is located with the protein bodies of aleurone layer (Tanaka et al. 1976B).

In a study of protein metabolism, the ability of the developing grain to incorporate ^{14}C-U-leucine into protein was maximum about 8 days after flowering and was higher in the high protein rice (Cruz et al. 1970). Protease and ribonuclease showed a similar trend as did ribonucleic acid and soluble amino nitrogen. Higher protein grain had more rough endo-plasmic reticulum in the endosperm cells than the lower protein grain (Harris and Juliano 1977).

In the mature grain, enzyme activity is relatively low as compared to developing and germinating grain (Palmiano and Juliano 1972B). Shinke *et al.* (1973) proved that the mature grain has no α-amylase, and only free and zymogen β-amylases. A limit dextranase, which hydrolyzes α-1, 6-D-glucosidic linkage, has been characterized from ungerminated rice (Dunn and Manners 1975). α-Glucosidases, which convert starch to glu-cose, have also been characterized (Takahashi and Shimomura 1973). Lipase activity is also low (Palmiano and Juliano 1973) and the lipase activity in rice bran is probably microbial in origin (Juliano 1977A). The lipolytic acyl-hydrolases in bran and milled rice have recently been purified and characterized (Hirayama and Matsuda 1975; Matsuda and Hirayama 1975). Lipase, phospholipase and galactolipase were detected in both fractions but lipase activity predominates in bran and phos-pholipase in milled rice.

Dormancy and Germination Changes

Dormancy in rice is not embryo dormancy as the embryo can already germinate even in the developing grain by 7 days after flowering (Nava-sero *et al.* 1975). Peroxidase activity is higher in dormant than in nondormant grains. Oxygen uptake is also higher in dormant grains, indicating that nongermination of dormant grain is due to depletion of oxygen supply to the embryo due to high peroxidase activity of aleurone layer. Pricking the aleurone layer over the embryo results in complete breaking of dormancy. Various treatments such as dry heat, acid, and hydrogen peroxide break dormancy. Recent results indicate that dor-mant seeds absorb water slower on soaking for 1—3 days than non-dormant seeds (Martires 1975).

Rice grains lose dormancy on storage for a few weeks (Navasero *et al.* 1975). Maintenance of viability of the germ requires storage below 14% moisture or 75% relative humidity. Storage of seed stock at low humidity and temperature insures prolonged viability.

During germination, the root protrudes on the second day, followed by the coleoptile on the third day, and the primary leaf on the fourth day (Palmiano and Juliano 1972B). Starch and protein reserves of the endo-sperm and protein and lipid of the aleurone are broken down and utilized by the growing embryo. Gibberellic acid from the embryo is translocated into the aleurone layer, which activates the synthesis of hydrolyzing enzymes. Oxygen uptake and adenosine triphosphate level increase dur-ing the first day of germination and phytase shows earliest increase among the enzymes (Palmiano and Juliano 1973). Phytase activity was shown to be associated with the globoids or phytin-containing inclusions of aleurone protein bodies (Yoshida *et al.* 1975).

Most of the other enzymes follow closely the increase in soluble protein by the fourth day of germination—phosphatases, protease, esterase, lipase, peroxidase and catalase, β-glucosidase, and α- and β-galactosidase. The third class of enzymes still increases in level after 7 days of germination, represented by ribonuclease, α-amylase, R-enzyme and β-1, 3-glucanase. The earlier synthesis of lipase to α-amylase indicates that aleurone lipids are the initial source of energy of germinating rice grain instead of endosperm starch. Low glycolate (lactate) oxidase activity was found in roots (Marwaha and Juliano 1976). The operation of glutamine synthetase and glutamate synthetase in primary nitrogen metabolism of rice seedling has been established (Miflin and Lea 1977).

The sequence of enzyme production of phytase, protease, and β-1,3-glucanase and α-amylase can be demonstrated in embryoless seed half in the presence of 0.12 μM gibberellin A$_3$, but the production of lipase is delayed (Palmiano and Juliano 1972B, 1973).

Changes in starch granules structure during germination have been studied (Kiribuchi and Nakamura 1974) and are similar to those of granules incubated *in vitro* in pancreatic α-amylase (Evers and Juliano 1976). Surface erosion is common resulting in a decrease in granule size, together with the presence of pits on the surface of the granule. Changes in scutellum structure during germination have been investigated (Wada and Maeda 1976). Hydrolysis of scutellum protein bodies commences with 6 hr of soaking.

Acknowledgement: W.G. Rockwood assisted in editing the manuscript.

REFERENCES

ACKER, L. and SCHMITZ, H.J. 1967. On the lipids of wheat starch. III. The remaining lipids of wheat starch and the lipids of other types of starches. Stärke *19,* 275-280. (German)

AKAZAWA, T. 1972. Enzymes of rice. *In* Rice Chemistry and Technology. D.F. Houston (Editor). Amer. Assoc. Cereal Chem., St. Paul, Minnesota.

ALI, S.Z. and BHATTACHARYA, K.R. 1972. An alkali reaction test for parboiled rice. Lebensmitt. Wissen. Technol. *5,* 216-218.

ANTONIO, A.A. and JULIANO, B.O. 1973. Amylose content and puffed volume of parboiled rice. J. Food Sci. *38,* 915-916.

ANTONIO, A.A. and JULIANO, B.O. 1974. Physicochemical properties of glutinous rices in relation to *pinipig* quality. Philip. Agric. *58,* 17-23.

ANTONIO, A.A., JULIANO, B.O. and DEL MUNDO, A.M. 1975. Physicochemical properties of glutinous rice in relation to "suman" quality. Philip. Agric. *58*, 351-355.

BALDI, G., FOSSATI, G. and FANTONE, G.C. 1976. By-products of rice. II. Protein fractions and amino acid composition. Riso *25*, 347-356, (Italian)

BALDI, G., MALAGONI, R., PELA, E. and RANGHINO, F. 1974. Chemical composition and quality of Italian rice varieties. Riso *23* (Spec. Is.), 3-20.

BANKS, W. and GREENWOOD, C.T. 1975. Starch and Its Fractions. Edinburgh Univ. Press, Edinburgh, Scotland.

BARBER, S. 1972. Milled rice and changes during aging. *In* Rice Chemistry and Technology. D.F. Houston (Editor). Amer. Assoc. Cereal Chem., St. Paul, Minnesota.

BARBER, S., NAVARRO, L. and TORTOSA, E. 1976. Histochemistry of rice embryo. II. Lipids, proteins, amino acids and minerals. Rev. Agroquim. Tecnol. Aliment. *16*, 516-529. (Spanish)

BAUN, L.C., PALMIANO, E.P., PEREZ, C.M. and JULIANO, B.O. 1970. Enzymes of starch metabolism in the developing rice grain. Plant Physiol. *46*, 429-434.

BEACHELL, H.M., KHUSH, G.S. and JULIANO, B.O. 1972. Breeding for high protein content in rice. *In* Rice Breeding. Int. Rice Res. Inst., Los Baños, Philippines.

BECHTEL, D.B. and POMERANZ, Y. 1977. Ultrastructure of the mature ungerminated rice *(Oryza sativa)* caryopsis. The caryopsis coat and aleurone cells. Amer. J. Bot. *64*, 966-973.

BECHTEL, D.B. and POMERANZ, Y. 1978A. Ultrastructure of the mature ungerminated rice *(Oryza sativa)* caryopsis. The germ. Amer. J. Bot. *65*, 75-85.

BECHTEL, D.B. and POMERANZ, Y. 1978B. Ultrastructure of the mature ungerminated rice *(Oryza sativa)* caryopsis. The starchy endosperm. Amer. J. Bot. *65*, 684-691.

BOLLING, H. and EL BAYA, A.W. 1975. Effects of lipids on the determination of amylose content in rice and wheat. Chem. Mikrobiol. Technol. Lebensm. *3*, 161-163. (German)

BRESSANI, R., ELIAS, L.G. and JULIANO, B.O. 1971. Evaluation of the protein quality of milled rices differing in protein content. J. Agric. Food Chem. *19*, 1028-1034.

BRIONES, V.P., MAGBANUA, L.G. and JULIANO, B.O. 1968. Changes in physicochemical properties of starch of developing rice grain. Cereal Chem. *45*, 351-357.

BUTTROSE, M.S. 1969. Personal communication. *In* 1969 Ann. Rep. Int. Rice Res. Inst., Los Baños, Philippines (1970), 31.

CAGAMPANG, G.B., CRUZ, L.J. and JULIANO, B.O. 1971. Free amino acids in the bleeding sap and developing grain of the rice plant. Cereal Chem. *48*, 533-539.

CAGAMPANG, G.B., PERDON, A.A. and JULIANO, B.O. 1976. Changes in salt-soluble proteins of rice during grain development. Phytochem. *15*, 1425-1430.

CAGAMPANG, G.B., PEREZ, C.M. and JULIANO, B.O. 1973. A gel consistency test for eating quality of rice. J. Sci. Food Agric. *24*, 1589-1594.

CAGAMPANG, G.B. *et al.* 1966. Studies on the extraction and composition of rice proteins. Cereal Chem. *43*, 145-155.

CARTAÑO, A.V. and JULIANO, B.O. 1970. Hemicelluloses of milled rice. J. Agric. Food Chem. *18*, 40-42.

CHIKUBU, S. 1970. Stale flavor of stored rice. Jpn. Agric. Res. Quart. *5*, 63-68.

CHIKUBU, S. 1975. Quality of rice in South-East Asia. *In* Rice in Asia. Tokyo Univ. Press, Tokyo.

CHO, J. 1956. Double fertilization in *Oryza sativa* L. and development of the endosperm with special reference to the aleurone layer. Bull Natl. Inst. Agric. Sci. *D6*, 61-101.

COFFMAN, W.R. and JULIANO, B.O. 1976. Current status of breeding high protein rice. *In* Improving the Nutrient Quality of Cereals. II. H.L. Wilcke (Editor). U.S. Agency Int. Dev., Washington, D.C.

CRUZ, L.J., CAGAMPANG, G.B. and JULIANO, B.O. 1970. Biochemical factors affecting protein accumulation in the rice grain. Plant Physiol. *46*, 743-747.

DEL ROSARIO, A.R., BRIONES, V.P., VIDAL, A.J. and JULIANO, B.O. 1968. Composition and endosperm structure of developing and mature rice kernel. Cereal Chem. *45*, 225-235.

DUNN, G. and MANNERS, D.J. 1975. The limit dextrinases from ungerminated oats *(Avena sativa* L.) and ungerminated rice *(Oryza sativa* L.). Carb. Res. *39*, 283-293.

EGGUM, B.O. and JULIANO, B.O. 1973. Nitrogen balance in rats fed rices differing in protein content. J. Sci. Food Agric. *24*, 921-927.

EGGUM, B.O. and JULIANO, B.O. 1975. Higher protein content from nitrogen fertiliser application and nutritive value of milled-rice protein. J. Sci. Food Agric. *26*, 425-427.

EGGUM, B.O., RESURRECCION, A.P. and JULIANO, B.O. 1977. Effect of cooking on nutritional value of milled rice in rats. Nutr. Rep. Int. *16*, 649-655.

ESMAMA, B.V. and JULIANO, B.O. 1976. The effect of subherbicidal application of Simetryne and Benzomarc on the nitrogen metabolism of rice seedlings. Kalikasan Philip. J. Biol. *5*, 315-324.

EVERS, A.D. and JULIANO, B.O. 1976. Varietal differences in surface ultrastructure of endosperm cells and starch granules of rice. Stärke *28*, 160-166.

FOOD AND AGRIC. ORGAN. U.N. (FAO). 1973. Energy and protein requirements. FAO Nutr. Meet. Rep. Ser. *52*, 63.

FOSSATI, G., BALDI, G. and RANGHINO, F. 1976. By-products of rice I.

Chemical composition and inorganic constituents. Riso *25*, 339-345. (Italian)

FRASER, J.R. and HOLMES, D.C. 1958. The proximate analysis of rice carbohydrates. J. Sci. Food Agric. *9*, 511-515.

FUJII, T. 1972. Studies on the purification and processing of starch. Purification of starch by surface active agents, and fatty substances in starch and its effects on physical properties of starch. J. Jpn. Soc. Starch Sci. *19*, 159-168. (Japanese)

FUJINO, Y. *et al.* 1976. Structural and metabolic relationships of lipids in rice. Ppr. presented at Jpn.–U.S. Sem. on Lipids in Higher Plants, Sept. 7–9. Tokyo.

GREMLI, H. and JULIANO, B.O. 1970. Studies on the alkali-soluble, rice-bran hemicelluloses. Carb. Res. *12*, 273-276.

HARRIS, N. and JULIANO, B.O. 1977. Ultrastructure of endosperm protein bodies in developing rice grains differing in protein content. Ann. Bot. *45*, 1-5.

HARTMAN, L. and LAGO, R.C.A. 1976. The composition of lipids from rice hulls and from the surface of rice caryopsis. J. Sci. Food Agric. *27*, 939-942.

HAYAKAWA, T. 1977. Variation of myo-inositol and myo-inositol phosphate contents and some starch properties in rice seed (Biochemical studies of myo-inositol on rice seed. VI.). Bull. Fac. Agric. Niigata Univ. *29*, 35-42. (Japanese)

HAYAKAWA, T., KOYAMA, S., KURASAWA, H. and IGAUE, I. 1977. Myo-inositol 1-phosphate synthase of rice seed in embryo, bran, endosperm, seedling and milky stages (Biochemical studies of myo-inositol on rice seed. V.). Bull. Fac. Agric. Niigata Univ. *29*, 27-34. (Japanese)

HEGSTED, D.M. and JULIANO, B.O. 1974. Difficulties in assessing the nutritional quality of rice protein. J. Nutr. *104*, 772-781.

HINTON, J.J.C. 1948. The distribution of vitamin B_1 in the rice grain. Brit. J. Nutr. *2*, 237-241.

HIRAYAMA, O. and MATSUDA, H. 1973. Lipid components and distribution in brown rice. J. Agr. Chem. Soc. Jpn. *47*, 371-377. (Japanese)

HIRAYAMA, O. and MATSUDA, H. 1975. Purification and characterization of lipolytic acyl-hydrolases from rice bran. J. Agric. Chem. Soc. Jpn. *49*, 569-576. (Japanese)

HORIKOSHI, M. and MORITA, Y. 1975. Localization of γ-globulin in rice seed and changes in γ-globulin content during seed development and germination. Agric. Biol. Chem. *39*, 2309-2314.

HOSHIKAWA, K. 1973. Morphogenesis of endosperm tissue in rice. Jpn Agric. Res. Quart. *7*, 153-159.

HOSHIKAWA, K. 1974. Histology of endosperm development and reserve substance accumulation in cereal grains. *In* Gamma-Field Symp. *13*, Inst. Radiat. Breed. Jpn. 1-15.

HOUSTON, D.F. and KOHLER, G.O. 1970. Nutritional Properties of Rice. Natl. Acad. Sci. Washington, D.C.

IGNACIO, C.C. and JULIANO, B.O. 1968. Physicochemical properties of brown rice from Oryza species and hybrids.   J. Agric. Food Chem. *16*, 125-127.

INT. RICE RES. INST. (IRRI). 1975. Annual report for 1974. Int. Rice Res. Inst., Los Baños, Philippines.

IRRI. 1976. Annual report for 1975. Int. Rice Res. Inst., Los Baños, Philippines.

IRRI. 1977A. Annual report for 1976. Int. Rice Res. Inst., Los Baños, Philippines.

IRRI. 1977B. Unpublished data. Chem. Dep., Int. Rice Res. Inst., Los Baños, Philippines.

JULIANO, B.O. 1966. Physicochemical data on the rice grain. Int. Rice Res. Inst. Tech. Bull. *6.*

JULIANO, B.O. 1967. Physicochemical studies of rice starch and protein. Int. Rice Comm. Newsl. (Spec. Is.), 93-105.

JULIANO, B.O. 1970. Relation of physicochemical properties to processing characteristics of rice. *In* Proc. 5th World Cereal Bread Congr. Dresden. Vol. 4.

JULIANO, B.O. 1971. A simplified assay for milled-rice amylose. Cereal Sci. Today *16*, 334-338, 340, 360.

JULIANO, B.O. 1972A. Physicochemical properties of starch and protein and their relation to grain quality and nutritional value of rice. *In* Rice Breeding. Int. Rice Res. Inst., Los Baños, Philippines.

JULIANO, B.O. 1972B. The rice caryopsis and its composition. *In* Rice Chemistry and Technology. D. F. Houston (Editor). Amer. Assoc. Cereal Chem., St. Paul, Minnesota.

JULIANO, B.O. 1972C. Studies on protein quality and quantity of rice. *In* Seed Proteins. G.E. Inglett (Editor). AVI Publishing Co., Westport, Conn.

JULIANO, B.O. 1973A. Quality of milled rice. Riso *22*, 171-184.

JULIANO, B.O. 1973B. Recent developments in rice grain research. *In* Reports 7th Working Discussion Meet., Int. Assoc. Cereal Chem., 1972. Vienna.

JULIANO, B.O. 1976. Biochemical studies. *In* Rice Postharvest Technology. E.V. Araullo, D.B. de Padua and M. Graham (Editors). Int. Dev. Res. Cntr., Ottawa, Canada.

JULIANO, B.O. 1977A. Rice lipids. Riso *26*, 3-21.

JULIANO, B.O. 1977B. Recent developments in rice research. Cereal Foods World *22*, 284-287.

JULIANO, B.O. 1978. Metabolic evaluation of rice protein. Food Chem. *3*, 251-264

JULIANO, B.O., ANTONIO, A.A. and ESMAMA, B.V. 1973. Effects of protein content on the distribution and properties of rice protein. J. Sci. Food Agric. *24*, 295-306.

JULIANO, B.O., BAUTISTA, G.M., LUGAY, J.C. and REYES, A.C. 1964. Studies on the physicochemical properties of rice. J. Agric. Food Chem. *12*, 131-138.

JULIANO, B.O. and BEACHELL, H.M. 1975. Status of rice protein improvement. *In* High-Quality Protein Maize. Int. Maize and Wheat Improvement Cntr./Purdue Univ. Dowden, Hutchinson and Ross, Stroudsburg, Pa.

JULIANO, B.O. and BOULTER, D. 1976. Extraction and composition of rice endosperm glutelin. Phytochem. *15*, 1601-1606.

JULIANO, B.O., NAZARENO, M.B. and RAMOS, N.B. 1969. Properties of waxy and isogenic nonwaxy rices differing in starch gelatinization temperature. J. Agric. Food Chem. *17*, 1364-1369.

JULIANO, B.O., OÑATE, L.U. and DEL MUNDO, A.M. 1965. Relation of starch composition, protein content, and gelatinization temperature to cooking and eating qualities of milled rice. Food Technol. *19*, 1006-1011.

JULIANO, B.O., OÑATE, L.U. and DEL MUNDO, A.M. 1972A. Note: Amylose and protein contents of milled rice as eating quality factors. Philipp. Agric. *56*, 44-47.

JULIANO, B.O. and PERDON, A.A. 1975. Gel and molecular properties of nonwaxy rice starch. Stärke *27*, 115-120.

JULIANO, B.O., PERDON, A.A., PEREZ, C.M. and CAGAMPANG, G.B. 1974. Molecular and gel properties of starch and texture of rice products. *In* Proc. 4th Int. Congr. Food Sci. Technol. Madrid, Vol. 1. E. Portela (Editor). Inst. Agroquim. Tecnol. Aliment., Valencia, Spain.

JULIANO, B.O., PEREZ, C.M. and GOMEZ, K.A. 1972B. Variability in protein content of rice. Kalikasan Philipp. J. Biol. *1*, 74-81.

KENNEDY, B.M. and SCHELSTRAETE, M. 1975. Chemical, physical and nutritional properties of high-protein flours and the residual kernel from the overmilling of uncoated milled rice. III. Iron, calcium, magnesium, phosphorus, sodium, potassium, and phytic acid. Cereal Chem. *52*, 173-182.

KENNEDY, B.M., SCHELSTRAETE, M. and TAMAI, K. 1975. Chemical, physical and nutritional properties of high-protein flours and residual kernel from the overmilling of uncoated milled rice. IV. Thiamin, riboflavin, niacin and pyridoxine. Cereal Chem. *52*, 182-188.

KIRIBUCHI, S. and NAKAMURA, M. 1974. Mechanism of decomposition of starch in germinating rice seeds. J. Jpn. Soc. Starch Sci. *21*, 299-306. (Japanese)

KONGSEREE, N. and JULIANO, B.O. 1972. Physicochemical properties of rice grain and starch from lines differing in amylose content and gelatinization temperature. J. Agric. Food Chem. *20*, 714-718.

KUNZE, O.R. and CHOUDHURY, M.S.U. 1972. Moisture adsorption related to the tensile strength of rice. Cereal Chem. *49*, 684-696.

KUNZE, O.R. and HALL, C.W. 1965. Relative humidity changes that cause brown rice to crack. Trans. Amer. Soc. Agric. Eng. (ASAE) *8*, 396-399, 405.

LEONZIO, M. 1967. Contents of pentosans in rice caryopsis and in principal by-products. Riso *16*, 313-320. (Italian)

LITTLE, R.R. and DAWSON, E.H. 1960. Histology and histochemistry of raw and cooked rice kernels. Food Res. *25*, 611-622.

LUGAY, J.C. and JULIANO, B.O. 1965. Crystallinity of rice starch and its fractions in relation to gelatinization and pasting characteristics. J. Appl. Polymer Sci. *9*, 3775-3790.

MANDAC, B.E. and JULIANO, B.O. 1978. Properties of prolamin of mature and developing rice grain. Phytochem. *17*, 611-614.

MANINGAT, C.C. and JULIANO, B.O. 1978. Alkali digestibility pattern, apparent solubility and gel consistency of milled rice. Stärke *30*, 125-127.

MARSHALL, J.J. and WHELAN, W.J. 1974. Multiple branching in glycogen and amylopectin. Arch. Biochem. Biophys. *161*, 234-238.

MARTIRES, G.A. 1975. A rapid method of determining percent germination of rice seeds *(Oryza sativa* L.) and the factors affecting the accuracy of the method. Ph.D. dissertation. University Sto. Tomas, Manila, Philippines.

MARWAHA, R.S. and JULIANO, B.O. 1976. Aspects of nitrogen metabolism in the rice seedling. Plant Physiol. *57*, 923-927.

MATSUDA, H. and HIRAYAMA, O. 1975. Purification and characterization of lipolytic acyl-hydrolases from rice endosperm. J. Agric. Chem. Soc. Jpn. *49*, 577-583. (Japanese)

MIFLIN, B.J. and LEA, P.J. 1977. Amino acid metabolism. Ann. Rev. Plant Physiol. *28*, 299-329.

MITSUDA, H. and MURAKAMI, K. 1969. Protein body isolated from rice endosperm. Physiol. Plants *8*, 1-5. (Japanese)

MITSUDA, H., MURAKAMI, K., KUSANO, T. and YASUMOTO, K. 1969. Fine structure of protein bodies isolated from rice endosperm. Arch. Biochem. Biophys. *130*, 678-680.

MITSUDA, H. *et al.* 1967. Studies on the proteinaceous subcellular particles in rice endosperm: electron-microscopy and isolation. Agric. Biol. Chem. *31*, 293-300.

MURATA, K., KITAGAWA, T. and JULIANO, B.O. 1978. Protein quality of a high protein rice in rats. Agric. Biol. Chem. *42*, 565-570.

NAGATO, K. and KONO, Y. 1963. Grain texture of rice. 1. Relations among hardness distribution, grain shape and structure of endosperm tissue of rice kernel. Proc. Crop Sci. Soc. Jpn. *32*, 181-189. (Japanese)

NAKAMURA, A., KÔNO, T. and FUNAHASHI, S. 1958A. Nature of lysolecithin in rice grains. I. Lysolecithin as a constituent of nonglutinous rice grains. Bull. Agric. Chem. Soc. Jpn. *22*, 320-324.

NAKAMURA, A., SHIMIZU, R., KOŃO, T. and FUNAHASHI, S. 1958B. Nature of lysolecithin in rice grains. II. Complex formation of lysolecithin with starch. Bull. Agric. Chem. Soc. Jpn. *22*, 324-329.

NAVASERO, E.P., BAUN, L.C. and JULIANO, B.O. 1975. Grain dormancy, peroxidase activity and oxygen uptake in *Oryza sativa*. Phytochem. *14*, 1899-1902.

NIKUNI, Z. *et al.* 1969. The effect of temperature during the maturation period on the physicochemical properties of potato and rice starches. Mem. Inst. Sci. Ind. Res. Osaka Univ. *26*, 1-27.

NISHIZAWA, N. *et al.* 1977. Protein quality of high protein rice obtained by spraying urea on leaves before harvest. Agric. Biol. Chem. *41*, 477-485.

OBATA, Y. and TANAKA, H. 1965. Studies on the photolysis of L-cysteine and L-cystine. Formation of the flavor of cooked rice from L-cysteine and L-cystine. Agric. Biol. Chem. *29*, 191-195.

OGAWA, M., TANAKA, K. and KASAI, Z. 1975. Isolation of high phytin containing particles from rice grains using an aqueous polymer two phase system. Agric. Biol. Chem. *39*, 695-700.

PADUA, A.B. and JULIANO, B.O. 1974. Effect of parboiling on thiamin, protein and fat of rice. J. Sci. Food Agric. *25*, 607-701.

PALMIANO, E.P., ALMAZAN, A.M. and JULIANO, B.O. 1968. Physicochemical properties of protein of developing and mature rice grain. Cereal Chem. *45*, 1-12.

PALMIANO, E.P. and JULIANO, B.O. 1972A. Physicochemical properties of Niigata waxy rices. Agric. Biol. Chem. *36*, 157-159.

PALMIANO, E.P. and JULIANO, B.O. 1972B. Biochemical changes in the rice grain during germination. Plant Physiol. *49*, 751-756.

PALMIANO, E.P. and JULIANO, B.O. 1973. Changes in the activities of some hydrolases, peroxidase and catalase in the rice grain during germination. Plant Physiol. *52*, 274-277.

PASCUAL, C.G., SINGH, R. and JULIANO, B.O. 1978. Free sugars of rice grain. Carb. Res. *62*, 381-385.

PERDON, A.A., DEL ROSARIO, E.J. and JULIANO, B.O. 1975. Solubilization of starch synthetase bound to *Oryza sativa* starch granules. Phytochem. *14*, 949-951.

PERDON, A.A. and JULIANO, B.O. 1975A. Gel and molecular properties of waxy rice starch. Stärke *27*, 69-71.

PERDON, A.A. and JULIANO, B.O. 1975B. Amylose content of rice and quality of fermented cake. Stärke *27*, 196-198.

PERDON, A.A. and JULIANO, B.O. 1978. Properties of a major α-globulin of rice endosperm. Phytochem. *17*, 351-353.

PEREZ, C.M. *et al.* 1973. Protein metabolism in leaves and developing grains of rices differing in grain protein content. Plant Physiol. *51*, 537-542.

PEREZ, C.M. *et al.* 1975. Enzymes of carbohydrate metabolism in the developing rice grain. Plant Physiol. *56*, 579-583.

PRIESTLEY, R.J. 1974. Physicochemical studies of rice with particular reference to the processing of rice. Ph.D. dissertation. University of Reading, Reading, England.

PRIESTLEY, R.J.. 1976. Studies on parboiled rice. I. Comparison of the characteristics of raw and parboiled rice. Food Chem. *1*, 5-14.

RAGHAVENDRA RAO, S.N. and JULIANO, B.O. 1970. Effect of parboiling on some physicochemical properties of rice. J. Agric. Food Chem. *18,* 289-294.

RESURRECCION, A.P., HARA, T., JULIANO, B.O. and YOSHIDA, S. 1977. Effect of temperature during ripening on grain quality of rice. Soil Sci. Plant Nutr. *23,* 109-112.

REYES, A.C., ALBANO, E.L., BRIONES, V.P. and JULIANO, B.O. 1965. Varietal differences in physicochemical properties of rice starch and its fractions. J. Agric. Food Chem. *13,* 438-442.

ROXAS, B.V., INTENGAN, C.LL. and JULIANO, B.O. 1975. Effect of protein content of milled rice on nitrogen retention of Filipino children fed a rice-fish diet. Nutr. Rep. Int. *11,* 393-398.

ROXAS, B.V., INTENGAN, C.LL. and JULIANO, B.O. 1976. Protein content of milled rice and nitrogen retention of preschool children fed rice-mung bean diets. Nutr. Rep. Int. *14,* 203-207.

SATO, K. and EHARA, Y. 1974. Studies on the starch contained in the tissues of rice plants. XIV. Electron microscopic observation of plastids of various organs and tissues. Proc. Crop Sci. Soc. Jpn. *43,* 111-122. (Japanese)

SCHOCH, T.J. 1967. Properties and uses of rice starch. *In* Starch: Chemistry and Technology, Vol. 2. R.L. Whistler and E.F. Paschall (Editors). Academic Press, New York.

SHINKE, R., NISHIRA, H. and MUGIBAYASHI, N. 1973. Types of amylase in rice grain. Agric. Biol. Chem. *37,* 2437-2438.

SILANO, V. 1976. Personal communication. Laboratori di Chimica Biologica, Instituto Superiore di Sanitá, Rome, Italy.

SINGH, R. and JULIANO, B.O. 1977. Free sugars in relation to starch accumulation in developing rice grain. Plant Physiol. *59,* 417-421.

SOWBHAGYA, C.M. and BHATTACHARYA, K.R. 1971. A simplified colorimetric method for determination of amylose content in rice. Stärke *23,* 53-56.

SUZUKI, H. and JULIANO, B.O. 1975. Alkaliviscogram and other properties of starch of tropical rice. Agric. Biol. Chem. *39,* 811-817.

TABATA, S., NAGATA, K. and HIZUKURI, S. 1975. Studies on starch phosphate. 3. On the esterified phosphate in some cereal starches. Stärke *27,* 333-335.

TAKAHASHI, N. and SHIMOMURA, T. 1973. Action of rice α-glucosidase on maltose and starch. Agric. Biol. Chem. *37,* 67-74.

TAMURA, S. and KENMOCHI, K. 1963. Studies on amino acid content of rice. III. Distribution of amino acid in rice grain. J. Agric. Chem. Soc. Jpn. *37,* 753-756. (Japanese)

TANAKA, K., OGAWA, M. and KASAI, Z. 1976A. The rice scutellum: studies by scanning electron microscopy and electron microprobe X-ray analysis. Cereal Chem. *53,* 643-649.

TANAKA, K., YOSHIDA, T., ASADA, K. and KASAI, Z. 1973. Subcellular particles isolated from aleurone layer of rice seeds. Arch. Biochem. Biophys. *155,* 136-143.

TANAKA, K., YOSHIDA, T. and KASAI, Z. 1974. Distribution of mineral elements in the outer layer of rice and wheat grains, using electron microprobe X-ray analysis. Soil Sci. Plant Nutr. *20,* 87-91.

TANAKA, K., YOSHIDA, T. and KASAI, Z. 1976B. Phosphorylation of myo-inositol by isolated aleurone particles of rice. Agric. Biol. Chem. *40,* 1319-1325.

TANAKA, Y., HAYASHIDA, S. and HONGO, M. 1975. The relationship of the feces protein particles to rice protein bodies. Agric. Biol. Chem. *39,* 515-518.

TASHIRO, T. and EBATA, M. 1975. Studies on white-belly rice kernel. IV. Opaque rice endosperm viewed with a scanning electron microscope. Proc. Crop Sci. Soc. Jpn. *44,* 205-214. (Japanese)

TECSON, E.M.S., ESMAMA, B.V., LONTOK, L.P. and JULIANO, B.O. 1971. Studies on the extraction and composition of rice endosperm glutelin and prolamin. Cereal Chem. *48,* 168-181.

UTSUNOMIYA, H., YAMAGATA, M. and DOI, Y. 1975A. Scanning electron microscopy of the endosperm of cereal crops. 3. Starch cell layer of white-core rice. Bull. Fac. Agric. Yamaguti Univ. *26,* 1-18. (Japanese)

UTSUNOMIYA, H., YAMAGATA, M. and DOI, Y. 1975B. Scanning electron microscopy of the endosperm of cereal crops. 4. Starch cell layer of imperfect grain of rice (non-glutinous) and glutinous rice. Bull. Fac. Agric. Yamaguti Univ. *26,* 19-44. (Japanese)

VIDAL, A.J. and JULIANO, B.O. 1967. Comparative composition of waxy and nonwaxy rice. Cereal Chem. *44,* 86-91.

VILLAREAL, R.M. and JULIANO, B.O. 1977. Some properties of 3-phos-phoglycerate phosphatase from developing rice grain. Plant Physiol. *59,* 134-138.

VILLAREAL, R.M. and JULIANO, B.O. 1978. Properties of glutelin of mature and developing rice grain. Phytochem. *17,* 177-182.

VILLAREAL, R.M., RESURRECCION, A.P., SUZUKI, L.B. and JULIANO, B.O. 1976. Changes in physicochemical properties of rice during storage. Stärke *28,* 88-94.

WADA, T. and MAEDA, E. 1976. Light microscopy of cell organelles in the scutellum of rice during germination. Proc. Crop Sci. Soc. Jpn. *45,* 582-590. (Japanese)

WANG, Y.L., YANG, K.C. and CHENG, S.M. 1950. Vitamin-B_1 in the rice grain and the distribution of nicotinic acid. Chem. Sci. *1* (1) 99-107. (Chinese)

WATABE, T. and OKAMOTO, H. 1960. Experiments on "Ryokka" phenomenon on glutinous rice grain. 3. Electronmicroscopic investigation on the surface structure of starch granules. Proc. Crop Sci. Soc. Jpn. *29,* 89-92. (Japanese)

WATSON, C.A. and DIKEMAN, E. 1977. Structure of the rice grain shown by scanning electron microscopy. Cereal Chem. *54,* 120-130.

WOLF, M.J. and KHOO, U. 1970. Mature cereal grain endosperm: rapid glass knife sectioning for examination of proteins. Stain Technol. *45*, 277-283.

YAMAGISHI, T., MATSUDA, K. and WATANABE, T. 1975. Studies on the proteoglycan from rice bran. I. Isolation and partial characterization of proteoglycans from rice bran. Carb. Res. *43*, 321-333.

YAMAGISHI, T., MATSUDA, K. and WATANABE, T. 1976. Characterization of the fragments obtained by enzymic and alkaline degradation of rice-bran proteoglycans. Carb. Res. *50*, 63-74.

YASUMATSU, K. and MORITAKA, S. 1964. Fatty acid composition of rice lipid and their changes during storage. Agric. Biol. Chem. *28*, 257-264.

YOSHIDA, T., TANAKA, K. and KASAI, Z. 1975. Phytase activity associated with isolated aleurone particles of rice grains. Agric. Biol. Chem. *39*, 289-290.

ZEE, S.Y. 1972. Vascular tissue and transfer cell distribution in the rice spikelet. Austr. J. Biol. Sci. *25*, 411-414.

11

Nutritional Quality of Rice Endosperm

Barbara M. Kennedy

The composition of the rice kernel has been studied extensively as has that of its major parts: endosperm, bran and germ (Juliano 1966). That these fractions differ considerably in certain constituents is well known. As early as 1900–1911 it was suggested that the consumption of rice from which the outer layers had been removed by milling was the cause of beriberi, a nutritional disease now known to be due to a deficiency of thiamin (Hinton and Shaw 1953).

Analyses of brown and white rices show that brown rice, as compared with unenriched white, has higher percentages of all constituents except for total carbohydrate (Table 11.1). Concentrations of thiamin and fat are about 5 times greater than in the white, while fiber, niacin, phosphorus, potassium, iron, sodium and riboflavin are about 2 to 3 times greater. In the milling of white rice, a variable amount of the outer layers of paddy rice is removed as bran and polish, from 5 to 9% of the weight of paddy milled as bran (commonly 8 to 9%) and an additional 2 to 3% as white bran or polish (Houston 1972). The amount of bran and aleurone layer removed affects the nutritive value of the milled rice. For a more detailed discussion of the composition of the rice caryopsis, see Chapter 10.

However, the distribution of nutrients within the endosperm, which constitutes from 80 to 90% of the edible kernel, has been given serious attention only since about 1960. Until then only a minimum of data were available.

Subrahmanyan *et al.* (1938) noted that both nitrogen and phosphorus progressively decreased as more of the polishings were removed. Hinton (1948) reported differences of vitamin B_1 determined by a microthiochrome method, in the outer and inner portions of endosperm, obtained by hand dissection from a sample of white rice. Concentration (in Inter-

TABLE 11.1. COMPOSITION OF BROWN AND WHITE RICE

	Brown rice	White rice[1]
Moisture (%)	12.0	12.0
Kcal (100 g)	360	363
Protein (%)	7.5	6.7
Fat (%)	1.9	0.4
Ash (%)	1.2	0.5
CHO (total, %)	77.4	80.4
Fiber (%)	0.9	0.3
Calcium, (mg/100 g)	32	24
Phosphorus (mg/100 g)	221	94
Iron (mg/100 g)	1.6	0.8
Sodium (mg/100 g)	9	5
Potassium (mg/100 g)	214	92
Thiamin (mg/100 g)	0.34	0.07
Riboflavin (mg/100 g)	0.05	0.03
Niacin (mg/100 g)	4.7	1.6

Source: Watt and Merrill (1963).
[1]Unenriched, fully milled.

national Units) in the outer 20% of the endosperm was 4.5 times as great as for the inner portion. Later Hinton and Shaw (1953) published data on the distribution of nicotinic acid (niacin) determined by microbiological assay, also on hand-dissected samples. They reported that the concentration in the outer 6% of the endosperm averaged 24 times as great as that in the inner portion so that about ⅔ of the nicotinic acid of the endosperm was found in the outer 6%.

Only after sufficient material of successive layers of the kernel was available has it been possible to make a thorough study of the distribution of nutrients within the endosperm.

DISTRIBUTION OF NUTRIENTS WITHIN THE ENDOSPERM

Techniques for Study or Separation of Fractions

Histological Studies.—Concentrations of nutrients in rice endosperm have been demonstrated visually using histological techniques for fat and especially for starch and protein (Little and Dawson 1960; Del Rosario *et al.* 1968; Barber 1969). Little and Dawson published photographs of stained sections of rice kernels. In particular they observed that within a section, protein was most concentrated in peripheral-lateral and peripheral-ventral cells where starch granules were fewest, was least concentrated in peripheral-dorsal cells, and was often rather sparse in ventral, dorsal and ventral cells (Fig. 11.1).

Hand Dissection.—Quantitative analyses of constituents of rice endosperm can be obtained from fractions separated by hand dissection as

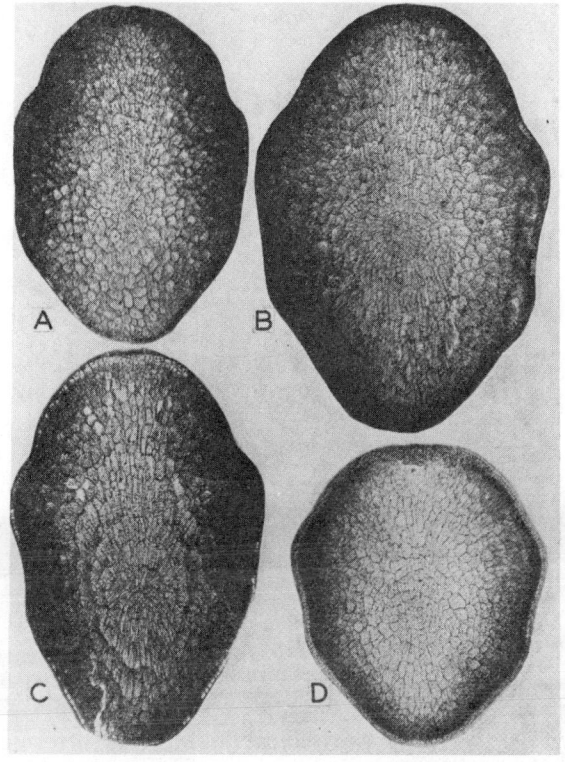

From Little and Dawson (1960)

FIG. 11.1. FREEHAND CROSS SECTIONS OF BROWN
RICE VARIETIES AFTER ENZYMATIC REMOVAL OF
STARCH AND STAINING WITH PROTEIN STAINS

(A) Fortuna—long; (B) Caloro—short; (C) Zenith—
medium; (D) Improved Bluebonnet—long. Magnification
× 262.5

was done by Hinton (1948). However the tedious and painstaking work required would be very time consuming—e.g., 500 kernels of rice weighing 20 mg each would be required to produce 600 mg of the outer 6% of the endosperm. Hence other methods have been used to separate the fractions.

Abrasive Milling.—The removal of successive peripheral layers of rice by abrasive milling, using experimental mills as well as existing commercial machinery, produces large amounts of material in a short time and has made possible the quantification of a number of constituents (Primo *et al.* 1963; Hogan *et al.* 1964; Houston *et al.* 1964). By adjusting the settings, from each mill pass 1 to 3% of the total weight of rice can be removed as abraded flour with the remainder as residual kernel. As many as 12 mill passes have been made on a single lot of rice, removing in all about 20% of the outer portions.

True anatomical separation of the bran is not made, however. In addition the flours are not always of uniform particle size. Pieces of the layers are sometimes chipped off producing larger particles of "chits" so that there is not a clean separation of the layers. Nevertheless, a fairly good separation of regions of the endosperm can be attained.

Analysis of Milled Flours

Following the earlier studies a number of reports were published in which the distribution of protein in abraded rice fractions was studied. In addition several papers included fat and ash while starch, sugars, fiber, B-vitamins, amylases, phosphorus, calcium and amino acid composition of the protein were each determined in 1 or 2 of the studies.

These reports indicated that the concentration of the constituents analyzed, except for starch, was greatest in the outer layers and decreased as the center of the kernel was approached while that for starch was the reverse. There was variation among varieties of rice and in the degree of gradation among the various constituents.

Since only 1 or 2 rices had been analyzed in any one study and only limited information obtained for many of the constituents, a comprehensive study of 12 lots of commercially milled rice, differing in variety, growing conditions and processing, was undertaken (Kennedy et al. 1974). Milling data, chemical determinations on about 20 constituents, acceptability, characterization for potential food use and effects of storage were studied, some of which will be referred to in the present discussion.

COMPOSITION OF FRACTIONS

Protein

Although among cereal proteins rice protein ranks high in nutritive quality, its content in the grain is low. Reported values for protein in brown rice range from 5 to 15.5% and for milled rice, 4.5 to 14.3% (Juliano 1966). Simpson et al. (1965) reported means of 8.3% for 156 samples of long grain rice, 7.8% for 147 samples of medium grain, and 7.3% for 129 samples of short grain with ranges of 6.4 to 10.0, 6.2 to 10.2 and 5.6 to 9.4%, respectively. The International Rice Research Institute (IRRI) currently has a program to develop strains of high protein rice and to improve the protein content.

Within the endosperm, the outer layers have been shown to have much higher concentrations of protein than that of the whole kernel. Quantitative data from reports of several investigators (Primo *et al.* 1963; Houston *et al.* 1964; Houston 1967; Hogan *et al.* 1964, 1968; Normand *et al.* 1966) show that, in agreement with histological findings, the protein concentration is highest in the peripheral layers of milled rice and decreases as the center of the kernel is approached. These data and that of other constituents including carbohydrates, enzymes, vitamins and minerals have been summarized graphically by Barber (1972).

In 1964 the Research Institute of Brewing, Japan, issued a report on the protein concentration in residual kernels from the milling of brown rice for saké. As proteins and fats are undesirable for saké quality, the proteins are gradually removed until the polishing rate reaches 50%, after which the protein remains practically constant (Nunokawa 1972).

The protein concentration of the outer 2 to 3% of milled rice is about twice that of the whole kernel with ratios ranging from 1.6 to 2.8. Normand *et al.* (1966) studied a commercially milled long grain rice, predominantly Bluebonnet 50, the flour through an 80-mesh screen from each of the 12 mill passes representing 1 to 2% of the original sample. From this rice with 7.9% protein, dry basis, the first pass flour had 21.8% protein, nearly 3 times that of the original; the 12th pass flour, 11.9%; and the residual kernel (53% by weight of the original rice), 5.1%.

In a later study (Kennedy *et al.* 1974) 12 different lots of commercially milled rice grown in the United States and of different varieties and treatments were analyzed for protein among other constituents. The percentage of protein in flours from 3 mill passes is shown in Table 11.2. Protein concentration decreased progressively from the outer layers to the center in all lots except for one, a long grain Belle Patna, for which the values were 19.2, 20.0 and 19.5% after the first, second and third passes, respectively.

Amino Acids

The amino acid composition of the protein in the various portions of the endosperm does not vary appreciably. Normand *et al.* (1966) milled off 12 layers of a sample of commercially milled Bluebonnet 50 and found that the amino acid composition was similar in all the layers except that tryptophan decreased from 1.51% of the protein to 0.91% in the 12th layer. Inexplicably tryptophan was the same in both original and residual kernels, 1.32%. In our study of 12 different samples of rice (Kennedy and Schelstraete 1974) we also found no appreciable differences in composition including tryptophan (Table 11.3). For cystine, methionine and tryptophan varietal differences were greater than differences among

TABLE 11.2. PROTEIN CONCENTRATION OF FRACTIONS OF OVERMILLED RICE, N = 12

Sample[1]	Percentage of original kernel	Protein % of dry matter		Ratio with respect to original kernel	
		Mean St.Dev.	Range	Mean St.Dev.	Range
Original kernel	100	7.5 ± 1.3	6.0–10.2	1.00	
First pass flour	1.97 ± 0.54	14.5 ± 2.7	10.9–19.2	1.92 ± 0.15	1.68–2.18
Second pass flour	2.19 ± 0.36	14.0 ± 3.0	10.2–20.0	1.84 ± 0.15	1.58–2.11
Third pass flour	1.98 ± 0.50	13.2 ± 3.1	9.9–19.5	1.74 ± 0.17	1.53–2.07
Residual	87.2 ± 2.9	7.0 ± 1.1	5.9– 9.4	0.93 ± 0.03	0.88–0.95

Source: Kennedy et al. (1974).

[1]All flours through a 40-mesh screen.

fractions. Cystine as cysteic acid and tryptophan, both of which were analyzed separately, showed very little variation among fractions within a given variety (coefficients of variation were 3% or less) while among varieties and for the same part of the kernel, coefficients were 10 to 17%.

TABLE 11.3. AMINO ACID COMPOSITION OF COMMERCIALLY MILLED RICES—MEAN AND STANDARD DEVIATION OF 12 LOTS

	Amino Acids (g/16 g N)	
Amino acid	Original kernel (n = 12)	Original rice, High-protein flours and Residual kernel (n = 60)
Lysine	4.0 ± 0.1	3.9 ± 0.2
Histidine	2.6 ± 0.1	2.7 ± 0.1
Arginine	9.2 ± 0.1	9.3 ± 0.2
Aspartic acid	9.2 ± 0.3	9.2 ± 0.4
Threonine	3.5 ± 0.2	3.5 ± 0.2
Serine	5.0 ± 0.2	5.1 ± 0.2
Glutamic acid	17.6 ± 0.5	18.0 ± 0.8
Proline	4.6 ± 0.4	4.7 ± 0.4
Glycine	4.6 ± 0.2	4.6 ± 0.3
Alanine	5.5 ± 0.2	5.6 ± 0.2
Cystine	1.7 ± 0.2	1.7 ± 0.2
Valine	5.8 ± 0.4	5.8 ± 0.3
Methionine	2.2 ± 0.3	2.2 ± 0.3
Isoleucine	4.1 ± 0.1	4.1 ± 0.2
Leucine	8.2 ± 0.3	8.2 ± 0.3
Tyrosine	5.2 ± 0.3	5.2 ± 0.3
Phenylalanine	5.1 ± 0.2	5.1 ± 0.2
Tryptophan	1.7 ± 0.3	1.7 ± 0.3

Protein Bodies

Much of the protein in numerous plant seeds is localized in subcellular storage particles which are commonly called "protein bodies." Mitsuda *et al.* (1967,1969) isolated such particles from rice polish and found that these rice protein bodies (RPB) were round in shape and ranged in diameter from 1.5 to 4 μm. Electron microscopy indicated that they were distributed throughout the endosperm but not uniformly. By analysis the isolated particles contained about 60% protein while lipid and carbohydrate were variable, 10 to 28% and 12 to 29%, respectively. The fractions of highest density contained less lipid and more carbohydrate than did those of lower density. Also found were small amounts of ash, RNA, phospholipid, phytic acid and niacin. No remarkable difference was apparent in amino acid composition between the protein bodies and rice polish.

Electron microscopy revealed electron-dense bodies, more than half of which had a limiting membrane and a distinct concentric strata structure. Electron-dense and electron-thin layers were arrayed alternately and the electron-dense layers appeared to be composed of minute granules (about 15 nm) of exceedingly high electron density which suggested that they may be a basic unit of structure of the protein bodies. Protein bodies having characteristic strata structures similar to those just described were also found in rice endosperm by K. Tanaka *et al.* (1973).

Recently Y. Tanaka *et al.* (1975) reported the presence of spherical or oval particles in fresh feces of the Japanese, reaching a density of more than 2×10^9/g. These particles, from 1 to 3.5 μm in diameter, were found to be present following the ingestion of meals containing cooked rice but not present after meals otherwise identical but containing sweet potatoes in place of rice. Electron micrograph studies revealed that these bodies, called feces protein particles (FPP), had a distinct concentric strata structure and diameters which agreed with the observations of Mitsuda *et al.* (1969) for rice protein bodies (RPB). Chemical compositions, size and shape of the FPP and the RPB were quite similar. Digestibility of the individual RPB differs, according to in vitro tests, since 15 to 20% of the RPB of rice polish was indigestible and excreted as FPP. An explanation for the indigestibility of these bodies is currently being sought.

Biological Evaluation of the Protein

Rice is among those cereals—oats, rye and buckwheat—which have relatively good amino acid patterns. The protein efficiency ratio (PER)—gain in weight in grams per gram of protein eaten using rats—for milled rice is about 2 although values range from 1.38 to 2.56 depending upon variety, the level of protein and experimental conditions (Juliano 1972). Maximum values for PER, 3.5 to 4, are obtained with egg protein which is often used as a standard for evaluating protein quality. Proteins with a PER of 2 or higher are considered to be of good nutritional quality and produce good growth in young animals.

High protein rice flours (first mill pass) from the overmilling of commercially milled rice gave a mean value with standard deviation for PER of 2.10 ± 0.20, adjusted to casein standards at 2.5 (Kennedy and Schelstraete 1974B). Protein concentrations of the flours from 12 different lots of milled rice ranged from 9 to 15.4%. Each flour was compared with a casein control at a comparable protein level in the diet.

For 2 lots of rice, Bluebonnet 50 and Colusa, PER was determined for the original and residual kernels and first, second and third pass flours. Mean PER for all 5 fractions from each of the 2 rices were 2 ± 0.10 and 1.9 ± 0.15, respectively. Within each rice variety there were no significant differences in PER among the 5 fractions.

Only one other value for PER has been reported for high-protein rice flours, 1.84 for a flour containing 14.2% protein which represented 8.1% weight fraction of a California Pearl rice with 7.8% protein (Milner

1965). This value is within the range of values found in the previous study.

CARBOHYDRATES

Plant carbohydrates include starch, sugars, cellulose, hemicellulose, pectin, gums, and sugar alcohols and acids. Noncarbohydrate compounds such as lignin and organic acids are often included when carbohydrate is determined "by difference" or as a nitrogen free extract. A sample of milled rice, Caloro, had a lower organic acid content than did other cereals, about 0.05%, somewhat less than half of which was citric acid (Houston *et al.* 1963). Values for other acids were, per 100 g, dry basis: malic, 9.7 mg; succinic, 5.3 mg; fumaric, 4.5 mg; oxalic, 4.3 mg; and acetic, 3.9 mg. Very few studies on the composition of hemicellulose and cellulose have been made (Juliano 1966). As starch comprises 90% of milled rice, most of the studies on carbohydrates in rice have been concerned with this constituent. The distribution of starch, sugars and fiber within the endosperm will be discussed below.

Starch

Starch is the only constituent for which the concentration is smallest in the periphery of the endosperm and increases as the center of the kernel is approached. From a commercially milled rice, predominantly Bluebonnet 50, with 90.7% starch, dry basis and 32% of the starch as amylose, Normand *et al.* (1966) found 60.3% starch in the outer 3% of the kernel. Concentrations increased as the center of the kernel was approached until the 12th fraction reached 90.1% Residual kernel had 94.3% starch. Amylose made up about 26% of the starch in the first 2 fractions and increased to about 34% in fractions 7 through 12.

Hogan *et al.* (1968) reported starch concentrations for fractions milled from a commercially milled Belle Patna rice (see Table 11.4). Fractions, each about 3% of the original rice, were sieved through an 80-mesh screen to give material passing through the screen, called flour, and material retained, termed chits. For comparable fractions, flours with smaller particles showed a greater decrease in starch than did the chits.

Kennedy *et al.* (1974) reported concentrations of starch in 6 varieties of rice with 90 to 93% starch, dry basis (Table 11.5). Trends were similar to those in the previous reports. Amylose and starch concentrations, as percentage of rice, were both lowest in the outer layers and increased toward the center of the kernel while residual rices had a higher percentage than did the original rice. Within the same variety only minor differences in amylose as percentage of starch were noted.

TABLE 11.4. PERCENTAGE OF STARCH IN FRACTIONS OF OVERMILLED BELLE PATNA RICE

	Original kernel	Fractions 1	2	3	4	5	Residual kernel
Flour		42.9	52.6	61.9	75.5	79.1	
	91.7						90.9
Chits		65.5	79.5	84.3	86.5	89.0	

Source: Hogan et al. (1968).

Sugars

Brown rice contains 0.83 to 1.36% total sugars as glucose with reducing sugars ranging from 0.09 to 0.13%; milled rice contains 0.37 to 0.53% total sugars with 0.05 to 0.08% reducing sugars (Juliano 1972). Percentage of total sugar varies with variety and in milled rice, with degree of milling. Reducing sugars increased and nonreducing sugars decreased with the aging of rice (Barber 1972); nonreducing sugars also decreased with increased temperatures. Sugar concentration is also affected by processing, 2 milled parboiled rices having 0.66 and 1.06% total sugars with 0.19 and 0.14% reducing sugars, about twice the values for milled rice (Williams and Bevenue 1953). The main nonreducing and reducing sugars are sucrose and glucose, respectively. Fructose, galactose, maltose, raffinose and other oligosaccharides have also been reported (Juliano 1966).

The distribution of total and nonreducing sugars in fractions of a Spanish rice, variety Balilla × Sollana, was studied by Barber et al. (1967). Total sugars in the external layer and residual kernel of milled rice were:

Description of sample	A	B	C
Year harvested	1965	1965	1966
Degree of milling	Under	Over	Intermediate
Percentage of bran removed	7.7	12.0	9.8
External layer (%)	10	10	5
Total sugars (% dry basis)			
Whole kernel	0.61	0.25	0.38
External layer	4.02	1.23	2.92
Residual kernel	0.17	0.13	0.18

Sugars in the external layer were about 6 times as concentrated as in the whole kernel and accounted for ½ to ⅔ of the sugar in the original rice. Little published data on sugars in rice bran is available. Saunders (1977), in 2 samples of bran, obtained values of 7.6 and 8.4% total sugars as glucose of which 80% was sucrose. In addition, oligosaccharides above tetra, raffinose, and traces of glucose and fructose were found. Total sugar concentrations in the external layers of the under, intermediate

TABLE 11.5. PROXIMATE COMPOSITION OF SAMPLES FROM THE OVERMILLING OF MILLED RICE

Constituent	Rices no.	Whole kernel	Flour through 40-mesh screen			Residual kernel
			First pass	Second pass	Third pass	
		(Mean and standard deviation, % dry basis)				
Starch	6	91.9 ± 1.3	60 ± 3	71 ± 7	82 ± 3	93.2 ± 0.6
Amylose	6	24 ± 5	15 ± 5	18 ± 4	20 ± 5	24 ± 7
Ash	10	0.54 ± 0.13	4.8 ± 2.3	2.7 ± 1.4	1.8 ± 1.2	0.30 ± 0.05
	2[1]	0.76 ± 0.05	5.8 ± 0.9	3.7 ± 0	2.3 ± 0	0.60 ± 0.05
Fat	12	0.49 ± 0.23	7.5 ± 2.9	4.4 ± 2.1	2.6 ± 1.4	0.12 ± 0.06
Phytic acid	6	0.13 ± 0.02	3.0 ± 0.8	1.7 ± 0.5	1.0 ± 0.4	0

Source: Kennedy et al. (1974).

[1]Parboiled rices.

and overmilled rices studied by Barber *et al.* (1967) were 4.02, 2.92 and 1.23%, respectively, which together with the data of Saunders show that the sugar concentration in both brown and milled rices decreases as the center of the kernel is approached.

Fiber

Fiber in brown, milled raw and milled parboiled rices are 0.9, 0.3 and 0.2%, respectively (Watt and Merrill 1963). Houston (1967) determined fiber concentration in milled portions of 2 samples of rice: (1) a laboratory milled, high protein rice, BPI-76; and (2) a commercially milled, short grain parboiled rice. From each rice 5 portions, each about 2% (1.5 to 3%) of the original rice, were milled off and the data obtained is listed in Table 11.6.

In both rices fiber in the outer 2 to 3% was 6 to 7 times as concentrated as in the original rice and accounted for 17% of the total fiber.

TABLE 11.6. PERCENTAGE OF FIBER IN MILLED RICE, DRY BASIS

Original Kernel	Fractions					Residue (89% of original)
	1	2	3	4	5	
0.54[1]	3.18	1.89	1.33	1.04	0.88	0.28
0.43[2]	1.94	2.17	1.70	1.48	1.45	0.24

[1]Laboratory milled, high protein rice, BPI-76.
[2]Commercially milled, short-grain parboiled rice.

FAT

Total fat concentrations (petroleum-ether extract) for 241 milled rices was 0.65% with a range of from 0.19 to 2.73% (Simpson *et al.* 1965). Fatty acids from milled rice extracted with petroleum ether consisted of about 20% polyunsaturated acids, linoleic and linolenic (Resurreccion and Juliano 1975). The ratio of oleic to linoleic was essentially 1.0:1.0.

Fat is unevenly distributed within the endosperm, the highest concentration being in the outer layer and the lowest in the central portion as reported by several investigators (Casas *et al.* 1963; Houston *et al.* 1964; Normand *et al.* 1966; Houston 1967; Hogan *et al.* 1968).

Mean values for 12 lots of rice (Table 11.4) show these trends (Kennedy *et al.* 1974). Fat in the whole kernel ranged from 0.20 to 0.92%; concentrations in the first pass flour through a 40-mesh screen were 17 times as much as in the whole kernel, with the highest value 11.6% and the lowest, 4.1%. The percentage of fat in flours over 40 mesh was 4 times that of the whole kernel and about ⅓ that of the flours through 40 mesh. Mean value for the residual kernel was 0.12% fat, ¼ that of the original rice.

The composition of the fat also differs within the endosperm. The percentage of total lipid in the outer 5% of a sample of Spanish rice (Variety Ballilla × Sollana) was 17.3% and the inner 80% of the kernel had 0.24% (Primo *et al.* 1965). Neutral fats accounted for 85 to 90% of total lipids in the outer layers and for only 60% in the center. Unsaturated acids oleic and linoleic showed, in general, an inverse pattern of distribution: oleic acid content decreased and linoleic acid content increased from the outer to the inner layers of the kernel, in both the free fatty acid and the neutral fat fractions.

MINERALS

Reported ash values for 239 samples of milled rice ranged from 0.26 to 1.95%, dry basis, with means of 0.69, 0.64 and 0.61% for long, medium and short grain rices, respectively (Simpson *et al.* 1965). Means for 17 long grain and 20 medium grain parboiled milled rices were higher, 0.75 and 0.78%. Several studies have been made on the percentage of ash in different portions of the endosperm of which the data in Table 11.7 is representative.

TABLE 11.7. ASH IN FRACTIONS OF OVERMILLED WHITE RICE

No. of samples	Type of rice	Whole kernel	1	2	Fractions 3	4	5	Residue
		(mean and standard deviation, dry basis)						
1[1]	Calrose	0.50	3.60	2.11	1.32	1.02	0.86	0.29
2[2]	Parboiled high- protein	1.04 ±0.02	6.7 ± 1.5	5.1 ±1.2	4.0 ±0.7	3.2 ±0.4	2.8 ±0	0.45 ±0.18
4[2]	Varied	0.52 ±0.05	4.2 ± 1.4	2.5 ±0.9	1.7 ±0.5	1.3 ±0.3	1.0 ±0.3	0.29 ±0.06
1[3]	Belle patna	0.77	16.00	8.97	5.52	2.85	1.59	0.16
1[3]	Parboiled	0.90	9.61	8.17	—	—	—	0.76

[1]Source: Houston *et al.* (1964).
[2]Source: Houston (1967).
[3]Source: Hogan *et al.* (1968).

Houston (1967) and Houston *et al.* (1964) analyzed 7 types of rice for ash for which the concentrations were about 0.5% except for a parboiled rice and 1 high in protein which had 1%. The first fractions abraded had from 4 to 7 times as much as the original rice with ratios decreasing for each successive fraction removed. Residual rices, 85 to 90% by weight of the whole kernel, accounted for about 50% of the ash, with only 27% for the parboiled rice. Data obtained by Hogan *et al.* (1968) showed similar trends. Ash concentration in the first fraction of the Belle Patna was considerably greater than in the whole kernel, about 21 times more, and the residual kernel, 85% of the original rice, accounted for about 18% of the ash in the whole kernel.

Kennedy *et al.* (1974) obtained ash concentrations with standard deviation of 0.58 ± 0.14% for 12 lots of milled rice. Concentrations in the first pass flours through a 40-mesh screen were 5% ranging from 3 for a Calrose to 10.6% for a Belle Patna. Ratios for the percentage of ash of the milled fractions to that of the original rice for 6 lots of rice of different varieties is shown graphically in Fig. 11.2. Residual kernels accounted for about 60% of the ash in the original rice—as little as 36% for the Belle Patna and as much as 75 to 80% for the short grain and parboiled rices.

Among the major mineral constituents of the ash are phosphorus, potassium, silicon, magnesium, calcium, sodium and iron, all of which have been studied with respect to their distribution within the endosperm.

Macrominerals

Phosphorus.—Phosphorus and potassium are the most abundant mineral elements in rice; reported values in milled rice range from 88 to 192 mg/100 g dry weight for phosphorus and 58 to 117 mg for potassium (Juliano 1972).

The distribution of phosphorus in the endosperm has been reported in 2 studies. Normand *et al.* (1966) found 1.24% P in the outer layer of milled rice, Bluebonnet 50, for which the whole kernel had 0.14% P—9 times as much as in the outer layer as in the whole kernel. Concentrations decreased as the center of the kernel was approached until the 12th layer had 0.085% and the residual kernel only 0.028%. The residual kernel, 53% of the original rice, accounted for only 10% of the phosphorus in the whole kernel.

Kennedy and Schelstraete (1975A) determined phosphorus in portions of 6 commercially milled rice of different varieties for which the mean value in the original rices was 0.126%. Concentrations of phosphorus in flours obtained by abrasive milling from 3 mill passes decreased from periphery to the center as in the Normand study (Fig. 11.2). First pass flours through a 40-mesh screen also had 9 times as much phosphorus as in the whole kernel while the residual kernel, 88% of the original rice, accounted for 57% of the phosphorus. Data from these 2 studies are in general agreement.

Potassium.—Data for potassium were very similar to those for phosphorus (Kennedy and Schelstraete 1975A), with a mean concentration of 128 mg/100 g in the original rice. Distribution of potassium within the endosperm was also similar to that of phosphorus (Fig. 11.2) except that the gradient was not so steep.

Phytic Acid.—Phytic acid, myoinositol 1,2,3,4,5,6-hexakis (dihydrogen phosphate), is present in seeds often thought to be in the form of salts of calcium and magnesium. Kennedy and Schelstraete (1975A) reported a mean value with standard deviation of 132 ± 17 mg/100 g of phytic acid for the whole kernel of 6 lots of milled rice of different varieties. Flours through a 40-mesh screen from the outer layers (4% by weight of the rice) had $2,972 \pm 830$ mg/100 g of phytic acid, 23 times that in the whole kernel, with values ranging from 1.9 to 4% of the flour. Phytic acid concentrations decreased from the outer to the center of the kernel; no phytic acid was detected in the residual rice (Fig. 11.2).

Phytate phosphorus made up 30% of the total phosphorus in the whole kernel and from 77 to 70% in the flours; none was present in the residual rice.

Phytic acid in rice has been associated with the aleurone layer which lies between the outer layers and the endosperm of the seed. Aleurone particles in the aleurone layer, as distinguished from protein bodies described earlier in this chapter, were found to have the following composition: moisture, 14.9%; protein, 11.7%; carbohydrate, 7.9%; acid-soluble organic phosphorus, 11.3%; other forms of phosphorus, 0.31%; total myoinositol, 9.4%; potassium, 9.45%; magnesium, 8.3%; calcium, 0.42%; and zinc, iron, copper and manganese, each less than 0.05% (Tanaka et al. 1973).

Scanning electron micrographs showed that sizes of the aleurone particles in the aleurone layer as well as those of isolated particles ranged between 1 and 3 μm. A transmission electron micrograph of the aleurone layer was characterized by electron-dense bodies embedded in the aleurone particles. The isolated aleurone particles exhibited no strata structure as was found in protein bodies but instead had the characteristic electron-dense bodies thought to be phytic acid. Part of the outer layer of the aleurone particles was lost during isolation.

High phytin containing particles have recently been isolated from rice grains by Ogawa et al. (1975). After discarding the bran—pericarp and seed coat, 2% of the whole grain by weight—another 2% by weight was collected and used after sieving to remove embryos. This portion roughly corresponded to the aleurone layer. Particles, 1 to 2 μm in diameter and embedded in aleurone particles of matured rice grains, were isolated using a combination of differential centrifugation and an aqueous 2 phase system.

Composition of these particles on a dry weight basis included 67.2% phytic acid, 1.26% protein, 0.47% carbohydrate, 18.9% potassium, 10.8% magnesium, 1.4% calcium and 0.13% iron. Manganese and zinc accounted for 0.1%. Over 90% of the compounds in the isolated particles were phytic acid, potassium and magnesium. Electron microscopic observation of the

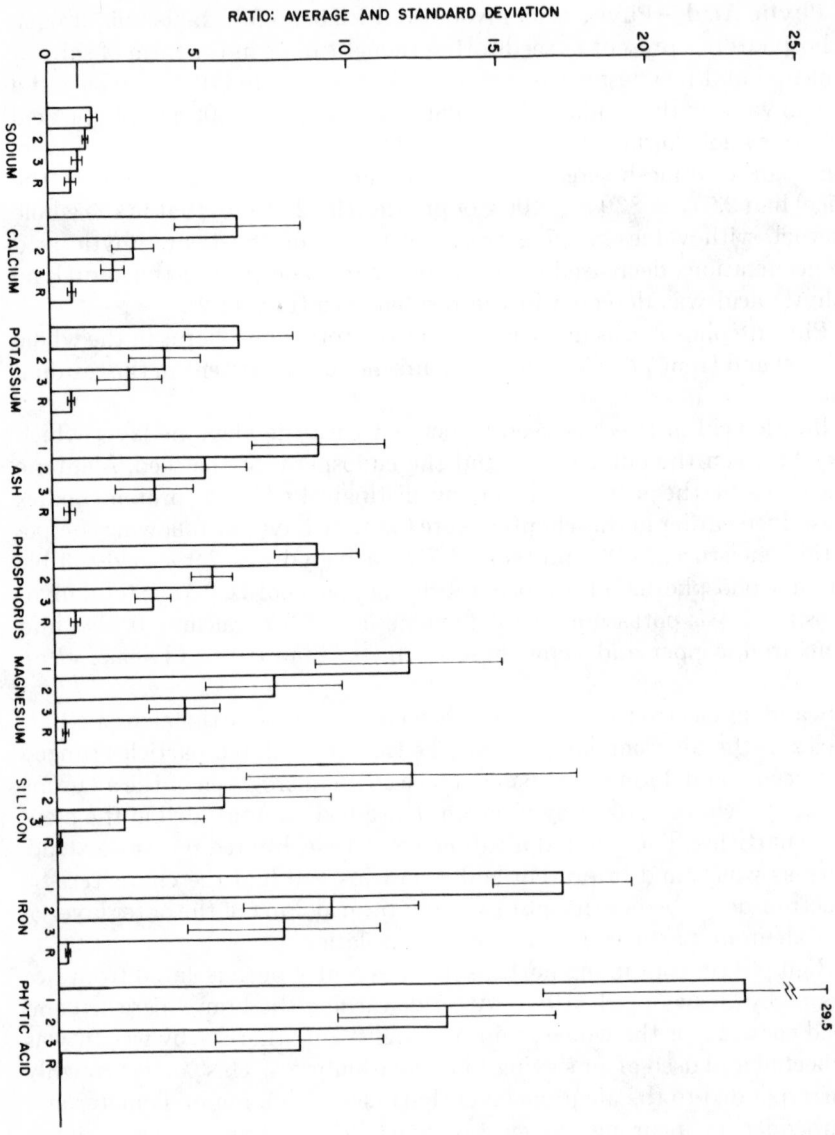

Revised from Kennedy and Schelstraete (1975A)

FIG. 11.2. RATIO AND STANDARD DEVIATION OF CONTENTS OF SODIUM, CALCIUM, POTASSIUM, ASH, PHOSPHORUS, MAGNESIUM, SILICON, IRON AND PHYTIC ACID IN FLOURS AND RESIDUAL KERNEL OF 6 VARIETIES OF RICE WITH RESPECT TO THOSE OF THE ORIGINAL WHOLE KERNEL RICE

(1) First pass flour; (2) Second pass flour; (3) Third pass flours; (R) Residual kernel.

isolated particles confirmed that the electron-dense material, which is embedded in aleurone particles, is composed of potassium and magnesium salts of phytic acid.

Ogawa *et al.* (1975) suggested that aleurone particles are composed of at least 2 major parts: 1 is the high phytin containing particle which forms the core of the aleurone particle and the other is the surrounding coat which is composed of protein and carbohydrate. The peripheral membrane appears to have important roles in the dephosphorylation of phytic acid during the germination of grains.

For some time phytates have been considered to prevent the absorption of mineral elements such as calcium and iron from the intestinal tract, particularly of human subjects. Within the last 2 to 3 years, however, work has appeared showing that certain forms of iron phytate have a high biological availability. Morris and Ellis (1976) reported monoferric phytate to have a high bioavailability for the rat, essentially equal to the reference salt, ferrous ammonium sulfate. This is the major form of iron in wheat, a water-soluble complex of iron isolated from wheat bran. In contrast, they found ferric phytate with 3 to 4 moles of iron/mole phytate to be insoluble and, as a dietary source, to have a low bioavailability. It thus seems that the form of the phytate salt is important in its bioavailability. Little work on the availability of mineral elements in rice appears to have been done.

Silicon

The rice plant is a member of the family Gramineae, the best known group belonging to silicicolus plants, those which selectively accumulate silicon. Silica (SiO_2) content of rice plants may range from 5 to 20% at harvest depending on the silicon-supplying power of soils (Yoshida *et al.* 1959). It is also possible to lower the silicon content of rice plants to about 0.1% using water culture techniques. Other than the fact that silicon does not appear to be essential for growth, the role of silicon in physiological processes is unknown. The greater part of the silicon absorbed by plants is accumulated in the aerial parts and is deposited in the cell walls, especially in the epidermis of the leaf. The content of silicon is greater in mature than in young plants. Yoshida *et al.* (1962) fractionated the silicon in rice plants into 3 forms: silica gel, the most prevalent form constituting 90% or more of the total silicon; silicate ion, ranging from 0.5 to 8%; and colloidal silicic acid, 0 to 3.3%. Silica gel is characterized as polysilicic acids derived from orthosilicic acid by nonenzymatic condensation polymerization.

Silicon is found in the root, stem, leaf-blade, leaf-sheath and husk (hull) of the plant. It has been determined to constitute, as silica, about 97% of the ash, or calculated as silicon, 11% in the total husk (Houston 1972). Silicon has not been studied so extensively in the rice seed. A few values, ranging from 32 to 291 mg/100 g, dry basis, for brown rice and from 5 to 292 mg for milled, have been reported (Juliano 1966). Recent work on silicon as a possible essential trace mineral for man suggests that more data on the silicon content in foods is desirable.

In the study by Kennedy and Schelstraete (1975B), the mean value of silicon in 6 varieties of milled rice was 46 ± 30 mg/100 g, dry basis, with the lowest value being 13 mg for a short grain Colusa and the highest, 100 for a long grain Belle Patna. Within the kernel, as compared with the whole kernel, silicon in first, second and third pass flour was, respectively, 14, 7 and 4 times more concentrated (Fig. 11.2). Mean value for the residual kernel was only 4 mg/100 g. Of the 7 mineral elements analyzed, silicon, as the percentage of total ash, was the third most abundant in all fractions except for the residual kernel in which it was sixth. The forms in which silicon appears in the rice kernel have not been studied but it is possible that they might be similar to those found in other parts of the plant.

Magnesium and Calcium.—Mean concentrations of magnesium and of calcium in the whole kernel of 6 lots of milled rice were 28 and 25 mg/100 g, respectively (Kennedy and Schelstraete 1975A). Both elements had the highest concentration in the periphery of the kernel, decreasing as the center of the kernel was approached. The gradient for magnesium, however, was steeper than that for calcium, the mean value for the first pass flours being about 11 times greater than that of the whole kernel for magnesium as compared with 7 times for calcium. The residual kernel accounted for 25% of the magnesium in the whole kernel but for 60% of the calcium.

Varietal differences were greater for magnesium than for calcium due largely to the long grain Belle Patna which had the greatest values for both but was particularly high in magnesium. Ranges for calcium were from 17.7 to 38.5 mg/100 g in the original rice for all 6 varieties but 17.7 to 29.7 mg omitting Belle Patna while magnesium ranges were 17.1 to 70.4 mg for all the varieties but only 17.1 to 21.2 mg without Belle Patna. Calcium and magnesium values for Belle Patna were higher than those of the other varieties for all milled fractions.

In the only other study on calcium distribution within the endosperm (Normand et al. 1966), the rice was milled deeper into the kernel. Starting with a milled rice, predominantly Bluebonnet 50, with a calcium concentration of 23 mg/100 g, calcium in the flour through an 80-mesh screen from the first 3.3% of the outer layers was 20 times as concentrated as that of the original rice. Calcium concentration in flours decreased, in the third pass being 7 times that of the original rice while that from the seventh pass, only 0.04. Flours from these 7 passes, from the outer 31% of the whole kernel, accounted for 94% of the total calcium. No calcium was reported for the remaining 5 flours or in the residual kernel.

Ca:P ratios in rice are much below the recommended 1:1. The ratio in the whole kernel milled rice (Normand et al. 1966) was 0.16:1, higher for the first pass flour, 0.37:1, and gradually decreased until the ratio for the seventh pass was only 0.006:1. In a study by Kennedy and Schelstraete (1975A), the whole kernel rices had a mean ratio of 0.18:1 with a range of 0.13 to 0.21. There was no consistent trend in ratios for the flours and residual kernel: values ranged from 0.07 to 0.35. Calcium concentrations were so low that even the highest ratio was much below the desirable 1:1.

Sodium.—Of the macrominerals sodium had the lowest concentration with a mean and standard deviation of 8.1 ± 2.2 mg/100 g for the original rices of 6 varieties. Distribution within the kernel was the most even of all the constituents analyzed, with sodium concentration in the first pass flour only 1.5 times greater than that of the whole kernel. Residual kernels accounted for about 63% of the sodium in the original rice.

Trace Minerals

Iron.—Concentrations of iron in brown and in white rice (12% moisture) are given as 1.6 and 0.8 mg/100 g, respectively (Table 11.1). Values of 0.7 to 4.6 mg for brown rice and 0.2 to 2.7 mg for milled rice have been reported (Juliano 1972). In the study by Kennedy and Schelstraete (1975A), the mean value with standard deviation for 12 lots of milled rice was 1.03 ± 0.65 mg/100 g, dry basis. The lowest values, 0.26 and 0.37 mg, were in 2 short grain rices while the highest, 1.17 to 2.34 mg, were from 5 different lots of long grain Belle Patna.

First pass flours had much higher concentrations ranging from 5.6 to 35.6 mg, with mean and standard deviation of 15.1 ± 8.7 mg/100 g— about 15 times greater than that of the whole kernel. Concentrations decreased as the center of the kernel was approached. Iron in residual rices accounted for about 27% of that in the whole kernels. Mean values for the 5 lots of Belle Patna were 2 to 3 times greater than those for all corresponding fractions of the other varieties; however, ratios of the concentrations in the flours and residual kernel with respect to the whole kernel were essentially the same. Values for parboiled rices were within the range of those for the raw rices.

Zinc

Mean values for 27 samples of unpolished and 100 of polished rices collected from 22 countries and areas were, not corrected for moisture, 16.4 and 13.7 $\mu g/g$ (1.64 and 1.37 mg/100 g) Zn, respectively (Masironi et al. 1977).

In a study by Franz (1977), relative availability of zinc in a lot of Calrose brown rice, from a slope-ratio assay of gains in body weights of rats, was 40 to 50% using dietary zinc sulfate as 100%. Availability in milled rice, from the same lot, was high—99% of the standard. Zinc sulfate is not completely absorbed, the amount being affected by the composition of the diet. In the absence of other data, assuming an absolute absorption of zinc from zinc sulfate as 40% for both rices, the absolute absorption of zinc from rice could be approximated by: concentration of zinc $\times$ % relative absorption $\times$ % absolute absorption. Concentrations in the brown and the milled raw rices were 1.74 and 1.21 mg Zn/100 g dry matter, respectively. Thus the relative and the absolute availabilities would be:

Rice	Zn (mg/100g)	Phytic acid	Neutral detergent fiber (% dry basis)	Availability of Zinc Relative determined (%)	Availability of Zinc Absolute calculated (mg/100 g)
Brown	1.74	836	3.63	48	0.33
Milled	1.21	135	0.91	99	0.48

From the above data 45% more zinc is available from milled rice than from brown. Milled rice contains much less phytic acid and dietary fiber than does the brown rice. Both of these substances have been considered to reduce the availability of mineral elements although some evidence points to phytate rather than to the fiber.

VITAMINS

Of the vitamins present in rice perhaps thiamin, riboflavin and niacin have been most studied. As is found in most cereal grains, concentrations of these nutrients, as compared with other portions of the kernel, is lowest in the endosperm (Houston and Kohler 1970). Distribution of the vitamins within the endosperm has been studied only for thiamin, riboflavin, niacin and pyridoxine.

Thiamin and niacin are probably the most important of the vitamins in rice. When compared with the Recommended Daily Allowance (RDA) on an energy basis, thiamin concentration in brown rice is more than sufficient, being 0.09 mg/100 Kcal (from Table 11.1) as compared with 0.05 mg/100 Kcal, the RDA for males, 19 to 22 years of age (FNB 1974). Milled rice, however, is low with only 0.02 mg/100 Kcal.

Thiamin

Several studies have been made on thiamin concentrations in various portions of milled rice. Houston *et al.* (1964) determined thiamin in high protein flours abraded in 6 successive mill passes from a milled rice, Caloro, while Normand *et al.* (1966) overmilled a sample of commercially milled rice, predominantly Bluebonnet 50, obtaining 12 fractions. Results from the flours obtained from the outer layers are as follows:

| | *Thiamin in milled rice* | | | |
	Original kernel (mg/100 g)	Outer 7% (mg/100 g)	Outer 7% as % of whole kernel	Residual kernel (mg/100 g)
Houston	0.074	0.42	39	0.007
Normand	0.081	0.86	43	0.014

In relation to the original kernel, thiamin in the outer layers was concentrated about 6 and 11 times, while concentrations in the residual kernels were only 9 and 17% that of the original rice.

Fractions of 6 lots of different varieties of raw milled rice and 2 of parboiled rice were analyzed for thiamin by Kennedy *et al.* (1975). Concentrations of thiamin in the whole kernel varied among the varieties, the sample of Saturn having twice the concentration of Colusa, 0.18 mg/100 g versus 0.09 mg. These differences were reflected in the abraded flours. About 12% of flour was removed from 3 mill passes.

Concentrations of thiamin are shown in Table 11.8. The flours through a 40-mesh screen from the outer 6.2% (5 to 8%) of the kernel accounted for 24% of the thiamin in the original kernel of the raw rices but only 9% for the parboiled rices. Thiamin in the flours of the raw rices was 7 times more concentrated than in the original kernel but in the parboiled rices, only 3 times more. The residual kernel after the third pass, about 88% of the rice, account for 32% of the thiamin in the original raw milled rice but for the 2 parboiled rices, 74%. Except for thiamin in the third pass flour of the parboiled rices, all flours had concentrations of thiamin greater than those for brown rice (Table 11.1).

It has been shown that, although the level of thiamin in milled parboiled rice is greater than that of the corresponding raw rice and the distribution more even within the kernel, the total amount of thiamin in the brown rice was less than in the raw brown rice indicating destruction of thiamin in the processing (Padua and Juliano 1974).

Niacin

Quantitative determinations of niacin in parts of the endosperm have been made in only a few studies. Hinton and Shaw (1953) determined

TABLE 11.8. VITAMINS IN FRACTIONS OF OVERMILLED RICE—MEAN AND STANDARD DEVIATION

Vitamins and number of rices analyzed	Original kernel	Flour through a 40-mesh screen			Residual kernel After third pass
		First pass	Second pass	Third pass	
		(mg/100 g, dry basis)			
Thiamin					
Raw (6)	0.13 ± 0.04	0.98 ± 0.28	0.62 ± 0.19	0.47 ± 0.21	0.074 ± 0.020
Parboiled (2)	0.15 ± 0.03	0.59 ± 0.38	0.37 ± 0.19	0.26 ± 0.11	0.124 ± 0.013
Niacin					
Raw (10)	1.54 ± 0.57	20.4 ± 6.8	10.0 ± 4.9	6.1 ± 3.1	0.73 ± 0.28
Parboiled (2)	3.20 ± 0.08	23.8 ± 1.2	55.6 ± 0.4	9.9 ± 1.3	2.08 ± 0.21
Pyridoxine					
Raw (6)	0.14 ± 0.02	1.08 ± 0.16	0.74 ± 0.15	0.53 ± 0.13	0.09 ± 0.02
Riboflavin					
Raw (6)	0.040 ± 0.003	0.20 ± 0.04	0.13 ± 0.02	0.10 ± 0.02	0.032 ± 0.002
Parboiled (2)	0.044 ± 0.006	0.19 ± 0.01	0.17 ± 0.005	0.10 ± 0.06	0.032 ± 0.002

nicotinic acid (niacin) by microbiological assay in various portions of two samples of hand dissected whole grain rice. One sample was a long, narrow, red skinned Indian variety, A.S.D.₃, and the other a short, broad, white Egyptian rice of unknown origin. Normand *et al.* (1966) reported niacin values for flours, sieved through an 80-mesh screen, from the outer 12 layers of a commercially milled rice, predominantly Bluebonnet 50, the 12 layers making up 47% by weight of the rice.

Data from these 2 studies are summarized in Table 11.9. From Hinton's data for fractions of brown rice, niacin concentration in the whole endosperm, excluding the aleurone layer, was determined by calculation.

Results of the 2 studies are in general agreement. Niacin in the outer endosperm was about 63% of that in the whole kernel from Hinton's work on hand dissected fractions while that of Normand *et al.* (1966) was only 32%. Hinton's fraction, however, contained all of the outer layer while Normand *et al.* analyzed only the flour through an 80-mesh screen, which was but 4.4% of the whole kernel.

TABLE 11.9. NIACIN IN RICE ENDOSPERM

Variety	Outer layers % by wt. of whole kernel	Concentration (mg/100 g, dry basis)			Ratio b/a	Niacin in fraction % of whole kernel
		Whole kernel a	Outer layers b	Residue		
Indian[1]	6.5	0.96	9.38	0.37	10.2	63.5
Egyption[1]	6.6	0.92	8.70	0.37	9.5	62.4
Blue-bonnet 50[2] (4.4% flour)	7.6	1.26	9.27	0.65	7.4	32.3

[1]Source: Hinton and Shaw (1953).
[2]Source: Normand *et al.* (1966).

Concentrations of niacin in the outer endosperm of 10 raw and 2 parboiled rices (Table 11.8), as compared with that of the original rices, were greater for niacin than for any of the other vitamins (Kennedy *et al.* 1975). First pass flours of the raw rices had concentrations 13 times that of the original kernel; parboiled rices, 7 times. Second pass flours of the parboiled rices, however, had the greatest increase in concentration, 17 times that of the original rice.

Niacin values of the second pass flour of parboiled rices were the only exception to the gradual decrease in concentration from periphery to center which was found with all of the constituents except for starch. Hinton and Shaw (1953) determined that the aleurone layer (4.7% of the whole grain of the Indian rice), with a concentration of 120 mg/100 g, contained 80% of the niacin of the whole grain. Consequently the amount of niacin in milled rice will depend a great deal on the amount of aleurone layer left adhering to the endosperm.

Pyridoxine

Reported concentrations for pyridoxine in milled rice range from 0.037 to 0.69 mg/100 g (Juliano 1966). Kennedy *et al.* (1975) found a mean concentration of 0.14 mg/100 g, dry basis, with values ranging from 0.11 to 0.17 mg for 6 milled rices of different varieties (Table 11.8). Values were in the same range as those for thiamin. Flours through a 40-mesh screen from the outer 5 to 8% of the kernel accounted for 25% of the pyridoxine in the original kernel and the degree of concentration was 8 times that of the whole kernel. Residual kernels after the third pass accounted for 57% of the pyridoxine in the original rice.

Riboflavin

Of the vitamins studied, riboflavin was the most evenly distributed throughout the endosperm. Two studies (Houston *et al.* 1964; Normand *et al.* 1966) on the same rices used in their determinations of thiamin gave the following:

		Riboflavin in milled rice		
	Original kernel mg/100 g	Flour (Outer 7%) mg/100 g	In outer 7% as % of whole kernel	Residual kernel mg/100 g
Houston	0.037	0.203	37	0.026
Normand	0.022	0.080	16	0

In relation to the original kernel, riboflavin in the outer layers was concentrated about 5.5 and 3.6 times. Concentration in the residual kernel (Houston *et al.* 1964) was 70% that of the original rice while Normand *et al.* (1966) did not detect any riboflavin in the residual kernel.

Concentrations of riboflavin in 6 milled rices and 2 parboiled rices are shown in Table 11.8 (Kennedy *et al.* 1975). Riboflavin in flours through a 40-mesh screen from the outer 5 to 8% of the kernels was about 4.5 times as concentrated as in the original rices and accounted for about 14% of the riboflavin in the whole kernel. Residual kernels accounted for about 70% of that in the corresponding original rices. There was no essential difference between raw and parboiled rices.

OVERALL NUTRITIONAL QUALITY

Brown vs Milled Rice

When compared on an energy basis with the recommended dietary allowance per day (RDA) for adult women and when rice is a large part of the diet (2000 Kcal), brown rice would give more than sufficient niacin, thiamin and phosphorus, 200, 190 and 154%, respectively (Table 11.10). Protein needs would be nearly enough—91%, zinc—60%, but calcium and riboflavin needs would be low, both about 23%. Milled rice, however, would be deficient in all of the nutrients with only protein, niacin and phosphorus supplying more than 50% of the RDA.

Since men have a greater energy requirement than do women, 3000 versus 2000 Kcal, they can more easily meet their dietary needs for some nutrients when eating the same types of food. For this reason brown rice would provide, in addition to phosphorus, niacin and thiamin, more than sufficient iron and protein, 230, 196, 189, 133 and 116% of the RDA, respectively, and 90% of zinc needs. Calcium and riboflavin are again low, 33 and 23%. Milled rice would provide sufficient amounts only of protein and phosphorus—each about 100% of the RDA, between 50 and 75% of zinc, niacin, iron, magnesium and pyridoxine needs, but less than 40% each of thiamin, calcium and riboflavin.

Nutritionally desirable features of whole grain brown rice in relation to its energy value are an adequate amount of good quality protein and sufficient niacin, thiamin, phosphorus, potassium and, for men, iron. Less desirable are the low concentrations of riboflavin and calcium, the low Ca/P ratio, and the presence of phytates and fiber, one or both of which may make certain mineral elements unavailable to the body.

Milled rice would provide sufficient protein and phosphorus for young men and nearly sufficient protein for adult women with insufficient amounts of all other nutrients—thiamin, calcium and riboflavin providing the least, less than 40% of the RDA.

The low sodium content makes rice a recommended food for persons on low-sodium diets. Also desirable for those on low-fat diets is the small amount of fat with a polyunsaturated to saturated fatty acid (P/S) ratio of 2.1:1.0.

Fractions of Overmilled Rice

Although all of the various components, except for starch, were most concentrated in the periphery of the kernel—decreasing as the center was approached, the constituents varied greatly in the degree of concentration. The most evenly distributed were sodium, protein and riboflavin which had ratios of concentration in the outer layers compared with the whole kernel of 1.5, 1.9 and 4.5, respectively, with the residual kernel accounting for 70 to 80% of the amount in the whole kernel (Kennedy *et al.* 1974, 1975; Kennedy and Schelstraete 1975A). Components with the steepest gradient from periphery to center were fat, iron and phytic acid with ratios of 15, 17 and 22.5, respectively, while the

TABLE 11.10. NUTRIENTS IN RICE PER 100 KCAL

Nutrient (RDA[1])	Male	Female	Brown[2] rice	Milled[2] rice	Wk	Overmilled Rice[3]			
Sex: / Years of age:	19–22	23–50				Fl	F2	F3	R
Protein (g)	1.5	2.3	2.1	1.8	1.85	3.7	3.5	3.1	1.68
Calcium (mg)	27	40	8.8	6.5	6.2	45	19.2	13.2	4.2
Phosphorus (mg)	27	40	61	26	31.5	294	174	103	20.2
Ca/P	1	1	0.14	0.25	0.20	0.15	0.11	0.13	0.21
Iron (mg)	0.33	0.9	0.44	0.22	0.16	3.1	1.8	1.3	0.05
Thiamin (mg)	0.05	0.05	0.10	0.02	0.03	0.26	0.16	0.11	0.02
Riboflavin (mg)	0.06	0.06	0.014	0.008	0.01	0.05	0.02	0.02	0.01
Niacin (mg)	0.67	0.65	1.30	0.44	0.38	5.4	2.6	1.5	0.18
					0.80[4]	6.3[4]	14.4[4]	2.4[4]	0.52[4]
Pyridoxine (mg)	0.067	0.10	—	0.034[4]	0.034	0.29	0.19	0.13	0.022
Magnesium (mg)	12	15	—	7.0[3]	7.0	79	51	30	2.0
Zinc (mg)	0.5	0.75	0.45	0.38	—	—	—	—	—

[1]RDA: Recommended daily allowance. Calculated from Food and Nutr. Board, Nat'l Res. Counc. (1974).
[2]Calculated from Table 11.1
[3]Calculated from Kennedy et al. (1974, 1975) and Kennedy and Schelstraete (1975A). Kcal per 100 g for WK, whole kernel; Fl, first pass flour; F2, second pass flour; F3, third pass flour; and R, residual kernel, estimated by calculation are 400, 375, 385, 410 and 400, respectively.
[4]Parboiled rice.

residual kernel accounted for only ⅓ or less of the amounts in the whole endosperm, none for phytic acid.

As a result of the concentration in the periphery, all flours contained more of the nutrients studied than did brown rice (Table 11.9), for the first pass flour from nearly twice as much protein to 4 or 5 times as much calcium, phosphorus and niacin and 7 times as much iron.

On an energy basis, the first pass flours provided more than sufficient nutrients to meet the RDA for both young men and adult women, except for riboflavin, for which the flours provided 83% of the allowance. For men amounts of niacin, iron and phosphorus were high—8, 9 and 11 times the RDA, respectively.

With the second and third pass flours, although the amounts of the nutrients studied progressively decreased, they were still several times greater than the RDA for both men and women with the exception of calcium and riboflavin neither of which met the allowance. Niacin was particularly high in the second pass flours from the parboiled rices, about 22 times the RDA.

Residual kernels were deficient in all nutrients with respect to dietary allowances except that protein for young men was sufficient. Phosphorus supplied 75% of the RDA and niacin in parboiled rice, 78%. All other nutrients supplied only 40% or less.

Varietal and Other Differences

In assessing the nutritive qualities, mean values of a variable number of samples of rice have been used. This may be misleading in some instances where variations in the whole grain occur due to environmental factors, varietal differences, or in the case of milled rice as a result of the amount of bran removed.

In our work with from 6 to 12 lots of rice the original raw milled rices differed in the amount of the constituent, ratios of highest to lowest, by factors ranging from as low as 1.03 in the case of starch to a high of 9 for iron. Starch, phytic acid, and riboflavin (ratios of high to low of 1.03 to 1.2) showed the least variation in concentration among lots. Ratios for 10 of the 18 constituents studied were from 1.2 to 2.2. Niacin, magnesium and fat, silicon and iron had the greatest variations, with ratios of 3, 4, 8 and 9, respectively. Consequently these variations should be kept in mind when using the generalizations on nutritive value as just presented.

Non-nutrients

In addition to increased concentrations of many nutrients, flours from the outer layers contained large amounts of silicon, phytic acid and fiber.

Two long grain rices contained 1.77 and 1.08% silicon, 17 and 19 times that of the original milled rices. Flours from the medium and short grain rices had smaller amounts, 0.14 to 0.46% silicon while the residual kernels had a mean value of only 0.004%. Whether the presence of the large amounts found in the flours is desirable or detrimental is at present undetermined.

Phytic acid was also found in the outer layers of the milled rices, 3% in the first pass flours, 1.7 and 1.0% in the second and third passes. Phytate phosphorus accounted for about 30% of the total phosphorus in the original milled rices, from 60 to 80% in the flours from all three passes, and none in the residual kernels. Whether phytic acid aids or prevents the utilization of mineral elements should be studied.

Fiber is also concentrated in the outer layers. Flours from the first, second and third fractions had about 3, 2 and 1.5% fiber as compared with 0.5 and 0.25% in the original residual kernels (Houston 1967). Dietary fiber is currently being studied by a number of investigators to elucidate its effect on the utilization of nutrients.

Evaluation and Uses of High Protein Rice Flours and Residual Kernel

Because of the high concentration of many of its valuable nutrients especially the B-vitamins and good quality protein, the high protein rice flours from the outer layers of the endosperm have been suggested for use in foods for infants, the aged and others requiring special diets (Houston 1967). Sodium is concentrated only slightly so that the flours would be useful in low-sodium diets. Before extensive use is made of these flours, however, long term feeding tests on animals are recommended to determine whether or not the relatively large amounts of phytic acid, silicon and fiber are harmful.

Residual kernels are low in nearly all the nutrients but this deficiency is balanced by the very low amounts of non-nutrients such as phytic acid and fiber. As a result the bioavailability of the smaller amounts of mineral elements in the residual kernel may compensate for the lower concentrations. Bioavailability studies would be helpful. Residual kernels should be fortified with the necessary nutrients, or the nutrients furnished by other components of the diet.

REFERENCES

BARBER, S. 1969. Basic studies on aging of milled rice and application to discriminating quality factors. U.S. Dep. Agric., Agric. Res. Serv., For. Res. and Tech. Prog. Div. Final Rep. Proj. No. *E-25-AMS-(9).*

BARBER, S. 1972. Milled rice and changes during aging. *In* Rice: Chemistry and Technology. D.F. Houston (Editor). Amer. Assoc. Cereal Chem., St. Paul, Minnesota.

BARBER, S., BENEDITO DE BARBER, C., GUARDIOLA, J.L. and ALBER-OLA, J. 1967. Chemical composition of rice. IV. Distribution of sugars in the milled kernel. Revista Agroquim. Tecnol. Aliment. 7, 346-353. (Spanish)

CASAS, A., BARBER, S. and CASTILLO, P. 1963. Quality factors of rice. X. Distribution of fat in the endosperm. Revista Agroquim. Tecnol. Aliment. 3, 241-244. (Spanish)

DEL ROSARIO, A.R., BRIONES, V.P., VIDAL, A.J. and JULIANO, B.O. 1968. Composition and endosperm structure of developing and mature rice kernel. Cereal Chem. 45, 225-235.

FOOD AND NUTR. BOARD, NATL. RES. COUNC. (FNB). 1974. Recommended Dietary Allowances, 8th edition. Natl. Acad. Sci., Washington, D.C.

FRANZ, K.B. 1977. Bioavailability of zinc in selected cereals and legumes. Ph.D. dissertation. University of California, Berkeley.

HINTON, J.J.C. 1948. The distribution of vitamin B_1 in the rice grain. Brit. J. Nutr. 2, 237-241.

HINTON, J.J.C. and SHAW, B. 1953. The distribution of nicotinic acid in the rice grain. Brit. J. Nutr. 8, 65-71.

HOGAN, J.T., NORMAND, F.L. and DEOBALD, H.J. 1964. Method for removal of successive surface layers from brown and milled rice. Rice J. 67 (4) 27-34.

HOGAN, J.T., DEOBALD, H.J., NORMAND, F.L., MOTTERN, H.H., LYNN, L. and HUNNELL, J.W. 1968. Production of high-protein rice flour. Rice J. 71 (11) 5-6, 8-9, 32.

HOUSTON, D.F. 1967. High-protein flour can be made from all types of milled rice. Rice J. 70 (9) 12-15.

HOUSTON, D.F. 1972. Rice bran and polish. *In* Rice: Chemistry and Technology. D.F. Houston (Editor). Amer. Assoc. Cereal Chem., St. Paul, Minnesota.

HOUSTON, D.F., HILL, B.E., GARRETT, V.H. and KESTER, E.B. 1963. Organic acids in rice and some other cereal seeds. J. Agric. Food Chem. 11, 512-517.

HOUSTON, D.F. and KOHLER, G.O. 1970. Nutritional Properties of Rice. Natl. Acad. Sci., Washington, D.C.

HOUSTON, D.F., MOHAMMAD, A., WASSERMAN, T. and KESTER, E.B. 1964. High-protein rice flours. Cereal Chem. 41, 514-523.

JULIANO, B.O. 1966. Physicochemical data on the rice grain. Int. Rice Res. Inst. Tech. Bull. 6, Los Baños, Laguna, Philippines.

JULIANO, B.O. 1972. The rice caryopsis and its composition. *In* Rice: Chemistry and Technology. D.F. Houston (Editor). Amer. Assoc. Cereal Chem., St. Paul, Minnesota.

KENNEDY, B.M. and SCHELSTRAETE, M. 1974. Chemical, physical, and nutritional properties of high-protein flours and residual kernel from the over-milling of uncoated milled rice. II. Amino acid composition and biological evaluation of the protein. Cereal Chem. *51*, 488-457.

KENNEDY, B.M. and SCHELSTRAETE, M. 1975A. Chemical, physical, and nutritional properties of high-protein flours and residual kernel from the over-milling of uncoated milled rice. III. Iron, calcium, magnesium, phosphorus, sodium, potassium and phytic acid. Cereal Chem. *52*, 173-182.

KENNEDY, B.M. and SCHELSTRAETE, M. 1975B. A note on silicon in rice endosperm. Cereal Chem. *52*, 854-856.

KENNEDY, B.M., SCHELSTRAETE, M. and DEL ROSARIO, A.R. 1974. Chemical, physical, and nutritional properties of high-protein flours and residual kernel from the overmilling of uncoated milled rice. I. Milling procedure and protein, fat, ash, amylose, and starch content. Cereal Chem. *51*, 435-448.

KENNEDY, B.M., SCHELSTRAETE, M. and TAMAI, K. 1975. Chemical, physical, and nutritional properties of high-protein flours and residual kernel from the overmilling of uncoated milled rice. IV. Thiamin, riboflavin, niacin, and pyridoxine. Cereal Chem. *52*, 182-188.

LITTLE, R.R. and DAWSON, E.H. 1960. Histology and histochemistry of raw and cooked rice kernels. Food Res. *25*, 611-622.

MASIRONI, R., KOIRTYOHANN, S.R. and PIERCE, J.O. 1977. Zinc, copper, cadmium and chromium in polished and unpolished rice. Sci. Total Environ. *7*, 27-43.

MILNER, M. 1965. High protein fractions from white rice and simple methods for their production. Food and Agric. Organ. U.N./World Health Organ./U.N. Children's Emerg. Fund Protein Adv. Group News Bull. *5*, 39.

MITSUDA, H., MURAKAMI, K., KUSANO, T. and YASUMOTO, K. 1969. Fine structure of protein bodies isolated from rice endosperm. Arch. Biochem. Biophys. *130*, 678-680.

MITSUDA, H., YASUMOTO, K., MURAKAMI, K., KUSANO, T. and KISHIDA, H. 1967. Studies on the proteinaceous subcellular particles in rice endosperm: Electron-microscopy and isolation. Agric. Biol. Chem. (Tokyo) *31*, 293-300.

MORRIS, E.R. and ELLIS, R. 1976. Isolation of monoferric phytate from wheat bran and its biological value as an iron source to the rat. J. Nutr. *106*, 753-760.

NORMAND, F.L., SOIGNET, D.M., HOGAN, J.T. and DEOBALD, H.J. 1966. Content of certain nutrients and amino acids pattern in high-protein rice flour. Rice J. *69* (9) 13-18.

NUNOKAWA, Y. 1972. Saké. *In* Rice: Chemistry and Technology. D.F. Houston (Editor). Amer. Assoc. Cereal Chem., St. Paul, Minnesota.

OGAWA, M., TANAKA, K. and KASAI, Z. 1975. Isolation of high phytin containing particles from rice grains using an aqueous polymer two phase system. Agric. Biol. Chem. (Tokyo) *39*, 695-700.

PADUA, A.B. and JULIANO, B.O. 1974. Effect of parboiling on thiamin, protein and fat of rice. J. Sci. Food. Agric. 25, 697-701.

PRIMO, E., BARBER, S. and TORTOSA, E. 1965. Chemical composition of rice. II. Free fatty acids, neutral fats, and phospholipids: Their distribution in the kernel and their fatty acid composition. Revista Agroquim. Tecnol. Aliment. 5 (4) 458-466. (Spanish)

PRIMO, E., CASAS, A., BARBER, S. and BENEDITO DE BARBER, C. 1963. Quality factors of rice. IV. Distribution of nitrogen in the endosperm. Revista Agroquim. Tecnol. Aliment. 3 (1) 22-26 (Spanish)

RESURRECCION, A.P. and JULIANO, B.O. 1975. Fatty acid composition of rice oils. J. Sci. Food. Agric. 26, 437-439.

SAUNDERS, R.M. 1977. Unpublished data. Western Reg. Res. Cntr., Berkeley, California.

SIMPSON, J.E. et al. 1965. Quality evaluation studies of foreign and domestic rices. U.S. Dep. Agric., Agric. Res. Serv. Tech. Bull. 1331.

SUBRAHMANYAN, V., SREENIVASAN, A. and DAS GUPTA, H.P. 1938. Studies on quality in rice. I. Effect of milling on the chemical composition and commercial qualities of raw and parboiled rices. Indian J. Agric. Sci. 8, 459-486.

TANAKA, K., YOSHIDA, T., ASADA, K. and KASAI, Z. 1973. Subcellular particles isolated from aleurone layer of rice seeds. Arch. Biochem. Biophys. 155, 136-143.

TANAKA, K., HAYASHIDA, S. and HONGO, M. 1975. The relationship of the feces protein particles to rice protein bodies. Agric. Biol. Chem. (Tokyo) 39, 515-518.

WATT, B.K. and MERRILL, A.L. 1963. Composition of foods—Raw, processed, prepared. U.S. Dep. Agric., Agr. Res. Serv., Agric. Handbk 8.

WILLIAMS, K.T. and BEVENUE, A. 1953. A note on the sugars in rice. Cereal Chem. 30, 267-269.

YOSHIDA, S., OHNISHI, Y. and KITAGISHI, K. 1959. The chemical nature of silicon in rice plant. Soil and Plant Food 5, 23-27.

YOSHIDA, S., OHNISHI, Y. and KITAGISHI, K. 1962. Chemical forms mobility and deposition of silicon in rice plants. Soil Sci. Plant Nutr. 8, 107-113.

12

Rice Flours in Baking

Bor S. Luh
Yuan-Kuang Liu

Presently there are 2 types of commercial rice flour available in the United States (Hogan 1977). The first is produced from waxy or glutinous rice, which is commercially grown in limited quantities in California. The waxy rice flour has superior qualities for use as a thickening agent for white sauces, gravies, puddings and in oriental snack foods. It can prevent liquid separation (syneresis) when these products are frozen, stored and subsequently thawed. A characteristic of waxy flour is that it has little or no amylose. Since the waxy rice starch is essentially amylopectin, flour prepared from it has this unique food use property. The other type of rice flour is prepared from broken grains of ordinary raw or parboiled rice. The flour prepared from parboiled rice is essentially a precooked flour. It differs from wheat flour in baking properties because it does not contain gluten, and its doughs do not readily retain gases generated during baking. There is, however, a steady basic demand for rice flours for use in baby foods, breakfast foods and meat products; for separating powders for refrigerated, preformed, unbaked biscuits, dusting powders, breading mixes; and for formulations for pancakes and waffles. These uses are sufficient to sustain a market for rice flour production.

Rice is largely consumed as a whole grain. Some breakage, however, is unavoidable. Matthews *et al.* (1970) have shown that some of the rice kernels are cracked while still in the husk and some breakage occurs during harvesting, handling, drying and milling. The amount of broken rice produced annually in the United States is 363 to 408 million kg, about 15% of the rice milled.

The larger pieces of broken rice sell for a little more than ½ the price of whole rice, whereas the smaller fragments sell for less than ½ of comparable whole grains. Therefore, these smaller pieces are used for grinding into rice flour or for brewing.

About ⅓ of the broken rice produced in the United States is used by brewers and a small portion by cereal manufacturers and baby food formulators. The amount of rice used in the brewing industry is generally a function of the price of broken rice (USDA 1970). The brewing industry would use more broken rice if the price were competitive with that of other grains, such as corn grits.

Some of the broken rice is ground into flour, which is listed in a statistical tabulation of the products of milling or processing (USDA 1968A). The amount used is determined by the price paid for the flour (which must be high enough to cover the cost of grinding and related charges), since there is a ready market for this fraction at a lower price in the brewing industry.

Incorporation of 30% rice flour in wheat flour is acceptable for bread baking (Mosqueda-Suarez 1958). Replacement of 5% wheat flour with rice flour can scarcely be detected. In countries where rice production is more favorable than wheat or corn because of climatic conditions, it is highly desirable to partially substitute rice flour for wheat flour in bread and bakery products.

There is also a market for packaged rice flour which is sold in specialty food stores to people with allergies to wheat and for use in special recipes. Recipes for such baked products have been developed (Goertz et al. 1965; USDA 1968B).

PROPERTIES OF RICE FLOURS

Rice flours made from long, medium, and short grain and waxy rice are available commercially. Since rice flours are made from broken milled rice, their chemical composition is the same as that of whole rice. There are, however, varietal differences in protein, lipid, starch contents, and the amylose and amylopectin ratios in the starch. The proximate analyses of some rice are presented in Table 12.1 (Ejlali et al. 1978). Compositional differences contribute to the diversity of chemical and physical properties of various rice flours, such as viscometric properties, starch gelatinization temperatures, water absorption and other characteristics. The rheological properties of starch slurries and gels have been reviewed by Matz (1962) and Kruger and Murray (1976).

Developers of new rice foods should recognize the wide diversity in composition and properties of various rice flours. Occasionally, discovery of a unique property of a variety has led to new uses for rice flours made from it. For example, the freeze resistance of waxy rice flour pastes led to their use in frozen gravies (Hanson et al. 1953). The viscosity behavior of

TABLE 12.1. PROXIMATE ANALYSES OF FOUR RICE VARIETIES GROWN IN CALIFORNIA IN 1976

Variety	Grain type	Amylose %	Amylopectin %	Protein %	Fat %	Ash %
S-6	Short	19.0	56.5	6.50	2.05	1.22
M-5	Medium	22.5	54.7	6.60	1.50	1.07
74-Y-52	Medium	24.0	56.2	6.75	1.82	1.55
72-3764	Long	27.8	54.2	7.07	1.88	1.65

Source: Ejlali *et al.* (1978)

rice flour pastes provides information for the application of rice flour to food formulation. The varieties used in pancake mixes and expanded breakfast cereals are examples. Rice flours from each variety, with the exception of the waxy type, have characteristic viscosity patterns during the heating and cooling cycles of their pastes. The changes in viscosity depend largely on the composition of the starch, and to a lesser extent on the protein and oil components. Usually long grain rice containing starch with an amylose content of over 22% have a relatively low peak viscosity and form a rigid gel on cooling (high set-back viscosity). Those with starches low in amylose have high peak and low set-back viscosities.

The Brabender Amylograph can be used to study rice flour rheology. The amylograph data on viscosities of various rices and their starches have been reviewed by Horiuchi (1967). Voisey *et al.* (1972) used the viscometer for studying starch slurry behavior during cooking. In general, the amylograms of flour and of starch from a given sample were indicative of its behavior as related to peak viscosity, breakdown and set-back viscosity. They exhibited the generally recognized relationships based on amylose content noted above. The α-amylase content of milled rice, unlike that of wheat, had a negligible effect on the amylogram, except in the case of waxy rice, which contains appreciable amounts of α-amylase in the endosperm. Water washing flours from waxy rices increased their peak viscosities by removing a large portion of the amylolytic activity.

Besides varietal differences in the properties of rice flours, other modifications can be induced by processing the rice before grinding. Rice flours are used in preparation of snack and convenience foods. Commercial pregelatinized flours made from parboiled rice and from various quick cooking rices are used in the meat industry because of their bland taste, high water absorption and improvement in texture.

Lynn and Anderson (1969) patented a process for making modified rice flours by partially or totally gelatinizing rice through cooking in an excess of water for various lengths of time, followed by water washing, drying and grinding. Flours made by controlled, limited gelatinization of rice may offer texturizing properties for baby foods and breakfast foods. The extent of steaming in the parboiling process can be estimated by the amount of reduction induced in the amylograph peak viscosity of the original sample (Ferrel and Pence 1964).

Waxy Rice Flour

Waxy rice flour, or sweet rice flour, has the usual viscosity characteristics of waxy type flours from corn or sorghum (Whistler and Paschall 1967). It contains less than 0.5% amylose in the starch and an appreciable amount of α-amylase. This flour has a lower peak viscosity than some of the short grain rices, probably because of its amylolytic activity, and it has practically no set-back viscosity.

Waxy rice flour is different from other rice flours or starches in its resistance to liquid separation (syneresis) during freezing and thawing. The remarkable stability of cooked waxy rice flour pastes after repeated freeze-thaw cycles was observed during a study of factors determining the stability of white sauces and gravies commonly used with frozen, precooked meat and vegetables (Hanson et al. 1953).

Waxy rice flours and starches are superior to other grain starches and flours because they are more stable under freeze-thaw treatment than any other flours or starches. This behavior may be attributable to the virtual lack of amylose starch. This flour, mixed with wheat flour even at 40 to 60% levels, stabilizes sauces and gravies held at $0°C$ ($32°F$) for 5 to 6 months. When waxy rice flours are used as a sole source of starch, stability can be maintained for a year or longer. It shows no retrogradation on cooking. However, this lack of retrogradation is equally evident in the other waxy grain flours and particularly in the purified starches from these flours. The unusual stability of waxy rice flour over other starches or flours may be due to its special chemical structure or to the small size of the starch granules. The lower stability of purified waxy rice starches, when compared to the flour, may be caused by the amylase activity or incipient gelatinization induced by the extensive alkali treatment employed in isolating the starch.

High-protein Rice Flour

Air classification was very satisfactory for the separation of high protein fractions from wheat (Pfeifer and Griffin 1960). However, when this process was tried on rice flour, there was no success (Stringfellow et al. 1961). The difference may be attributed to the vitreous nature of rice, its very small starch granules, and the intimate dispersions of the protein bodies throughout the starch matrix in the endosperm. Some separation of protein was achieved from flour that had been ground very finely, but segregation was not sufficient to be practicable.

Primo *et al.* (1963) made intensive efforts to separate a high protein flour by peripheral abrasion of the milled rice kernel. They analyzed a peripheral fraction of the kernel and demonstrated that the outer layer was higher in protein. Hogan *et al.* (1964) constructed a rice mill consisting of a knurled disc, about 12.7 cm in diameter, which rotated in a horizontal plane. A 7.62 cm length of 5.08 cm diameter glass tubing was positioned vertically above and to one side of the disc so that the lower end of the tubing was almost in contact with the knurled surface. The milled rice sample was put into the tube, and the circulating action caused by rotation of the disc resulted in an even, mild abrasion of each kernel, with almost no breakage or temperature change. The flour produced by the abrasion was discharged into a hood that surrounded the entire apparatus. The protein content of the outer layer was more than twice that of the original kernels. This high protein layer extended into the grain to a depth equivalent to more than 20% by weight of the milled kernel. Flours having a protein content as high as 20 to 22% were obtained from milled rices having a protein content of 8.5%. In these experiments, removing 6.8% of the peripheral material as high protein flour improved the appearance and cooking characteristics of the residual kernels.

Houston *et al.* (1964) compared the air classification and peripheral abrasion methods. They used a CeCoCo mill, a commercial pearling machine of Japanese origin, to produce high-protein rice flour by surface abrasion. The protein content of the outer layers was more than twice that of the original milled kernel. They concluded that the abrasion process was economically more feasible than the air classification method.

Houston *et al.* (1968) reported on analyses of successively removed layers from milled rice. The flours obtained from the outer layer were very rich in protein and other important nutrients such as calcium, phosphorus, lipids, thiamin, niacin and riboflavin. Starch was correspondingly lower in these peripheral layers. The concentrations of the various nutrients extended to various depths in the kernel. Forty percent of the kernel was removed before the protein content of the innermost layer was equivalent to that of the whole kernel.

The amino acid distribution in the rice protein was about the same in all layers and in the residual kernel (Normand *et al.* 1966). High protein flour was also produced from rice processed in a commercial Satake rice whitening machine (Hogan *et al.* 1968). All flours obtained from the various abraded layers were similar to the initial laboratory and pilot scale products in composition and properties. By a similar process, a low oil, high-protein rice flour can be made from the product of the solvent milling process.

Rice Bran Flour

The solvent process for simultaneously milling and removing rice oil has renewed interest in the possibility of obtaining a high protein food flour from rice bran. After oil extraction, rice bran is much more stable and easier to grind and fractionate than that which is not defatted. Rice bran is higher in protein content than any other portion of the rice kernel. Houston and Mohammad (1966) tried to upgrade extracted bran and white bran (polish) by air classification and by sieving. The 25 to 30% fraction of the extracted rice bran passing through a 100-mesh screen was a light tan flour having a protein content 50% greater than the original bran. The crude fiber content—5 to 7% compared with the original 11 to 12%—was further reduced to 4.5% by regrinding and passing through a 140-mesh screen.

A new approach to obtaining a high protein food from rice bran is to extract the protein from defatted bran. Chen and Houston (1970) reported that up to 80% of the protein in bran could be extracted with sodium hydroxide solutions. They recovered 50% of the protein at pH 11, obtaining a 40% protein product by neutralizing and drying the extract, and an 85% concentrate representing 37% of the bran protein by precipitating the dissolved proteins at pH 5.5. These protein concentrates from rice bran are potential sources of soluble high protein, low fiber products from rice.

Barber and Barber (1974) made a status report on utilization of rice bran as food and feed. They quoted Houston's report (1972) that little use is presently made in most countries of the nutritional qualities of rice bran for human foods, although such use has long been suggested. Since then, the situation has hardly changed. Solution of the problems of stabilization of bran on an industrial scale, and of fiber removal from bran and the extraction of its proteins, introduces a new stage in the utilization of bran in foods. But rice bran foods are not yet on the shelves of the supermarkets, nor do they cover even a minimal part of the protein and calorie deficiencies which they are capable of covering. In many regions, people eat what they like regardless of the nutritional value of the food. Some prejudice against rice bran products is to be expected. Rice bran has been traditionally considered as an animal feed.

There is a potential for the use of bran in baked goods. It has been claimed that X-M bran increases dough yields because of increased water absorption, contributes to an attractive tan crumb and crust, does not disturb fermentation or mixing tolerances of the dough, causes baked products to remain fresher and more moist due to moisture retention, and adds significant essential amino acids, minerals and vitamins to the baked goods (Lynn 1969).

Rice Starch

Rice starch is a special product used as a major component of face powder. The fine particle size of rice starch makes it especially suitable for cosmetic use. According to Matz (1970), simple washing methods will not release rice starch granules from their protein matrix. Chemical treatment is necessary to disperse the rice proteins. For example, broken rice may be steeped at ambient temperature for 24 hr in 5 times its weight of a 0.3% caustic soda solution as the first step in rice starch production. The soaking solution may be heated to 48.8°C (120°F) to speed up the extraction process.

The caustic treated granules are washed and then dried before being ground into flour. The flour is then mixed with 10 times its weight of 0.3% caustic soda solution and stirred for 24 hr. The starch is allowed to settle and the supernatant solution, which contains most of the rice protein, is removed. Washing with water, settling and decanting are employed to remove most of the soluble materials from the starch granules. The washed starch is dewatered by filtering or centrifuging; complete removal of residual alkali is a very important step which must be carefully controlled. The washed starch is dried in ovens or rotary drum driers; the rice starch cake is ground to the desired particle size and sieved.

The rice protein in the combined effluents is precipitated by the addition of hydrochloric acid. The supernatant fluid is discarded, and the precipitated material is partially dewatered in a filter press and finally dried in a rotary drier. The product can be used as a protein supplement for cattle feed (Matz 1970).

RICE FLOUR PRODUCTS

Rice Snack Foods

In the Oriental countries, sweet rice flour is a very important raw material for making snack foods. The details of this subject are covered in Chapter 20.

Rice Cookies

Przybyla and Luh (1977) used rice flours made from medium grain Calrose rice in making rice-oatmeal and rice-peanut butter cookies. The results are summarized as follows:

Rice-oatmeal Cookies.—Two typical recipes for rice-oatmeal cookies are presented in Table 12.2. The shortening and sugar were mixed for 1

TABLE 12.2. RECIPES FOR RICE-OATMEAL COOKIES

Ingredients	16.4% rice flour Weight (g)	%	26.4% rice flour Weight (g)	%
Shortening	80	13.2	80	13.2
Sugar (granulated/brown)	90/90	29.7	60/60	19.8
Eggs	50	8.3	50	8.3
Vanilla extract	3	0.5	3	0.5
Water	20	3.3	20	3.3
Oats	100	16.5	100	16.5
Rice flour	100	16.5	160	26.4
Nonfat dry milk	40	6.6	40	6.6
Cinnamon	2	0.3	2	0.3
Baking soda	3	0.5	3	0.5
Walnuts (chopped)	28	4.6	28	4.6

Source: Pryzybyla and Luh (1977).

min at medium speed in a Hobart electric mixer. Then eggs, vanilla and rolled oatmeal were added. The mixture was mixed in the Hobart mixer for 2 min at medium speed and then for 1 min at high speed. The rice flour, nonfat dry milk, cinnamon, baking soda and chopped walnuts were then added. The product was mixed for 1 min at low speed and then rolled out with a cloth covered rolling pin to a thickness of 7 mm. Wooden guides were used to ensure even height. The dough was cut with a cookie cutter in circular pieces 60 mm in diameter, and the product was baked on a greased sheet in a rotary oven at 177°C (350°F) for 10 min and then cooled to ambient temperature.

Rice-peanut butter Cookies.—The recipe for rice-peanut butter cookies is presented in Table 12.3. The margarine, shortening, peanut butter, sugar, eggs and water were beaten together for 2 min on medium speed and for 1 min on high speed. Flour, baking soda and nonfat dry milk were added, and the mixture was beaten for another minute on medium speed. The cookies were formed in the same way as the rice-oatmeal cookies mentioned above and baked for 9 min at 177°C (350°F).

Spread of Cookies.—The dough handled well when the AACC (1962) Spread-factor Test for baking cookies was followed (Przybyla and Luh 1977). Nonfat dry milk was added to increase protein content and to improve crumb color. In 2 of the formulas, the percentage of flour was increased to make it the major ingredient so that differences in the cookies due to rice versus wheat flour would be emphasized. The rice flour cookies exhibited more spread, as shown in Table 12.4 where the spread factors are recorded.

The spread mechanism is known to be a function of the total availability of water and other factors subordinate to it. Flour, sugar and water are the major components of cookie dough. The less hydrophilic the flour is, the more water will be available for the sugar. Spread increases

TABLE 12.3. RICE-PEANUT BUTTER COOKIES (27.6% FLOUR)—A MODIFICATION OF THE CROCKER'S RECIPE (1970)

Ingredients	Weight (g)	%
Margarine	55	7.6
Shortening	40	5.5
Peanut butter	150	20.6
Sugar (granulated/brown)	45/130	24.1
Eggs	50	6.9
Water	5	0.7
Flour	200	27.6
Baking soda	3	0.4
Salt	3	0.4
Nonfat dry mild	45	6.2

Source: Przybyla and Luh (1977).

when the amount of sugar is increased. Milk solids, egg white and whole egg tend to decrease spread while shortening, leavening agents and egg yolk usually increase spread. The greater spread of the rice flour cookies shows that the rice flour is less hydrophilic than the wheat flour.

The greater spread of rice flour may also be partly attributed to the lack of gluten in rice flour. Gluten forms an elastic dough when mixed with water, and is capable of holding air, resulting in a spongy structure when baked.

TABLE 12.4. SPREAD RATIO OF RICE AND WHEAT FLOUR COOKIES

	Spread ratio width/thickness
Rice oatmeal 16.4% flour	10.73 ± 0.26
Wheat oatmeal 16.4% flour	8.07 ± 0.18
Rice oatmeal 26.4% flour	5.58 ± 0.12
Wheat oatmeal 26.4% flour	5.13 ± 0.12
Rice peanut butter 27.4% flour	6.93 ± 0.14
Wheat peanut butter 27.4% flour	6.66 ± 0.13

Source: Przybyla and Luh (1977).

Proximate Analyses.—Proximate analyses of the rice and wheat flour cookies are presented in Table 12.5. The peanut butter cookies contained more protein than the others.

Wheat flour cookies had a slightly higher lipid content than the rice flour cookies. Although rice is slightly higher in lipid content than wheat, milled rice has a lower lipid content than wheat flour. Lipid content

TABLE 12.5. PROXIMATE ANALYSES OF RICE AND WHEAT FLOUR COOKIES

Cookies	Moisture (%)	Protein (%)	Lipids (%)	Ash (%)
Rice oatmeal 16.4% flour	4.72	8.77	16.79	1.71
Wheat oatmeal 16.4% flour	5.07	8.83	17.03	1.79
Rice oatmeal 26.4% flour	5.06	8.64	16.87	1.76
Wheat oatmeal 26.4% flour	5.21	9.48	19.63	1.88
Rice peanut butter 27.6% flour	4.11	10.58	22.09	2.55
Wheat peanut flour 27.6% flour	4.22	11.64	23.81	3.00

Source: Przybyla and Luh (1977).

decreases as the center of the rice kernel is approached (Kennedy *et al.* 1974).

Ash content was slightly higher in the wheat flour cookies than the rice flour cookies. For the oatmeal cookies, ash content varied from 1.71 to 1.88%, whereas for the peanut butter cookies it increased to 2.55 and 3%.

Riboflavin and Thiamin Contents.—Riboflavin and thiamin contents of the cookies, and of the ingredients highest in these vitamins (the flours and oatmeal), were reported by Przybyla and Luh (1977). The results are presented in Table 12.6. Enriched wheat flour was used, whereas the rice

TABLE 12.6. RIBOFLAVIN AND THIAMIN CONTENT OF RICE FLOUR, WHEAT FLOUR, OATMEAL, AND RICE AND WHEAT FLOUR COOKIES

Cookies	Riboflavin (mg/100 g)	Thiamin (mg/100 g)
Rice flour	0.04	0.28
Wheat flour	0.88	1.98
Oatmeal	0.22	0.54
Rice oatmeal 16.4% flour	0.04	0.12
Wheat oatmeal 16.4% flour	0.15	0.27
Rice oatmeal 26.4% flour	0.04	0.10
Wheat oatmeal 26.4% flour	0.24	0.34
Rice peanut butter 27.6% flour	0.01	0.05
Wheat peanut butter 27.6% flour	0.25	0.38

Source: Przybyla and Luh (1977).

flour was unenriched. The oatmeal cookies lost approximately 10% ribo-flavin during baking, but the peanut butter cookies had a slight increase in riboflavin, probably due to the amount contributed by the peanut butter. The thiamin content of the cookies decreased by approximately 30%. The result agrees with data on loss of riboflavin and thiamin in baking.

Sensory Analysis.—A panel compared the flavor, color, texture and acceptability of the rice and wheat cookies on a 9-point scale. Two by two (with interaction) analysis of variance tables was made to compare the rice versus wheat cookies for each of the four attributes.

Color.—There was no significant difference in color between the cookie samples. Using rice flour instead of wheat flour did not affect the color of the cookies.

Texture.—The cookies' texture was rated from very crispy to very soft. Judges' scores for the peanut butter cookies and oatmeal cookies, with 26.4% flour, did not differ significantly. Texture of the rice and wheat oatmeal cookies, with 16.4% flour, differed and was significant at the 99% probability level, with the rice cookies being more crisp. When the amount of flour was decreased to 16.4%, the rice flour produced crispier cookies.

Acceptability.—The cookies were rated as highly acceptable. Accep-tability did not differ significantly between the rice and wheat flour peanut butter cookies. The rice oatmeal cookies were rated more ac-ceptable than the wheat flour ones, significant at the 95% probability level.

Rice Bread

The problem associated with rice bread formulation is due to the absence of gluten in rice flour. The manufacture of rice bread without gluten presents considerable technological difficulties, because gluten is the important structure-forming protein. Research has been done using gum or suitable surfactants such as glyceryl monostearate (GMS) as the binding agent. Kim and deRuiter (1968) studied the influence of various surfactants on the loaf volume at different water levels when making nonwheat bread (Table 12.7). Nishita *et al.* (1976) described the devel-opment of a yeast-leavened rice bread formula which consists of mod-ifying a typical wheat bread formula in which wheat flour is completely replaced by rice flour. They compared the effects of hydroxypropyl-methylcellulose, locust bean and guar, sodium carboxymethylcellulose, carrageenan and xanthan gum on the loaf volume of rice bread.

TABLE 12.7. EFFECT OF KIND AND LEVEL OF VARIOUS BAKING AIDS AND OF QUANTITY OF WATER ON BREAD LOAF VOLUME

Baking test no.	Tailoring agent	Quantity % on flour	Water % on flour	Loaf volume ml/kg flour	Loaf volume ml/g bread
1	—	—	85	2600	1.5
2	carboxymethylcellulose	1	70	2400	1.5
3		1	85	2980	1.7
4		1	100	3540	1.9
5	methylethylcellulose	1	70	2400	1.5
6		1	85	2890	1.7
7		1	100	3440	1.9
8	methylcellulose	1	70	2360	1.5
9		1	85	2900	1.7
10		1	100	3380	1.8
11	methocel	1	70	2420	1.5
12		1	85	2800	1.6
13		1	100	3220	1.8
14	visutal 1201 alginate	2	85	3040	1.8
15		2	100	4120	2.2
16		4	120	3440	1.7
17		8	140	2880	1.3
18	manucol SS LL alginate	2	85	3200	1.9
19		2	100	3800	2.1
20		4	120	4000	2.0
21		8	140	4200	1.9
22	guar gum	1	100	3360	1.8
23		4	120	4140	2.1
24		8	140	4600	2.1
25	locust bean gum	1	100	3290	1.8
26		2	100	3400	1.9
27		4	120	4060	2.0
28		8	140	4640	2.1
29	pregelat, potato starch	5	100	3700	2.0
30		10	130	4360	2.1
31		20	140	4200	1.9
32	Pregelat. tapioca starch	10	130	4220	2.0
33	Pregelat. corn starch	10	130	3680	1.7
34	Pregelat. wheat starch	10	130	4360	2.1
35	potato starch phosphate	2	100	3260	1.8
36		8	140	4200	1.9
37	soya lecithin	1	100	3440	1.9
38		4	100	3920	2.1
39	soya lecithin + manucol	4+4	140	5300	2.4
40		4+4	120	4400	2.2
41		2+4	120	4400	2.2
42		2+2	100	3120	1.7
43	glyceryl monostearate	0.5	85	3380	2.0
44		0.5	100	3920	2.1
45		1	85	3600	2.1
46		1	100	3800	2.1
47		1	120	4300	2.1

Source: Kim and deRuiter (1968).

Investigation was carried out by Delgado (1977) to improve bread formulations. By adjusting the levels of sucrose, yeast, water and methocel, an improved recipe was developed. An enriched bread was obtained by introducing nonfat dry milk (Table 12.8). Comparisons of the specific

TABLE 12.8. COMPARISON OF THREE FORMULAS USED FOR BAKING RICE BREADS

	Basic formula	Improved formula	Enriched formula	Commercial white wheat bread	LSD (p=0.05)	LSD (p=0.01)	LSD (p=0.01)
Rice flour (g)	100.0	100.0	100.0	—			
Compressed yeast (g)	3.0	5.0	5.0	—			
Sucrose (g)	7.5	12.5	12.5	—			
Salt (g)	2.0	2.0	2.0	—			
Vegetable oil (g)	6.0	6.0	6.0	—			
Methocel (g)	3.0	4.0	4.0	—			
Water (ml)	75.0	90.0	90.0	—			
Nonfat dry milk (g)	—	—	3.0	—			
Wheat gluten (g)	—	—	1.0	—			
Specific volume (ml/gm)	2.0 ± 0.3	4.5 ± 0.4	4.1 ± 0.1	6.5 ± 0.1	0.5	0.9	1.9
Moisture loss on baking %	15.9 ± 0.3	21.5 ± 1.4	20.9 ± 0.8	—	4.0	NS	NS
Total moisture in bread %	—	25.7 ± 0.4	24.8 ± 0.8	35.6 ± 0.6	2.2	5.1	16.3
Protein content (D.B.)%	—	7.4 ± 0.1	11.0 ± 0.2	13.5 ± 0.1	0.4	0.8	2.6
Sensory score average (scale 1-10)	6.3 ± 0.7	7.3 ± 0.1	7.3 ± 0.4	7.0 ± 0.9	NS	NS	NS

Source: Delgado (1977).

volume, protein and moisture contents were made on the improved, enriched and commercial wheat bread. The primary results showed that rice bread made from these improved and enriched formulas were acceptable to the sensory evaluation panel.

REFERENCES

AMER. ASSOC. OF CEREAL CHEM. (AACC). 1962. Cereal Laboratory Methods. AACC, St. Paul, Minn.

BARBER, S. and BARBER, C.B. 1974. Basic and applied research needs for optimizing utilization of rice bran as food and feed. *In* Proc. Int. Conf. on Rice By-Products Utilization, Vol. 4 Sept. 30–Oct. 2. Valencia, Spain. *In* Rice Bran Utilization: Food and Feed. 1977. Inst. Agron. Technol. Alim., Valencia 10, Spain.

CHEN, L. and HOUSTON, D.F. 1970. Solubilization and recovery of protein from defatted rice bran. Cereal Chem. *47*, 72-79.

DELGADO, C. 1977. Improvement of rice bread. M.S. thesis. University of California, Davis.

EJLALI, M., LUH, B.S. and MICKUS, R.R. 1978. Physicochemical and cooking characteristics of four rice varieties. Ppr. presented to the Rice Tech. Working Group Meeting, Feb. 14–16. College Station, Texas.

FERREL, R.E. and PENCE, J.W. 1964. Use of the amylograph to determine extent of cooking in steamed rice. Cereal Chem. *41*, 1-9.

GOERTZ, G.E., HOOPER, A.S. and ROGERS, P.A. 1965. Effect of variations on rice flour cake. J. Amer. Dietet. Assoc. *46*, 207-209.

HANSON, H.L., NISHITA, K.D. and LINEWEAVER, H. 1953. Preparation of stable frozen puddings. Food Technol. *11*, 462-465.

HOGAN, J.T. 1967. The manufacture of rice starch. *In* Starch, Chemistry and Technology. R.L. Whistler and E.F. Paschall (Editors). Academic Press, New York.

HOGAN, J.T. 1977. Rice and rice by-products. *In* Elements of Food Technology. N.W. Desrosier (Editor). AVI Publishing Co., Westport, Conn.

HOGAN, J.T., NORMAND, F.L. and DEOBALD, H.J. 1964. Method for removal of successive surface layers from brown and milled rice. Rice J. *67* (4) 27-34.

HOGAN, J.T., DEOBALD, H.J., NORMAND, F.L. and HUNNELL, J.W. 1968. Production of high-protein rice flour. Rice J. *71* (11) 5-9.

HORIUCHI, H. 1967. Studies on cereal starches. Part 7. Correlation of rice starch and flour. Agric. Biol. Chem. *31*, 1003. (Japanese)

HOUSTON, D.F. 1972. Rice bran and polish. *In* Rice, Chemistry and Technology. D.F. Houston (Editor). Amer. Assoc. Cereal Chem. St. Paul, Minnesota.

HOUSTON, D.F., MOHAMMAD, A., WASSERMAN, T. and KESTER, E.B. 1964. High-protein rice flours. Cereal Chem. *41*, 514-523.

HOUSTON, D.F. and MOHAMMAD, A. 1966. Air classification and sieving of rice bran and polish. Rice J. *69* (8) 20-21.

HOUSTON, D.F., IWASAKI, T., MOHAMMAD, A. and CHEN, L. 1968. Radial distribution of proteins by solubility classes in the milled rice kernel. J. Agric. Food Chem. *16*, 720-724.

KENNEDY, B.M., SCHELSTRAETE, M. and DEL ROSARIO, A.R. 1974. Chemical, physical and nutritional properties of high-protein flours and residual kernel from the overmilling of uncoated milled rice. Part I. Cereal Chem. *51* (4) 435-448.

KIM, J.C. and DERUITER, D. 1968. Bread from non-wheat flours. Food Technol. *22*, 867-874.

KRUGER, L.H. and MURRAY, R. 1976. Starch texture. *In* Rheology and Texture in Food Quality. J.M. deMan, P.W. Volsey, V.R. Rasper, D.W. Stanley (Editors). AVI Publishing Co., Westport, Conn.

LYNN, L. 1969. Edible rice bran foods. *In* Protein-enriched cereal foods for world needs. M. Milner (Editor). Amer. Assoc. Cereal Chem., St. Paul, Minnesota.

LYNN, L. and ANDERSON, R.M. 1969. Method for preparing precooked rice flour. U.S. Pat. 3,432,309. March 11.

MATTHEWS, J. *et al.* 1970. Relation between head rice yields and defective kernels in rough rice. Rice J. *73* (10) 6-12.

MATZ, S.A. 1962. Food Texture. AVI Publishing Co. Westport, Conn.

MATZ, S.A. 1970. Starch and oil production from cereals. *In* Cereal Technology. S.A. Matz (Editor). AVI Publishing Co., Westport, Conn.

MOSQUEDA-SUAREZ, A. 1958. A new type of bread: Wheat and rice bread. Food Technol. *12*, 15-17.

NISHITA, K.D., ROBERTS, R.L. and BEAN, M.M. 1976. Development of a yeast-leavened rice-bread formula. Cereal Chem. *53*, 626-635.

NORMAND, F.L., SOIGNET, D.M., HOGAN, J.T. and DEOBALD, H.J. 1966. Content of certain nutrients and amino acids pattern in high-protein rice flour. Rice J. *69* (9) 13-18.

PFEIFER, V.F. and GRIFFIN, E.L., Jr. 1960. Fractionation of soft and hard wheat flours by fine grinding and air classification. Amer. Miller Proc. *88* (2) 15-18.

PRIMO, E., CASAS, A., BARBER, S. and BENEDITO DE BARBER, C. 1963. Quality factors in rice. IV. Nitrogen distribution in the endosperm. Rev. Agroquim. Tecnol. Alimentos. *3*, 22−24.

PRZYBYLA, A. and LUH, B.S. 1977. Nutritive value and sensory properties of rice and wheat flour cookies. Ppr. presented to Amer. Assoc. Cereal Chem. 62nd Ann. Meeting. Oct. 24. San Francisco.

STRINGFELLOW, A.C., PFEIFER, V.F. and GRIFFIN, E.L., Jr. 1961. Air classification of rice flours. Rice J. *64* (7) 30-32.

U.S. DEP. AGRIC. (USDA). 1968A. Consumer and marketing summary. Grain division. Rice Annual Market Summary. USDA, Washington, D.C.

USDA. 1968B. Baking for people with food allergy. Home and Garden. Human Nutr. Bull. *147.* USDA, Washington, D.C.

USDA. 1970. Rice Situation. U.S. Dep. Agric. Res. Serv. *RS-15*, 14. USDA, Washington, D.C.

VOISEY, P.W., MURRAY, R. and KEIGHTLY, G. 1972. Viscometer for studying starch slurry behavior during cooking. Canad. Inst. Food Technol. J. *5,* 129-133.

WHISTLER, R.L. and PASCHALL, E.G. 1967. Starch Chemistry and Technology. II. Academic Press, New York.

13

Rice Enrichment with Vitamins and Amino Acids

Robert R. Mickus
Bor S. Luh

Rice is an important food for more than half the world's population; however, it is deficient in some nutrients. Diseases such as beriberi are common in areas where large amounts of rice are consumed. This is partially due to the practice of polishing the grain to remove the bran layer and germ which contain the B complex vitamins. Table 13.1 shows how the nutritional value of rice is altered by milling. The levels of thiamin, riboflavin, nicotinic acid. pantothenic acid, pyridoxine, biotin and iron decrease markedly during the polishing process.

The nutritional value of rice may be further diminished because of the method used to prepare the grain. In many countries rice is packaged in cloth bags which allow the rice to become dirty and infested with insect eggs. The grain must therefore be washed before cooking. Rice is usually cooked in excess water which is subsequently discarded.

Malakar and Banerjee (1959) studied the effects of washing and cooking on the vitamin and mineral content of rice. It was found that ⅓ of the minerals and almost ½ of the water-soluble vitamins are lost upon washing and cooking the rice in 8 vol. of water. Table 13.2 shows the loss of thiamin, riboflavin and niacin by various cooking methods. Loss of vitamins is greatly minimized when just enough water is added so that it will be totally absorbed by the rice.

Kik and Williams (1945) studied the methods for cooking rice in various countries and noted that the amount of water used in washing and cooking the rice has a tremendous impact on vitamin and mineral retention. They reported "people who persist in drastic rinsings and in discarding of cooking water will of necessity sacrifice a large part of any nutritional improvement that can be achieved with present known meth-

TABLE 13.1. EFFECT OF MILLING ON B VITAMINS AND IRON CONTENTS OF RICE

Milled product	Thiamin	Riboflavin	mg/453.6 g of dry weight				
			Nicotinic acid	Pantothenic acid	Pyridoxine	Biotin	Iron
Brown rice	1.1 – 2.3	0.21–0.34	18– 29	7.2– 8.4	4.3– 5.1	0.052–0.057	7.0–8.0
First break	0.4– 0.8	0.09–0.22	10– 15	3.7– 4.8	2.0– 4.0	0.028–0.037	–
Second break	0.3– 0.7	0.09–0.22	9– 14	3.4– 3.6	1.2– 3.5	0.023–0.027	–
Brushed	0.2– 0.5	0.08–0.18	5– 10	3.2– 4.4	1.2– 3.3	0.022–0.024	–
Finished head	0.2– 0.5	0.08–0.14	5– 9	2.9– 3.0	0.9– 2.8	0.015–0.023	3.0–4.0
Bran	5.4–15.0	0.7 –1.5	117–221	30 –37	11 –18	0.17 –0.25	–
Polish	6.8–13.0	0.5 –1.4	94–195	33 –50	13 –50	0.24 –0.38	–

Source: Furter and Lauter (1946).

TABLE 13.2. LOSS OF VITAMINS BY COOKING RICE IN EXCESS WATER

Type of rice and method of cooking	Thiamin Content (μg/g)	Loss (%)	Riboflavin Content μg/g	Loss (%)	Niacin Content (μg/g)	Loss (%)
Brown						
Double boiler	4.40	9.0	0.81	6.2	54.0	4.0
Open vessel	4.40	32.2	0.81	26.0	54.0	31.0
Earle undermilled						
Double boiler	2.94	1.4	0.38	5.2	50.0	3.0
Open vessel	2.94	42.2	0.38	36.0	50.0	37.7
White						
Double boiler	0.65	1.3	0.27	7.4	20.6	3.4
Open vessel	0.65	54.0	0.27	18.2	20.6	41.0
Malekized parboiled						
Double boiler	2.01	6.5	0.40	7.5	40.2	2.2
Open vessel	2.01	57.2	0.40	50.0	40.2	37.8
Converted parboiled						
Double boiler	3.02	5.3	0.41	7.3	49.0	2.0
Open vessel	3.02	43.7	0.41	29.4	49.0	37.6
Package recipe	3.02	19.2	0.41	26.8	49.0	25.1
White enriched						
Double boiler	1.40	2.8	0.29	6.8	19.5	4.5
Double boiler	3.20	4.7	0.32	6.2	19.2	3.6
Open vessel	1.40	53.6	0.29	37.9	19.5	41.0
Open vessel	3.20	50.0	0.32	37.5	19.2	47.9
Brand enriched						
Double boiler	3.00	3.3	0.25	8.0	20.0	4.0
Open vessel	3.00	45.7	0.25	36.0	20.0	40.0
Package recipe	3.00	63.7	0.25	32.0	20.0	47.0

Source: Houston and Kohler (1970).

ods. The customary discards of cooking waters from these foods are so large as to cause a loss of nearly half the nutrients added. No measure other than reeducation in cooking methods will supply an adequate remedy for this defect."

The Food and Nutritional Board of the National Research Council in 1943 expressed the desirability of developing physiologically active and rinse-resistant vitamin derivatives. These derivatives can be added to grains, thus increasing their nutritional value.

ENRICHMENT PROCESSES

Premix Kernels

Several processes have been developed for the production of rice premix kernels. One of these, the HLR-Mickus process, was developed by Hoffman-LaRoche in 1946 and later described by Mickus (1955). This enriched rice was fed to a population in Bataan Province in the Philippines during 1948–50 and resulted in a significant decrease in mortality from beriberi.

This process involves the application of an acidic solution containing thiamin and niacin to the surface of the milled rice kernel. After air drying 2 coats of an alcoholic solution consisting of stearic acid, zein and abietic acid are applied. Finally, a ferric pyrophosphate and talc mixture is dusted over the grains. The product, called the premix, is added to milled rice in a ratio of 1:200.

Table 13.3 shows the composition of a premix prepared by Cort *et al.* (1976) and Table 13.4 shows its stability. After 6 months at room temperature there was a 4% loss of vitamin A and 10% loss of folic acid and pyridoxine. No significant loss in vitamin E was found.

Merck and Company developed a similar process for making a rice premix several years later. A zein solution containing vitamins and iron is applied in thin layers. Several coats of shellac and tricalcium phosphate, a whitening agent, are applied next. A high gloss finish is provided in the final coat by the addition of talc. Finally, 199 parts of enriched rice are added to 1 part of fortified kernels.

Cort *et al.* (1976) have produced a premix by this method referred to as the Wright procedure. Table 13.5 shows the results of storage of Wright rice premix for 3 months at room temperature. The retention of the various nutrients is quite high.

The addition of riboflavin by the above methods presents a problem. The vitamin-covered kernels are yellow and upon cooking the color spreads to the surrounding grains of rice. The colored grains are quite often discarded by the consumer, thus defeating the purpose of enrichment. Lease *et al.* (1962) have devised a process that permits inclusion of riboflavin. Kernels are first coated with vitamins and then soaked in a solution of polyethylene glycol or methyl cellulose and riboflavin-5′-phosphate sodium. The solvent gives rapid disperson of the riboflavin derivative which is quite soluble. The grains are then coated with iron pyrophosphate, a color masking agent. The premix is added to milled rice in a 1:100 ratio. Rice enriched by this process gives good resistance to washing for niacin and thiamin, but not for riboflavin. The yellow color is not observed in the raw rice nor are spots produced during cooking.

Powdered Premix

A powdered premix has been produced which is added to milled rice at a ratio of 0.5–1.0 part premix to 16,000 parts rice. Ordinary handling of the rice does not cause the powder to separate from the kernels. However, the enriched product will not withstand any washing procedures.

TABLE 13.3. HLR–MICKUS RICE PREMIX FORMULATION

Additives	Amount added per kg of rice	Label claim per kg of premix	Level found	
			Per kg premix	Per kg rice after 1:200 dilution
Thiamin (mg)	1,852	1,279	1,213	6.06
Pyridoxine-HCl (mg)	1,367	794	882	4.63
Niacin (mg)	13,228	10,582	Not run	53
Vitamin E (IU)	8,378	6,614	7,143	35
Vitamin A (IU)	(12.8 g)	3,218,766	2,645,561	13,228
Folic Acid (mg)	181	132	101	0.51
Iron (ferric orthophosphate) (mg)	83,776	17,637	11,177	55
Calcium phosphate (mg)	119,050	—	Not run	132
Zinc oxide (mg)	5,512	4,409	Not run	22
Talc (mg)	83,776	—	Not run	(Some mg)

Source: Cort et al. (1976).

TABLE 13.4. STABILITY OF HLR–MICKUS PREMIX

		Level found			
		Per kg of premix			
Nutrients	Initial	6 months at r.t.	4 weeks at 45°C (113°F)	Per kg of uncooked rice with premix diluted 1:200	Per 1382 g of cooked fortified rice[1]
Vitamin A (IU)	2,645,561	2,314,865	2,314,865	13,228	5,800
Pyridoxine (mg)	882	882	860	4.6	2.5
Folic acid (mg)	101	97	90	0.5	0.27
Vitamin E (IU)	7,143	7,077	7,055	35	15.9
Thiamin (mg)	1,213	1,168	Not run	5.7	Not run

Source: Cort et al. (1976).
[1] 454 g of uncooked rice yield 1380 g of cooked rice.

TABLE 13.5. WRIGHT RICE PREMIX FORMULATION

Additives	Amount added per kg of rice	Label claim per kg of premix	Level found per kg premix		% loss on cooking in rice at 1:200 dilution
			Initial	3 months at r.t.	
Thiamin (mg)	1,852	1,279	1,499	1,499	Not run
Pyridoxine-Hcl (mg)	1,367	794	311	309	0
Niacin (mg)	13,338	10,582	Not run	Not run	Not run
Iron (ferric orthophosphate) (mg)	83,776	17,637	Not run	Not run	Not run
Vitamin E (tocopheryl acetate) (mg)	8,378	6,614	6,614	6,614	0
Vitamin A (250 SD) (IU)	4,299,037	3,218,766	3,196,720	3,174,673	1
Folic Acid (mg)	181	132	88	88	0
Zinc (oxide) (mg)	18,276	4,409	Not run	Not run	Not run
Calcium (phosphate) (mg)	110,231	33,070	Not run	Not run	Not run
Talc (mg)	44,092	—	Not run	Not run	Not run

Source: Cort et al. (1976).

Simulated Rice

A totally different method for enriching rice involves the manufacture of an artificial kernel premix (Bauernfeind and Cort 1974). A dough is prepared by adding water to wheat flour which has previously been blended with the desired vitamins and minerals. The mixture is put through a device under pressure and cut and dried to form the simulated kernels. The process is very similar to that used to make pasta (spaghetti, noodles, etc.). The kernels contain 200 times the amount of vitamin present in unpolished rice, therefore 1 part of simulated grains is added to 200 parts of milled rice to obtain the final product. One advantage of using this method is that the composition of the premix can be altered, depending on the nutrient deficiencies of a particular country. Also, rice flour or meal may be used instead of wheat flour. The major drawback of the simulated rice premix is that there is a high loss of added nutrients during washing and cooking. In addition, the simulated premix tends to spoil the palatability of boiled rice. Water-insoluble vitamin derivatives are now being used to minimize washing losses.

Acid Parboiling

A soaking method for producing vitamin B_1 enriched rice was the result of the finding made in Japan in 1948—that the vitamin contained in the bran and embryo of brown rice could be transferred to the endosperm by soaking in aqueous acetic acid. The product of this process is called acid parboiled rice.

In the original method, polished rice was soaked in a 1% aqueous acetic solution to which vitamin B_1 was added. The rice was then drained, steamed and dried to minimize vitamin loss in washing and cooking. The resulting enriched rice was found to be digestible, tasty and, when mixed in a 1:100 ratio with milled rice, to provide adequate amounts of vitamin B_1.

A major disadvantage of this procedure is that cracks in the grain result. The grains appear to be of inferior quality and the nutrients are more readily lost in washing. Soaking in organic solvents such as ethanol, acetone and chloroform does not result in crack formation. This is true even when the water content of the solvent is 40% by volume. By using these organic compounds the process of steaming, which was introduced to prevent cracking, may be eliminated.

Vitamin B_1 derivatives have been produced which are soluble in acetone, ethanol and chloroform, but not in water (Table 13.6). By using these derivatives along with organic solvents, vitamin loss during washing and cooking can be significantly reduced.

TABLE 13.6. SOLUBILITY OF VITAMIN B_1 DERIVATIVES IN ORGANIC SOLVENTS AT 30°C (G/100 ML)

Solvent	Concentration	B_1-HCl	DBT[1]	B_1 derivatives Cetyl-B^2_1	B_1-naph.[3]
Acetone	100%	0.00	1.08	0.16	0.00
	90	1.76	2.18	0.96	0.004
	80	3.20	1.76	1.24	0.047
	70	10.81	1.03	0.97	0.16
	60	22.07	0.57	0.58	0.362
Ethanol	99	0.22	0.35	1.82	0.044
	90	2.14	0.68	2.20	0.025
	80	8.59	0.70	2.37	0.105
	70	20.40	0.61	1.65	0.181
	60	34.27	0.36	0.85	0.31
Chloroform	100	0.00	27.67	2.12	0.00
	90	0.00		3.05	0.00
	80	0.00		1.52	0.00
	70	0.00		0.85	0.00
	60	0.00		0.51	0.00
5% chloroform in 70% acetone		0.00	1.79	3.31	0.13

Source: Mitsuda (1962)
[1] Dibenzoylthiamin
[2] Thiamin dicetylsulphate
[3] Thiamin naphthalene-2, 6-disulphonic acid

The widespread use of this acid-soaking technique in Japan has replaced the old method of undermilling rice to retain more of the vitamins.

Amino Acid Fortification

Since rice is naturally deficient in lysine and threonine, many studies have been carried out to determine the action of these amino acids. Pecora and Hundley (1951), followed by Harper and coworker (1955), have shown that a combination of lysine and threonine has a marked effect on growth of weanling rats. Administration of either amino acid by itself does not show this effect. Sure (1955) has reported that 0.4% L-lysine plus 0.3% DL-threonine as a supplement to milled rice and processed milled rice (Minute Rice) at the 5% protein level increases the protein efficiency ratio (PER) by 64% and 75% respectively. Rosenberg and Culik (1957) found that the total diet lysine-to-threonine was optimal at 1.4, in contrast to the 1.66 or 2.0 as suggested earlier by Rose *et al.* (1949). The supplementary value of the lysine-threonine combination is well known.

Although the protein quality of rice can be enhanced by supplementation with lysine, it is technically difficult to apply a large amount of this amino acid to the kernels. As a consequence no premix formula has of yet been commercialized.

Mitsuda reported in 1969 that the soaking method designed for the production of vitamin enriched rice can be modified to include lysine. Kernels soaked in a 1% acetic acid solution containing lysine can be enriched with up to 66 mg of the amino acid per gram of kernel. Figure 13.1 shows the time course for the impregnation of rice kernels with lysine. Washing losses of the amino acid may be drastically minimized by employing the steaming process, as can be seen in Fig. 13.2. Threonine, another amino acid found in low quantities in rice, may be introduced along with lysine. The product obtained by the process diagrammed in Fig. 13.3 is wash resistant, appears to be very similar to plain white rice and neither discoloration nor cracking is evident.

The Institute of Nutrition of Central America and Panama reported in 1975 (Brenes *et al.* 1975) the results of a study designed to determine the

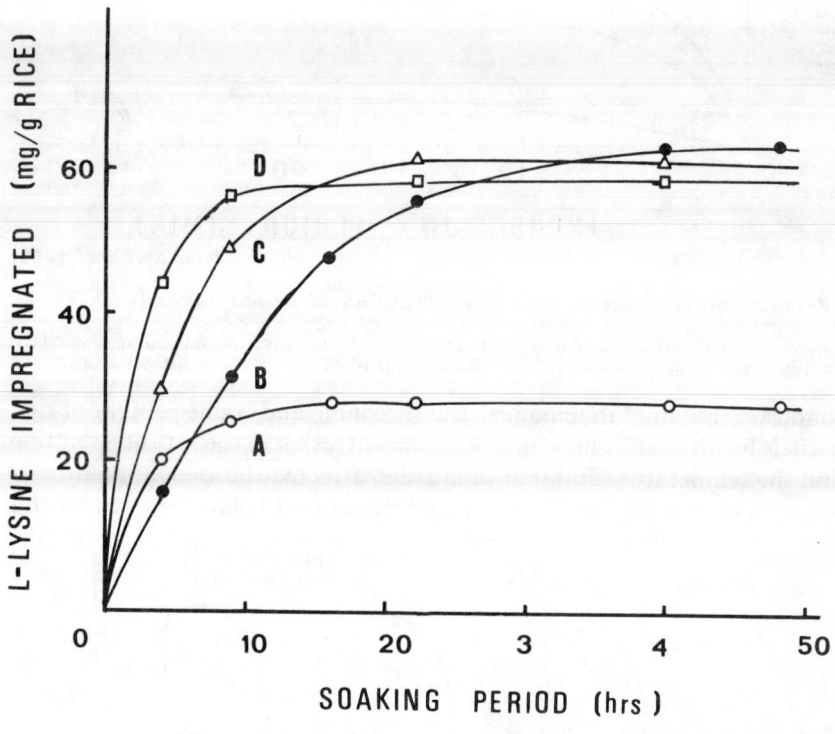

From Mitsuda (1969)

FIG. 13.1. TIME COURSE FOR THE IMPREGNATION OF RICE KERNELS WITH LYSINE

Polished rice, unit of 100 g, was soaked in 1% acetic acid solution (unit of 200 ml) containing L-lysine (1.1 M) at 23°C (73.4°F) (A) or by 2.1 M at 23°C (73.4°F), (B) 37°C (98.6°F), (C) and 50°C (122°F) D, respectively.

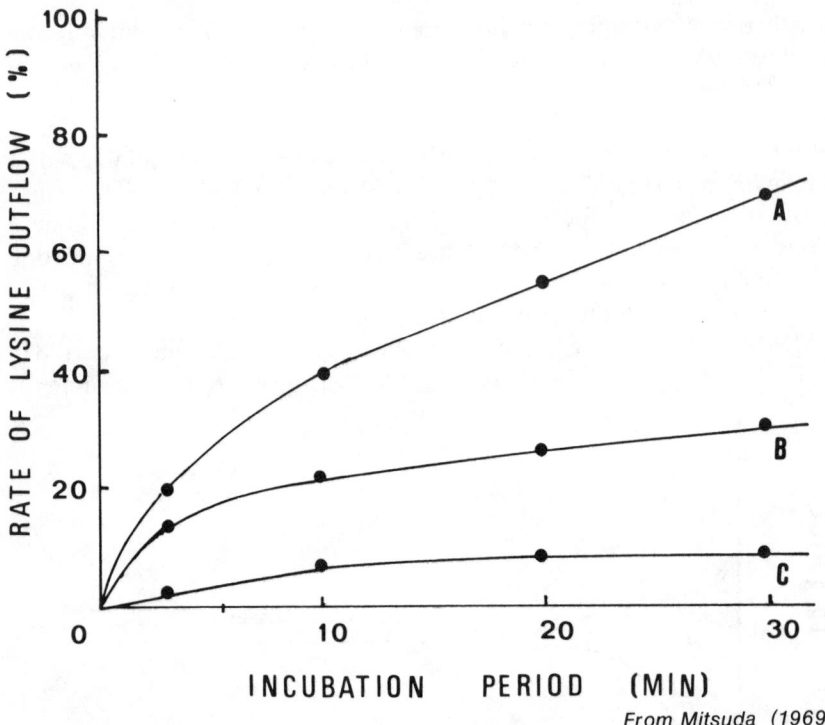

From Mitsuda (1969)

FIG. 13.2. PREVENTION BY STEAMING PROCESS OF LYSINE LOSS IN WASHING

The enriched rice, unit of 10 g, obtained by steaming for zero (A), 3 (B) and 10 min (C), respectively, was rinsed with fresh water, unit of 50 ml at room temperature.

conditions required to maximize the threonine and lysine content of rice kernels by infusion. The amino acid concentration in the infusion solution and the temperature and time of infusion affected the amount of infused amino acid in the kernel. Initial moisture content below 18% caused the

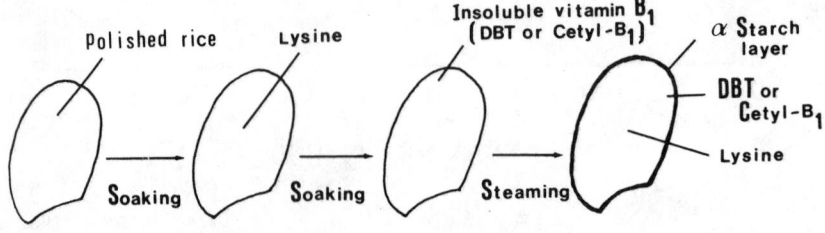

From Mitsuda (1969)

FIG. 13.3. PRODUCTION OF LYSINE AND VITAMIN B_1 ENRICHED RICE BY THE DOUBLE SOAKING METHOD

kernels to crack after drying. The best infusion conditions for lysine are: a 10% concentration, neutral pH of the solution and 4 hr of infusion at a temperature of 40°C (104°F) for kernels having an initial moisture content of 18% or more.

L-threonine has the same optimum infusion conditions with the exception of temperature of infusion which is 18°C (64.4°F).

To determine biological availability of the infused amino acid in rice, infused kernels were added to milled rice to provide 0.25% L-lysine and 0.10% L-threonine. Table 13.7 shows the effect of various diets on weight gain, food intake and PER. The results indicate the infused amino acids were totally available to animals. The results also serve to reemphasize the fact that the first limiting amino acid in rice is lysine. Addition of this amino acid to the diet caused a marked increase in the PER. This increase was not seen when threonine alone was added. In combination, the 2 amino acids improved the protein quality significantly.

TABLE 13.7. WEIGHT GAIN, FEED INTAKE AND PROTEIN EFFICIENCY RATIO OF RICE FORTIFIED WITH AMINO ACIDS IN THE SYNTHETIC FORM AND FROM INFUSED RICE KERNELS

Treatment	Average weight gain (g)	Average food intake (g)	PER
Rice	45.0 ± 5.9[3]	302.0 ± 38.3	1.90 ± 0.10
Rice + Lysine[1]	61.1 ± 7.9	307.0 ± 19.1	2.59 ± 0.26
Rice + L-Lysine[2]	60.0 ± 5.8	309.0 ± 42.6	2.55 ± 0.20
Rice + L-Threonine[1]	41.0 ± 7.4	277.0 ± 53.4	1.94 ± 0.13
Rice + L-Threonine[2]	47.0 ± 12.0	264.0 ± 39.3	2.06 ± 0.36
Rice + L-Lysine + L-Threonine[1]	99.0 ± 13.5	375.0 ± 46.7	3.47 ± 0.26
Rice + L-Lysine + L-Threonine[2]	102.0 ± 9.1	379.0 ± 46.7	3.42 ± 0.32
Casein	92.0 ± 14.2	365.0 ± 39.8	2.93 ± 0.45

Source: Brenes et al. (1975).
[1]Fortification by addition of infused kernels.
[2]Fortification by addition of synthetic amino acids.
[3]Standard deviation.

FORTIFICATION STANDARDS

On February 27, 1958, the U.S. Standard Identity for Enriched Rice went into effect. This order, issued by the U.S. Food and Drug Administration, prescribed the minimum and and maximum levels of thiamin, riboflavin, niacin and iron that should be added to rice. Calcium and vitamin D levels were also given, although these ingredients were optional.

The proposed addition of riboflavin met with a large amount of opposition. Riboflavin gives a yellow tint to raw rice and upon cooking causes yellow splotches to develop around the premix kernels. Millers and processors feared that consumers would object to the colored rice. The requirement for riboflavin has been stayed until a public hearing is held.

U.S. Food and Drug Administration's currently effective enriched rice standard of identity excludes coated rice. However, the territories of Puerto Rico and Guam were specifically exempted from this requirement by letter from George P. Larrick, Commissioner of Food and Drugs, dated April 24, 1964.

State cereal enrichment legislation almost always adopts verbatim the federal cereal enrichment standards. These standards cover all commodities that are shipped in interstate commerce. Federal standards for enriched staple cereals which are not mandatory thus become mandatory via state laws for selected staple cereals in the 36 states which currently have cereal enrichment laws. The exclusion of coated rice from the currently effective enriched rice standard was, therefore, automatically adopted in the 6 state cereal enrichment laws which require enrichment of all milled rice (Arizona, California, Connecticut, Florida, New York and South Carolina).

TABLE 13.8. ORIGINAL (1958) AND REVISED (1972) LEVELS OF NUTRIENTS IN ENRICHED RICE

Nutrients	Original		Revised
Thiamin (mg/kg)	4.4–	8.8	6.4
Riboflavin (mg/kg)[1]	2.6–	5.3	4.0
Niacin (mg/kg)	35 –	71	53
Iron (mg/kg)	29 –	57	88
Calcium (mg/kg)	1102	−2205	2116
Vitamin D (USP Units/kg)	551	−2204	–

Source: Brooke (1972)
[1]The level oriboflavin in enriched rice is fixed at the suggested present amount until final action taken.

A new Standard of Identity was issued in 1972. Table 13.8 compared the updated values with those published in 1958. Vitamin D was dropped from the list and riboflavin remained.

In 1974 the Food and Nutrition Board of the National Academy of Science proposed fortification of cereal products with 11 nutrients, 5 of which are being included for the first time. Table 13.9 shows the current nutrient levels and the proposed ones. The low values in the first column are the currently used minima for baked goods; the high values are the maxima for macaroni and rice. It should be noted that fortifying rice with the proposed levels of calcium and magnesium is not possible (Cort et al. 1976). Eight hundred grams of calcium phosphate would have to be added to 0.45 kg of premix so that after a 1:200 dilution the final product would contain the appropriate levels of calcium. Even if all the powder were to adhere to the kernels, they would be 2 to 4 times their usual size.

TABLE 13.9. FORTIFICATION LEVELS FOR CEREAL PRODUCTS IN MG /KG

Nutrients	Presently used[1]	Proposed[2]
Thiamin	2.4 – 11	6.4
Riboflavin	1.5 – 5.3	4.0
Niacin	22 – 75	53
Iron	18 – 57	88
Calcium	1102 –3307	1984
Vitamin A	–	(16,094 IU) 4.9
Pyridoxine	–	4.4
Folic Acid	–	0.7
Magnesium	–	441
Zinc	–	22

Source: Cort et al. (1976).
[1]Range of minima to maxima listed in Anon. (1973).
[2]Nat. Acad. of Sci. (1974).

South Carolina passed a law in 1956 which prohibits the sale of plain white rice. Since then other states have enacted similar laws stating that milled rice must be enriched to the nutrient levels outlined in the Federal Standard of Identity.

Two types of enrichment are allowed under law. Packaged rice may be treated with nonrinse resistant ingredients provided it bears the following label: "To retain vitamins do not rinse before or drain after cooking." Rice that contains premix kernels must not lose more than 15% of the added nutrients.

REFERENCES

ANON. 1973. Enrichment levels of baked products, floura, farina, etc. Code of Federal Regulations Title 21, 15.525. U.S. Govt. Print. Off., Washington, D.C.

BAUERNFEIND, J.C. and CORT, W.M. 1974. Nutrification of food with added vitamin A. Crit. Rev. Food Tech. 4, 337-375.

BRENES, R.G., ELIAS, L.G., RUILOBA, M.H. and BRESSANI, R. 1975. Report from Institute of Nutrition of Central America and Panama (INCAP). In Rice Report 1975. S. Baker, H. Mitsuda and H.S.R. Desikachar (Editors). Intl. Union of Food Sci. Tech. Instituto de Agroquimicay Tecnologia de Alimentos, Valencia, Spain.

BROOKE, C.L. 1972. Rice enrichment. In Rice Chemistry and Technology. D.F. Houston (Editor). Amer. Assoc. Cereal Chemists. St. Paul, Minn.

CORT, W.M. et al. 1976. Nutrient stability of fortified cereal products. Food Tech. 30, 52-58.

FURTER, M.F. and LAUTER, W.M. 1946 Enrichment of rice with synthetic vitamins and iron. Ind. Eng. Chem. 38, 486-492.

HARPER, A.E., WINJE, M.E., BENTON, D.A. and ELVEHJEM, C.A. 1955.

Effect of amino acid supplements on growth and fat deposition in the livers of rats fed polished rice. J. Nutr. *56*, 187-198.

HOUSTON, D.F. and KOHLER, G.O. 1970. Nutritional properties of rice. Natl. Acad. Sci., Washington, D.C.

KIK, M.C. and WILLIAMS, R.R. 1945. The nutritional improvement of white rice. Natl. Acad. Sci., Natl. Res. Counc. Bull. *112.* Washington, D.C.

LEASE, E.J., WHITE, H. and LEASE, J.G. 1962. Enrichment of rice with riboflavin. Food Tech. *16,* 146-158.

MALAKAR, M.C. and BANERJEE, S. 1959. Effect of cooking rice with different volumes of water on the loss of nutrients and digestibility of rice *in vitro.* Food Res. *24,* 751-756.

MICKUS, R.R. 1955. Seals enriching additives on white rice. Food Eng. *27,* 91-93.

MITSUDA, H. 1962. Enrichment of rice by soaking method. *In* proceedings of 1st Int. Congr. Food Sci. and Tech., London.

MITSUDA, H. 1969. New approaches to amino acid and vitamin enrichment in Japan. *In* Protein Enriched Cereal Foods for World Needs. M. Milner (Editor). Amer. Assoc. Cereal Chemists. St. Paul, Minn.

NATL. ACAD. SCI. 1974. Proposed fortification policy for cereal grain products. Publ. No. *2232.* Natl. Acad. Sci., Washington, D.C.

NATL. RES. COUNC. 1958. Cereal enrichment in perspective 1958. Committee on Cereals. Food and Nutrition Board. Washington, D.C.

PECORA, L.J. and HUNDLEY, J.M. 1951. Nutritional improvement of white polished rice by the adding of lysine and threonine. J. Nutr. *44,* 101-112.

ROSE, N.C. *et al.* 1949. The utilization of the nitrogen of ammonium salts, urea, and certain other compounds in the synthesis of nonessential amino acids *in vivo.* J. Biol. Chem. *181,* 307-316.

ROSENBERG, H.R. and CULIK, P. 1957. The improvement of the protein quality of white rice by lysine supplementation. J. Nutr. 63, 447-487.

SURE, B. 1955. Effect of amino acid and vitamin B_{12} supplements on the biological value of protein in rice and wheat. J. Amer. Dietet. Assoc. *31,* 1232-1234.

U.S. FOOD AND DRUG ADMIN. 1958. Standard of identity for enriched rice. *In* Code of Federal Regulations. Title 21. Part 15. Sec. 15.525. U.S. Food and Drug Administration, Washington, D.C.

14

Parboiled Rice

Bor S. Luh
Robert R. Mickus

Cultivated rice varieties are grouped according to their morphological characteristics, such as Indica, Japonica and Javanica (Adair 1972). The rice caryopsis varies widely among cultivars in shape and size, length and width. The Food and Agriculture Organization of the United Nations (FAO) classifies milled rice by length into sizes of extra long, more than 7 mm; long, 6.0 to 7.0 mm; medium or middling, 5.0 to 5.9 mm; and short, less than 5 mm (Chang and Bardenas 1965). The caryopsis of the Indica group is usually long, slender and rather flat. These varieties are used more often for parboiling because the operations of steeping and heating are quicker and easier as water and heat rapidly reach the center of the endosperm. Long grain rice is quite distinct from medium and short grain rice in cooking and processing characteristics. These differences are described by Webb (1967): "Raw milled kernels of high quality domestic long grain varieties, frequently called "hard-rice," usually cook dry and flaky with a minimum of splitting, and the cooked grains tend to remain separate. High quality short and medium grain varieties, referred to as "soft-rice," are more moist and firm when cooked and the grains tend to stick together." The long grain varieties are characterized by a comparatively high amylose content and a medium high gelatinizing temperature. Medium and short grain varieties have lower amylose content and lower gelatinizing temperature than long grain varieties. Long grain rices are used for canned soups and quick cooking products. Medium and short grain varieties generally are used for making dry breakfast cereals and baby foods, and as an adjunct in brewing.

Parboiling is a hydrothermal process in which the crystalline form of starch is changed into an amorphous one, due to the irreversible swelling and fusion of starch. This is accomplished by soaking, steaming, drying and milling the rice. The parboiling process is to produce physical, chemical and organoleptic modifications in the rice with economic and nutritional advantages (Gariboldi 1972, 1974). The major objectives of parboiling are to: (1) increase the total and head yield of paddy, (2) prevent the loss of nutrients during milling, (3) salvage wet or damaged paddy, and (4) prepare the rice according to the requirements of consumers (Ali and Ojha 1976).

The changes occurring in the parboiling process are as follows:

(1) The water-soluble vitamins and mineral salts are spread throughout the grain, thus altering their distribution and concentration among its various parts. The riboflavin and thiamin contents are 4 times higher in parboiled rice than in whole rice. The thiamin is more evenly distributed in the parboiled rice and the niacin level in this rice is 8 times greater (Kennedy *et al.* 1975). Vitti *et al.* (1975) have also noted a considerable increase in thiamin levels in parboiled rice.

(2) The moisture content is reduced to 10−11% for better storage.

(3) The starch grains embedded in a proteinaceous matrix are gelatinized and expanded until they fill up the surrounding air spaces.

(4) The protein substances are separated and sink into the compact mass of gelatinized starch becoming less liable to extraction. Dimopoulos and Muller (1972) have reported that the parboiling process alters the solubility of rice protein in various solvents (Table 14.1).

(5) The enzymes present in the rice kernel are partially or entirely inactivated. Shaheen and coworkers (1975) found a reduction in the free fatty acid content of parboiled rice that has been stored for 10 months and presumably due to the inactivation of lipase.

(6) Proliferation of fungus spores, growth of eggs, larvae or insects, etc., are prevented.

(7) The solids leached into the cooking water and the extent of solubilization of the kernels on cooking are considerably reduced.

After parboiling, the milling yield is higher and the quality improved as there are fewer broken grains. The grain structure becomes compact, translucent and shiny. The milled parboiled rice keep longer and better than in the raw state, as germination is no longer possible. The endosperm has a compact texture making it more resistant to attacks by insects. The grains remain firmer during cooking and are less likely to become sticky. The nutritional value of parboiled rice is greater because of the higher content of vitamins and mineral salts which have spread into the endosperm. The starchy endosperm of parboiled rice has a greater resistance to milling and therefore the bran and germ are more effectively removed (Padua and Juliano 1974).

TABLE 14.1. SOLUBILITY FRACTIONATION OF PROTEINS OF COMMERCIAL RICE SAMPLES

Grain type	Treatment	Source	Fractionation by percolation % of total protein soluble in:						Extraction efficiencies of 2 solvents on rice proteins by shaking	
			Water	NaCl 5%	Alcohol 60%	Detergent 3%	Alkaline detergent containing bisulfite	Extraction efficiency	Detergent 3%	Alkaline detergent containing bisulfite
Long	Raw	U.S.	11.0	6.7	1.9	34.9	52.4	96.9	73.2	96.8
Long	Raw	Thailand	2.4	1.7	1.6	—	—	—	67.1	96.8
Short	Raw	China (Shoonan)	—	—	—	—	—	—	77.9	100.0
Long	Parboiled	U.S.	2.4	0.3	0.3	5.9	69.2	78.2	19.5	88.1
Long	Parboiled	Thailand	0.8	0.3	0.4	—	—	—	16.5	85.2
Long	Parboiled	"Uncle Ben's"	1.8	0.5	0.4	9.2	70.7	82.5	20.0	93.7
Long	Parboiled	"Whitworth's"	—	—	—	—	—	—	21.3	84.2
Long (waxy)	Parboiled	Thailand	—	—	—	—	—	—	21.6	92.0
Medium (waxy)	Parboiled	China	2.7	0.5	0.3	7.5	60.2	71.2	18.0	73.9

Source: Dimopoulos and Muller (1972).

Several workers have studied the improvement in rice kernel characteristics following parboiling. Rao and Juliano (1970) offer retrogradation as an explanation for the effect of parboiling on iodine staining, amylograph characteristics, resistance to breakdown during cooking and reduced solubility of the starch. Studies of amylose solubility and hydration characteristics of parboiled rice have led Ali and Bhattacharya (1972A, 1972B, 1976) and Bhattacharya et al. (1975A,B) to a similar conclusion.

Priestley (1976) has suggested that the resistance of parboiled rice paste to swelling and solubilization may be attributed to the presence of associative bonding in the starch. To test whether this bonding is a result of retrogradation or the formation of an insoluble amylose complex, he studied the X-ray diffraction pattern of cereal starch with peaks at 3.84Å and 5.85Å. Following parboiling, the A-type pattern is replaced by a V-type pattern with diffraction lines at 6.80Å and 4.42Å. These peaks confirm the presence of a helical amylose complex. There was no evidence of a B-type pattern characteristic of retrograded starch which would be expected to give peaks at 15.8Å and 5.2Å (Zobel 1964).

It is well known that amylose can form complexes with molecules capable of entering and stabilizing the helix, such as fatty acid and lysolecithin which are both in the rice grain. The complex is insoluble but stable at relatively high temperatures. The formation of such complexes is reported to restrict swelling and solubilization of starch by stabilizing the granule structure (Gray and Schoch 1962).

The changes parboiling brings about in the rice are closely related to the techniques used. Lack of experience may nullify the advantages described, even reducing the food value of the cereal originally possessed. If the paddy is allowed to ferment during or after steeping, the sensory quality of the rice will be unacceptable for consumption. There are problems hindering the more extensive consumption of parboiled rice. If improperly dried, milled and prepared, the product may have poor color, odor, taste and texture. Better technical knowledge in processing, packaging and storage is needed to improve the quality of parboiled rice.

CHARACTERISTICS OF PADDY RICE FOR PARBOILING

The rice varieties used for parboiling are those that are more brittle because of the soft structure of their endosperm. Long grain paddy rice that gives a low output on milling is preferred for parboiling. The long and slender rice varieties are usually parboiled because they are fragile as compared with the short or medium length grains. Varieties which have good milling quality are generally not parboiled (Ali and Ojha 1976).

Some characteristics of paddy rice which affect the yield and quality of parboiled rice are: (1) the presence of partially or fully shelled grains, (2) the awn and hairiness of the husk may make the soaking operation difficult because of the tendency of the grains to float on the surface of the water, (3) the pigments of the husk and pericarp may be dissolved during the soaking and steaming operations causing discoloration of the endosperm, (4) microbial infestation may cause a partial or total darkening of the endosperm, and (5) some injuries on the seed caused by mechanical impact or by insects may lead to partial discoloration of the parboiled rice (Gariboldi 1974).

Awned Rice and Empty Glumes

Awned rice and empty glumes tend to float on the surface of the steeping water. They tend to break loose during processing and obstruct the flow and movement of the paddy, water and steam. Awned rice has a low bulk weight which greatly reduces the output of the processing plant.

Shelled Grains

The grains that are not protected by hulls absorb more water and heat and this may alter their shape. They stand out in the milled product because of their darker color resulting from the greater quantity of heat absorbed. The grains which are not shelled, but with the hulls loosened or partly pulled off during threshing, are likely to deteriorate in shape and color.

Mold Infestations

Mold infestation leaves spots on the paddy, shelled and milled rice. The grains become colored or stained. The severity of the disease depends both on the species of microorganism and its concentration. It may affect the outer surface of the hull with spots visible on the paddy, or the outer layers of the caryopsis with spots visible on the shelled rice. In severe cases the mold may reach the starchy part of the endosperm. Discoloration is more or less inevitable when the mycelium has penetrated to the layers below the hull, the endosperm of the caryopsis and the germ.

The paddy must be shelled in order to see whether the surface of the caryopsis is stained or discolored. Polishing the grain will also reveal the presence of any caryopses with yellow or amber colored endosperms formed by enzymic amylolysis and lipasic action of the enzymes in the fungi on the grain. Parboiling can further accentuate the color change in the grains to form a dark or reddish shade.

Injuries Caused by Insects or Threshing

Infestation by insects as well as threshing may injure the caryopsis. The caryopsis will appear pitted and stains will be seen after shelling or milling. The injury will lead to darkening of the grains during steaming. After shelling and polishing the paddy, the depth of the injury can be estimated and its effect after parboiling can then be gauged.

Chalky, Green and Red Grains

Chalky or green grains are caused by imperfect ripening due to morphological reasons or by certain growth conditions. The chalky kernels present in a glassy or waxy textured variety of paddy assume importance in relation to parboiling operations. In order to gelatinize them completely, steeping and steaming conditions must be altered. Chalky or green grains are generally not immune to attacks by fungi which can cause partial or complete darkening of the milled rice grains. Green or red grains turn dark due to the effects of the parboiling process. After parboiling, those with a red pericarp and white endosperm show a deep or dark red pericarp with an endosperm which is slightly darker than normal. Thorough polishing is necessary to completely remove the colored pericarp and to avoid any traces of remaining color which would give the milled grain a streaky appearance.

Mechanical grading will remove the unripe (chalky or green) grains and sometimes part of the red ones. When stored for some time, the unripe grains shrink considerably in size so that they can be easily sorted out by mechanical sizing according to thickness.

To make a qualitative examination of the paddy and the parboiled rice, a shelling and polishing machine must be available together with sorting trays. The operations are as follows:

(1) The raw paddy is visually examined to ascertain the amount of awn and empty glumes present and the color and hairiness of the hulls. Notice should be taken of any spots which would denote fungal infections or damage. The percentage of shelled grains (whole or broken) and of partly cracked hulls can be determined by hand sorting;

(2) After shelling part or all of the paddy, the shelled sample is closely examined to see whether fungi or lesions have damaged the caryopsis or the germ. If this has happened the affected grains will be discolored after parboiling. Having been picked out by hand, the quantity by weight of shelled caryopses showing the characteristic colored spots can be determined;

(3) Another unshelled sample is polished to see the texture of the caryopsis and to find discolored grains and chalky ones. Any defects in regard to texture and color due to fungal attack and deep-seated injuries must be recorded;

(4) The yield of milling including total yield and percentage of broken grains must be ascertained. This is of importance when calculating the cost of the process. First and foremost, the miller considers what the milling yield will be since parboiling raises the milling yield.

CLEANING

The impurities present in paddy rice are varied in nature. Weeds, lodging in the fields, animals used for threshing and natural drying all account for the extraneous materials found in the paddy. Impurities and seeds other than rice are usually removed during milling. Some are removed before shelling, and others after polishing along with the broken and damaged grains. The impurities considerably affect the results of parboiling.

To ensure high grading, several machines are needed with each operating on a different principle. One of the typical cleaners is shown in Fig. 14.1. Mechanical cleaning may be completed by washing and floating the paddy in water before it is put into the steeping tanks. This operation called water floatation is used to remove stones and sand. Straw and defective grains (chalky, empty, unripe, etc.) are light enough to float away in the process.

The equipment used for cleaning is similar to that employed in flour mills for removing stones and for dampening the wheat before grinding it. If paddy grains (kernels) have long awns, separation by water floatation may be impractical because even ripe grains will float on the surface of the water together with the lighter impurities.

In some modern processes, separation by water floatation is preceded by mixing the rice with water and whirling it to free the surface of the paddy grain from air bubbles.

GRADING

The thickness of the grain is very important to the parboiling process, as the necessary period of steeping and steaming increases with grain thickness. If the grains are of different thicknesses, gelatinization of the starch will be uneven. If steeping and steaming times are prolonged and temperatures raised so as to gelatinize the thicker grains completely, the thinner ones will be gelatinized to a greater degree. The thinner grains will be darker, more compact and harder. The resulting product will look uneven in color.

From Ferrell Ross Co. (Saginan, Mich.)

FIG. 14.1. HIGH CAPACITY CLEANER-SUPER K-2248 BDW

Sorting on the basis of kernel thickness is essential for good quality parboiled rice. This is done by means of grading reels fitted with a steel sheet with rectangular slots or with wire netting. Further grading may complete the selection according to the length and bulk weight of the grain to obtain a final product of improved and uniform quality. Sorting according to grain length is done by means of disc separators. Sorting by bulk weight, if necessary, is done by specific gravity separators (Fig. 14.2).

The paddy is divided into lots. Each lot with grains of similar size is parboiled separately with different steeping and steaming times and temperatures. Drying and milling will thus be facilitated and there will

From Aeroglide Corp., Raleigh, N.C.

FIG. 14.2. AEROGLIDE PADDY SEPARATOR

be less broken grains. The thinner grains obtained from sorting are usually those which are unripe or naturally misshapen. It may sometimes be best to use them for the production of low quality raw rice rather than to parboil them. If one lot is formed of mixed varieties, sorting is essential to separate them according to length, thickness and bulk weight.

Sorting according to thickness also separates part of the shelled grains present in the mass of the paddy. It is, however, preferable to do this in "compartment separators" which are used to separate the shelled grains from the unshelled. The compartment separator utilizes the difference in the bulk weight between shelled and unshelled grains. To obtain the best results it is advisable to feed the separator with lots of paddy which have been previously graded. The paddy should first be sorted and then passed in separate lots through the compartment separator. The specific gravity separator, which occupies less space than the compartment separator but uses more electricity, can also be used for separating the shelled grains.

Extraction of shelled grains with the compartment separator also means sorting out stones. Because of their bulk weight, stones are extracted together with shelled grains. Thus a high percentage of stones which must be removed on a densimetric table are removed here.

STEEPING

Different varieties of paddy rice have their steeping characteristics. An efficient and properly controlled steeping process is used for the production of parboiled rice. The treatment should be done quickly to avoid fermentation which would adversely affect the color, taste and smell of the product. To achieve effective and uniform results, the grain size must be uniform and the caryopsis must be entirely covered by the hulls. If the caryopsis is exposed, the shape and color of the parboiled rice will be unsatisfactory.

The methods used to achieve steeping include: (1) the use of high and medium temperature water, (2) application of vacuum and/or hydrostatic pressure, and (3) the addition of wetting agents to the steeping water. These systems have been used either singularly or in conjunction to increase water penetration and to reduce steeping time. Steeping is needed to provide the starch with a sufficient amount of water for gelatinization. A moisture content of not less than 30% is required to fully gelatinize the starch in the caryopsis. Water absorption by the grain involves spreading the water-soluble substances evenly. Steeping also facilitates the transmission of heat from the surface of the hull to the middle of the endosperm.

Bandyopadhyay and Roy (1975) have developed an equation to describe the relationship of water uptake of paddy to the time and temperature of soaking:

$$\overline{X} - X_o = \frac{2}{\sqrt{\pi}}(X_s - X_o)\left(\frac{S}{V}\right)\sqrt{D_m \cdot \theta + X_i} \tag{1}$$

$$\overline{X} - X_o = K_m\sqrt{\theta + X_i} \tag{2}$$

$$K_m = K\sqrt{e - E/RT} \tag{3}$$

Where:
$$K_m = \frac{2}{\sqrt{\pi}}(X_s - X_o)\left(\frac{S}{V}\right)\sqrt{D_m}$$

$$K = \frac{2}{\sqrt{\pi}}(X_m - X_o)\left(\frac{S}{V}\right)\sqrt{D_o}$$

Where:
X_o = Initial, uniform moisture content. (g/g, dry basis)

$\overline{X}$ = Average moisture content at a given hydration time. (g/g, dry basis)

X_s = Effective surface moisture content at the bounding surface at times, greater than 0. (g/g, dry basis)

X_i = Moisture content corresponding to initial hydration. (g/g, dry basis)

θ = Time. (sec)

$\left(\dfrac{S}{V}\right)$ = Surface to vol. ratio of paddy.

D_m = Diffusion coefficient. (cm²/sec.)

D_o = Diffusion constant (in Arrhenius equation), cm^2/sec.
E = energy of activation (cal/mol)
H = gas constant (cal/mol °K)
T = absolute temperature (°K)
K_m, K = constants.

The above diffusion equations give fair predictions of the hydration only if the activation energy is accurate enough.

If steeping is prolonged, enzymatic action in the paddy is activated. Fermentation of the grain results in removal of the adhering organic impurities.

The use of very hot water for steeping has been advocated as a means of reducing processing times. Within certain limits, the quantity of liquid absorbed in the time unit is in proportion to the temperature of the water. The use of water at a temperature slightly below that of starch gelatinization (typical for each single variety, roughly about 70°C, 158°F), increases the speed of absorption. If the water temperature exceeds that of starch gelatinization, the absorption time is reduced but more water is absorbed than is necessary for moistening the inner part of the kernel.

In Fig. 14.3, water absorption by 1 variety of paddy (Indian) at different temperatures is presented. A moisture content of about 30% appears to be the very lowest for water to reach the core. This can be achieved without increasing the quantity of water absorbed with temperatures below 50°C (122°F) and steeping times varying from 12−60 hr. If more water than necessary is absorbed, the caryopsis swells considerably, cracking open the hull and becoming exposed. Cracking of the hull is a serious drawback as many of the substances contained in the grains will diffuse into the steeping water.

Although hot water accelerates absorption, the milled product is more likely to be discolored by it. The discoloration of the parboiled milled rice increases with the duration of steeping and the temperature of the water, subsequent steaming being the same in both cases (Fig. 14.4 and 14.5). The color becomes much deeper once the limit of 70°C (158°F) is exceeded. In Fig. 14.4, it is shown that the maximum discoloration at 70°C (158°F) appears about 5 hr after the paddy has been in water.

The time and temperature of steeping are related to the darkness of the parboiled milled rice. When the amylase in the rice is activated, sugars such as maltose and glucose are formed. Research has shown that 60°C (140°F) is the ideal temperature for amylase activity (Refai et al. 1967).

The color of the parboiled rice varies with the pH of the steeping water. If the pH is close to 5, coloring is at a minimum. The color deepens as the pH rises.

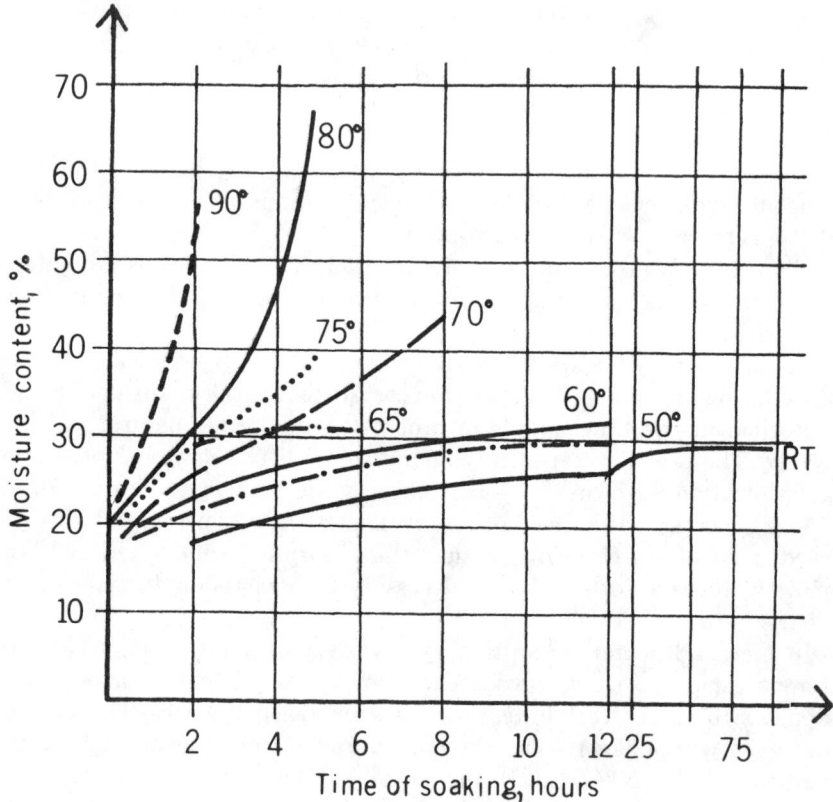

From Bhattacharya et al. (1966)

FIG. 14.3. WATER ABSORPTION BY PADDY RICE IN RELATION TO TEMPERATURE OF STEEPING WATER (RT = ROOM TEMPERATURE)

The smell and flavor of parboiled rice has been shown to be related to the condition of the steeping process. Steeping time and water temperature affect solubilization of the albuminoids contained in the outer layers of the caryopsis. By hydrolysis, the albuminoids are split up into amino acids.

Heating the sulphur-containing amino acids will split them to form hydrogen sulphide and organic sulphides of low molecular weights. These compounds will combine with the alcohols produced by decomposition of the lignin in the paddy hull, producing odorous products such as thioalcohols and thioethers, which give some characteristic odor and flavor to the parboiled rice.

Occasionally, both vacuum and hydrostatic pressure methods are used to reduce steeping time, keeping the temperature of the water within

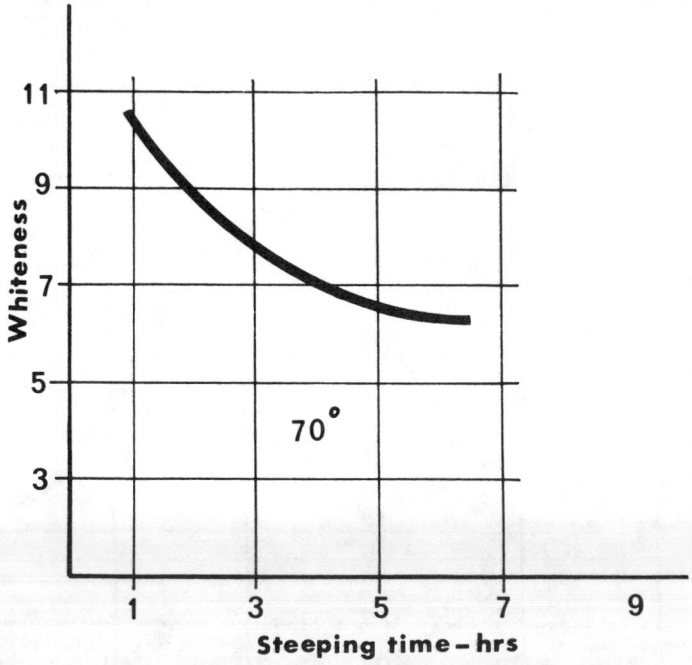

From Jayanarayanan (1964)

FIG. 14.4. DISCOLORATION OF PARBOILED RICE AS RELATED TO
STEEPING TIME IN WATER AT 70°C (158°F)

limits which will not adversely affect the quality of the final product. By
removing interstitial air and by applying hydrostatic pressure to the
steeping water, the steeping time can be reduced.

STEAMING

The purpose of steaming is to increase the milling yield and to improve
storage characteristics and eating quality. It improves the firmness after
cooking, and achieves better vitamin and salt retention in the milled rice.
These advantages, however, are offset by practical and economic disadvantages which sometime make it necessary to reach a compromise
between the 2 tendencies. For example, while complete starch gelatinization results in a high milling yield, the color of the final product will
be deeper.

Heating the steeped paddy with steam causes the following changes in
the paddy: (1) the moisture content of the paddy rice increases because of
the extra water formed by condensation; (2) water soluble substances

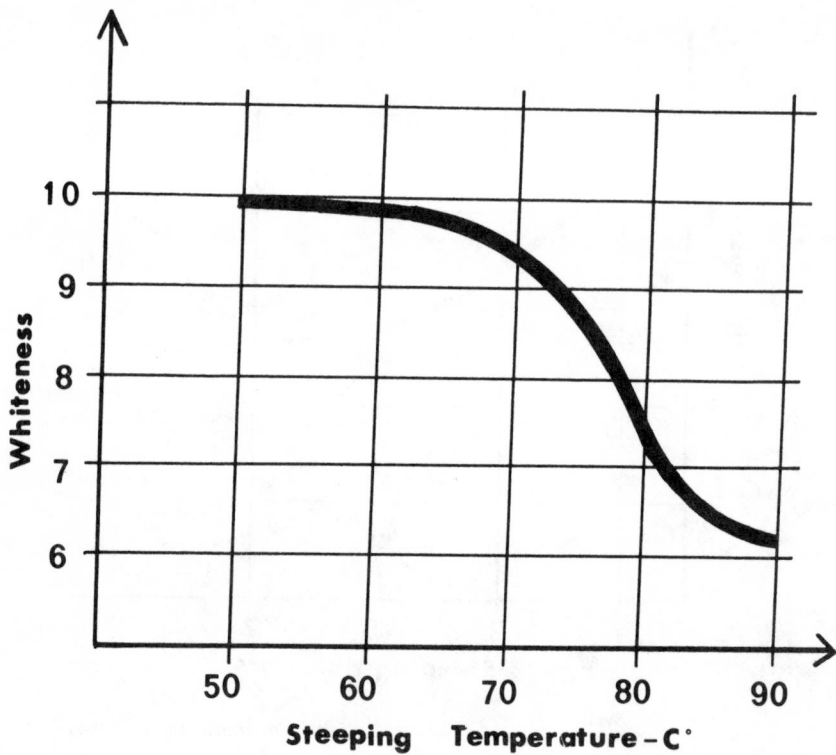

From Jayanarayanan (1964)

FIG. 14.5. LOSS OF WHITENESS BY PARBOILED RICE IN RELATION TO STEEPING WATER TEMPERATURES

spread inside the paddy grain; (3) the granular texture of the endosperm becomes pasty due to gelatinization of starch; (4) the cracks in the caryopsis become sealed and the texture of the endosperm becomes more compact; (5) the single parts making up the pericarp and the perisperm, the aleuronic cells of the endosperm and part of the germ (scutellum) become partially softened and embedded in the amylaceous endosperm of the caryopsis; and (6) germination, fungus spores, insects, their eggs and larvae and enzymes are annihilated or inactivated.

The fusion of the outer perisperm layers and the scutellum of the germ with the endosperm makes these parts difficult to remove during milling if it is desired to obtain a highly milled product which keeps well without becoming rancid during storage. Steam heating must therefore be applied uniformly to have an even exposure of the rice. The heating time and temperature must be controlled precisely in relation to the paddy variety being processed.

Steam heating may be done at atmospheric pressure by injecting steam onto the product contained in an open vessel or in an autoclave at a pressure higher than atmospheric. Continuous steam injection devices are also used at the Sacramento plant of the California Rice Growers Association.

When steam is injected on the rice in an open vessel, the excess water from steam condensation is easily taken off. But it is more difficult to distribute the steam evenly. When heating is done under pressure, the temperature can be varied easily and heat distribution is more uniform. The equipment, however, is more expensive. There are more expenditures for mechanical installations to remove excess condensate and for loading and unloading the rice.

In order to gelatinize the starch the paddy must absorb sufficient quantities of water and must reach the minimum gelatinizing temperature. Steam temperature at atmospheric pressure is always higher than that needed for gelatinization. To make sure that all starch in the rice grain is gelatinized, sufficient moisture and heat must be applied.

If the starch in the endosperm is not fully gelatinized, there will be white cores present in the parboiled product. The time exposed to steam must therefore be long enough so that the whole kernel is completely gelatinized.

The quantity of water to be absorbed, the time of exposure to steam and the temperature or pressure of the steam itself provide the parameters which will decide the quality of the parboiled rice.

Through a variation of these factors, parboiled rice possessing particular characteristics and degrees of gelatinization can be obtained. "Fully parboiled rice" means that the starch has been gelatinized right through the middle of the grain and "partially or surface parboiled rice" means that gelatinization is only surface deep and the product has typical white cores. "Light parboiled rice" is obtained by steaming it for a minimum time and at the lowest temperature needed. "Dark parboiled rice" is steamed for a long time at a high temperature.

Roberts *et al.* (1954) found that within defined limits of temperatures and pressure of the steam used the milled parboiled rice shows differences in: (1) color, (2) volume after exposure to air heated to 121°C (249.8°F), and (3) soluble starch content.

A graph showing the increase in volume (as an ascertainable ratio) of milled parboiled rice made at different steam temperatures is presented in Fig. 14.6. An increase in expanded volume is observed when the steam temperature during parboiling is increased from 100° to 120°C (212° to 248°F).

The soluble starch contents of parboiled rice made by steaming at various temperatures are presented in Fig. 14.7.

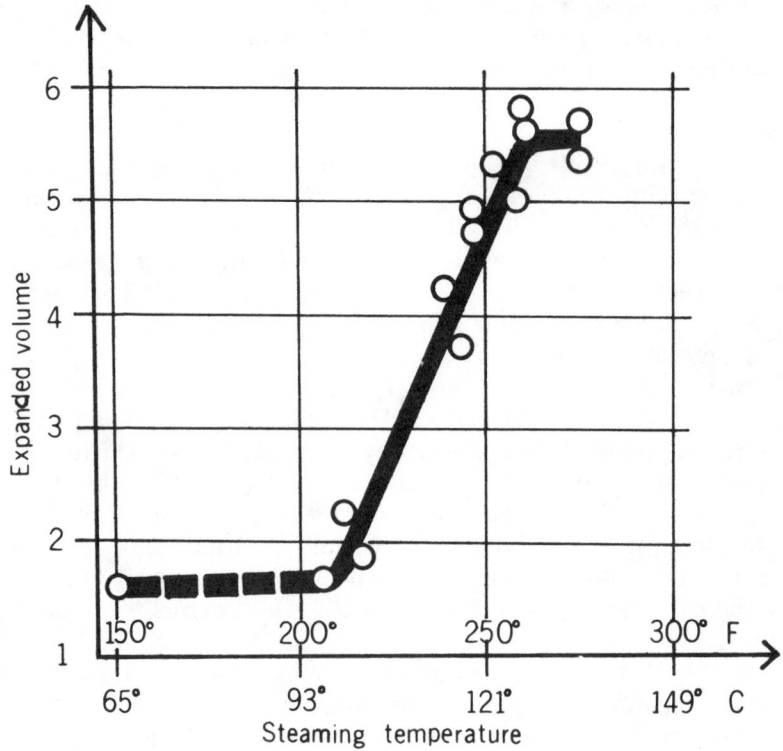

From Roberts et al. (1954)

FIG. 14.6. INCREASE IN VOLUME OF MILLED PARBOILED RICE STEAMED AT VARIOUS TEMPERAUTRES AND AFTER EXPOSURE TO A STREAM OF AIR HEATED TO 121°C (249.8°F)

The relationship between the color of milled parboiled rice and steaming parameters has been given by Bhattacharya *et al.* (1966). Color changes are caused by chemical and physical transformations induced by heat. The absorbed water can dissolve the coloring pigments in the hulls and the parboiling process drives them inward to the endosperm. Through the parboiling process the starch assumes a different refraction to light which alters the appearance and color of the product. During steeping, activation of various enzyme actions leads to the formation of reducing sugars which are responsible for Maillard type browning reactions with amino acids and protein when heat is applied during drying. Activation of the enzyme during steeping was shown to occur at about 60°C (140°F) (Gariboldi 1974).

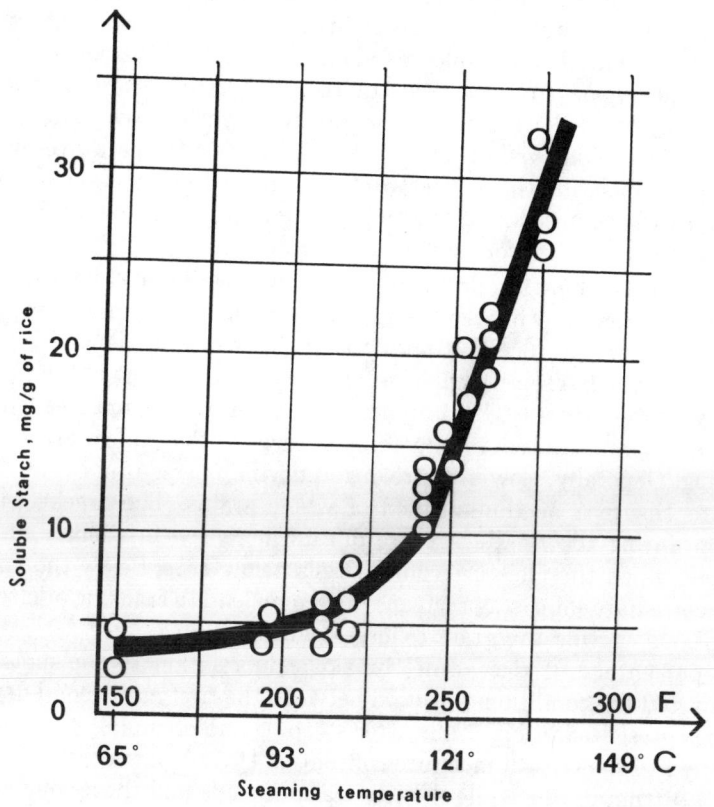

From Roberts et al. (1954)

FIG. 14.7. SOLUBLE STARCH CONTENT OF MILLED PARBOILED RICE
AFTER STEAMING AT DIFFERENT TEMPERATURES

DRYING

Objectives and Methods

The objectives of drying parboiled rice are to reduce the moisture content to an optimum level for milling and subsequent storage, and to obtain the maximum milling yield. Drying also affects the texture and color of the final product.

The conditions required for drying parboiled rice differ considerably from the processes normally used on threshed paddy direct from the field because the moisture content of parboiled paddy is higher than that of harvested paddy. The grain texture is also different as the starch has been gelatinized to form a compact grain. At the beginning of the drying process the temperature of the parboiled paddy approaches 100°C (212°F), while that of the threshed paddy is at ambient temperature. In many cases the hulls of parboiled paddy are cracked open to some extent. Thus the drying of parboiled rice requires a different process. Threshed rice must be dried slowly at air temperatures slightly above that of the environment. Milling yield is not affected by drying temperatures when reducing the moisture content to 16–18%. More attention is given to the temperature used in reducing the moisture content to 14% or lower which is considered optimum for storage.

To prevent cracking the drying process is stopped for a while when the moisture content reaches 16% and then drying is resumed using the appropriate temperature and drying time. This interval is called "conditioning," which varies somewhat with the variety of rice and the severity of the process. The optimum temperature and time needed for final drying are related to the temperature of the paddy after conditioning. Generally, slow and prolonged drying is essential in the final stage to ensure a maximum yield of whole grains. The cracks which develop during the final stage are due to increased brittleness of the caryopsis when the moisture content falls below 16%. Below this moisture level, a hardening stage sets in which may lead to cracking or else set up tensions causing the grain to break during milling.

The percentage of rice kernel breakage in relation to the moisture content and the conditioning period between the first and second drying stages is presented in Fig. 14.8. After steeping and steaming, the sample was dried to the critical moisture content of 16%. A part of it was then given a further drying while the rest was put aside and the second stage begun after periods varying from 2–48 hr. The percentage of broken grains after milling decreased with the increase of tempering time and with a reduction in moisture content.

Drying in the shade with the rice spread out on roofed over floors would yield excellent results, but the long time and large space requirements have made this system rather unpopular. In some parts of Asia before the rice is spread to dry naturally, while still hot from the steaming process, it is heaped up and left for several hours. In this way the gelatinizing action is prolonged, making the milled product harder and darker in color.

Dryers

Various types of vertical column dryers and horizontal rotating cylindrical dryers have been developed in modern plants (Gariboldi 1974). The vertical cylindrical dryers are preferred where low temperature drying air is used. The rice is exposed to the drying air for a long time

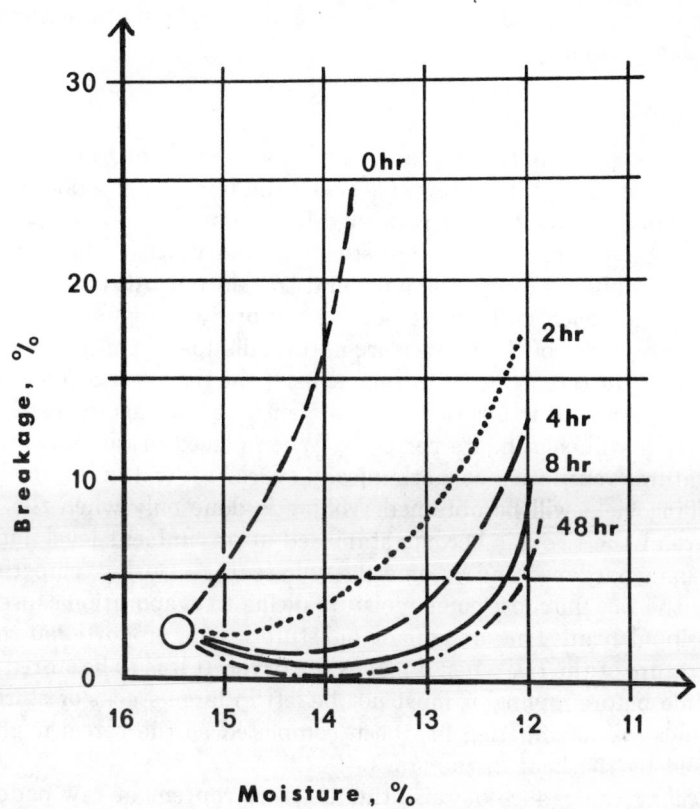

From Bhattacharya et al. (1966)

FIG. 14.8. EFFECT OF GRAIN MOISTURE AND CONDITIONING ON
BREAKAGE IN MILLED PARBOILED RICE

while the horizontal rotating cylindrical dryers are normally used when
rapid drying at high temperatures is required.

In rotary cylindrical dryers hot air is used and heat is also applied to the
cereal by fitting an external steam jacket to the dryer and a tube nest
inside. High temperature drying with a horizontal cylindrical dryer is
used for reducing the moisture content to 16–18%. It is followed by
further drying at lower temperatures in a conventional column dryer.

The vacuum dryer offers savings in fuel and removal of unpleasant
smelling substances formed when the rice is steeped and steamed. Vacuum drying is usually done inside the container in which the rice was
steamed.

In the hot air dryer, the drying time is determined by the temperature
and the amount of air. In the vacuum type, the determining factors are

the temperature, the surface heating area of the rice in contact and the attainable vacuum.

TEMPERING

After drying, the parboiled paddy must be allowed to rest for a time before milling. This time interval is called the tempering period. A tempering period of about 48 hr is needed for the product to dissipate the heat it received during steeping, steaming and drying. The caryopsis needs several hours to become hard and translucent. Also the moisture content inside each grain must become uniform throughout.

Tempering must be done to ensure natural dissipation of heat without speeding up the cooling by artificial means. If the rice is arranged in small heaps or spread out in a closed but well ventilated storage space, 20–30 cm thick, it will consolidate perfectly. When placed in tall narrow silos with natural ventilation or mechanically stirred several times, the highest milling yield will be obtained. Milling is done only when the temperature of the rice has become stabilized at an ambient level and the grains have hardened and become glassy in texture. During tempering or cooling the rice may lose some moisture owing to evaporation caused by the residual heat. The amount of moisture lost is proportional to the temperature of the rice when it leaves the dryer. If it is to be stored for a long time before milling, it must not be left in large stacks or stored in large silos until tempering has been completed as the cereal might be damaged by the heat in the stack.

Parboiling can reduce or raise the moisture content of raw paddy to levels which are most appropriate for milling and storage. The moisture content of parboiled milled rice may be brought up to 12–14%, even if that of the raw paddy used for the process is below these percentages.

MILLING

When properly prepared and milled, parboiled rice will give the maximum yield of edible rice with a minimum amount of broken grains. Parboiling gives hardness and resistance to grains and seals any cracks in the caryopsis. Any breakages will only be caused by mechanical action of the milling machines. Even if steeping and steaming have given the grains a satisfactory degree of gelatinization, they may become brittle and cracked during drying. Good results from the treatment depend to a great extent on the drying process.

Before parboiling the paddy must be properly cleaned and graded according to thickness, length and weight. Milling will be easier and the chance of breakage in the machines is minimized, if the machine is correctly adjusted.

The part of the germ known as the scutellum contains oil and protein. It is partly combined with the caryopsis so that the embryo is held in place and is usually not detached during shelling.

It is necessary to pass the product through a cone type whitening machine or a horizontal cylinder covered with abrasive material in order to remove the pericarp, the perisperm and the layer of aleuronic cells. Polishing is done in a friction machine.

In many cases parboiled rice is undermilled and still carries most of the aleuronic cells and traces of the perisperm as well as the germ at one end.

When comparing raw milled rice with another lot of an identical variety that has gone through parboiling, the respective milling degrees must first be defined in terms of the quantity of bran removed during whitening. Milling parboiled paddy becomes a difficult operation not only because the process has hardened and merged the outer layers with the endosperm, but also because the fatty substances, especially those contained in the germ, have been dissolved and distributed throughout the caryopsis. These substances make the grains slippery during the process of mechanical erosion and tend to cause the bran to cake. To avoid this, the whitening machines must be thoroughly air cooled by means of a central aspiration system. Between one whitening operation and the next, the rice is allowed to stand for some time in feeding bins. When pearling parboiled rice, the cones are usually made to turn at a 10% higher rpm than for raw paddy. At least 4 whitening machines are used to get parboiled rice completely polished.

When the paddy is put directly into the huller without prior shelling, the hull which came off the caryopsis during the first stage acts as an abrasive and at the same time absorbs some of the fatty substances, thus facilitating polishing.

The bran and polish are darker in color and contain more fatty substances compared with those obtained after milling raw paddy rice due to the spread of the fats in the germ toward the perispermic layers and the aleuronic cells. The bran from parboiled rice has prolonged resistance to the formation of free fatty acids. This makes it better and easier to use for the extraction of edible oil. The bran obtained from raw paddy has a fat content of between 12 and 14%, and that from parboiled rice may contain 16−22%. The bran obtained from processing raw paddy may show an increase of fatty acids by about 1% per hour during the first 12 hr after milling. In bran from parboiled rice, there is hardly any increase of free fatty acids during the first 15−20 days following milling. The latter can therefore be easily collected, transported and stored for subsequent extraction of its oil content. The action of heat during the parboiling process has a stabilizing effect and inhibits lipasic action which causes the fats to hydrolyze.

COLOR SORTING

The parboiled rice must be sorted to remove discolored grains. A flat conveyor belt about 0.9 m wide is used. The speed of the belt is adjustable as desired by the operator. The rice is spread on the belt in a thin layer and inspected as it moves along by sorters who pick out the discolored grains by suction, using a plastic or rubber tube connected to a centrifugal air pump. The grains thus sucked up are deposited inside a cyclone separator through which the flow of air passes before reaching the pump. The average speed a sorter can reach is about 1 grain/sec. The cost of sorting depends largely on the percentage of discolored grains present.

Recently automatic machines based on photoelectric devices have been built to sort the rice by color. The existence of such machines enables rapid sorting of parboiled rice.

The automatic sorting machines have the following advantages over the manual sorting belts: (1) the speed is faster and the rice passes through the machine at the same speed irrespective of its content of discolored grains, and (2) sorting is more efficient as the grains are checked from all angles.

The photoelectrical cell consists of a metal plate covered by a layer of oxidized selenium. Its electrical conductivity varies according to the amount of light striking its surface. The principles upon which these machines work are essentially that of presenting the grains of rice to a uniformly lit chamber so that they can be scanned simultaneously on 2 sides by 2 photocells. The light reflected by the grain passes through a filter and by means of a system of lenses, its intensity is metered by the photocell. The color is electronically compared by reference to a standard background. The sorting is based on the ability of the photocells to determine if the light of a certain hue reflected from a given object is more or less than that reflected from the background selected. Any minute change of light intensity reaching the photocells causes an electrical change within them. This change, which is due to the photoelectric effect, causes an output voltage to swing above or below the constant background signal level. Whenever the output signal level for the grain being viewed is equal to or above the background standard signal level, the grain is acceptable. An ejection circuit is actuated whenever the output signal level for the grain in view swings below the background level.

In modern sorting machines, the rice presented for scanning is made to slide in line down a straight slope from which it reaches the scanning area at a previously calculated speed and curve (Fig. 14.9). The scanning unit, the photo detectors and the impulse amplifier are the essential components of the machine. Transistors have completely replaced electron tubes and plug in circuit boards are now commonly used. Many improve-

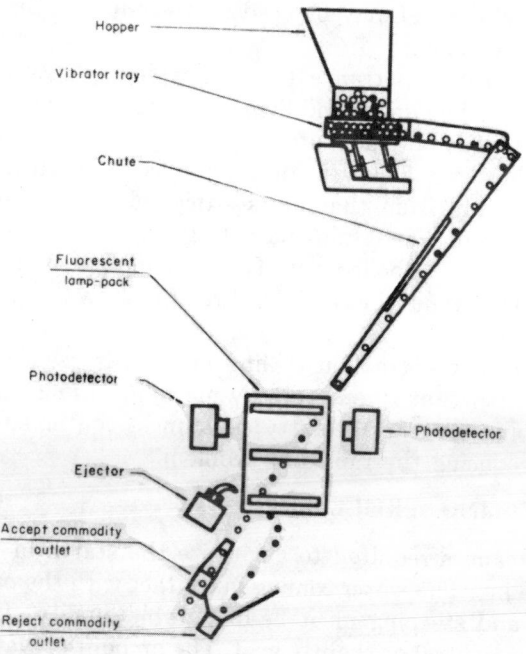

Hopper

Vibrator tray

Chute

Fluorescent
lamp-pack

Photodetector

Photodetector

Ejector

Accept commodity
outlet

Reject commodity
outlet

From Gariboldi (1974)

FIG. 14.9. SORTING DEVICE OF A SOLID STATE ELEC-
TRONIC RICE SORTER

ments have been made in extracting the discolored grains from the rest
of the rice so that the selection between rejected and acceptable grains
has become quicker and more accurate.

Recently built machines use a stream of pressurized air which knocks
the grain aside from its path. Improvements made on these machines
have brought about a continuous rise in sorting speeds and in output.
Running costs are low both in regards to power consumption and maintenance.

STABILITY OF MILLED RICE

Parboiled paddy can be shelled where it is produced then stored and
carried to other destinations where it is consumed after whitening.

After milling, the smell and flavor of the rice may undergo changes if
the fatty substances in it become rancid. The reason why parboiled rice
turns rancid is that the antioxidants, especially the tocopherols (vitamin
E), are inactivated by the hydrothermic treatment. This is especially
true when it is packaged in paper, cardboard or transparent materials.

Antioxidants such as BHA (butylated hydroxy anisole) and BHT (butylated hydroxy toluene) may be used to prevent deterioration in smell and flavor after milling.

A factor of great importance to the keeping quality of rice is the moisture content. The lipase and lipoxidase are more active if the moisture content is high. The moisture content of parboiled rice both before and after milling is not easy to determine as the texture of the grain differs considerably from that of raw rice. Most moisture measuring instruments used for raw grains do not give a correct indication of the moisture content of parboiled rice. To avoid any possibility of error the classic method of drying the rice in an oven for a few hours has proved to be the most reliable.

When the solvent extraction milling method is used, the rice keeps much better than that given ordinary mechanical milling. The organic solvents dissolve most of the fatty substances contained in the endosperm, thus reducing the rancidity problem.

THE PARBOILING PROCESSES

Water or steam is required to gelatinize the starch in the rice grain during parboiling. The water coming in contact with the paddy in washing, floating and steeping must be of potable quality. The pH of the water must be neutral or slightly acid. The amount of water needed for washing and floating the paddy usually varies from 2 to 3 times the weight of the cleaned paddy. Some of the washing water may be regenerated.

About 600 kg of water are needed to steep 1000 kg of rice paddy. During steeping, 20% of the water is absorbed by the paddy. The steeping water left behind contains the impurities in suspension, and is either thrown away or regenerated for further use.

The amount of steam needed to heat 1 MT of paddy varies with the degree to which the starch in the caryopsis must be gelatinized and with the method of heating. In a modern plant, about 80 kg of steam are required for heating 1000 kg of paddy. If steaming is done in open containers by injection at ambient pressure, the quantity used will be considerably more.

The quantity of steam needed to parboil paddy rice is a mere fraction of that required for the whole process because steam is also needed for heating the steeping water and the air where the paddy is artificially dried. The steam required to produce parboiled rice in modern plants is supplied by high pressure boilers sent to the various points where steam is needed at a low pressure. In some plants, steam is produced at high pressure for power in turbines or engines before it enters the heating system of the parboiling plant at a low pressure.

The parboiling process includes cleaning and grading the paddy, parboiling, steam production, drying, milling, color sorting and packaging. Between stages, bins are used for storing the products or byproducts so that the various operations are kept flexible.

There have been developments in industrial production of parboiled rice and further changes are taking place as a result of research. Numerous technological improvements have been accomplished to increase the yield, improve the quality of edible rice and to save labor costs.

The plants may operate under continuous or batch type processes. Some use a long steeping and steaming cycle using low temperatures and others use short cycles, but with high temperatures and pressures.

The technique to be used depends on the variety and quality of the paddy to be treated and the quality of the final product desired. The various operations may be automated to reduce running costs and to ensure more constant quality.

Some principles can be used as a guide in building a parboiling plant. A continuous production system is preferable if the paddy consists of only a few varieties grown on a large scale. A batch production system is more suitable where the paddy is of many different varieties and characteristics, or if the usable properties of the edible rice are to be different. The steeped rice should be moved from the soaking tanks to the steaming autoclave and from there to the dryers by gravity.

Steeping the paddy in water at certain temperatures, with or without the vacuum or pressure, is an essential step. The steeping water must be kept at an even temperature and the whole lot must be steeped for the same length of time. The same principal applies to the steaming process. It is important that saturated steam be used to avoid too high a temperature. Care must be taken to prevent the rice from absorbing impure condensate which would adversely affect the flavor and color of the product.

Different techniques may be used to provide the best possible milling yield and a moisture level that will ensure good steeping qualities. The use of dryers constructed for drying threshed paddy should be avoided unless their design and materials have been suitably altered.

A highly mechanized plant with automated processes may be suitable in a country where labor costs are high, but unsuitable for countries where labor costs are cheap and maintenance facilities for sophisticated equipment are not available.

Examples of Parboiling Systems

Schule Process.—The Schule process (Fig. 14.10) was originated by a German rice machinery manufacturer (Gariboldi 1974). In this system,

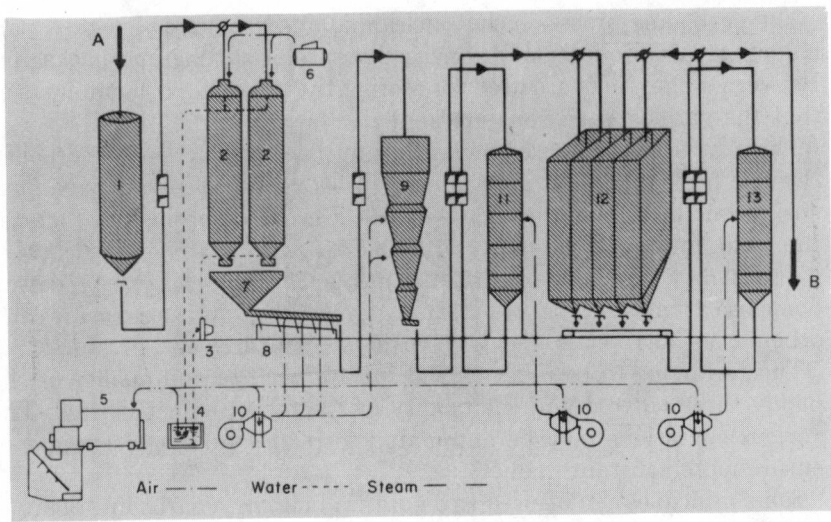

From Gariboldi (1974)

FIG. 14.10. THE SCHULE PROCESS: A. RAW PADDY; B. PROCESSED PADDY

(1) Feeding bin, (2) parboiling pressure tanks, (3) preheated water circulation pump, (4) water heater, (5) steam boiler utilizing husk as fuel, (6) air compressor, (7) wet parboiled rice receiving hopper, (8) vibratory conveyor, (9) predryer, (10) drying air heater and blower, (11) column dryer, (12) tempering bins, (13) column dryer.

steam is not directly applied to the paddy but is used to heat the water. The paddy is put into a pressure tank and is first soaked for about 120–160 min in water at medium temperature while the water is kept in circulation. When the rice has reached the temperature of the soaking water, the water supply is turned off and hydrostatic pressure (4–6 kg/cm²) is applied by admitting compressed air. The second cooking period starts by lowering pressure and readmitting the water that has been heated to a very high temperature to ensure that the starch gelatinizes completely. The water is then drained away and the paddy, with a moisture content of about 40%, is carried by a vibratory conveyor to a predryer designed to take a full batch of wet paddy. Here the moisture content is reduced and the product is then unloaded into 2 or more column dryers where drying is continued until a moisture content of 13% is reached. The milled parboiled paddy is pale in color.

The Central Food Technological Research Institute (CFTRI) Process.—The parboiling system developed by the Central Food Technological Research Institute (CFTRI) of Mysore, India, has improved the quality of the rice, shortened the processing times and reduced the equipment costs (Gariboldi 1974).

A boiler supplies steam under pressure to the steeping and steaming containers. The steam enters through perforated pipes running the length of the cylindrical container. At the bottom there are perforated pipes, arranged radially to provide the best possible distribution. The base of the steeping and steaming cylinder is cone shaped and is closed at the bottom by a water-tight hatch. At the side of the hatch there is a valve for draining off the steeping water.

The steeping and steaming cylinders are raised about 1 m above ground level. The cylinders are fed with water, which is heated by steam injection to 85°C (185°F). The paddy is poured manually into the soaking and steaming tank. The temperature of the water drops from 85°C (185°F) to 70°–75°C (158–167°F) when all the paddy has been poured in. After 2–3.5 hr of steeping, the water is drained off. Pressurized steam is then admitted and heating is continued until the hull begins to crack open. The condensate is drained off at the bottom of the tank by opening the drain valve.

The wet rice is unloaded by opening the bottom hatch and transported to the drying floor where it is spread out. In order to prevent the steeping water from fermenting, a pump may be provided to regenerate the water continuously through a filter. After filtering, the water is kept at high temperature by continued steam injection.

The batch parboiling process is similar to the CFTRI method but with some improvements. A bucket elevator and a screw conveyor raise the paddy from ground level to the steeping and steaming cylinders of the closed autoclave type. Steeping water and steam are fed from the bottom of the container. A valve at the top acts as a water overflow and as an air outlet when pressurized steam is admitted. Steeping water and steam condensate are drained off by a valve at the bottom of the container. The wet parboiled paddy falls out by itself when the hatch is open.

The Jadavpur University Process.—In the parboiling process developed at the University of Jadavpur, India, the operations are fully automatic. The average processing time is about 5 to 6 hr (Gariboldi 1974).

Steeping is completed in water 60°–70°C (140°–158°F) within 2.5–3 hr while steaming requires only from 3–5 min. The high temperature of the water and the short steeping time contribute to the production of a good quality parboiled rice.

After steaming the paddy is rapidly cooled. Drying takes place in a rotary steam-jacketed, high temperature air dryer to 13% moisture. The rice is then milled. In this process 2 different systems may be applied. With the first, the steeping and steaming take place in the same tank, whereas with the second, these 2 operations are performed separately in a horizontal apparatus. In both cases saturated steam is used. The steamed paddy is rapidly cooled in a draught of cold air.

The Avorio Process.—The Avorio was developed in Italy (Gariboldi 1974). It is a mechanized and automatic process. Steeping is done by mechanically submerging baskets filled with paddy in a tank of medium temperature water. A chain conveyor passes the baskets through the tank. The water is kept in continuous circulation and aerated by compressed air. Steeping time is controlled by the speed at which the baskets pass through the tank, and varies from 50–120 min, depending on the variety of paddy used. After steeping it is steamed in autoclaves containing rotating perforated cylinders through which the steam spreads throughout the paddy. The paddy is loaded and unloaded through special valves which function alternately. Steaming times and pressures may be varied according to the variety of the paddy. Pressure may be raised up to 1 kg/cm^2. The steamed paddy is cooled by a stream of cold air and then transferred to a series of vertical column dryers utilizing air at 45°–50°C (113°–122°F).

This process is well controlled at all stages and produces a parboiled rice of high head yields with excellent properties.

The Crystal Rice Process.—The Crystal Rice process was developed in Italy (Gariboldi 1974). The paddy is first washed in cold water to remove impurities and the lighter grains. Steeping takes place in a stationary autoclave applying first vacuum and then high hydrostatic pressure using water at a controlled temperature. Steaming and drying are carried out in a rotary autoclave which is fitted with a steam jacket and coils for heating the paddy. The process permits variations of time, temperature and pressure, and can produce various types of parboiled rice.

The Rice Conversion Process.—This process was the first parboiling process adopted in the United States in 1941–42 (Fig. 14.11). The entire rice kernel is completely gelatinized following a number of operations. The parboiled product is pale in color and completely vitreous. There are no grains with white starchy centers. Steeping is done in an autoclave where the paddy is deaerated under a vacuum to facilitate water saturation. A pressure treatment is applied to the steeping water in such a way that the combination of the vacuum and pressure processes reduces steeping time to less than 3 hr.

Steaming is done in a rotating, steam-jacketed autoclave. The pressure is kept at less than 1 kg/cm^2 for about 1 hr, after which a vacuum is applied to free the grain of excess water. The drying takes place while the paddy is still in the autoclave, by applying a vacuum and keeping the paddy hot through contact with steam-heated surfaces. The drying pro-

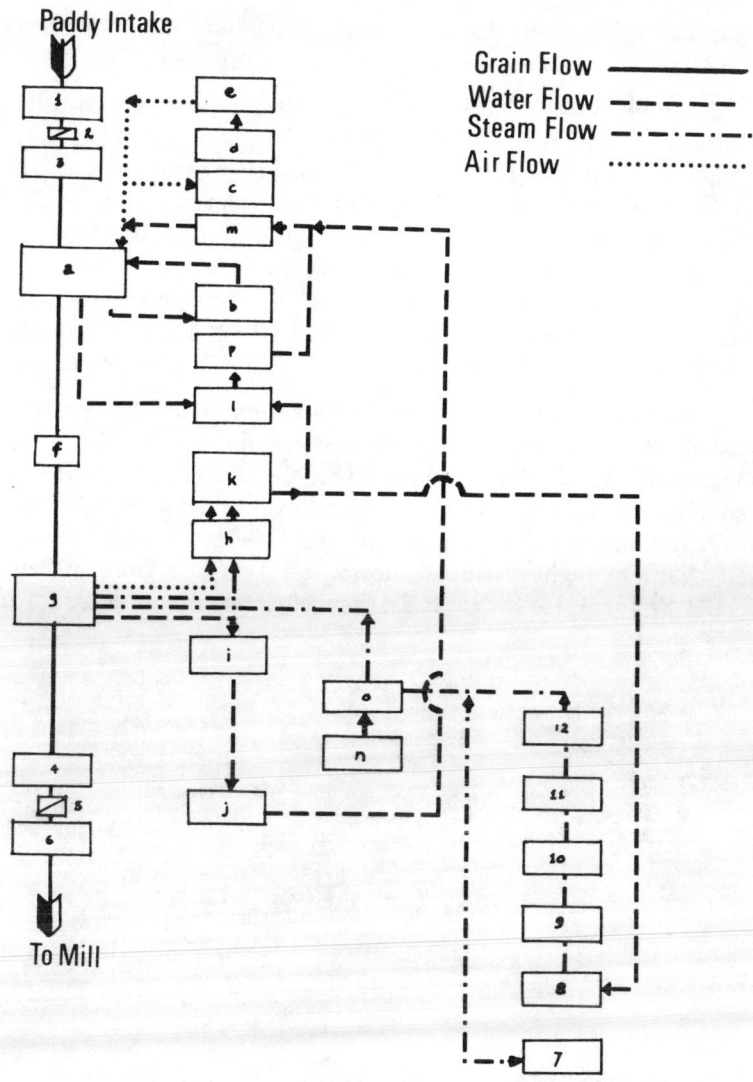

Paddy Intake

Grain Flow ——————
Water Flow – – – – –
Steam Flow –··–··–··–
Air Flow ·················

To Mill

From Gariboldi (1974)

FIG. 4.11. RICE CONVERSION PROCESS

(1) Paddy floater (to float off shrunken and dead grains); (2) elevator; (3) bin for paddy; (4) dryer discharge hopper with elevator feed regulator; (5) grain elevator; (6) mill feed bins; (7) exhaust from existing steam engine; (8) economizer; (9) boiler feed water tank; (10) boiler; (11) steam accumulator; (12) steam oil separator; (a) stationary pressure steeping tank; (b) circulating water pump; (c) dry vacuum pump; (d) air compressor; (e) air compressor tank; (f) double outlet hopper with wire mesh and drain; (g) combined steaming vessel and vacuum dryer; (h) balanced nonreturn valve; (i) condensate collecting tank; (j) hot water pump; (k) wet vacuum pump; (l) water settling tanks; (m) hot water tank, adjustable ball valve, level indicator; (n) reducing valve cold water tank; (o) reducing valve.

cess is completed in a rotating dryer utilizing medium-temperature air to carry away the moisture.

The Malek Process.—This process produces an amber colored and fully gelatinized parboiled rice (Gariboldi 1974). The paddy is steeped in tanks with high temperature water for 3 to 6 hr. It is then steamed by injecting steam into a vertical, cylindrical autoclave which has a truncated conical base and is fitted with inlet and outlet valves. The paddy is let in and taken out of the autoclave by gravity.

Drying is done in 2 stages. In the first stage, air is blown into a steam heated, rotary cylindrical dryer; in the second, air is forced through a vertical dryer at lower temperatures.

The California Rice Growers Association (CRGA) Parboiling Process.—The California Rice Growers Association has developed a parboiling process at its Sacramento plant (Fig. 14.12). The raw paddy rice is soaked in moderately warm water, depending on the variety, for several hours. After that the rice paddy is moved to another tank where it is soaked again in higher temperature water (varying from 40°–90°C, 104°–194°F) for 1 to 10 hr. It is then steamed under pressure in a

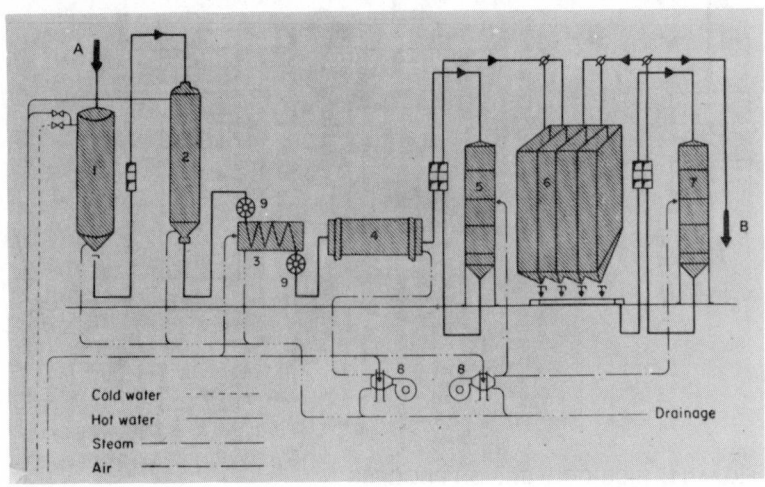

From Gariboldi (1974)

FIG. 14.12. RICE GROWER'S ASSOCIATION OF CALIFORNIA PROCESS:
A. RAW PADDY; B. PARBOILED PADDY

(1) Cold or hot water soaking tank, (2) hot water soaking tank, (3) steaming autoclave (pressure steamer), (4) hot air dryer, (5) column dryer, (6) tempering bins, (7) column dryer, (8) air heaters and blowers, (9) steaming autoclave rotary valves.

continuous cooker for a very short time, ranging from 15 sec to 3 min. The steam pressure varies between 1 and 5 kg/cm².

After draining, the parboiled paddy is dried first with hot air, succeeded by several passages through a column dryer stage. The drying process includes long intervals of tempering. The main features of the process are the long soaking time and the extremely short exposure to steam at high pressure; the resulting product is very pale. The process gives a good milling yield. The parboiled rice so obtained possesses good characteristics. It is usually applied to rice with a short, round grain typical of the varieties grown in California. The milling and packaging processes are the same as those described by Gariboldi (1974).

Miscellaneous Processes.—Barber *et al.* (1975) has developed a new hydrothermal process, similar to parboiling, and has tested it for the retention of nutrients in comparison to the normal parboiling process.

Three processes were used. Process A involves soaking the paddy for 1 min at room temperature and then autoclaving (130°C, 266°F) for 3 min; process B, for 8 min; process C, soaking for 2 hr at 70°C (158°F) and autoclaving at 121°C (249°F) for 15 min. In all cases drying was carried out at room temperature.

All processes caused slight increases in protein concentration in the outermost layer. They did not appreciably effect the distribution of crude grain fiber in the grain. The pattern of fat distribution remained the same.

The processes caused a loss of available lysine, tryptophan and methionine. The losses were larger in the outermost layer than in the intermediate layers. All the processes had significant decreases in thiamin content in the outermost layers and an increase in the center portion.

The effects of the process on fat, protein and fiber contents were limited to the outermost layers; the contents of these constituents in the well milled kernel remained unchanged. Available lysine, tryptophan and methionine contents of milled rice from process A were similar to those of the raw rice, but there were large losses in processes B and C. All treatments increased the thiamin content of milled rice.

The outer bran from all the processes were richer in fat, and somewhat higher in protein than the raw rice, but lower in available lysine, tryptophan, methionine and thiamin. In general, all treatments decreased the nutritive value of the brans.

Summary of Rice Parboiling Processes

A summary of parboiling processes for rice is presented in Table 14.2 (Gariboldi 1974).

TABLE 14.2.   SUMMARY OF PARBOILING PROCESSES

Process	Soaking	Steaming	Drying
Schule	Batch system in medium temperature water followed by a second stage in high temperature water under pressure in the same tank.	Steaming not required. Starch gelatinization obtained by soaking in high temperature water under pressure.	In high temperature air followed by medium temperature air.
CFTRI	Batch system in high temperature water. Continuous circulation of high temperature filtered water.	Batch system in the open tank used for soaking. Steam is pressure-injected through perforated pipes.	Sun drying or mechanical drying by medium temperature air.
Jadavpur Univ.	Batch system in high temperature water.	Batch system in the open tank used for soaking. Steam is pressure-injected through perforated pipes. Alternatively, continuous system with steam at ambient pressure in an autoclave equipped with a screw conveyor.	Cooling before drying. This is done by using high temperature followed by medium temperature air.
Avorio	Continuous system in medium temperature water.	Continuous steaming under pressure in an autoclave equipped with mechanical conveyors.	Cooling before drying. This is done by using medium temperature air.
Crystal rice	Batch system in high temperature water under vacuum followed by hydrostatic pressure.	Batch system in a rotary autoclave under steam pressure.	Under vacuum in the same autoclave. Final drying may be done after milling.
Malek	Batch system in high temperature water.	Continuous steaming under pressure in a vertical stationary autoclave.	By high temperature air followed by medium temperature air.
CRGA Parboiling	Batch system in medium temperature water followed by higher temperature water.	Continuous system in a horizontal cooker under high steam pressure for a short time.	In high temperature air followed by medium temperature air.

Source: Garibaldi (1974).

QUICK COOKING PARBOILED RICE

One disadvantage of untreated parboiled rice is that the parboiling process extends the required cooking time. While milled rice may require 15 min of cooking, the same rice when parboiled must be cooked for 30 min. Mickus and Brewer (1957) have developed a patented process which eliminates this problem. Milled parboiled rice is heated in a dry condition after the rice has reached its normal moisture content of 14%. Such heating has the effect of accelerating the breakdown of the starch structure of the rice which is started by the parboiling step.

The dry heating step may be carried out in several ways. One method is to circulate rapidly moving hot air around the rice kernels at 149°–260°C (300°–500°F). If the exposure of the rice to the hot air at the above temperature is continued for a period of 30–60 sec, the resultant product may be completely cooked in a minute or less. If this dry heating step is performed on ordinary rice which has not been parboiled, no reduction in the subsequent cooking time is effected.

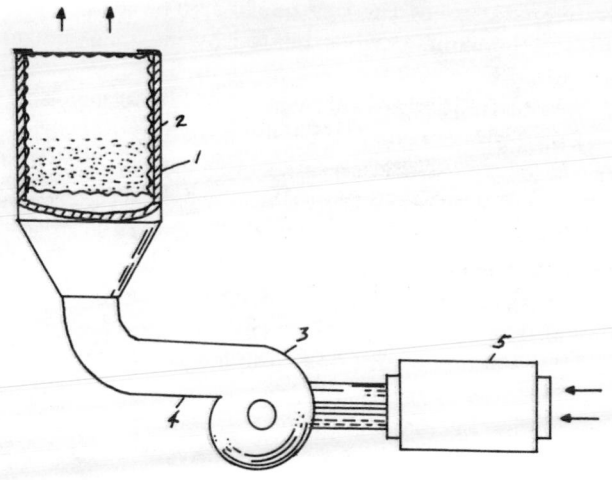

From Daniels (1970)

FIG. 14.13. APPARATUS FOR HEATING QUICK COOKING PARBOILED RICE DEVELOPED BY MICKUS AND BREWER (1957)

The apparatus for carrying out the above step is illustrated in Fig. 14.13. A vertically extending conduit was designed *1*, in which a basket *2* of wire mesh may be inserted for supporting the rice within the path of the air flow through the conduit *1*. A blower *3* is provided for conducting air through a pipe *4* into the lower end of the conduit *1*. Before reaching the blower 3 the air may be heated to the required temperature by means of any conventional type of heater 5.

The dry heating step must be done quickly. For this reason it is preferable not to insert the rice into the air stream until the air from the heater 5 has been raised to the desired temperature. Naturally the time required depends on the temperature of the air, but as stated above, the time may be made as short as 30 sec if the air temperature is around 260°C (500°F), and if the temperature is reduced to 149°C (300°F), the time may run as long as 60 sec.

Another method of carrying out the dry heating step is to use dielectric heating whereby the rice is made the dielectric between a pair of opposedly facing plates acting as a capacitor. The rice is placed in bags supported on a conveyor belt for movement between a pair of plates. The plates are connected to a source of high frequency alternating current such as a vacuum tube oscillator so that a dielectric heating effect is imposed on the rice.

The dielectric heating process heats the interior of the rice first, hence the final temperature need not be as high as that used in hot air heating. When the temperature of the rice reaches 65.6°−82.2°C (150°−180°F), its condition is the same as when heated by means of hot air at 149°−260°C (300°−500°F).

By using various frequencies between 1 and 30 megacycles with a power input of 1.8 kw, an exposure of 30 sec is required to break down the starch structure so that the rice may be cooked in less than 1 min.

The above description assumes a relatively light weight package such as cellophane. With heavier containers made of cardboard or the like, a greater amount of energy must be employed or the exposure time lengthened. This process is particularly applicable to the processing of rice for making breakfast cereals and the like. Such rice cereals have been puffed or exploded by various operations to make the kernels edible without further cooking.

Puffed rice may be made with less expensive equipment by using parboiled rice because the dry heating step returns the rice kernel to its original whiteness. To puff the rice it is merely necessary to continue the heating step beyond that described above.

The dry heated rice may be eaten without any additional cooking and makes a particularly tasty breakfast cereal especially if sugar and cream are added. Thus, whether whole kernels or broken kernels are used, the above described process produces either a quick cooking rice if prepared in the conventional manner or a ready-to-eat product having a high nutritive value.

Serbia and Benett (1968) have developed a method for producing quick-cooking parboiled rice. The parboiled rice is soaked in water at

71.1°C (160°F) for 10—15 min until the moisture content is 40%. It is next steamed for 18—25 min at 100°C (212°F), resulting in a 5% increase in moisture content.

The steamed rice is immersed in water at 15.6°C (60°F) to cool the rice, thereby causing the grains to toughen and the cooking to be arrested.

The rice, now containing 50% moisture, is retreated with steam and soaked 2 more times. Following this treatment the moisture content increases to 65%.

The soaked, cooked rice is then drained of free water and fed to a 3-section, through circulation, continuous belt dryer. Air in the dryer varies from 104.4°—121.1°C (220°—250°F) in the first section to 82.2°—110°C (180°—230°F) in the third section. The dried product has a terminal moisture content of 10—13% and shows no evidence of puffing.

Rice treated in this manner can be immersed in gently boiling water and allowed to simmer for 7 min in an open or covered pot. Alternately, the rice may be added to boiling water, removed from heat and allowed to stand for 10 min. The cooked product will have a better overall eating quality from the standpoint of texture, flavor and aroma than untreated parboiled rice.

RECENT DEVELOPMENTS IN EQUIPMENT

Transfer of Materials

McPhail (1975) has devised a method for introducing materials into or removing materials from a hermetically sealed vessel. The apparatus consists of a plurality of sealed vessels in series, with means for transferring the contents from one vessel to the next without breaking the seal. This device is useful in the process of parboiling rice, where soaked rice maintained at ambient pressure is transferred to a cooking vessel which is at a higher pressure.

Rubber Rollers

Sarda (1975) reported that rice milling was carried out in India until about 10 years ago by the traditional disc sheller type or huller type machines. Rubber rollers were not used for commercial milling of paddy. The advantages of the use of rubber roller shellers were demonstrated during a pilot study of 7 modern mills installed by the government of India. Based on these results, modern rubber roller shellers were introduced into Indian rice mills. Sarda Rice and Oil Mills, a rice processing company in Ahmadpur, West Bengal, was one of the first to introduce this technology into commercial milling of parboiled paddy.

Difficulties with the poor life of rubber rolls, black coloration of rice and low capacity of shelling were experienced in early trials. However, these were found to be due to faulty adjustment and operation and improper cooling of the rollers.

After putting an air circulation system for continuous cooling of the rubber rolls, an output of about 250 MT in winter and about 200–250 MT of parboiled paddy in summer were obtained per pair of rubber rolls. The moisture content of the parboiled paddy before milling was kept at 15%. Cleaning of chaff and immature grains prior to shelling was found to increase milling capacity and increase the life of the rubber rolls.

Black color in the rice could be due to insufficient cooling of the rolls or improper working of paddy separator, thus returning a high proportion of brown rice to the rubber rolls.

The rubber rolls can prevent removal of bran during shelling and can increase output of rice by 0.5%. The oil content in the bran from the polishers has been found to be slightly higher also. The use of rubber roller shellers in the milling of parboiled paddy has been found to be beneficial. However, a disadvantage faced by the millers using a rubber sheller is the presence of dark colored or brownish colored grain in the milled product, caused by infested or immature grains. These are normally removed with the hull in conventional milling of rice, but are recovered and retained when rubber rollers are used. Methods of eliminating this problem in milling rice with rubber rollers are therefore necessary.

ECONOMICS

Shivanna (1972) compared the cost of modern and traditional methods of parboiling rice in India. Three types of mills were compared: (1) a huller rice mill, (2) an emery cone sheller-polisher rice mill, and (3) a modern rice mill.

Modern methods of parboiling involve a huge capital investment for machinery. The processing cost is also higher as fuel oil or steam is used for heating the air in the drying process. These modern mills, however, produce a clean, wholesome product with higher rice yields. Although not clearly delineated the modern mill is generally held to be more economical.

The modern mill complex obtains freshly harvested paddy and handles sequentially the operations of cleaning and drying, silo storage, parboiling by the hot soaking method, mechanical drying of the parboiled paddy, milling by modern rubber roller type mill (Schule make), and extraction of purified bran with solvent hexane by a batch extraction method. All the processes are largely mechanized and controlled. The

TABLE 14.3. COST OF PROCESSING PER MT OF PADDY (ALL DATA BASED ON CLEANED PADDY)

Item	Huller rice mill[1]				Emery cone sheller-polisher rice mill[1]				Modern rice mill[1,2]			
	Fixed	Recurring	Processing	Total	Fixed	Recurring	Processing	Total	Fixed	Recurring	Processing	Total
					Rupees							
Expenditure												
Precleaning and drying	0.39	8.24	1.07	9.70	0.39	8.24	1.07	9.70	0.79	2.28	5.70	8.77
storage	—	—	—	—	—	—	—	—	1.94	5.16	0.40	7.50
Parboiling and drying	1.58	1.87	5.25	8.70	1.58	1.87	5.25	8.70	4.26 (3.47)	8.89 (7.63)	11.76 (8.44)	24.91 (19.54)
Milling	0.39	2.80	4.37	7.56	2.08	2.54	1.12	5.74	4.58	12.45	4.68	21.71
Total cost of processing	2.36	12.91	10.69	25.96	4.05	12.65	7.44	24.14	11.57 (10.78)	28.78 (27.52)	22.54 (18.22)	62.89 (57.52)
Cost of paddy	—	—	—	485.00	—	—	—	485.00	—	—	—	485.00
Net expenditure	—	—	—	510.96	—	—	—	509.14	—	—	—	547.89 (542.52)
Value of raw byproducts	—	—	—	10.80	—	—	—	27.00	—	—	—	39.50
Value after processing of byproducts	—	—	—	10.63	—	—	—	36.36	—	—	—	55.22
Realization by sale of polished rice	—	—	—	516.80	—	—	—	524.40	—	—	—	532.00

TABLE 14.3. (CONTINUED)

Net realization							
By sale of rice and byproducts	—	—	527.60	—	551.40	—	571.50
By sale of rice and byproducts after processing	—	—	533.43	—	560.76	—	587.22
Net profit							
By sale of rice and byproducts	—	—	16.64	—	42.26	—	23.61
By sale of rice and byproducts after processing	—	—	22.47	—	51.62	—	39.33

Source: Shivanna (1972).

[1] Yield of polished rice from the huller, emery cone sheller-polisher and modern types of mills are 68, 69 and 70%, respectively, valued at rupees 760 per MT (aver. government price).

[2] The figures in brackets are parboiled by hot soaking, rinsing with brine solution and steaming. If the polished rice is sold in the open market @ rupees 1050 per MT an extra net profit of rupees 197.20, 200.10 and 209.00 is recoverable from huller, disc sheller-polisher and modern rice mill, respectively.

steam for the parboiling is obtained from a Schule boiler with capacity of 1100 kg steam/hr. The drier is one of the LSU type and the air is heated by mixing with burnt gases after burning fuel oil in a combustion furnace.

The traditional mill is an old disc type sheller mill of the same Schule make and it runs on a steam engine from a boiler. Paddy hull is used as fuel. Parboiling is done by the traditional method followed by sun drying. A simple huller type mill also operates on the same premises.

Table 14.3 provides summarized data of item particulars for the 3 types of mills. In addition to the rice itself the table provides information on the value of the byproducts (brans, brokens/germs) and the potential value, after solvent extraction, of bran oil and bran meal for the 3 types of mills. The huller mill gives the least return and the modern mill the best return; the disc-sheller gives intermediate benefits.

The net returns are approximately rupees 22.50, 51.60 and 44.70, respectively, for the huller, disc sheller and modern type mills. It is an interesting observation that the modern mill should give lower returns than the disc sheller mill in spite of the former giving more rice yield (with less brokens) and better quality bran. The higher processing costs of the modern mill more than offset the advantages of higher yield. The processing cost could be brought down by (1) the use of hull instead of furnace oil for heating the air in the driers, (2) use of rubber rollers that will have lower wear and tear and longer life, (3) urging and realizing better market value for the processed rice as it has a lower percentage of brokens and refractions than the normal commercial rice, and (4) reducing high investment costs on silo storage. As a result of these methods it should be possible to make the economics of processing in the modern mill more attractive than the customary methods of storage and processing.

A method of pressure parboiling has been developed by Shivanna (1974). The paddy is washed in water in a pressure vessel, kept under saturated steam and then steamed at high pressure to ensure full gelatinization of the starch.

The comparative costs of the parboiling methods are presented in Table 14.4. The total parboiling and drying costs are 50% lower for the pressure parboiling process.

Additional advantages of the pressure parboiling process are that: (1) the grain becomes quite hard due to the pressure steaming, resulting in reduced breakage during milling; (2) total processing time is reduced by about 50% and therefore production capacity can be much increased; and (3) the oil content in the bran is much higher than that obtained from the hot soaking method.

TABLE 14.4. COMPARATIVE COST ESTIMATES OF PARBOILING UNDER PRESSURE AND HOT SOAKING AND STEAMING

Sample	Particulars	Parboiling Under pressure	Hot soaking and steaming
		Rupees	
A.	Nonrecurring expenditure Parboiling tanks, raw paddy elevator, belt conveyor, mechanical dryers with 1 elevator for 2 dryers and boiler (2.0 MT/hr capacity).	3,25,000	4,70,000
B.	Recurring expenditure per month Supervisory staff, spares, repairs, etc.	5,300	8,850
C.	Processing expenditure (Parboiling and mechanical drying) per MT of paddy		
	Electricity	1.30	2.95
	Fuel charges (furnace oil)	8.40	16.80
	Total expenditure on the process	9.70	19.75
	Recurring expenditure per MT of paddy	2.20	3.70
	Nonrecurring expenditure per MT of paddy (depreciation on machinery @ 10%)	1.15	1.60
	Total cost of parboiling and mechanical drying per MT of paddy	13.05	25.05

Source: Shivanna (1974).
Note: Steam is obtained by using husk as fuel to the boiler. The above figures are subject to variation with fluctuation in cost of material and units.

There are, however, several drawbacks to this new process. The rice obtained has a deep yellow-brown color which may be objected to in some markets. Also the higher oil content in the bran tends to clog the polishing sieves.

REFERENCES

ADAIR, C.R. 1972. Production and utilization of rice. *In* Rice Chemistry and Technology. D.F. Houston (Editor). Amer. Assoc. Cereal Chem., St. Paul, Minn.

ALI, N. and OJHA, T.P. 1976. Parboiling. *In* Rice Postharvest Technology. E.V. Araullo, D.B. Depadua and M. Graham (Editors). Int. Dev. Res. Ctnr., Ottawa, Canada.

ALI, S.Z. and BHATTACHARYA, K.R. 1972A. An alkali reaction test for parboiled rice. Lebensmittel-Wissenschaft und Tech. 5 (6) 216–218.

ALI, S.Z. and BHATTACHARYA, K.R. 1972B. Hydration and amylose-solubility behaviour of parboiled rice. Lebensmittel-Wissenschaft und Tech. 5 (6) 207-212.

ALI, S.Z. and BHATTACHARYA, K.R. 1976. Comparative properties of beaten rice and parboiled rice. Lebensmittel-Wissenschaft und Technolog. 9, 11-13.

BANDYOPADHYAY, S. and ROY, N.C. 1975. Prediction of time of hydration during parboiling of paddy from activation energy. J. Food Sci. and Tech. *12*, 197-199.

BARBER, S., BARBER, C. and TORTOSA, E. 1975. Effects on parboiling processes on the chemical composition and nutritional characteristics of rice and rice bran. *In* Rice Report. S. Barber, H. Mitsuda and H.S.R. Desikachar (Editors). Inst. Agric. Chem. and Food Technol., Valencia, Spain.

BHATTACHARYA, K.R., SUBBA RAO, P.V. and SWAMY, Y.M.I. 1966. Processing and quality factors in parboiling of rice. Central Food Technol. Res. Inst., Mysore, India.

BHATTACHARYA, K.R., ALI, S.Z., SOWBHAGYA, C.M., SWAMY, Y.M. and INDUDHARA SWAMY, Y.M. 1975A. Physicochemical properties of Indian rice and changes during parboiling. *In* Rice Report. S. Barber, H. Mitsuda and H.S.R. Desikachar (Editors). Inst. Agric. Tech. Food, Valencia, Spain.

BHATTACHARYA, K.R., ALI, S. and ZAKIUDDIN, A. 1975B. A sedimentation test for pregelatinized rice products. Lebensmittel-Wissenschaft und Tech. *9*, 36-37.

CHANG, T.T. and BARDENAS, E.A. 1965. The morphology and varietal characteristics of the rice plant. Int. Rice Res. Inst. Tech. Bull. *4*.

DANIELS, R. 1970. Rice and bulgar quick-cooking process. Brown and parboiled rice. *In* Food Processing Review *16*, 166.

DIMOPOULOS, J.S. and MULLER, J.G. 1972. Effect of processing conditions on protein extraction and composition on some other physicochemical characteristics of parboiled rice. Cereal Chem. *49* (1) 54-62.

GARIBOLDI, F. 1972. Parboiled rice. *In* Rice Chemistry and Technology. D.H. Houston (Editor). Amer. Assoc. Cereal Chem., St. Paul, Minn.

GARIBOLDI, F. 1974. Rice parboiling. FAO Agricultural Development Paper *97*. Food and Agric. Organ. U.N., Via delle Terme de Caracalla, Rome.

GRAY, V.M. and SCHOCH, T.J. 1962. Effects of surfactants and fatty adjuncts on the swelling and solubilization on granular starches. Starke *14*, 239-245.

JAYANARAYANAN, E.K. 1964. The influence of working conditions on the browning value of parboiled rice. Nahrung. *8*, 129-137.

KENNEDY, B.M., SCHELSTRAETE, M. and TAMAI, K. 1975. Chemical, physical, and nutritional properties of high-protein flours and residual kernel from the overmilling of uncoated milled rice. IV. Thiamin, riboflavin, niacin, and pyridoxine. Cereal Chem. *52* (2) 182-188.

MCPHAIL, J.L. 1975. Method for introducing materials into or removing materials from a hermetically sealed vessel. U.S. Pat. 3,914,499. October 21, 1975.

MICKUS, R.R. and BREWER, G.W. 1957. Assigned to Rice Growers Assoc. of California. U.S. Pat. 2,808,333. October 1, 1957.

PADUA, A.B. and JULIANO, B.O. 1974. Effect of parboiling on thiamin, protein and fat of rice. J. Sci. Food Agric. *25*, 697-701.

PRIESTLEY, R.J. 1976A. Studies on parboiled rice. Part 1. Comparison of the characteristics of raw and parboiled rice. Food Chem. *1*, 5-14.

PRIESTLEY, R.J. 1976B. Studies on parboiled rice. Part 2. Quantitative study of the effects of steaming on various properties of parboiled rice. Food Chem. *1*, 139-148.

RAO, S.N. and JULIANO, B.O. 1970. Effect of parboiling on some physico-chemical properties of rice. J. Agric. Food Chem. *18*, 298.

REFAI, F.Y., KAMAI, M.A. and AHMED, S.A. 1967. Biochemical changes in rice upon parboiling. From 3rd Conf. Int. Probl. of Modern Cereal Processing and Chem. Bergholz-Rehbruecke, Germany.

ROBERTS, R.L., POTTER, A.L., KESTER, E.B. and KENEASTER, K.K. 1954. Effect of processing conditions on the expanded volume, color, and soluble starch of parboiled rice. Cereal Chem. *31*, 121−129.

SARDA, P.S. 1975. Suitability of rubber rollers for milling of parboiled rice. *In* Rice Report, 1975. S. Barber, H. Mitsuda and H.S.R. Desikachar (Editors). Inst. Agric. Tech. Food, Valencia, Spain.

SERBIA, G. and BENETT, I. 1968. Quick-cooking rice. U.S. Pat. 3,408,202. October 29, 1968.

SHAHEEN, A.B., EL-DASH, A.A. and EL-SHIRBEENY, A.E. 1975. Effect of parboiling of rice on the rate of lipid hydrolysis and deterioration of rice bran. Cereal Chem. *52* (1) 1−8.

SHIVANNA, C.F. 1972. Comparative costing of improved parboiling process in a modern mill as compared with traditional methods. J. Food Sci. and Tech., Mysore, India. *9* (1) 7-9.

SHIVANNA, C.S. 1974. Economics of pressure parboiling of paddy. J. Food Sci. and Tech. India. *6*, 286-287.

VITTI, P., LEITAO, R.F.F. and DIZZINATO, A. 1975. Parboiling of rice varieties. Bull. Inst. Food Technol. *6* (1) 103−199.

WEBB, B.D. 1967. Cooking and processing qualities required of rice varieties in the United States; the evaluation in rice breeding programs. Int. Rice Comm. Newsl. (spec. issue), Food and Agric. Organ. U.N., Bangkok.

ZOBEL, H.F. 1964. Methods in carbohydrate chemistry. *In* Starch. R.L. Whistler (Editor). Academic Press, New York.

15

Rice Quality and Grades[1]

B.D. Webb

Quality in rice *(Oryza sativa* L.) may be categorized into 4 broad areas: (1) milling quality; (2) cooking, eating and processing quality; (3) nutritive quality; and (4) specific standards for cleanliness, soundness and purity. All of these categories are important, collectively, in judging the suitability of rice for a particular use. However, rice has many different uses so the quality characteristics desired vary considerably, being ultimately related to the final consumer acceptance of each rice product or rice-containing food.

Essentially all rice is used as food for human consumption in one or more of its many forms. Boiled rice prepared in the home or institution constitutes by far the greatest consumption pattern. Other important uses include: parboiled rice, quick cooking rice, dry breakfast cereals, canned rice, soups, baby foods, frozen dishes, rice flour and brewing. Consequently it is necessary that rice quality be judged on the basis of its suitability for its intended use and that it meets established requirements regarding wholesomeness.

Characteristics that influence rice quality include those that are under genetic control and those independent of genetic control such as purity and cleanliness. These latter characteristics are primarily a function of handling and storage and as such are described thoroughly in the United States grading standards for rough rice, brown rice for processing and milled rice (see Appendix).

The genetic makeup of the grain is a major factor influencing the quality of rice. Modern rice breeding programs continually strive to refine and improve the genetic characteristics that influence quality in order to

---

[1]Cooperative investigations: U.S. Department of Agriculture (USDA) Science and Education Administration, Agricultural Research, Texas Agricultural Experiment Station and the Texas Rice Improvement Association. Mention of a trade name does not constitute a guarantee or warranty of the product by the USDA or an endorsement by the Department over other products that are also available.

obtain the most desirable product. In the United States all new rice cultivars are developed through intensive genetic selection of all important quality attributes (Adair *et al.* 1973). Selecting for desirable milling, cooking, eating and processing of hybrid selections, breeding lines and new varieties of rice is an essential part of responsible rice breeding programs conducted by the U.S. Department of Agriculture (USDA) and the State Agricultural Experiment Stations in Arkansas, California, Louisiana, Mississippi and Texas. New varieties developed in these programs must meet established standards for the milling, cooking, eating and processing qualities required of their particular grain type before they are released for commercial production (Adair *et al.* 1973; Webb *et al.* 1972; Webb 1975).

Another major factor that influences rice quality is the environment under which the plant is grown. Once a new variety is released for commercial production it will be used wherever it can be produced advantageously in comparison with currently grown varieties. Consequently, before release, each new variety is extensively tested agronomically and for quality for its likely production area. In the United States tests are carried out in the Uniform Rice Performance Nurseries and other trials in each of the rice producing states. These trials provide the means for evaluating the quality characteristics of each new variety within environmental and modifying influences such as soil, climate and cultural conditions.

Rice quality characteristics range in level of heritability from very low to very high (Beachell and Halick 1957; Ghosh and Govindaswamy 1972; Bollich and Webb 1973; Anon. 1975; Rao and Siddiq 1976). The response of certain characteristics to environmental and cultural conditions also varies. Some rice characteristics are highly affected while others rarely are influenced by certain environmental or cultural factors.

Specific characteristics influencing the quality of rice in the United States include: (1) hull and pericarp color; (2) grain size, shape, weight, uniformity and general appearance; (3) milling outturn; (4) kernel chalkiness, translucency and color; (5) cooking, eating and processing characteristics; and (6) cleanliness, soundness and purity. Also, since rice is consumed and processed mainly in whole kernel form, the physical attributes of the intact endosperm are always of foremost importance.

The purpose of this chapter is to identify the components of quality considered to be the most important in rice for specific uses and to present data and reference methods which may be useful as guidelines in judging the quality and grade of rice. Most of the discussion will be confined to practices used in evaluating rice quality in the United States. These concepts and principles involved, however, are used in varying degrees throughout the world.

Most likely anyone who is concerned with rice quality will at one time or another use most or all of the quality factors and component measurements discussed and presented in this chapter. It is therefore difficult to categorically class the significance of the different quality factors because all the components are cumulatively involved in judging the quality of rice. Furthermore, the values for the various quality criteria presented in tables used in this chapter should not be considered as iron-clad parameters, but rather as indicating reasonable ranges in line with current concepts of rice quality.

It is recognized that the nutritional qualities of rice are necessarily involved in all areas of rice quality. This subject is beyond the scope of this chapter and the reader should consult Chapter 11 for a discussion of the nutritive value of the rice endosperm. Also, the reader should refer to previous reviews (Kester 1959; Beachell 1959; Adair *et al.* 1962; Matz and Beachell 1969; Barber 1972; Juliano 1972A, B, 1973; Webb and Stermer 1972; Chikubu 1975; Webb 1975) for additional information on the many aspects of rice quality.

INTERRELATIONSHIPS OF VARIETY, GRAIN TYPE AND QUALITY

Virtually all of the United States rice crop is produced from varieties developed by the rice research centers and experiment stations located at Stuttgart, Arkansas; Biggs, California; Crowley, Louisiana; and Beaumont, Texas (Bollich *et al.* 1972; Carnahan *et al.* 1978; Jodon *et al.* 1971; Johnston *et al.* 1967). These centers have had active breeding programs for many years and are operated cooperatively by the 4 state agricultural experiment stations, Science and Education Administration, USDA and state and local producers organizations.

Traditionally, rice varieties in the United States are classed as long, medium and short grain types (Fig. 15.1) and now, through responsible planned breeding programs, varieties of each grain type are associated with specific cooking, eating and processing qualities. High quality U.S. long grain varieties cook dry and fluffy, and the cooked grains tend to remain separate. Cooked kernels of high quality medium and short grain varieties, on the other hand, are more moist and chewy than long grain types and the grains tend to clump together. All 3 grain types with their characteristic cooked textural qualities are in widespread demand by both the domestic and foreign trade (Shafer and Grant 1970).

Different cultural groups prefer specific as well as varied textures in home boiled rice. Processors of rice require a variety of grain types and textural qualities for use in various kinds of prepared and convenience type food products. In the United States, a substantial and ever increas-

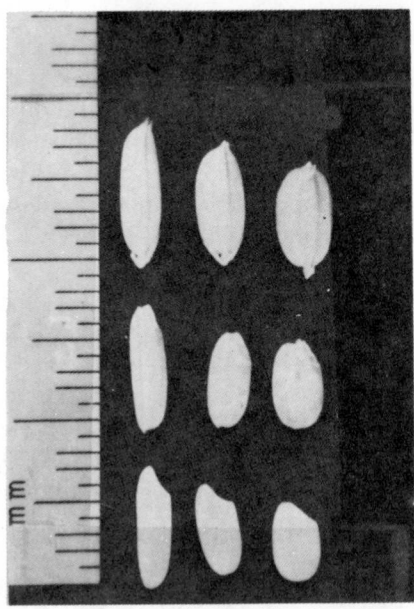

FIG. 15.1. ROUGH, BROWN AND MILLED
FORMS OF TYPICAL U.S. LONG, MEDIUM
AND SHORT GRAIN TYPES

ing amount of the domestic rice crop is processed and reprocessed into
numerous kinds of prepared products such as parboiled rice, quick cooking rice, dry breakfast cereals, canned products, baby foods and frozen
dishes. There exists a strong demand for rice in brewing and rice flour is
used in various prepared mixes.

In many rice products the qualities and grain types of U.S. long grain
varieties are preferred, whereas in others, the short and medium grain
types with their characteristic textural properties are in demand. Thus,
since the domestic and world trade (about ⅔ of the annual U.S. rice crop
is exported) associates U.S. long, medium and short grain varieties with
specific cooking, eating and processing qualities, it is essential for new
varieties of each grain type to have the same or improved milling, cooking
and processing qualities as the varieties they replace.

It should be noted that it is principally in the United States that grain
type is associated with specific cooking and processing characteristics as
a result of planned breeding. Worldwide there are short grain varieties
that have the cooking, eating and processing characteristics of U.S. long
grain types and long grain varieties possessing the properties of U.S.
short and medium grain types (Anon. 1965; Webb *et al.* 1968). Varieties
of a given grain type possessing nontypical cooking and processing qual-

ities pose serious identity, drying, handling, storage and blending problems to the rice industry.

Virtually all commercially produced rice varieties are classed as straw hulled, nonpigmented pericarp, translucent, nonscented, nonwaxy types containing varying ratios of amylose and amylopectin starch and are considered to have a mild flavor. There is also a very limited production of so-called specialty rice varieties. These include a highly aromatic long grain variety possessing a "nutty" aroma and having a taste similar to the highly prized "Basmati" types of varieties in India (Jodon *et al.* 1971) and a glutinous short grained variety, also called "sweet rice," characterized by an opaque endosperm containing almost 100% amylopectin type starch.

In any crop year a number of rice varieties of each grain type are produced commercially and new varieties must constantly be developed to keep abreast of the ever-changing needs and requirements of the rice industry. However, regardless of whether a new variety is developed for greater field yielding ability, greater nitrogen responsiveness, improved plant type, disease resistance or specific end use, the requisite standards for high quality rice of each grain type must be maintained.

COMPONENTS OF QUALITY

Hull and Bran (Pericarp) Color

Color and anthocyanin pigmentation in the apiculus of the rice hull are factors influencing different aspects of rice quality. Rice varieties produced in the United States are classed as either light (straw) or dark (gold) hulled. While hull color is not of major importance in the production of regular white milled rice, it is of considerable importance in the manufacture of parboiled rice. Varieties with light colored hulls are generally preferred by parboilers because they tend to produce a lighter colored parboiled product than dark hulled varieties processed under similar parboiling conditions (Gariboldi 1974). Most users prefer very light parboiled products, thus genetic selection for light hulled varieties is an important consideration in developing new long grain varieties suitable for the parboiling industry.

Similarly, bran (pericarp) color affects the quality of rice. This is a factor in parboiling where dark bran colors may be imparted to the parboiled endosperm and in milling of both parboiled and regular white milled rice where increased milling pressure, resulting in lower milling yields, must be applied in order to completely remove the colored bran streaks. Varieties in the United States traditionally have light brown, nonpigmented bran and this, too, is an important criteria for selection in varietal development. Selection for both hull and bran color is accomplished subjectively by close visual examination.

Grain Dimensions, Weight and Uniformity

Since rice in the United States is produced and marketed according to 3 grain size and shape classes known as long, medium and short grain, kernel dimensions are primary quality factors in many areas of processing, drying, handling equipment, breeding, marketing and grading (Kramer 1951; Anon. 1977; Adair *et al.* 1973). There are established size and shape requirements for each grain type and varieties must conform to these specifications. In developing new commercial varieties, grain size and shape is one of the first quality characteristics considered. In early generation breeding material, close visual comparisons are made by rice breeders and their associates to make sure that grain size and shape conforms to the established requirements of each grain type. Also, in the same early developmental stages intensive genetic selection is carried out to eliminate other heritable kernel abnormalities that would detract from milling quality and general appearance of the grain. These inherent grain defects include deep creases which tend to leave bran streaks on milling, irregularly shaped kernels, sharp pointed extremities that break easily in milling and oversized germs.

Visual classification of grain type is suitable for early generation comparisons but more exact measurements are needed for final classification. The various grain types are objectively classified according to length, width, length/width ratio, thickness and grain weight. Grain dimensions are defined as follows (Adair *et al.* 1973; Chang and Parker 1976; Chang and Bardenas 1965):

(1) Length of awnless rough (paddy) rice is the straight line distance (mm) from the point of disarticulation of the grain, which is below the outer glumes to the tip of the apiculus (Fig. 15.2B). Length for brown and milled rice is the distance between the most distant tips of the kernel including the embryo of the brown rice kernel (Fig. 15.2A);

(2) Width (dorsiventral diameter) for rough rice is the distance (mm) across the lemma and palea at the widest point (Fig. 15.2B). Width for brown and milled rice is the distance across the kernel at the widest point (Fig. 15.2A);

(3) Thickness (lateral diameter) for rough rice is the distance (mm) from one outside surface of the lemma to its opposite at the thickest point (Fig. 15.2D). Thickness for brown and milled rice is the distance from one side of the kernel to the opposite side at the thickest point (Fig. 15.2C).

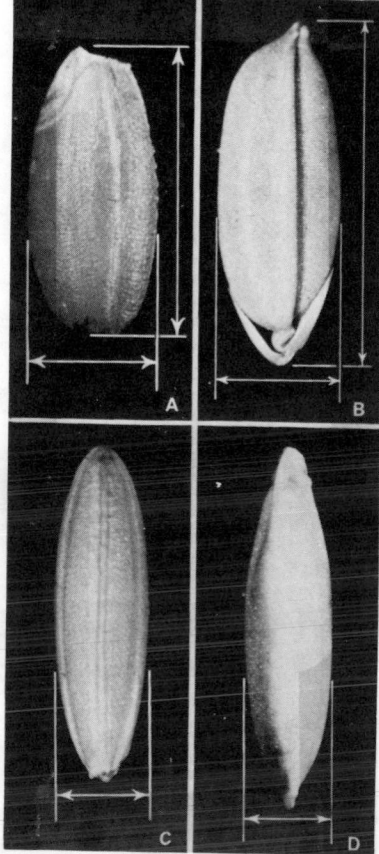

From Adair et al. (1973)

FIG. 15.2. POINTS FROM WHICH
RICE GRAIN AND KERNEL MEAS-
UREMENTS SHOULD BE MADE

Methods for measuring rice kernels include the use of photographic
enlargers to magnify the kernel or simply measuring with a transparent
ruler the length, width and thickness of several kernels placed adjacent
in the desired position for a particular measurement (Adair *et al.* 1973).
Probably the simplest method for length, width and thickness determination is to use a grain shape tester designed by the Satake Engineering Company in Tokyo specifically for obtaining kernel dimensions.

Uniformity of grain size, shape and weight is determined by calculating
the coefficient of variation for each measurement on randomly selected
kernels from a representative sample. Grain weight (size) is expressed in
g/1000 kernels.

The range of average values for grain size and shape of rough, brown

TABLE 15.1. RANGE OF AVERAGE GRAIN SIZE AND SHAPE MEASUREMENTS AMONG TYPICAL U.S. COMMERCIAL LONG, MEDIUM AND SHORT GRAIN TYPES

Grain type	Grain form	Average length (mm)	Average width (mm)	Aver. length/width Ratio	Average thickness (mm)	Average 1000 grain wt. (gms)
Long	Milled[1]	6.7 to 7.0	1.9 to 2.0	3.4:1 to 3.6:1	1.5 to 1.7	15 to 18
Medium		5.5 to 5.8	2.4 to 2.7	2.1:1 to 2.3:1	1.7 to 1.8	17 to 21
Short		5.2 to 5.4	2.7 to 3.1	1.7:1 to 2.0:1	1.9 to 2.0	20 to 23
Long	Brown[2]	7.0 to 7.5	2.0 to 2.1	3.4:1 to 3.6:1	1.6 to 1.8	16 to 20
Medium		5.9 to 6.1	2.5 to 2.8	2.2:1 to 2.4:1	1.8 to 2.0	18 to 22
Short		5.4 to 5.5	2.8 to 3.0	1.8:1 to 2.0:1	2.0 to 2.1	22 to 24
Long	Rough[3] (Paddy)	8.9 to 9.6	2.3 to 2.5	3.8:1 to 3.9:1	1.8 to 1.9	21 to 24
Medium		7.9 to 8.2	3.0 to 3.2	2.5:1 to 2.6:1	1.9 to 2.1	23 to 25
Short		7.4 to 7.5	3.1 to 3.6	2.1:1 to 2.4:1	2.1 to 2.3	26 to 30

Source: Adapted in part from Adair et al. (1973) and Webb (1975). Data based on measurements of fully developed mature kernels of typical varieties within each grain type.
[1]Whole milled kernals with hull, bran and germ removed.
[2]Grain with hull removed.
[3]Unhulled grain.

and milled forms of traditional U.S. commercial long, medium and short grain types are given in Table 15.1.

Weight per Unit Volume

Test weight is an important quality characteristic for many segments of the rice industry. It is particularly useful as a comparative indication of total milled rice yields. Test weight also provides a relative measure of dockage and/or foreign material present, and the proportion of unfilled, shriveled and immature kernels. It is important in rice storage and handling as it can be used to estimate the weight contained in holding bins of known volume. Test weight of rice is simply the weight of a known volume and is one of the easiest measurements to obtain. Procedures and equipment of determining test weight have been described in detail (Anon. 1953; Anon. 1976A). In the United States test, weight is expressed in pounds per Winchester bushel. To convert to kilograms per hectoliter, multiply by the factor 1.287. The average test weight of U.S. rough rice is 58 kg/hl (45 lb/bu) but this varies widely, affected by factors such as pubescence, amount of dockage, unfilled and immature kernels, and grain type. Average test weight for typical U.S. long, medium and short grain types are tabulated in Table 15.2.

TABLE 15.2. RANGE OF AVERAGE TEST WEIGHT AMONG TYPICAL U.S. COMMERCIAL LONG, MEDIUM AND SHORT GRAIN TYPES[1]

Grain type	Average bushel weight (lb)	Average kilograms/ hectoliter
Long	42 to 45	54 to 58
Medium	44 to 47	57 to 60
Short	45 to 48	58 to 62

Source: Adapted in part from the 1970–76 Uniform Rice Performance Nurseries Yield and Quality Report. Data based on measurements of fully developed mature kernels of varieties within each grain type.

General Appearance

Since virtually all rice is consumed in the form of whole intact kernels, the overall appearance of the processed rice kernel is extremely important in judging the quality of rice. Many factors constitute general appearance in rice. Some, including grain size, shape and uniformity have been discussed in previous sections of this chapter. Other factors such as vitreousness, translucency, chalkiness, color, damaged and imperfect kernels are equally important contributors to general appearance. There is, of course, no instrument available to objectively measure general appearance. This important quality characteristic is judged subjectively with the human eye which has the capacity for integrating all of the factors contributing to general appearance into a very important index of quality.

Translucency.—Clear, vitreous, translucent kernels are demanded by practically all segments of the rice industry (Brockington 1967; Hagberg 1967; Littlejohn 1967; Smith 1967). Typical nonwaxy U.S. varieties are required to possess these traits to a high degree. Consequently rice breeders practice intensive genetic selection for bright, clear, translucent kernels in new varieties at all stages of varietal development. Exceptions to this are the specialty waxy varieties which are characterized by completely opaque endosperms.

Chalkiness.—Chalkiness is an undesirable trait for practically all forms of rice. Chalkiness detracts from general appearance and usually results in lower milling yields because chalky kernels tend to be weak and break up more during milling. Excessive chalkiness is undesirable for many processed products because of the nonuniformity produced by overprocessing of chalky kernels under usual processing conditions. Most processors specify the amount and type of chalky kernels permissable in rice used for each manufactured product (Brockington 1967; Smith 1967; Littlejohn 1967). Chalkiness in rice results from many factors. It occurs when rice is harvested at too high a moisture level or in varieties of nonuniform maturity where excessive numbers of immature kernels, referred to as immature chalk, exist. Adverse weather conditions and cultural practices also influence the incidence of chalkiness in rice. In many instances both the type and amount of chalk are highly heritable and intensive selection is carried out in developing varieties as free of chalk as the genetic and environmental conditions allow. Kernels inherently free of chalkiness is one of the first quality characteristics the breeder must select for in new U.S. rice varieties. The position of chalky areas on the endosperm, as well as the amount, is also an important consideration for processors of rice. Chalkiness in rice is most often referred to as "white belly," "white core," "white back," "germ tip" or "immature" depending upon its location on or within the endosperm. Close visual examination is the method currently used in determining the type and amount of chalkiness in rice. A recently developed translucency instrument offers promise as an objective measure of transparency, including chalkiness, in rice (Kushibuchi 1973).

Milling Yields

It is sufficient to say that no rice variety would make it commercially unless it possessed high whole kernel (head) and total milled rice turnout. The whole kernel (head) yield is the quantity of intact whole kernels of well milled rice that can be obtained from a given quantity of rough rice. Total milled rice yields include the whole kernel (head) and all sizes of broken kernels obtainable from a specified amount of rough rice.

The objective of rice milling is the removal of hulls, bran and germ with a minimum breakage of the endosperm. Generally, 4 basic operations (reviewed by Witte 1972; Adair *et al.* 1973; Anon. 1976B) are involved in the conventional milling process:

(1) cleaning the field run rough rice to remove foreign materials such as mud lumps, rice stems and leaves, weed seeds and stems and other extraneous matter;

(2) shelling the cleaned rice to remove the hulls;

(3) milling the brown rice to remove the outer and inner bran layers, aleurone layers and germs; and .

(4) grading the mixture of whole and broken milled kernels according to size classes known as whole kernel (head) rice, second head (larger pieces of broken milled kernels), screenings (smaller pieces of broken milled kernels), and brewers rice (very small pieces of broken milled kernels).

Milling quality in rice is based on the yield of whole kernel (head) rice because this is the milled product of greatest economic value. Thus, the accurate determination of milling quality is extremely important from a marketing standpoint. Yields of whole kernel (head) rice vary widely, depending on many factors such as: variety, grain type, chalkiness, cultural practices, other environmental factors, and the drying, storing and milling conditions (reviewed by Wasserman and Calderwood 1972; Adair *et al.* 1973). The yield of total milled rice is important as well, and this yield is influenced by the proportion of hulls and the amount of fine endosperm particles unavoidably included in the bran fraction during the milling process.

Because many factors affect milling yields (Autrey *et al.* 1955), rigid laboratory milling tests are required to insure that new varieties released for commercial production will consistently produce high yields of whole (head) and total milled rice. Thus, milling yield is one of the most important components of quality. Intensive genetic selection for this characteristic is practiced in all stages of varietal development.

Several laboratory instruments have been used to determine the milling yields in rice. Adair (1952) reviewed the more important laboratory rice milling methods used in the United States. The methods most commonly used are: (1) the official grading method for determining milling quality of rough rice by U.S. standards (Appendix; Anon. 1976A) which requires a 1000 g rough rice sample, and (2) a modification of the official method requiring only a 125 g rough rice sample for analysis. Both methods involve the use of the McGill type millers. The modified method is particularly applicable for use in rice breeding programs.

Whole kernels are separated from the broken kernels in a milled sample with a sizing device developed by the Grain Division, USDA. This device uses indented plates, with flat bottomed holes which are tilted at a slight angle and mechanically shaken in an eccentric motion. During shaking, broken kernels fall into the indents while whole kernels travel the length of the plate. This device, as well as the McGill milling methods are described in detail in a series of articles by Smith (1955 A,B,C,D) with more recent modifications outlined in the USDA Rice Inspection Manual (Anon. 1976A; Adair et al. 1973).

Milling quality results are usually reported as the percentage of whole kernel (head) rice and total milled rice. Average milling yields for typical U.S. long, medium and short grain types are shown in Table 15.3.

TABLE 15.3. RANGE OF AVERAGE MILLING YIELDS AMONG TYPICAL U.S. COMMERCIAL LONG, MEDIUM AND SHORT GRAIN TYPES

| Grain type | Average Milling Yields (%) | |
	Whole kernel (head rice)	Total milled rice
Long	56−61	68−71
Medium	65−68	71−72
Short	63−68	73−74

Source: Adapted in part from Adair et al. (1973) and Webb (1975). Data based on clean, mature, rough rice samples of varieties within each grain type.

Cooking and Processing Behavior

Historically, the cooking and processing characteristics of rice have always been factors of primary importance in rice eating areas of the world. Cooking and processing quality, along with milling quality, are the fundamental components of quality that determine and establish the economic value of the rice grain.

Concepts of rice cooking and processing quality may be described in several ways. In the United States, one of the most successful has been to characterize cooking and processing quality on the basis of chemical and physical terms which serve as indices of specific rice qualities (Beachell and Halick 1957; Webb et al. 1972; Adair et al. 1973).

Prior to the mid 1950s, rice quality in the United States was judged solely on the basis of its milling yields, factors affecting milling yields and cleanliness and purity of the rice. However, in the 1950s, an agronomically superior long grain variety was developed and released for commercial production which later proved to be completely unacceptable for traditional long grain cooking and processing uses. This event emphasized the immediate need for developing and producing only those varieties with the inherent cooking and processing characteristics required by the ultimate user. To meet this need, representatives of the USDA, the State Agricultural Experiment Station and various segments of the rice industry met to develop plans for a rice quality laboratory. The purpose of the laboratory was to perform specific chemical and physical tests to serve as guides for rice breeders in developing new varieties with the desired qualities in combination with the agronomic features required by the rice industry.

Consequently, in 1955 the Regional Rice Quality Laboratory was established at the Texas A & M University Agricultural Research and Extension Center, Beaumont, Texas. The Laboratory is operated by the USDA in cooperation with the Texas Agricultural Experiment Station and the Texas Rice Improvement Association. The laboratory serves the national Federal-State varietal improvement programs conducted cooperatively by the USDA and the State Agricultural Experiment Stations in Arkansas, California, Louisiana, Mississippi and Texas. It has the support of all phases of the United States rice industry.

In the United States, specific chemical and physical criteria are used to describe the cooking and processing qualities desired in new varieties of each grain type. These criteria are based on a series of physicochemical tests which, when taken together, serve as indices of rice cooking and processing behavior. Standard and new varieties being developed are systematically tested for amylose content (Williams *et al.* 1958; Juliano 1971; Webb 1972); alkali reaction of whole kernel milled rice in contact with dilute alkali, a measure of gelatinization temperature (Little *et al.* 1958); amylographic gelatinization and pasting characteristics (Halick and Kelly 1959); water uptake capacity at 77°C (170.6°F) (Halick and Kelly 1959); birefringence end-point temperature, BEPT (Halick *et al.* 1960); protein content (Anon. 1962); and parboil canning stability (Webb and Adair 1970). The average and range of test values for these characteristics have been established for all commercially acceptable varieties (Webb *et al.* 1972; Adair *et al.* 1973). Hence, the characteristics of prospective new varieties can be directly compared with those of the typical standard varieties. The chemical and physical characteristics of new varieties are always compared with those of similarly grown leading commercial varieties of the appropriate grain type according to the format illustrated in Fig. 15.3. If after a number of years at several locations, the properties of the new varieties are similar or superior to those of the traditional standard varieties, then the new varieties are considered to have satisfactory or superior cooking and processing quality; if not, they are considered to have undesirable or unknown quality.

Amylose content is considered to be the single most important char-

Rice Cooking and Processing Quality Evaluation Form

Chemical and Physical endosperm characters	Standard Variety	Prospective new variety
Amylose content-(%)		
Alkali spreading value-avg. no.		
Amylographic paste viscosity:		
Peak-Brabender Units-(BU)		
Cooked 10 min at 95°C-(BU)		
Cooled to 50°C-(BU)		
Water uptake at 77°C-(ml / 100 g)		
Gelatinization temperature (BEPT)[1]-°C		
Protein (N × 5.95)-(%)		
Parboil-canning stability:		
Dry matter (Solids Loss)-%		
Cookability of rice with malt:		
Viscosity flow time-(Sec / 150 ml)		

[1](BEPT) = Birefringence-end-point temperature.

FIG. 15.3. RICE COOKING AND PROCESSING QUALITY EVALUATION FORM

Prospective new varieties are compared with those of typical varieties each grown at several locations for a number of years.

acteristic for predicting rice cooking and processing behavior (Rao *et al.* 1952; Williams *et al.* 1958; Halick and Keneaster 1956; Juliano *et al.* 1972A). The alkali spreading value (Warth and Darabsett 1914; Little *et al.* 1958) used to measure gelatinization characteristics also is of prime importance. Since 1955 virtually all U.S. breeding material, new and standard varieties have been tested for these 2 characteristics. Now, these 2 tests are used universally in breeding programs in most rice producing countries to determine and describe rice cooking and processing behavior.

Although the chemical and physical properties of rice outlined have been invaluable in characterizing and evaluating the cooking and processing qualities of rice in the United States, they do not always explain the fundamental cause or reason for the observed differences in rice cooking and processing behavior. For example, why do certain rice varieties of similar amylose content and other measured characteristics have substantially different cooked kernel stability, textural quality and mouth appeal? The gel consistency test described by Cagampang *et al.* (1973) reportedly measures these differences to some extent. Also, why do varieties of similarly measured characteristics differ markedly in their suitability for specific brewing needs? To answer these and other questions, need for basic research on the factors and constituents responsible for observed differences in rice cooking and processing behavior is indicated now and for the future. This information would accelerate breeding rice for special uses. Also, as new uses and improved techniques for processing rice are developed the quality characteristics now needed may change or require revision. Thus, constant attention must be given to all aspects of the characteristics and reaction of the rice kernel.

Average values for some comparative chemical and physical (quality) characteristics of typical cooking and processing long, medium and short grain U.S. rice types are given in Table 15.4. Environmental and other factors influence these characteristics to some extent; however, within a limited range, the values shown are representative of each grain type.

TABLE 15.4. RANGE OF AVERAGE CHEMICAL AND PHYSICAL (QUALITY) CHARACTERISTICS AMONG TYPICAL U.S. COMMERCIAL LONG, MEDIUM AND SHORT GRAIN TYPES

Endosperm characteristic	Grain type		
	Long	Medium	Short
Amylose content (%)	23 to 26	15 to 20	18 to 20
Alkali spreading value (aver.)	3 to 5	6 to 7	6 to 7
Gelatinization temperature (BEPT)	71 to 74	65 to 68	65 to 67
Gelatinization temperature (class)	Intermediate	Low	Low
Water uptake at 77°C (ml/100g)	121 to 136	300 to 340	310 to 360
Protein (N × 5.95) (%)	6 to 7.5	6 to 7	6 to 6.5
Parboil-canning stability: solids loss %	18 to 21	31 to 36	30 to 33
Amylographic paste viscosity: peak-brabender units (B.U.); cooked 10 min at 95°C (B.U.)	765 to 840	890 to 980	820 to 870
	400 to 500	370 to 420	370 to 400
Cooled to 50°C (B.U.)	770 to 880	680 to 760	680 to 690

Source: Adapted in part from Adair et al. (1973) and Webb (1975). Data based on measurements of fully developed mature kernels of typical varieties within each grain type.

The chemical and physical characteristics associated with traditional cooking and processing of U.S. long grain types (Table 15.4) are: a relatively high amylose content, a slight to moderate reaction of whole kernel milled rice in contact with dilute alkali, a moderate water uptake capacity at 77°C (170°F), and an intermediate gelatinization temperature (BEPT). Amylographic pasting characteristics of the typical long grain varieties usually show an intermediate peak viscosity and a relatively high viscosity of the cooled paste on cooling to 50°C (122°F). Parboil canning characteristics of the typical long grain varieties in terms of percentage of solids lost during processing are relatively low, and the canned kernels show a minimum amount of splitting and fraying of edges and ends.

In contrast, the typical cooking and processing of U.S. medium and short grain varieties are characterized by a relatively low amylose content, a pronounced extensive reaction of whole kernel rice in contact with dilute alkali, a relatively low gelatinization temperature (BEPT), and a relatively high water uptake capacity at 77°C (170°F). Amylograms of the typical medium and short grain varieties usually show relatively low cooked paste viscosities on cooling to 50°C (122°F). The parboil canning stability of the traditional medium and short grain varieties, compared directly with that of the typical long grain varieties, show relatively high losses of solids during processing, and the canned kernels show extensive splitting and fraying of edges and ends.

Inheritance studies (Bollich and Webb 1973; Anon. 1975; Ghosh and Govindeswamy 1972; Beachell and Halick 1957) indicate that amylose and alkali spreading value—a measure of gelatinization temperature—have relatively high levels of heritability. Thus, in rice breeding programs in the United States, these characteristics are selected for and "fixed" in very early stages of varietal development. Hence rice breeders in the United States are now selecting for the chemical and physical indices of rice cooking and processing quality that will appear in new varieties to be released for commercial production in the next 5 to 10 years.

Moisture Content

Moisture content affects rice quality in several different ways. Of great significance is its effect on the keeping quality of all forms of rice. Sound dry rice can be maintained for years if properly stored but only a few days are required for wet rice to spoil. Rough rice moisture content of 13% is commonly accepted as a safe level for storage for less than 6 months, whereas 12% or less moisture is recommended for long term storage (reviewed by Wasserman and Calderwood 1972; Johnston and Miller 1973; Bolling et al. 1977.) Moisture contents of rough rice in excess of 14% is designated as sample grade under the U.S. standards for rice (see Appendix).

Of equal significance is the influence of moisture content on milling yields in rice. To gain and maintain optimum milling quality, rice must be harvested at the proper moisture content, dried carefully to safe storage moisture levels and stored and milled under moisture conditions recommended for maximum milling yields. Excellent reviews on this subject have been published (Wasserman and Calderwood 1972; Johnston and Miller 1973).

Excessive moisture levels in rice may also adversely affect various cooking and processing quality characteristics in rice as changes in these characteristics during storage occur more readily in high than in low moisture rice. Also, various limitations on moisture levels for raw rice manufactured into certain processed products are specified (Brockington 1967; Hardwick 1967; Hays 1967; Kelly 1961; Littlejohn 1967; Smith 1967). The moisture content of rice, like that of wheat (Zeleny 1972), is also of direct economic importance since the amount of dry matter in rice is inversely related to the amount of moisture it contains.

Methods for measuring moisture content vary widely. Zeleny (1972) has reviewed a number of these which may be broadly classified as air oven, solvent extraction, electrical conductance and electrical capacitance. The air oven or vacuum oven procedures are usually the basic methods for determining moisture in rice, but when rapid results are needed the properly calibrated electric moisture meters are often used and are sufficiently accurate for most control work.

RICE GRADES

United States Standards for Rice

The U.S. standards for rough rice, brown rice for processing and milled rice provide a useful and necessary way for the orderly marketing of rice by grades. The U.S. official grain standards for rice is reproduced in the Appendix and the reader should refer to these standards for detailed information regarding the complexities involved in determining the various grades of rice. Although the grades are a useful tool, they do not provide a sufficiently refined classification of rice to satisfy all the quality requirements of all segments of the rice industry. Thus, many processors and buyers purchase rice on the basis of grade plus additional quality factors specific for their individual product needs. For detailed step by step instructions and procedures for establishing a grade of a particular lot of rice by U.S. standards, the reader should consult the U.S. Standards for Rice (Appendix) and the Rice Inspection Manual (Anon. 1976A) covering the sampling, inspection, certification and grading of rice.

Some of the factors affecting the grade of a particular lot of rice, including grain type, moisture, chalkiness and milling yields are discussed in previous sections of this chapter. Other equally significant factors involved in establishing rice grades include degree of milling, color, dockage (impurities), damaged kernels, red rice, odors, and seeds or kernels of any plant other than rice. Since these latter grading factors are adequately defined and described in the U.S. Rice Standards (Appendix) and the Rice Inspection Manual (Anon. 1976A) only a brief discussion will be presented here.

Color and Milling Requirements.—The U.S. Standards (Appendix) for whole kernel milled rice specify that "U.S. No. 1 grade shall be white or creamy, and shall be well milled. U.S. No. 2 may be slightly gray, and shall be well milled. U.S. No. 3 may be light gray, and shall be at least reasonably well milled. U.S. No. 4 may be gray of slightly rosy, and shall be at least reasonably well milled. U.S. No. 5 and No. 6 may be dark gray or rosy and shall be at least lightly milled." Colors of raw milled rice range from white to dark gray or rosy whereas, parboiled rice is usually graded from "parboiled light" to "parboiled dark." As stated previously, color and general appearance in rice are used somewhat interchangeably and are usually determined subjectively by close visual examination. Numerous attempts (reviewed by Webb and Stermer 1972) have been made to use optical devices for measuring color and degree of milling in rice but none has so far proven entirely satisfactory.

Degree of Milling.—The extent to which the bran layers and germs have been removed from the endosperm is referred to as the degree of milling. The U.S. Standards (Appendix) recognize 4 degrees of milling: well milled, reasonably well milled, lightly milled and under milled. At the present time degree of milling is determined subjectively by visual inspection by trained technicians. There is a great need for a simple, rapid, reliable and objective measure of the milling degree (Stermer 1968) but so far none of the many proposed have been acceptable. Hogan and Deobald (1961, 1965) and Webb and Stermer (1972) reviewed many of the proposed chemical and optical methods. More recently Nuclear-Magnetic-Resonance (NMR) (Pomeranz et al. 1975) has been investigated as a measure of degree of milling in rice; and Barber and Barber (1976) and Bhattacharya and Sowbhagya (1976) propose approaches based on staining techniques.

Dockage.—According to the U.S. Standards (Appendix) "dockage shall be any matter other than rice which can be removed readily from the rough rice by the use of appropriate sieves and cleaning devices, and underdeveloped, shriveled, and small pieces of kernels of rough rice which are removed in properly separating the dockage and which cannot be recovered by properly rescreening and recleaning." Other impurities which are difficult to remove because they are similar in size, shape and density of the rice are classed as objectionable materials. These include metal and glass fragments and certain weed seeds. The recommended instrument for removing dockage in the United States is the Carter Dockage Tester. Its proper operation and description are described by Smith (1955A) and in the Rice Inspection Manual (Anon. 1976A).

Damaged Kernels.—The U.S. Standards defines damaged kernels as "kernels and pieces of kernels of rice which are distinctly discolored or damaged by water, insects, heat, or any other means. Kernels and pieces of kernels of parboiled rice when found in nonparboiled rice shall function as damaged kernels." Kernels damaged by heat are classified as heat damaged because this form of kernel damage is considered to be more serious than other kinds of damage. Visual inspection by trained inspectors is the only reliable method available for determining the amount and type of damaged kernels in rice.

Distinctly discolored kernels, regardless of the source of discoloration, are considered by parboilers and other processors as "peckiness" or "pecky" rice. In parboiled rice, "peck" is a serious problem because the parboiling process tends to intensify kernel discoloration of the damaged kernel which must ultimately be removed from the finished product. Usually this is accomplished by electronic color sorters which remove most of the peck kernels from normal kernels in parboiled rice.

Odors.—Off odors in rice severely affects the grade. The U.S. Standards (Appendix) specify that rice which is musty or sour, or which has any commercially objectionable foreign odor, shall be graded U.S. sample grade. According to the Rice Inspection Manual (Anon. 1976A), musty or sour odors include earthy, moldy ground odors, insect odors, rancid odors and sharp acrid odors. Commercially objectional foreign odors include odors of fertilizers, hides, oil products, skunk, smoke and decaying animal and vegetable matter. Trained inspectors subjectively determine the type and severity of odors in rice, preferably at the time of sampling.

Red Rice.—By definition the U.S. Standards (Appendix) classify red rice as kernels and pieces of kernels of rice which are distinctly red in color or which have an appreciable amount of red bran within. Red rice (*Oryza sativa* L.) is one of the more severe problems facing the U.S. rice industry today. It is a serious weed pest on large acreages of the rice growing states. Its growing habits are similar to the cultivated varieties but readily shatters after maturity. The seeds have the ability to retain their viability buried in the soil for many years. Red rice is objectionable because the red bran is not completely removed in regular milling and this detracts from the general appearance and market value of the rice. The current U.S. Standards specify that a rice kernel be classified as red if it has a streak of red bran ½ or more of the kernel length or 2 or more streaks which total ½ or more of the length of the kernel. Determinations of red rice for grading purposes are accomplished visually by trained rice inspectors.

REFERENCES

ADAIR, C.R. 1952. The McGill miller method for determining the milling quality of small samples of rice. Rice J. *55* (2) 21.

ADAIR, C.R., MILLER, M.D. and BEACHELL, H.M. 1962. Rice improvement and culture in the United States. Advances in Agron. *14,* 61−108.

ADAIR, C.R. *et al.* 1973. Rice breeding and testing methods in the United

States. *In* Rice in the United States: Varieties and production. U.S. Dep. Agric.—Agric. Handbk. *289.* (rev)

ANON. 1953. The test weight per bushel of grain: Methods of use and calibration of the apparatus. U.S. Dep. Agric.—Agric. Circ. *921,* 11.

ANON. 1962. Cereal laboratory methods. Method 46—13. Am. Assoc. Cereal Chem., Inc., St. Paul, Minn.

ANON. 1965. Quality evaluation studies of foreign and domestic rices. U.S. Dep. Agric. Tech. Bull. *1331.*

ANON. 1975. Grain quality. *In* The IRRI Report for 1975. Int. Rice Res. Inst., Los Baños, Philippines.

ANON. 1976A. Inspection handbook for the sampling, inspection, grading and certification of rice. U.S. Dep. Agric.—Agric. Marketing Ser. *HB 918—11.*

ANON. 1976B. Rice postharvest technology. International Development Research Center—IDRC-053e. E.V. Araullo, D.B. De Padua and M. Graham (Editors). Int. Dev. Res. Cntr., Ottawa, Canada.

ANON. 1977. United States standards for rough rice, brown rice for processing, milled rice. U.S. Dep. Agric.—Agric. Mktg. Serv. (rev)

AUTREY, H.S., GRIGORIEFF, W.W., ALTSCHUL, A.M. and HOGAN, J.T. 1955. Effects of milling conditions on breakage of rice grain. J. Agric. Food Chem. *3* (7) 593.

BARBER, S. 1972. Milled rice and changes during aging. *In* Rice Chemistry and Technology. D.F. Houston (Editor). Amer. Assoc. Cereal Chem., St. Paul, Minn.

BARBER, S. and BARBER, C. 1976. An approach to the objective measurement of the degree of milling. Rice Process. Engin. Cntr. Rept. *2* (2) 1.

BEACHELL, H.M. 1959. Rice. *In* The Chemistry and Technology of Cereals as Food and Feed. S.A. Matz (Editor). AVI Publishing Co., Westport, Conn.

BEACHELL, H.M. and HALICK, J.V. 1957. Breeding for improved milling, processing and cooking characteristics of rice. Int. Rice Comm. Newsl. *6* (2) 1.

BHATTACHARYA, K.R. and SOWBHAGYA, C.M. 1976. An alkali degradation test and an alcoholic alkali bran-staining test for determining the approximate degree of milling of rice. J. Food Tech. *11,* 309.

BOLLICH, C.N., ATKINS, J.G., SCOTT, J.E. and WEBB, B.D. 1972. Labelle-A blast resistant, very early maturing, long-grain rice variety released in Texas. Rice J. *75* (3) 28.

BOLLICH, C.N. and WEBB, B.D. 1973. Inheritance of amylose in two hybrid populations of rice. Cereal Chem. *50* (6) 631.

BOLLING, H., HAMPEL, H. and EL BAYA, A.W. 1977. Changes in physical and chemical characteristics of rice during prolonged storage. Il Riso *26* (1) 65.

BROCKINGTON, S.F. 1967. Puffed rice products. Proc. Natl. Rice Util. Conf., April 5—6, 1966. New Orleans, La. U.S. Dep. Agric., Agric. Res. Serv. ARS 72—53. USDA, Washington, D.C.

CAGAMPANG, G.B., PEREZ, C.M. and JULIANO, B.O. 1973. A gel consistency test for eating quality of rice. J. Sci. Food. Agric. *24,* 1589.

CARNAHAN, H.L., JOHNSON, C.W., TSENG, S.T. and MASTENBROEK, J.J. 1978. Registration of M9 Rice. Crop Sci. *18,* 357.

CHANG, T.T. and BARDENAS, E. 1965. The morphology and varietal characteristics of the rice plant. Int. Rice Res. Inst. Tech. Bull. *4.*

CHANG, T.T. and PARKER, M.B. 1976. Characteristics of rice cultivars. Il Riso *25* (3) 195.

CHIKUBU, S. 1975. Quality of rice in Southeast Asia. *In* Rice in Asia. Assoc. Jpn. Agric. Soc. (Editors). Univ. of Tokyo Press, Tokyo.

GARIBOLDI, F. 1974. Rice parboiling. Food and Agric. Organ. U.N. Agric. Dev. Ppr. *97.* FAO, Rome.

GHOSH, A.K. and GOVINDASWAMY, S. 1972. Inheritance of starch iodine blue value and alkali digestion value in rice and their genetic association. Il Riso *21* (2), 123.

HAGBERG, E.C. 1967. Canned rice products. Proc. Natl. Rice Util. Conf. New Orleans, La. USDA, ARS 72-53, Washington, D.C.

HALICK, J.V., BEACHELL, H.M., STANSEL, J.W. and KRAMER, H.H. 1960. A note on the determination of gelatinization temperature of rice varieties. Cereal Chem. *37,* 670.

HALICK, J.V. and KELLY, V.J. 1959. Gelatinization and pasting characteristics of rice varieties as related to cooking behavior. Cereal Chem. *36,* 91.

HALICK, J.V. and KENEASTER, K.K. 1956. The use of a starch iodine-blue test as a quality indicator of white milled rice. Cereal Chem. *33,* 315.

HARDWICK, W.A. 1967. Anheuser-Busch's use of rice as a brewing adjunct. Proc. Natl. Rice Util. Conf. New Orleans, La. USDA, ARS 72-53, Washington, D.C.

HAYS, W.E. 1967. Characteristics desired in rice for Adolph Coors Company. Proc. Natl. Rice Utiliz. Conf. New Orleans, La. USDA, ARS 72-53, Washington, D.C.

HOGAN, J.T. and DEOBALD, H.J. 1961. Note on a method of determining the degree of milling of whole milled rice. Cereal Chem. *38,* 291.

HOGAN, J.T. and DEOBALD, H.J. 1965. Measurement of the degree of milling of rice. Rice J. *68* (10) 10.

JODON, N.E., SONNIER, E.A. and McILRATH, W.O. 1971. Two new rice varieties developed at Crowley station. La. Agric. *14* (3) 4.

JOHNSTON, T.H. 1967. Breeding rice for milling, cooking and processing characteristics in a comprehensive varietal improvement program. Int. Rice Comm. Newsl. (spec. iss.) 166.

JOHNSTON, T.H. and MILLER, M.D. 1973. Culture. *In* Rice in the United States: Varieties and production. U.S. Dep. Agric.—Agric. Handbk. 289. (rev)

JONES, J.W. 1938. The "alkali test" as a quality indicator of milled rice. Amer. Soc. Agron. *30,* 960.

JULIANO, B.O. 1971. A simplified assay for milled-rice amylose. Cereal Sci. Today *16* (10) 334.

JULIANO, B.O. 1972A. Physicochemical properties of starch and protein in relation to grain quality and nutritional value of rice. *In* Rice Breeding. Int. Rice Res. Inst., Los Baños, Philippines.

JULIANO, B.O. 1972B. The rice caryopsis and its composition. *In* Rice Chemistry and Technology. D.F. Houston (Editor). Amer. Assoc. Cereal Chem., St. Paul, Minn.

JULIANO, B.O. 1973. Quality of milled rice. Il Riso *22* (2) 171.

KELLY, V. 1961. Properties of rice products desirable for baby food formulations. Proc. 2nd Rice Util. Conf., Western Reg. Res. Cntr., Albany, Calif.

KESTER, E.B. 1959. Rice processing. *In* The chemistry and technology of cereals as food and feed. S.A. Matz (Editor). AVI Publishing Co., Westport, Conn.

KRAMER, H.A. 1951. Physical dimensions of rice. Agric. Eng. *32* (10) 544.

KUSHIBUCHI, K. 1973. Introducing a testing instrument for determining the transparency of rice grain (Jap). Jpn. J. Breed. *23* (3) 35.

LITTLE, R.R., HILDER, G.B., DAWSON, E.H. and ELSIE, H. 1958. Differential effect of dilute alkali on 25 varieties of milled white rice. Cereal Chem. *35*, 111.

LITTLEJOHN, J.P. 1967. Production of and characteristics desired in rice for Rice Krispies and Special K. Proc. Natl. Rice Util. Conf. New Orleans, La. USDA, ARS 72-53, Washington, D.C.

MATZ, S.A. and BEACHELL, H.M. 1969. Rice. *In* Cereal Science. S.A. Matz (Editor). AVI Publishing Co., Westport, Conn.

POMERANZ, Y., STERMER, R.A. and DIKEMAN, E. 1975. NMR-OIL Content as an index of degree of rice milling. Cereal Chem. *52*, 849.

RAO, B.S., MURTHY, A.R.V. and SUBRAHMANYA, R.S. 1952. The amylose and amylopectin content of rice and their influence on the cooking quality of the cereal. Proc. Indian Acad. Sci. *368*, 70.

RAO, G.M. and SIDDIQ, E.A. 1976. Studies on induced variability for amylose content with reference to yield components and protein characteristics in rice. Environ. and Exp. Bot. *16*, 177.

SHAFER, C.E. and GRANT, W.R. 1970. World production, exports, prices and consumption. *In* The Southern Rice Industry. M.R. Godwin and Lonnie L. Jones (Editors). Texas A & M Univ. Press, College Station, Texas.

SMITH, R.G. 1967. Desirable rice characteristics for Ralston Purina Company. Proc. Natl. Rice Utiliz. Conf. New Orleans, La. USDA, ARS 72-53, Washington, D.C.

SMITH, W.D. 1955A. The use of the Carter dockage tester to remove weed seeds and other foreign material from rough rice. Rice J. *58* (9) 26.

SMITH, W.D. 1955B. The use of the McGill sheller for removing hulls from rough rice. Rice J. *58* (10) 20.

SMITH, W.D. 1955C. The use of the McGill miller for milling samples of rice. Rice J. *58* (11) 20.

SMITH, W.D. 1955D. The determination of the estimate of head rice and of total yield with the use of the sizing device.   Rice J. *58* (12) 9.

STERMER, R.A.   1968.   An instrument for objective measurement of degree of milling and color of milled rice.   Cereal Chem. *45,* 358.

WARTH, F.J. and DARABSETT, D.B.   1914.   Disintegration of rice grains by means of alkali.   Agric. Inst. Pusa Bull. *38.* (Indian)

WASSERMAN, T. and CALDERWOOD, D.L.   1972.   Rough rice drying. *In* Rice Chemistry and Technology.   D.F.Houston (Editor).   Amer. Assoc. Cereal Chem., St. Paul, Minn.

WEBB, B.D.   1972.   A totally automated system of amylose analysis in whole kernel milled rice (ABST).   Cereal Sci. Today *17* (9) 141.

WEBB, B.D.   1975.   Cooking, processing and milling qualities of rice. *In* Six Decades of Rice Research in Texas.   Texas Agric. Exp. Sta. Res. Monog. *4.*

WEBB, B.D. and ADAIR, C.R.   1970.   Laboratory parboiling apparatus and methods of evaluating parboil-canning stability of rice.   Cereal Chem. *47,* 708.

WEBB, B.D., BOLLICH, C.N., ADAIR, C.R. and JOHNSTON, T.H.   1968. Characteristics of rice varieties in the U.S. Department of Agriculture Collection.   Crop Sci. *8,* 361.

WEBB, B.D. and STERMER, R.A.   1972.   Criteria of rice quality. *In* Rice Chemistry and Technology.   D.F. Houston (Editor).   Amer. Assoc. Cereal Chem., St. Paul, Minn.

WEBB, B.D. *et al.*   1972.   Evaluating the milling, cooking and processing characteristics required of rice varieties in the United States.   U.S. Dep. Agric., ARS-S-1, Washington, D.C.

WILLIAMS, V.R., WU, W.T., TSAI, H.Y. and BATES, H.G.   1958.   Varietal differences in amylose content of rice starch.   J. Agric. Food Chem. *6,* 47.

WITTE, G.C.   1972.   Conventional rice milling in the United States. *In* Rice Chemistry and Technology.   D.F. Houston (Editor).   Amer. Assoc. Cereal Chem., St. Paul, Minn.

ZELENY, L.   1972.   Criteria of wheat quality. *In* Wheat Chemistry and Technology.   Y. Pomeranz (Editor).   Amer. Assoc. Cereal Chem., St. Paul, Minn.

16

Quick Cooking Rice

Bor S. Luh
Robert L. Roberts
Chin-Fung Li

Ordinary rice requires 20–30 min to cook to a culinary acceptability. In some instances the rice is soaked, washed and steamed, requiring a total attention time of about 1 hr. The relatively long preparation time has restricted rice consumption in the United States. Thus, effort has been directed toward developing a quick cooking rice to increase rice consumption. With the development of the automatic electrical rice cooker, the attention time has been reduced for household rice cooking.

Various rice varieties require different cooling times and yield cooked rice of different textural characteristics. Variations in recipes also have a significant effect upon the texture, flavor and acceptability of the cooked rice.

The quality of quick cooking rice developed over the past few years has varied considerably. For example, instant rice has been utilized by reconstituting dehydrated rice as an emergency food in Japan. However, several disadvantages have been encountered with this rice: (1) the cooking method is not popular and the cooking time is more than 20 min, (2) the cooked rice grains tend to crumble and so the rice does not taste like that which is conventionally cooked, and (3) the cost per meal is higher.

Quick cooking rice should be cooked within 3 to 5 min and the cooking method should be simple. After cooking, the product should match the characteristic flavor, taste and texture of conventionally cooked rice. It should be rich in nutrients, well balanced in composition and easily processed in mass quantities.

Recently, the Nissin Food Company in Osaka, Japan, has developed an instant "Cup Rice" (Fig. 16.1) which can meet most of the conditions

Courtesy of Nissin Food Co.,
Osaka, Japan

FIG. 16.1.   CUP RICE

mentioned above. The rice is precooked under high pressure and temperature and then dehydrated. The product can be reconstituted with
boiling water within 5 min in a polystyrene cup. Today there are many
kinds of instant rice products available on the market (Fig. 16.2–16.3).

Quick cooking rice is precooked and gelatinized to some extent in water,
steam or both. The cooked or partially cooked rice is usually dried in such
a manner as to retain the rice grains in a porous and open structured
condition. The finished product should consist of dry, individual kernels,
substantially free of lumps or aggregates, and should have approximately
1.5–3.0 times the bulk volume of the raw rice. The boiling water used in
the final preparation of the rice should penetrate the rice grains in a
relatively short time.

Many quick cooking rice products, although varying in texture, bulk
volume, appearance, taste and performance qualities, are designed spe-

Courtesy of Food Industry Research and Development Institute,
Hsinchu, Taiwan

FIG. 16.2. VARIETIES OF INSTANT RICE AVAILABLE

Courtesy of Dragon Gate Food Corp., Taiwan

FIG. 16.3. INSTANT RICE PORRIDGE

cifically for certain consumer uses. Some quick cooking rice for special applications, such as in dry soup mixes, casseroles or other dry food mixtures which have certain rehydration time requirements, are designed to be compatible with the other ingredients in the mix.

TYPES OF QUICK COOKING RICE

Differences in moisture levels, precooking times and temperatures, drying conditions and other processing variables can produce various types of quick cooking rice. They range from relatively undercooked rice requiring 10–15 min of "cooking," or a good quality "table" rice requiring a 5 min preparation time, such as "Minute Rice," to a variety which can be hydrated in several seconds to 1 or 2 min. The last type yields a fairly mushy, soft product when boiled. Some of these are marketed as ready-to-eat breakfast cereals.

Some consumers prefer long grain, light, fluffy or slightly dry individual kernels of rice with typical cooked rice flavor, having essentially no gritty or hard, uncooked centers. This has been the target for most quick cooking rice developments over the past 20–30 years. A notable exception to this is that in Japan and China, people prefer short grain rice which is somewhat pasty and sticky when cooked. In fact, as rice gradually becomes "drier" (when cooked) with time in storage after harvesting and milling, the short grain type may become too dry and nonpasty for Japanese textural preference, to the extent that in some cases "waxy" or "sweet" rice is added in small amounts to increase the pastiness. Short grain rice merits commercial interest in developing and marketing a quick cooking rice.

Practically all processes described in patents primarily emphasize the treatment of the rice. Effects have been made to improve milling characteristics and yields, to remove surface fats, to improve storage stability and to enhance flavor by parching the grain. Some of these processes improve nutritional quality by infusion of the surface vitamins from the bran and aleurone layers into the endosperm (Mickus and Brewer 1957; Roberts *et al.* 1951; Roberts 1952A). This latter treatment has been developed to form a product now commonly known as parboiled rice which is discussed in Chapter 14.

QUICK COOKING PROCESSES

Many quick cooking rice products and processes have been developed and patented during the past 30 years. More than 10 different approaches have been used, as well as several combinations of these and numerous equipment design modifications. Among the several processes and products developed in the past, the following are the commercially useful ones reviewed by Roberts (1972):

(1) Raw-milled white rice is soaked to 30% moisture and cooked in hot water to 50–60% moisture with or without steaming. The product is further boiled or steamed to increase the moisture content to 60–70% and then dried carefully to 8–14% moisture to maintain a porous structure. A significant modification of the procedure is a dry heat pretreatment to fissure the grains prior to cooking and drying.

(2) Rice is soaked, boiled, steamed or pressure-cooked to thoroughly gelatinize the grain, dried at a low temperature to yield fairly dense glassy grains, then expanded or puffed at a high temperature to produce the desired porous structure.

(3) Rice is pregelatinized, rolled or "bumped" to flatten the grains and dried to a relatively hard and glassy product.

(4) Rice is treated in a blast of hot air at 65.6°–315.6°C (150°–600°F) to dextrinize, fissure or expand the grains. No boiling or steaming is applied.

(5) Rice is precooked, then frozen, thawed and dried. This procedure combines the hydration and gelatinization steps of 1, 2 and 3, as well as the critical steps of freezing and thawing before drying.

(6) Gun puffing is a combination of some preconditioning of rice plus high temperature pressure, followed by explosive puffing to atmospheric pressure or into a vacuum.

(7) Freeze-drying cooked rice.

(8) Chemical treatments.

(9) Combinations of 2 or more of the above.

(10) Miscellaneous procedures.

THE "SOAK-BOIL-STEAM-DRY" METHODS

The first process for a quick cooking rice was that of Ozai-Durrani (1948). This process was used by General Foods Corporation to make "Minute Rice" which was marketed as the first convenience rice of this type.

First, the milled white rice is soaked in water at room temperature (Table 16.1). The moisture content of the rice is increased to 30%. The rice is next boiled 8–10 min until the moisture content has increased to 65–70%. After draining, cooling and washing in cold water for 1–2 min, the rice is spread on screens to be dried. A chamber with forced air at an inlet temperature of 140°C (284°F) and an air velocity through the grain of 61 m/min is used to bring the final moisture content of the rice to 8–14%. The drying conditions are critical in that the temperature should be fairly high, at least initially, so that moisture is removed from the surfaces of the grain at a rate sufficiently faster than it can diffuse from their interiors. This process sets a porous structure into the grain.

The precooked rice is dried in 2 or more successive steps, usually at gradually decreasing temperatures. When treated according to the Ozai-Durrani process, the dry, precooked rice grains are enlarged to about twice their original volume. The products, referred to as "Minute Rice 1," require 10−13 min preparation time.

Roberts (1952A) developed another process of this type (Table 16.1). The rice is soaked in water at room temperature and then boiled in water for 1−3 min, so that the moisture is increased to about 45−55%. By limiting the boiling time and the moisture uptake, the rice grains remain intact, with little or no sloughing of surface starch. Drying is done in 2 stages. Initially, the air temperature is set at about 200°C (390°F). The rice is further gelatinized by the high heat, and the surface of the grains are quickly dried and hardened so that the grains remain in a porous condition. Some puffing occurs simultaneously forming a number of small, uniform voids throughout the rice grains. The precooked rice is placed in a drying chamber through which hot air is forced at a velocity high enough to suspend the rice in the air over a screen or porous plate. The drying at this high temperature is held for only 1−3 min to prevent scorching the grain while still doing the job of puffing and setting the structure. The temperature is next reduced to about 100°C (212°F) to complete the drying of the rice to about 10−15% moisture. The product can be prepared for serving in about 5 min.

Various researchers at General Foods Corporation have devoted their efforts toward developing new and improved quick cooking rice products. Two patents were granted to Campbell and Hollis (1954A,B) and one to Shuman and Stanley (1954). These patents pretreat the raw rice grains to develop numerous small cracks or "fissures" throughout the kernels prior to any soaking or cooking treatment. This "fissuring" effect is believed to facilitate a subsequent cooking operation by allowing a more rapid penetration of moisture to the interior of the kernels. Thus, the soaking time and the boiling or steaming times required for proper gelatinization are decreased, with a consequent increase in yield. The dry volume of the finished product is increased and the product requires less time to prepare for serving.

Flynn and Hollis (1955) described an improved process in which rice is hydrated in stages and gelatinized to various degrees at an intermediate stage of hydration by heating to or above the gelatinization temperature in the absence of excess water. For example, rice is soaked in water to a moisture content of about 25−30%, steamed at atmospheric pressure to increase the moisture to about 30−35%, then resoaked in water at

TABLE 16.1. SOME PROCEDURES USED IN PREPARATION OF QUICK COOKING RICE

Method	Soak			Boiling H₂O		Steam	Freeze	1st Stage Drying Temp.	2nd Stage Drying Temp.	Final Moist.
	Time	Temp.	Moist.	Time	Moist.					
Ozai-Durrani (1948)	***[1]	25°–60°C	30%	8–10 min	65–70%	—[2]	—	140°C	—	8–14%
Roberts (1952A)	***	25°–60°C	30%	1–3 min	45–55%	—	—	200°C 1–3 min	100°C	10–15%
Roberts (1955)	30–60 min	20°–65°C	25–35%	—	—	10–15 psig 5–20 min	—	35°–100°C to 8–14%	200°–260°C for 1 min	8–14%
Keneaster & Newlin (1957)	16 hr	27°C	***	50–60 min	***	—	2°–3°C then –18°C for 4 hr	37°–40°C	—	6%
Yasumatsu et al. (1971)	0.5–16 hr 1–10 min	10–50°C 20–70°C	25–35% 45–60%	—	—	atm. 5 min –1 hr atm. 15 min– 2 hr	—	30–100°C	200–400°C 5–30 sec	8–20%

Source: Lethem (1976).
[1]Information not given.
[2]Step not included in process.

38°−88°C (100°−190°F) to a moisture content of 60−70%. The rice is drained, rinsed and suitably dried to maintain a light porous structure. A compression or "bumping" step is suggested as an intermediate step between the steaming and resoaking steps to further modify the structure and to reduce the resoaking time. A patent was issued to Hollis *et al.* (1958) for a quick cooking process which embodied most of the significant and practical quality improvements and the processing steps disclosed in the earlier patents. The discussion reveals that the fissuring process, followed by either steam cooking or water cooking, has several advantages over the initial Ozai-Durrani patent. The soaking time prior to gelatinization is decreased and possibly eliminated. The precooking time is decreased, with a consequent increase in yield and the dry volume of the product is increased. Most important, the time required for preparation of the finished product for serving is significantly reduced. A specific example of processing steps is as follows: Raw milled white Blue Bonnet rice is heated in dry air (usually forced draft) at about 93°C (200°F) for about 15 min. The fissured rice grains are then immersed in water at 92°C (197°F) and cooked for about 11 min. The moisture content increases to about 60% at this stage. The water-cooked rice is then steamed at atmospheric pressure for 10 min. At the end of this period, the rice has undergone uniform and substantially complete gelatinization and has a moisture content of about 70%. The rice is next washed in cold water for about 2 min to halt the cooking process and to remove any foreign material. It is then drained and placed on a continuous conveyor belt in a layer of about 2.5 cm thick. The rice is passed through a forced air drier, where air at 121°C (250°F) is forced upward and downward through the bed of rice at a velocity of 53.34 m/min. This process is believed to be the one used for "Minute Rice II," an improved and quicker cooking rice, which can be prepared for serving in about 5 min.

In another of Ozai-Durrani's methods (1956A,B) for making quick cooking rice, white, brown or parboiled rice, the rice is soaked to about 30% moisture and then cooked by "blasting" with high pressure steam. Pressure in the chamber is being maintained at 5 to 10 psig for about 5−15 min. The precooked rice, at about 35% moisture, is dried at 60°C (140°F) to about 10−14% moisture. The product is said to be recooked by boiling for 5−10 min, followed by standing in a covered pan for 10 min.

Carcassonne-Leduc (1963) obtained a patent for precooking rice, beans, lentils, tapioca, semolina and corn. The process appears to be useful in cooking large quantities of farinaceous foods for commercial use. The quick cooking rice treatment is quite similar to earlier disclosures—soak, boil, steam and dry.

Autrey and Lynn (1965) patented a process in which white, brown or parboiled rice is hydrated, cooked and dried to yield quick cooking rice.

The main departure from prior work is the thermal cycling treatment. This involves alternately heating dry, raw rice, cooking it, and reheating it again through 3 or more cycles. The process creates rifts and crevices during heating and compression and cracks during cooling. It enlarges the crevices during the second heating, whereby the rice grains are rendered highly permeable and absorptive. The remaining steps of soaking, cooking and drying are similar to the previously discussed patents.

Yasumatsu *et al.* (1971) used a 2 step soak-steam method to gelatinize the starch (Table 16.1). After the rice is soaked and steamed at atmospheric pressure, it is soaked a second time in water containing an edible oil and surfactant. The product is then steamed to completely gelatinize the starch. The oil is used to ensure grain separation after steaming and during the drying operation. The rice is dried at $30°-100°C$ ($86°-212°F$) until the moisture content is between 8 and 20%. Then the temperature is raised to $200°-400°C$ ($392°-752°F$) for 5–30 sec.

The soaking step provides an inexpensive method for increasing the moisture content of the raw rice. The speed of water penetration and the amount of water absorbed by the soaked rice may differ with the varieties used, extent of milling, storage duration after milling and temperature of the soaking water. In general, soft rice, highly milled rice and higher water temperature enhance water absorption. Li *et al.* (1976) have measured water absorption in rice. Table 16.2 shows that a long soaking time facilitates the uptake of water. For a given soaking time, an increase in temperature results in an increase in hydration. Thirty to sixty minutes of soaking appears to be sufficient to reach the equilibrium state.

TABLE 16.2. RATE OF WATER ABSORPTION BY TAIWAN RICE

Soaking Time (min)	Water Absorption %		
	26°C	36°C	52°C
15	21.3	22.7	25.3
30	26.7	26.7	28.0
60	27.3	27.3	28.0

Source: Li *et al.* (1976).

Gelatinization of starch granules is one of the 2 key steps in the quick cooking process. Fully cooked and ready-to-serve white rice is completely gelatinized. Since gelatinization of the starch is irreversible, it is only necessary to rehydrate fully gelatinized dried rice to prepare it for serving. In terms of preparation time for serving, fully gelatinized dried rice takes about 5 min, whereas partially gelatinized rice takes 10–15 min to finish cooking.

Drying conditions are also very critical. The porous structure must be maintained during drying to facilitate the penetration of water upon rehydration.

THE EXPANDED AND PREGELATINIZED RICE

Roberts (1955) patented a process (Table 16.1) in which raw milled white rice is soaked to increase the moisture to about 30%. The product is steamed under pressure at about 10–15 psig for 5–20 min and then dried at a fairly low temperature to a moisture content of 8–14%. The grains shrink to a dense glassy structure in contrast to the light porous structure obtained by the high temperature drying process. Finally, the rice is expanded or puffed in a hot air stream at 200°–260°C (300°–500°F). The velocity of the air stream must be sufficient to tumble the rice grains and suspend them while puffing takes place. In one example, pregelatinized white rice dried to about 10% moisture was expanded in 12 sec in an air blast at 250°C (480°F) to a volume about 4 times that of the raw rice. This light porous product can be prepared for serving in 2–3 min. The process has been commercialized and marketed as a flavored rice dish or casserole type product.

Wayne (1963B) heated dry parboiled rice or dry pregelatinized white or brown rice in a fluidizing gas stream at temperatures between 150°–455°C (300°–850°F). It is a continuous process in which previously gelatinized dry rice is moved through a fluidized or air suspension gas stream. The expanded kernels are separated from the line through a cyclone.

Mickus and Brewer (1957) described a method for tenderizing parboiled rice. The dry parboiled rice is heated in a hot air blast or an electric field to expand or puff the grains to yield either a quick cooking rice or a breakfast cereal.

THE ROLLING OR "BUMPING" TREATMENT

Another process for making quick cooking rice involves rolling, compressing or "bumping" to produce a thin cross section or cracks with a resultant increase in the rate of moisture absorption during preparation for serving.

In Ozai-Durrani's patent (1956A), either parboiled rice or pregelatinized rice is soaked in water, steamed or both to increase the moisture to a predetermined level. Pregelatinized rice at a moisture content of 40% or less is then compressed to 30–80% of its original volume and dried to 10–14% moisture at a relatively high temperature so that the grains are somewhat porous. In some cases the rice is soaked before compression and in others the rice is soaked after compression.

A Unilever Ltd. patent (1957) describes the preparation of quick cooking rice for use in dry soup mixes. The steps of soaking, gelatinizing, compressing and drying are similar to those disclosed by Ozai-Durrani. However, the drying temperature is quite low, so that the grains collapse to a hard glassy texture rather than a porous one. The relatively quick cooking quality of this product is due to the thin cross section.

Seltzer's patent (1959A,B) describes a compression treatment in detail. The first patent involves a pressure cooking and pregelatinizing treatment while the second involves utilization of parboiled rice. Both steps essentially complete gelatinization of the starch. These processes yield a specialized product for use in dry soup mixes.

DRY HEAT TREATMENTS

Alexander (1954) described a dry heating process for both raw white rice and raw brown rice in which 3−4% of water is removed from the rice in its natural state by circulating air at 57°−82°C (135°−180°F) at 0.85−1.70 m³/min for 10−30 min. The process creates transverse striations in the rice grains, thus yielding a quick cooking product. This product has been marketed as a quick cooking rice which will cook in about 10 min. The effect of this dry heat treatment is similar to the "fissuring" treatment described previously which facilitates cooking by allowing more rapid penetration of moisture.

Bardet and Giesse (1961) described a dry heat treatment of raw brown rice at very high temperatures (230°−315°C, 450°−600°F) to yield a quick cooking product. Selected processing conditions involve passing heated air at about 272°C (522°F) at a velocity of about 762 m/min through a bed of raw brown rice for about 17.5 sec to fracture the bran layer. Immediately thereafter cool air is passed through the rice to terminate any further processing treatments. The endosperm becomes quite opaque or chalky as a result of this high temperature treatment. Some degree of swelling also occurs as well as fracturing of the surface. This product can be prepared for serving in about 15 min.

THE FREEZE-THAW PROCESS

A "freeze-thaw" treatment has been incorporated into the previously developed series of steps (soaking, boiling, steaming and drying) by Keneaster and Newlin (1957). Dry milled parboiled rice is steeped to a point approaching saturation by holding the rice under water for 16 hr at 27°C (80°F). The steeping time can be reduced to about 1.5 hr by increasing the water temperature. The rice is completely gelatinized without undue sloughing or ragged surfaces. The hot, hydrated and gelatinized rice is then cooled in water, then drained and frozen. The rice is first refrigerated to the point of freezing (0°C, 32°F) and then held at this temperature for 1–3 hr. During this period, large ice crystals are formed which breakdown the colloidal starchy structure and produce a porous kernel. The product can thus absorb water readily during the recooking step. Following this purposefully prolonged chill-freeze step, the rice is subjected to a rapid temperature drop sufficient to insure complete freezing. The frozen rice is then thawed under nondrying conditions to prevent the grains from sticking together and to obtain the desired characteristics in the final product. Thawing is accomplished by holding the frozen rice at room temperature or in warm air for 5 hr or longer. The thawed product is then dried to about 43°C (110°F) for not less than 3.5 hr. Conditions may vary somewhat, especially in the accelerated thawing and drying process. To avoid formation of vitrified rice particles in the final product, it is necessary to carefully control the temperature and humidity during thawing, especially when the rice is dried at temperatures ranging up to 121°C (250°F).

In another procedure, milled white rice is steeped for 2 hr at 49°C (120°F) and then boiled for 10 min to reach a moisture content of about 70%. The cooked rice is quickly drenched in cold water to prevent further self-cooking, drained and carefully frozen as described above. The frozen product is thawed in a forced air draft at 29°C (85°F) for 30 min and dried at 110°C (230°F) for 2 hr to a final moisture content of 8%. The finished product resembles whole grain rice, with well-rounded and unbroken kernels of good color. The product is prepared for serving by boiling in water for 5 min.

The "freeze-thaw" treatment was the first effective method for producing quick cooking rice on a commercial scale. "Uncle Ben's Quick Rice" is made by this method. It has been marketed successfully in the United States.

Ozai-Durrani's patent (1965) describes a similar freeze-thaw process in which raw white rice is soaked and boiled or steamed (or both) to gelatinize the starch (70% moisture) and then frozen, thawed and dried. Except for the compression step and the fast freezing treatment, this method does not differ greatly from that described by Keneaster and Newlin (1957).

GUN PUFFING

Gun puffing has been practiced for many years to produce breakfast cereals such as puffed rice, oats, corn and wheat. In a process patented by Carman and Allison (1953A,B), properly conditioned grain is placed in a puffing gun. After removal of air, the pressure is increased with steam to a moderate degree. The grain is quickly released into an expansion chamber maintained under high vacuum. The 2 patents are quite similar in the general processing steps with the first one concerned with details of processing several other cereals in addition to rice. Raw white rice is conditioned to a moisture content of 18–26%, preferably to 20–22%. The tempered rice, at a temperature not less than 33°C (91°F), is placed in the steaming chamber of the puffing gun which is then tightly sealed. The pressure is reduced to about 3.8 cm of mercury (absolute) or less, to remove the air and noncondensable gases from the rice kernels in about 2 min. Steam is introduced and the pressure maintained until the rice is substantially cooked without loss of cellular identity. Best results are obtained by slowly increasing the pressure from about 3.8 cm Hg (absolute) to a final pressure of 3.49–5.60 kg/cm² at a uniform rate, sometimes followed by a holding period at the final steam pressure. Examples are given which include variations in initial steam pressure from 1.74–3.49 kg/cm² for about 5–35 min, followed by an increase to about 3.49–5.60 kg/cm² with a holding time of about 15–60 sec before firing. In some cases, the pressure is increased to 3.49–3.85 kg/cm² in a single stage and held there for 5–10 min before firing. The firing step is for the purpose of releasing the pressure very quickly, i.e., to fire the charge through a triggered door into the expansion chamber which has a reduced pressure of about 0.25 cm Hg. The vacuum in the expansion chamber is maintained by continuing the evacuation during and after the expansion of the rice. In some instances the pressure in the expansion chamber may rise to as high as 8.9–17.8 cm Hg, but is reduced to below 2.5 cm before releasing the vacuum for discharging the product. The puffed product is further dried to below 15% moisture to maintain desired storage stability. The finished product can be recooked in about 5 min.

Grains of gun puffed rice may not be uniform in texture. This effect is believed to be caused by the noninstantaneous release of pressure from one end of the pressure chamber to the other as the product discharges through the triggered door or breach. A process to produce quick cooking rice by gun puffing the rice into an expansion chamber at atmospheric pressure was patented by Flynn and Ricker (1961). The advantages of this process are high yields, simple manipulations, small capital investments and readily available equipment.

Quick cooking rice can be produced by gun puffing, provided that the degree of enlargement is no more than 2 to 3 times the original size of the raw rice grains. In making puffed, ready-to-eat breakfast cereals, the final temperatures in the puffing chamber are distinctly lower and the moisture contents of the rice are considerably higher than usual. These final conditions just prior to firing are referred to as the "terminal" temperature and moisture conditions. The rice is subjected to a water vapor pressure equal to the pressure of saturated steam at the particular terminal temperature selected. Additional pressure achieved by introducing air or an inert gas will serve no useful purpose. Upon reaching the terminal conditions, it is desirable to release the pressure to atmospheric pressure.

The terminal conditions of rice temperature and moisture employed in carrying out the process are shown in Fig. 16.4 in which the terminal moisture content of the rice is plotted against its terminal temperature. Regardless of the moisture content, no puffing of any useful degree is obtained at terminal rice temperatures below 163°C (325°F). The conditions in the areas designated by A, B, D and E are not suitable. The conditions designated by area B provide an excessively sticky product. In the area designated by D, the product obtained is a mixture of quick cooking rice and highly puffed ready-to-eat breakfast cereal. Conditions in the area designated by E, characterized by relatively lower terminal rice moisture content and relatively higher terminal rice temperatures, yield a highly enlarged puffed product for breakfast cereals. Particularly useful conditions are found in areas C_1 and C_2. The conditions designated by area C_1 are most desirable. The product provided by the conditions under area C_2 is generally soft in texture but is a satisfactory product. The general public is believed to prefer the somewhat firmer product made under conditions of area C_1.

The terminal moisture is established by soaking the raw white rice in water followed by a tempering step, condensation of steam on the rice within the puffing chamber, or by a combination of both. The preferred terminal moisture levels are about 25−30%. Similarly, the terminal temperature is achieved either by introduction of steam (preferred method), by external heating of the puffing gun through applying a gas flame to the outer surface or by a combination of both. Rice produced under the preferred conditions is ready for serving in 5 min after adding excess boiling water.

FREEZE-DRYING

Preparation of quick cooking rice by freeze-drying fully cooked rice has been explored. Generally the process has been considered more expensive than the product made by hot air drying or puffing. An improved and somewhat more economical freeze-drying approach was patented by Wayne (1963A,B). His method combines an initial freeze-drying step with a finish drying in hot air at atmospheric pressure. Variations in freezing treatment and freeze-drying at different or "dynamic" pressure and temperature conditions are given with or without a final drying at atmospheric pressure.

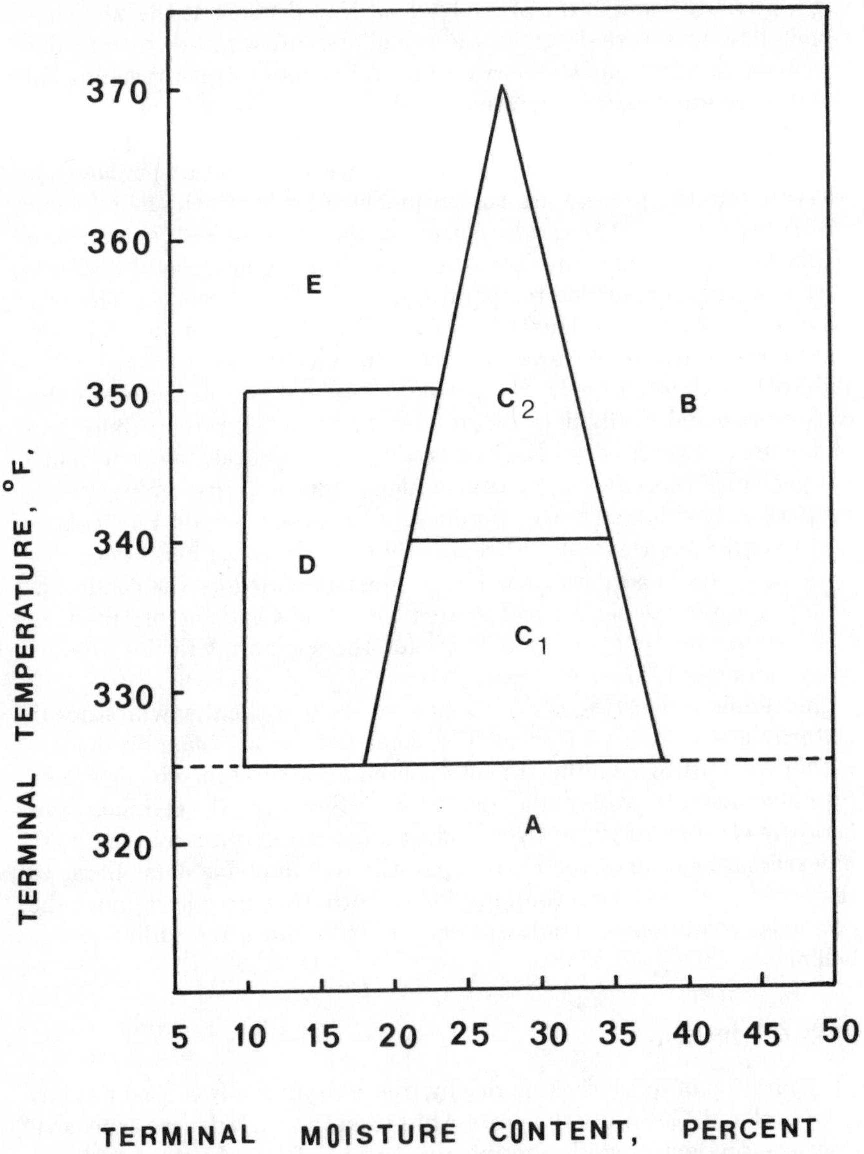

From Roberts (1972)

FIG. 16.4. RELATIONSHIP OF TERMINAL TEMPERATURE AND TERMINAL MOISTURE CONTENT TO THE QUALITY OF GUN-PUFFED RICE

In the processes disclosed by Wayne, rice is first cooked to a "desired extent." The product is frozen by contact with chilled plates within the vacuum chamber or by direct contact with a refrigerant, such as liquid nitrogen, prior to charging into the freeze-drying chamber. Efficiency in time and energy is achieved by operating the freeze-drying process at a temperature slightly below the freezing point of the product so that moisture is sublimed at a faster rate than is usually experienced in conventional freeze-drying. A suggested variation is that of drying the frozen precooked rice at atmospheric pressure by circulating dehumidified cold air through the rice, so as to remove moisture while keeping the rice in the frozen state. After drying under frozen conditions to 10–20%, moisture, it is further dried in a hot air blast at 150°–315°C (300°–600°F) which provides further voids or pores within the product.

Although there are some novel steps in Wayne's patents, as well as some previously disclosed treatments for quick cooking rice, it is believed that the overall economics could not compare favorably with those of currently successful commercial retail and institutional products.

CHEMICAL TREATMENTS

The principal physical modification of rice produced by most of the methods developed thus far is that of gelatinizing the starch by application of heat with a gradual increase in moisture content to about 70%. Hot water, steam, dielectric, infrared or microwave heating methods have been applied. Certain chemical treatments may also be used to produce quick cooking rice.

Lewis and Lewis (1965) described a sodium chloride treatment method to yield quick cooking rice. The rice is impregnated with a saturated solution of sodium chloride at about 80°C (175°F) whereby the starch is partially gelatinized. The weight of sodium chloride solution absorbed is 25–100% of that of the rice. After pasteurization, the product is resistant to attack by vermin and microorganisms. The cooking time of the dry product is substantially reduced. Cooking in a large volume of water reduces the salt content to a palatable level. This method is not as good as those described earlier.

Li et al. (1976) have described an enzyme treatment method by which the outer layer of unhulled rice is removed while leaving the germ intact. The rice is washed and soaked in 0.3% NaOH at 40°C (104°F) for 2 hr. After the base is neutralized with HCl, the rice is again washed with water. Hemicellulase (10 μ/g) is next added to the treatment tank. After 2 hr at 40°C (104°F), the softened and swollen skin may be removed from the grains. The enzyme-treated rice, after steaming or cooking by conventional methods, has the appearance of rice refined by the mechanical method, but its taste is superior. They also used glycerine fatty acid ester, sorbitan fatty acid ester, propylene glycol and fatty acid sugar ester in preparing quick cooking rice. These compounds serve as wetting, permeating, emulsifying, solubilizing, cleaning and foaming agents. They are also involved in complex formation with starch, water-keeping and gel formation at higher temperatures.

Addition of 0.3% by weight of fatty acid glyceride powder prevents clump formation when steaming the rice (Li et al. 1976). It also increases aeration during dehydration. The level of surface active agent added should be less than 1% to avoid foaming during cooking.

Tanaka and Yukami (1969) received a patent for making quick cooking rice by chemical treatment. The rice is first soaked at 20°−30°C (68°−86°F) in a solution of phosphates or polyphosphates at pH 7.6−8.2. After draining, the soaked rice is cooked in a solution containing phosphates (0.05−0.5%), saccharide (0.3−10%) and a surfactant (0.1−0.5%) until the rice is about 70% gelatinized and the moisture content is 50−70%. Further steaming is applied until the rice reaches 100% gelatinization. The product is then dried rapidly. As an example, white rice is soaked for 1 hr at 30°C (86°F) in a solution of disodium phosphate (DSP) at pH 7.6. The drained rice is cooked for 15 min in a solution of 0.3% DSP, 0.5% lactose and 0.25% glyceryl monostearate, then steamed for 20 min at 100°−120°C (212°−248°F) and finally dried. The finished product re-cooks in 5 min.

Ozai-Durrani (1960) obtained a patent for making quick cooking rice. In this process, gelatinization of the rice was accomplished by applying a series of sprays. The temperature of the rice was increased gradually until the moisture of the rice reached 70%. The grain was thoroughly cooked by then. Other desirable steps include a compression or bumping step to facilitate moisture absorption and freezing the completely cooked rice before drying in a hot air blast. This patent is essentially a combination of the 5 methods described earlier. It may be desirable to treat the rice with a fat solvent to remove residual oil.

MISCELLANEOUS PROCESSES

Several "miscellaneous" quick cooking rice processes have been developed during the past decades. These include fabricating quick cooking rice from broken grades of rice, products for specialized applications, minor modifications and multiple step treatments applied to previously known methods, microwave heating, and adaptation of quick cooking white rice processes to brown rice.

Gorozpe's patents (1959,1963) for preparing quick cooking rice present a novel method of utilizing the less expensive broken rice to yield either a precooked meal or flour, or glomerates of small granular particles shaped to resemble whole grain, quick cooking rice. Broken rice grains are heated in an air blast at 50°−70°C (122°−158°F) for about 30 min, which checks the rice and enables it to absorb water more quickly and thoroughly. The heat-treated rice is soaked in warm water to a moisture content of about 40%. The drained rice is passed through a roller mill having differential speed rolls to produce porous clusters. The product is subjected to a wet steaming treatment in 1 to 3 stages to further gelatinize the rice to about 90−95% completion. There is a milling or crushing step between the steaming treatments. The product is finally dried with hot air.

For granular products, the clusters of lumps from a single-stage gelatinization treatment are dried at 80°−140°C (178°−284°F) to about 15−25% moisture, broken up on rolls, dried again to about 10−14% moisture and then ground and screened.

If a whole grain, quick cooking rice product is desired, the partially hydrated clusters may be gelatinized in steam in 2 or 3 stages. These stages will accomplish gelatinization to 30−40%, then 50−70% and finally 80−95%. There is a milling or crushing action between the steaming steps. The clusters, which have been gelatinized to about 80−95% in a single stage, are extruded, with or without further shaping, to form rod-like segments resembling regular long grain rice and dried to 10−14% moisture. In the 3-stage method, the rice from the first stage is extruded and shaped, further gelatinized in 2 steps and finally dried.

Although there are economic advantages in using broken rice, the processing steps and equipment costs make it difficult to estimate whether whole grain, quick cooking rice made by this process could compete favorably with presently marketed products.

Gorozpe (1964) described a process of fissuring, multiple soaking and steaming steps. It is a combination of steps similar to the various developments previously cited. Either white or brown rice is fissured by a hot air treatment at 50°−120°C (122°−248°F). The product is hydrated in water below the gelatinization temperature, briefly immersed in boiling water, and then treated with steam in several stages in which the moisture is increased 4−6% in each step. The rice may be treated first with hot water sprays at 15 sec intervals for a total of 3−10 min. It is then subjected to wet steam from below, with cold water sprays from above at 30 sec intervals for 6−20 min, cooled by a blast of cold air and finally dried.

A specialized product was developed by Willock (1965) for rice pudding. The rice is purposely under-gelatinized and the conditions adjusted to result in the desired tenderness and thickening qualities when the rice pudding product is prepared in milk. Willock (1968) patented a process in which rice is soaked to increase the moisture to about 28%. The product is coated with a mucilage of an edible binder such as starch, vegetable gum or cellulose derivative, flavored and then dried. In both processes the rice is essentially not precooked.

Kester and Ferrel (1957) describe a process for making quick cooking brown rice in which the raw brown rice is soaked, cooked and dried at a relatively low temperature, then expanded in a hot air blast at 250°–300°C (480°–570°F). Miller (1963) expanded the established technology of quick cooking white rice processes to apply it to a quick cooking brown rice.

Utilization of microwave energy for producing quick cooking rice was presented by Huxsoll and Morgan (1968). The process and equipment illustrate the advantages in using a "system approach" to designing microwave applications. Basically, microwave energy is used to replace other forms of heat energy in cooking and drying the rice. Raw rice is soaked in water to 25–30% moisture, then heated with either steam or microwaves and further hydrated in water at about 66°C (150°F). An adequate microwave cooking is attained for long grain Patna rice at about 50% moisture. The hydrated and cooked rice is then dried by air to about 15% or less moisture and subjected to microwave energy which leads to extensive fissuring and slight puffing of the grains. The process yields a product which can be prepared for serving in about 5 min. Microwave energy is said to be advantageous due to the "volume heating" effect which cooks the rice with minimum clumping. It produces a rapid heating effect which can impart a porous structure to the dried precooked rice grains.

Serbia and Benett (1968) described a quick cooking rice made from parboiled rice. The dry, milled parboiled rice is soaked in excess water at 70°C (160°F) for 14 min, whereby the moisture content reaches 47%. The rice is drained, steamed for 13–25 min, washed in water, steamed and washed twice until the moisture reaches about 69%. The hydrated parboiled rice is then dried at 115°–160°C (240°–320°F) in 3 stages at gradually decreasing temperatures.

RECENT DEVELOPMENTS IN EQUIPMENT

The long grain rice varieties have been the only variety in common use in the quick cooking products. There is less application of the medium and short grain varieties due to the difficulty in handling and removal of water. Carlson et al. (1976) describe a process in which all varieties may be used successfully (Fig. 16.5). The process involves the use of the centrifugal fluidizing bed (CFB) in the critical drying process (Fig. 16.6). The CFB drier removes the moisture at a higher air velocity, lower

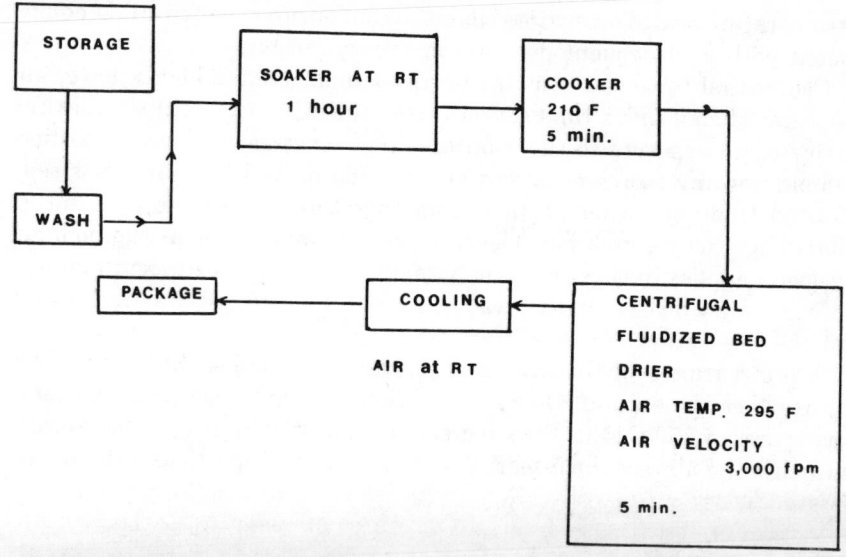

From Carlson et al. (1976)

FIG. 16.5. FLOW DIAGRAM FOR PREPARATION OF QUICK COOKING MEDIUM GRAIN WHITE RICE

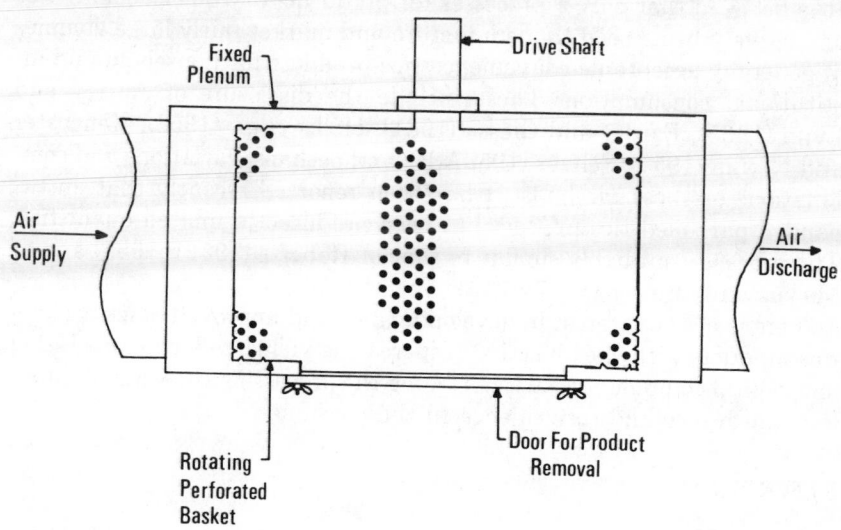

From Carlson et al. (1976)

FIG. 16.6. CENTRIFUGAL FLUIDIZING BED DRIER

temperature and shorter time than the conventional dehydration equipment with a consequent decrease in energy output.

Centrifugal force restrains the particles fluidized in a high velocity air stream. Homogenous fluidization occurs, insuring that all particles receive equal exposure to the countervailing air stream. Constant motion eliminates any scorching or surface heat damage. Each particle is separated from the other particles and therefore there is no problem in handling sticky starch particles. Clumping is minimized, as the high air velocity applies the heat extremely rapidly. There is an efficient removal of moisture which is carried away as soon as it is diffused into the interior of the rice kernel.

A pretreatment of the rice kernels, by soaking and slight cooking, will cause them to expand. Quick application of heat using air above temperatures of 250°C (482°F)) will create a void structure in the rice kernel, which will maintain the integrity and the size of the particle in its cooked state.

CONCLUSION

Among the quick cooking rice processes and combinations of steps which have been reported over the past 35 years, only 9 products have been produced and marketed in the United States in significant volume. It is believed that only 5 processes for producing dry quick cooking rice are actively being used to manufacture and market fairly large volumes of generally acceptable convenience rice products for household and institutional consumption. These include the disclosure of Autrey and Lynn (1965), Bardet and Giesse (1961), Hollis *et al.* (1958), Keneaster and Newlin (1957), Seltzer (1959A,B), and perhaps, variations and combinations of these methods. It has been reported recently that an expanded parboiled rice product has been produced in limited quantities. The process is probably similar to that of Roberts (1952B,1955) and of Mickus and Brewer (1957).

There is active interest in developing new and improved quick cooking rice products with the objective of increasing yields, reducing processing and capital equipment costs, shortening the processing time, and improving appearance and convenience to the consumer.

REFERENCES

ALEXANDER, W.P. 1954. Process of preparing quick cooking rice. U.S. Pat. 507,242. November 9.

AUTREY, H.S. and LYNN, L. 1965. Process for the preparation of precooked rice. U.S. Pat. 3,189,462. June 15.

BARDET, G.V. and GIESSE, R.C. 1961. Processing of brown rice. U.S Pat. 2,992,921. July 18.

CAMPBELL, H.A. and HOLLIS, F., Jr. 1954A. Process of preparing quick-cooking rice. U.S. Pat. 2,696,156. December 7.

CAMPBELL, H.A. and HOLLIS, F., Jr. 1954B. Method of preparing quick-cooking rice. U.S. Pat. 2,696,157. December 7.

CARCASSONNE-LEDUC, R.P.C. 1963. Process for pre-cooking rice. U.S. Pat 3,083,102. March 26.

CARLSON, A.R., ROBERTS, R.L. and FARKAS, D.F. 1976. Preparation of quick cooking rice products using centrifugal fluidized bed. J. Food Sci. *41*, 1177-1179.

CARMAN, C.R. and ALLISON, J.E. 1953A. Quick cooking cereal and method of making same. U.S. Pat. 2,653,099. September 22.

CARMAN, C.R. and ALLISON, J.E. 1953B. Pre-cooked rice. US. Pat. 2,653, 100. September 22.

FLYNN, C.E. and HOLLIS, F., Jr. 1955. Production of quick-cooking rice. U.S. Pat. 2,720,460. October 11.

FLYNN, C.E. and RICKER, M.O. 1961. Method for preparing quick-cooking rice. U.S. Pat. 2,696,288. January 24.

GOROZPE, R.D. 1959. Apparatus for preparing a quick-cooking rice product from broken rice. U.S. Pat. 2,914,005. November 24.

GOROZPE, R.D. 1963. Process for preparing a quick-cooking rice product. U.S. Pat. 3,071,471. January 1.

GOROZPE, R.D. 1964. Method for preparing quick-cooking rice. U.S. Pat. 3, 157,514. November 17.

HOLLIS, F., Jr., MILLER, F.G. and MILLER, F.J. 1958. Process of preparing a quick-cooking rice. U.S. Pat. 2,828,209. March 25.

HUXSOLL, C.C. and MORGAN, A.L., Jr. 1968. Microwaves for quick-cooking rice. Cereal Sci. Today *13* (53) 203.

KENEASTER, K.K. and NEWLIN, H.E. 1957. Process for producing a quick-cooking product of rice or other starchy vegetable. U.S. Pat. 2,813,796. November 19.

KESTER, E.B. and FERREL, R.E. 1957. Method of preparing precooked puffed brown rice cereal. U.S. Pat. 2,785,070. March 12.

LETHEM, L.F. 1976. Effect of antioxidant, packaging material and storage temperature on the stability of quick-cooking calrose rice. M.S. thesis. University of California, Davis, California.

LEWIS, D.A. and LEWIS, J.M. 1965. Quick cooking rice. Australian Pat. 262, 788. October 18.

LI, C.F., CHANG, P.Y. and CHANG, J.L. 1976. Instant rice. Food Ind. Res. and Dev. Inst. Rept. *79.* Hsinchu, Taiwan.

MICKUS, R.R. and BREWER,B.W. 1957. Rice treating process. U.S. Pat. 2, 808,333. October 1.

MILLER, F.J. 1963. Process of preparing a quick-cooking brown rice. U.S. Pat. 3,086,867. April 23.

OZAI-DURRANI, A.K. 1948. Quick-cooking rice and process for making same. U.S. Pat. 2,438,939. April 6.

OZAI-DURRANI, A.K. 1956A Quick-cooking rice and process. U.S. Pat. 2,733,147. January 31.

OZAI-DURRANI, A.K. 1956B. Quick-cooking rice and process. U.S. Pat. 2,740,719. April 3.

OZAI-DURRANI, A.K. 1960. Process for preparing quick-cooking rice. U.S. Pat. 2,937,946. May 24.

OZAI-DURRANI, A.K. 1965. Process for producing quick-cooking rice. U.S. Pat. 3,189,461. June 15.

ROBERTS, R.L. 1952A. Production of quick-cooking rice. U.S. Pat. 2,610,124. September 9.

ROBERTS, R.L. 1952B. Production of expanded rice products. U.S. Pat. 2,616,808. November 4.

ROBERTS, R.L. 1955. Preparation of pre-cooked rice. U.S. Pat. 2,715,579. August 16.

ROBERTS, R.L. 1972. Quick-cooking rice. In Rice Chemistry and Technology. D.F. Houston (Editor). Amer. Assoc. Cereal Chem. St. Paul, Minn.

ROBERTS, R.L., HOUSTON, D.F. and KESTER, E.B. 1951. Expanded rice product. A new use for parboiled rice. Food Tech. 5, 361–363.

SELTZER, E. 1959A. Process for preparing quick-cooking rice. U.S. Pat. 2,890,957. June 16.

SELTZER, E. 1959B. Method of making quick-cooking cereals from parboiled grains. U.S. Pat. 2,903,360. September 8.

SERBIA, G.W. and BENETT, I. 1968. Process for preparing a quick-cooking rice. U.S. Pat. 3,408,202. October 29.

SHUMAN, A.S. and STANLEY, C.H. 1954. Method of preparing quick-cooking rice. U.S. Pat. 2,696,158. December 7.

TANAKA, M. and YUKAMI, S. 1969. Method of preparing precooked dry rice. U.S. Pat. 3,484,249. December 16.

UNILEVER, LTD. 1957. Improvements in the preparation of quick-cooking cereal products. Brit. Pat. 771,378. April 3.

WAYNE, T.B. 1963A. Process of preparing a rice product. U.S. Pat. 3,085,012. April 9.

WAYNE, T.B. 1963B. Process of preparing a rice product. U.S. Pat. 3,085,013. April 9.

WAYNE, T.B. 1963C. Process of preparing a rice product. U.S. Pat. 3,113,032. December 3.

WILLOCK, J.T. 1965. Rice process. U.S. Pat. 3,164,475. January 5.

WILLOCK, J.T. 1968. Rice product and process. U.S. Pat. 3,364,299. January 23.

YASUMATSU, K., SAWADA, T., SAWADA, K. and MORITAKA, S. 1971. Preparation of quick-cooking rice. U.S. Pat. 3,582,352. June 1.

17

Canning, Freezing and Freeze Drying

Bor S. Luh
Yuan-Kuang Liu

Rice *(Oryza sativa)* is a staple grain consumed by more than half of the world's population (Hogan 1977). It is the chief source of calories in Asia and has become more popular in western countries in recent years (Burns 1972). Common objections to using rice in western countries include difficulty and length of preparation and established eating habits. To increase consumption of rice, the industry has tried to reduce cooking time by developing easily prepared rice products such as dried precooked rice, frozen precooked rice and canned rice.

The types of canned rice products on the market include soups with rice, meat and rice dinners, casseroles, Spanish rice, unflavored cooked rice, fried rice and rice pudding (Burns 1972). Some desirable characteristics for canned rice are white color, separate noncohesive kernels, a minimum amount of splitting and fraying, and clear canning liquor (Burns 1972). Varietal characteristics, age of the grain and parboiling all affect the texture of canned rice. Varietal differences have been attributed to variations in amylose content (Kester 1959; Williams *et al.* 1958). Shibuya *et al.* (1977A,B) reported that the changes in rheological properties of cooked rice and its paste, during storage of rice grains at 4° and 23°C (39.2° and 73.4°F), should be explained by the changes in some structural components such as proteins and cell wall constituents, but not the free fatty acids and lipids.

High amylose rice is flaky, dry and bland, while varieties with low amylose tend to be sticky, moist and better tasting. Rice with a high protein content takes long to cook due to the physical barrier to water absorption created by the protein matrix around starch granules. Low protein rice is more tender, cohesive and sweeter than high protein rice (Juliano *et al.* 1965). Aging differences are probably caused by altered colloidal properties of the grain and cell wall during storage (Desikachar *et al.* 1960). Parboiled rice is often used in preparing canned rice products because of the stability of the kernel and the retention of its shape without disintegration under rigid retorting and heating conditions. Kruger and Murray (1976) reviewed the literature on starch texture and instruments used in the measurement of starch properties.

Other factors which may affect the quality of canned rice include pH, fat content, salt concentration and blanching time (Ferrel *et al.* 1960; Ghosh 1959). For example, alkaline solutions cause rice to develop a yellow color. Addition of 0.01% acetic acid slightly improves the color and flavor. A boiling water with a pH close to 7 has been found to be optimal (Roberts *et al.* 1953).

VARIETIES

Short, medium and long grain rice varieties are grown commercially in the United States. Long grain rice is dry, fluffy and not sticky upon cooking and therefore considered suitable for canning. Table 17.1 shows the adaptation of varieties for particular uses. Variety preference is primarily based on flavor, appearance and cooking characteristics, and secondarily on grain shape and size (Beachell and Halick 1956).

EFFECTS OF AMYLOSE CONTENT AND PARBOILING CONDITIONS ON CANNED RICE CHARACTERISTICS

The potential market for canned rice in soups, salads, rice dishes, desserts and baby foods is high, but difficulties in rice canning processes slow down the increase in consumption. In addition to the canning process, characteristics of the canned rice may also be influenced by the properties of the raw material.

Feillet and Alary (1975) examined 48 rice varieties, with amylose content ranging from 20–30%. Samples were parboiled in a fully automatic laboratory-scale apparatus. Standard parboiling conditions included: vacuum for 10 min, steeping for 30 min at 65°C (149°F) under 3.5 kg/cm², and steaming for 20 min at 112°C (233.6°F) under 2.1 kg/cm². After milling, parboiled rice was canned in excess of water. Eleven grams of parboiled rice and 110 ml of water at 80°C (176°F) were transferred to a tin can (⅙ flat French standard). The can was then sealed, stirred for 3 min and retorted for 20 min at 120°C (248°F). In some cases, rice was first cooked in boiling water until the moisture content reached 65% (16–19 min), then transferred to a can and retorted.

A high correlation was observed between amylose content and water absorption (r= − 0.74) or firmness (r= + 0.76). Table 17.2 shows the results in regard to characteristics of canned rice processed from par-

TABLE 17.1. UTILIZATION OF RICE VARIETIES

Variety	Grain type	Table quality	Canning quality (parboiled)	Product use
Bluebonnet 50	Long	Good	Good	General
Rexoro	Long	Excellent	Excellent	General and canned soups
TP 49	Long	Excellent	Excellent	General and canned soups
Texas Patna	Long	Excellent	Excellent	General and canned soups
Improved Bluebonnet	Long	Good	Good	General
Sunbonnet	Long	Good	Good	General
Zenith	Medium	Good	Unsuitable	General, breakfast cereal, baby food
Nato	Medium	Good	Unsuitable	General and breakfast cereal
Nova 66	Medium	Good	Unsuitable	General
Magnolia	Medium	Good	Unsuitable	General, breakfast cereal, baby food
Saturn	Medium	Good	Unsuitable	General
Belle Patna	Long	Excellent	Good	General
Bluebelle	Long	Excellent	Good	General
Starbonnet	Long	Excellent	Excellent	General

Source: Burns (1972).

TABLE 17.2. CHARACTERISTICS OF CANNED RICE PROCESSED FROM PARBOILED FRENCH VARIETIES

Variety of rice	Type of grain	Amylose (% d.b.)	Firmness (g)	Water absorption (g water per 100 g rice)	Appear- ance	Solid losses (%)
Delta	Long	23.4	540	500	3	8.6
Cesariot	Long	23.6	610	470	4	7.7
Ciglon	Short	24.6	740	510	3	8.7
Cristal	Short	25.5	450	620	2.5	8.0
Balilla	Short	25.5	890	480	3	8.2
Duribe	Medium	26.4	560	550	3	7.2
Cesariot	Long	27.0	1060	480	3	7.8
Arlesienne	Medium	28.0	1350	420	3.5	6.8

Source: Feillet and Alary (1975).

boiled French varieties. Comparison of French rice varieties with a different amylose content showed that the Arlesienne variety was the most suitable for canning.

Steeping conditions during the parboiling process slightly affect the properties of canned parboiled rice, but steaming temperature and steaming time have a tremendous effect on the quality of the canned products (Table 17.3).

TABLE 17.3. EFFECTS OF STEAMING CONDITIONS ON SOME CHARACTERISTICS OF PARBOILED CANNED RICE

Steaming temperature (°C)	Steaming time (min)	Appearance	Water absorption (g water/100 g)	Firmness (g)	Solid losses (%)
105	20	2.4	512	530	8.5
112	10	2.1	606	429	10.3
112	20	3.1	504	780	7.9
112	30	3.2	545	879	8.1
120	20	2.9	532	920	7.8

Source: Feillet and Alary (1975).

An increase in firmness and formation of a chalky texture were observed during post-canning storage. These changes in textural characteristics developed faster at 4°C (39°F) than at room temperature. The difference disappeared when the product was heated before consumption (Table 17.4). Changes during storage of the canned rice appear to be similar to hardening of bread due to retrogradation of gelatinized starch.

CANNING

Various methods have been studied for making canned rice more ac-

TABLE 17.4. EFFECT OF HEATING ON FIRMNESS OF STORED CANNED RICE (1 WEEK AT 4°C)

Varieties	Amylose content (% dry basis)	Firmness (kg) of canned rice after storage[1]	
		Before heating	After heating
Delta	23.4	4.1	1.3
Arlesienne	28.0	11.4	7.0

Source: Feillet and Alary (1975).
[1]Moisture content of canned rice = 65%.

ceptable (Hara 1950; Gallenkamp 1952; Ferrel and Kester 1959; Altares and Luh 1976). These fall into 2 categories: wet pack and dry pack.

A product in which there is an excess of liquid, such as in soup media, is termed wet pack. Proper density is the prime objective with these types of products. The rice is precooked or blanched sufficiently to promote buoyancy in the product and prevent settling and matting, but not to the point that kernel texture is degraded. The parboiled rice is cooked slowly in an excess of water, followed by draining and washing in cold water. This washing process removes excess surface starch and stops the cooking process. The rice is filled into cans together with the sauce. The cans are sealed and the product is retorted.

A canned product, such as Chinese style fried rice, in which the grains are separate and devoid of free or excess moisture, is called the dry pack. The prime objective is to provide enough moisture for gelatinization of the starch during retorting without causing pastiness or cohesiveness in the kernels. Cooking oil as an ingredient helps to minimize grain cohesion. The usual procedure involves slowly precooking parboiled rice in an excess of water. The rice is subsequently washed in cold water and then mixed with the other ingredients. After filling and sealing, the canned product is slowly heated and then retorted. An example of dry pack canning of short grain rice has been developed by Ferrel et al. (1960). The process is outlined in Fig. 17.1. The acidified soaking water is claimed to produce white rice while kernel stickness is reduced by the emulsion rinse step.

Roberts et al. (1953) developed a process for canning white rice. The rice is washed and soaked in cold water for 30−45 min and boiled for 2−4 min, or until the moisture content is approximately 55%. Limiting the moisture to this level minimizes kernel disintegration. The partially cooked rice is filled into cans, sealed under 71.12 cm of vacuum and then retorted. The canned product is prepared for serving after being heated in boiling water. The grains remain white and well separated and become distorted or mushy in appearance.

An inherent disadvantage of canned rice is the clumping of the kernels

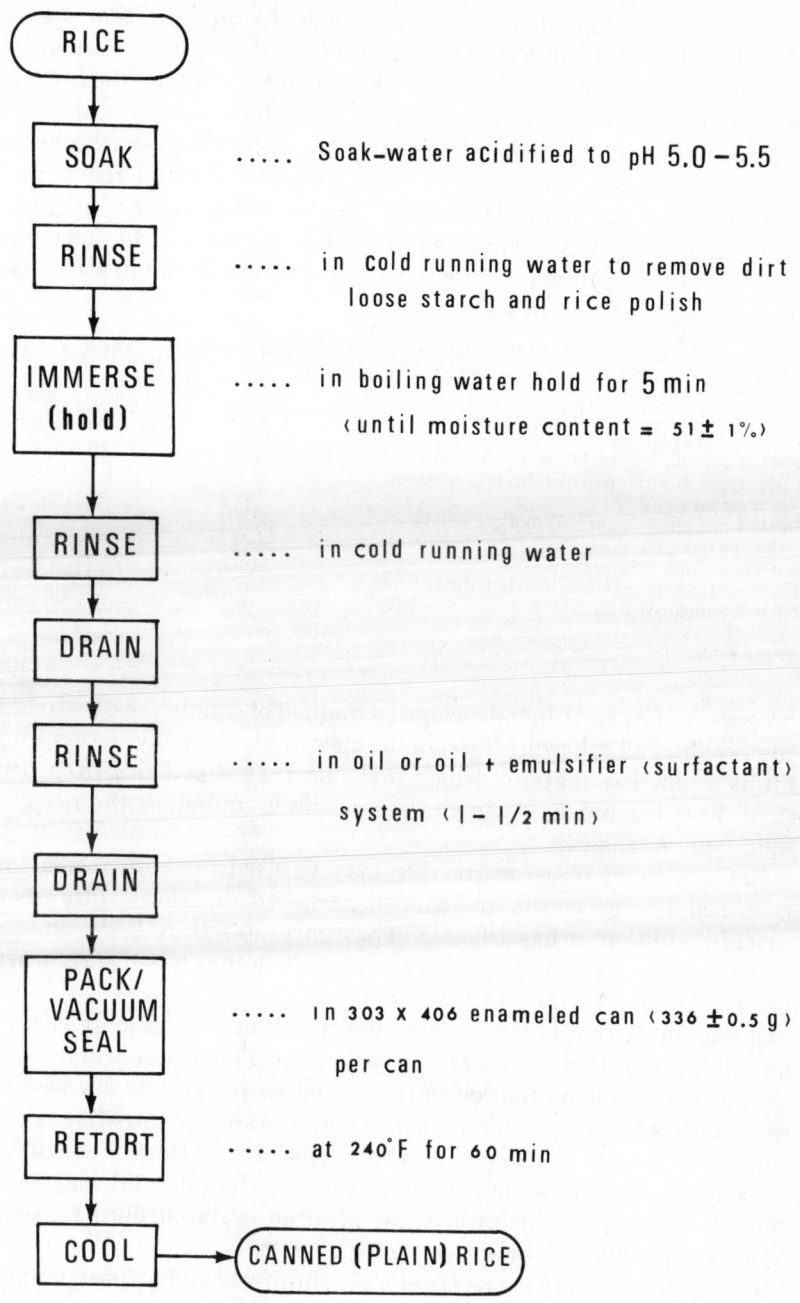

FIG. 17.1. RICE CANNING PROCESS

From Ferrel et al. (1960)

and the concomitant difficulty in removing the product from the can. Ferrel *et al.* (1960) and Ferrel and Kester (1959) have used oil emulsions and surfactants to minimize this problem. Table 17.5 shows the results of their study. Oil emulsions applied in the rinse step following soaking and cooking of the rice caused a significant reduction in the kernel cohesion. Only a few of the surfactants tested gave a substantial reduction in clumping; sorbitan mono-oleate was found to be effective. Sripathy *et al.* (1960) reported that similar use of buffalo butter and hydrogenated vegetable fat helped to promote the use of canned rice and to avoid these difficulties for its use in India.

TABLE 17.5. EFFECT OF EDIBLE OIL EMULSIONS AND SURFACTANTS ON REDUCTION OF COHESION IN CANNED PEARL RICE

Sample treatment	% Passing screen
Control (no treatment)	25
Cottonseed oil, 5%; sorbitan mono-oleate, 0.5%	89
Cottonseed oil, 5%; polyoxyethylene sorbitan monostearate, 0.5%	93
Cottonseed oil, 5%; carboxymethyl cellulose (15 cps.), 0.25%	91
Sorbitan mono-oleate, 0.5%	75
Polyoxyethylene sorbitan monostearate, 0.5%	42
Sorbitan monostearate, 0.5%	23

Source: Ferrel and Kester (1959).

Verity and Allen (1964) developed a method of canning rice by freezing the rice product following the canning operation. Normally, excess starch forms a glue-like material which prevents the canned rice from being fluid. Freezing breaks down the excess starch and stops the rice from solidifying in the can.

The process of post-can-freezing may be used for both regular and parboiled rice. However, the canned parboiled rice flows with greater ease. In addition, it has a better appearance, stands up better in soups and casseroles, and has better keeping quality in the unused portion after being taken out of the can.

Ghosh and Sarbar (1959) studied the effects of various inorganic salts on water absorption by rice during cooking. Sodium chloride and sodium sulfate, depending on the concentration, either increase or decrease water uptake. Magnesium sulfate increases water absorption, while calcium chloride decreases it; magnesium chloride has little effect. The addition of sodium chloride, sodium sulfate, calcium chloride and magnesium chloride result in a diminished loss of grain material during cooking. Magnesium sulfate does not create this effect.

Kastorykh *et al.* (1971) received a patent in which 0.5% sodium pyrophosphate, 0.002% citric acid, 0.5% sodium glutamate and extracts of juniper and black pepper (10 ml/250 g product) were added to canned rice and meat products. The process improved shelf life, organoleptic properties and food value of the canned product.

Tollefson and Bice (1972) developed a process in which raw rice was cooked in acidified water to cause gelatinization. After canning and sterilization, the cans were allowed to age for several days before consumption of the product.

There are relatively few commercially canned rice products on the market primarily due to the lack of stability of the rice grain and the high cost of agitated retorting equipment. Most canned products which could use rice in the formulation require processing for approximately 60 min at 115.6°C (240°F) or a shorter process at 121.1°C (250°F) in conventional retorting equipment. In the course of processing, a point is reached where the hydrogen bonds responsible for starch integrity are weakened and irreversible swelling occurs. If processing continues, the starch granules will eventually rupture, resulting in the leaching of solids and grain distortion. White rice is less resistant to thermal degradation than parboiled rice in this respect.

The process of promoting cross-linkages in rice starch consists of 3 steps: activation, cross-linking and neutralization. The treated rice will be able to withstand the processing conditions encountered in still retorting while still maintaining the desirable organoleptic properties associated with rice. Rutledge *et al.* (1972, 1974) and Rutledge and Islam (1973, 1976A,B) made cross-linked white rice by treating the kernels with epichlorohydrin. Parboiled rice may be cross-linked with epichlorohydrin, sodium trimetaphosphate or phosphorous oxychloride.

Cross-linking the starch in the intact rice kernel appears to greatly increase the stability of the product during thermal processing. Cross-linked samples show approximately 68% less leaching at pH 7 and approximately 82% less at pH 5, as compared with the untreated samples. The difference in properties between the cross-linked and untreated samples is related to the increased starch granular stability toward thermal degradation and pH extremes. The results are tabulated in Table 17.6. Figure 17.2 illustrates the well defined grains of the cross-

TABLE 17.6. EFFECT OF TREATMENT WITH EPICHLOROHYDRIN ON SOLID LOSS AND ORGANOLEPTIC EVALUATION OF PARBOILED RICE

	Treated		Control	
	pH 7	pH 5	pH 7	pH 5
% solid loss	7.32	6.62	23.14	36.06
Color	4.88	4.78	3.72	3.84
Cohesiveness	4.96	5.00	1.70	1.60
Flavor	4.96	4.78	4.30	4.40
Doneness[1]	3.16	3.18	1.14	1.20

Source: Rutledge and Islam (1973).
[1]In the case of doneness, a score of 3 was excellent, whereas a score of 5 or 1 was considered underdone or mushy, respectively.

linked sample as compared to the control, which shows considerable leaching with longitudinal splitting and fraying of edges and ends of the grains. The control samples also show considerable clumping at the bottom of the can. No clumping occurs in the cross-linked samples. Taste panel evaluations involving color, cohesiveness, flavor, doneness and general appearance indicate that the cross-linked rice is superior in all attributes tested. In addition, the modified rice samples are considerably more stable than the control samples after storage at 25°C (77°F) for 6 months. Hogan (1977) stated that canned rice may be considered a type of quick cooking or convenience product. A brief procedure of evaluating the quality of the products was given for canning milled rice.

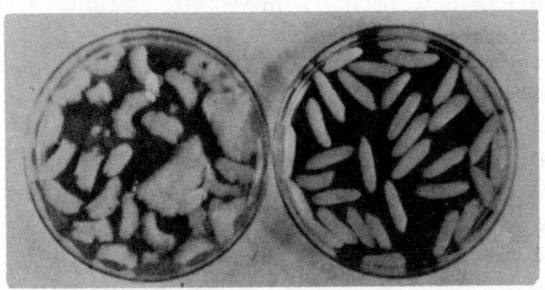

From Rutledge and Islam (1973)

FIG. 17.2. CANNING STABILITY OF EPICHLORO-HYDRIN-TREATED PARBOILED RICE (CONTROL—LEFT; TREATED—RIGHT)

QUALITY EVALUATION OF RICE

The methods and techniques for evaluating rice quality have been published in the U.S. Department of Agriculture Technical Bulletin No. 1331, entitled "Quality evaluation studies of foreign and domestic rices" (USDA 1965). The subject is also discussed in detail by Dr. Webb in this book (Chapter 15).

Texture

A negative correlation was reported to exist between the water uptake ratio (cooked weight divided by weight before cooking) and cohesiveness (Dawson *et al.* 1960). Long grain rice absorbed more water and was less cohesive than either short or medium grain rice. The alteration of rice starch during heat treatment was found to be directly related to the

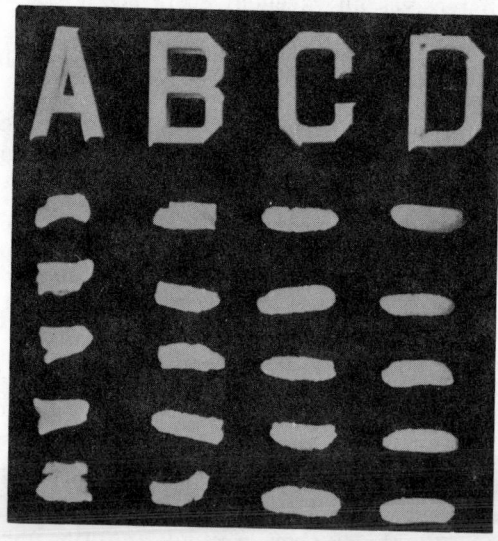

From Demont and Burns (1968)

FIG. 17.3. SUBJECTIVE RATING OF THERMAL PROCESSING QUALITY

A—Poor: B—Fair:C—Good:D—Excellent.

cohesiveness of cooked rice samples (Little and Hilder 1960). The rice with the highest heat alteration values was found to be the most cohesive by a taste panel. Ejlali *et al.* (1978) reported that varietal characteristics and the water-rice ratio are important factors influencing texture of cooked rice.

Wigman *et al.* (1956) utilized the phase contrast microscope to examine rice starch for granule size, stage of hydration during cooking and granule disruption. Their technique has been used to determine the effect of cooking treatment. Similarly, microscopic studies by Little and Dawson (1960) revealed that different patterns of swelling and disruption may be caused by the delaying or limiting effects of nonstarch components.

A subjective rating for kernel breakdown due to the thermal processing was developed by Demont and Burns (1968). The results are shown in Fig. 17.3. Rice kernels which exhibit severe end and longitudinal splitting or more than 5 lateral fissures are considered poor. A score of fair is given when the product is slightly mushy and the grains have more than 2 lateral fissures, including nonsevere longitudinal fissures and split ends. To be rated good the rice should be only slightly mushy with no more than 2 lateral fissures. Rice which responds to the canning process with no more than occasional cracking and no sloughing is scored excellent.

Only rice which receives an excellent score should be considered acceptable for canning purposes.

The L.E.E. Kramer shear press has been proven a valuable tool for objective measurement of rice firmness and of the textural qualities of many foods (Kester 1959; Kramer 1972; Szczesniak and Shinner 1973). At the present time, several large food manufacturers are using the shear press to measure the texture of rice products (Kramer 1972). Voisey (1976) and Voisey and deMan (1976) presented comprehensive reviews on instrumental measurement of food texture.

Composition

Batcher *et al.* (1956) found that the amount of starch in the cooking water is independent of grain type. However, the total solids are significantly lower in the liquid in which long grain varieties have been cooked. Matz and Beachell (1969) reviewed literature on rice production and composition. Halick and Keneaster (1956) reported a positive correlation between soluble amylose content and the quality of rice in cooking tests. The iodine-blue reaction with amylose was used to obtain a spectrophotometric index which relates to the soluble amylose fraction (Juliano 1971). Roberts *et al.* (1954) used the iodine test as a method for determining the severity of treatment of parboiled rice. They theorized that the more severe parboiling treatments enchanced the breakdown of starch granule structure, resulting in formation of more soluble amylose which would give higher iodine values.

A method for evaluating quality of small amounts of breeding material is the alkali digestion test (Little *et al.* 1958). Quality is indicated where kernels exhibit resistance to spreading or when the spread exhibits clear rather than opaque masses.

Color

The Gardner tristimulus color and color difference meter accurately relates differences in color between samples and a calibrated standard plate. The "L" scale measures lightness, the "a_L" scale measures redness when positive and greeness when negative, and the "b_L" scale measures yellowness when positive and blueness when negative. This instrument has been applied to many products including paints, leather, textiles and foods (Mackinney and Little 1962).

Obviously, diverse techniques have been employed in the preparation and evaluation of rice material. Researchers are in agreement that long grain rice generally exhibits more desirable soup canning qualities than the other grain types grown in the United States.

Equipment

The development of new and improved varieties having the parboil canning stability required for use in heat processed formulation is an important part of rice breeding programs (Webb 1976). A laboratory scale parboiling apparatus has been described by Webb and Adair (1970) to aid in evaluating new rice varieties and hybrid selections.

The parboiling apparatus is shown in Fig. 17.4. It consists of a retort chamber, a vacuum pump system, an air compressor, a steam generator and a water heater. The retort chamber is connected directly to the other components and equipped with 3 thermometers to measure water temperature during steeping. Constant steeping temperature is maintained

From Webb and Adair (1970)

FIG. 17.4. LABORATORY RICE PARBOIL-CANNING APPARATUS

1—Water heater. 2—Air line to portable compressor. 3—Vacuum system. 4—Retort chamber. 5—Sample rack and baskets. 6—Steam generator, the horizontal autoclave. 7—Facilitates canning of parboiled rice.

by 3 flexible heating tapes connected to powerstats wound around the chamber. Automatic air and steam pressure controls on the chamber maintain constant conditions during parboiling. Asbestos sheeting is used for insulating the chamber.

RICE IN CANNED SOUPS

Canned condensed soup is one of the products in which rice stability is most important to overall product quality (Hagberg 1966). The rice process (Jones *et al.* 1946) includes, first of all, a precooking or blanching treatment of the rice in boiling water for 15–18 min. The rice, almost completely cooked after the blanching treatment, is blended with the other ingredients and the soup is canned and heat processed. During heat processing the cans may be heated to 121.1°C (250°F) and held for 30 min or longer. During both heat treatments, it is important to remember that there is an excess of water present. The water uptake and expansion of the rice kernels is not restricted by the lack of water. Rice showing kernel instability is not suitable for use in food. Upon heating, the ends of the kernels are split, the surfaces are ragged and a sloughing of fragments increases the turbidity of the surrounding liquid. The desired type of canning stability needed for heat processed, condensed soups will swell to a larger size while maintaining kernel identity. The kernel surfaces are smooth and noncohesive during the processing treatment.

CANNED FRIED RICE

According to Casimir (1970), the preparation of fried rice is a process requiring moisture uptake, heating and a uniform blending of a number of particulate materials. In a true Chinese style fried rice, it is essential that the vegetable ingredients do not receive a thermal process which is so severe that their natural crisp texture is destroyed. However, for the preparation of a product which is to be stored frozen, the thermal process should be adequate to destroy the naturally occurring enzymes which would result in the formation of off-flavor during the storage period.

A Conical Dryer Blancher is a unit in which the following operations may be carried out: (1) blending or mixing, (2) moisture addition, (3) heating or cooking, (4) vacuum steam blanching, and (5) vacuum cooling. The use of a CDB permits all the processes required for the production of fried rice to be carried out in the same vessel.

Equipment

A Pfaudler 61 cm Jacketted Conical Dryer Blancher (Model No. 24–45 CD-S B) with a volumetric capacity of 0.113 m³ may be used (Fig. 17.5).

Procedure

(1) The CDB is preheated by allowing the steam pressure in the jacket to rise to 1.38 kg/cm².
(2) Six kilograms of Calrose rice direct from the bag and having a moisture content of 11.4%, 30 g of distilled monoglyceride (glycerol

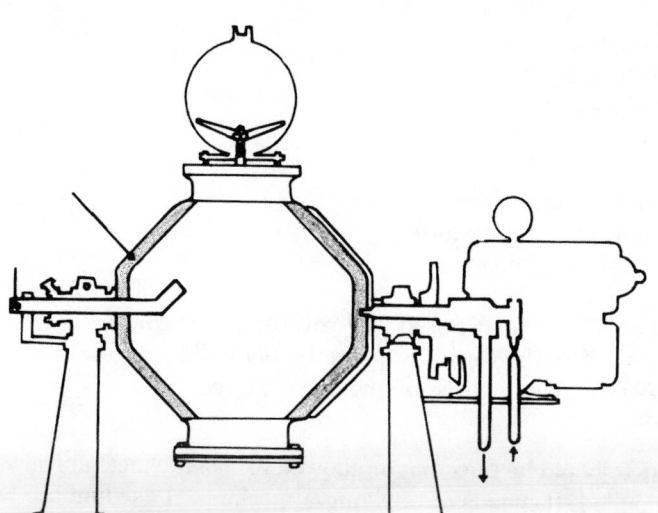

From Casimir (1970)

FIG. 17.5. PFAUDLER 61 CM JACKETED CONICAL DRYER BLAN-
CHER

monostearate), Riken Type T, and 50 g of common salt are preheated and blended in the CDB which is revolving at 5 rpm while
maintaining the steam pressure in the jacket at 1.38 kg/cm².

(3) Rotation of the CDB is stopped and 0.7 kg of maize oil is added and
then mixed in by further rotation to ensure that all rice grains are
covered with a layer of oil.

(4) The steam pressure in the jacket is then released and a vacuum of
71.12 cm Hg is applied to the oil covered rice.

(5) The vacuum is broken by the addition of steam to the vessel. Eighty
seconds are required to change the pressure from a 71.12 cm vacuum
to a positive 1.74 kg/cm² steam pressure. The steam pressure is then
maintained at 1.74 kg/cm². Rotation rate of the CDB is held at 5
rpm during the period.

(6) At the completion of this hydration process the moisture content
has reached 52% and the grains are discrete and free flowing. The
following frozen prepared materials, including lightly blanched vegetables, are then added to the rice in the CDB:

Cooked egg	0.45 kg
Lightly fried bacon	0.34 kg
Cooked prawns	0.11 kg
White onion	0.11 kg
Shallots	0.11 kg
Celery	0.34 kg
Soy sauce	60 g
Saromex celery (dry)	60 g
Monosodium glutamate	150 g
Worcestershire sauce	55 g

(8) The CDB is then closed and rotated at 10 rpm for 3 min to thoroughly mix the contents, partially blanch the vegetables and give a partial pasteurization of the other ingredients.

Filling

The rice should be filled hot or allowed to cool if a mechanical vacuum is to be applied. If means of applying a mechanical vacuum are not available, the rice should be filled as hot as possible immediately after blanching and draining.

The fill-in weight for 301 × 411 cans is 340 g of rice with a moisture content of 55−60%.

Closing

A high vacuum level is required to prevent oxidative browning of the products, and hence a mechanical vacuum of approximately 66 cm Hg is desirable. Hot filling and closing without applied vacuum should give a vacuum level of 25.4 to 38.1 cm Hg after processing and cooling. Discoloration at this vacuum level is present but it is not unduly objectionable, particularly in fried rice packs.

Processing

Can vacuum level, moisture content of the rice and fill-in weight affect the heating characteristics of the rice. The effect of fill-in weight of rice with a moisture content of 55−60% on heating characteristics in a 301 × 411 can is shown in Table 17.7.

The effect of moisture content of the rice in the canning range 54−60% on the heating characteristics of a 301 × 411 can with a 340 g fill-in weight may be expressed as follows:

$$F_h \text{ (min)} = 34 - 0.75 \text{ (Moisture \%}-54)$$

TABLE 17.7. EFFECT OF FILL-IN WEIGHT OF RICE ON HEATING CHARACTERISTICS IN A 301 × 411 CAN

Fill-in weight (g)	F_h from graph of the best fit (min)
255	42.7
284	39.0
312	35.3
340	31.5

Source: Casimir (1970).

This equation can only be applied to this particular can size at this particular fill weight.

The vacuum level does not appreciably affect the F_h value when the fill-in weight and moisture content are held constant.

A commercial process has been evaluated for canned white rice packed under the following conditions:

Can size	301 × 411
Fill-in weight	340 g
Moisture content of rice	55−60%
Initial temperature	15.6°−37.8°C (60°−100°F)

A come-up time of 10−15 min is recommended in order to avoid panelling or collapse of the high vacuum cans. Table 17.8 shows lethal ratio and F_o values for various processing times at 118.3°C (245°F). The recommended processing time is 55 min at this temperature.

TABLE 17.8. THE LETHAL RATIO AND F_o VALUES FOR VARIOUS PROCESSING TIMES AT 118.3°C (245°F)

Time at 118.3°C (245°F) (min)	Lethal ratio	F_o Value
30	0.30	0.82
35	0.78	2.14
40	1.35	3.70
45	2.80	7.67
50	3.60	9.86
55	4.85	13.29

Source: Casimir (1970).

Cooling

Cooling in the retort with cooling water to 37.8°C (100°F) requires approximately 25 min. It should be noted that the cans float in water. Tolson (1973) has described a method in which raw rice is fed into an oil bath at 350°−420°C for 5 to 20 sec, then drained, cooled to room temperature and packed.

RICE PUDDINGS

There is a growing trend toward using the aseptic canning process for rice puddings. The pudding is sterilized and cooled separately from the container, thus avoiding the slow heat penetration problems inherent in the in-container canning process. The sterilized and cooled product is filled into presterilized containers and sealed in a sterile atmosphere with a sterile cover.

The 2 components of the rice pudding, the rice kernels in a small amount of liquid and the sauce, are sterilized individually and then combined in the can (Kester and Matz 1970). This step is necessary because of the different sterilization treatments required by the 2 components. The sauce can be quickly sterilized by swept wall heat exchangers, Spiratherms, Uperizers, or triple-tube heat exchangers. Sterilization time of the rice is much longer, due to the greater time interval required for the heat to completely penetrate the kernels. If the sauce and grains are heated together until the rice is sterilized, there is a tendency for the sauce to become overheated, with excess browning and off-flavors as the result. The pudding may be sterilized at 137.8°C (280°F) for 30–60 sec. At this temperature, a F_o of 20–30 is reached.

Casimir and Lewis (1972) have described the process of flame sterilization which has been highly successful for canned milk rice puddings, as well as canned white and fried rice. The pudding mix is packed into the can, which is immediately closed and run into the heating section. Rapid heating brings the can to a temperature at which a microbiologically adequate (137.8°C, 280°F, for 30–60 sec) thermal process is achieved before rehydration and thickening are completed.

HIGH PROTEIN RICE VEGETABLE MIXTURE

Consumer demand for convenience foods of good quality, low cost and high nutritional value is increasing. Canned plain rice is convenient but less attractive to consumers. It is also known that rice has a low protein content. This low nutritional value is aggravated by the low lysine content of rice proteins.

The nutritive value of a protein eaten alone may be markedly different when included as part of a mixed diet. The value of the protein in a mixed diet depends, in part, not only on its limiting amino acids but also on excesses of other amino acids which may supplement differences existing in other dietary proteins.

The protein content and amino acid composition of rice, beans and textured soybean protein are listed in Table 17.9 (FAO 1970; Cagampang *et al.* 1976). Lysine is the limiting amino acid of the protein in rice and methionine is limiting in beans and textured soybean protein.

Legumes are the most important source of protein for those who do not choose or can not afford protein rich animal foods. Legume proteins can provide certain essential amino acids in which cereal proteins are deficient and thereby enhance the overall nutritive value of the protein in the mixed diet.

Altares and Luh (1976) have formulated and canned a product containing beans and tomato sauce. The product contains vegetable proteins from dry beans and rice, and is high in protein (Table 17.10). It is a low cost convenience food.

Dry beans and textured soybean protein are utilized to complement the rice protein. The use of the spiced sauce makes the product more attractive in flavor acceptance. Starch that leaches from the rice during canning serves as a thickening agent for the sauce and contributes to the body of the product. The textured soybean protein also adds a meat-like texture to the product.

Figure 17.6 is a flow diagram showing the canning procedures used. Calrose rice was soaked in water for 1 hr at room temperature, and then blanched in either boiling water or by steaming for 2 or 3 min. Subsequently, the rice was drained, rinsed and combined with red kidney and garbanzo beans and textured soy protein which had also been soaked previously. The solid ingredients were packed into enameled cans with water or the sauce mix, sealed and then heat processed in still or rotary retorts.

A taste panel indicated that the rice mixture with tomato sauce was more desirable than the product canned with plain water. Unblanched rice kernels tended to break down during processing, producing an unsatisfactory texture. Steam blanching the soaked rice for 3 min produced a better texture than blanching in boiling water.

As compared with the rotary retort processing the still retort produced an inferior product. During processing, the rice sank to the bottom of the can, producing a nonuniform product. The rotary retort yielded a uniform and fully cooked product.

The canned rice product consisted of Calrose rice (medium grain), kidney beans, garbanzo beans, textured vegetable protein and a sauce which was added to improve the nutritive value of the rice. Table 17.11 shows the quantity of prepared raw materials added to each 303 × 406 can. The addition of beans and textured soybean protein to the rice product not only increased the protein content of the canned rice product but also balanced the amino acid profile of the product.

TABLE 17.9. AMINO ACID COMPOSITION OF RICE, BEANS AND TEXTURED SOYBEAN PROTEINS

Essential amino acid (% total protein)	Rice[1] (Oryza sativa)	Kidney beans (Phaseolus vulgaris)	Garbanzo beans (Cicer arietinum)	Textured soybean (Glycine max) protein
% Protein	(9.1)	(22.1)	(20.1)	(50)
Valine	7.0	4.6	4.5	4.6
Leucine	8.5	7.6	7.5	7.7
Isoleucine	4.7	4.2	4.4	4.6
Threonine	3.8	4.0	3.8	4.0
Lysine	4.0	7.2	6.8	5.7
Methionine	2.2	1.0	1.0	1.3
Cystine	2.2	0.8	1.2	—
Cystine + Methionine	4.4	1.8	2.2	2.8
Phenylalanine	5.4	5.2	5.7	5.1
Tyrosine	4.9	2.5	2.9	2.1
Tryptophane	1.2	—	—	1.3

Source: FAO (1970).
[1]Cagampang et al. (1976).

TABLE 17.10. COMPOSITION OF RICE, KIDNEY BEANS, GARBANZO BEANS AND TEXTURED SOYBEAN PROTEIN

Nutrient	Polished rice	Kidney beans	Garbanzo beans	Textured vegetable protein
Moisture	12.0%	10.4%	10.7%	7%
Protein	6.7%	22.5%	20.5%	50%
Carbohydrates	80.4%	61.9%	61.0%	32%
Fat	0.4%	1.5%	4.8%	1%
Ash	0.5%	3.7%	3.0%	7%
Thiamin	4.4 μg/g	5.1 μg/g	3.1 μg/g	4.0 μg/g

Source: Altares and Luh (1976).

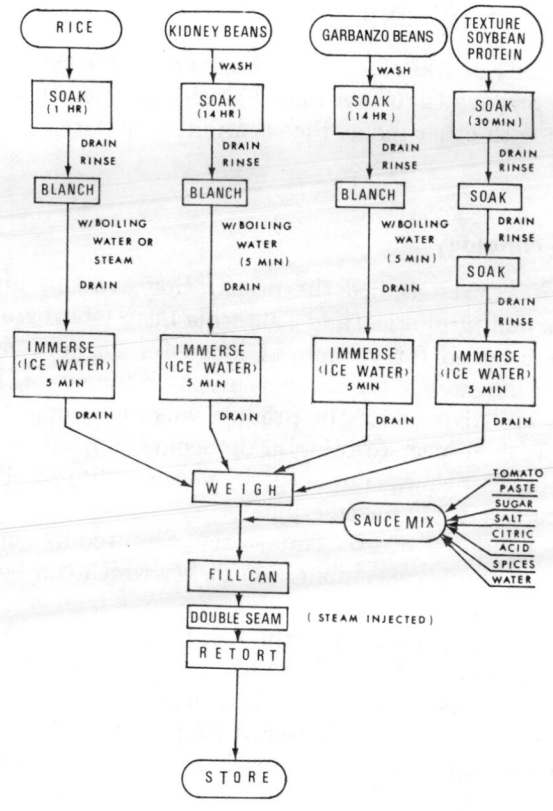

From Altares and Luh (1976)

FIG. 17.6. CANNING FLOW DIAGRAM

TABLE 17.11. QUALITY OF PREPARED RAW MATERIALS ADDED TO EACH 303 × 406 CAN

Prepared material	Moisture (%)	Material wet (g)	Weight dry (g)	Dry basis (%)	Protein Weight (g)
Rice	55.7	70	31.0	7.6	2.4
Kidney beans	63.5	40	14.6	25.1	3.7
Garbanzo beans	58.6	30	12.4	23.0	2.9
Textured soybean protein (Miratex 200)	78.5	40	8.6	60.2	5.2

Source: Altares and Luh (1976).

THE FREEZING OF RICE

Frozen cooked rice, like canned rice, is convenient to use since it requires less time to prepare than raw rice. The rice may be frozen plain or in combination with other foods. Rice is an integral part of Chinese frozen dinners.

Freezing Technology

Boggs et al. (1951) studied the preparation, freezing and storage of Texas Patna (long grain rice) and California Pearl (short grain rice). Both varieties were boiled for 10 min, steamed for 25 min, air cooled and frozen. Following storage the rice was prepared for serving in 10 min. As can be seen in Table 17.12, the product was fully equal to the freshly cooked rice with respect to color, grain separation, flavor and texture. The Texas Patna rice was not as sticky as the California Pearl and was therefore more suitable for freezing.

There are a number of excellent frozen rice products on the market today. Some of these are combination dishes which can be reheated by the boil-in-bag method.

Tressler et al. (1968) reviewed the literature relating to freezing and frozen storage of cooked rice. They quote the following suggestions for commercial processing of frozen cooked rice:

(1) Place rice in an excess of water at 54.4°–60°C (130°–140°F) which contains enough citric acid to reach a pH of 4.0–5.5. Enough water should be used to cover the rice after it has soaked for 2 hr.

(2) After 2 hr, drain off the soak water and rinse with more of the same pH adjusted water to remove fines.

(3) Drain thoroughly, tapping the screen to shake loose the adhering water or by blowing the rice layer with air.

(4) Place a small volume of water in the bottom of a pressure cooker and heat to boiling with the cover on to heat up the apparatus. The

TABLE 17.12. QUALITY OF FROZEN COOKED RICE EVALUATED AT DIFFERENT STORAGE PERIODS

| Rice | Storage period | Average scores[1] | | | | | Least significant difference at 1% level |
| | | Dry rice stored at | | Frozen cooked rice stored at | | | |
		r.t.	−30°F	−10°F	0°F	10°F	
				Color			
Patna	1 wk	6.3	6.4	6.7	6.6	6.7	0.2[2]
Patna	2 mo	5.7	5.9	6.6	6.5	6.7	0.3
Patna	8 mo	6.2	6.7	7.0	7.0	7.0	0.2
Pearl	1 wk	6.2	6.2	6.4	6.5	6.4	0.3
Pearl	2 mo	5.9	6.1	6.2	6.2	6.2	0.3
Pearl	8 mo	6.5	6.6	6.7	6.7	6.7	0.2
				Grain Separation			
Patna	1 wk	6.1	5.8	6.4	6.6	6.4	0.3
Patna	2 mo	6.0	5.8	6.0	6.2	6.1	—
Patna	8 mo	6.2	5.9	6.2	6.2	6.2	—
Pearl	1 wk	5.2	5.2	5.4	5.5	5.2	—
Pearl	2 mo	4.8	4.8	5.5	5.3	5.5	0.3
Pearl	8 mo	5.2	5.1	5.6	5.4	5.2	0.4
				Flavor			
Patna	1 wk	6.7	6.7	6.7	6.7	6.7	6.7
Patna	2 mo	6.6	6.6	6.6	6.7	6.5	—
Patna	8 mo	6.4	6.6	6.5	6.5	6.3	—
Pearl	1 wk	6.6	6.6	6.5	6.7	6.6	0.2
Pearl	2 mo	6.4	6.3	6.5	6.4	6.5	—
Pearl	8 mo	6.6	6.7	6.7	6.7	6.5	—
				Texture			
Patna	1 wk	5.7	5.7	6.3	6.3	6.2	0.4
Patna	2 mo	5.6	5.8	6.4	6.4	6.4	0.4
Patna	8 mo	5.8	6.0	6.2	6.4	6.3	0.5
Pearl	1 wk	5.0	5.1	5.5	5.6	5.3	0.6
Pearl	2 mo	5.3	5.3	5.8	5.7	5.6	0.3
Pearl	8 mo	5.5	5.4	5.8	5.9	5.5	0.4

Source: Boggs *et al.* (1951).
[1]Highest scores indicate best quality, with 7 as maximum.
[2]This least significant difference value of 0.2 means that a difference of 0.2 between 2 scores in this row is probably real and not due to chance.

soaked drained rice is placed in layers 5 cm deep or less over screens which are supported above the water in the vessel. Close the vessel and heat with the vent open until steam is emitted, then close off the vent and raise the pressure to 2.09 kg/cm² and hold for 12 to 15 min. Then blow off steam gradually enough to prevent violent boiling and flashing of the hot water.

(5) Place the hot steamed rice in an excess of hot water at 93.3°−98.9°C (200°−210°F) without stirring. The rice will imbibe water until the

grains are large, tender and quite free. Stirring will cause the rice to become sticky. The rice should be held in a perforated vessel so that water may circulate freely through it. Do not boil the rice.

(6) Cooking should require only 10 to 15 min following the step described in (4). Drain off the hot water and rinse twice with cold water which has the pH adjustment described in (1).

(7) Tap and shake to remove free water or suck off the free water over a vacuum filter.

(8) Convey the cooked rice on a stainless steel mesh belt through an air blast cooler to reduce it to room temperature and then package in cartons or plastic pouches. Freeze the rice in air blast freezers. The rice may be frozen as individually quick frozen (IQF) products prior to packaging in a fluidized bed freezer.

Boiled and steamed white rice which has been frozen and reheated are virtually indistinguishable from their unfrozen counterparts. Frozen storage at $-18.8°C$ ($0°F$) up to 1 year appears to have no deleterious effects on quality (Boggs et al. 1951)

Boggs et al. (1952) did similar studies on brown rice. The long grain (Brown Patna) and short grain (Brown Pearl) rice were boiled for 15 min, then steamed for 50 min, air cooled, packaged and the packaged product frozen at $-23.3°C$ ($-10°F$). After storage for periods of 1 week, 2 months, 6 months and 12 months, the reheated product was scarcely distinguishable from freshly cooked brown rice in any respect.

The methods recommended for preparation, cooking, packaging and freezing on a commercial scale are the same as those suggested by Boggs et al. (1951) for polished rice, except longer periods of time are required both for the boiling and steaming operations.

Miller (1960) developed a process for freezing rice by which small quantities of the product can be removed from the package easily without thawing it entirely. The rice is first cooked by any conventional method until it is ready to be served and then allowed to cool. Good results are obtained if the product is chilled to $0.6°C$ ($33°F$).

Prechilling the rice results in removing most of the surface moisture that may be present after cooking and at the same time permits quicker freezing. Before freezing takes place, the individual grains are separated and maintained out of contact with one another during the freezing part of the process. The product is then frozen solid. Any appropriate freezing temperature may be used but good results have been obtained by subjecting the rice to a moving air blast at $-34.4°C$ ($-20°F$). After the separate grains are solidly frozen, they can be brought together in a mass and packed in any desirable manner. It is of course necessary to maintain the product in a frozen condition until used.

The process lends itself ideally to continuous operation and freezing can be done in a single stage, without the necessity of glazing, agitation or other means to prevent adherence of the separate particles during the freezing operation.

Ragab (1971) has developed a method for producing frozen table rice. The raw rice is roasted for a short time in fat, then fully cooked with the correct quantity of water, packed in polyethylene bags and frozen. The rice is heated before serving.

Frozen Fried Rice

Casimir (1970) has utilized a conical dryer blancher for preparing frozen fried rice. The initial procedure has been described in the "Canning" section of this chapter. After step 7, the CDB is cooled by flowing water through the jacket and reapplying the vacuum to cool the rice. During the vacuum cooling, water is sprayed in at approximately 300 ml per min to partially replace the water lost by evaporation. Vacuum cooling reduces the temperature from 98.9° to 34.4°C (210° to 94°F) in 2 min under 71.12 cm vacuum.

Characteristics of Frozen Rice

Raw rice consists of firm, dense grains comprised of tightly packed cells filled largely with granules of raw starch. At the low moisture content (8 to 13%) at which the grains are stored, they are hard and vitreous in appearance and able to absorb water only slowly. When the kernels are put in water, their generally solid nature serves to retard the penetration of water into their inner areas. Although cracks or fissures may be present, they are insufficient in making the grains more hydratable. Roseman (1958) conducted a study to determine whether an increase in the porosity and ability to absorb water could be brought about by subjecting rice to freezing. Raw Rexoro rice was hydrated by 1 of 2 methods: (1) rice was hydrated in excess water at 86°C (184.8°F) and 100°C (212°F) for 5−20 min; and (2) rice was soaked for 10 min, steamed in an autoclave at 2.09 kg/cm^2 121°C (249°8°F) for 15 min, then resoaked in water for 50 min. Freezing was done in an alcohol bath at −25°C (−13°F). Drying was done in a forced air oven. For water uptake measurements, the dried samples were placed in a water bath at 100°C (212°F) for the desired length of time, followed by drying in a forced air oven at 104°C (219.2°F). Table 17.13 shows the effect of the processing conditions on the apparent specific volume of the rice.

Rice grains that had been cooked, frozen and dried were chalky, much larger than the unfrozen control and uniformly spongy. The change in structure was induced when the rice was hydrated to above 60% water

TABLE 17.13. APPARENT SPECIFIC VOLUME OF DRIED UNCOOKED AND COOKED MILLED RICES

Sample	Vol[1] (ml)	Vol of 1 g (ml)
Uncooked (Rexoro)	3.8	1.3
Parboiled[2]	3.8	1.3
Cooked[3]	4.6	1.5
Quick cooking, sample A[2]	6.8	2.3
Cooked and frozen	10.0	3.3
Quick cooking, sample B[2]	10.4	3.5

Source: Roseman (1958).
[1]To obtain the vol, 3 g of sample was placed in a 25 ml graduated cylinder (Corning No. 3022) and tapped until the sample was free of large voids.
[2]Commercial long grain products; variety of rice unknown.
[3]Soaked 30 min; treated 10 min at 15 psig; resoaked 60 min.

content by the conditions shown in Table 17.14, frozen, held at 1.7°C (35°F) and dried. Samples with less than this moisture content during freezing, appeared progressively less chalky, less spongy and more vitreous as the moisture content was decreased. All chalky samples took up water rapidly upon immersion, but only those which were completely gelatinized tasted fully cooked when reconstituted. The other samples had an unpleasant grittiness and tasted slightly raw or starchy. The method and degree of hydration prior to freezing also affected the volume and appearance of the product.

A study of conditions of freezing necessary to produce optimum conversion of structure indicates that chalky or spongy grains were obtained when the rice was frozen to −6.7°C (20°F) or lower, followed by slow

TABLE 17.14. EFFECT OF HYDRATION CONDITIONS ON PROPERTIES OF FREEZING RICE

Sample No.	Hydration conditions		Moisture content (wet basis) (%)	Chalky grains (%)
	Time (min)	Temp. (°C)		
1	20	100	77	90
2	15	100	72	95
3	10	100	65	60
4	5	100	56	30
5	20	86	68	95
6	15	86	65	99
7	10	86	61	99
8	5	86	51	99
9[1]	10	25		
	15	121 (steam 2.09 kg/cm²)		
	50	25	61	95

Source: Roseman (1958).
[1]Sample 9 received 3 successive treatments as indicated.

thawing (16 hr) at 1.7°C (35°F). The appearance and rehydration characteristics of rice treated by this method may be controlled by varying the conditions of freezing and/or rehydration.

Freezing rice in the presence of calcium nitrate, a swelling agent, caused an inhibition in the alteration of the properties mentioned above. Different rice varieties containing various amounts of amylose showed different changes after freezing. A waxy rice showed the least change.

Further studies by Roseman and Deobald (1959) indicate that retrogradation is induced by freezing the rice. Ungelatinized starch (raw rice) is resistant to amylase activity and gelatinization increases its susceptibility. The behavior of raw and cooked rice is in agreement with this concept. However, the freeze processed rice, which has also been gelatinized, is resistant to β-amylosis. Freeze processed rice also gives a β-type x-ray diffraction pattern, which is characteristic of retrograded starches. The raw rice and cooked rice give A- and V-type x-ray diffraction patterns, respectively.

FREEZE–DRYING

The freeze-drying process consists of removing moisture from foods at the frozen state by sublimation under high vacuum. The low temperature used in the process inhibits undesirable chemical and biochemical reactions and minimizes the loss of volatile aromatic compounds. The dried product is light in weight and can be stored in airtight containers for long periods without refrigeration. For institutional feeding and for markets where low temperature facilities are absent, freeze-dried foods are even more desirable than frozen or dehydrated foods.

Bird (1963, 1964) estimated the yearly growth of freeze-drying industry in the United States from 1962 through 1970. In 1967, only 20 freeze-drying plants were in operation. The production increased from 40.8 million kg of freeze-dried products in 1967 to 113.4 million kg in 1970. Because of the high vacuum and low temperature requirements, the freezing-drying process must have high vacuum equipment freezing facilities. It is therefore an expensive method of dehydration. But for highly valued items, freeze-drying is a desirable process yielding high quality products.

Literature on freeze-drying of foods has been given by Harper and Tappel (1957), Fisher (1962), Luh (1962), Cotson and Smith (1963), Burke and Decareau (1964), King (1970), Rey (1966, 1972), Goldblith *et al.* (1963), Warman and Reichel (1970) and Ziemba (1960).

General Principles

In freeze-drying, the food is frozen first. The temperature is maintained below the triple point of the constituent aqueous solution so that water vapor can be sublimated from the frozen solution. There is a direct transformation from the solid to a vapor state without passing through the liquid phase. The heat required to sublime a given quantity of ice at a temperature is equivalent to the heat of fusion of ice and the heat of vaporization of water plus the heat necessary to raise the temperature of the ice to its melting point. The quantity of heat required is the same whether the process is carried out slowly at ordinary pressures or rapidly under a high vacuum. The vapor pressure of ice increases as the temperature is raised. The higher the temperature is, the faster the drying process and the lower the cost will be. Obviously, the food should not reach 0°C (32°F) until nearly all of the water has been removed because ice melts at this temperature. It is necessary to remove the vapor which evolves when drying at very low pressure. This can be done either by condensation, pumping or by absorption with a desiccant. In order to condense the water vapor, the temperature of the condensing medium must be below that of the frozen product being dried. This is costly because of the amount of heat to be extracted from the vapor and the relatively low efficiency of the refrigeration machines when operating at low temperatures.

The apparent activation energy of water-removal rates during the falling-rate stages of drying is in the range of 7–12 kcal/g mole, corresponding closely to the latent heat of vaporization of water (10 kcal/g mole) which characterizes the dependence of vapor pressure upon temperature. The apparent activation energies of nonenzymatic browning reactions are close to 30 kcal/g mole. Because of the higher activation energy for the browning reaction, the amount of browning for a given amount of dehydration is much less at lower temperatures. For example, lowering the temperature from 40° to 10°C (104° to 50°F) reduces the rate of water removal by a factor of 5.6, whereas the rate of browning is lowered by a factor of 176 (King 1970).

At sufficiently low temperature and low partial pressure of water vapor, drying will occur by sublimation rather than evaporation. Freeze-drying causes very little shrinkage of the product and consequently allows nearly complete product rehydration. This lack of shrinkage results from the absence of a rigid rice structure at the location where sublimation occurs. The absence of free liquid water during freeze-drying is a definite advantage because there will be no migration of dissolved solutes carried by capillary liquid flow, and no spattering or undesired frothing resulting from liquid entrainment in the water vapor.

Application to Rice

Blanchaud (1972) has developed a method for making instant dishes based on starch products; this is applicable to rice. Pasta, rice or a flour-based food is cooked with water and then kept at 40°–70°C (104°–167°F) for 5–60 min. While cooking, the product is mixed with 15–100% by weight of a hydrogenated fatty substance. After absorbing the fatty substances, the starchy food is then freeze-dried. The product is easily rehydratable.

Huber (1972) has gelatinized rice, subjected it to 2 or more freezing-thawing cycles and then freeze-dried it. Freeze-dried rice and curry mixes have been prepared in India by Bhatia and Nath (1971). Because of its higher cost of production, freeze-dried rice may not be able to compete with quick cooking rice.

REFERENCES

ALTARES, R.A. and LUH, B.S. 1976. Thermo-processing of canned mixture of rice, beans, and textured vegetable protein. Ppr. presented at the Rice Tech. Working Group Meeting. Lake Charles, Louisiana.

BATCHER, O.M., HELMINTOLLER, K.F. and DAWSON, E.H. 1956. Development and application of methods for evaluating cooking and eating quality of rice. Rice J. *59* (13) 4–8.

BEACHELL, H.M. and HALICK, J.V. 1956. Research being done to improve milling, processing and cooking qualities of rice. Rice J. *59* (9) 20–22.

BHATIA, B.S. and NATH, H. 1971. Some technological aspects of defense food research in India during the sixties. Indian Food Packer *25* (1) 59–65.

BIRD, K.M. 1963. Freeze-dried foods. Palatability tests. U.S. Dep. Agric.— Mrktg Res. Rept. No. *617.*

BIRD, K.M. 1964. Selected writings on freeze-drying. U.S. Dep. Agric. *ERS-147.*

BLANCHAUD, M. 1972. Manufacture of instant dishes based on starch products. French Patent 2,088,695. June 22.

BOGGS, M.M., SINNOTT, C.E., VASAK, O.R. and KESTER, E.B. 1951. Frozen cooked rice. Food Tech. *5* (6) 230-232.

BOGGS, M.M., WARD, A.C., SINNOTT, C.N. and KESTER, E.B. 1952. Frozen cooked rice. II. Brown rice. Food Tech. *6* (8) 53–54.

BURKE, R.F. and DECAREAU, R.V. 1964. Recent advances in the freeze-drying of food products. *In* Advances in Food Research. Vol 13. C.O. Chichester, G.F. Stewart and E.M. Mrak (Editors). Academic Press, New York, London.

BURNS, E.E. 1972. Canned rice foods. *In* Rice: Chemistry and Technology. D.F. Houston (Editor). Amer. Assoc. Cereal Chem. St. Paul, Minnesota.

CAGAMPANG, G.B., PERDON, A.A. and JULIANO, B.O. 1976. Changes in salt-soluble protein of rice during grain development. Phytochem. *15*, 1425–1429.

CASIMIR, D.J. 1970. Frozen fried rice. *In* Food Preservation Rept No. *25:* Fried Rice Preparation in Conical Dryer Blancher. Commonwealth Sci. and Ind. Res. Organ., Australia.

CASIMIR, D.J. and LEWIS, P.S. 1972. Product formulation specifically for flame sterilization. *In* Specialist Courses for the Food Industry, No. *2.* Australia.

COTSON, C. and SMITH, D.B. 1963. Freeze-drying of food stuffs. Columbus Press, Manchester and London.

DAWSON, E.H., BATCHER, O.M. and LITTLE, R.R. 1960. Cooking quality of rice. Rice J. *63* (5) 16–22.

DEMONT, J.I. and BURNS, E.E. 1968. Effects of certain variables on canned rice quality. Food Tech. *22* (9) 1186–1188.

DESIKACHAR, H.S.R. and SUBRAHMANYAN, V. 1960. The relative effects of enzymatic and physical changes during storage on the culinary properties of rice. Cereal Chem. *37,* 1–8.

EJLALI, M., LUH, B.S. and MICKUS, R.R. 1978. Physicochemical and cooking characteristics of flour rice varieties. Ppr. presented at the Rice Tech. Working Group Meeting. Feb. 14–16. College Station, Texas.

FEILLET, P. and ALARY, R. 1975. Effects of amylose content and parboiling conditions on canned rice characteristics. Rice Rept., 1975. Int. Union of Food Sci. and Tech., Spain.

FERREL, R.E. and KESTER, E.B. 1959. Reduction of cohesion in canned pearl rice by use of edible oil emulsion and surfactants. Food Tech. *13* (8) 473–474.

FERREL, R.E., KESTER, E.B. and PENCE, J.W. 1960. Use of emulsifiers and emulsified oils to reduce cohesion in canned white rice. Food Tech. *14* (2) 102–105.

FISHER, F.R. 1962. Freeze-drying of foods. Proc. Conf. on Freeze-drying of foods, April 12–14, 1961. Chicago, Illinois. Natl. Acad. Sci., Natl. Res. Counc., Advis. Board on Quartermaster Res. and Develop., Washington D.C.

FOOD AND AGRIC. ORGAN. U.N. (FAO). 1970. Amino acid content of foods and biological data on proteins. Nutritional Studies No. *24.* FAO, United Nations, Rome.

GALLENKAMP, N.B. 1952. Process for canning rice. U.S. Pat. 2,616,810 November 4.

GHOSH, B.P. and SARBAR, N. 1959. Effect of some inorganic salts on water absorption by rice during cooking. Ann. Biochem. Exp. Med. Calcutta, *19,* 83.

GOLDBLITH, A.S., KAREL, M. and LUSK, G. 1963. The role of food science and technology in the freeze-dehydration of foods. Food Tech. *17,* 139–144.

HAGBERG, E.C. 1966. Canned rice products. Proceedings Natl. Rice Util. Conf. New Orleans, Louisiana. U.S. Dep. Agric *ARS 72–53,* Jan., 1967.

HALICK, J.V. and KENEASTER, K.K. 1956. The use of a starch iodine-blue test as a quality indicator of white milled rice. Cereal Chem. *33,* 315–320.

HARA, T. 1950. U.S. Pat. 2,495,001. January 17.

HARPER, J.C. and TAPPEL, A.L. 1957. Freeze-drying of food products. In: Advances in Food Research. Vol. 7. E.M. Mrak and G.F. Stewart (Editors). Academic Press, New York and London.

HOGAN, J.T. 1977. Rice and rice products. *In* Elements of Food Technology. N.W. Desrosier (Editor). AVI Publishing Co., Westport, Conn.

HUBER, C.S. 1972. Freeze-dried rice. U.S. Pat. 3,692,533. September 19.

JONES, J.W., ZELENY, L. and TAYLOR, J.W. 1946. Effect of parboiling and related treatments on the milling nutritional and cooking quality of rice. U.S. Dep. Agric. Circ. *752.* Washington, D.C.

JULIANO, B.O. 1971. A simplified assay for milled-rice amylose. Cereal Sci. Today. *16* (10) 334–340, 360.

JULIANO, B.O., BAUTISTA, G.M., LUGAY, J.D. and REYERS, A.C. 1964. Studies on the physiochemical properties of rice. J. Agric. Food Chem. *12* (2) 131–138.

JULIANO, B.O., ONATE, L.J. and DEL MUNDO, A.M. 1965. Relation of starch composition, protein content, and gelatinization temperature to cooking and eating qualities of milled rice. Food Technol. *19,* 116.

KASTORYKH, M.S. *et al.* 1971. Method for production of canned meat-and-vegetable products. USSR Pat. 310, 639. August 9.

KESTER, E.B. 1959. Rice processing. *In* the Chemistry and Technology of Cereals as Food and Feed. S.A. Matz (Editor). AVI Publishing Co., Westport, Ct.

KESTER, E.B. and MATZ, S.A. 1970. Rice processing. *In*: Cereal Technology. S.A. Matz (Editor). AVI Publishing Co. Westport, Conn.

KING, C.J. 1970. Freeze-drying of food stuffs. CRC Critical Rev. Food Tech. *1,* 379–451.

KRAMER, A. 1972. Texture—its definition, measurement and relation to other attributes of food quality. Food Tech. J. *26,* 34–36, 38–39.

KRUGER, L.H. and MURRAY, R. 1976. Starch texture. *In* Rheology and Texture in Food Quality. J.M. deMan *et al.* (Editor). AVI Publishing Co., Westport, Conn.

LITTLE, R.R. and DAWSON, E.H. 1960. Histology and histochemistry of raw and cooked rice kernels. Food Res. *25,* 611.

LITTLE, R.R. and HILDER, G.B. 1960. Differential response of rice starch granules to heating in water at 62°C. Cereal Chem. *37,* 456–463.

LITTLE, R.R., HILDER, G.B. and DAWSON, E.H. 1958. Differential effect of dilute alkali on 25 varieties of milled white rice. Cereal Chem. *35,* 111–126.

LUH, B.S. 1962. Freeze-drying of foods. Fruchtsaft-Ind. *7,* 310–316. (German)

MACKINNEY, G. and LITTLE, A.C. 1962. Color of Foods. AVI Publishing Co., Westport, Conn.

MATZ, S.A. and BEACHELL, H.M. 1969. Rice. *In*: Cereal Science. S.A. Matz (Editor). AVI Publishing Co., Westport, Conn.

MILLER, C.R. 1960. Freeze-drying of rice. U.S. Pat. 2,938,802. May 31.

RAGAB, J.T. 1971. Frozen table rice. German Pat. 1,767,917. June 2.

REY, L. 1966. Advances in Freeze-drying. Hermann, St. Germain, Paris France.

REY, L. 1972. Recent developments in freeze-drying and cryogenics. *In* Nestle Research News. Nestle Products Tech. Assistance Co., Lausanne, Switzerland.

ROBERTS, R.L., HOUSTON, D.F. and KESTER, E.B. 1953. Process for canning white rice. Food Tech. 7 (2) 78–80.

ROBERTS, R.L., POTTER, A.L., KESTER, E.B. and KENEASTER, K.K. 1954. Effect of processing conditions on the expanded volume, color, and soluble starch of parboiled rice. Cereal Chem. 31, 121–129.

ROSEMAN, A.S. 1958. The effect of freezing on the hydration characteristics of rice. Food Tech. 12, 464–468.

ROSEMAN, A.S. and DEOBALD, H.J. 1959. Effect of freeze-processing on amyloclastic susceptibility, crystallinity and hydration characteristics of rice. Agric. Food Chem. 7 (11) 774–778.

RUTLEDGE, J.E. and ISLAM, M.N. 1973. Canning and pH stability of epichlorophydrin-treated parboiled rice. J. Agric. Food Chem. 21 (3) 458–460.

RUTLEDGE, J.E. and ISLAM, M.N. 1976A. Improved shelf-life at room temperature of canned rice modified by cross-linking. Cereal Chem. 53 (5) 862–866.

RUTLEDGE, J.E. and ISLAM, M.N. 1976B. Limited moisture canning of rice—effects of cross-linking. Cereal Chem. 53 (5) 982–987.

RUTLEDGE, J.E., ISLAM, M.N. and JAMES, W.H. 1972. Improving canning stability by chemical modification. Cereal Chem. 49 (4) 430–436.

RUTLEDGE, J.E., ISLAM, M.N. and JAMES, W.H. 1974. Improved canning stability of parboiled rice through cross-linking. Cereal Chem. 51 (1) 46-51.

SHIBUYA, N., IWASAKI, T. and CHIKUBU, S. 1977A. Role of the fatty acids in the changes of rheological properties of cooked rice and its paste during storage of rice. J. Jpn. Soc. Starch Sci. 24 (3) 67–68.

SHIBUYA, N., IWASAKI, T. and CHIKUBU, S. 1977B. On the changes of rice starch during storage of rice. J. Jpn. Soc. Starch Sci. 24 (3) 55–58.

SRIPATHY, N.V., BALIGA, B.R. and LABIRY, B.L. 1960. A note on the canning of rice. Ind. Food Packer. 14 (9) 11.

SZCZESNIAK, A.S. and SHINNER. 1973. Meaning of texture words to the consumer. J. Texture Stud. 4, 378–384.

TOLLEFSON, C.I. and BICE, C.W. 1972. Canned rice. U.S. Pat. 3,647,486. March 7.

TOLSON, R.C., Sr. and TOLSON, R.C., Jr. 1973. Fried rice product. Canadian Pat. 936,040. June 8.

TRESSLER, D.K., VAN ARSDEL, W.B. and COPLEY, M.J. 1968. The Freezing Preservations of Foods. 4th edition. Vol. 4. AVI Publishing Co., Westport, Conn.

U.S. DEP. AGRIC. (USDA). 1965. Quality evaluation studies of foreign and domestic rices. Agric. Res. Serv. Tech. Bull. No. *1331.*

VERITY, N.S. and ALLEN, R.C. 1964. Method of canning rice. U.S. Pat. 3,132,030. May 5.

VOISEY, P.W. 1976. Instrumental measurement of food texture. *In*: Rheology and Texture in Food Quality. J.M. deMan *et al.* (Editor). AVI Publishing Co., Westport, Conn.

VOISEY, P.W. and DEMAN, J.M. 1976. Application of instruments for measuring food texture. *In*: Rheology and Texture in Food Quality. J.M deMan *et al.* (Editor). AVI Publishing Co., Westport, Conn.

WARMAN, K.G. and REICHEL, A.J. 1970. Development of a freeze-drying process for food. Chem. Eng. *283,* 134−139.

WEBB, B.D. 1976. Evaluation of rice milling, cooking and processing qualities. *In* Six Decades of Rice Research in Texas. Texas Exper. Stat. Res. Monograph No. *4.*

WEBB, B.D. and ADAIR, C.R. 1970. Laboratory parboiling apparatus and methods of evaluating parboil-canning stability of rice. Cereal Chem. *47,* 708−714.

WIGMAN, H.B., LEATHEN, W.W. and BRACKENBAYER, M.J. 1956. Phase contrast microscopy in examination of starch granules. Food Tech. *10* (4) 179−184.

WILLIAMS, V.R., WEI-TING, W.A. and TSAI, H. 1958. Varietal differences in amylose content of rice starch. J. Agric. Food Chem. *6,* 47−48.

ZIEMBA, J.V. 1960. Freeze-drying. Food Eng. *32* (12) 57−64.

18

Breakfast Rice Cereals and Baby Foods

Bor S. Luh
Amara Bhumiratana

Rice is one of the most important cereals in the world. Most people in Asia and tropical and subtropical countries use rice as a major staple food. Almost all cultivated rice plants belong to *Oryza sativa* L. which was originated in Asia. They are divided roughly into 2 subspecies, Indica and Japonica. Between these two typical subspecies rice has differentiated into many kinds of ecotypes according to local conditions. Each ecotype has a different response to day length, temperature, soil fertility and water supply. Accordingly, different ecotypes of rice are cultivated in a manner corresponding to the soil properties of the paddy fields, growth seasons, climatic conditions and the methods of cultivation.

In general, Indica rice is mainly grown in tropical and subtropical zones, and Japonica rice in temperate zones and mountainous regions. However, high yielding varieties adapted to the tropics can be obtained in both the Indica and Japonica groups (Oka 1975).

In this chapter, methods for the preparation and utilization of rice as breakfast cereals and baby foods are presented.

CHARACTERISTICS OF RICE FOR PROCESSING

The rice varieties grown in the United States are classified as short (Pearl), medium and long grain types which associate with specific cooking and processing characteristics (Webb *et al.* 1972). Raw milled kernels of long grain varieties, frequently called "hard rice," usually cook dry and fluffy and the cooked grains tend to remain separate. The long grain rice is used more for canning and freezing of precooked rice. On the other hand, high quality short and medium grain varieties, called "soft rice," cook moist and firm and the cooked grains tend to stick or clump together. They are used for making puffed rice and parboiled rice.

All 3 grain types with their characteristic textural qualities are in widespread demand by the domestic and foreign trade because the different ethnic groups prefer the various textures in home cooked rice. Processors of rice also require all grain types and textural qualities for use in various kinds of prepared and convenient type rice containing products such as dry breakfast rice cereals, parboiled rice, quick cooking rice, canned rice, canned soups, dry rice soup mixes, baby foods and frozen dishes. Annually in the United States a substantial and increasing amount of the domestic rice crop is processed and reprocessed into numerous kinds of prepared products. There is a strong demand for the broken grade of rice in brewing and rice flour is used in various prepared mixes. In many of these processed and convenience foods, the textural qualities and grain size of long grain varieties are preferred, while in other foods, the qualities of the short and medium grain types are required for their specific uses. Hence, the domestic and world trade associates United States long, medium, and short grain rice with certain specific cooking and processing characteristics. For this reason, new rice varieties must have the same (or improved) milling, cooking and processing characteristics as the established varieties they replace.

Some comparative chemical and physical characteristics of selected rice varieties in the United States are presented in Table 18.1 (Webb *et al.* 1972).

The amylose/amylopectin ratio in milled rice greatly affects the texture and fluffiness of cooked rice. The ranges of amylose, amylopectin and the amylose/amylopectin ratios of the starch in milled rice in the United States are presented in Table 18.2 (Simpson *et al.* 1965). The long grain rice has a higher amylose/amylopectin ratio. The grains remain separate after cooking, and have a firmer texture than the medium and short grain rice.

Variation in puffing, popping and expansion properties of short, medium and long grain rices may also be a matter of the shape and physical structure of the kernels. Juliano (1970) indicated that the physical structure of the rice kernel, such as the "hardness" characteristic, will also influence expansion properties of rice during processing. Hardness appears to be related to protein content. There is presumptive evidence that a level of amylopectin similar to that in short grain rice (Pearl) is necessary for optimum expansion, with medium grain rice (Calrose) in an intermediary position. High amylose appears to result in less expansion.

Mottern *et al.* (1967) studied the differences in popping characteristics of rice varieties. The yield of popped cereal from long grain rice (Blue Bonnet) was ¾ that of Mochi-Gome, a short grain waxy rice which is the acceptable popping variety in the Orient. Rice with low amylose/amylo-

TABLE 18.1. COMPARATIVE CHEMICAL AND PHYSICAL CHARACTERISTICS OF SELECTED RICE VARIETIES IN THE UNITED STATES[1]

Variety	C.I.[2] No. 2	Milled Rice[3]								Parboiled Rice
		Amylose content (%)	Alkali reaction (avg.)	Water uptake at 77°C (ml/100g)	Gelatinization temperature (BEPT) (°C)	Amylograph viscosity			Crude protein (N × 5.95) (%)	Parboil canning dry matter (solids) loss (%)
						Peak	After 10 min at 95°C	Cool to 50°C		
						← Brabender Units →				
LONG GRAIN										
Typical cooking varieties:										
Belle Patna (standard)	9433	23.8	3.6	130	73	820	410	790	6.9	19
Bluebelle	9544	24.1	3.7	124	74	820	430	860	6.7	18
Bluebonnet 50	8990	23.7	4.1	136	72	815	400	770	6.8	20
Dawn	9534	25.1	3.9	130	73	765	415	830	6.9	21
Labelle	9708	24.8	3.5	125	73	820	410	810	6.3	19
Starbonnet	9584	25.0	3.4	122	72	840	400	780	6.1	20
Vegold	9385	24.6	3.7	127	74	850	430	820	6.7	20
Della (aromatic type)	9485	23.6	3.9	128	72	800	410	780	6.9	21
Nontypical cooking varieties:										
Century Patna 231	8893	16.2	2.2	97	79	1040	430	740	6.0	36
Toro	9013	16.5	6.7	399	66	915	380	660	6.3	35
MEDIUM GRAIN										
Typical cooking varieties:										
Calrose	8988	19.0	7.0	340	65	960	400	680	6.3	33
CM-M3	9675	18.4	6.8	332	67	980	420	750	6.2	33
Earlirose	9672	18.6	6.4	350	66	840	430	730	6.8	32

Kokaho Rose	9673	18.2	7.0	310	68	880	430	710	6.6	34
Nato (standard)	8998	15.3	6.3	316	67	940	410	720	6.4	34
Nova 66	9481	15.8	6.3	320	66	960	390	720	6.3	36
Saturn	9540	15.2	6.2	300	68	970	420	760	6.5	31
Vita	9628-2	17.0	6.5	340	65	890	370	740	6.4	32
Nontypical cooking variety:										
Early prolific	5883	14.8	2.6	92	78	1020	390	700	6.2	34
SHORT GRAIN										
Typical cooking varieties:										
Caloro (standard)	1561-1	19.0	7.0	360	66	850	390	690	6.0	32
Colusa	1600	18.0	6.5	350	66	840	400	690	5.9	30
CS-S4	9835	18.4	7.0	320	67	820	370	680	6.0	33
Nortai	9836	18.8	6.7	310	67	870	380	680	6.2	31

Source: Webb et al. (1972).

[1] Results of tests conducted at the Regional Rice Quality Laboratory, Beaumont, Texas. Values are averages of representative samples grown in the Uniform Rice Performance Nurseries and other trials in Arkansas, Louisiana, Mississippi and Texas. Data on samples grown in California are not included. Data taken in part from Adair et al. (1966); Bollich et al. (1966, 1972); Johnston et al. (1963, 1966, 1967); and Webb (1967A, 1967B).

[2] C.I. refers to the accession number used in the Plant Genetics Research Laboratory, Plant Genetics and Germplasm Institute, Agricultural Research Service, USDA.

[3] Milled rice moisture content was 11.5%.

TABLE 18.2. STARCH CONTENT OF MILLED RICE IN THE UNITED STATES

	No. of Samples	Amylose		Amylopectin		Amylose:Amylopectin Ratio	
		Mean	Range	Mean	Range	Mean	Range
Long grain	6	22.65	17.90−25.18	65.67	62.30−71.55	0.347	0.250−0.402
Medium grain	5	18.66	14.16−23.77	70.49	65.94−73.35	0.267	0.193−0.360
Short grain	2	20.94	20.70−21.17	69.86	69.28−70.45	0.300	0.294−0.305

Source: Simpson *et al.* (1965).

pectin ratios generally has lower gelatinization temperatures than those with higher ratios, and on cooling, their starch pastes show considerably less retrogradation than rices with high ratios.

Waxy or glutinous rice has been used in large quantities for making snack foods in Taiwan and Japan. The product is called Arare or Senbei (larger disc form). The details for making such products are presented in Chapter 20 of this book. The waxy rice is useful in the production of oven-expanded cereals. Waxy rice differs from common rice in that the grains contain essentially no amylose, are opaque and tend to disintegrate on conventional cooking. It is preferred in the Orient for preparation of desserts. In the Philippines, waxy rice is used for making "pinnipig." The rice is first cooked until the starch is partially gelatinized and the cooked grains are flattened in a wooden mortar-like bowl with a wooden pestle. The flattened grains are then heated in a pan with a small amount of vegetable oil until the flakes become puffed into a product that closely resembles oven-expanded cereal. Waxy rice has been used in the United States as a thickener for sauces and gravies and as a tenderizing agent in frozen foods.

RICE BREAKFAST CEREALS

Rice breakfast cereals may be divided into 2 classes: those that require cooking before serving and those that are ready to eat directly from the package. Rice cereals on the market that require cooking are the "farina" type made from granulated white milled rice. The ready-to-eat rice cereals may be made from the entire rice grain or its milled products as well as fabricated rice products in the form of doughs cooked in various ways.

The use of rice in prepared cereals, both alone and in combination with products of other cereal grains, has assumed considerable importance in recent years. Rice not only imparts its special flavor, but also contributes to the texture-modifying properties in formulation and processing. Hogan (1967, 1977) reviewed the literature on the use of rice in cereals in which rice, alone or in combination with other cereal grain products, is precooked, dried, flaked or formed into doughs, then expanded or puffed and toasted. The cooking time, steam pressure, temperature of the basic materials and toasting procedure greatly influence the quality of the final product. Flavoring materials, vitamin B complexes, minerals and protein containing ingredients may be added to improve the nutritive value. There are many kinds of rice breakfast cereals on the market. Many variations of the manufacturing processes are applied to make more nutritious and attractive ready-to-eat breakfast cereals.

Puffed Rice

The puffing processes may be divided into 2 types: (1) atmospheric pressure procedures which rely upon the sudden application of heat to obtain the necessary rapid vaporization of water, and (2) pressure-drop processes which involve suddenly transferring superheated moist particles into a space at lower pressure. In the latter case, the pressure-drop may be achieved by releasing the seal on a vessel containing a product which has been equilibrated with high temperature steam, or it may be secured by transferring the hot material from the atmosphere into an evacuated chamber. The former process is much more widely used (Matz 1970; Hogan 1977).

The puffing phenomenon results from the sudden expansion of water vapor (steam) in the interstices of the granule. The particle is fixed in its expanded state by the dehydration resulting from the rapid diffusion of the water vapor out of it. Gun puffing may result in an increase of apparent volume (bulk density decrease) of sixfold to eightfold. Oven puffing causes a lesser increase, about threefold to fourfold.

Puffed products must be maintained at about 3% moisture in order to achieve the desired crispness. The moisture level is very critical and must be maintained to insure a good keeping quality.

Oven-puffed Rice.—Oven-puffed rice is prepared from whole kernels of California Pearl (short grain) rice. Frequently the rice is parboiled. Each batch consists of 635 kg of rice and 202.5 l of sugar syrup. Some salt may be added. The mixture is cooked in a retort for 5 hr under 21.1 kg/cm^2 steam pressure. Sometimes nondiastatic malt syrup and enriching ingredients are added before cooking (Matz 1970; Hogan 1977).

The cooked rice is broken up and dried to approximately 25 to 30% moisture content in a rotating louver drier. The partially dried product is stored in a stainless steel bins for about 15 hr to equilibrate the moisture. Lumps may form during the tempering process. They must be broken up before being sent to the flaking rolls.

The individual kernels are separated and again dried so that a moisture content of 18 to 20% is reached. They are passed under a radiant heater which brings the external layers of the rice to 82.2°C (180°F). The outside layers of the kernel are plasticized by the heat so they do not split when the grain is run through the flaking rolls. The rolls used in preparation of oven-puffed rice are set relatively far apart so that they contact only the central part of the rice kernel. The "bumped" grains are again tempered, this time for about 24 hr.

To secure the puffed effect, the cooled and tempered rice is passed through toasting ovens at 232.2° to 301.7°C (450° to 575°F). Transit time is about 30 to 45 sec.

A cereal called "Special K," manufactured by the Kellogg Co., is a rice kernel cooked and then coated while in a moistened condition with wheat gluten, wheat germ meal, dried skim milk, debittered brewers' yeast and other nutritional adjuncts. Finally the material is oven-puffed (Hogan 1977).

Oven-puffed rice cereal seems to be preferred to rice flakes by consumers. This is believed to be due to the superior tenderness developed by oven-puffing. Oven-puffed rice also maintains crispness for a longer period when served with milk, whereas rice flakes tend to soften more quickly.

Rice cereals may be coated with flavored sugar syrup, making sweetened products for greater appeal, especially for children. Many kinds of flavoring, coatings and colorings are used for a variety of finished products. However, the consumer groups are against the consumption of cereals that are high in sugar content. This is largely due to the lack of vitamins, minerals and fibers in sugar-coated cereals.

Gun-puffed Rice.—According to Hogan (1977), short grain rice is preferred for gun-puffing. California Pearl rice is generally used. The rice contains 13% moisture. It is preheated with air at 521°C (970°F) to 638°C (1180°F) before puffing. The preheated rice is fed to the gun in which pressure is built up by superheated steam and then suddenly released. The gun can be stationary or rotating and can be either heated or unheated. Superheated steam pressure up to 15.1 kg/cm² is applied at temperatures up to 241.6°C (475°F) at the gun. After a short cooking time, the gun is suddenly opened and the puffed grain is caught in metal hoppers. Unpuffed and clumped grains, along with fine material, are removed in a Bauer-type separator. The puffed rice is dried to 3% moisture before packaging.

The puffing process is related to the moisture present in the original rice and that involved in the process of steaming within the gun. If the surfaces of the kernels become too wet during the puffing stage, poor expansion will result. Satisfactory puffing depends on attaining grain temperatures where starch exhibits plastic flow characteristics under pressure. The time and temperature at which the rice is preheated are critically important. In general, the required temperature should be reached as quickly as possible without scorching the grain. If the grain temperature in the gun is too low, the grains will clump together. If the grain temperature is too high, it will not expand as much and will likely have hollow centers, and losses due to breakage and abrasion will increase. The rate of steam flow to the puffing gun is very important and must be controlled precisely.

Puffed parboiled rice has a darker color than puffed milled rice. The former lacks the characteristic sheen and has a pitted surface in the individual kernels which imparts a less pleasing appearance. Puffed parboiled rice tends to develop a scale off-flavor during storage probably due to oxidation of small amounts of rice oil in the original parboiled rice. Houston and Kohler (1970) point out that parboiled rice does not store well due to development of a rancid flavor. A maximum of 60 days has been specified for storage of parboiled rice prior to processing. Houston and Kohler (1970) suggested that the parboiling process may destroy the natural antioxidants present, since raw milled rice which has not been submitted to the parboiling process has longer storage shelf-life.

Puffing by Extrusion.—Puffed breakfast cereals, as well as snacks, are being made by extruding superheated and pressurized doughs through an orifice into the atmosphere. The sudden expansion of water vapor as the excess pressure is released increases the volume several times. Apparent specific volumes can reach or exceed those attained by gun-puffing and the process seems to have several advantages over gun-puffing (Matz 1970).

The rice premix containing 60 to 75% expandable starch base is moisturized with water or steam. The resultant mash is compacted by a screw revolving inside a barrel which may be heated by steam. The thread of the screw has a progressively closer pitch as it approaches the discharge. The pressurizing and steam heating bring the dough to a temperature of around 148.9°C (300°F) to 176.7°C (350°F) and a pressure of 2.46 to 35.2 kg/cm^2 at the die head. Under these conditions the dough is quite flexible and easily adapts to complex orifice configurations.

The die head may contain several orifices and pieces of correct size are sliced off by revolving blades resting on the exterior die surface. Adjustment of the speed of rotation of the knife assembly controls the piece size. Figure 18.1 illustrates a typical extrusion device.

The dough pieces expand very rapidly as they leave the die orifice and the expansion may even continue for a few seconds because the dough is hot and still flexible, and water continues to boil off. Even so, the moisture content of 24 to 27% is too high for satisfactory stability and

From Wenger Manufacturing, Sabetha, Kansas

FIG. 18.1. WENGER X-200 EXTRUSION UNIT

the pieces are further dried on vibrating screens in hot air ovens. Fines and agglomerates are removed at this time and the products cooled and packaged.

Rice Flakes

Short grain rice such as California Pearl is used for making rice flakes (Hogan 1977). The rice is first cooked in a rotary cooker at 8.2 to 8.6 kg pressure. After an initial cooking period of 20 min, the steam pressure is bled off to eliminate gases. Then cooking is continued for 1 to 2 hr, depending on the kind of rice used. Completeness of cooking is determined by the presence of uncooked centers in the individual grains. In cooking rice, it is necessary to add an agent that will prevent the cooked rice from sticking and forming lumps. Finely ground wheat bran is usually added up to about 5% of the cooking formula. Moisture of the rice after cooking should be about 33%. It is desirable to apply a vacuum to the cooker after cooking has been completed to surface-dry the cooked grains and thus improve handling in subsequent parts of the process. After being dried in steam-tube or lourver driers to about 17% moisture, the cooked rice is then tempered for several hours to equalize moisture between grains. After tempering the rice is flaked on large, smooth rolls operating with a slight differential to exert a stretching action and to improve flaking. Flaking-roll pressure is adjusted so that the rice flakes are well blistered after toasting. Blistering is accomplished at the head end of the toasting oven and is necessary to produce tender crisp flakes. At this point, moisture in the rolled flakes and oven temperature are critical. Moisture in the flakes is quickly converted to steam as they enter the oven. Thereafter, the blistered flakes are dried to 3% moisture and should be a light tan color. A typical dryer is shown in Fig. 18.2.

Puffed Rice Cereal with Fruit Filling

McKown *et al.* (1969) received a Canadian patent which was granted to

From the National Drying Machinery Co., Philadelphia, Pa.

FIG. 18.2. DRYER FOR RICE CEREALS

General Mills, Inc., Minneapolis, Minnesota. Dried, puffed and toasted cereal dough shells are given an expanded fruit filling to improve the savoriness, nutritional value and other qualities of ready-to-eat breakfast cereals.

Previous attempts to incorporate fruit into cereal, by adding fruit in the form of pastes and juices to cereal doughs which are then made into cereal products, have encountered difficulties. When the dough with the fruit product is subjected to elevated temperatures involved in cooking or other processing, the flavor quality of the fruit product is degraded.

The McKown patent (1969) inserts the fruit material into the cereal shells or biscuits only after the biscuits have been made. The fruit material is not subjected to the elevated processing temperatures required to form a finished cereal product. The process by which this product is achieved can be described quite broadly as first making a finished cereal shell or biscuit, and subsequently filling the same with a fruit material that is expanded and dried in the cavity of the biscuit. The process involves: (1) making a workable cooked dough; (2) forming the dough into closed, generally flattened pellets having a sealed perimeter and a closed inside surface; (3) drying, puffing and toasting the pellets to form finished cereal shells or biscuits, each with a large closed cavity; (4) inserting by means of an injection needle of small diameter into the cavity of each biscuit a syrup-like filling material; and (5) expanding and drying the filler, this being advantageously done by vacuum drying the biscuits to form a stable foam structure. This patent alleviates the difficulties of quality loss in the fruit and fruit products. However, the process requires special equipment and the operation is relatively slow due to the injection and vacuum drying processes. The processor may encounter financial difficulties due to the higher cost and slower speed of production.

Shredded Rice Cereal

Shredded rice cereal is a very popular item. The basic ingredients are milled rice, sugar, salt and malt. It contains 3% moisture when packaged. The product is usually enriched with sodium ascorbate, thiamin, niacin, pyridoxine, folic acid, vitamin B_{12} and iron. White milled rice is cooked in a rotary cooker with sugar, malt syrup and salt in a manner similar to that described under rice flakes. The cooked rice receives a preliminary drying in lower ovens. The cooked and partially dried rice is transferred to stainless steel bins and tempered before it goes to the shredding rolls. The tempered rice kernels, while in a plastic condition, are passed through shredding rolls similar to those used in making shredded wheat cereals (Matz 1970). The shredding rolls are from 15.2 to 20.3 cm in diameter and as wide as the finished chex is to be, and thus are much smaller than flaking rolls. On 1 of the pair of rolls is a series of 20 shallow corrugations running around the periphery. In cross section, these corrugations may be square, rectangular, triangular or a combination of these shapes. The most popular one is 1.9 cm square with double layers. Soft cooked rice is drawn between these rolls as they rotate and issued as continuous strands of dough. The individual rice grains are shredded by passing through rolls, with a cutting roll operating under pressure against a smooth roll. The shredded dough sheets are scored on another set of rolls forming the characteristic 1.9 cm squares which are placed on a metal belt moving through a high temperature gas fired oven. After 10–15 min, the outside of the product is dry and toasted while the interior is still wet. Then the product enters a different section of the same oven where it is dried at 121 °C (250°F) for 30 to 60 min, depending upon the size and air flow.

Bite-sized rice breakfast cereals are made with a triple shredding mill. The rice dough is fed to long, water cooled shredding rolls. These rolls deposit a shredded dough sheet onto a constant speed conveyor to form a wide, 3-layer ribbon. The rolls on the first and third shredding mills extrude dough sheets with a laced pattern due to the presence of smooth and grooved rolls. The middle set of rolls revolves at a higher speed than the other 2 sets. As a result, the dough sheet in the middle folds as it falls onto the relatively slow moving conveyor belt covered by the first sheet. Sugar is sprinkled over the middle dough sheet and the top sheet is added. The combined structure passes between scoring rolls and the baked cereal is finally broken along scored lines to form individual bite sized pieces (Anon. 1965).

The process for preparation of a biscuit having a lattice like network of shreds was patented by Hale and Carpenter (1956). The product may be puffed and therefore resembles not only shredded wheat biscuits but also the product patented by Huber (1955).

Rancid odors tend to accumulate if shredded rice is stored in sealed containers. For this reason, the product is sold in "breather" boxes without outer or inner linings. When so packaged, the product is just as stable to storage deterioration as any other prepared cereals except that moisture adsorption may occur in atmospheres of high relative humidity with a consequent loss of crispness.

Many variations of the basic rice cereals are made possible by the use of coatings or sprays containing different flavors and vitamins, along with syrup-enrobing, to achieve better consumer acceptance.

Rice in Multi-grain Cereals

In multi-grain cereals, rice is generally used in the form of flour mixed with cereal grain flours from corn, oats, wheat and soybeans. Since rice flour is relatively expensive, its use is restricted to about 10 to 15% by weight in the formula. There are many brands of multi-grain cereals on the market today. As consumer demand for variety has increased, cereal production has moved from the relatively simple metods of preparing flaked cereals to multi-grain formulations involving extrusion-cooking of doughs which can be more easily shaped by the cutting heads in a multiplicity of designs and forms.

A patent for making multi-grain type cereal was issued to Clausi *et al.* (1967). The dry ingredients are first thoroughly mixed together and moistened to about 30% with water. The mix is passed through an extruder-cooker and the dough cooked at temperatures ranging from 65.6°C (150°F) to 121.1°C (250°F). Cooked dough pieces are cut as they pass from the extruder into a variety of shapes and sizes. The pieces are then toasted usually in a traveling-belt type of oven at temperatures around 260°C (500°F). Moisture is reduced to about 3% before packaging. Further product variety is obtained by syrup coating and "frosting" containing flavors and colors.

Enrichment of Rice Cereal

Because of the consumer awareness of nutrition in foods, most cereal manufacturers add vitamins and minerals to the products they produce. The consumers also like to read the manufacturers' statements of nutritive quality expressed as percentages of the established minimum daily requirement.

Usually, solutions of thiamin, riboflavin, niacin and iron salts are sprayed on the product after toasting and before the final packaging. Minerals such as calcium salts may be added in the basic dough formula in the case of multi-grain cereals, or in the cooking formula in the case of single grain cereals. Nutrients can also be added to the syrups used in coating the final products. Care is needed in the case of thiamin addition to cereal doughs. High temperatures used in cooking may be destructive to a portion of the thiamin. For this reason, it should be added separately, e.g., in the syrup if sugar-coated, or as a separate spray. Amounts generally used are those necessary to bring nutritional levels up to those of the natural grains in terms of the minimum daily requirement as represented by an average serving (usually 28.3 g) for ready-to-eat cereals. The chemical composition and B vitamin content of rice cereals are presented in Table 18.3 (Adams 1975). Feldman (1959) suggests amounts of nutrients necessary to restore cereals to whole grain levels. For rice

TABLE 18.3. PROXIMATE ANALYSES AND VITAMIN B CONTENTS OF RICE CEREALS

	Oven popped rice[1]	Puffed rice[2]	Shredded rice[1]	Rice flakes	Parboiled rice[3] (long)	Instant rice
Moisture (%)	3.2	3.7	3.0	3.2	10.3	9.6
Protein (%)	6.0	6.0	5.2	5.9	7.4	7.5
Fat (%)	0.3	0.66	0.4	0.3	0.3	0.2
Calcium (mg/100 g)	20.0	20.0	16.0	29.0	60.0	5.0
Phosphorus (mg/100 g)	93.3	93.3	96.0	132.0	200.0	65.0
Iron (mg/100 g)	2.67	2.0	2.0	1.6	2.92	2.9
Sodium (mg/100 g)	943.3	trace	916.0	987.0	9.19	1.0
Thiamin (mg/100 g)	1.16	0.47	0.40	0.35	0.44	0.44
Riboflavin (mg/100 g)	1.4	0.066	0.08	0.04	0.04	—
Niacin (mg/100 g)	11.6	4.66	7.2	5.4	3.51	3.5

Source: Adams (1975); Watt and Merrill (1963).
[1]Added sugar, salt, iron and vitamins.
[2]Without salt and sugar.
[3]Added salt, iron and vitamins.

cereals he suggests thiamin, 1.5 to 2.5; niacin, 20 to 30; and iron, 22 to 44 mg per kg of cereal as packaged. The vitamin contents of rice and its milled products are presented in Table 18.4 (Houston and Kohler 1970).

TABLE 18.4.   VITAMIN CONTENTS OF RICE AND ITS BYPRODUCTS (Mg/100 G)

	Brown rice	Milled rice	Rice bran	Rice polish	Rice germ
Thiamin	0.34	0.07	2.26	1.84	6.5
Riboflavin	0.05	0.03	0.25	0.18	0.5
Niacin	4.7	0.6	29.80	28.20	3.3
Pyridoxine	1.03	0.45	2.50	2.00	1.6
Pantothenic acid	1.5	0.75	2.8	3.3	3.0
Folic acid	0.02	0.016	0.15	0.19	0.43
Inositol	119.0	10.0	463.0	454.0	373.0
Choline	112.0	59.0	170.0	102.0	300.0
Biotin	0.012	0.005	0.06	0.057	0.058

Source: Houston and Kohler (1970).

As the milling process proceeds from whole rice (bran, polish and germ) to milled rice, significant losses of essential vitamins occur because the vitamins are contained in the outer layers of the kernels. Niacin, pyridoxine and inositol are higher in the bran than in the germ.

The fiber content of rice flakes is 0.6%, rice cereals, 0.2% and instant rice 0.4% (Brockington and Kelly 1972). The ash content of rice flakes is 2.9%, rice cereal, 2.7%, and instant rice 0.2% (Brockington and Kelly 1972).

Packaging of Breakfast Cereals

Packaging of breakfast cereals has been discussed by Luh and Zee (1976A). Food packaging in general has been reviewed by Brody (1970), Chen et al. (1975), Hanlon (1971), Heiss (1970), Luh and Zee (1976B), Paine (1967), Pinner (1967), and Sacharow and Griffin (1970).

Packaging of rice cereals is a very important entity in rice processing. Many packaged ready-to-eat cereals are stored for 6 months or longer before they are actually consumed. Thus, it is necessary to use special packaging materials to protect the crispness and storage stability of rice cereals.

The cereals must be acceptable in flavor, aroma and texture at the time of consumption. For this reason, many cereal manufacturers adjust their production schedule to meet the demand of the market so that the products will not stay longer than 6 months on the shelves of supermarkets and stores.

The most important qualities of ready-to-eat breakfast rice cereals are crispness and flavor. Color is also a factor contributing to the attractiveness of the product to the consumer.

Crispness of breakfast cereals is dependent on the moisture content of the products. Puffed rice, for example, is more hygroscopic than other rice cereals. It will rapidly lose its crispness unless a good moisture barrier is provided in the package. Optimum crispness can be kept if the moisture content of the puffed rice is 3 to 4%.

Packaging Requirements.—Except for the bowl-ready instant products, cereals do not have exact packaging requirements. They are dry and thus are not subject to microbiological deterioration. Hot cereals absorb moisture only with difficulty and so require little moisture protection. Fat contents of other than the whole grain cereals have been reduced considerably and so rancidity from exposure to air is not a major problem. However, to preserve freshness, antioxidants are added either in the formula or to the package material. Most cereal manufacturers apply butylated hydroxyanisole, butylated hydroxytoluene and propyl gallate to the packaging material as antioxidant (Stuckey 1955; Caldwell *et al.* 1964). Many flavoring agents are stable and little protection is required to retain them. Vitamins used for enrichment do not present any problems if the cereals are kept reasonably dry.

Some cereals are subject to insect infestation and extra protection is required to keep moths and beetles from penetrating the packages at weak spots, such as corners.

Bowl-ready precooked cereals contain hydrated starch which is designed to accept moisture readily and which, therefore, requires water vapor protection. The incorporation of flavors, such as cocoa or cinnamon, leads to the need for an odor barrier to supplement the moisture barrier.

In contrast, ready-to-eat cereals are hard and often fragile. They can crack and be abraded and crushed if seriously abused. The light weight and density allow them to flow rather easily and move inside containers with only minimum surface powdering and breaking.

Unsweetened, ready-to-eat cereals require some moisture protection because slow absorption of water can lead to loss of crispness and a toughening of the texture.

Ready-to-eat cereals may be produced from a variety of grains and grain flours with varying quantities of fat remaining. There is a long-term possibility of rancidity and short-term potential from fat flow into packaging as a result of temperature fluctuations. The need to protect these cereals from odor contamination of the cylinder board carton in which they are contained also exists. As with crackers and cookies, ready-to-eat cereal products often need to breathe through the package.

Sweetened cereals present a more difficult moisture problem because dried sugar syrup is an active moisture absorber when the vapor pressure exceeds the equilibrium. Moisture absorption leads to partial liquefaction and surface stickiness.

All of this points to a requirement for moisture protection. Some advocate total sealing and others a good barrier without the need for hermetic sealing. Regardless, there is the need for enclosure in the larger sizes.

Synthetically sweetened dry cereals do not contain additives which readily absorb moisture and so water vapor protection requirements for these would be about the same as for unsweetened cereals. Those with added nutrients, however, (and most cereals are fortified with certain vitamins and minerals) may contaminate the products with the flavor of the vitamin.

Cereals with inclusions such as raisins, nuts and candies do not have major problems of moisture uptake.

Cereals with freeze-dried fruit (all have been withdrawn from the market) experienced just the reverse. The porosity and low moisture content of the dried fruit structure led to extremely rapid rates of water absorption by the fruit. If the moisture available was insufficient for total reconstitution, the moisture distributed itself throughout the fruit and led to toughening. Moisture absorption in freeze-dried fruit was not reversible.

Packaging Materials.—Cereals not requiring such protection are packaged in linerless chipboard cartons. The traditional paperboard cylinder with a lithographed paper label is used by the Quaker Oatmeal Company and others. Farina and whole wheat hot cereals in rectangular paperboard cartons with printed double-wound overlabels (to protect against insect infestation) are close packaging relatives of the cylinder. These packages are inexpensive, readily decorated and they stack well.

Instant cereals which require moisture protection are packed in a paper/polyethylene pouch. Polyethylene extrusion on the interior provides an inexpensive moisture barrier which also acts as a heat sealant.

Ready-to-eat cereals in family-size packages are almost all packed in DPM, lined, rectangular paperboard cartons.

Liners for unsweetened cereals belong to the same family of waxed and laminated glassines as the cookie and cracker package liners. Wax, of course, acts to retard moisture passage. The glassine acts as the fat and odor barrier. Generally, 2-side bleached or amber waxed glassine is used, amber being slightly less expensive.

Waxed glassine liners are heat-sealed on the long and bottom seams and the foldover which is closed on the top after filling. Waxed glassine has sufficient stiffness for foldover closure.

Shredded wheat is now individually packaged (a reuse convenience as well as moisture protection) in paper/plastic that is heat sealed face-to-face.

For sweetened cereals, some employ wax laminated overwaxed glassine with double foldover closure. Wax laminated, overwaxed, glassine liners may also be heat sealed and folded over for additional protection.

Other barriers are employed for dry, ready-to-eat cereal protection. Aluminum foil wax laminated to tissue is used on some individual portion packages. Wax on the interior acts as a sealant because a total seal is required to assure against leakage. The wax laminant is a moisture barrier.

Although a similar material could be employed for larger packages, one construction which is employed is a triplex, glassine/wax/aluminum foil/ wax/glassine. The glassine can be inexpensive cereal grade because its principal function is to protect the foil, which provides the major barrier. Aluminum foil is now used in 0.0076 cm and even 0.0064 cm gages instead of the former 0.0009 cm. The small pinhole level in these thicknesses is readily overcome by the microcrystalline laminating wax and provides an effective moisture barrier.

Glassine on the inside protects the foil against abrasion from cereal, is a fat and odor barrier, and also separates the product from the wax, reducing the possibility of waxy flavors being imparted to the cereal.

Into the area of sugar-sweetened cereal liners has come what is sometimes referred to as an intermediate barrier: overwaxed one-side saran-coated glassine or paper. Saran is not as impermeable as aluminum foil but is the best of the commercially available barrier coatings.

Portion packaging of ready-to-eat cereals has been in both lined paperboard cartons and in sealed thermoformed tubs.

DPM portion pack cartons (Hanlon 1971) are multipacked in 10s and other multiples for retail sale. For hotel, restaurant, institution and some retail use the cartons are bulk cased. Multipacking is by overwrapping in printed cellophane or polyethylene, with or without a printed paperboard tray. The package is conventionally overwrap bundled, using transparent film and printing to show the contents or a paperboard to try to express the message of many packages in a single unit.

Equipment.—The major equipment used in breakfast cereal packaging is the double-package maker by Pneumatic Scale Corp. (Quincy, Mass.).

The latest major commercial cereal packaging method is the pouch packing of instant hot cereals. Bartelt continuous or intermittent motion equipment is probably the most widely used in the United States. Bartelt equipment operates at speeds above 60 pouches per min for intermittent motion and up to 400 per min for continuous motion. Bartelt equipment can continuously couple pouch making with single or multiple cartoning, using sleeve-type paper board cartons. The Canadian Delamere and Williams equipment is probably the second most popular.

RICE IN BABY FOODS

Strained baby foods were first introduced to parents as "convenience foods." Convenience ranks high as a motivating force in baby food purchases. Modern commercial canning procedures result in a higher retention of nutrients than many home cooking methods.

Rice in the form of rice flour or as granulated rice is used in the formulation of many of these products, particularly in meat and vegetable combinations (Luh and Woodroff 1977). Rice flour, waxy rice flour, parboiled rice, rice polishings and rice oil are used in baby foods. The largest use of rice in the baby food industry is in the manufacture of precooked infant rice cereal.

Inspection of Raw Material and Finished Goods

The baby food manufacturers have imposed rigid quality standards on their products. To produce quality products, quality ingredients must be employed. All ingredients used by baby food manufacturers are carefully inspected for rodent and insect contamination, foreign materials and pesticide residues. Ingredients in processed foods must conform to strict bacteriological standards. It is not unusual to inspect a supplier's operation prior to purchase. According to Brockington and Kelly (1972), a typical specification for rice flour is as follows:

Definition:	The flour obtained by grinding and bolting clean, sound rice of short or medium grain varieties.
Materials:	Rice shall be clean, sound, and scoured, and shall be free from smut, weed seeds or other foreign matter.
Workmanship:	The rice shall be properly ground and bolted, and otherwise processed in accordance with good commercial principles and under strict sanitary conditions.
Containers:	Sound, multi-wall paper bags meeting consolidated freight classifications, adequately sealed by sewing or taping, and free from dirt, foreign matter, or adulterating substances constitute the containers.
Net weight:	Net weight is set at 45.36 kg.
Labeling:	Each container shall be clearly and properly labeled, indicating the name of the manufacturer and the net weight, and preferably using the ingredient name, "Rice Flour."
Shipping:	The product is to be shipped in first-class transportation equipment which has been thoroughly cleaned

Requirements:

to prevent any possible contamination of the product or its containers.

Deliveries shall comply in every respect with the Federal Food, Drug and Cosmetic Act, with state laws and with requirements given in this specification.

Samples:

Upon the buyer's request before purchase, the seller shall furnish representative samples for inspection. Such samples shall be drawn from actual lots which the buyer may purchase, and should be no smaller than 2.3 kg.

Testing:

Acceptance will be subject to a sampling of the material actually delivered and to testing of these samples according to the laboratory instructions in this specification.

Methods:

Methods of analysis will be in accordance with the methods of the Association of Official Analytical Chemists or any other standard method deemed advisable by the buyer.

Sampling:

Representative samples will be obtained from 6 bags selected by random from the lot as delivered, and composited into 3.8 l containers. If resampling is required, representative samples from an additional 12 bags may be tested individually or in any composite form.

Besides the microphysical and bacteriological examinations, rice flour is subjected to some essential laboratory tests. An amylograph for paste viscosity is run to determine peak viscosity and gelatinization temperature. Usually, flour milled from short and medium grain rice varieties are used. Color, odor, flavor and granulation are also evaluated. Since the rice flour could be used both in cereals and in processed foods, total thermophilic spores, flat sour spores and thermophilic anaerobes must meet established standards. The rice flour is tested for protein and ash contents.

The finished goods are inspected for flavor, consistency, color and other established quality standards. Samples are incubated to ascertain that sterilization has been achieved. The thermometer charts on the reports are carefully checked to make certain that the time and temperatures required for commercial sterilization have been met.

Precooked Rice Cereal

Precooked rice cereal is an easily digested rice product frequently prescribed as the infant's first solid food. Precooked cereals are excellent vehicles for the introduction of essential minerals and vitamins into the infant's diet. The cereal must be easily reconstituted with milk or formula with a minimum of lumps. The quantity of liquid required to reconstitute a given weight or volume of cereal must be uniform. This is particularly important when cereal is reconstituted with formula from a nursing bottle.

The process for making precooked cereals has been reported by Hogan (1967, 1977). It consists of preparing and cooking a cereal slurry. The slurry is dried on a double-drum atmospheric drier and the dried cereal is flaked and packaged. Each baby food manufacturer has his own formulation and process for the manufacture of rice cereal. The ingredients used in the formulation of precooked rice cereal are: rice flour, rice polishings, sugar, dibasic calcium phosphate, iodized salt, sodium iron pyrophosphate, glyceryl monosterate (emulsifier), rice oil, thiamin hydrochloride, riboflavin, niacin or niacinamide. The composite analyses of various rice cereal products available in the United States are presented in Table 18.5.

TABLE 18.5. COMPOSITE ANALYSIS OF VARIOUS PRECOOKED RICE CEREAL PRODUCTS AVAILABLE IN THE UNITED STATES

Content	A	B	C	D
Protein (N × 6.25), %	6.7	5.5	7.51	6.0
Fat (acid hydrolysis), %	5.5	5.6	1.20	1.0
Carbohydrate (by difference), %	74.9	74.7	80.21	80.9
Crude fiber, %	1.0	2.0	1.09	0.3
Minerals (ash), %	5.5	4.7	3.94	3.8
Calcium, %	0.663	0.8	0.92	0.78
Phosphorus, %	0.67	0.6	0.72	0.53
Iron, %	0.05	0.053	0.0565	0.03
Moisture, %	6.4	7.5	7.46	8.0
Thiamin, mg/g	0.028	0.028	0.035	0.014
Riboflavin, mg/g	0.021	0.032	0.018	0.011
Niacin, mg/g	0.141	0.176	0.367	0.141
Calories, per g	3.81	3.63	3.60	3.53

Source: Brockington and Kelly (1972).

Rice cereal is more difficult to manufacture than other cereals. The precooked cereal slurry is dried on an atmospheric drum drier. The thickness of the film on the drier surface, the spacing between the drums, the temperature of the drum surface, the drum speed and the flowing properties of the slurry influence the drying time greatly. The rheological properties of the slurry are most difficult to control and influence the drying operation to a much greater degree than any other factor involved. The bulk density of the cereal is related to the thickness of the sheet coming off the drier and the size distribution of the flakes. Dry rice cereal does not flake well and if the cereal process is not controlled, an excess weight of cereal must be packaged in order to meet the package headspace requirement. Because of the high starch content in rice, the apparent viscosity of the cooked cereal slurry is markedly affected by slight variations in the solids content. The solids, drum speed, drum temperature, etc., are adjusted to obtain a finished product of good quality.

Kelly *et al.* (1972) patented a process for preparing a dried rice cereal product by incorporating an ester-containing organic releasing agent (emulsifier), containing at least 1 phosphatide linkage, in a rice cereal slurry prior to dehydration of the slurry surface. The precooked, dehydrated rice cereal product so obtained is rapidly reconstitutable with liquid to form a rice cereal with a homogeneous smooth texture suitable for infant feeding.

An example of the patented process is as follows (Kelly *et al.* 1972):

A rice cereal slurry was prepared approximately in the following proportions:

Ingredients	Quantity parts by weight
Rice flour (medium or short grain)	800
Rice polish	300
Dicalcium phosphate	36
Sodium chloride	20
Lecithin	8
Vitamins, seasoning, acid etc.	1

Sufficient water was added to reduce the overall solids content to about 22.6%.

The slurry, having a pH of about 5.0, was pumped through a line strainer with a screen size of 0.15 cm to a holding tank at a temperature of 71.1° to 82.2°C (160°–180°F), from where it was transferred to an agitating heater at about 96.1°C (205°F). The heated slurry was then fed to a conventional double drum dryer, each drum operating at an internal pressure of 80 psig. The drums were rotated at 5 revolutions per min and the dried sheet was removed with doctor blades. The resulting sheet was nonplastic, continuous and had a film thickness of about 5 mm. The doctor blades were loosely held to the drums, yet the sheet was readily removed. Once equilibrium of the operation was reached, no loss of dried puree was incurred on the dryer surface. The production rate was increased by 50% when drying the lecithin-containing slurry, while steam requirements were reduced from 7.5 kg/kg of dry cereal to 6.0 kg/kg of dry cereal.

The dried rice cereal sheet was continuously conveyed to a flaker employing a No. 5 (U.S. standards) screen size. The resulting flaked product had a moisture content of 2%.

A process for the manufacture of precooked cereal with fruit has been patented by Billerbeck et al. (1970). A precooked rice cereal product with strawberries appears to be gaining favorable consumer acceptance. These products are prepared in a manner similar to that employed for the regular precooked infant cereals. Cereal ingredients, fruit, sugar, oil, vitamins and minerals are cooked, dried on an atmospheric drum drier, flaked and packaged. Because of the hygroscopicity related to the fruit and sugar, fruit cereals require moisture-proof packages. An antioxidant is incorporated in the packaging material to prolong shelf life.

Sometimes, a diastatic enzyme is used to control the quality of the dried cereal. The diastase lowers the liquid requirement for reconstitution by hydrolyzing part of the starch. This is a very sophisticated process. The temperature, time of digestion, enzyme activity and solids concentrations must be closely controlled to obtain a satisfactory product.

Prior to 1970, most infant cereals were prepared with salt or iodized salt. The baby food manufacturers have discontinued its use due to the recommendation by the National Academy of Sciences for reduced salt intake by infants.

Rice cereal products contain little sugar, so caking is usually not a problem, particularly when they are subjected to storage at relative humidities less than 75%. Rice cereal, however, will become rancid if packaged in a hermetically sealed container. The most suitable package material for rice cereal is one which will allow both moisture vapor and gas transmission. Most precooked infant cereals are packaged in paperboard cartons. A bleached manila liner on the interior of the carton is occasionally used. The carton is overwrapped with a glue-mounted, preprinted paper label. The tight wrap offers sifting and insect protection to the package. Some manufacturers package rice cereals in preprinted cartons.

Rice cereals are packaged in 1, 8 and 16 oz (28.4, 226.8 and 453.6 g) packages and sold in units of 6 with several varieties of cereals.

Dry cereal products other than precooked cereals are used in infant feeding. These are manufactured by dry blending cereal grains, including rice with sugar and milk powder. These blends require cooking and are fed as a gruel or pap.

Formulated Baby Foods

Meats, fruits and vegetables are puréed for making baby foods (Luh and Kean 1975; Luh and Woodroof 1975). In vegetable and meat combinations or in soups, not only are the meat and vegetable particles comminuted but the products are also thickened to a desirable consistency so that they can be conveniently fed to a baby. Rice cereal products are customarily used in the preparation of soups and casserole dishes. Similar products designed for infant feeding also contain these ingredients. Not only is rice a food ingredient, but its use in baby foods has a significant role in the consistency of the product. This feature is probably associated with the amylose content of rice starch. The consistency of a product containing rice flour develops upon cooling. The variety of rice used in these products is very important. Long grain rices, because of their amylose content, cause the product to thicken during storage (retrograde) and to eventually produce a very rigid gel and water separation. The presence of free liquid in a product packed in a glass container is a serious defect. The strong gel associated with water separation is also undesirable.

The consistency of a baby food or puree is evaluated by measuring the distance in centimeters traveled by a given volume of material in 5 sec. The lower the Bostwick measurement, the thicker the product. Spoonability is a very important characteristic in baby foods, since a cohesive product is very difficult to feed babies.

A formula for the split-pea-and-ham baby food was reported by Brockington and Kelly (1972) as follows:

Ingredients	lbs/100 gal.	kg/100 liter
Precooked split peas	70	8.39
Carrots	50	5.99
Ham	25	2.99
Bacon	20	2.40
Rice flour	15	1.80
Modified food starch	12	1.44
Powdered skim milk	10	1.20
Salt	4	0.48
Onion powder	2	0.24
Celery powder	1	0.12

The split peas were precooked in 132.5 l of water for 1.5 hr at 95°C (203°F). The meat was ground in a meat grinder. The vegetables were comminuted in a Robinson cutter through a 0.48 cm screen. The dry materials were blended together and slurried in 113.6 l water. All ingredients were mixed together and the volume adjusted to approximately 359.6 l. The mixture was cooked at 95°C (203°F) for 10 min. The consistency was adjusted to a Bostwick of 6.0 with the remaining 18.9 l of water. The product was packaged in 212.6 g junior baby food jars and processed for 47 min at 121°C (250°F).

The samples were stored at 4°C (39.2°F). The amount of water separation or syneresis was measured, the greatest amount being noted in samples containing long grain rice varieties. Waxy rice flour is a good stabilizer for canned and frozen food products. The stability was possibly due to a reduction in the amylose:amylopectin ratio.

"Junior" baby foods have a coarser texture. They help the baby acquire the "mouth feel." To produce the "junior particle," granulated rice is incorporated into the formulation of many junior vegetable and meat items. Care must be taken in the formulation of junior baby foods because the consistency can thin upon cooking and the particle may settle out and form a mat in the bottom of the jar. To insure uniform distribution of the junior particle, modified waxy maize starch is frequently incorporated into the product.

Kaset Infant Foods

Bhumiratana (1975) reported on a new low cost infant food at the 10th International Congress on Nutrition in Kyoto, Japan. It is a precooked rice based product, high in protein content, designed as a complete food for infants. The product is also suitable as a snack food for young children. The raw materials are rice flour, full fat soy flour, sugar, amino acid and minerals. The components are blended together and then processed in a Wenger cooker extruder model X-25. The product then passes through a belt drier, after which the product has the appearance of a puffed snack. In this form it is suitable for use as a snack food for infants. A typical formula is shown in Table 18.6 (Kaset infant food, Bhumiratana 1975).

TABLE 18.6. INGREDIENTS USED FOR A NEW INFANT FOOD

Rice flour (g)	717.4
Sugar (g)	150.0
Full fat soy flour (g)	125.0
Calcium carbonate (g)	5.0
Salt (g)	2.0
Potassium iodide (mg)	0.4
DL-Methionine (mg)	500.0
Vitamin A acetate, type 500 (mg)	40.0
Dry vitamin E adsorbate, 50% (mg)	30.0
Nicotinic acid (mg)	10.0
Riboflavin (mg)	4.0
Vitamin D_2, type 850 (mg)	3.0
Thiamin HCl (mg)	2.0
Vitamin B_{12} (μg)	10.0

Source: Bhumiratana (1975).

The vitamin-mineral premix used for making Kaset infant food is presented in Table 18.7. One part of Kaset infant food is mixed with 3 parts of boiling water in a bowl. The product is stirred to form a thick porridge, allowed to cook and then fed to the baby with a spoon. Emphasis on the use of boiling water in diluting is to insure better sanitation. They suggest that the bowl and spoon are easier to clean than a feeding bottle. The possibility of over-dilution is minimized when the Kaset infant food is served as a thick porridge.

TABLE 18.7. VITAMIN–MINERAL PREMIX FOR KASET INFANT FOOD

Calcium carbonate (kg)	25.0
Salt (kg)	10.0
DL-Methionine (kg)	2.5
Rice flour (kg)	12.0
Dry vitamin A acetate, type 500 (g)	200.0
Dry vitamin E adsorbate, 50% (g)	150.0
Nicotinic acid (g)	50.0
Riboflavin (g)	20.0
Potassium iodide (g)	2.0
Thiamin HCl (g)	10.0
Vitamin B_{12} (mg)	50.0
Dry vitamin D_2, type 850 (g)	15.0
Approx. total (kg)	50.0
(for use in 5000 kg infant food)	

Source: Bhumiratana (1975).

The proximate analysis of the Kaset infant food was reported by Bhumiratana (1975). It contained 78.4% carbohydrate, 5.2% moisture, 11% protein, 3% fat, 1.4% ash and 1% fiber.

REFERENCES

ADAIR, C.R., BEACHELL, H.M. and JODON, N.E. 1966. Rice breeding and testing methods in the United States. *In* Rice in the United States: Varieties and Production. U.S. Dept. Agric., Agric. Handbk. *289*, 19–64.

ADAMS, C.F. 1975. Nutritive Value of American Foods in Common Units. U.S. Dep. Agric. Res. Serv. Agric. Handbk. *456.*

ANON. 1965. Let rolls work for you. Food Eng. *37* (2) 60–64.

BHUMIRATANA, A. 1975. A new infant food for Thailand. Abstracts from Symp. and Comm. from 10th Int. Congr. on Nutrition. August 3–9. Kyoto, Japan.

BILLERBECK, F.W. *et al.* 1970. Method for preparing a fruit cereal product. U.S. Patent 3,506,447. April 14.

BOLLICH, C.N., ATKINS, J.G., SCOTT, J.E. and WEBB, B.D. 1966. Dawn— a blast resistant, early maturing, long-grain rice variety. Rice J. *69* (4) 14–20.

BOLLICH, C.N., ATKINS, J.G., SCOTT, J.E. and WEBB, B.D. 1972. Labelle-a blast resistant, very early maturing, long-grain rice variety released in Texas. Rice J. *75* (3) 28–30.

BROCKINGTON, S.F. and KELLY, V.J. 1972. Rice breakfast cereals and infant foods. *In* Rice Chemistry and Technology. D.F. Houston (Editor). Amer. Assoc. Cereal Chem., St. Paul, Minnesota.

BRODY, A.C. 1970. Flexible packaging of foods. Chem. Rubber Co. Press, Cleveland, Ohio.

CALDWELL, E.F. *et al.* 1964. Package treatment versus direct application as a means of incorporating BHT in shredded breakfast cereals. Food Tech. *18,* 383–386.

CHEN, H.C. *et al.* 1975. Tin, glass and plastic containers. *In* Commercial Vegetable Processing. B.S. Luh and J.G. Woodroof (Editors). AVI Publishing Co., Westport, Conn.

CLAUSI, A.S. *et al.* 1967. Breakfast cereal process. U.S. Patent 3,381.705. May 9.

FELDMAN, C. 1959. Adequacy of processed cereals. The Chemistry and Technology of Cereals as Food and Feed. S.A. Matz (Editor). AVI Pub. Co., Westport, Conn.

HALE, D. and CARPENTER, E.J. 1956. Apparatus for manufacturing a cereal food product. U.S. Patent 2,743,685. May 1.

HANLON, J.F. 1971. Handbook of Packaging Engineering. McGraw-Hill Book Co., New York.

HEISS, R. 1970. Principles of Food Packaging. Published by arrangement with the Food and Agric. Organ. of the P. Keppler Verlagkg, Germany.

HOGAN, J.T. 1967. Processed rice products. Rice J. *70* (11) 25–31.

HOGAN, J.T. 1977. Rice and rice products. *In* Elements of Food Technology. N.W. Desrosier (Editor). AVI Publishing Co., Westport, Conn.

HOUSTON, D.F. and KOHLER, G.O. 1970. Nutritional properties of rice. Food and Nutrition Board, Natl. Res. Council—Natl. Acad. Sci., Washington, D.C.

HUBER, L.J. 1955. Process of preparing a puffed cereal product and the resulting product. U.S. Patent 2,701,200. February 1.

JOHNSTON, T.H., ADAIR, C.R. and TEMPLETON, G.E. 1963. Nova and Vegold—New rice varieties. Ark. Agric. Expt. Sta. Bull. *675.*

JOHNSTON, T.H., TEMPLETON, G.E. and SIMS, J.L. 1966. Nova 66 and other medium-grain varieties—Performance in Arkansas, 1960–1965. Ark. Agric. Expt. Sta. Bull. *148.*

JOHNSTON, T.H., TEMPLETON, G.E. and WEBB, B.D. 1967. Performance in Arkansas of Starbonnett and other long-grain rice varieties, 1962–1966. Ark. Agric. Expt. Sta. Bull. *160.*

JULIANO, B.O. 1970. Relation of physico-chemical properties to processing characteristics of rice. Ppr. presented at the 5th World Cereal and Bread Congr. May 24–29. Dresden, East Germany.

KELLY, V.J. *et al.* 1972. Process for preparing a dried rice cereal product. U.S. Patent 3,690,894. September 12.

LUH, B.S. and KEAN, C.E. 1975. Canning of vegetables. *In* Commercial Vegetable Processing. B.S. Luh and J.G. Woodroof (Editors). AVI Publishing Co., Westport, Conn.

LUH, B.S. and WOODROOF, J.G. 1975. Commercial Vegetable Processing. AVI Publishing Co., Westport, Conn.

LUH, B.S. and WOODROOF, J.G. 1977. Baby Foods. *In* Elements of Food Technology. N.W. Desrosier (Editor). AVI Publishing Co., Westport, Conn.

LUH, B.S. and ZEE, W.M. 1976A. Food packaging with flexible laminates. Proc. Mod. Eng. and Tech. Sem. Chinese Inst. of Eng. VI. Taichung, Taiwan.

LUH, B.S. and ZEE, W.M. 1976B. Properties of plastic films for food packaging. Proc. Mod. Eng. and Tech. Sem. Chinese Inst. of Eng., VI. Taichung, Taiwan.

MATZ, S.A. 1970. Manufacture of breakfast cereals. *In* Cereal Technology. S.A. Matz (Editor). AVI Publishing Co., Westport, Conn.

McKOWN, W.L. *et al.* 1969. Fruit product and process. Canadian Patent 803,960. January 14. (Granted to General Mills, Inc., Minneapolis, Minnesota.)

MOTTERN, H.H. *et al.* 1967. Popping characteristics of rice. Rice J. *70* (8) 9–31.

OKA, H.I. 1975. The origin of cultured rice and its adaptive evaluation. *In* Rice in Asia. Assoc. of Jpn. Agric. Sci. Soc. (Editor). Univ. of Tokyo Press, Tokyo.

PAINE, F. 1967. Packaging Materials and Containers. Blackie and Sons, London.

PINNER, S.H. 1967. Modern Packaging Films. Plenum Press, New York.

SACHAROW, S. and GRIFFIN, R. 1970. Food Packaging. AVI Publishing Co., Westport, Conn.

SIMPSON, J.E. *et al.* 1965. Quality Evaluation Studies of Foreign and Domestic Rices. U.S. Dep. Agric.—Agric. Res. Serv. Bull. *1331.*

STUCKEY, B.N. 1955. Increasing shelf life of cereals with phenolic antioxidants. Food Tech. *9,* 585–587.

WATT, B.K. and MERRILL, A.L. 1963. Composition of Foods. U.S. Dep. Agric.—Agric. Handbk. *8.* Washington, D.C.

WEBB, B.D. 1967A. Cooking and processing qualities required of rice varieties in the United States—Their evaluation in rice breeding programs. Int. Rice Com. Newsl. (Special issue) 115–125.

WEBB, B.D. 1967B. New rice varieties—Their cooking and processing characteristics. Proc. of Natl. Rice Utilization Conf., New Orleans. U.S. Dept. Agric., Agric. Res. Serv., ARS 72–53. USDA, Washington, D.C.

WEBB, B.D. *et al.* 1972. Evaluating the milling cooking and cooking characteristics required of rice varieties in the United States. Ppr. presented at First Australian Rice Res. Conf., Leeton, New South Wales, Australia.

19

Fermented Rice Products

Hsi-Hua Wang

Starch is the major constituent of milled rice and makes up 90% of the milled rice in dry weight. Milled rice contains 0.37–0.53% total free sugars (Juliano 1972). One may consider utilization of rice in biomass formation, saccharification, alcoholic or acetic fermentation. Hesseltine (1965) discussed the use of fungi in Oriental traditional foods including rice products. Van Veen and Steinkraus (1970) briefly summarized the nutritive value of fermented foods derived from rice. It is interesting that saccharification of starch in the west is characterized by the use of malt, while in the Orient, fungi are widely used (Table 19.1).

In terms of the nature of products, fermented rice foods are classified into 3 categories in this chapter: solid, paste and gel or liquid. The solid state fermented products include the starters such as p'eka or pekka, angka (China), ragi (Indonesia), koji (Japan), predigested "yellow rice" (Ecuador) (Van Veen and Steinkraus 1970) and those that are ready to eat such as Tieng-chiou-niang (China). Characteristic paste products are miso (Japan) and chiang (China) for seasoning of foods. Shao-hsing wine (China), sake (Japan) and rice vinegar are well known fermented rice foods in liquid form (Hesseltine 1965; Yamasaki 1945). Those being discussed in this chapter are summarized in Table 19.1 with regard to the nature of products and processing.

SOLID STATE PREDIGESTED FOODS

Tieng-chiou-niang

The method for making Tieng-chiou-niang (fermented sweet rice) in Taiwan is described in Fig. 19.1. Chiou-yao (Yamasaki 1945; Liu *et al.* 1959A,B,C, 1960) contains *Rhizopus, Mucor, Monilia, Aspergillus,*

TABLE 19.1. THE NATURE OF PRODUCTS AND PROCESSING OF FERMENTED RICE FOODS

Name	Nature of product	State of rice to be processed	Other raw materials	Microbes	Area	Reference
Predigested						
Yellow rice Arros fermantado Amarillo Reauemado Sierra rice	predigested solid	unhusked undried		*Bacillus subtilis* and others	Ecuador	9
Tieng-chiou niang	predigested solid in saccharified liquid	steamed milled		fungi and yeasts	China	3
Idli	precooked	sometimes parboiled	black gram	*Leuconostoc meaenteriodes*	India	9
Dosci	souring partially, the same as Idli		less black gram		India	9
Tape ketan Tape ketella		glutinous rice			Indonesia	2
Starter						
Ragi or Java yeast Peh-khak or	dried solid	raw grained rice		*Mucor, Rhizopus Mucoraceae*	Indonesia	2,9
Pekka (Chinese) or Shirokoji (Japanese)	dried solid	raw grained rice		*Rhizopus* yeasts	China	2
Bakhar murcha	dried solid	rice	plant material ginger	*Mucor* yeasts	Northern India	2

TABLE 19.1. *(CONTINUED)*

Name	Nature of product	State of rice to be processed	Other raw materials	Microbes	Area	Reference
Koji	moist solid	steamed		*Asp. oryzae*	Japan	
Chiu-chu Chou Kyoku-shi Tane-koji	air dried solid	raw, sometimes steamed			China	2,8
Saraimandie			barks	*Mucor*	India	2
Angkak (Chinese) Akakoji Benikoji (Japanese) Chinese red rice Red rice	dried solid	steamed		*Monascus purpureus*	China	7,8
Seasoning						
Miso	paste	steamed	soybean	*Asp. oryzae*	Japan	2,5,7
Chiang Kanji	paste	steamed cooked rice		Yeast	China India	2,8 2
Mirin	liquid	steamed glutinous rice	koji		Japan	2
Alcoholic beverages						
Shao-hsing wine	liquid, alcohol	polished rice, steamed		*Rhizopus* yeasts	China	3
Sake (Japanese)	liquid, alcohol	polished rice		*Asp. oryzae*	Japan	4,6,8

TABLE 19.1. (CONTINUED)

Name	Nature of product	State of rice to be processed	Other raw materials	Microbes	Area	Reference
Chin-chiou (Chinese)		steamed				
Anchu Hong-ru Hong-chiou	liquid, alcohol	polished rice steamed	angkak or red koji		China	2
Arak	liquid, alcohol	rice	ragi, juice and molasses		East India	2,9

[1]Hayashida et al. (1972).
[2]Hesseltine (1965).
[3]Liu et al. (1959A,B,C, 1960).
[4]Murakami (1972).
[5]Nakano et al. (1976).
[6]Nunokawa (1972).
[7]Shibasaki and Hesseltine (1962).
[8]Yamasaki (1945).
[9]Van Veen and Steinkraus (1970).

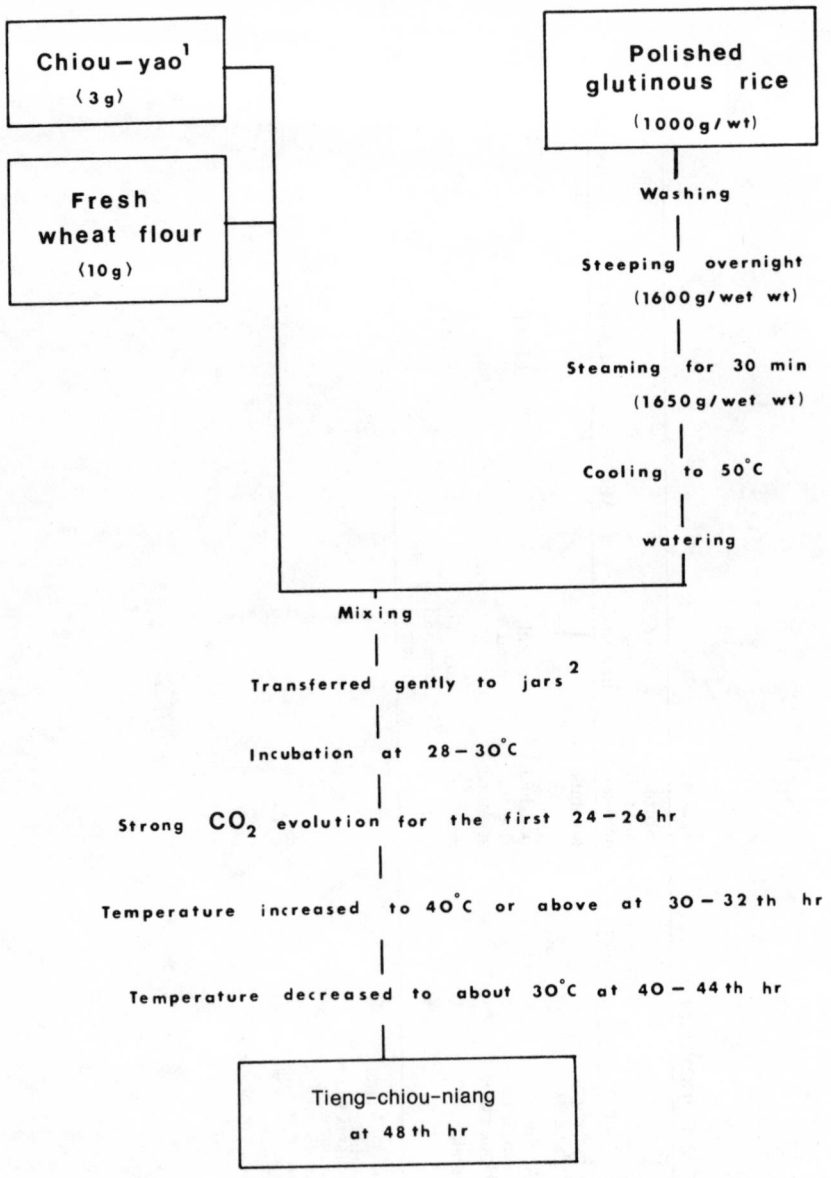

From Liu and Chen (1976)

FIG. 19.1. FLOW SHEET OF THE PREPARATION OF TIENG-CHIOU-NIANG DISTRIB-UTED IN TAIWAN

[1]Refers to Starter; [2]Refers to Fig. 19.2

yeasts and bacteria. If pure culture of *Rhizopus* spp. isolated from chiou-yao is used instead of chiou-yao in the above process, the product will show an increase in sweetness but a decrease in alcoholic content. Tieng-chiou-niang is a mixture of rice grains and saccharified liquid, which contains 50–53% sugar, 42–45% reducing sugar and 1.5–2.0% alcohol with an acidity of 0.5–0.6% (as shown by lactic acid). For preparing desserts, canned or fresh fruits may be added to the product (Fig. 19.2). The product can be stored at 10°C (50°F) for a time, but those kept at 25°C (77°F) or above will be further fermented. Hence the acidity will be increased to 1% or more and the alcohol content more than 5%.

FIG. 19.2. TIENG–CHIOU–NIANG

(A) Commercial product, (B) Product added with some water, (C) Fresh fruit, (D) canned fruit.

Yellow Rice

The procedure for making yellow rice is described by Van Veen and Steinkraus (1970) as follows. The unhusked, undried rice is fermented by the microorganisms with which it is contaminated. During fermentation the grain acquires a yellowish brownish color. After drying and milling, the resulting product is still colored. After cooking it develops a faint, specific flavor. Arroz fermando, amarillo, requemado, or simply Sierra rice are the synonyms of "yellow rice." The adjectives amarillo and requemado are used because of the color which may remind one of yellow toasted rice. The product is consumed mainly in the Sierras and is used in the preparation of "dry rice" which is considered indispensable for all meals by the mountain population. It is prepared by cooking the kernels until they separate evenly. Fermentation has already "precooked" the rice because it has been heated during fermentation at $50°-70°C$ ($122°-158°F$) and sometimes even higher. Moreover, the microorganisms break down some of the nutrients, at least partly, thus effecting a kind of "predigestion." The product therefore requires less cooking time, which is important at the Andean altitudes where fuel is scarce and water may boil below $90°C$ ($194°F$).

Van Veen and others were interested mainly in the nutritive value of fermented rice and in the possible occurrence of microtoxins which might present health hazards. They isolated several fungi and bacteria from Ecuadorian fermented rice samples, but it was difficult, if not impossible, to state which microorganisms were most important for the fermentation process. In general all samples showed more or less the same population of microorganisms. All bacteria were gram positive spore formers, which is not surprising because the fermenting rice heats itself considerably. The most prominent bacteria appeared to be *Bacillus subtilis* with strong proteolytic and amylolytic activity. Two other less important bacilli resembled *B. pumilis* and *B. cereus,* although both were not quite identical with these strains in connection with some of the biochemical properties as described in *Bergey's Manual.* Among the fungi, the *Aspergillus* were prominent, especially *A. flavus* L.K. var. *flavus; A. flavus* var. *columnaris* Raper and Fennel; *A. candidus* L. K.; and *A. fumigates. Rhizopus rhizopodiformis, Absidia corymbifers* and an *Actinomycetes* species were also isolated. It appeared that fermented grains retained their consistency much better than unfermented, which is in accordance with Herzfeld's observations (1957). Judging from the increase in soluble nitrogen, which may double from $1.5-3\%$ to $3-6\%$, a small portion of the protein was broken down by microorganisms. As a result of this and other microbiological effects, the cooking properties of rice and the consistency of the product were changed.

In a rat feeding experiment, the nutritive value of the protein (protein efficiency ratio, PER) had not improved. On the same level of protein intake, the unfermented product showed a PER of 1.90; the fermented product, 1.63; and casein as a control, 2.45. This is an abservation made with many fermented products (Van Veen and Steinkraus 1970). The apparent digestibility in rats even decreased from 86.8 to 72.6. Of the vitamins investigated, the thiamin content was practically unchanged, but the riboflavin content increased considerably (three to fourfold). This is also a common observation with fermented foods (Van Veen and Steinkraus 1970). It may be important for rice eating populations which usually have a rather low riboflavin intake. The fermented rice samples did not contain aflatoxin and the 2 strains of *Aspergillus flavus* were not able to form aflatoxin in the usual wheat medium.

Idli and Dosci

Idli is a food prepared in South India. It is made by the fermentation of rice and black legume (Hesseltine 1965). It is believed that yeasts and lactic bacteria are involved (Hesseltine 1965; Joseph *et al.* 1961). Dosci is practically the same as idli, but it contains somewhat less black legumes (Hesseltine 1965; Rao 1961).

Tape, Tape Ketan, Tape Ketella and Tape Nasi

Tape ketan is an Indonesian name for a product made by adding ragi to cooked glutinous rice. This mixture is wrapped in banana leaves at room temperature. In 2 or 3 days the rice becomes soft, moist, sweet and alcoholic (Hesseltine 1965). The difference between tape ketan and tape ketella is that the ketella is made of smashed cassava instead of the glutinous rice used in tape ketan.

STARTERS

The making of the starter (chu) involves a solid state fermentation occurring in an open system in which the moisture may be lost by evaporation (Wang and Chiou 1977). Crushed or milled raw grains may select or enrich the naturally inhabiting microbes. The size and the texture of the starter determine the distance the gases move, the volume of gas and the water holding capacity. The size and texture of the starter may also affect the thermodiffusibility of gas in the starter (Wang *et al.* 1968) and the germination or growth of naturally occurring microbes. The saccharifying agents are fungi. Therefore nuclear selection in heterokaryotic mycelium of some fungi may happen under extreme conditions. For example, the effect of high partial pressure of respiratory CO_2 inside the starter is found in the solid state mash of Kao-liang liquor brewing (Wang *et al.* 1970,1974), or in the mycelium of mushroom (Wang *et al.* 1976). The properties of the grains are as important as the size and texture of the starter used. Historically, the Japanese made their starter from steamed rice and the Chinese made theirs from raw crushed or milled grains. It should be emphasized that *Aspergillus oryzae* cannot saccharify raw starch, but when starch is steamed or dextrinized, it can be saccharified.

Chu

In ancient China, there were 3 kinds of saccharifying agents, malt, chu and tsao-chu. The latter 2 contained microbes, particularly molds. An old Chinese book *Chi Ming Yao Su* (late Wei Dynasty, A.D. 220), notes that nin-chu (maiden starter) is made from cooked or raw rice in the shape of a cake, which is packed with leaves (Fig. 19.3). Ping-chu has been made in northern China, whose raw material is wheat, but pekka, noted by Hesseltine (1965), is one of the chu. Another book *Nang Fang Tsao Mu Chung* (*Tsin* Dynasty, A.D. 265) describes that tsao-chu (grass starter) is in the form of balls packed with leaves and fermented for a month and

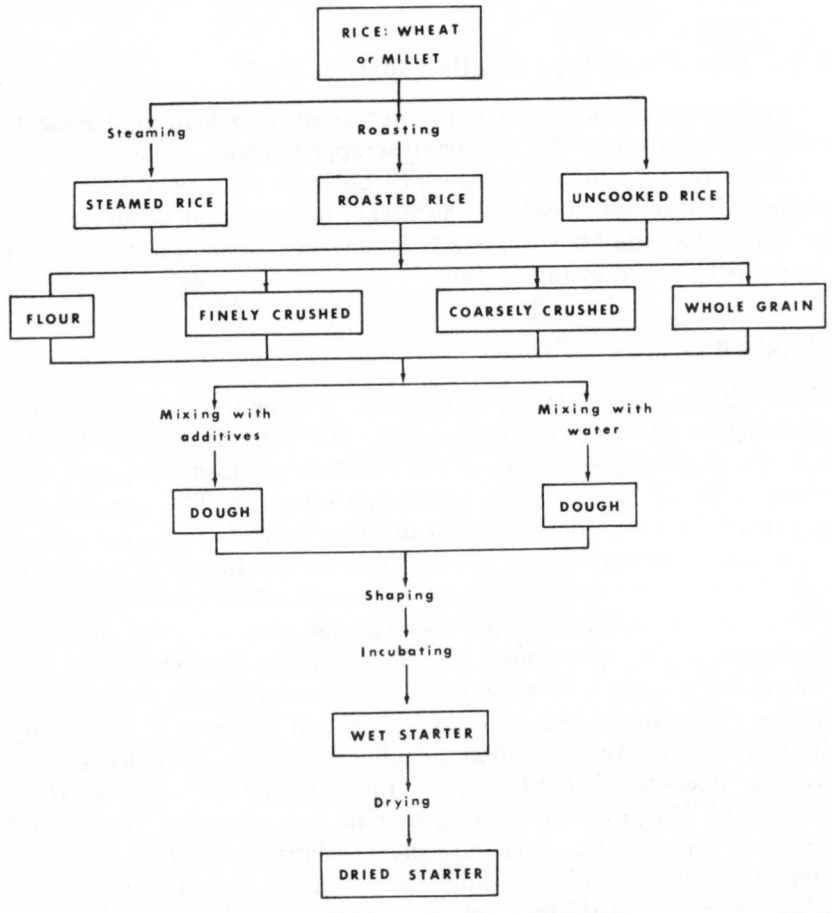

From The Methods of Farming (1968)

FIG. 19.3. FLOW SHEET OF OLD CHINESE PROCESS FOR MAKING STARTERS DESCRIBED IN *CHI MING YAO SU*

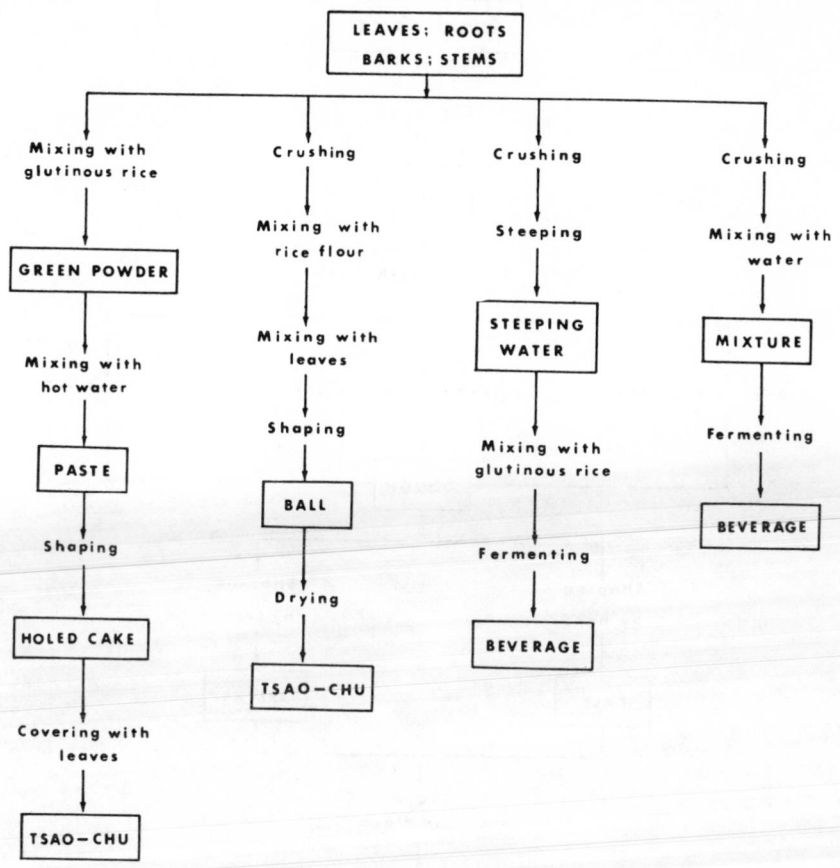

From Yamasaki (1945)

FIG. 19.4. FLOW SHEET OF MAKING TSAO-CHU

the balls are made from ground rice and plant juice (Fig. 19.4). It seems from the processes shown in Fig. 19.3 and 19.4 that tsao-chu is more primitive than chu. There is an intermediate starter, chiou-yao, between chu and tsao-chu as shown in Fig. 19.5. Chiou-chu, chiou-niang and chou, described by Hesseltine (1965), seem to be chu described here and equivalent to the koji of Japan. But the microbes of those starters in China are mucoraceous fungi and yeasts, while those of koji are *Aspergilli*. Various enzyme activities during the making of chu in the laboratory are shown in Table 19.2.

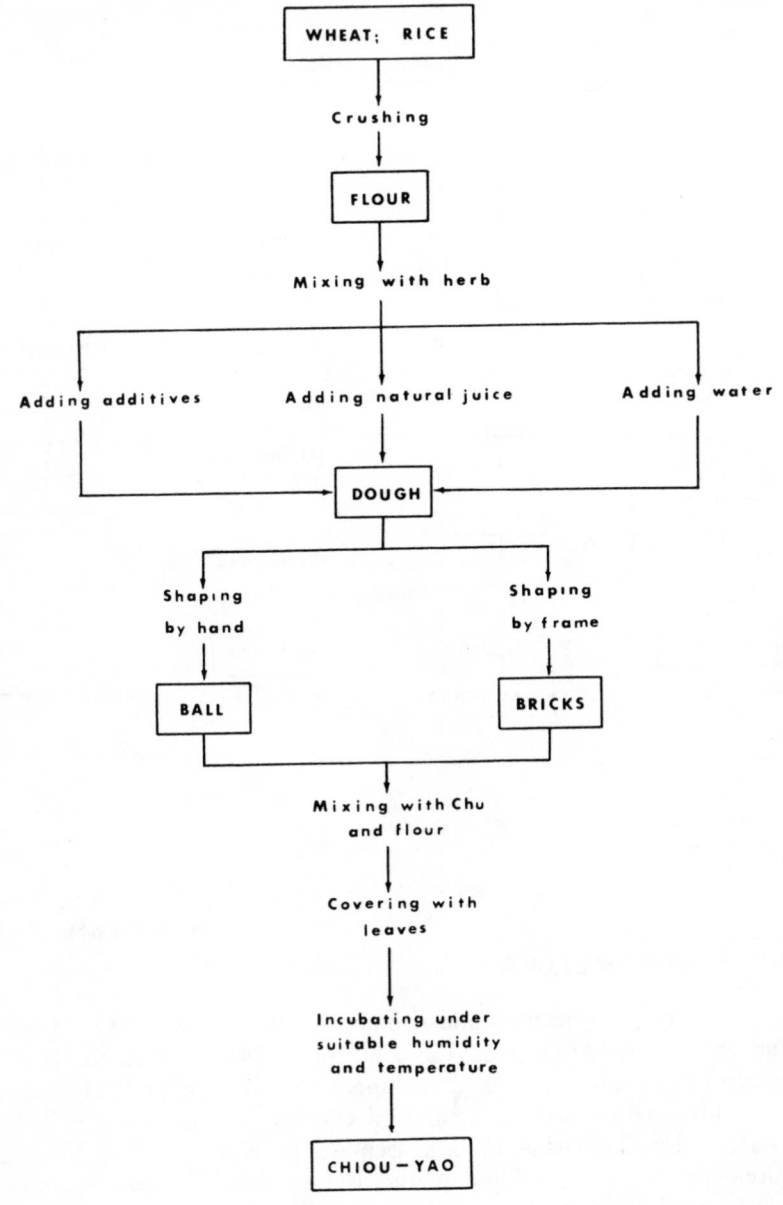

From Yamasaki (1945)

FIG. 19.5. FLOW SHEET OF MAKING CHIOU-YAO

TABLE 19.2. ENZYME ACTIVITIES FOUND IN THE STARTERS OF CHINESE RICE WINE PRODUCED IN TAIWAN

Wines	Microbes	Enzymes	References
Hua-tiao	*Saccharomyces cerevisiae* Hansen	glucoamylase ribonuclease	Lin (1972A, 1972B) Lin *et al.* (1974)
	Aspergillus sp. *Rhizopus* sp.	lipase	Mantani (1968)
Shao-hsing	*S. shaohsing* No. 1,2,3 and 4	amylase $\left\{\begin{array}{l}\alpha\text{-amylase}\\ \beta\text{-amylase}\end{array}\right.$	Yeh and Chiuch (1972)
	Asp. oryzae shaohsing No. 1 and 2 *Rhizopus* sp.	protease	
Hong-ru	*S. peka* Takeda *Rhizopus delemer* *S. formosanesis* Nakazawa *Monascus anka*	amylase $\left\{\begin{array}{l}\alpha\text{-amylase}\\ \text{glucoamylase}\\ \text{isoamylase}\end{array}\right.$ maltase protease invertase emulsin peroxidase	Su *et al.* (1970) Cheng (1970)

Ragi

Indonesian ragi is a starter grown on rice and sugar cane. It is very similar to Chinese pekka. The descriptions of chu (Fig. 19.3 and 19.4) are almost the same as those described by Hesseltine (1965) for the Indonesian ragi which is similar to "levure Chinoise." A commercial starter from Taiwan, when plated out, shows a mixed culture of 2 mucoraceous fungi and a yeast. Ragi is prepared from rice flour to which sugar cane and rhizomes of *Alpinia galanga* are added. These 2 materials are cut into small pieces and mixed with rice flour, dried in the sun and then ground into pulp with the addition of the juice of *Citrus limonellus.* After 3 days the large plant parts are removed, the liquid is decanted and round balls are made from the rest of the materials which are then dried in the sun or laid between rice straw. The ragi organisms may come from the rice straw. However, Hesseltine (1965) has corresponded with Mr. K.S. Djien and learned that the ragi is inoculated with old ragi. One of the yeasts was named *Monilia javanica* and a strain of this yeast still exists in the Dutch Yeast Collection and in the Agricultural Research Service Culture Collection NRRL Y-6703. Actually it is *Hansenula anomala.* A second yeast isolated from ragi and named *Saccharomyces vordermannii* is actually *S. cerevisiae.* The fungi in ragi are the mucors, *R. oryzae* and *Chlamydomucor oryzae.* The modernized process for the making of the starter in Taiwan and Japan will be noted later in this chapter.

Koji and Tane koji

Koji is a typical starter of Japanese fermented rice products. Tane koji is the spore inoculum for preparing koji whether it is used for miso, sake, shoyu or any other mold product (Fig. 19.6) (Hesseltine 1965; Nunokawa 1972). Koji is a general term for molded masses of various cereals or soybeans. The molds used are strains of *Aspergillus oryzae* or closely related species. The molded material serves primarily as a source of enzymes, although in some instances it is used as a source of inoculum for the second stage of fermentation. Each type of the koji used is from a different strain. Thus certain mixtures of strains are used for shoyu, others for miso and still others for sake. For sake brewing, more active amylolytic koji are required, but for miso or shoyu fermentation, more active proteolytic ones are needed. For a further detailed description see the sections of this chapter on Miso and Sake. The difference between chu and koji is the genus of the fungi involved, *Rhizopus* spp. for chu and *Aspergillus* spp., especially *A. oryzae* for koji.

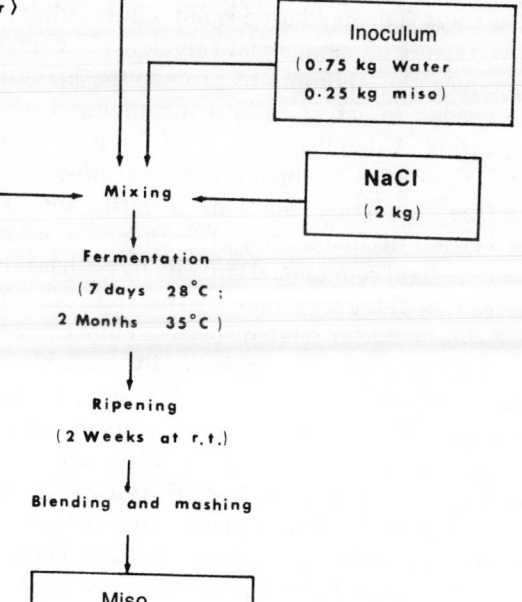

From Hesseltine (1965)

FIG. 19.6. FLOW SHEET FOR THE PRODUCTION OF MISO FROM GRITS

PASTE STATE FERMENTED RICE FOODS

Miso

Miso is a traditional Japanese pasty seasoning used in food. The raw materials for fermentation are rice, together with a suitable inoculum, soybean and salt (Fig. 19.6) (Hesseltine 1965; Yamasaki 1945; Nakano et al. 1976). Miso is used by the Japanese principally to make soup every morning. The consumption per capita per day is approximately 0.24 g according to national statistics in Japan (Nakano et al. 1976). The process developed by the Northern Regional Research Laboratory (Hesseltine 1965) is simplified in Fig. 19.7. It seems that there are some minor differences between the processes used, either in industry or at home. For example, Nakano et al. (1976) have described the mixing ratio of rice, soybean and salt as shown in the following equation: Rice = 5 × (soybean − 2 × salt).

The concentration of salt selects microflora during fermentation (Nakano et al. 1976). As carried out in Japan, the miso fermentation consists of the following steps: polished rice is washed, soaked, steamed and inoculated with Aspergillus oryzae (Ahlburg) Cohn and placed in shallow, wooden trays. The mold used is not a pure culture but a mixture of pure culture starters in commercial tane koji. The mixed strains used in the koji produce the appropriate proteolytic, lipolytic and amylolytic enzymes needed to act in various substances in the second phase of the fermentation. A detailed account of the preparation of tane koji can be found in the review of the miso fermentation (Shibasaki and Hesseltine 1962). One interesting aspect in preparing the spores is the addition of ash of certain deciduous trees to enhance spore production. The observation of several commercial starters by Shibasaki and Hesseltine (1962) indicates that these materials are free from other microorganisms except Aspergillus oryzae or related species. They agree very favorably both in purity and viability with commercially available inoculum in the United States. Although accounts in the literature by the Japanese state that other Aspergillus species besides A. oryzae and A. soyae are present, these are still species of Aspergillus which are closely related to A. oryzae.

The microbes other than Aspergillus involved in the miso process were also found by Shibasaki and Hesseltine (1962) as lactic bacteria, osmophilic and halophilic bacteria and yeasts. The yeast is Saccharomyces rouxii Boutroux, a heterothallic yeast. Synonyms of this species are S. soya Saito, Zygosaccharomyces soya (Saito) Takahashi and Yukawa, Z. japonica var. soya (Saito) Dekker, and Zygopichia japonica (Saito) Klöcker (Hesseltine 1965).

Chiang

Chiang is a Chinese seasoning food, equivalent to miso, but the Chinese do not dissolve chiang in water to make soup in the morning as the Japanese do. The major uses of chiang are seasoning for dishes and preservation for the storage of vegetables, meat and fishes as used by the Japanese. The quality of steamed rice added to chiang or miso during the fermentation affects the sweetness of the processed product.

LIQUID STATE FERMENTED RICE FOODS

In the Orient, rice is widely used to make wine and vinegar, although in the west it is only used as an adjunct in the production of beer (Promeranz 1972). Rice wine in the Orient is produced according to recipes handed down through generations (Anon. 1968,1974A, B). Modifications and improvements have been described by Liu et al. (1959A,B,C, 1960), Wang et al. (1970), Lin et al. (1975), Murakami (1972), Hayashida et al. (1972) and Nunokawa (1972). The Chinese prefer a natural starter, i.e., chu. In Taiwan, pure cultures, either *Rhizopus* spp. or *Aspergillus oryzae,* are used. In Japan, *A. oryzae* is used.

Shao-hsing Wine

In Taiwan, there are 2 kinds of shao-hsing wine, shao-hsing and hua-tiao. A diagram for the brewing of shao-hsing wine is given in Fig. 19.7. The difference between shao-hsing and hua-tiao lies in the type of cultures used: *Asp. oryzae* shao-hsing for the former and *Rhizopus* sp. for the latter (Liu et al. 1959A,B,C, 1960; Lin 1976). In consequence, shao-hsing resembles Japanese sake and hua-tiao a true Chinese rice wine.

Raw Materials.—The materials used in making shao-hsing wine include glutinous rice, pon-lai rice (a variety of Japonica short grain), wheat and water.

Water.—The ideal water for the brewing of shao-hsing wine should be colorless, tasteless and odorless. For example, when the water is warmed up to 50°−60°C (122°−140°F), a bad smell should not be present. The optimal ranges of water hardness, pH, residues and minerals required for this purpose are summarized in Table 19.3.

If the water does not contain enough effective components, they can be artificially supplemented. The brewing water of shao-hsing wine used by the Puh-li Winery (Taiwan) is spring water taken from a well nearby. Its components almost fit the characteristics of brewing water, although supplements are added to increase the hardness and sometimes chloride

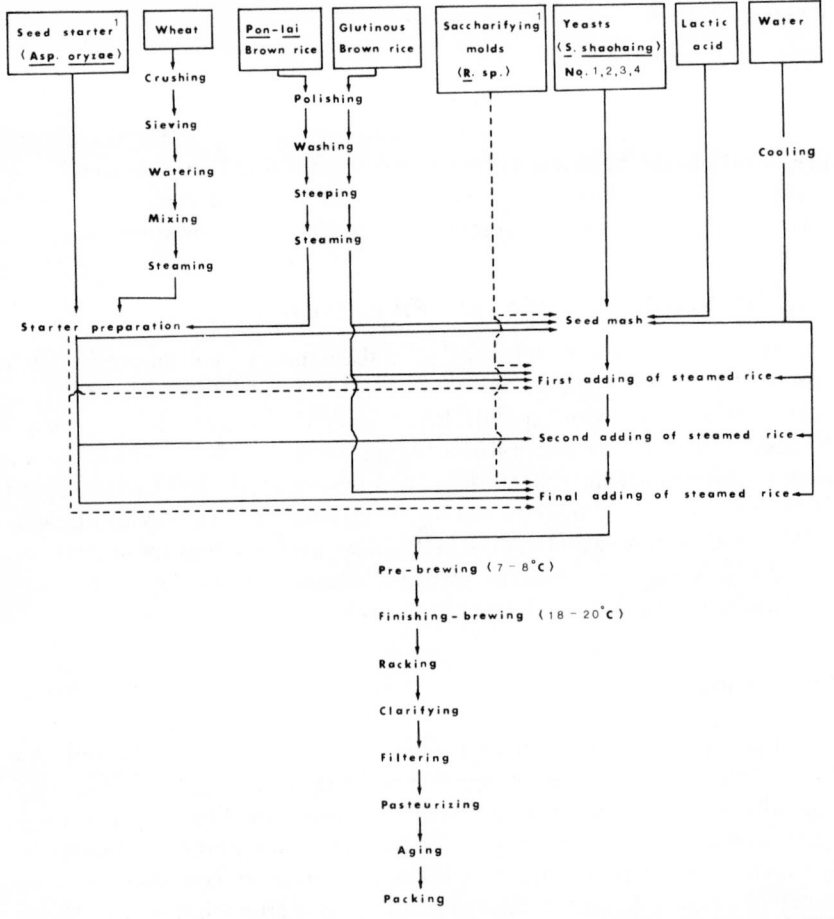

From Lin et al. (1975); Liu et al. (1959); Nunokawa (1972)

FIG. 19.7. FLOW SHEET OF SHAO-HSING WINES (SHAO-HSING AND HAU-TIAO) IN TAIWAN

[1]The different step in making Hua-tiao and Shao-hsing—*Asp. oryzae* for Shao-hsing and *R. sp.* for Hua-tiao.

is added, 100–200 mg/1 for seed mash water and 50–100 mg/1 for main mash water, respectively, to reach the optimum levels. Before it is used, the brewing water must be cooled for 1 day for 2 important reasons. One is to liberate the carbonate in the water and to allow the residues to settle out. During this period, the water will become slightly alkaline which is the best condition for fermentation. The other reason is to achieve the low temperature needed for fermentation.

Rice and Wheat.—The principal raw materials of shao-hsing wine are

TABLE 19.3. THE CHARACTERISTICS OF BREWING WATER FOR SHAO-HSING WINE

Items	Range
pH	6–7
Hardness	50–70 ppm
Residues after evaporation	100 ppm
Iron	<0.1 ppm
Organic material[1]	<7.0 ppm
NO$^-_3$	trace
NO$^-_2$ and NH$_3$	negative

Source: Puh-li Winery (1969, 1974)
[1]Organic materials are indicated as the titre of 0.01 N KMNO$_4$.

rice and wheat, the qualities of which greatly influence the brewing of shao-hsing wine. To select rice and wheat, the physical and chemical properties described in Tables 19.4, 19.5 and 19.6 are considered.

TABLE 19.4. PHYSICAL PROPERTIES OF GRAINS USED FOR SHAO-HSING WINE

Items	Pon-lai rice[1]	Glutinous rice[2]	Wheat
Size of grain	large and full	large and full	large and full
Appearance	waxy, shining luster in light gray or light brown, no adulteration	waxy, shining luster, no adulteration	waxy, shining luster, no adulteration, recent crop
Moisture	<15%	<15%	<14%
Starch value	>67%	>67%	>60%
Mixture	<8.5%	<20%	<0.5%

Source: Puh-li Winery (1967, 1974)
[1]*Oryza sativa* L. (japonica type)
[2]Indica type.

Preparation of Raw Materials.—There are more proteins, fats, fibers and minerals in the peripheral layers of rice grains than in the inner endosperm. The color of the peripheral layer is undesirable for the quality of shao-hsing wine. Therefore, both the glutinous rice used for brewing and the pon-lai rice used in the preparation of the starter should be polished. The rate of polishing is defined as the percentage of weight of white rice obtained from the original brown rice:

$$\text{the rate of polishing} = \frac{\text{white rice obtained (kg)}}{\text{original brown rice (kg)}} \times 100$$

The rate of polishing of the raw materials is 80% for shao-hsing, while

TABLE 19.5. CHEMICAL PROPERTIES OF GRAINS USED FOR SHAO-HSING WINE

Items	Pon-lai[1] Rice (%)	Glutinous rice[1] (%)	Wheat (%)
Starch	75	74	64
Protein	5.8	6.2	13
Fat	0.35	1.6	1.5
Acid	0.6	0.6	2.5
Moisture	13	13	13.5
Ash	0.25	0.28	2.2

Source: Puh-li Winery (1967,1974)
[1]The raw materials of 80% polishing rate.

TABLE 19.6. THE RATIO OF RAW MATERIALS USED FOR BREWING 5270 LITER SHAO-HSING WINE

Items / Raw materials	Seed Fermentation Mash Steamed rice (kg)	Main Fermentation Mash Initial addition (kg)	Final addition (kg)	Total (kg)	Remark
80% polished glutinous rice	150	600	1200	1950	added in the form of steamed rice
80% polished Pon-lai rice	45	144	231	420	for rice starter
wheat	—	18	129	210	for wheat starter

Source: Puh-li Winery (1967,1974)

sake is 70−80%. For superior sake, the rate is only 50−60% (Nunokawa 1972).

The purpose of washing the rice grains is to remove the bran and dust which adhere to their polished surfaces. The steeping process follows after washing in order to enhance the penetration of water into the grain via the interstices of the endosperm and to promote the penetration of heat into the grain during steaming. In general, water is kept at room temperature during steeping. The process usually takes 10−12 hr to allow the rice to absorb enough water. In the case of hard grain, steeping in 30°C (86°F) water is proper. If steeping in 30°C (86°F) water still does not prepare the rice properly for steaming, one may use the so-called "alternately steeping method," that is, steeping first in room temperature water, then in 10°C (50°F) water and finally in room temperature water again. In that case, swelling and shrinking will occur in the grain and enough water will be absorbed. Water absorption for pon-lai rice is usually 25−30% and for glutinous rice 35−40%.

In making the starter of shao-hsing wine, steaming of the rice is necessary for the modification of the structure of rice starch. In brewing, it is necessary not only for sterilization but also for the starch to become susceptible to the enzymatic action. An ideal steamed rice must have the proper moisture content, smell, texture, taste and color. It is noteworthy that after steaming the glutinous rice gains about 50% water in weight and pon-lai rice 32%.

According to its different usages, the steamed rice is cooled to different temperatures; for example, 48°−52°C (118.4°−125.6°F) when it is used for seed mash (35 min), 34°−36°C (93.2°−96.8°F) when used for the first addition of steamed rice during main mash (80 min), 26°−27°C (78.8°−80.6°F) for the final addition during the main mash (130 min) and 35°−38°C (95°−100.4°F) when it is used for rice starter preparation.

Preparation of Wheat.—The treatments of wheat include milling, water addition and steaming. The purpose of milling is not only to increase the contact area of microorganisms and hence promote the reproduction of microorganisms in the starter, but also to increase the rates of water absorption, steaming and fermentation. With the speed of 3400 rpm, the mill is able to process 250 kg of wheat per hour. Since the fine particles of milled wheat grains will increase the difficulty of handling during water absorption, screening must be performed before the preparation of the starter. After screening, the coarse particles of wheat are placed in an aluminum chamber of 450 cm × 100 cm × 16 cm and 30% water (by comparative weight) is added. The added water will be completely absorbed in 1 hr. In addition to accelerating the retrogradation of wheat starch and increasing the susceptibility to the enzymatic action of amylase and protease, steaming has another important purpose, that is, sterilization. In the sterilized wheat, the proper microorganisms are then able to easily reproduce. The steamed wheat must, however, be cooled to a temperature of about 32°C (89.6°F) before inoculation.

Starter Preparation.—There are 2 kinds of starters for making shao-hsing wine. One is rice starter and the other is wheat starter. The purpose of the preparation of the starter is to grow mold on the rice and/or wheat grains to produce various kinds of enzymes useful in the production of shao-hsing wine. The difference between rice starter and wheat starter is that there is more saccharifying amylase in the former and more protease in the latter.

Equipment.—The equipment needed to make the starters is a starter cultivation room and trays for starter incubation (Fig.19.8 and 19.9).

The necessary conditions of the starter cultivation room are: (1) a large enough area for incubating at least 20 kg rice/0.033 acre, (2) programmed temperature which can be exactly controlled and easily adjustable, (3) available methods of controlled gas exchange, (4) convenience for working, and (5) a large area outside the room.

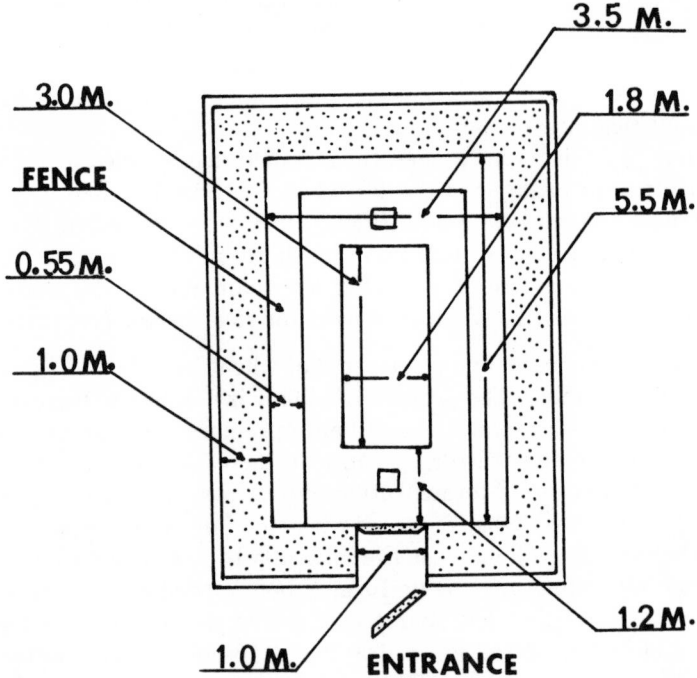

From Puh-li Winery (1967, 1974)

FIG. 19.8. PROFILE OF STARTER OF CULTIVATION ROOM

There are 2 kinds of wooden trays, 12 kg (65 cm × 95 cm) and 1.4–1.5 kg (30 cm × 45 cm). Usually, the trays are covered with a clean white cloth to keep a constant fermentation temperature.

Raw Materials.—The raw materials used for making starter are pon-lai rice (rate of polishing 80%) and wheat. The seed inocula are the cultures of *Asp. oryzae* shao-hsing No. 1 and No. 2.

Process of the Preparation of Starter.—(1) When making rice starter and wheat starter, a few differences occur in the steps. When preparing the former, the first step consists of transferring steamed rice to the starter cultivation room and cooling it to 37°–38°C (98.6°–100.4°F). It is then piled and covered by a cheese cloth for 4 hr to achieve uniform temperature 25°–30°C (77.0°–86.0°F) and humidity of 80–90%. In making wheat starter, steamed wheat is cooled to 30°–32°C (86.0°–89.6°F) and need not be covered with a cheese cloth.

(2) Inoculation with mold spores. The moist rice is spread thinly over the bed using a sieve inoculated with spores of *Asp. oryzae* shao-hsing.

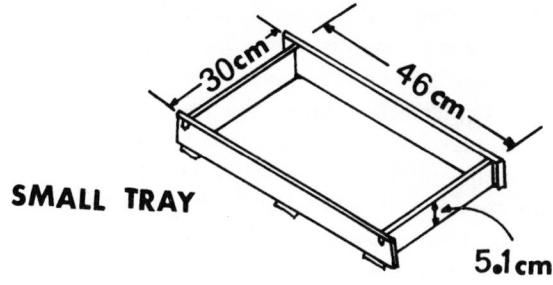

SMALL TRAY

30cm

46cm

5.1cm

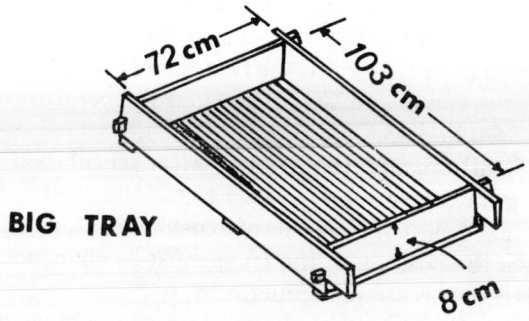

BIG TRAY

72 cm

103 cm

8 cm

From Puh-li Winery (1967, 1974)

FIG. 19.9. TRAY CONSTRUCTION

The density of inoculation allows each grain to acquire 2000 spores of mold. After inoculation, the moist rice should still be mixed, piled and covered with cheese cloth in order to maintain the correct temperature.

(3) Breaking up the grain heap. The inoculated rice becomes wet and loses it original smell due to germination of the spores. If it stands for a long time, a big difference in the temperature and moisture between the inner and outer layers of the heap will result. Thus the purpose of breaking up the grain heap is to expel the heat and CO_2 produced inside the heap. During the breaking up of the heap, the temperature will decrease by 1° or 2°C (1.8° or 3.6°F). It is important however that the temperature should not fall beneath 30°C (86°F). The dispersed grains are again heaped on the bed and covered with a cloth.

(4) Separation of piled grains onto trays. The following step is to separate the piled grains onto trays. Otherwise the temperature of the heap and the growth of microorganisms in the heap will increase rapidly and influence the flavor of shao-hsing wine. The time for separation of the piled grains to trays is 15−17 hr after the inoculation. At this time the temperature is about 33°−34°C (91.4°−93.2°F). The piled grains must be

separated quickly so that the temperature will not decrease too much. The capacity of the large trays is about 7−10 kg and that of the small trays about 1.4−1.5 kg. If 3 or 4 hr later the temperature of the grains in the upper trays is higher than that of the grains in the lower trays, the trays must be exchanged in position in order to balance the temperature differences. Since the microbial growth in wheat starter is more rapid than that in rice starter, only the small trays (about 1 liter) are used in making the wheat starter. Owing to the fast growth of microbes, the temperature of the wheat starter quickly rises to 36°−38°C (96.8°−100.4°F). Therefore, the first and second (or final) mixing needed in the preparation of rice starter is not necessary.

(5) The first mixing. As the growth of mold progresses, the temperature of the grain mass increases rapidly. It reaches 35°−38°C (95°−100.4°F) 24 hr after inoculation. At this point, the first mixing takes place in order to expel heat, water vapor and CO_2. The final temperature after this process is 35°−37°C (95°−98.6°F).

(6) Second (or final) mixing. About 29 hr after inoculation, the temperature of the grain mass reaches 39°−40°C (102.2°−104.0°F). The final mixing also takes place for reasons of expelling heat, vapor and CO_2. During this period, the position of the trays must be changed 3 times in order to prevent the increase in temperature.

(7) Finishing. After the change of the position of the trays during the final mixing, the molds grow heavily towards the inner part of the grains, the hyphae becomes tangled and the starter acquires a nutty flavor. The length of time needed for the preparation of rice starter is usually 45 hr while for wheat starter, 55 hr. Temperatures start at 35°−38°C (95°−100.4°F) and rise to 39°−40°C (102.2°−104.0°F). Once finished, the starter's surface is white in color, greenish white indicating top quality. The hyphae are short and stump-like in form and the flavor nutty and quite sweet. Some differences between wheat and rice starters are given in Table 19.7.

TABLE 19.7. CHARACTERISTICS OF RICE AND WHEAT STARTERS FOR SHAO-HSING WINE

	Rice starter	Wheat starter
Moisture (%)	26.5−31.0	21.8−28.5
Acidity	low	high
Glucoamylase[1]	low	high
Acid protease	low	high

Source: Puh-li Winery (1967,1974)
[1]Little difference.

Seed Mash.—The method of making seed mash for the brewing of shao-hsing wine is a quick process, lactic acid being added to enhance the growth of yeasts. The yeasts used in seed mash are a mixed culture of *Saccharomyces* shao-hsing No. 1, 2, 3 and 4.

It takes about 15 days to make a seed mash. The procedure includes the steeping of the starter, the addition of steamed rice, stirring and settling. The rice starter should be steeped for $1-2$ hr in 5°C (41°F) brewing water with a suitable amount of lactic acid, $KHPO_4$, $CaHPO_4$ and NaCl. This solution of water starter contains the enzymes for the saccharification of steamed rice. At first yeasts are added to the water solution and then the steamed rice which has been cooled to $48°-52°C$ ($118.4°-125.6°F$). When the temperature of the mixture has cooled to $27°-29°C$ ($80.6°-84.2°F$), the lid of the tank is put on and a cloth is used to cover it in order to maintain the temperature. Twenty to twenty one hours after the addition of the steamed rice, the mixture should be stirred in order to achieve a uniform temperature, to enhance the saccharification of grains and to promote the growth of yeasts. After the first stirring, the temperature will drop to $25°-27°C$ ($77°-80.6°F$). The mixture should be stirred every 4 hr until the 12th day, at which time the temperature has cooled to $10°-11°C$ ($50°-52°F$). Despite the low temperature, fermentation reactions still continue. The mixture is then stirred every 8 hr until the 15th day, at which time the temperature has dropped to 10°C (50°F) and the making of seed mash completed. Usually the seed mash contains 15.2% alcohol, 0.17% total acid, 1.6% sugar and has a yeast count of 3 billion/ml. The aqueous phase of seed mash has a pale color and is not viscous.

Main Mash.—The main mash consists of seed mash, steamed rice, rice starter, wheat starter and water. During the fermentation of the main mash, saccharification and alcohol fermentation occur simultaneously, although saccharification is more active in the early stage and alcohol fermentation in the later stage. The main product of this process is alcohol. In addition there are traces of aldehyde, glycerol, acids and esters which all contribute to the flavor of shao-hsing wine.

It takes about 30 days to finish the main mash fermentation. The processes can be divided into two stages: a prebrewing that takes about 12 days, and a finishing brewing that takes about 18 days.

Prebrewing Stage (first stage of the brewing process).—At first the seed mash, water, rice starter and wheat starter should be mixed together in the fermentation tank, so that the enzymes in the starter are dissolved into the solution. The mixture of starters is steeped for $1-2$ hr, and then the cooled, steamed rice at $34°-36°C$ ($93.2°-96.8°F$) is added. Mixing and stirring are performed as mentioned in the seed mash preparation. On the third day steamed rice is once again added, although this time the alcohol content of the mixture is about 4.15%, total acid 0.26% and sugar

10.8%. The mixture is stirred and mixed every 2−4 hr to adjust the temperature and to promote the reactions of saccharification and alcohol fermentation. By the 10th day, the materials in the main mash have settled and the temperature has dropped to 14°−16°C (58°−61°F). The prebrewing process is complete.

Finishing Brewing Stage (second stage of brewing process).—The main purpose of the finishing brewing is to increase the flavor of shao-hsing wine. The main mash should be transferred to a room with higher temperature (18°−20°C, 64°−68°F) in order to continue fermentation. With increased amounts of alcohol, total acid and amino acid contents, the flavor of the wine becomes stronger and the quality becomes better. It takes about 18 days to finish this process. During this period, stirring, heating and/or cooling should be done as needed in order to adjust the temperature, to expel the excess CO_2 and to promote the growth of yeasts.

Deterioration of Mash.—During the process of brewing, wild acid-producing microorganisms may occur. In an attempt to prevent contamination, sterilized equipment is used and low temperature should be maintained. If contamination occurs during the prebrewing stage, the main mash is not transferred to the warmer room where the finishing brewing stage proceeds, for the higher temperatures will increase the acidity of the fermentation. The main mash is then put through the filtration procedure. If contamination occurs in the finishing brewing stage, the process should be halted in order to inhibit further increase in the acidity of the wine.

Filtration.—In order to protect the quality of the product from microbial contamination, the room for filtration and settling should be kept at the temperature of 10°C (50°F). A suitable mash for filtration consists of 17.5−19.0% alcohol, 0.52% acid (i.e., calculated as succinic acid) and no free sugar. Twenty hours before filtration, the main mash (20°C, 68°F) is transferred into a tank (6 kl) to cool down to 13°−15°C (55°−59°F). Then the mash is placed into small (6−7 liter capacity) bags made of synthetic fiber. The bags are piled in a rectangular box. The mash is drained naturally for the first 4−5 hr and then filtered with a gradual increase of hydraulic pressure. The maximum pressure (of 15 kg/m²) should be maintained for 6−8 hr. After the removal of the pressure, the bags of greater moisture are placed in the center part of the box to obtain better filtration efficiency. The piling may be repeated once again. Recently, new ways of filtration have been devised. The drainage from the mash is kept in a vessel (6 kl) in the settling room (10°C, 50°F) for about 1 week to allow the yeast cells, fibers and insoluble protein to settle out.

After the settling, the supernatant is pumped without pressure to another vessel through a cotton filter. The depth of the cotton in the filter is 15 cm. The filtrate is aged gradually during settling.

Adjustment and Blending.—The room for blending must be kept at a temperature of 10°C (50°F) or less. The process of blending starts 16 days before bottling. The liquid in the blending bowl is mixed thoroughly with air, and then the components and color are analyzed. To adjust the proper color of the liquid, caramel is added. The quality of shao-hsing wine after blending is 14.3–15.0% of alcohol, 0.36–0.5% of total acid (i.e., using lactic acid) and less than 0.5% of sugar. The blended wine should be pasteurized at 65°C (149°F), then transferred to a closed tank to cool, and finally packed before it is sold.

Sake

The rice wine in Japan, similar to shao-hsing wine in China, is sake which is transparent and pale yellow in color with a specific gravity of 1.0, alcohol content of 15–16%, little acidity and slight sweetness. About 1.6 million kl of sake were produced in 1970 by more than 3000 breweries, but only about ⅛ of the breweries are large ones, each of which produces more than 500 kl of sake in a year (Murakami 1972). The brewing of the sake is characterized by (1) using rice as the raw material and koji from a culture of mold to digest the rice enzymatically, and (2) by a natural mixed culture of various microbes utilizing some interactions between them and their fermentation products at low temperature. In the brewing of sake the saccharification by mold enzymes and the fermentation by sake yeast proceed simultaneously. If a high proportion of solid matter (rice grain) and high proportions of sake yeast are present, these characteristics have been considered to be the cause of high alcohol content (over 20%) in the fermentation mash.

Treatments of water or rice grain in the brewing of sake have been reviewed exclusively (Nunokawa 1972). The differences between the treatments in the brewing of sake and shao-hsing wine are summarized in Table 19.8. The characteristics of microorganisms in the brewing of sake described by Murakami (1972) are summarized in Table 19.9.

In the culture of seed mash (yeast inoculum), as the enzymatic digestion of rice by koji advances, some nitrate-reducing bacteria start to grow and produce a certain amount of nitric acid by which the over-propagated yeasts can be temporarily depressed. Various lactic acid bacteria gradually grow and produce lactic acid, the nitric acid completely disappears and then the *sake* yeast begins to grow. These yeasts show high

TABLE 19.8. COMPARISONS BETWEEN THE BREWING PROCESSES OF SHAO–HSING WINE (HAU–TIAO AND SHAO–HSING) IN TAIWAN AND SAKE IN JAPAN

	Hua-tiao	Shao-hsing	Sake
1. Raw materials	glutinous rice, wheat, lactic acid	glutinous rice, wheat, pon-lai rice (Japonica type), lactic acid, sulfate salts, phosphate salts	rice (Japonica type)
2. Steamed rice	steaming in a liquid form by pressure and steamed rice	steamed rice	steamed rice
3. Seed mash	using liquid seed starter, by aeration with pure culture (63 hr)	batch culture incubated and low temperature (10°C (50°F), 15 days)	batch culture incubated under low temperature (15°C (59°F), 21 days)
4. Rice starter	liquid state starter (glutinous rice) $Rhizopus$ sp.	solid state rice starter (pon-lai rice) $Asp.$ $oryzae$	solid state rice starter $Asp.$ $oryzae$
5. Wheat starter	solid state wheat starter $Asp.$ sp.	solid state wheat starter $Asp.$ $oryzae$ Shaohsing	—
6. Pre- and finishing brewing	pre-brewing (7–8°C, 44.6°–46.4°F), 12 days finishing brewing (18–20°C, 64–68°F), 18 days	same as left	—

Source: Lin et al. (1975); Liu et al. (1959A); Nunokawa (1972)

TABLE 19.9. PRINCIPAL MICROORGANISMS AND THEIR ROLES IN SAKE BREWING

Microbes	Functions	Process	Origin
Molds (A. oryzae)	enzymatic digestion of rice and flavor- ing the sake	koji making	koji
Bacteria (Pseudomonas)	production of nitric acid from nitrates in water	culture of the seed mash	brewing water rice
(Leuconostoc) (Lactobacillus)	production of lactic acid followed by elimination of nitric acid		
Yeast (Sake yeast)	production of alcohols, flavor, and aroma	main mash	natural and/or pure culture

Source: Murakami (1972).

resistance to the acidity of lactic acid and propagate to the extent of cell numbers of $3-4 \times 10^8$ per ml of the mash with purity at 95−99% of all the microbes in the mash. The sake yeasts are somewhat different from S. cerevisiae (Murakami 1972). They usually show:

(1) poor growth at 35°C (95.0°F) in Berkholder's medium containing β-alanine instead of pantothenic acid;
(2) they have different characteristics with respect to biotin require-ment and assimilability of sodium during potassium deficiency;
(3) the yeasts agglutinate with Lactobacillus plantayum but not with L. casei; and
(4) the yeasts are absorbed on the surface of gas bubbles in the solution. The nonfoaming sake yeasts selected by bubbling make it possible to use a smaller fermentation tank in many factories (Murakami 1972).

During the storage in bottles, sake is sometimes damaged or spoiled by the invasion of certain kinds of lactic acid bacteria called hiochic bac-teria. The hiochic bacteria are mostly found to show certain biological characteristics different from those of the common lactic acid bacteria (Table 19.10). Most of the hiochic bacteria essentially require the pres-ence of hiochic acid (mevalonic acid) for their growth (Tamura 1958A,B; Tamura and Nagura 1958; Tamura and Suzuki 1958). They can grow well in a solution with a high alcohol content, reaching over 20%, and at a pH ranging from 4.0 to 5.5. The hiochic bacteria have been classi-fied by Kitahara et al. (1957) into 4 types according to their sugar fer-

TABLE 19.10. CLASSIFICATION OF HIOCHIC BACTERIA (64 STRAINS) [1]

Fermentation scheme	Growth on nutrient broth (pH 9.0)	Requirement of Hiochic acid	Classification
Hetero-fermentative	−	E	*L. heterohiochii* True hiochic bacteria (44 strains)
	+	N	*L. fermentum* Hiochi natured lactobacilli (1 strain
Homo-fermentative	−	variable	*L. homohiochii* True hiochic bacteria (14 strains)
	+	N	*L. acidophilus* (1 strain) *L. plantrum* (4 strains) Hiochi natured lactobacilli

Source: Kitahara *et al.* (1957).
[1] Common characteristics: Gram positive, catalase negative, rod shaped, optimum growth temperature 28°–33°C (82.4°–91.4°F).
− = no growth
+ = growth
E = essential
N = independent

mentation schemes and hiochic acid requirements. Recently, some other physiological characters have been investigated. It showed that most hiochic bacteria belong to the true hiochic bacteria. The kind of carbohydrates fermentable by *Lactobacillus homohiochic* seem to be somewhat different from that reported formerly.

Murakami (1972) briefly described the procedures for the brewing of sake. The raw rice is polished to remove the outer layers of raw kernel. The weight ratio of polished rice (white rice) is 70−80% and sometimes 50−60%, according to the quality of the sake. Some parts of the white rice are steamed and, after being cooled to about 40°C (104°F), are transferred into a warm room of 28°−30°C (82.4°−86°F). The spores of *A. oryzae,* called *tane-koji* (moldy starter), are sprayed on the rice followed by culturing for about 40 hr. The koji is thus completed.

The seed mash is cultured either with a classic method or a modern one. The classic method is to culture the sake yeast in a mixture of steamed white rice, koji and brewing water at a low temperature for 3−4 weeks. The modern method is to culture at a comparatively high temperature for 1−2 weeks with the addition of lactic acid and the cultured yeasts. After the seed mash is completed, it is transferred into a large tank and the addition of steamed rice, koji, and brewing water is carried out in 3 steps for 3 days by distributing these materials. Then the main mash begins. The temperature is initially 7°−8°C (45°−47°F), which gradually rises to a maximum of 15°−16°C (59°−61°F), finally lowering to room temperature in about 20−25 days. The mash is then filtered to remove solid matter such as yeast cells, and undigested rice which is called sake cake and the sake is obtained by the usual treatments shown in Fig. 19.10.

Based on the long history of brewing experience and the results of recent studies, it has been found that the rice bran should be polished to remove more than 25−30% of the original weight of the raw kernel. The maximum temperature of the main mash should be lower than 14°−16°C (57°−61°F) and the brewing water should not contain more than 0.02 ppm of iron. After heat pasteurization, sake sometimes becomes turbid with many small white particles produced by the coagulation of the protein of saccharogenic enzymes of A. oryzae after heating. These particles could be excluded by the addition of certain precipitation reagents. Under UV-light, sake usually shows blue fluorescence which is caused by a photochemical reaction between acetaldehyde and tryptophan under sunlight. Any presence of iron in water has been considered undesirable as a cause of deep coloration of sake. Several kinds of ferrichrome mainly consisting of white deferri-ferrichrysin are found in sake and koji. The deferri-ferrichrysin in koji change quickly to red ferrichrysin by chelating with iron in the brewing water, resulting in the deep coloration of sake. The koji for the brewing of sake is usually white but sometimes turns brown with age in the air. This browning is caused by the tyrosinase of A. oryzae. L-Dopa was found in koji as a precursor of melanoidines. Browning was prevented by the addition of SO_2 to koji. The coloring of sake is caused by 3-deoxyglucosone. A compound in rice consisting of glucose-glutamic acid-ferulic acid-calcium is highly influenced by the growth of koji mold (Yoshizawa et al. 1970). In order to exclude iron from the brewing water, various types of water treatment have been devised and set up in most of the factories. Brewers become able to exclude a trace of iron in water by settling up an apparatus suitable for each type of water.

Semicontinuous sake brewing has been worked on by Hayashida et al. (1972) (Table 19.11 and Fig. 19.11). The first characteristic of semi-continuous brewing of sake is the 4 main stainless steel vessels arranged in a stair-step fashion. The second characteristic is that the seed mash process of the batch system is omitted and the total period for the production is reduced by the addition of a preliminary vessel (¼ the capacity of the main vessels) from which the unsterilized feed with a controlled microbial population is steadily drained. The third characteristic is the use of vertical partition plate for transferring heterogeneous mash without any stirring that makes it possible to maintain the continuity of the brewing process.

The koji molds have been long considered to belong to yellow green

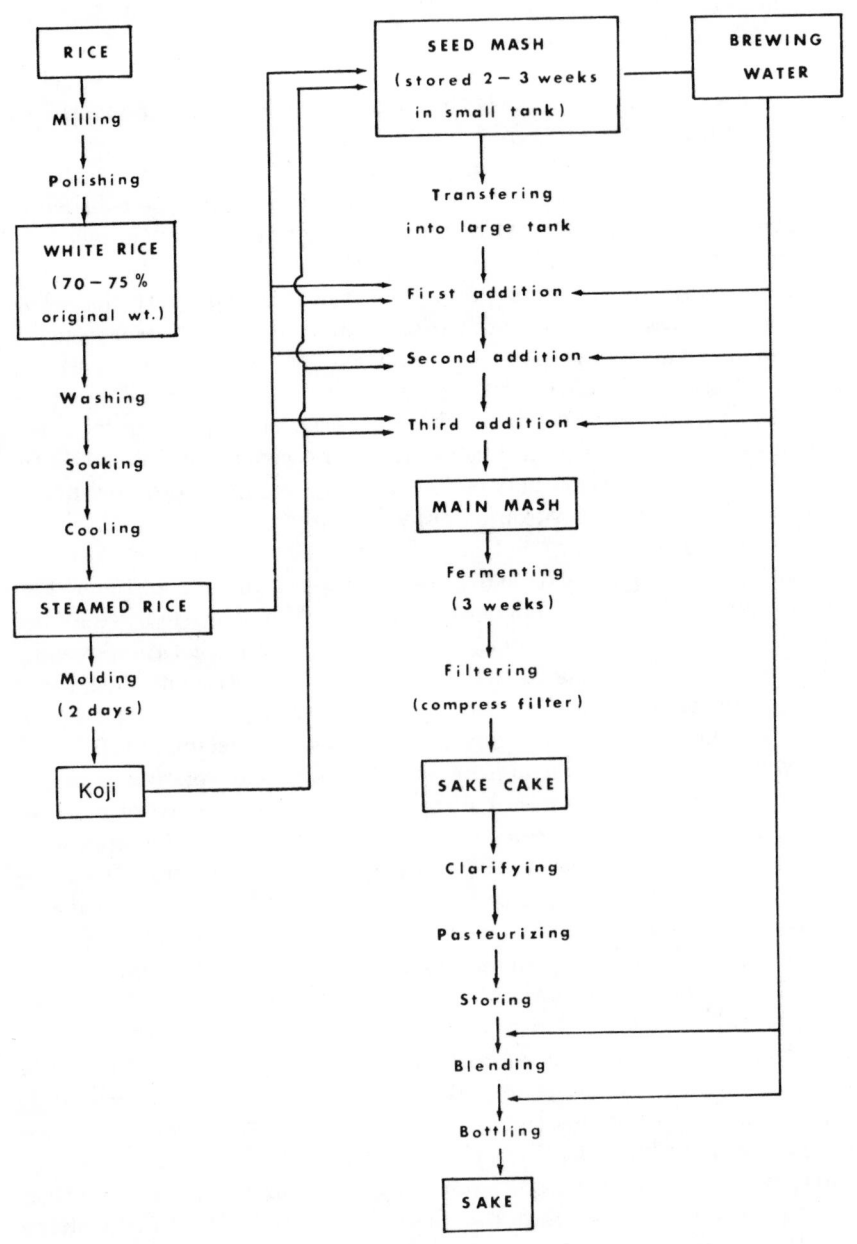

From Puh-li Winery (1967, 1974)

FIG. 19.10. FLOW DIAGRAM OF SAKE BREWING

TABLE 19.11. MASH CONDITIONS FOR SEMICONTINUOUS SAKE BREWING

Total weight of rice in each vessel: 12,000 kg.
Dilution rate: 0.25/day
Temperature: 13°–15°C (55.4°–59°F)
Composition of feed:
    Total weight of rice: 300 kg
    Ratio of koji to total weight of rice: 24%
    Ratio of mash concentration: 13%
    Lactic acid (75%): 200 ml

Source: Hayashida *et al.* (1972).

aspergillus, *A. oryzae.* By multivariate statistical analysis, Murakami (1972) claimed that the koji molds are definitely different from the other group of yellow green aspergilli, aflatoxin-producing *A. flavus.* In this connection, it is worthwhile to note that Wang *et al.* (1968) could not find aflatoxin in the semifinished and finished products of rice wines produced in Taiwan.

Yellow Wine

The so called yellow wine of Shang-hai is actually shao-hsing wine or a likeness to shao-hsing wine. In the province of Shang-tung, yellow wine is made from yellow rice (i.e., glutinous corn) and wheat starter, etc., whereas in Taiwan yellow wine is made from polished pon-lai rice, rice starter, wheat starter and seed mash. The latter is a wine similar to brewed shao-hsing wine and has an aroma similar to that of distilled rice liquor.

The brewing takes place in a cooled (6°–7°C, 43°–45°F) room especially designed for the process of fermentation. It takes about 9 months to brew yellow wine. The process is faster than the brewing of shao-hsing wine and slower than the brewing of rice liquor. Hence, the quality and cost of yellow wine lie between those of shao-hsing wine and rice liquor.

The flow sheet of yellow wine brewing is given in Fig. 19.12.

Rice Liquor

Rice liquor is a traditionally distilled liquor in Taiwan. It is colorless, transparent and equivalent to shao-chiou in Japan. The flow sheet of making rice liquor is given in Fig. 19.13. The process of making rice liquor is a modified Amylo-process. Starch hydrolysis and fermentation take place simultaneously when *Rhizopus javanicus* Takeda and *Saccharomyces peka* Takeda are added at the same time. The yeast is suitable for the fermentation at high temperatures (35°–38°C, 95.0°–100.4°F). The alcohol content and acidity (using acetic acid) of the rice liquor are 23% and 0.06%, respectively.

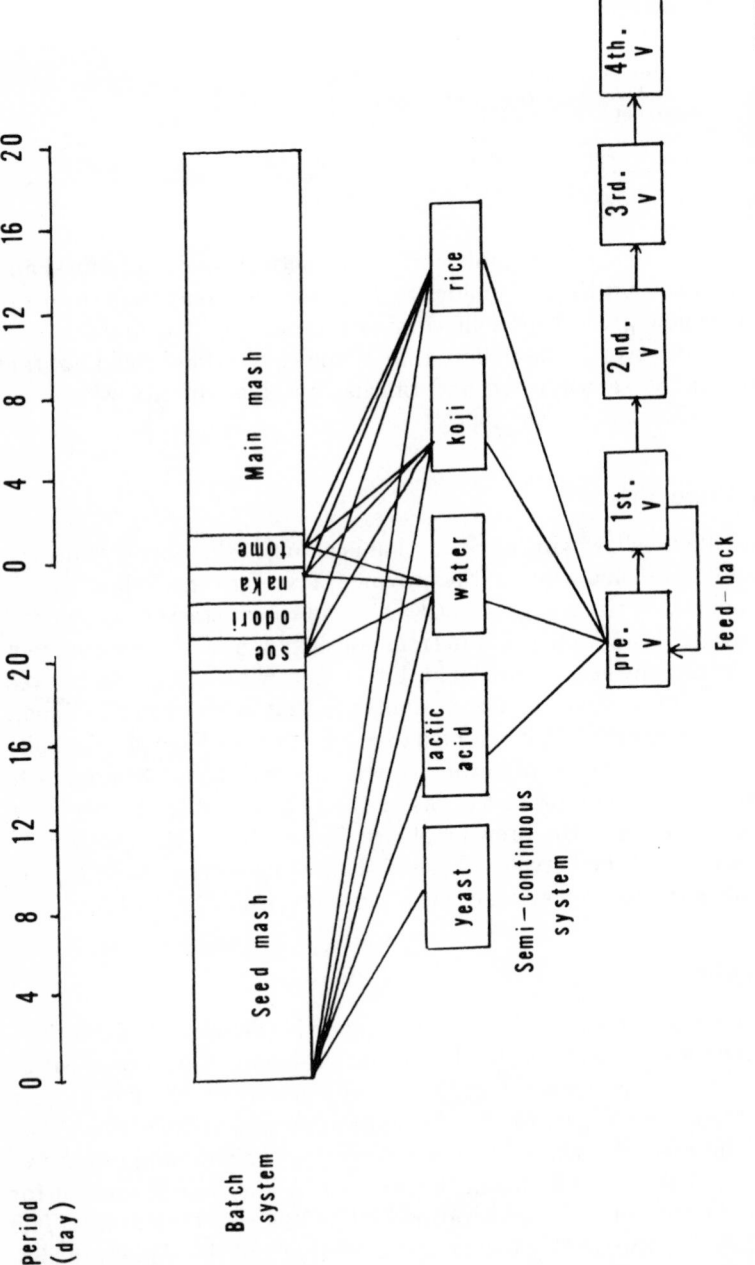

From Nunokawa (1972)

FIG. 19.11. A COMPARISON BETWEEN THE BATCH AND THE SEMICONTINUOUS SYSTEM OF SAKE BREWING

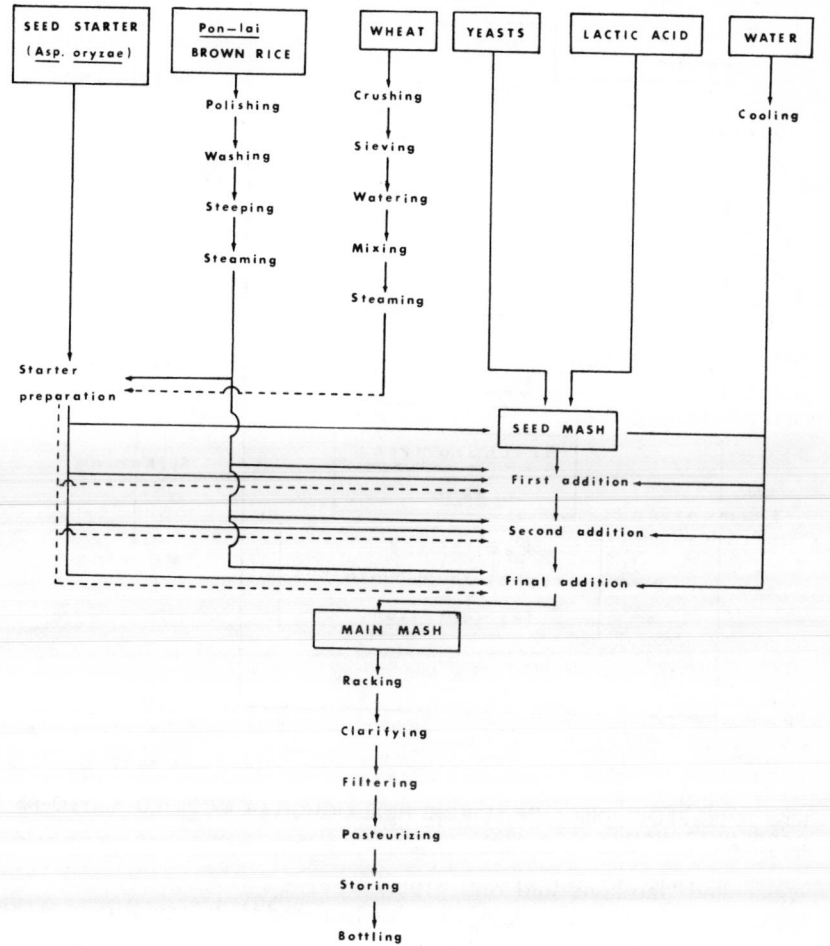

From Chung-Hsiung Brewer, Taiwan

FIG. 19.12. FLOW SHEET OF YELLOW WINE BREWING

Hong-ru Wine

Hong-ru wine has been one of the favorite wines in Taiwan. It originates from lao-hong wine of the Province of Fu-chen in mainland China. A hundred years ago, lao-hong wine was brought from the mainland of China to Taiwan and 30 years ago the Monopoly Bureau of Wine of Taiwan Province changed its name to Hong-ru wine. Owing to the deep red color which indicates the presence of the *Monascus* pigments of the freshly brewed hong-ru wine, the name was called "hong (red) wine" at first. The product should be stored for 2 years before marketing. Then, it

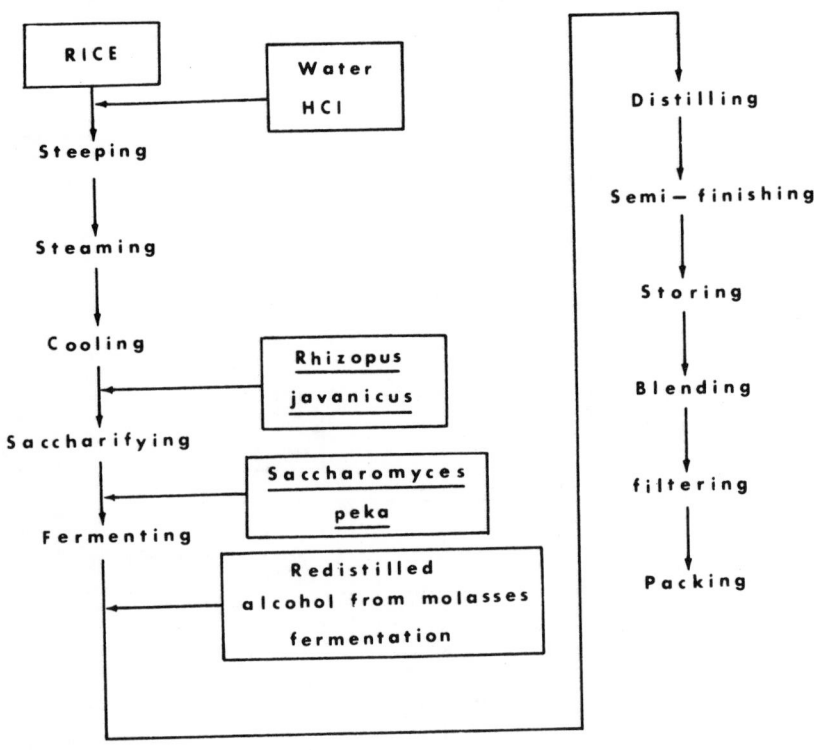

From Taichong Winery (1969)

FIG. 19.13. FLOW SHEET OF MAKING RICE LIQUOR (A MODIFIED AMYLO-PRO-CESS) IN TAIWAN

was called "lao-hong (old red) wine."

The process of making hong-ru wine is a modified Amylo-process (Fig. 19.14). At first, the saccharifying mold *(Rhizopus javanicus* Takeda) is added into the steamed glutinous rice in a closed, sterilized fermentation tank. Thirty-five hours later, after the inoculation of mold *(Rhizopus javanicus* Takeda), the yeast *(Saccharomyces formosanesis)* is inoculated under aseptic conditions to progress the fermentation and 20–24 hr later, the "red starter" (Fig. 19.15) is added for further fermentation. After fermentation, redistilled alcohol is also added. In such a case, the freshly brewed hong-ru wine is deep red in color, and has a fluorescence of purplish-green. The color will fade after storage and will become pale yellow after it has been stored for a year. At this time the fluorescence will also be lost and a special flavor produced.

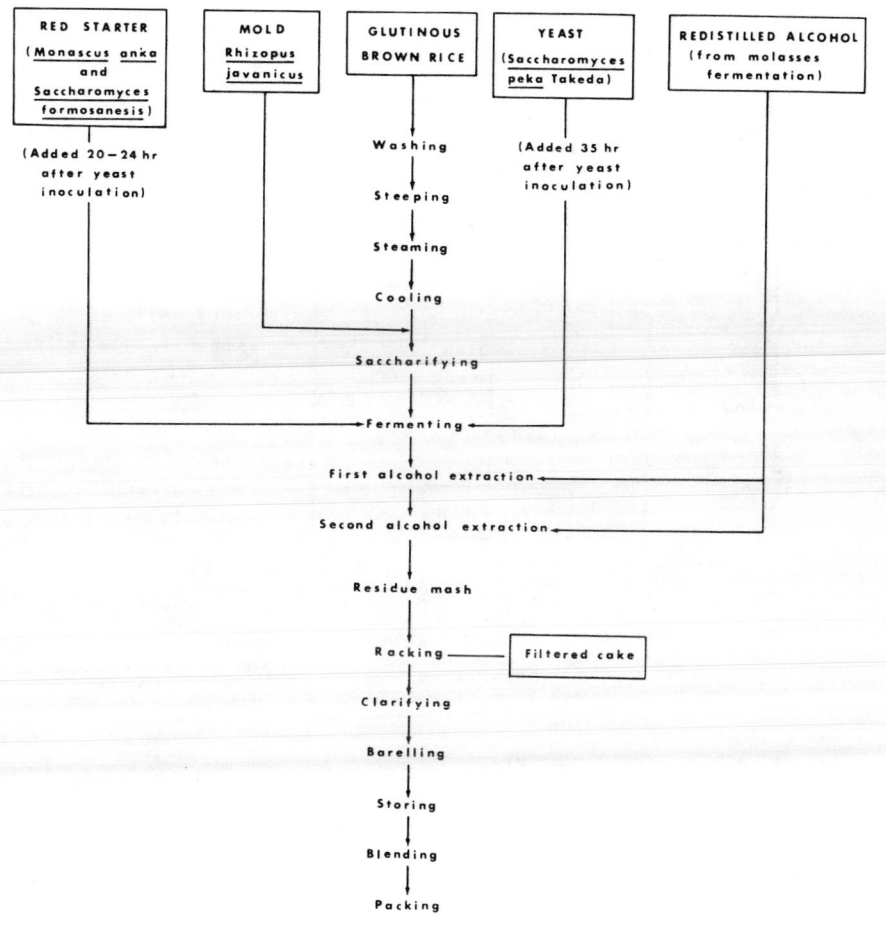

FIG. 19.14. FLOW SHEET OF HONG-RU WINE PRODUCTION (MODIFIED AMYLO-PROCESS)

From Su et al. (1970)

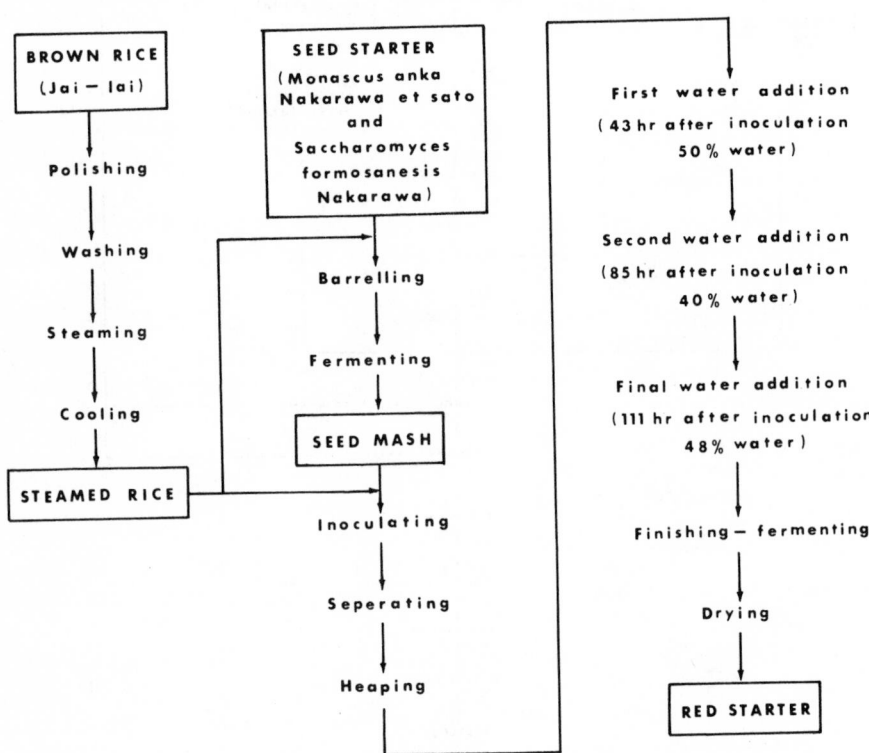

FIG. 19.15. FLOW SHEET OF MAKING RED STARTER

'Cementic barrel

REFERENCES

ANON. 1968. *In* The Methods of Farming. (Chi Ming Yao Su, the oldest farming book of China). Reprinted by the Commercial Press, Taipei, Taiwan. (Chinese)

ANON. 1974A. *In* The Outline of Herbs. (Peng Tsao Kang Mu). Reprinted by the Horng Yeh Book Co., Taipei, Taiwan. (Chinese)

ANON. 1974B. *In* The Outline of Herbs. (Nang Fang Tsao Mu Chung). Reprinted by Yee Wee Publishing Co., Taipei, Taiwan. (Chinese)

CHENG, Y.S. 1970. Studies on amylase from *Monascus anka.* M.S. Thesis. National Taiwan University, Taiwan.

HAYASHIDA, S., KAMACHI, T. and HONGO, M. 1972. Semi-continuous sake fermentation. Fermentation Technology Today. Proc. of the 4th Intl. Ferment. Symp. Kyoto.

HERZFELD, H.C. 1957. Rice fermentation in Ecuador. Econ. Bot. *11* (3) 267.

HESSELTINE, C.W. 1965. A millennium of fungi, food and fermentation. Mycologia *57* (2) 149-197.

JOSEPH, K. *et al.* 1961. *In* A millennium of fungi, food and fermentation. C.W. Hesseltine. 1965. Mycologia *57* (2) 149-177.

JULIANO, B.O. 1972. The rice caryopsis and its composition. *In* Rice Chemistry and Technology. D.F. Houston (Editor). Amer. Assoc. Cereal Chem., St. Paul, Minnesota.

KITAHARA, K., KANEKO, T. and GOTO, O. 1957. Taxonomical studies on the hiochi-bacteria. I. Isolation and grouping of bacteria strains. J. Agric. Chem. Soc. *31,* 556-564.

LIN, C.F. 1972A. Production of glucoamylase by using temperature-sensitive mutant of *Rhizopus formosaensis* R 13-5 and its properties. J. Ferment. Tech. *50,* 371.

LIN, C.F. 1972B. Study on the production of hydrolytic enzymes of *Rhizopus thailandensis.* J. Chinese Agric. Chem. Soc. *10,* (3–4) 77–83.

LIN, C.F., CHANG, C.C. and IISUKA, H. 1974. Taxonomic studies on the yeasts used in shao-hsing wine manufacture. J. Ferment. Tech. *52* (3) 196–200.

LIN, C.F., CHANG, C.C. and LU, S.C. 1975. Taxonomy study on the *Rhizopus* sp. isolated from starter of shao-hsing wine. J. Chinese Agric. Chem. Soc. *13* (1–2) 81-87.

LIU, P.W. and CHEN, S.H. 1976. Personal Communication. Taipai, Taiwan.

LIU, P.W., CHEN, S.H., CHIU, T.H. and CHEN, C.C. 1959A. Studies on *shao-hsing* wine. I. Taxonomical studies of saccharifying mold from *chiu-chu* of *shao-hsing* wine. Chem. *48* (3) 118-188. (Chinese)

LIU, P.W., CHEN, S.H., LEE, T.H. and CHANG, K.T. 1960. Studies on *shao-hsing* wine. IV. On the pilot-plant experiment of *Lin-fan* wine. Chem. *49* (3) 155–159. (Chinese)

LIU, P.W. *et al.* 1959B. Studies on *shao-hsing* wine. II. On the manufacture of *tien-chiu-niang* and *lin-fan* wine. Chem. *48* (3) 189-195. (Chinese)

LIU, P.W. *et al.* 1959C. Studies on *shao-hsing* wine. III. On the pilot-plant experiment of *lin-fan* wine. Chem. *40* (4) 241-244. (Chinese)

MANTANI, S. 1968. Studies on ribonuclease *Rhizopus* fungi. Amino Acids and Nucleic Acids *18*, 64−70. (Japanese)

MURAKAMI, H. 1972. Some problems in sake brewing. Ferment. Tech. Today. Proc. 4th Intl. Ferment. Symp. Kyoto.

NAKANO, M., EBINE, H. and ITO, H. 1976. On the miso, a Japanese traditional fermented food, in view of ecological system of fermentation microorganisms depending on the types of miso productions. Proc. of 5th Int. Ferment. Symp., July 3, Berlin.

NUNOKAWA, Y. 1972. Sake. *In* Rice Chemistry and Technology. D.F. Houston (Editor). Amer. Assoc. Cereal Chem., St. Paul, Minnesota.

POMERANZ, Y. 1972. Rice in brewing. *In* Rice Chemistry and Technology. D.F. Houston (Editor). Amer. Assoc. Cereal Chem., St. Paul, Minnesota.

PUH-LI WINERY. 1969. The effect of different ratios of rice and wheat starters on the aroma of Shao-Hsing Wine. *In* Annual Report of Wine Factories (1967−1968). Taiwan Tobacco and Wine Monopoly Bureau, Taiwan, ROC.

PUH-LI WINERY. 1974. The use of a native variety of paddy rice in Shao-Shing brewing. *In* Annual Report of Wine Factories (1967−1968). Taiwan Tobacco and Wine Monopoly Bureau, Taiwan, ROC.

RAO, M.V.R. 1961. In meeting protein needs of infants and preschool children. Nat. Acad. Sci. Nat. Res. Counc. Pub. *843*, 291-293. (in process)

SHIBASAKI, S.K. and HESSELTINE, C.W. 1962. Miso fermentation. Econ. Bot. *16* (3) 180−195.

SU, Y.C. *et al.* 1970. Mycological study of *Monascus anka.* J. Chinese Agric. Chem. Sc. *8* (1−2) 40−54.

TAICHONG WINERY. 1969. The use of concentrated mash for rice-liquor brewing. *In* Annual Report of Wine Factories (1967−1968). Taiwan Tobacco and Wine Monopoly Bureau, Taiwan, ROC.

TAMURA, G. 1958A. Hiochic acid, a new growth factor for hiochi-bacteria. II. Isolation of hiochic acid. J. Agric. Chem. Soc. *32*, 707−712.

TAMURA, G. 1958B. Hiochic acid, a new growth factor for hiochi-bacteria. IV. Determination of structure of hiochic acid. J. Agric. Chem. Soc. *32*, 783−790.

TAMURA, G. and NAGURA, A. 1958. Hiochic acid, a new growth factor for hiochi-bacteria. I. Distribution of hiochic acid. J. Agric. Chem. Soc. *32*, 701-706.

TAMURA, G. and SUZUKI, Y. 1958. Hiochic acid, a new growth factor for hiochi-bacteria. III. Nutritional requirements of hiochi bacteria. J. Agric. Chem. Soc. *32*, 778-783.

TAWANATANI, T. and SHIBASAKI, I. 1972. Effect of moisture content on the microbicidal activity of propylene oxide and the residue in foodstuffs. J. Ferment. Tech. *50*, 351.

VAN VEEN, A.G. and STEINKRAUS, K.H. 1970. Nutritive value and wholesomeness of fermented foods. J. Agric. Food Chem. *18*, 576.

WANG, H.H. and CHIOU, Y.Y. 1977. The application of solid state fermentation in agriculture. Taiwan Mushrooms *1*, 25–33.

WANG, H.H. *et al.* 1968. A short method for the determination of aflatoxin in wine products. J. Chinese Agric. Chem. Soc. *6* (3–4) 69–75.

WANG, H.H. *et al.* 1970. Some properties of mutants *Rhizopus* spp. by UV irradiation. Chinese J. Microbiol *3*, 105–110.

WANG, H.H. *et al.* 1974. CO_2 level inside the composts and fructification of *Agaricus biosporus.* Mushroom Sci. *9*, 20–22.

WANG, H.H. *et al.* 1976. The role of steeping water of mung bean starch manufacture on environmental control in Chinese traditional wet process. Proc. from 5th Intl. Ferment. Symp., June 28–July 3. Berlin.

YAMASAKI, M. 1945. Notes on Far East fermentation chemistry. (Toa-Hakkokagaku-Ronko). Dai-chi Pub., Tokyo. (Japanese)

YEH, C.I. and CHIUCH, S.H. 1972. Study on protease and amylase produced in wheat koji used for the fermentation of *shao-hsing* wine. *In* Annual Report of Research Institute for Wines (1971–1972). Taiwan Tobacco and Wine Monopoly Bureau, Taiwan, ROC.

YOSHIZAWA, K., ITOH, M., AIDA, K. and OISUKA, K. 1970. On growth-inhibition against *Aspergillus oryzae* No. 40 by the compound III-3 in rice grain. Agric. Biol. Chem. *34*, 1262–1264.

YOSHIZAWA, K., KOMATSU, S., TAKAHASHI, I. and OISUKA, K. 1970. Phenolic compounds in the fermented products. I. Origin of ferulic acid in sake. Agric. Biol. Chem. *34*, 170–180.

20

Rice Snack Foods

Chin-Fung Li
Bor S. Luh

In the Orient rice is not only eaten as the main food but also as a snack food. Some of the rice snack foods are made from either glutinous rice or nonglutinous rice only, while others are made from both types. The reason for this difference in rice preparation is that the sticky characteristic of glutinous rice is necessary in some cases. Another reason for using glutinous rice in baked or popped snacks is that glutinous rice expands readily and produces a more porous texture.

The rice snack foods consumed in Japan, China and some other countries are introduced in this chapter.

RICE CRACKER

Rice cracker is a Japanese baked snack food made from rice. Arare and Senbei are the major and traditional rice crackers in Japan. The consumption ratio of Arare to Senbei is 100 to 40–45. The flavor and the soft taste of such products are quite different from western snacks which are rich in the flavor of butter or cheese. Although in past years Japanese traditional snacks and cakes have competed with western style cakes and snacks, rice crackers are now increasing sales each year. Rice cracker, chocolate and chewing gum are the 3 largest confectionary industries in Japan. In 1968, the annual production of rice cracker in Japan was 148,000 MT which amounts to 12% of the total snacks and bakery products (1,219,000 MT).

Many problems exist in the rice cracker industry. These problems are mainly due to the techniques of manufacturing which are kept secret and traditionally man-performed. Moreover, government control of rice distribution and the increasing price of rice in the past few years have limited the accessibility of this raw material. Although the future of this industry is thought to be promising, no large manufacturer would be interested in this product. In Japan, there are 2800 rice cracker factories, 95% of which have only 50 workers or less.

Although studies have been done on the refining of rice for brewing sake (rice wine) and the inorganic constituents and starches of rice, little research has been done on the manufacturing of rice crackers.

Classification of Rice Crackers

The method of processing rice crackers varies with the kinds of rice used. The quality of the products obtained from each method is not the same.

Rice crackers made from glutinous rice are generally called "Arare" or "Okaki" (Fig. 20.1). They are soft in texture and can be easily dissolved

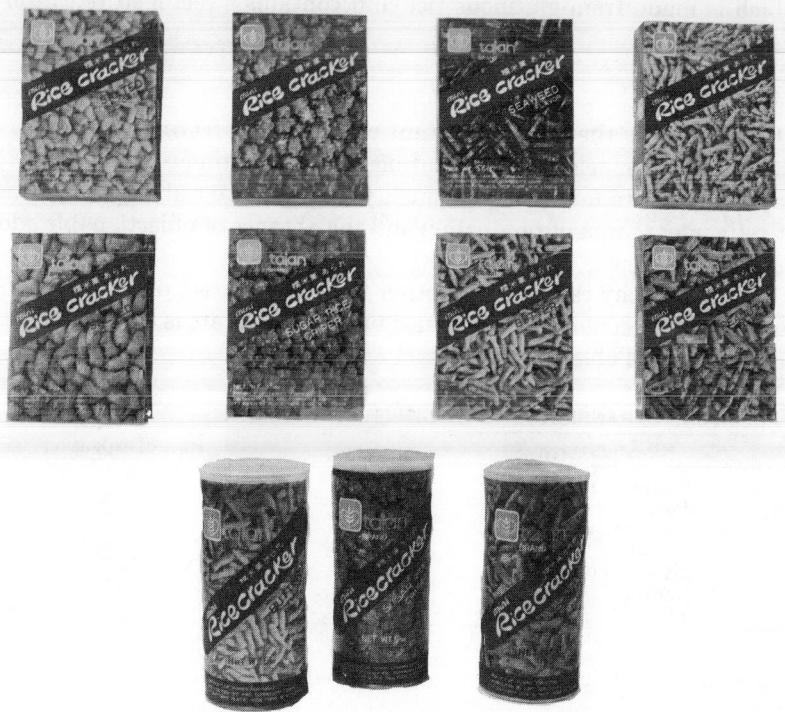

From Tai An Food Co.

FIG. 20.1. PRODUCT OF RICE CRACKERS

in the mouth. Rice crackers made from nonglutinous rice are called "Senbei," which has a hard and rough texture. The classification of rice crackers is as follows:

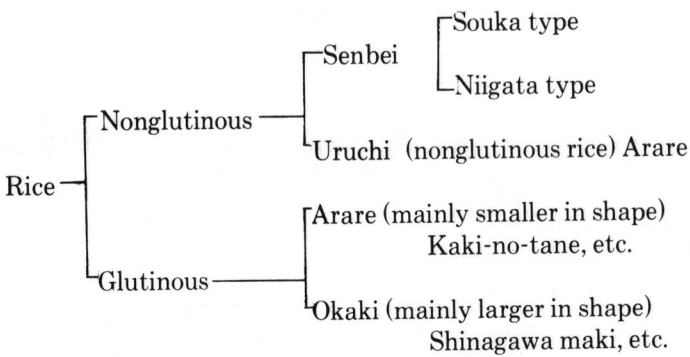

There are also modified rice crackers in Japan. One type is called Aghe (fried) Arare which is fried in oil; another type is called Monaka Shell which is made from glutinous rice and contains sweetened red bean.

Raw Materials

Rice.—Rice is the most important component affecting the quality of the crackers. In selecting the glutinous or nonglutinous rice, one has to pay attention to uniformity in quality, rate of water absorption, extent of refinement, area of production and the absence of objectionable odors and taste.

When a starchy raw material other than rice is used, potato starch is a suitable replacement. Consideration of expansion rate is essential for the replacement (Table 20.1).

TABLE 20.1. EXPANSION RATE OF DIFFERENT STARCHES

Kinds of starch	Rate of expansion
Glutinous rice	100
Potato	17.0
Sweet potato	15.5
Wheat	10.8
Nonglutinous rice	10.8
Corn	9.2
Pea	7.0

Soy Sauce.—Soy sauce is used for flavoring.

Other Raw Materials.—In order to improve the flavor and taste (preference), sea weeds, sesame, red peppers, sugar, pigments and spices may be used. For oil fried rice crackers the oil used must be of good quality, well refined and highly stabilized.

Processing Method

Glutinous Rice Crackers.—As shown in the flow sheet (Fig. 20.2) glutinous rice is 91% refined. It is washed by rice washing machines and soaked for 16−20 hr (usually 16 hr) in water at temperatures below 20°C (68°F). After draining the rice of 38% moisture, it is crushed by rollers into fine powder, passed through an 80-mesh sieve and steamed for 15−30 min. After cooling for 2−3 min it is kneaded 3 times. This kneaded cake is put in a cake vessel and quick-frozen to 2°−5°C (35.6°− 41.0°F) for 2−3 days for hardening. The hard cake is cut into various shapes and dried by hot air at 45°−75°C (113°−167°F) to a final moisture content of 20%. This cake is coated with soy sauce, spice and other

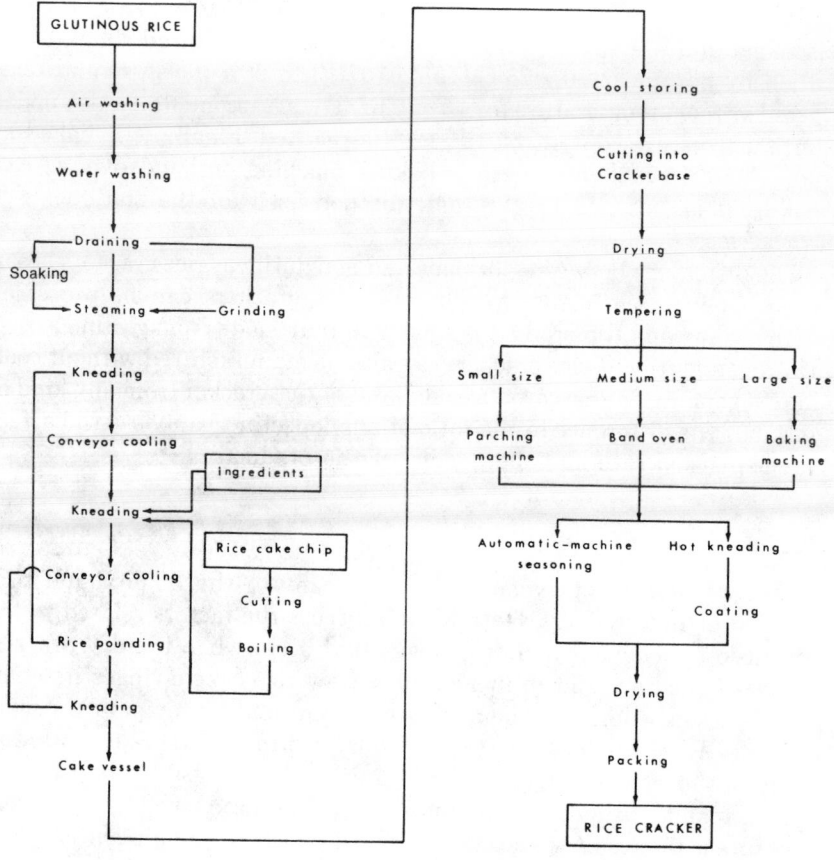

From Tai An Food Co.

FIG. 20.2. FLOW CHART OF RICE CRACKERS (ARARE)

seasoning materials and placed in a continuous baking machine or band oven. After baking it is dried through a finishing dryer at 90°C (194°F) for 30 min.

Currently the machines used for manufacturing rice crackers are almost all continuous. With the same machines we can manufacture both glutinous and nonglutinous rice crackers. The process of washing the cake chips is done by a machine that can treat 750–1000 kg of rice per hour. The conventional method takes 3–4 days to manufacture nonglutinous rice crackers, but now it only takes 3–4 hours by continuous processing.

Nonglutinous Rice Crackers.—After milling, the rice is washed, soaked in water to a moisture content of 20–30% and ground into powder. After adding some water it is placed into a kneading machine where it will be steamed for 5–10 min. After cooling to 60°–65°C (140°–149°F) the rice is rolled and pressed into thin layers and cut into desired shapes. It is dried by hot air at 70°–75°C (158°–167°F) to 20% moisture and tempered at room temperature for 10–20 hr. Then a second drying is applied until a moisture content of 10–12% is reached. Finally it is baked at 200°–260°C (392°–500°F) in a baking machine or band oven. After baking it is seasoned by the same method used for the glutinous rice crackers.

The difference between glutinous and nonglutinous rice crackers lies in their cooling treatment. Nonglutinous rice crackers can be processed with cooling and the product is similar to that made from glutinous rice. However, glutinous rice crackers can also be manufactured without cooling. In general, one can produce any kind of rice cracker from any kind of rice, but the difference in the ratio of amylopectin to amylose will affect the expansion rate and the quality of the products.

Puffing of Rice Cake

The mechanism of expansion is important to the quality and production of rice crackers. Here we will discuss the factors related to the expansion phenomenon. The changes that occur when the dry raw rice cake is baked are shown in Fig. 20.3. Raw rice cake changes its characteristics depending on the moisture content and air temperature upon heating. Raw rice cake softens gradually into a glass-like state and increases in extensibility. Due to the expansion pressure, the raw rice cake expands, dries and hardens to form the product. The expansion pressure is the result of the changes in the moisture and air content with rising temperatures. The raw rice cake is different from bread dough. Bread dough is in a glass state but has no extensibility at the first stage of heating and therefore cannot form an effective expansion pressure. In raw rice cake, moisture is sealed, i.e., in a tightly sealed container. If the

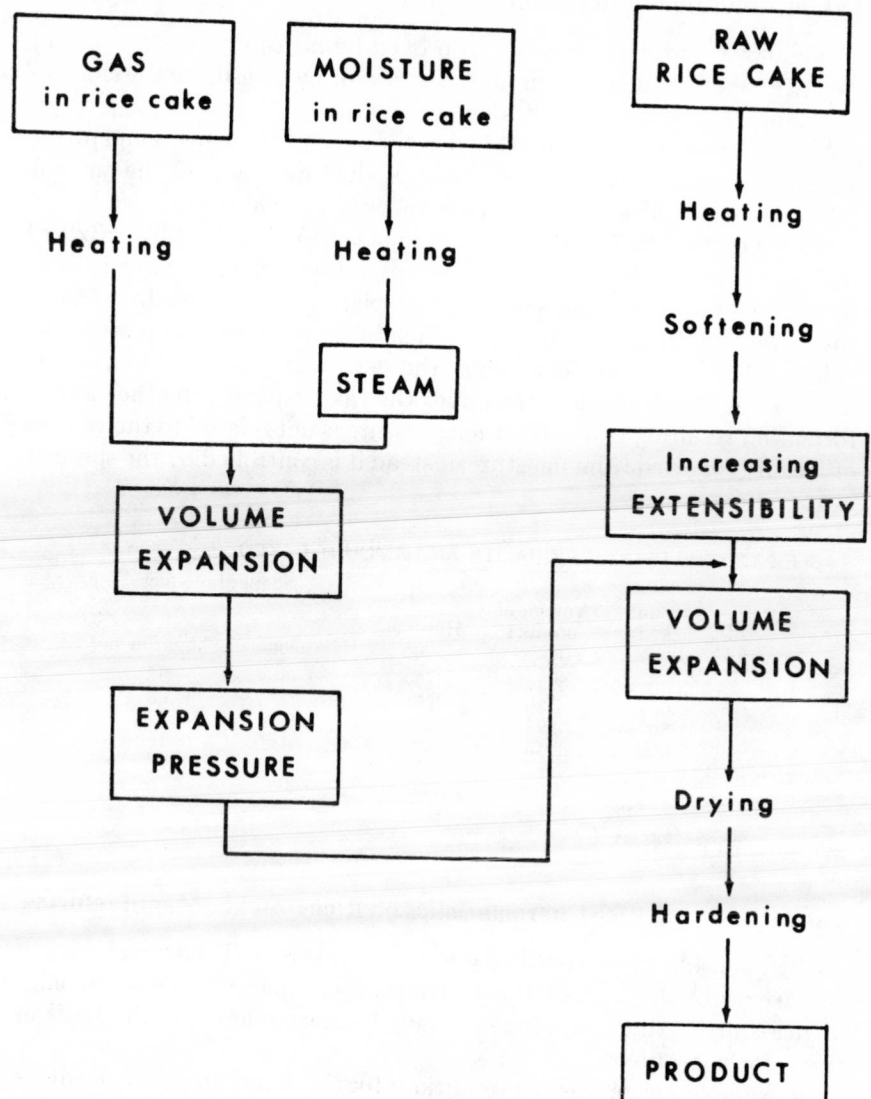

FIG. 20.3. CHANGES OF RAW RICE DURING BAKING

extensibility of the cake is strong after being heated to 100°C (212°F), an outer pressure forms and the evaporating temperature will increase. It will continue to expand until the steam pressure and extensibility are balanced. For this reason the expansion phenomenon of raw rice cakes is considered to be controlled by the evaporation of moisture and the physical state of the rice cake.

Baking Conditions and Product Quality

The above consideration may be proved by measuring the amount of raw rice cake baked and its product volume, or by measuring the changes in temperature during the baking processes.

When the raw rice cakes are baked in the oven at the same temperature as shown in Table 20.2, the volume of product decreases as the amount baked increases. The temperature changes of rice cake, as shown in Fig. 20.4, have the transition points at about $145°-165°C$ $(293°-329°F)$ where the temperature rate is lowered. By observing the baking process we understand that expansion is taking place at this point. In addition the period of this transition point has a close relationship with the volume of the product. The longer the period, the smaller the volume, and the shorter the period, the larger the cake expands. In other words, formation resulting from expansion pressure is not related to the volume of steam generated from moisture; instead it is controlled by the speed of evaporation.

TABLE 20.2. THE PRODUCT QUALITY AND AMOUNT BAKED

No.	Amount baked (kg)	Volume of product (ml/g)	Hardness (kg)	Shape of product	
				Ukimono[1]	Shimarimono[2]
1	3.00	2.44	0.73	3	97
2	2.00	2.59	0.58	7	93
3	1.50	2.98	0.81	21	79
4	1.00	3.45	0.58	84	16
5	0.75	3.87	0.34	100	0
6	0.50	4.04	0.38	100	0

[1]Ukimono: like Fubuki-Arare (loosely expanded product).
[2]Shimarimono: like Kakinotane (tightly expanded product).

The Influence of Rice Characteristics on Rice Cracker Manufacturing

When considering extensibility of rice crackers and characteristics of rice, we might think about the dextrinization, expansion and wettability of rice starch. Many experiments have been conducted on the basis of these considerations.

For example, the processing conditions for the manufacture of glutinous rice crackers are different from those for rice grown on paddy or dry land. It was impossible to manufacture the rice crackers made from dry land rice by the techniques used for paddy land rice. This difference can not be explained by the physical-chemical properties of starches only. As shown in Table 20.3, almost no differences exist in the results measured by the amylogram. No differences exist in dextrinization, expansion and wettability, which all might be related to extensibility. However, the table shows that there are differences in the characteristics of rice crackers using different methods of kneading and milling.

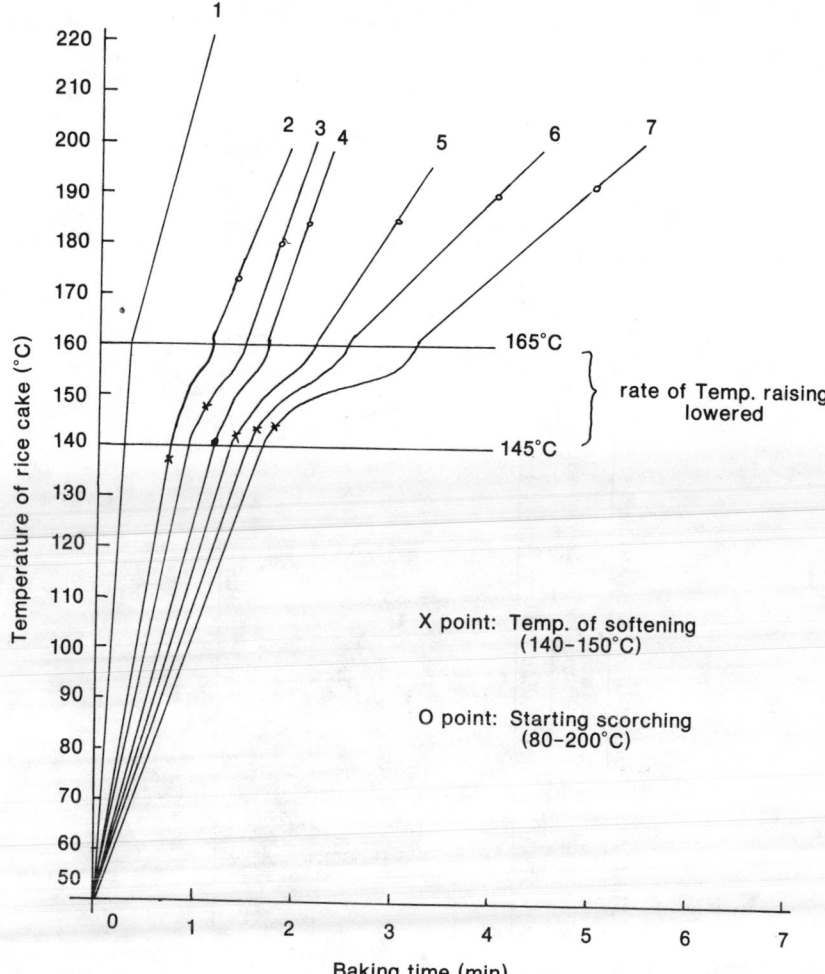

FIG. 20.4. CHANGES OF TEMPERATURE IN RICE CAKE

Amount baked: 1—control; 2—0.5 kg; 3—0.75 kg; 4—1.0 kg; 5—1.5 kg; 6—2.0 kg; 7—3.0 kg.

The Effect of Kneading on the Quality of the Products.—Rice flour is steamed and kneaded by a machine for different periods of time and is made into rice crackers by the ordinary method. The results show that there is an optimum time for kneading; those kneaded for shorter or longer periods have a nonuniform texture.

Kneading affects the physical properties of rice flour. These properties

TABLE 20.3. RESULTS FROM MEASURING AMYLOGRAM OF RICE STARCHES (TEMPERATURE °C, VISCOSITY BU)

Species	Starting dextrinization temp.	Maximum Viscosity		Visc. at 97.5°C	97.5°C, 10 min.		Minimum Viscosity		Viscosity lowering rate		Viscosity at 30°C (B)	B/A
		Temp.	Visc. (A)		Start	End	Temp.	Visc. (C)	A-C	A-C/A		
Paddy land grown												
Shimuzu glutinous	62.3	59.0	945	495	495	475	95.7	475	470	0.497	800	0.85
Chiukou No.2	61.8	66.8	880	427	427	427	94.5	379	501	0.569	680	0.77
Echigo nebari	60.5	65.3	905	445	445	415	95.1	410	495	0.547	670	0.74
Dry land grown												
(New rice)	60.0	67.5	1,045	511	511	465	93.0	445	600	0.574	615	0.59
(Old rice)	59.3	66.0	1,090	515	515	455	94.8	435	653	0.601	740	0.68

in turn affect the uniformity of the products made from rice flour.

The effect of kneading is more distinct for glutinous rice than for nonglutinous rice. The rice flour kneaded for a relatively longer period has a greater expansion rate than that which is kneaded for a shorter period.

Uniformity of Rice Cake and Characteristics of Rice.—As mentioned previously, the uniformity of rice cake is very significant, and here the different characteristics of rice which might affect the uniformity are considered.

Structure of Rice Grains.—Rice is not the assembly of chemical compounds such as starch, protein, etc. It is a plant seed and therefore has a plant cell structure. Rice grains have a definite arrangement in their cell structures and their compactness differs according to the positions of these cell structures. Ordinarily the outer portion of the rice grain is more rigid than the inner portion which is related to the physical properties of starch in the rice.

Distribution of Water Absorption.—The compactness of rice soaked in water varies with the distribution of absorbed water. The outer layer of rice grain, especially the layer of rice bran, is mostly composed of protein and therefore the water absorption rate in this portion is the highest. The layer next to the rice bran, the outer layer of the milled rice, shows the lowest water absorption rate. However the rate of water absorption increases as the core of the rice grain is approached.

Uniformity of Steamed Rice.—Differences in the water absorption rate give rise to a lack of uniformity in steamed rice. Upon observing the profile of steamed rice, one sees that the starch granules of the outer portion of the soft cells have retained their original shape, whereas the starch granules in the core portion no longer exist. Because the outside portion is hard, it is difficult to destroy it by kneading. It is this hardness that is responsible for the shape of raw rice cakes.

Granular Size After Grinding.—The water absorption rates also affect the granular size after grinding.

If the moisture content is higher, the rice can be ground further to a finer consistency. The outer portion of the rice grain contains less water and as a result the granules are coarser on grinding while those in the inner portion become finer.

Rice at 25% moisture content is ground, dried and separated by sieves according to different granular sizes. The fine powder of 250-mesh apparently has a lower dextrinizing temperature than that of 60-mesh. Generally, the finer the powder, the earlier it starts dextrinization. It also possesses the property of good expansion.

Application of Microwave Heating in Rice Cracker Manufacturing

A series of experiments concerning the application of microwave heating in the manufacturing of snack foods has been made (Li *et al.* 1973, 1976). One experiment dealt with a kind of Arare flavored with fish meats. It employed the method of preparing 2 parts of raw material. The first part was to soak 1 kg of nonglutinous rice (coarsely ground) and 1 kg of waxy corn (ground to small granules) in 30°C (86°F) water for 2 hr. The product was cooked for 15 min, cooled down to 40°C (104°F), ground and kneaded in the machine for 10 min. The second part involved blending 300 kg of frozen fish meat, 15 g of salt, 30 g of sugar, 80 g of dried egg white powder and 150 ml of water in a blender. The prepared raw materials were mixed together, kneaded for more than 5 min and then extended to the size of 300 mm $\times$ 400 mm $\times$ 120 mm dough. The product was frozen to a temperature of -5°C (23°F), cut into pieces of 5 mm $\times$ 25 mm $\times$ 3 mm (375 mm^3) and dried at 50°C (122°F) by blowing dried air (1.5 m/sec) for 1 hr. Then it was steamed for 2 hr, dried for 30 min and steamed in a tightly sealed container for 2 hr. At this point, the moisture content of the dough reached about 25% and the volume shrank to 3.5 mm $\times$ 22 mm $\times$ 2 mm (154 mm^3). One kilogram of dough was then mixed with salt heated to 150°C (203°F) and treated with microwave energy at an average rate of 12 kg of mixture in 45 sec. After treatment, the dough was dried and formed into a rice cracker having a moisture content of 7.5%, an average size of 8 mm $\times$ 35 mm $\times$ 7 mm (1960 mm^3) and a yield of 9.2 kg. The expansion rate of frozen dough to a product is 5.2 times greater in volume and that of dried dough to product is 12.7 times greater.

Li *et al.* (1973, 1976) used a microwave conveyor (Genesys Model 4003), capable of generating 5 kw of power at 2450 MHz instead of the conventional baking used in manufacturing Arare. During the baking process as in the conventional method the dough was not piled up. Otherwise the product did not expand uniformly. Only that which was directly illuminated with infrared light, or heated by flame could be baked. The dried dough was obtained from a rice manufacturing plant and treated by microwave heating. The results showed that 5 kw was more suitable. Within minutes it expanded to a volume almost 5 times its original volume (Fig. 20.5). It was observed that when the dried dough was treated, the larger pieces expanded with greater ease than the smaller pieces. When the dried dough was buried in salt, they found that it expanded more uniformly. A comparison of the product baked by flame with one treated with microwave energy showed that the former is more uniform than the latter. Although the former had a brown color and looked more attractive, the latter saved time and energy.

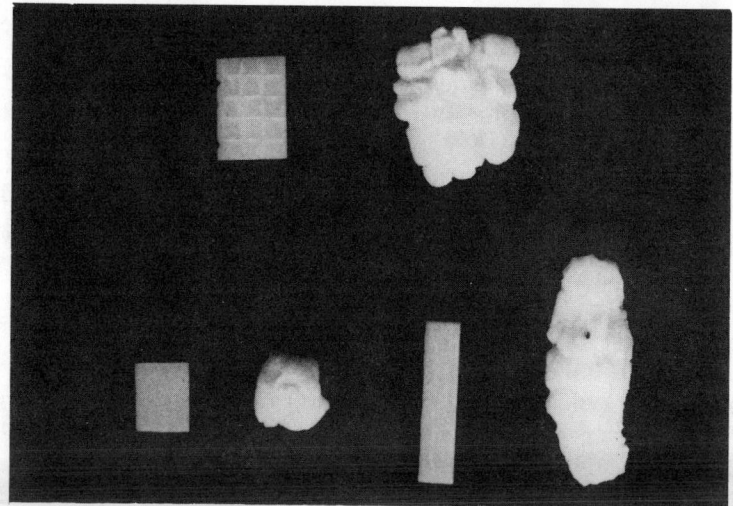

FIG. 20.5. COMPARISON OF CRACKER BASE AND RICE CRACKER
AFTER MICROWAVE TREATMENT

INSTANT RICE NOODLES

In China, rice noodles are called Mii-fen and in Japan, Harusame.
There is a difference between rice noodles (Mii-fen) and Harusame.
Mii-fen is made from rice only. Harusame may be made from Mung bean,
starches, rice or a mixture of them.

In China or Japan, rice noodles are consumed as main foods, soups or
snacks. Recently in Taiwan rice noodles have been manufactured as
instant food which can be easily reconstituted in hot water. Some consumers prefer instant noodles because of its convenience. Also the price
of such a product is very reasonable.

Manufacturing Process of Instant Rice Noodles

There are few references concerning rice noodles. Lin (1972) introduced a conventional rice noodle manufacturing method which is widely
used in Taiwan. Sakurai *et al.* (1975) also described rice noodles as
having been widely consumed in China, Japan and southeast Asian countries since ancient times. A raw material for rice noodle is nonglutinous
rice, which is highly elastic and sticky. The use of nonglutinous rice
enables easy extrusion of the dough. Like noodles, this rice should not
stick together or break. The yield from the rice is about 95%.

The conventional processing method is limited largely to sun drying.
There are a few exceptions using hot air drying. In order to improve the
quality of the rice noodle, Chen *et al.* (1971) tested the infrared drying
methods. They compared the final products by infrared drying, hot air

drying and sun drying. The results showed that the color and rehydration properties of the infrared dried rice noodle is superior to the hot air dried and sun dried noodles. They observed the cross section of the infrared dried noodle under microscopes and discussed the reasons contributing to the superiority of the infrared dried noodle.

As mentioned previously, rice noodles in Japan are sometimes called Harusame. In China the noodle made from Mung bean or starch other than rice is called Fung-Shu (or Tong-Fun). The differences between these 2 kinds of noodles are the raw materials and the transparency. The noodle after extrusion should be frozen and then dried. It also can be cooked for a longer time without becoming pasty.

Yamamura (1969) discussed modernization of the manufacturing techniques of Harusame. The production power and automation of the process are limited by the freezing process. Harusame should be frozen for 24 hr, thawed and then dried.

Instant rice noodles are a specialty developed in Taiwan. The processing method of instant rice noodles is almost identical with the conventional rice noodle, except for the fact that it can be easily reconstituted in hot water in just a few minutes. In order to attain the quick reconstitution characteristic, the steamed rice should be dried at a high temperature to prevent the gelatinous (α) starch from reverting to raw (β) starch. Normally hot air at a temperature of 80°C (176°F) is used for

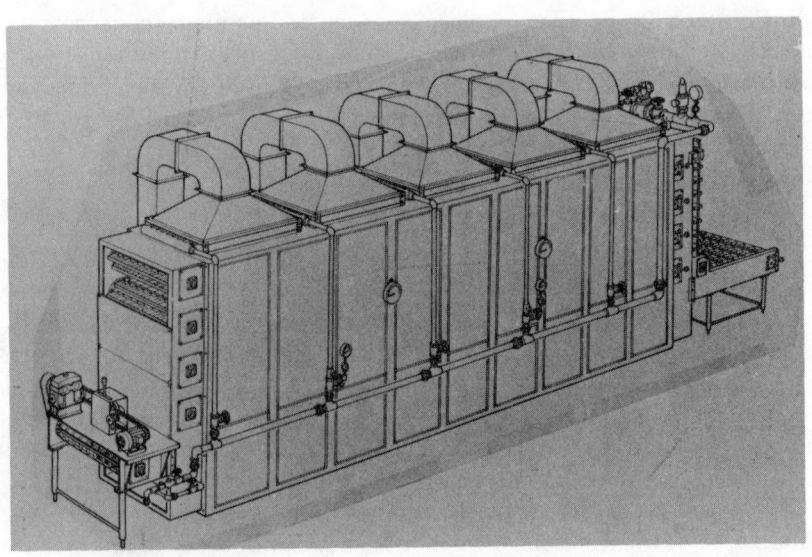

From Presidential Enterprise Co.

FIG. 20.6. DRYER FOR INSTANT RICE NOODLES

drying, but some plants are using frying instead of hot air drying in a tunnel drier (Fig. 20.6). Following is a short description of this process.

Nonglutinous rice (preferably 92% milled rice or nonglutinous Taichung No. 1) is soaked in water for 2–4 hr, ground and mashed into a rice paste which is then pressed to force out the excess water. This product (rice cake) is steamed for 80 min (80% gelatinization is the optimum condition), then kneaded and shaped into a column from which raw rice noodle is extruded. The raw rice noodle is steamed for 30 min, dipped into a seasoning solution, cut, put onto racks for hot air drying or fried in oil at a temperature higher than 80°C (176°F). The final product (instant rice noodle) is then cooled and packaged.

Generally rice noodle is manufactured to the dry state and reconstituted when it is cooked. There is another kind of snack called Bitaibah; it is made into a wet type of noodle, coarser (bigger in diameter) and shorter. This snack is eaten with syrup and ice water, therefore it is normally consumed in the summer. Some people in Taiwan also consume this kind of rice noodle with meat, green onion or leek. Salt, monosodium glutamate and pepper are always added as condiments.

RICE CAKE

In Japan a very popular rice cake is called Mochi, but in China it is called Nenkau. The method of manufacturing rice cake is similar to that of rice cracker, except it is not frozen and dried. In Japan the traditional method of pounding boiled rice into rice cake with pounders and a mortar is rarely used now. Instead rice cake is made by mechanical and automatic methods.

Manufacturing Process for Rice Cake

Glutinous or nonglutinous rice is milled, refined and washed. It is then soaked, allowed to drip until the moisture content is lowered considerably and then steamed. After this it is kneaded, packed and made into various forms. The rice cake is then heat pasteurized and cooled to give the final product, Mochi (Sakurai et al. 1975).

As a means of preventing rice cake from molds and spoilage, the above process is designed to eliminate contamination from micro-organisms. The packing material used is a plastic film such as the vinylidene chloride series. After packing and sealing, the rice cake is pasteurized. A high temperature during pasteurization will affect the quality of the product. Therefore heating is limited to 20 min at 80°C (176°F). It is very important to keep the environment under aseptic conditions. Rapid cooling after packing is desirable and very important.

Physical Properties

The physical properties of rice cake are also closely related to the quality and characteristics of rice from which the rice cake is made. This has already been mentioned in the section concerning rice cracker (Arare).

Little research has been done on the chemistry of rice cake. The physical properties of rice cake can be better understood by studying the main constituents of rice, starch and its major component, amylopectin.

Starch

Starch of glutinous rice is more easily dextrinized than that of nonglutinous rice. Even in its deteriorated state, starch of glutinous rice is still easily dextrinized. Actually, the physical properties of rice cake can not be explained satisfactorily by the characteristics of amylopectin. Kiribuchi (1976) observed the micro-texture of rice cake through an electron microscope and compared the structure of rice cake made by different methods. From these observations he found that in order to get good texture (with the exception of the pasty condition of dextrinized starch), there should be some grain-like structure composing the rice.

Traditional Manufacturing Process

The rice cake made by the traditional method involves pounding by pounders and mortars. The hard part is crushed easily. During this process, air can not enter the rice cake and hence a uniform texture can be obtained. When the rice cake is made by mechanical kneading, the soft part is easily crushed, whereas the hard portion remains uncrushed. In the latter case air bubbles easily get into the rice cake and give rise to a nonuniform product. If there are many air bubbles in the rice cake, it looks whiter because of the random reflection of light. The rice cake can easily be deformed during cooking and baking. (Yomiuri-Shinbun-Sha 1976).

Taiwan Manufacturing Process

In Taiwan, rice cake is made by a somewhat different method. The glutinous or nonglutinous rice is soaked overnight and ground with a stone mill with water to form a slurry. This slurry is transferred into a cotton cloth bag and sealed with a string. As a means of draining the water out, heavy stones are placed on the bag. The raw rice cake obtained is kneaded with water. For sweetened rice cake, sugar is added. The product is transferred into a vessel and steamed (Fig. 20.7). For other types of rice cake, ingredients such as salt, monosodium glutamate, crushed raddish and crushed taro may be added before steaming.

FIG. 20.7. SWEETENED RICE CAKE (NENKAU)

There is a particular kind of fermented rice cake made in Taiwan called Fakau (Fig. 20.8). It is made by adding sugar and a leavening agent to a ground, nonglutinous rice slurry. After fermentation, it is steamed and then consumed. In the Philippines there is a similar food called Bibingka, which is made by the same method. However, salted egg yolk is added on top of this rice cake before steaming.

FIG. 20.8. FERMENTED RICE CAKE (FAKAU)

Rice Cake Ingredients

Radish Rice Cake (Huang 1974)
Nonglutinous rice (600 g)
Water (750 ml)
Radish (1200 g)
Salt (20 g)
Pepper (3 g)
Monosodium glutamate (5 g)

Sweetened Rice Cake (Nenkau)
Glutinous rice (600 g)
White or brown sugar (300 g)
Water (1000 ml)

Fermented Rice Cake (Fakau)
Nonglutinous rice (300 g)
Water (500 ml)
Wheat flour (36 g)
White or brown sugar (240 g)
Leavening (10 g)

Ground Rice Powder Cake.—Glutinous rice is washed, drained, fried in a pan and ground to fine powder. Some maltose is added to this rice powder and pressed into the desired mold. If brown coloring and special flavor are desired, brown sugar is used. In some products, Mung bean may be added to the rice. In this case the Mung bean is washed and cooked in water until the skin has swollen. After removing the skin, the bean is cooked to dryness and used as an ingredient. The product made from glutinous rice and Mung bean is called Rhutou Kau in China.

In the above products, shortening, lard or cooking oil may be added before pressing into the desired mold. This kind of product tastes creamy and does not feel as dry as the product without oil.

BAMBOO LEAF WRAPPED RICE

This is a very popular food consumed in China and Japan. In China, it is called Tsongtsu, but in Japan it is called Chimaki.

In China, there are 3 kinds of Tsongtsu. Their ingredients and methods of preparation are shown as follows.

Chien Tsong (Bamboo Leaf Wrapped Alkaline Rice)

Six hundred grams of glutinous rice (round grain species) is washed and soaked for 1 hr, drained and has 12.5 g soda ash mixed with it. A

60 g portion of this rice mixture is wrapped with bamboo leaves to form a tetrahedron and bound with a string. This is simmered in water for 1.5 hr. After cooking, the bamboo leaves are removed and the Chein Tsong is served with honey or sugar. From the above ingredients, 20 bamboo leaf wrapped alkaline rice products can be made (Fig. 20.9).

FIG. 20.9.   CHIEN TSONG (BAMBOO-LEAF-WRAPPED ALKA-LINE RICE)

Zoutsong (Bamboo Leaf Wrapped Rice With Meat)

Six hundred grams of glutinous rice are washed, soaked in water for 1 hr and drained. The bamboo leaves are washed, boiled in water for about 5 min, cooled and taken out. Three hundred grams of pork (ham) are cut into 20 small cubes and 3 Shiitake mushrooms are cut into 10 pieces. These pieces are mixed with the following seasonings: 45 ml of soy sauce, 1.25 g of monosodium glutamate, 1.25 g of sugar, 0.75 g of black pepper, 2.5 ml of sherry wine, 18 g of fried red garlic. An amount of 60 ml of cooking oil is heated in a frying pan and 250 g of dried shrimp meat is fried until the flavor comes out. This is added to the glutinous rice and the special seasonings are mixed well before frying for several minutes. The special seasonings are 6 g salt, 1.5 g sodium glutamate, 7.5 ml soy sauce and 1.25 g black pepper. Two leaves are taken and bent to form a trigon in which the rice is placed. Some seasoned meat is also added and covered with a little rice. Finally the rice is wrapped with bamboo leaves to form a tetrahedron and bound with string. The wrapped rice is placed in hot water and boiled for about 20 min. The mentioned ingredients yield about 20 Zoutsong. The product is served hot with or without soy sauce or chilli sauce after removal of the bamboo leaves.

POPPED OR FRIED RICE SNACKS

Mishiian

Popped rice is mixed with melted sugar, malted sugar, cooking oil, flavoring and roasted peanuts (optional). It is then pressed into mold and cut into squares after cooling. This is called Mishiian in China.

Malau

A kind of snack called Malau is also very popular in Taiwan. The method of preparation is as follows. The glutinous rice is washed, soaked in water, ground and pressed to dryness. Then it is mixed with taro, kneaded and cut into stick-like shapes 4 cm long and 2—3 cm wide. This is dried under the sun and fried. Frying expands the stick and makes the inside porous. The stick is coated with maltose and dipped into popped rice or roasted sesame seeds. The product coated with sesame is called Malau.

Extruded Rice Snacks

The high temperature/short time extrusion-cooker (extruder) is versatile and thermodynamically efficient (Fig. 20.10). It has become a prime processing unit. The system produces a wide variety of products with different shapes, textures and rehydration characteristics. Smith (1975) has discussed the equipment, processing technology and factors that control production efficiency and product quality.

In this process, moistened, expansible, starchy and/or proteinaceous products are plasticized in a tube by a combination of heat, pressure and mechanical shear.

In Taiwan, several kinds of crispies, which are made from rice by an extruder, are very popular. There are different flavored products such as milk flavored, 5 spice flavored and curry flavored. The product can be made by feeding rice powder into the extruder. Then it is sprayed with flavoring solution and dried.

Rice Krispies

In the United States, there is a kind of breakfast cereal produced by the Kellogg Company which is an oven toasted rice fortified with 8 important vitamins and iron.

The ingredients for this product are milled rice, sugar, salt and malt flavoring. The vitamins added for fortification are sodium ascorbate (C), Vitamin A palmitate, niacinamide, ascorbic acid (C), pyridoxine hydro-

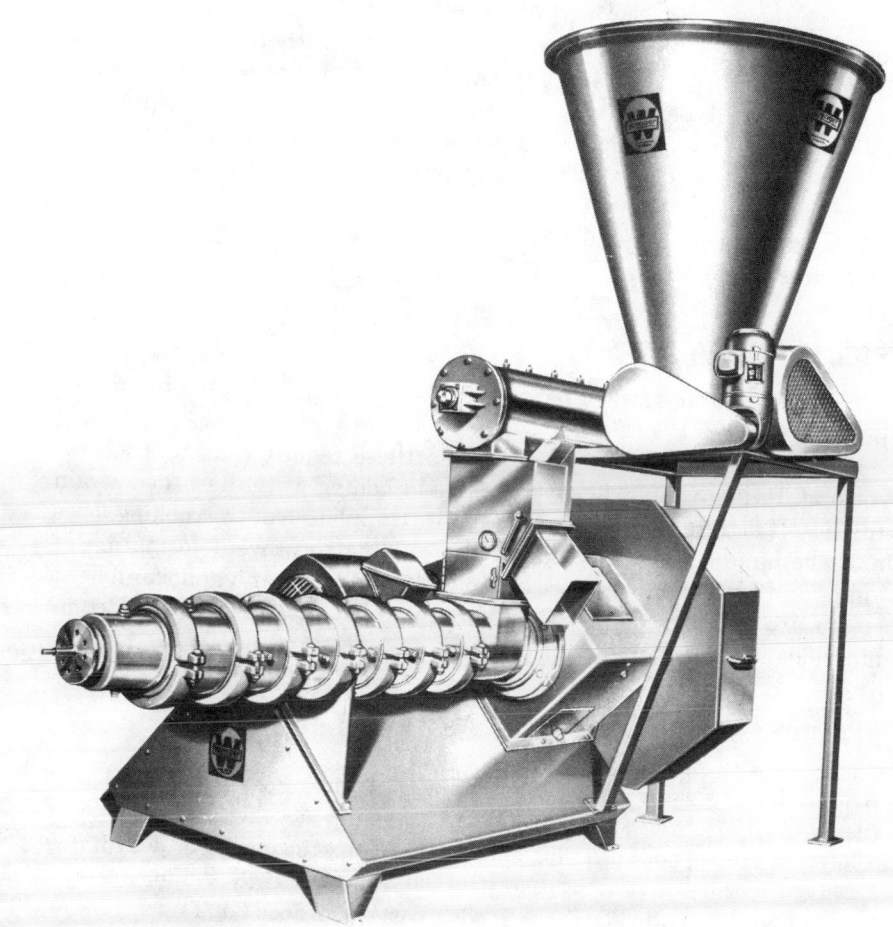

FIG. 20.10. EXTRUDER

chloride (B_6), thiamin hydrochloride (B_1), riboflavin (B_2), folic acid, and Vitamin D_2. For preserving product freshness, up to 0.02% BHA (butylated hydroxyanisole) and BHT (butylated hydroxytoluene or 2,6-ditert-butyl-4 methylphenol) are also added.

Rice Krispies Marshmallow Treats

The ingredients for this product are Rice Krispies, marshmallows and margarine or butter.

The product can be made by melting 62.5 ml of margarine or butter in a large saucepan over low heat. Add 284 g marshmallows and stir until melted and well blended. Cook 2 min longer with constant stirring. Remove from heat. Add 142 g of the Rice Krispies cereal and stir until the cereal is well coated. Add 184 g of roasted cocktail peanuts (optional). Using a buttered spatula or wax paper, press the mixture evenly and firmly in a buttered 33 × 23 × 5 cm pan. Cut the product into squares when cool.

The addition of roasted cocktail peanuts makes the product more attractive in texture, aroma and flavor.

RICE PUDDING

In Europe and in the United States, rice is frequently used in making pudding (Sultan 1977).

There are a variety of rice types and these require consideration for cooking. Pastry chefs will cook the rice in boiling water and then strain the rice. It is then mixed with milk before completion of the cooking. Rice must be handled carefully during the cooking to prevent formation of lumps and breaking of the rice kernels. Egg yolks, sugar, vanilla and light cream are the ingredients. Rice pudding with a variety of fruit combinations serves as a popular dessert.

REFERENCES

ANON. 1975. Quality, storage and utilization of rice. Food Res. Inst., Japan.

CHEN, C.M., HUNG, F.M., WANG, C.Y. and WANG, I.K. 1971. Effect of infrared drying, hot air drying and sun drying on the quality of dehydrated Mii-fen (rice noodles). Food Ind. Res. and Dev. Inst. Rept. 9 (Chinese)

HUANG, S. 1974. Chinese Snack Foods. Wei-Chuan Food Co., Taipei, Taiwan. (Chinese)

KIRIBUCHI, S. 1976. Observation of microtexture in rice cake by scanning electronic microscope. Food Dev. 11 (12) 35-39. (Japanese)

LI, C.F. and CHANG, J.L. 1973. Manufacturing of snack foods. Food Ind. Res. and Dev. Inst. Report 54. (Chinese)

LI, C.F. 1976. Rice snack foods have splendid future. Harvest 26 (24) 22−23. (Chinese)

LIN, K.N. 1972. Agriculture Products Processing. Fu Wen Publishing Co., Taiwan. (Chinese)

SAKURAI, Y. et al. 1975. General Food Industries. Koseikaku Publishing Co., Koseisha, Japan. (Japanese)

SMITH, O.B. 1975. Extrusion and forming: creating new foods. Food Eng. 47 (7) 48−50.

SULTAN, W.S. 1977. Modern Pastry Chef. AVI Publishing Co., Westport, Conn.

YAMAMURA, F. 1969. Modernization of the manufacturing techniques of Harusame (frozen starch noodle). Food Ind. *1–2*, 29–40. (Japanese)

YOMIURI-SHINBUN-SHA. 1976. Profiles of Foods. Yomiuri-Shinbun-Sha Publishing Co., Tokyo. (Japanese)

21

Rice Vinegar Fermentation

Min-Han Lai
William Tien Hung Chang
Bor S. Luh

Rice vinegar is made from rice after a series of saccharification, alcohol fermentation and acetic acid fermentation processes. In ancient China rice was one of the major crops, and wines and vinegars were mostly produced from rice. As early as 1200 B.C. in the Chou Dynasty, the record of the Emperor's chef (Chou Li, Tien Kwan Shan Fu) indicated the management guidance for rice vinegar production (Itagaki 1973). There was a detailed description of the methods of rice vinegar fermentation in *Chi Ming Yau Shu* or *Ways to Prosper People's Life* which was published in the Pei Wei Dynasty (368–534 A.D.). The method described was generally similar to the ones used presently (Yamashita 1976). This technique was introduced into Japan along with wine making techniques. Thus, rice vinegar became one of the important features in the lives of the Orientals.

In the last decade, rice vinegar has almost disappeared from the market because of strong competition from low-cost synthetic vinegar. However, in the last 3 years, the question of food safety has received much attention and the general fear of synthetic materials has resulted in a preference for products made from natural sources.

Thus the consumption of rice vinegar has increased substantially. At this moment, there are more brands of rice vinegar being sold on Taiwanese markets than at any other time. Recently, Kenkoigakusha Co., Ltd. of Japan produced rice vinegar for the Japanese market. A large investment has been put into the newly built rice vinegar factories.

The process of rice vinegar fermentation is shown in Fig. 21.1. The variety of rice, water quality, substrate composition, species of microorganisms and methods of fermentation are variable from one man-

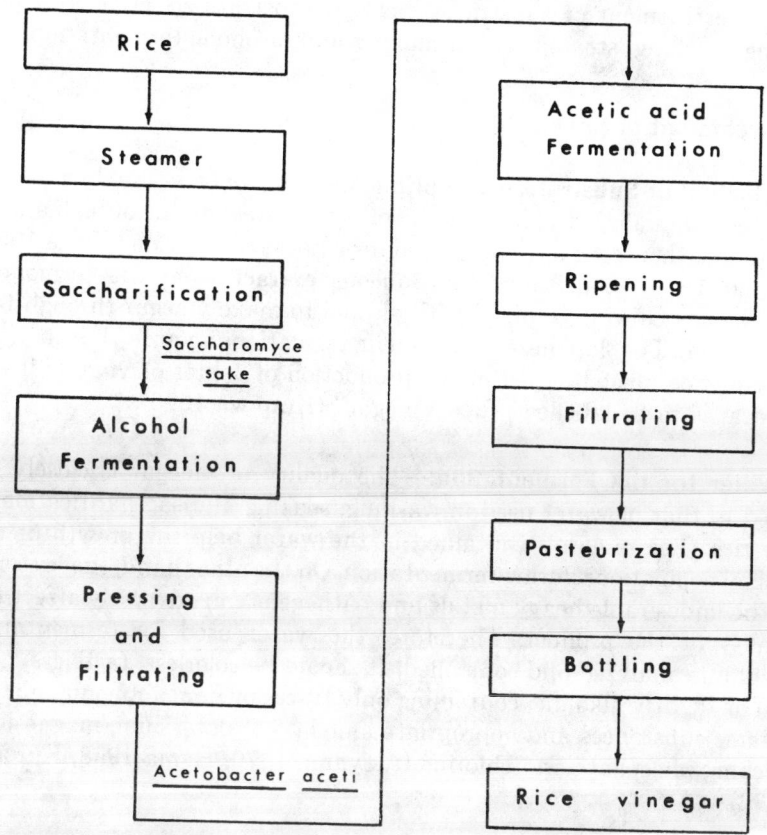

From Ito (1975) and Kaga (1977)

FIG. 21.1. GENERAL FLOW CHART OF RICE VINEGAR FERMENTATION

ufacturer to another. Therefore, the quality of the products produced by each manufacturer is quite different. This chapter will describe processes that are used in most commercial operations and also some modifications of the traditional techniques. The composition of rice vinegar and ways to distinguish rice vinegar from synthetic vinegar will be discussed.

ORDINARY METHOD OF RICE VINEGAR FERMENTATION

The problems in the rice vinegar fermentation process include pretreatment of substrates, fermentation, formation of acetic acid and processing of the final product via precipitation, separation, pasteurization and bottling. Many of these processes have been modernized by techniques common in the wine making, brewing and soft drink industries.

The pretreatment of substrates includes substrate selection, rice polishing, washing, steaming, Koji-making and alcoholic fermentation.

Pretreatment of Substrates

Selection of Substrate.—The primary substrate is definitely rice. Recently, low quality rice, broken rice or other cereal grains of high starch content such as corn have been used for vinegar production (Hama 1958; Itagaki 1970). Sometimes the aqueous extract from the fermented residue of rice wine is mixed with alcohol to make vinegar through fermentation. The Japanese government regulations require that at least 40 g of rice must be used in the production of 1 liter of vinegar if the vinegar is to be labelled "Rice Vinegar" (Kuroiwa 1976).

Water for the Fermentation.—The quality of vinegar is closely related to that of water used in washing, soaking and saccharification of the rice. The minerals contained in the water help the growth of microbial populations during fermentation. On the other hand, the presence of the undesirable heavy metals and pathogens can adversely affect the quality of the product. Therefore, the water used for fermentation should be analyzed and controlled. It should be colorless, tasteless, neutral or slightly alkaline, containing only traces of iron, ammonia, nitrite, organic substances and appropriate amounts of potassium, magnesium, calcium, phosphate and chlorine (Akiyama 1970; Hama 1958; Kuroiwa 1976).

Polishing the Rice.—The purpose of polishing the rice is to remove the hull and embryo of the rice so that the mycelia of the fungal cultures can penetrate into the rice easily and complete saccharification. The polishing is usually done with a polisher which controls the frictional force during this process (Akiyama 1970). The rice grains should be damaged as little as possible. The less mature or other unqualified grains are easily crushed. Therefore, the polisher's grinding force should be very light. The degree of polishing varies with the stages of the vinegar production (Hama 1958): for Koji inoculum, 98%; for Koji, 80–85%; and steamed rice, 90%. The rice used for the production of top quality wine usually requires the highest polishing rate in order to produce a colorless product (Akiyama 1970; Nunokawa 1976). However, the color of the rice is not important to the quality of the rice vinegar. Some commercial processes (Kuroiwa 1976) intentionally use rice polished at a low rate in order to increase the amino acid content of the vinegar (Kenkoigakusha Co., Japan). Since the outer layer of rice grain is rich in protein, fat, vitamins and ash, the polishing rate determines the level of these nutrients in the final product. Table 21.1 shows the relationship between the polishing rate and the chemical composition of rice (Akiyama 1970).

TABLE 21.1. CHANGES IN THE COMPONENTS OF RICE AFTER POLISHING

Polishing rates (% residue)	Moisture (%)	Crude protein (%)	Crude fat (%)	Ash (%)	Starch value
100 (whole rice)	13.3	7.95	1.90	1.06	69.63
90	13.3	7.24	0.55	0.39	72.75
80	12.9	6.36	0.10	0.25	74.62
70	11.8	5.83	0.07	0.20	75.75
60	11.0	5.47	0.05	0.18	76.88
50	11.0	5.12	0.03	0.19	78.34

Source: Akiyama (1970).

Washing and Soaking the Rice.—The purpose of washing the rice is to remove the residual hulls and other debris after polishing. Special washing machines are used for this purpose. The polished rice grains are usually very brittle. Therefore, the washing should be done carefully to keep the grains from being broken apart. The washed rice is then soaked in buckets in order to absorb water before steaming. The soaked rice should be 28–30% heavier than the unsoaked rice (Hama 1958; Nunokawa 1972). The time required for the soaking varies with the temperature, water quality, rice variety and polishing rate. Small scale soaking tests should be performed to find out the necessary conditions. Generally speaking, when warm water, soft rice and a high polishing rate are applied, the soaking time is shorter.

The mineral content of rice, such as K, P, Mg, etc., as well as some vitamins may decrease after washing and soaking. On the other hand, minerals such as Fe and Ca in the water may attach themselves to the rice. When the potassium content of the rice grains decrease by more than 50%, the growth of Koji fungi, yeast and acetic acid bacteria is affected. In this case, potassium phosphate should be added (Hama 1958).

Steaming and Cooling.—The purpose of steaming is to convert the starch into α-amylose and to denature the protein in the rice. *Aspergillus* spp. can then grow easily and its enzymes can react with the substrate rapidly. Steaming also destroys the contaminating microorganisms. It is one of the critical steps during vinegar fermentation. Two different techniques are currently used in steaming: the traditional steaming method and the continuous steaming method. The equipment used in the traditional steaming method consists of an iron or aluminum pot of 1.5–2 kl volume with the steamer baskets placed on top of it. Steam is injected into the pot until boiling occurs. The rice in the steamers is cooked for 1 hr (Nunokawa 1972). One type of continuous steaming method involves putting the rice on a conveyer and letting the conveyer pass through a steam bath continuously (Fig. 21.2). The time required for steaming by such a method is 20–30 min. Another type of continuous method is to pass the rice through a funnel on top of a cylindrical steamer while steam is injected from the bottom. The steamed rice is then removed from 2 doors at the bottom of the cylinder. The time required for the rice to move from the top to the bottom of the cylinder is approximately 20 min. (Fig. 21.3).

The steam cooked rice used for koji making should be cooled to 35°C (95°F), whereas the rice for yeast fermentation should be cooled to 20°–25°C (68°–77°F). The traditional method of cooling is to spread the rice on cloths and allow them to be air cooled. This method is time consuming. It also needs more space. Therefore, most of the processes now use cooling machines (Fig. 21.4).

Koji Making.—"Koji making" generally refers to the inoculation of steamed rice with A. oryzae to produce enzymes and other metabolites for wine or vinegar fermentation. The Koji contains approximately 50 enzymes, the most important ones being the amylases and the proteases. The amylases include α-amylase and glucoamylase (amyloglucosidase) (Nunokawa 1972). Because α-amylase degrades the starch into dextrin, glucose and other polysaccharides, β-amylase can thus reduce the viscosity of the starch solution. Glucoamylase splits glucose molecules off the nonreducing ends of the starch polymers. The more glucose there is available in the rice, the more alcohol is produced by fermentation.

Both acid and alkaline proteases are found in Koji. However, acid proteases are involved in most of the proteolytic activity. They react at low pH values (3–4) and remove amino acids and peptides from the ends of protein molecules.

Koji Inoculum Preparation.—The Koji inoculum is usually prepared by thoroughly mixing steamed rice (polishing rate of 98%) with 2% (w/w) of wood ash before inoculating with fungal spores (Akiyama 1970; Nunokawa 1972). The mixture is then incubated at 30°C (86°F) for 5 days. After sporulation is completed (0.4×10^9 spores/g of mixture), the mixture is dried until the moisture content reaches 10–12%. It is then wrapped in paper bags. The purposes of adding wood ash are to increase the pH of the media, to supplement the media with mineral contents and to decrease the contamination with undesirable organisms. The amount of Koji inoculum used is usually 60–100 g for each 100 kg of the original white rice.

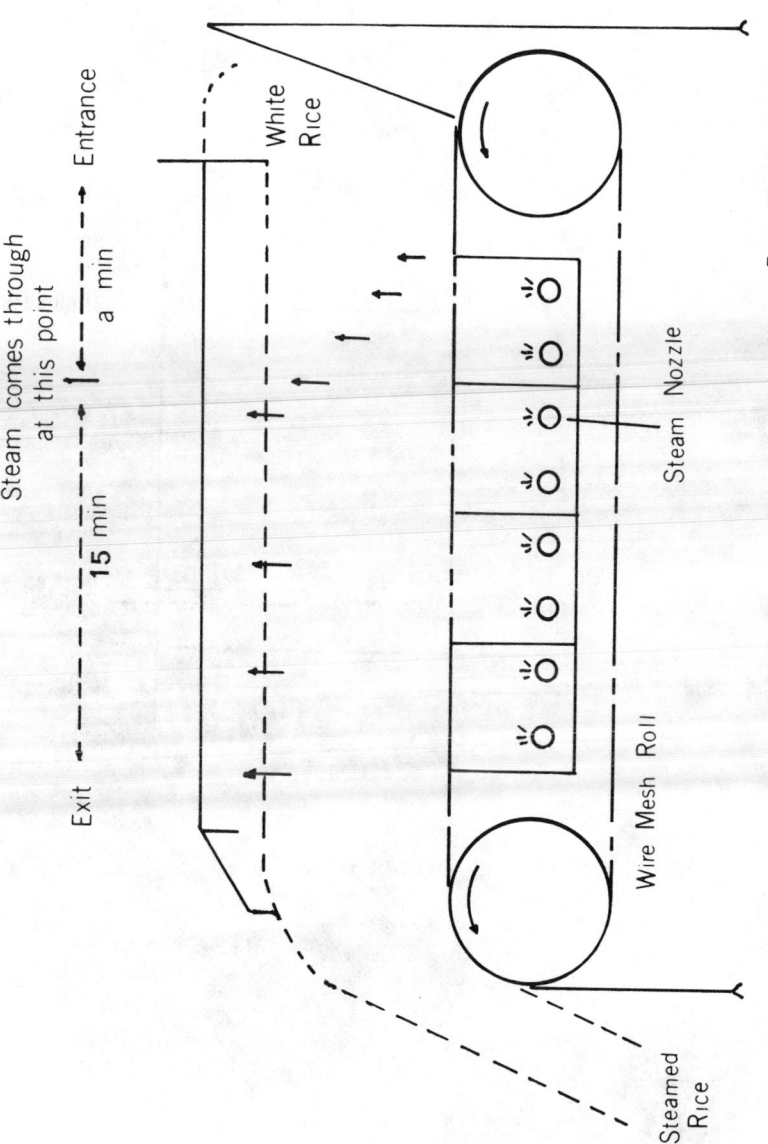

FIG. 21.2. STEAMING APPARATUS: BELT CONVEYOR SYSTEM

From Nunokawa (1972)

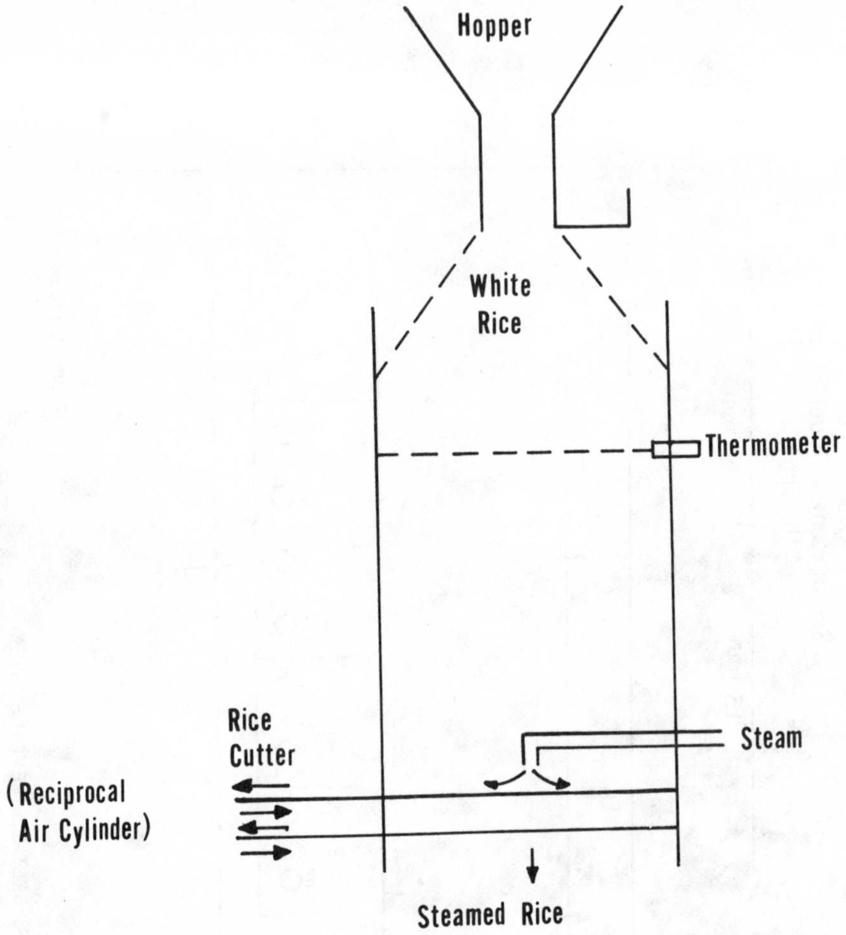

Hopper

White Rice

Thermometer

Rice Cutter

(Reciprocal Air Cylinder)

Steam

Steamed Rice

From Nunokawa (1972)

FIG. 21.3. STEAMING APPARATUS: CYLINDER SYSTEM

FIG. 21.4. STEAMING APPARATUS AND COOLING APPARATUS

The fungal strain used in making Koji belongs to *A. oryzae*, which does not produce aflatoxin. The color of its conidiophores is yellowish green and hence the name "yellow green mold" may be used. The morphology of the fungus is shown in Fig. 21.5.

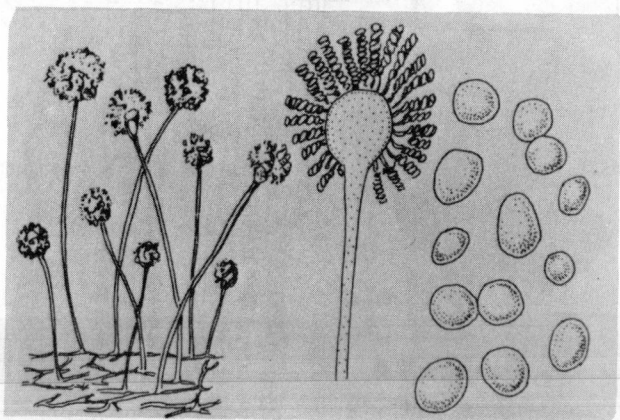

FIG. 21.5. THE MORPHOLOGY OF *ASPERGILLUS ORYZAE*

The following mycological characteristics have been found empirically to be of major importance in Koji molds: (1) rapid growth on and into the kernels of steamed rice, (2) production of abundant amylases and little proteases, (3) production of good fragrance and accumulation of flavor compounds, and (4) limited production of colored substances.

Production of Koji.—The conditions necessary for good growth of the *Aspergillus* spp. are the correct temperature, humidity and aeration (supply of oxygen and removal of CO_2). These conditions should be monitored constantly during Koji production. Several means for Koji production are the small-tray method, big-tray method, floor method and mechanized automatic method. The first 3 methods require heavy labor during fermentation to turn over the mass in order to control the temperature of fermentation. Therefore, most commercial processes presently use the mechanized automatic method. The equipment used is simple and does not require previous experience by the operator once the factors of operation are determined. The size of the Koji equipment can be designed to fit the needs of the factory. Many kinds of Koji equipment are available on the market and most of them operate under the same principle. A representative model is shown in Fig. 21.6 (Nunokawa 1972; Akiyama 1970). Humidified sterile air is passed through the rice bed. The temperature of the bed is regulated by the temperature of the water used for humidification and the aeration rate is controlled by the damper.

The general process of Koji production is described as follows (Akiyama

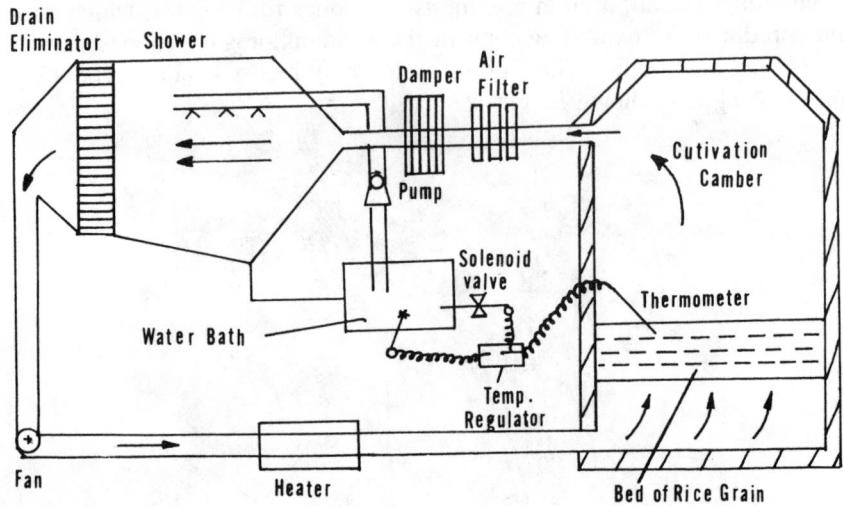

From Nunokawa (1972)

FIG. 21.6. A SCHEMATIC REPRESENTATION OF A KOJI-MAKING DEVICE

1970). After the steamed rice is cooled to 30°−32°C (86°−89.6°F), the Koji inoculum is added uniformly and is thoroughly mixed. The mixture is then spread out on the rice bed in a thin layer (30 cm). The air is passed in at a temperature of 32°C (89.6°F) and a relative humidity of 90−94%. After 20 hr, the mixture is turned over. The temperature of the Koji should rise to 35°C (95°F) after 24 hr, and 38°C (100.4°F) after 29 hr, and the humidity should be decreased to 80% after 29 hr. To complete the process, the temperature should be kept at 38°−40°C (100.4°−104°F) for 36−40 hr. For vinegar production, the Koji is usually incubated at 40°−42°C (104°−107.6°F) in the latter part of fermentation. The time necessary for fermentation is normally longer than 40 hr. The product is believed to contain higher enzymatic activity (Hama 1958; Itagaki 1970). The alcohol content will be higher in yeast fermentation if this kind of Koji is used.

Alcoholic Fermentation.—The general method of alcoholic fermentation is similar to that of sake brewing. The yeast inoculum is prepared in a large quantity. Both Koji and steamed rice are added 3 times during the process. Initially, a small quantity of rice is added and then a larger volume of rice is added at each additional time. The yeast used for alcoholic fermentation is *Saccharomyces sake*.

Yeast Inoculum (Moto).—The alcohol inoculum in vinegar production is as important as the acetic acid inoculum. Since the process is not a closed system, heavy inoculum of the yeast and large amounts of lactic acid are needed to control contaminating organisms. There are several kinds of alcohol inocula: Ki-moto (Yamahai-moto), rapidly processed moto and hot mashed moto. The latter 2 types are frequently chosen because less labor and time is needed to produce them.

For rapidly processed moto, the formula of the substrate for the inoculum is:

Steamed rice	75 kg
Koji	30 kg
Water	108 l
Lactic acid (75%)	756 ml
Phosphates	15−45 g

First potassium phosphate and lactic acid are dissolved in water and thoroughly mixed. Koji is then added and soaked for 1−2 hr to extract the amylase (Hama 1958). Next steamed rice is added slowly while stirring. The temperature should be in the range of 15°−20°C (59°−68°F). This mixture is stirred frequently in the first 2 days to saccharify the rice rapidly and to lower the temperature to 10°C (50°F) or below. On the third or fourth day, the temperature is raised by heating, then the yeast inoculum (100 ml) is added. The temperature is increased slowly until it reaches 13°−15°C (55.4°−59.0°F) on the ninth or eleventh day. During this period, the mixture bubbles substantially due to heavy yeast growth. On the twelfth day, the bubbling is so heavy that the solids float on the surface. This stage is called Fukure (swelling). Two days later, the solids sink back down and the surface of the liquid appears to be boiling. This stage is called Wakitsuki (gusting out). Heating is stopped at this time, when the temperature is around 15°−16°C (59.0°−60.8°F) and Baumé reaches 8−10. At this time the moto is divided in half. Each half is stirred, cooled to 10°−15°C (50.0°−59.0°F) and returned to the original container. The moto is stirred once a day until the Baumé reading reaches 6−8, at which time the moto is ready. This entire process takes 15−20 days. The yeast inoculum (moto) can be used immediately, or stored for 2−4 days to decrease the number of contaminating bacteria. However, storage for more than 7 days results in weakened yeast. The moto's chemical composition is as follows. Total acidity (in terms of lactic acid) 0.85−0.7%, alcohol 10−13%, glucose 0.5−0.8%, protein 1−1.2%, glycerin 0.8−1.0%, and ash 0.08−0.1%. One milliliter of yeast inoculum contains 4×10^8 yeast cells.

For hot mashed moto the Koji is mixed with warm water in an insulated container at a ratio of 1:1.6. Steamed rice is cooled to 60°C (140°F) and then added to the mixture. The saccharification process is continued for 6−8 hr at 55°−60°C (131.0°−140°F). After this time, the temperature is decreased rapidly to 25°−30°C (77.0°−86.0°F). The yeast and lactic acid (700 ml 75% lactic acid to 100 l of water) are then added so that alcoholic fermentation can begin (Akiyama 1970; Hama 1958). The fermentation process is identical to that of rapidly processed moto. It is safer to use this method during the warm seasons. The high temperature during saccharification also helps to kill most of the contaminating organisms.

Alcoholic Fermentation with Pure Yeast Inoculum.—The yeast inoculum can also be prepared by incubating yeast in a medium containing molasses, lactic or malic acid at a low pH (Akiyama 1970; Nunokawa 1972). The culture is aerated for 30 hr at 30°C (86.0°F), then the cell mass is harvested and washed. The yield of pressed cell mass from 100 g of molasses is 3−4 g. For alcoholic fermentation, 400−500 g of cell mass (70% water content) is added to 1 MT of rice. The alcohol product has the same quality as the product made from the "moto" method described above. Therefore, this product is adopted in many commercial processes.

Brewing Rice Wine with "Moto.".—The ratios of substrate ingredients at different stages of the rice wine fermentation are as follows (Hama 1958):

	Moto (yeast inoculum)	*Addition of the ingredients*			
		First	Second	Third	Total
Steamed rice (kg)	75	150	226	331	782
Koji (kg)	30	60	75	105	270
Water (l)	108	171	469	893	1641

The original substrates are added to the yeast inoculum in a 1260−1440 l barrel. The "moto" goes in first, then water, phosphates (100−200 g), Koji, and steamed rice are added and thoroughly mixed. The barrel is then covered and insulated to keep the temperature at 10°−12°C (50.0°−53.6°F). On the second day, the rice grains are saccharified rapidly and the yeast begins to grow in its log phase. On the third day, half of the mixture is transferred into a separate barrel. The substrates for the second addition are also divided in half and added to the 2 barrels. These materials are thoroughly mixed and the temperature is kept at 9°−10°C (48.2°−50.0°F). The third addition of substrate is on the fourth day. After the second addition, the mixture from each of the barrels is transferred into a large (3808 l) barrel and the fresh ingredients are added, mixed and allowed to ferment at 7°−8°C (44.6°−46.4°F). In the first few days after the last addition of substrate, the mixture should be stirred gently to allow uniform distribution of the material. On the third day when Baumé reaches 8, air bubbles start to form and soon appear on the surface. On the fifth day, foaming is very heavy (thick foam). On the seventh to the tenth day, the foam becomes slippery and swells to the edge of the barrel (high foam). At this time the temperature rises to $10°-13°C$ ($50.0°-55.4°F$) and saccharification and alcohol formation occur rapidly. Over-foaming should be controlled. On the tenth or eleventh day, the foam becomes "brittle" and breaks into "bead-like foam." The degree of Baumé reaches $4-5$ and alcohol concentration is $8-10\%$. On the twelfth or thirteenth day, the foam increases in size ("fat foam") and the temperature rises to $15°-20°C$ ($59.0°-68.0°F$). Alcohol is still being formed rapidly at this stage and the alcohol concentration soon reaches $13-15\%$. On the sixteenth to eighteenth day, the foam is at the meniscus. This shows that saccharification and fermentation have reached their final stage. At this stage in the process, Koji and steamed rice float to the surface. The container is closed after thorough stirring and the floating material is allowed to sink back down to the bottom. Twenty five days after the last addition of substrate, the wine with an alcohol concentration of $16-18\%$ can be pressed out.

The fermentation process described above is characteristically performed at a low temperature (Itagaki 1970). There is also a high temperature fermentation method where steamed rice is mixed with 30% Koji rice. Water is added at a weight $2-3$ times that of the total rice. The mixture is saccharified sufficiently at $60°C$ ($140.0°F$), then cooled to $22°-23°C$ ($71.6°-73.4°F$) before yeast inoculum is added. This method is superior to the low temperature fermentation because: (1) the high temperature during saccharification kills off the contaminating microorganisms, (2) more rice is saccharified and therefore more alcohol can be produced, and (3) the protein in the rice is completely hydrolyzed and thus the product has lower turbidity.

Both of the 2 fermentation methods mentioned above require the use of Koji for saccharification. Recently, inexpensive enzymes have become available and methods using enzymes for the saccharification have been developed. Using these methods, most of the labor costs and capital investment for Koji production can be saved and the quality of the vinegar can be more uniform. In recent years, consumers have changed preference to a light tasting wine. Takeuchi et al. (1972) used amylases in place of Koji to saccharify the steamed rice. They produced a wine with a lighter aroma, but in all other respects identical to the traditional product. The alcohol concentration was 17% and the acetic acid concentration reached 6%. Shimizu et al. (1969) used enzymes instead of Koji for the third addition of substrates during alcoholic fermentation. The alcohol content of the fermentation mixture reached 21% after $17-20$ days and the acidity was 1.9−2.0%. The product was identical to the wine brewed by the traditional method except that its aroma was lighter. Therefore, although Koji and its intermediate metabolic products are believed to provide nutrients for the growth of yeast and acetic acid bacteria, the amount of Koji used really does not affect the activity of fermentation of the steamed rice.

The saccharification of rice requires α-amylase, glucoamylase and acid protease (Iwano and Nunokawa 1977). These enzymes are found in bacteria and fungi. The production of glucose is closely related to the activity of glucoamylase. However, if acid protease is absent, the dissolution and saccharification of rice by either glucoamylase alone or in conjunction with α-amylase is slower (Iwano and Nunokawa 1977; Nunokawa 1976). The ratio of enzymes to the rice as well as to each other should be appropriate. For 1 g of rice, 500 μg of acid protease, 40 μg of glucoamylase and 200 μg of α-amylase should be used (Iwano and Nunokawa 1977). The fermentation rate is lowered when too much α-amylase is used.

Pressing.—The wine for vinegar production is pressed and filtered before acetic acid fermentation. The traditional method is to layer the bags containing the wine mixture on a hydraulic pressing machine. This method is both costly and time consuming. Therefore, many automatic pressing apparatus have been developed lately (Masai and Yamada 1973, e.g. Fig. 21.7). In these machines, mixture is filled automatically,

FIG. 21.7. AUTOMATIC PRESS

and the residue is removed continuously after pressing. Some vinegar is added to the mixture during pressing to prevent denaturation. Therefore, the material used to build the press should be acid-tolerant.

Acetic Acid Fermentation

The pressed and filtered wine is fermented to produce acetic acid. Two methods commonly used are "static fermentation" and "aeration process" (submerged acetic acid fermentation). The former is more popular for producing rice vinegar, because the supernatant vinegar is easier to filter and the aroma of the vinegar remains better (Itagaki 1970). The tank for steady state fermentation is made of wood, acid-tolerant metal or synthetic polymers. The usual volume of the tank varies from 1800–5400 l. For the deep circular type tank, the time required for fermentation is longer and the tank occupies a large space (Masai and Yamada 1973; Yanagida 1976). Therefore, shallow and angular tanks have gradually taken the place of deep circular tanks (Masai and Yamada 1973; Yanagida 1976). In the winter, insulation is needed between the inner and the outer wall of the tanks.

To start the process, seed vinegar is added to the tank and mixed thoroughly with the wine. The tank is covered with either a punctured lid or the lid is raised slightly with grills so that sufficient air is allowed into the barrel. Three to four days later, a layer of acetic acid bacteria film forms on the surface and acetic acid fermentation is being started. The time required for complete fermentation depends upon the size and the depth of the tank. Fermentation can take from 1 to 3 months to be completed (Itagaki 1970). The criteria for the readiness of the product are acidity and alcohol concentration. When the residual alcohol concentration reaches 0.3–0.4%, fermentation is complete. Half of the liquid is retained in the barrel as the seed vinegar for the next fermentation. The rest is transferred into a storage tank for ripening.

In order to maintain high fermentation activity and to shorten the time requirement, continuous steady state fermentation has been developed recently (Masai and Yamada 1973). In this method, the bacterial film on the surface of the liquid is undisturbed, but the liquid underneath is continuously renewed. Accurate alcohol monitoring equipment and controlling systems for inflowing wine are needed for this kind of process.

The bacteria for rice vinegar fermentation are: *Acetobacter aceti* and *Acetobacter rancens* (Yanagida *et al.* 1971; Yanagida *et al.* 1973). The optimum temperature for the fermentation is around 30°C (86.0°F). Naturally it is more economical to use less seed vinegar for fermentation. However, if the inoculum is too small, film-producing yeast grow rapidly and cover the surface of the liquid so completely that they inhibit the conversion of alcohol into acetic acid. Therefore, the inoculum needed for acetic acid fermentation must raise the acidity of the wine to 1.5—2% (Hama 1958; Itagaki 1970). The wine should be diluted before the inoculum is added. If the wine originally contains 16% alcohol and 1% acidity, and the desired acidity of the vinegar product is 5.2%, then the wine should be diluted 3.2 times. To this diluted wine, 30% by volume of seed vinegar should be added. The acidity of this mixture will be 1.718%.

Ripening

After the acetic acid fermentation process is complete, the liquid is transferred into a storage tank (usually made of stainless steel) and stored at a low temperature. Usually 2—3 months are needed for ripening (Itagaki 1970; Masai and Yamada 1973; Yanagida 1976). The stainless steel tank reduces evaporation and prevents the growth of vinegar eels.

Filtration

Diatoms and celite are usually used for pressurized filtration of the ripened vinegar. Contaminating material and bacterial cells are removed during filtration.

Pasteurization

The usual condition for pasteurization is 30 min at 60°C (140.0°F). The remaining acetic acid bacteria are killed by this process. Pasteurization is performed either before or after bottling. In large-scale production, pasteurization at higher temperature (<100°C) is usually done by disc-shaped stainless steel heat exchangers before bottling.

Bottling

There are many kinds of bottling machines available on the market, either semi-automatic or fully-automatic. The latter can fill 300—400 bottles per minute.

MODIFIED TRADITIONAL METHODS OF VINEGAR FERMENTATION

The traditional vessel used for producing vinegar is similar to that of wine making, i.e., earthenware jars covered with a nonporous glazed surface. The size and shape of the jars differ from place to place. Three sizes are most commonly used: 200 l (large size), 100 l (medium size) and

50 l (small size). The jars are either wide mouthed or narrow mouthed. The traditional methods have been modified in recent years to 2 major lines: the Japanese modification and the Chinese modification. Both methods allow saccharification, alcoholic fermentation and acetic acid fermentation to be performed in the same jar. The flavor of the product thus obtained is stronger and the amino acid content as well as other ingredients is higher (Kuroiwa 1976; Yamashita 1976).

Japanese Modification

The method of rice vinegar production was introduced into Japan from China in the Emperor Ojin's Era (369—404 A.D.). The most famous rice vinegars made in the traditional style are believed to be those produced in Fukuyama of Kagashima. This area has a fine climate and therefore is especially suitable for rice vinegar production by natural conditions.

The dimensions of the jars used in the Japanese modification include a diameter of 43 cm, a height of 62 cm and a mouth diameter of 14 cm (Higashi et al. 1973). The volume is 52 l. The substrates are mixed in the following ratio (Higashi et al. 1973; Higashi et al. 1974):

Rice koji	4.1— 5.4 kg
Upper koji	0.5— 0.9 kg
Lower koji	3.6— 4.5 kg
Steamed rice	9 — 9.9 kg
Water	28.9—29.8 l

According to Kenkoigakusha Co., Ltd. the process of production is as follows (Kuroiwa 1976):

The inside walls of the jars are wiped clean and sprayed with seed vinegar containing *Acetobacter aceti* and *A. acetosum*. Lower Koji, steamed rice and water are added, then the upper Koji is spread on the liquid surface. The jars are covered with lids and allowed to ferment either in a room or in an open field. The optimum temperature for the growth of these acetic acid bacteria is 28°C (82.4°F) or slightly above. Room temperature is usually maintained between 20°—28°C (68.0°—82.4°F). Under these conditions, the time required to complete fermentation is approximately 3 months. However, if the jars are located in an open field, the time required for fermentation takes more than 6 months. In order to shorten the fermentation time, the jars should be arranged in files that lie north to southwards (Fig. 21.8). In this way, the jars are exposed to the maximum sunlight and the temperature in the jars is higher.

Rice Koji can be separated into upper Koji and lower Koji. The former

FIG. 21.8. RICE VINEGAR FERMENTATION IN PORCELAIN JARS

is composed of old, dried fungal cultures containing mainly fungal spores. Upper Koji is spread on the surface of the fermentation liquid to overcome contamination. The lower Koji is composed of a younger culture which is high in enzymatic activity and can saccharify the steamed rice rapidly.

The supernatant from the rice vinegar is decanted after ripening. The residual precipitate is extracted by pressing and the liquid is collected and pooled with the first supernatant. The pooled fractions are heated for pasteurization and then bottled. The product usually contains a total acidity of 3–6% and amino nitrogen of 152.4 mg % or higher.

The traditional method of vinegar fermentation, which utilizes natural flora of yeasts and acetic acid bacteria, requires longer time for fermentation. If pure cultures of microorganisms are used, then the fermentation time and the amount of rice required are reduced. Higashi *et al.* (1973) used 3.3 kg of rice Koji, 4.95 kg of steamed rice, and 29 l of water to start the fermentation process. The inoculation schedule is shown in Table 21.2.

TABLE 21.2. DIFFERENT SCHEDULES OF INOCULATION FOR VINEGAR FERMENTATION

Treatment	No. jars	Inocula yeast (ml)	Acetic acid bacteria (ml)	Upper Koji
A	3	60	200	added
B	3	60	200	not added
C	5	0	0	added

Source: Higashi *et al.* (1974).

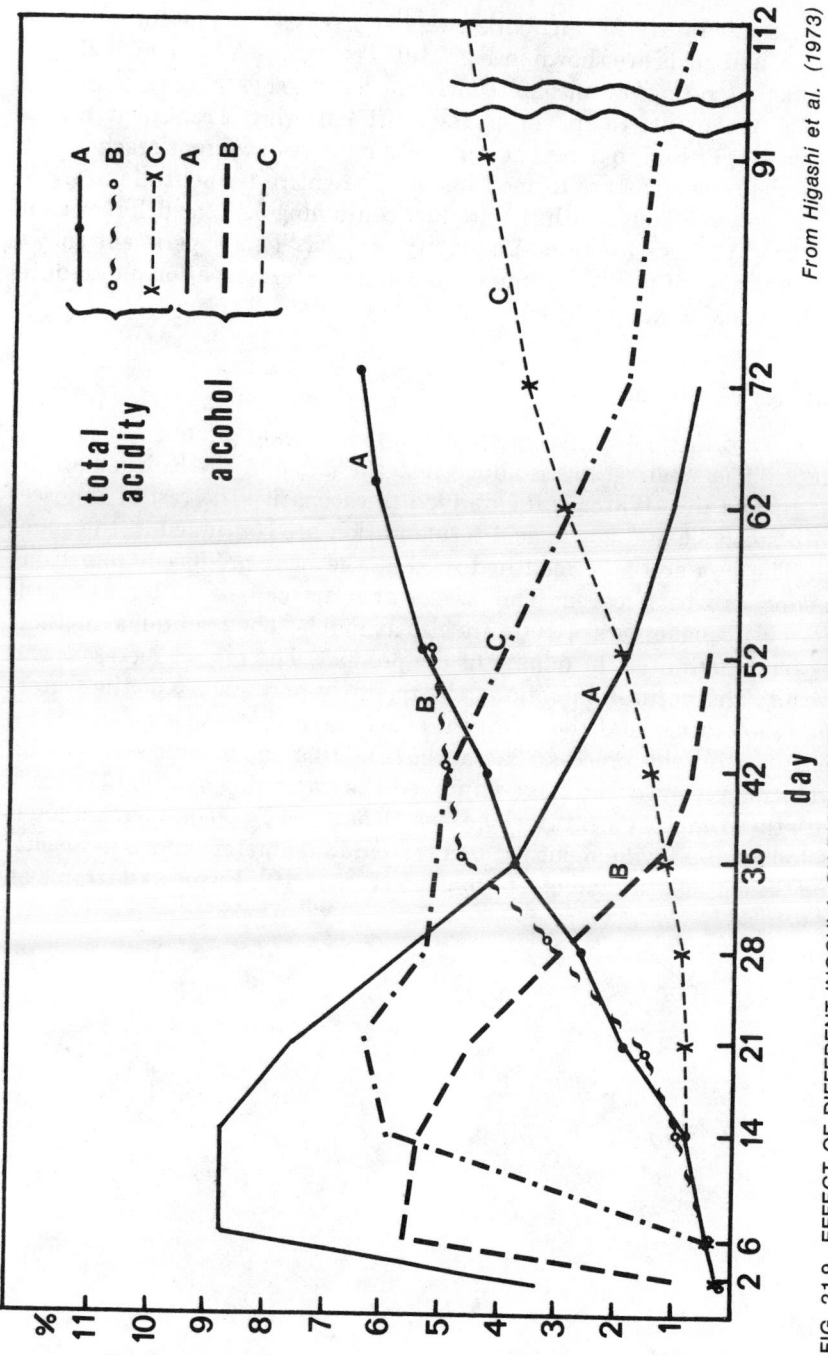

FIG. 21.9.   EFFECT OF DIFFERENT INOCULA ON RICE VINEGAR FERMENTATION

From Higashi et al. (1973)

The temperature for incubation was 23°C (73.4°F). The formation of alcohol and acids are shown in Fig. 21.9. Treatment A has a high alcohol production rate. The alcohol concentration reaches 5.8–8.8% in 1–2 weeks. Acids start to appear on the fourteenth day. Treatment B does not use upper Koji to cover the surface. The alcohol content reaches only 5.8%, but the acids are formed faster. The entire fermentation can be completed in 52 days with the product containing 5.1% acid. Treatment C takes 112 days to obtain an acidity of 4.7%. This experiment shows that addition of pure inoculum can hasten the fermentation and reduce the amount of rice required.

Chinese Modification

The traditional Chinese method for rice vinegar production is very simple. The steamed rice is added into the jars and the Koji is made in the cooked rice. Water is then added to saccharify the rest of the rice. Both alcoholic and acetic acid fermentation are continued in the same jar. The jars are wide mouthed so that the heat produced from Koji-making can be released. The size of the jars can be as big as 200 l. However, smaller jars of 100 l are also available. The traditional method does not guarantee the quality of the product. The rate of success is low because the natural microflora, the quantity and quality of the water, the temperature and the acidity are the major factors that can not be controlled in the process. The authors of this paper have studied the problems involved and have improved the method. The rate of success with these improvements may be as high as 95%. The modified method is as follows: 90 l wide mouthed jars (48 cm in diameter, 49 cm in height, and 38 cm mouth diameter) are used (Fig. 21.10). Fifteen kilograms of

FIG. 21.10.   RICE VINEGAR FERMENTATION IN WIDE MOUTH PORCELAIN JARS

steamed rice are cooled and added to each jar. The Koji inoculum of *A. oryzae* is added at 60 g per jar. After thorough mixing, the rice is pressed tightly against the inside wall. The jar is covered with paper and incubated at 20°–25°C (68.0°–77.0°F). On the second day, the rice grains are covered with white mycelia (Fig. 21.11). On the third day, the temperature in the mixture reaches 42°C (107.6°F). Water is then added for saccharification and *Saccharomyces cerevisiae* is inoculated. The jars are tightly covered with lids so that alcohol fermentation may start. A total of 15 kg of water is added in 2 portions. The concentration of alcohol is 10% after the first week. The jar is sealed for another 10 days to complete the fermentation process at which time total acidity is 1.2%. Finally, 25 l of water and 20–25 l of good quality seed vinegar is added and the mixture is incubated at 28°–31°C (82.4°–87.8°F). One month later, a rice vinegar containing 5% total acidity is obtained. This product has a higher amino nitrogen (89–98 mg %) content than rice vinegar produced by the traditional method and the flavor is believed to be better. Complete hydrolysis of the rice protein is believed to be the major reason for the improvement.

FIG. 21.11. THE GROWTH OF *ASPERGILLUS ORYZAE* ON KOJI RICE

COMPOSITION OF RICE VINEGAR

The composition of rice vinegar varies according to the substrates and the techniques used in production. The general composition of fermented vinegar is shown in Table 21.3. Just like other fermented vinegars, rice vinegar contains volatile and nonvolatile organic acids, sugars, amino

TABLE 21.3. COMPOSITION OF VARIOUS VINEGARS (%)

	Rice vinegar	Alcohol slop vinegar	Apple vinegar	Malt vinegar	Grape vinegar	Alcohol vinegar	Lemon vinegar
Specific gravity	1.018	1.016	1.021	1.024	1.020	1.012	1.033
Total acidity	4.51	4.52	5.04	5.06	5.03	4.21	5.83
Volatile acid	4.290	4.374	4.914	4.938	4.866	4.170	1.096
Nonvolatile acid	0.330	0.204	0.141	0.183	0.205	0.060	5.050
Total nitrogen	0.025	0.020	0.010	0.021	0.014	0.009	
Amino nitrogen	0.016	0.013	0.005	0.006	0.011	0.005	
Total sugar	1.808	1.269	2.888	3.369	3.231	0.295	2.909
Reducing sugar	1.770	1.265	2.865	2.812	3.124	0.278	2.847

Source: Kaga (1977).

acids, and traces of alcohols and esters. There are 60–70 kinds of organic acids present in fermented vinegar (Yamashita 1976). Those acids which exist in easily assayable amounts include: acetic acid, lactic acid, succinic acid, formic acid, propionic acid, fumaric acid, α-ketoglutaric acid, malic acid, citric acid and tartaric acid. Ordinary fermented vinegar has an acidity of around 4–5%, but concentrated vinegar can have an acidity as high as 10%. Acetic acid is the major organic acid in vinegar; it comprises 97% of the total organic acid. Rice vinegar contains more nonvolatile organic acids than other fermented vinegars. The major components are lactic acid and succinic acid.

The amino nitrogen content of rice vinegar is also very high. The reported concentration of amino nitrogen in the rice vinegar produced by Kenkoikusha Co. by the modified Japanese method is 152.4 mg % (Kuroiwa 1976). Reduced rice polishing, less water used and a longer fermentation time, which result in complete hydrolysis of the rice protein are the main causes for high amino nitrogen content in the vinegar. The high amino nitrogen contributes to creating a special mild flavor in the vinegar and makes it very popular.

DIFFERENTIATION BETWEEN FERMENTED VINEGAR AND SYNTHETIC VINEGAR

In order to correctly label the vinegar products with respect to their method of manufacture, the Japanese government ruled in 1970 (Yanagida 1976) that if a vinegar product is to be labeled as fermented vinegar it cannot contain glacial acetic acid. Thus, the sale of synthetic vinegar has been decreasing ever since. The fermented vinegar has a complicated composition; however, the producers try to reduce the cost of production by adding various known ingredients to the vinegar instead of allowing them to be produced by natural fermentation. It is rather difficult to distinguish between such vinegar and the truly fermented vinegar. Recently several methods have been developed to effectively identify the difference between the two. A liquid scintillation counter has been brought in to test for the content of ^{14}C in the products (Masai et al. 1973). The synthetic vinegar usually has its acetic acid derived from petrochemical sources and, therefore, does not have as high a ^{14}C content as fermented vinegar. Gas chromatography is also used to test for trace amounts of neutral components such as ethyl alcohol, acetoin, ethyl acetate, acetaldehyde, octyl alcohol and 2,3-butylene glycol in the fermented vinegar (Ito and Ebine 1975). Fermented vinegar has smaller amounts of n-butyl acetate, n-amylacetate, iso-amylacetate, isobutyl alcohol and furfural. Synthetic vinegar contains acetone, iso-butyl alcohol, n-heptyl alcohol, octyl alcohol, phenylacetaldehyde, ethyl alcohol, n-ethyl acetate, furfural and 2,3-butylene glycol in substantial amounts, but contains no traces of diacetyl, acetoin, iso-amylacetate, n-amylacetate, n-propylacetate and iso-amylalcohol. Except for the above mentioned instrumental analysis, the synthetic vinegar could also be distinguished from fermented vinegar by sensory tests (Table 21.4).

TABLE 21.4. DIFFERENCES BETWEEN SYNTHETIC AND FERMENTED VINEGAR

Items	Fermented vinegar	Synthetic vinegar
Flavor	Gentle.	Strong.
Taste	Sweet after-taste, not stimulating.	Bitter after-taste because of artificial sweeteners, stimulatingly sour.
Penetrating ability	Red fish turned white in the center of the meat after 30′ of soaking.	The fish meat turned white only on the outside after 30′ of soaking.
Effect of heating	Less change in flavor because of higher nonvolatile acids content.	Smells strongly at first, the sour aroma then evaporates rapidly.

REFERENCES

AKIYAMA, H. 1970. Sake. *In* General Food Industry (Sogo-Syokuryo-Kogyo). Kosei-Sha Publishing Co., Japan.

HAMA, M. 1958. Vinegar. Shikoku Ferment. Vinegar Soc., Japan.

HIGASHI, K., MIZUMOTO, H., MINAMI, H., MORI, T. and MAEDA, F. 1974. Studies on Fukuyama vinegar fermentation and improvement of brewing technique. Kagoshima-Ken Tech. Res. Lab. Rept. *21,* 61-66. (Japanese)

HIGASHI, K., MIZUMOTO, K., MORI, T. and MAEDA, F. 1973. Studies on Fukuyama vinegar fermentation and improvement of brewing technique. Kagoshihma-Ken Tech. Res. Lab. Rept. *20,* 58—77. (Japanese)

ITAGAKI, T. 1970. Vinegar. *In* General Food Industry (Sogo-Syokuryo-Kogyo). Kosei-Sha Publishing Co., Japan.

ITAGAKI, T. 1973. Traditional brewing of vinegar. Food Chem. *9,* 68—75.

ITO, H. 1975. On the flavor component of vinegars. Perfumery and Flavouring, No. *112,* 103. (Japanese)

ITO, H. and EBINE, H. 1975. Discrimination between commercial vinegar and synthetic acetate. III. J. Soc. Brew. Jpn. *70* (4) 271—276.

IWANO, K. and NUNOKAWA, Y. 1977. The effect of various enzymes on sake brewing. J. Soc. Brew. Jpn. *72* (1) 78—81.

KAGA, S. 1977. Vinegar and Dressing Sciences. Soc. for Sci. and Tech. Educ. Tokyo. (Japanese)

KUROIWA, T. 1976. Utilization of Pure Rice Vinegar. Kenkoigakusha Publishing Co., Japan. (Japanese)

MASAI, H. and YAMADA, K. 1973. Improvement of vinegar manufacture and its application. Food Ind. *16* (18) 26—33. (Japanese)

MASAI, H., OHMORI, S., KANEKO, T. and EBINE, H. 1973. Differentiation of synthetic from biogenic raw materials in various food with a liquid scintillation counter. Agric. Biol. Chem. *37* (6) 1321—1325.

NUNOKAWA, Y. 1972. Sake. *In* Rice: Chemistry and Technology. Amer. Assoc. Cer. Chem., St. Paul, Minnesota.

NUNOKAWA, Y. 1976. Technical problems concerning enzyme application in sake brewing. Food Ind. *19* (22) 45—54. (Japanese)

SHIMIZU, T., INOBE, K. and SHIMADA, S. 1969. The application of various enzymes on sake brewing. J. Brew. Soc. Jpn. *64* (5) 419—422.

TAKEUCHI, N., AMAMO, T., OKADA, Y., HOSOKAWA, N., YOSHIDA, M., NANBA, J. and YOKOO, Y. 1972. The improvement of qualities on rice vinegar. Aichi-Ken Res. Inst. of Food Tech. Ann. Rev. No. *13,* 95—100.

YAMASHITA, T. 1976. Vinegar Science. Kenkoigakusha *11* (277) 8—37. (Japanese)

YANAGIDA, T. 1976. Vinegar. J. Brew. Soc. Jpn. *71* (8) 608—610.

YANAGIDA, F., TAKASHIMA, K., YAMAMOTO, Y., NISHIZIMA, H. and SUMINOE, K. 1971. Studies on acetic acid bacteria and their utilization. II. J. Soc. Brew. Jpn. *66* (12) 1185—1189.

YANAGIDA, F., TAKAHASHI, K., TAKASHIMA, K., YAMAMOTO, Y., KANEKO, N. and SUMINOE, K. 1973. Studies on the acetic acid bacteria and their utilization. IX. J. Soc. Brew. Jpn. *66* (22) 130—134.

22

Rice Hulls

Wen-Hwei Hsu
Bor S. Luh

According to the statistics supplied by the Food and Agriculture Organization of the United Nations (FAO) (1974–1975), the world's annual paddy rice *(Oryza sativa)* production is estimated to be 341,000,000 MT. Most of this tonnage is produced in Southeast Asia. A major derivative of the rice crop is the hull, a fibrous, nondigestible commodity, representing some 20% of the dried paddy on-stalk. According to FAO-United Nations Industrial Development Organization (FAO-UNIDO 1973), dried paddy on-stalk yields 52% by weight of white rice, 20% hull, 15% stalk and 10% bran. The remaining 3% is lost in the conversion process. If all the paddy rice available were commercially milled, 68,000,000 MT of hulls would be produced. They would occupy 600 million cubic meters of storage space (Mehta and Pitt 1976). Due to their abrasive character, poor nutritive value, low bulk density and high ash content, only a little of the hulls can be disposed of for certain low value applications, such as chicken litter, juice pressing aid and animal roughage. The remaining rice hulls are transported to fields for disposal, usually by open field burning. The practice of open field burning of rice hulls is now strongly opposed. If not properly utilized, rice hulls will create a growing problem of space and pollution in the environment. It is likely that hull utilization will increase in the light of the high cost of fuel and the energy crisis confronting the world population.

MORPHOLOGY

Research has been done on the structure of the rice hulls by Weatherwax (1929); Winton and Winton (1932); Santos (1933); Juliano and Aldama (1937); Grist (1959); and Watson and Dikeman (1977).

The exterior surface of the rice hulls is composed of dentate rectangular elements. Thomas and Jones (1970) and Thomas *et al.* (1972) stated that silica is highly concentrated in the outer layer which is coated with a thick cuticle and surface hairs. The inner region of the hulls is fibrous and striated and is composed of elongated hypodermal fibers (Watson and Dikeman 1977). Thomas *et al.* (1972) showed that the midregion contains little silica. Houston (1972) had given a detailed description of the rice hulls' structural layer: (1) the outer epidermis coated with a thick cuticle of highly silicified sinuous cells among which the surface hairs are found, (2) sclerenchyma of hypoderm fibers also with a thick and somewhat lignified and silicified wall, (3) spongy parenchyma cells both elongated with rather wavy outline and short or quadrilateral, and (4) inner epidermis generally of isodiametric cells.

PHYSICAL PROPERTIES

Most rice hulls are straw or gold in color. Some rice hulls may be white, russet, reddish brown, shades of purple or sooty black (Houston 1972). The length of rice hulls is about 5–10 mm and the width varies from 2.5–5 mm.

According to Fieger *et al.* (1947), the bulk density of rice hulls is 0.100 g/ml. Other reports indicate bulk density values ranging from 96–160 kg/m^3 (Karon and Adams 1949; Lathrop 1952) and that the hulls can be compressed to 400 kg/m^3 (Lathrop 1952). Grinding can raise the bulk density from 192 or 208 to 384 or 400 kg/m^3 (Karon and Adams 1949).

The high concentration of opaline silica present in the outer layer of rice hulls results in an effective hardness of approximately 5½ to 6½ (Mohs Scale) reported for opal. Therefore, rice hulls can be used as an abrasive. When crushed into angular fragments, the abrasive characteristics of rice hulls are intensified. Regular whole rice hulls are reported (Arkansas Rice Growers' Co-op. Assoc. 1968) to have an angle of repose of 35°. When ground to 80- and 160-mesh, their angle is 43° to 45°; for material passing 80-mesh, it lowers again to 40°.

CHEMICAL PROPERTIES

Carbohydrate

The major carbohydrates of hulls are cellulose and hemicellulose. Hemicellulose, chiefly pentosan, is a glucoxylan which can be hydrolyzed to xylose (Buston 1947). According to Houston (1972), starch is apparently absent although small amounts are noted in some commercial products. The composition of rice in hulls is given in Table 22.1.

Crude Protein

The crude protein content of rice hulls is about 3%. A higher protein value undoubtedly reflects some bran contamination in it. The amino acid composition of protein in hulls is presented in Table 22.2.

Lipid

The lipid content of rice hulls ranges from 0.39 (Primo *et al.* 1970) to 2.98% (Limcango-Lopez *et al.* 1962). According to Houston (1972), true hulls generally would not have over 1.0% lipid and the average, somewhat less. Higher values reported in the literature are probably due to the presence of bran in the sample. According to Hartman and Lago (1976), the lipids from rice hulls contained unsaponifiable matter and free fatty acids 4 times higher than those from rice bran and rice caryopsis. There were also differences in the fatty acid composition, as evidenced by the presence of 2–3% of saturated C_{22} and C_{24} acids and a lower proportion of unsaturated acids in the lipids of rice hulls. Chromatographic analysis of the unsaponifiable matter of lipids from rice hulls showed that the sterols consist of about 51.95% β-sitosterol, 22.32% campesterol, 20.13% stigmasterol and 2.92% cholesterol.

Lignin and Cutin

There is evidence that a large part of the lignin is chemically combined with the hemicellulose, and that the middle lamella of the cell walls may contain 70% of the lignin associated with pentosans and a little cellulose (Neish 1965). Therefore, lignin values depend greatly on the extraction process used. Leonzio (1966) reported the presence of 19.20% to 24.47% purified lignin in rice hulls.

Cutin is a water repellent material covering the outer layers of the rice hulls. It appears to be a polymer of long chain hydroxymonocarboxylic acids. The cutin content of rice hull is 2.2% (Nelson *et al.* 1950).

Vitamins and Organic Acids

According to the reports (Kik 1943,1945; Kik and Vanlandingham 1943A,B, 1944; Kik and Williams 1945; Natl. Acad. of Sci. 1971), the levels of thiamin, riboflavin and niacin in rice hulls are 0.84–2.40 γ/g, 0.62–0.93 γ/g, and 14.0–39.5 γ/g, respectively.

TABLE 22.1. COMPOSITIONAL DATA ON RICE HULLS

Moisture (%)	Crude protein (%)	Crude fat (%)	Nitrogen-free extract (%)	Crude fiber (%)	Ash (%)	Pentosans (%)	Cellulose (%)	Other	Year
8.3	2.88	3.5	—	—	22.6	26.0	42.2	19.2 lignin	1950
9.23	6.38	0.65	31.58	31.30	—	—	39.05		1952
0.00	—	—	—	—	19.03–29.04	—	—		1953
10.0	2.00	1.20	—	—	17.60	21.95	41.22	32.88 lignin	1953
10.23	1.94	0.52	29.6	—	18.33	—	39.39		1954
8.0	3.0	0.8	28.4	40.7	—	—	—		1956
—	1.75	2.98	26.05	44.81	24.16	—	—		1962
8–11	2–3	0.5–1	25–30	36–45	22–24	—	—		1964
0.0	1.8–2.6	0.56,0.81	29.9,30.8	44.5,46.3	21.2–24.0	—	—		1969
0.0	2.18–4.84	0.38–0.78	26.0–34.1	47.28–49.92	15.27–20.32	—	—		1970
7.6	2.80	0.80	29.20	41.10	18.40	19.80	39.0	19.8 lignin	1971
9.02	2.63	0.72	37.84	35.74	14.50	—	35.04	1.35 starch	1972
9.60	3.70	0.80	37.70	41.80	16.00	—	—		1975
0.0	2.81–3.75		—	—	21.7–22.6	—	34.2		1976

Source: Houston (1972), Hsu et al. (1976).

TABLE 22.2. AMINO ACID CONTENT OF CALIFORNIA RICE HULLS (G AMINO ACIDS PER 16.0 G N)

Amino acid	Hulls (Calrose)	Amino acid	Hulls (Calrose)
Lysine	3.82	Cystine	1.90
Histidine	1.22	Valine	5.69
Ammonia	3.42	Methionine	1.76
Arginine	4.30	Isoleucine	3.66
Aspartic acid	8.60	Leucine	6.47
Threonine	4.20	Tyrosine	2.16
Serine	4.65	Phenylalanine	4.40
Glutamic acid	10.42	% N recovered	82.9
Glycine	5.43	% N in sample (d.b.)	0.32
Alanine	6.13		

Source: Houston et al. (1969).

Total acid extracted from rice hulls is 57.8 ± 1.4 meq/kg of which 30.3 meq is organic acid (Houston 1972). Houston et al. (1963) showed that oxalic and citric acid are the major organic acids in rice hulls. The other acids are acetic, fumaric, malic, succinic and an aromatic acid. The aromatic acids are ferulic, vanillic, p-coumaric, sinapic, p-hydroxybenzoic, salicylic and indolacetic acids (Mikkelsen 1967; Mikkelsen and Sinah 1961). Some of these phenolic acids and related substances act as germination inhibitors.

Inorganic Component

The major inorganic component of rice hull is ash. It varies from 13.2−29.0% of the weight of rice hull (Houston 1972). The silica content of the ash is around 94−96%. A value near or below 90% may indicate a mixture of bran or other low silica material in the hull sample (Table 22.3). X-ray diffraction studies have shown that pink ash consists essentially of tridymite and cristobalite (Jones 1953,1954). The other components of the ash are K_2O, CaO, Fe_2O_3, P_2O_5, SO_3, Na_2O, MgO and Cl. The rather wide range of values shown for the elements determined, indicates variation in purity of the samples and the accuracy of the analytical procedures used (Table 22.4).

HULL UTILIZATION

Agricultural Uses

Animal and Poultry Feeding.—Numerous efforts have been made to use rice hulls as cattle feed (Wahed 1965; Hsu et al. 1976; Hamad et al.

TABLE 22.3. INORGANIC COMPONENTS OF RICE HULL

Location	Ash	K	Na	Ca	Mg	Fe	P	Cu	Mn	Zn	S
					Components as % of hull						
U.S.[1]	21.5	0.73	0.02	0.08	0.04	—	0.04	—	—	—	—
U.S.[1]	19.9	0.34	—	0.09	0.03	—	0.08	—	—	—	—
Spain[2]	20.3	0.18	0.01	0.15	0.04	0.01	—	0.006	0.001	0.001	—
Australia[2]	22.6	0.275	0.078	—	—	—	—	—	—	—	0.002

Source: Wildman and Brandon (1968); Primo *et al.* (1970); Natl. Acad. of Sci. (1971); Choung and McManus (1976).
[1]On as-is basis; moisture not reported.
[2]On dry basis.

TABLE 22.4. ANALYSIS OF RICE-HULL ASH (% DRY BASIS)

SiO₂	K₂O	Na₂O	CaO	MgO	Fe₂O₃	P₂O₅	SO₃	Cl	Year
93.21	1.53	0.30	0.51	0.07	0.45	2.69	0.42	0.15	1870
87.71	1.60	1.58	1.01	1.96	0.54	1.86	0.92	—	1871
96.97	0.58	—	0.57	0.12	—	0.57	—	—	1916
—	—	1.75	—	—	0.38[1]	1.44	—	—	1917
97.3	—	—	0.43	—	—	2.78	—	—	1925
—	—	—	1.38	—	tr	0.53	1.13	tr	1928
94.50	1.10	0.78	0.25	0.23	—	0.23	—	—	1928
—	0.75	—	1.3	—	—	—	—	—	1929
95.4	—	—	—	—	0.94[1]	—	—	—	1929
95.49	1.88[2]	—	0.86	0.28	—	0.36	—	—	1930
—	1.8–2.5	—	1.0–1.5	—	—	0.5–2.0	—	—	1933
—	1.00	0.4	0.25	0.25	tr	0.3	1.00	tr	1941
96.5	1.59	0.0	0.32	0.76	—	—	0.40	0.42	1952
96.62[3]	—	—	0.52	—	—[3]	—	—	—	1953
—	—	—	—	—	—	0.32	—	—	1962
96.20	0.79	—	0.24	0.24	—	0.46	—	—	1966
92.47–95.04	1.70	—	0.25	0.23	—	0.67	—	—	1966
92.6	—	—	0.2	—	—	0.2	—	—	1966
86.9–96.2	—	—	—	—	—	—	—	—	1966
93.96	—	—	0.88	0.76	—	2.85	0.10	—	1968
91.16	4.75[2]	—	0.65	0.99	0.21	—	—	—	1970
86.9–97.3	0.58–2.5	0.0–1.75	0.2–1.5	0.12–1.96	tr–0.54	0.2–2.85	0.10–1.13	tr–0.42	Range

Source: Houston (1972).

[1] Total of iron plus aluminum as oxides.

[2] Total of potassium plus sodium as oxides.

[3] Trace elements present include aluminum, copper, iron and manganese as well as detectable amounts of barium, boron and zinc. Tin was not found.

1976; Choung 1976). Rice hulls are low in digestibility and nutritive value, and have sometimes caused harmful effects (Fieger *et al.* 1947). This characteristic may be related to the mineral nature of the rice hulls, rather than the massive encrusting silica sheath and the lignin that exist in the rice hulls (McManus and Choung 1976; Guggolz *et al.* 1971). If the integrity of the silica shield and the lignin existing in the hulls were the barrier to microbial attack on the sequestered organic matrix, it would be reasonable to expect that prior grinding of rice hulls to expose the organic matrix, would enhance the efficiency of the microbial attack.

According to Fraps (1904), the nutritive value of rice hulls is: digestion coefficient (for poultry), 0.17; nitrogen-free extract, 0.17; other extract, 0.41; protein, 0. The National Academy of Science (1971) showed that the energy of hulls is: 0.48 DE (Digestible Energy) Mcal/kg or 0.39 ME (Metabolizable Energy) Mcal/kg for cattle; 0.68 DE Mcal/kg or 0.5 ME Mcal/kg for sheep. The TDN (Total Digestible Nutrients) of hulls for cattle and sheep are 10.8% or 15.4%, respectively. Guggolz *et al.* (1971) reported that hulls contained 11% digestible dry matter. Morrison (1956) reported that rice hulls contain 9.9% total digestible nutrient. Hsu *et al.* (1976) found 8.75% total solubles in rice hulls after enzyme treatment (TSAE).

Animals will not voluntarily consume a diet of raw rice hulls alone. Substitutions from 5—50% have been reported with little economic or nutritional benefit (Noland and Gainer 1953; Noland and Ford 1954; Ray and Child 1963; White *et al.* 1969; Choung and McManus 1976).

Presently, utilization of rice hulls as poultry and ruminant feeds has been documented. Success in the utilization was largely due to economic reasons. There is also greater confidence now in rice hulls and rice hull derivatives from the mixers and feeders (Houston 1972). The legal obstacles in rice hull utilization as feeds have been gradually overcome (Beagle 1974).

There is a need for developing more feeding and nutrition data. This should include whole, ground, cracked, ammoniated and parboiled rice hulls from long, medium and short grain varieties. The optimum conditions for cooking, pulping, preparation in various solutions, pelleting and the use of furfural residue also need investigation. Combinations of rice mill byproducts as feeds offer some advantages.

Chemical Treatment.—Most workers consider the benefits of the sodium hydroxide treatment to arise from removal of the silica, or modification of the silica and other components of rice hulls. Treatment with increasing amounts of sodium hydroxide results in removal of more silica and lignin (McManus and Choung 1976). Acid treatment following the alkali treatment removed silica to a lesser extent than when the materials were not neutralized (Fig. 22.1). The greatest removal of lignin occurred in alkali and alkali plus acid treatment. Huntanuwatr *et al.* (1974) found that treatment of rice hulls with 8% NaOH, 12% NaOH and 16% NaOH for 24 hr followed by centrifugation increased the rice hull solubility in water and reduced silica content from 21.9% to 10.79%, 4.58% and 3.02%, respectively, with a little change in lignin, hemicellulose or cellulose content. Choung and McManus (1976) found that the digestibility of lignin and cellulose of the rice hulls was not significantly increased by alkali treatment and suggested that the function of this treatment was to free the hemicellulose of the cell wall of rice hulls for digestive attack.

Rice hulls treated with NaOH will increase the digestibility coefficient of dry matter from 5 to 20, of fiber from 12 to 28, and of extract matter from 5 to 38 (Lindsey 1922). Guggolz *et al.* (1971), using an enzymatic digestion test, showed that when rice hulls are treated with 28 kg/cm² of steam, there is an increase in DMD (dry matter digestibles) to 22%, and 28 kg/cm² of steam combined with alkali will raise the DMD to 38%. Hsu *et al.* (1976) used the enzymatic digestion test to determine the *in vitro* digestibility of alkali-treated rice hulls. They found that when the rice hulls were soaked at ambient temperature in NaOH solutions of various concentrations at a ratio of 1:8 (w/v), the enzymatic digestibility was raised from 8.81% to 40.79% as the concentrations of NaOH increased from 0% to 30% (w/v), Fig. 22.2.

An often cited limitation in the alkali soaking process involves its use of large volumes of both treatment solution and washing water, and the resulting loss of solubilized materials.

Proposed modifications have thus been directed toward overcoming these deficiencies. Hsu *et al.* (1976) described a "dry process" in which a much reduced volume of concentrated NaOH solution was used, resulting in an *in vitro* dry matter digestibility of 35.32% for treated rice hulls (Fig. 22.3). Considering the *in vitro* nutritive value, cost of treatment and pollution abatement, the optimal process may be achieved by spraying the rice hulls with 30% NaOH solution until a total of 5% NaOH (w/w) is obtained.

Choung and McManus (1976) tested and treated rice hulls in feeding trials with sheep against lucerne (alfalfa). Slow growth was shown by animals receiving 5.0 and 10.0% alkali-treated diets. Animals tolerated

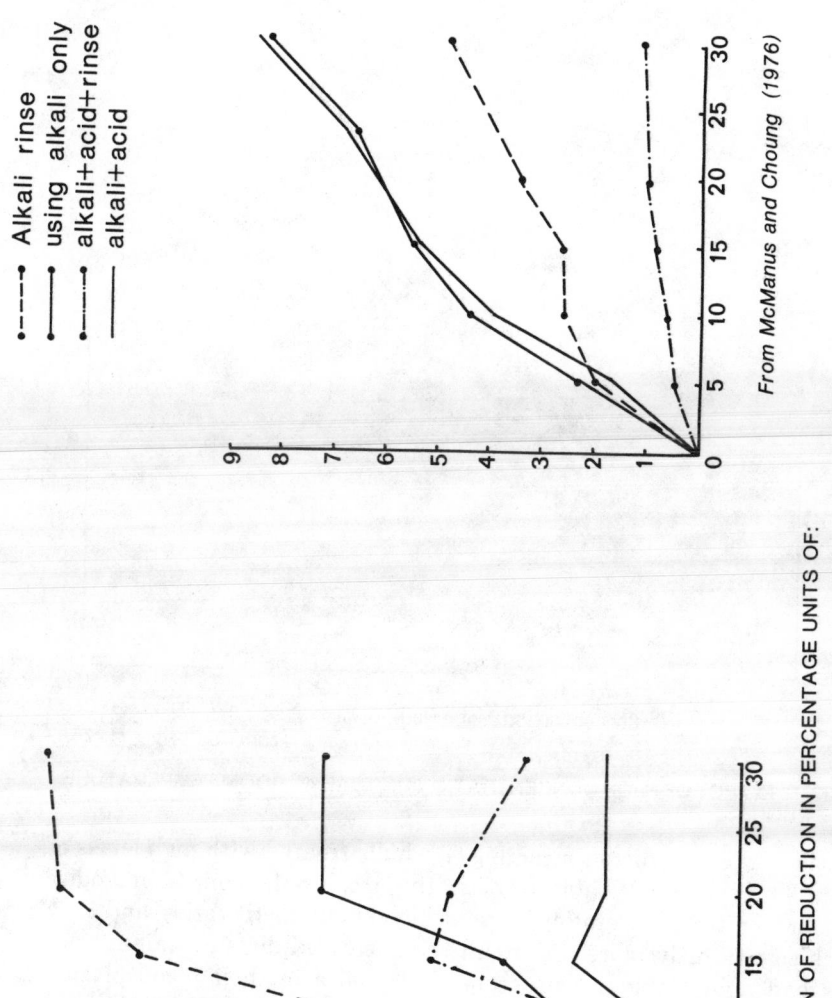

FIG. 22.1. THE PATTERN OF REDUCTION IN PERCENTAGE UNITS OF:

(1) silica content, and (2) lignin content of raw whole rice hull D.M. after treatment with different levels of alkali (0, 5, 10, 15, 20 and 30 g NaOH/100 g D.M.) using alkali only, alkali + acid, alkali + acid + rinse, and alkali + rinse. In both instances D.M. has been adjusted for ash added during treatment. Mean silica content prior to treatment was 17.2%. Mean lignin content prior to treatment was 22.9%.

From McManus and Choung (1976)

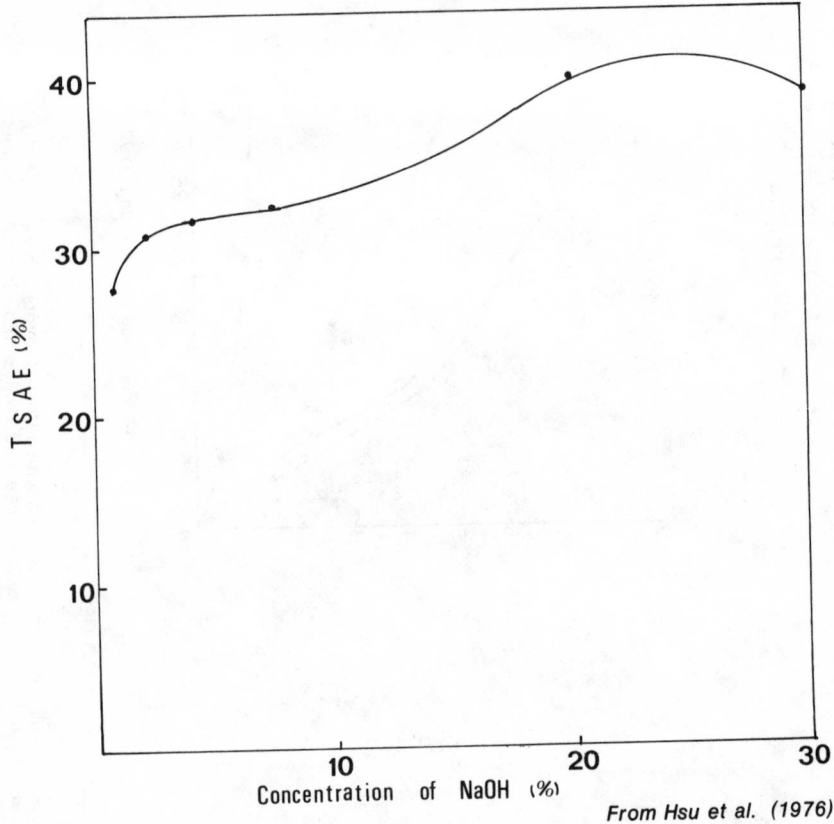

From Hsu et al. (1976)

FIG. 22.2. EFFECT OF NaOH CONCENTRATION ON TSAE AT ROOM TEMPERATURE FOR 24 HOURS

the salt content of diets containing rice hulls treated with high levels of alkali. The sheep were able to clear the excess salt from their bodies without significant alteration of their blood hematocrit ratios, and they drank significantly more water. All the sheep fed diets containing rice hulls had appreciable amounts of soluble silica in their blood plasma ($> 100 \mu g/1$) and urine. It was concluded that the alkali treatment of rice hulls to enhance their value as feed for ruminants is worthy of active consideration and that the process requires further study.

   Interest in the use of ammonia is related to its potential to increase both digestibility and the nitrogen content of the rice hulls. Hiroshi *et al.* (1975) found that when the rice hulls were treated with ammonia (10% by wt.) and water (30% by wt.) for 12 months at ambient temperature, the crude protein content of rice hulls was tripled through this treatment. Most of the increased nitrogen was occupied with nonprotein nitrogen. After ammonia treatment, the cell wall constituent of rice hulls was

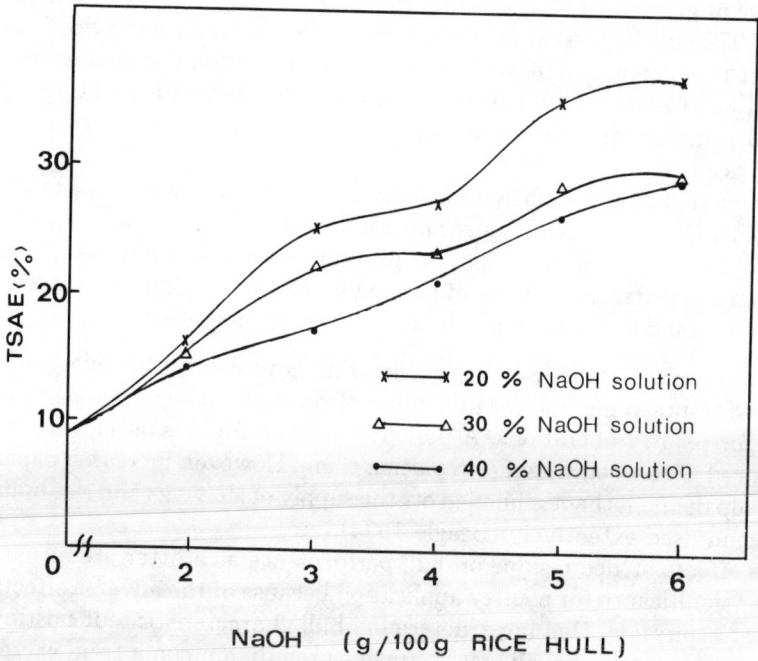

From Hsu et al. (1976)

FIG. 22.3. THE CHANGE IN TSAE VALUE OF RICE HULLS SPRAYED WITH DIFFERENT AMOUNTS OF NaOH AT VARIOUS CONCENTRATIONS

decreased by 6% and all of the noncell wall material was increased. This treatment significantly raised the nutritive value of rice hulls.

Another process used to produce ammoniated rice hulls was developed by Ulrey (1966). In this method, the rice hulls were treated in the presence of catalysts with heat and pressure in an atmosphere of ammonia. Presently, a number of plants have been set up throughout the world. After ammoniation, the fiber is softened and it provides a feed acceptable to sheep or cattle. This feed, when fed to steers, produced average daily gains of 1.25 kg on the 10% ration, and 1.16 kg on the 20% ration. When fed at a level of 20% of the total feed mixture, ammoniated rice hulls have been shown to cause toxicosis. Twenty percent ammoniated rice hulls in the feed is considered the maximum level which can be tolerated by cattle without depressing weight gain or feed intake (Eng 1964). The U.S. Food and Drug Administration (1966) established the limit for ammoniated rice hulls as a feed supplement not to exceed 20%.

Physical Treatment.—Physical means of enhancing the nutritional value of low quality forage include grinding (Minson 1963; Weston 1967) and use of gamma irradiations (McManus and Choung 1976; McManus *et al.* 1972A,B,C). Neither means are very satisfactory. Ground rice hulls are not injurious to cattle when the level in the ration does not exceed 30% (Rusoff *et al.* 1956). However, according to McManus and Choung (1976), grinding treatment could not increase the dry matter digestibility of rice hulls.

Gamma irradiation slightly increased the solubility in water of ground rice hulls (Fig. 22.4, McManus and Choung 1976). Application of 5 g NaOH/100 g dry matter prior to grinding markedly increased their solubility in water at all levels of gamma irradiation with no evidence of major interaction between alkali and the irradiation effects.

Bedding and Litter.—Bedding or littering is probably the oldest and most widespread method for utilization of rice hulls. They are especially good for poultry operations. A period of decline for this usage was experienced with wood shavings replacing them. However, increased paper and pulp demands have siphoned off the supply of shavings and rice hulls are again used extensively (Beagle 1974).

The effects of rice residue on hull performance as a litter are of particular significance for poultry application because of the adverse effects of dust aspiration. One must determine hull characteristics for hosting organisms, parasitic involvement, fungus growth, ammonia from excretion and long-term heating effects.

There is another area of utilization of the litter after it has been used, perhaps by pelleting it and feeding it as a high protein supplement. The litter might also be used as a fertilizer, but information must be developed as to beneficial effects obtainable with various combinations of litters available.

Soil Treatment and Fertilizers.—Incorporating hull into the soil normally occurs as an adjunct to disposal operations. Past efforts largely have been confined to demonstrating that the hull is not harmful to the soil, but slightly beneficial as a fertilizer. Some farmers apply the rice hulls or the ash to soils and increase the uptake of silica by the plants (Ishibashi 1954,1956). Browne (1904) indicated that decomposing hulls make soil phosphorus more available. Such benefits appear confined to tight clay or sandy soil in order to offset silica deficiencies.

In some instances, the benefit stems from the effect of hulls as a mulch instead of a fertilizer (Beagle 1974). Heavy soil such as claypan soil that is incorporated with rice hulls, which have open texture and slow decomposition characteristics, may sometimes result in a good yield of crops (Wildman and Brandon 1968). One must be sure that anaerobic decomposition effects are not promoted by excessive compaction or wetness of the treated areas.

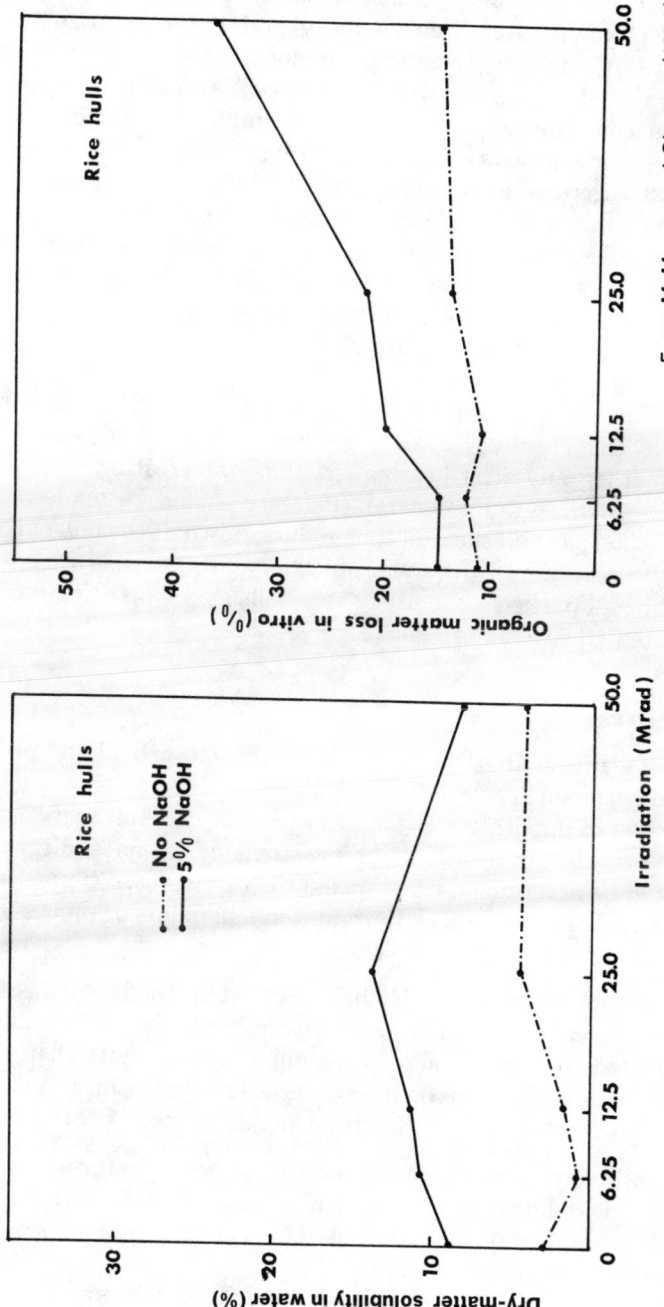

FIG. 22.4. DISAPPEARANCE OF SUBSTANCE FROM CONTROL AND 5% NaOH TREATED RICE HULLS AND RICE STRAW WHICH WERE TREATED WITH ALKALI AFTER IRRADIATION WITH DIFFERENT LEVELS OF GAMMA RAYS

Ammoniated rice hulls may be used in soil for the seedling of passion fruit, papaya, China ester, English daisy and French marigold (Fan 1976). In the soil mixture containing ammoniated rice hulls, vegetable growth and root development will be increased. Ammoniated rice hulls can also be used in the production of leaf vegetables. Kao (1977) showed that 1000 kg ammoniated rice hulls per 0.1 ha treatment, was very good for most leaf vegetable production and 2000 kg ammoniated rice hulls per 0.1 ha treatment was the best for head lettuce.

The cost of cultivation increases apace with the cost of transportation and the technique of soil incorporation as a hull usage factor will not prosper, except in instances of desperate disposal. The cost per acre for effective application is too high, except when most of such cost is borne by the disposal operation.

Rice hulls have been used successfully as a support medium for growing vegetables hydroponically (Hough and Barr 1956). The hulls were found to be satisfactory only after heating, grinding and treating with a synthetic detergent. Determination for their use should be made on the effect of rice hull ash and silica on the product grown. If the hull particles interfere with root movement and if objectionable materials are leached out from the hull particles, the use of rice hulls as a supporting medium will be limited (Beagle 1974).

Industrial Uses

Absorbent.—Rice hull ash has been utilized to absorb oil and provide an antiskid surface (Beagle 1974).

Another area of application is the utilization of rice hull or its derivative as a media for physically dispersing surrounding materials. Its behavior should be determined when mixed with a viscous, cohesive, semisolid material. This would allow for additional handling or processing of the mass.

Building Board.—Vasisth (1971) has reported on the use of rice hulls in the development of a water-resistant composite board for building. There is a need for development of techniques and plants that would provide beneficial utilizations on a large scale. Rice hull utilization can be pursued in areas where rice hulls are available (Beagle 1974).

Calcium Silicide.—The current practice for manufacturing calcium silicide is to heat lime, sand and carbon in an electric furnace. The product is a powerful reducing agent used in explosives and a deoxidizing agent in steel manufacturing.

Raw rice hulls contain approximately 20% silica in a very finely dispersed form. Pyrolyzed hull can be chlorinated at 1000°C (1832°F) to give a nearly quantitative yield of silicon tetrachloride, free of most other inorganic chlorides. Solid, organic chlorine contaminating species have been found to be reaction intermediates. Commercial application of such silicon derivatives appears economically attractive (Basu 1974).

Activated Carbon.—Considerable confusion stems from the use of the term "activated carbon" to describe both rice hull carbon containing silica and carbon with silica removed. There is an obvious significant difference in characteristics between the two. Similarly, much diverse activity with confusing results has been undertaken in connection with the production of activated carbon from rice hulls. Many market opportunities appear to exist for rice hull carbon, but have been somewhat deferred because most existing processes for char production result in a very fine material while the demand is for larger and more uniform particles (Beagle 1974).

Carrier.—Rice hull has been used as a carrier in many products. Both the rice mill byproduct and the ground hull have found markets as carriers for vitamins, pharmaceuticals, biologicals, toxicants and seeds. Most rice hull utilizations now are based on an empirical reaction to a need. There is a need to study the effect of particle size, shape and distribution on adsorption of various agents and carrier-concentrate ratios as they affect shelf life and appearance.

Cement.—A hydraulic cement can be made by grinding together 20–30% lime with rice hull ash (Mehta and Pitt 1974,1976; Pitt 1976). The setting and hardening characteristics of rice hull ash cement are similar to normal portland cement. Typical compressive strengths of rice hull ash cement mortars, when tested in accordance with the ASTM C109, were 175 kg/cm^2 (2500 psi) at 3 days, 320 kg/cm^2 (4500 psi) at 7 days, and 450 kg/cm^2 (6400 psi) at 28 days. The tensile strength and elastic modules of concretes made with rice hull ash cement are satisfactory. The water demand of the cement is generally higher than portland cements, hence mortars made from such cements tended to exhibit higher shrinkage.

The hydraulic cements made from rice hull ash can be treated as a premium product. The unique characteristic, a permanent black color in portland cement, is usually produced by blending with black pigments. Up to 10% carbon black or black iron oxide by weight of portland cement is used for this purpose. The black color produced by these pigmenting agents does not have good long time stability due to the discoloring effect of $Ca(OH)_2$ produced by portland cement hydration. The setting and strength characteristics of cements containing 10% carbon black or iron-oxide are adversely affected. The rice hull ash cement or a portland cement containing 10–20% rice hull ash as the black pigment, produced mortars and concretes of permanent black color. Moreover, when the rice hull ash was used as a pigmenting agent, the strength of the portland cement showed improvement (Fig. 22.5).

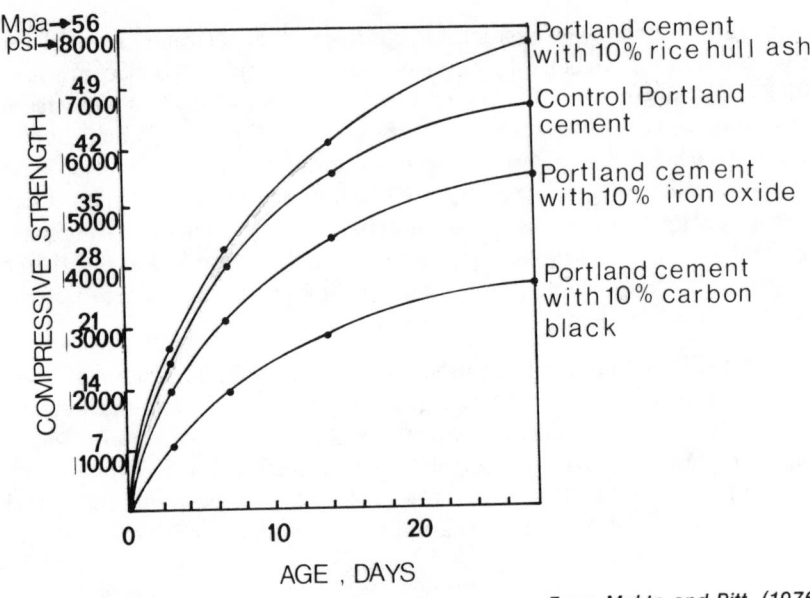

From Mehta and Pitt (1976)

FIG. 22.5. COMPARATIVE STRENGTH OF PORTLAND CEMENT PIGMENTED WITH RICE HULL ASH, IRON OXIDE OR CARBON BLACK

The mortars and concrete specimens made with a rice hull ash cement containing 20% CaO were far superior in resistance to acidic attack to companion specimens made with an ordinary portland cement. It may be noted that portland cement contains 60−65% CaO, and that upon cement hydration a considerable proportion of it is invariably present as free $Ca(OH)_2$. This accounts for the relatively poor performance of portland cement concretes under acidic environments.

The acid resisting nature of the rice hull ash cement is of great interest to the food industry. Many food products contain organic acids such as acetic acid and lactic acid. The former is present in vinegar and pickles, and the latter in milk, curds, cheese, green fodder, etc. Fruit juices and soft drinks are also acidic in nature. Since the $Ca(OH)_2$ constituent of hydrated portland cement is readily attacked by acidic materials, the floors and other structural parts made with portland cement concretes do not last when exposed to such environments.

Colloidal Usage.—Rice hull produces an ash with a high silica content made up of extremely fine particles with a high surface area.

For utilization as a catalyst carrier, it must have a high surface area as well as the capability for conversion into pellets.

For use as a thickening agent, it must have good swelling properties. Where it will act as a dehydrating agent, it must offer efficient water adsorption and provide a high water holding capacity (Beagle 1974).

Furfural.—Adequate furfural technology exists to permit its utilization on a plant size basis.

Cooking rice hulls in the presence of an aqueous acid, such as sulfuric acid, with steam distillation, will produce furfural. The Quaker Oats Chemical Company has used 13,000 to 15,000 MT of rice hulls per year in their Memphis, Tennessee plant (Lathrop 1952). According to Beagle and Beagle (1971), 2% of the United States rice hull supply was used for furfural production.

Motive Power.—Hawkey (1974) has made a brief review on the utilization of rice hulls. According to Grist (1975), the fuel value of rice hull is approximately 3200 kcals/kg. One can produce 2.5 kilos of steam from 1 kilo of hull (Schule 1971). Steam consumption is around 7 kg per hp/hr. The volume of hull ejected by a rice mill is as large as the volume of paddy fed to the mill.

A mill with capacity of 1 MT or more per hour can yield sufficient rice hulls to produce necessary steam for milling with a 20% surplus for drying and parboiling.

Owing to their high ash content, hulls are not easy to burn. It is essential that a special stepped grate furnace, preferably with an automatic feeding device, be supplied. A well designed combustion chamber can prevent the excessive formation of furnace ash. Automatic ash collectors are used to draw off the incinerated residue. The ash should be cooled, bagged up and packed for usage as a fertilizer base, thus eliminating hull residue.

Pressing Aids.—Rice hulls have been utilized by the food industry as pressing (or flow) aids for fruit juices, beverages, wines, etc. The more important areas of concern are yields, contamination problems and sterilization. In connection with yields, performance (yield comparisons-cake moisture) data should be developed for a given variety of fruit, using particle size and ripeness as parameters. The contamination problems are particularly critical when filtering wines. Methods of effective cleaning in preparation of the hull must conform to legal standards. The procedures for sterilization and cleaning must be developed which are cost effective in relation to yields. Rice hull utilization may be expanded to areas other than fruits and vegetables.

Water Purification.—Rice hull char and ash can be used as a processing aid in filtration, adsorption or coagulation of impurities in water. It can compete with diatomaceous earth as a filtering aid.

Rice hull char can compete with active carbon as an absorption medium. The char produced must have maximum absorption. Theoretically, rice hull char contains a large ash fraction which is relatively inert. It may have a lower efficiency when judged on a comparative weight basis. Actually, the performance of rice hull char ranges from poor to excellent in adsorbing some organic compounds.

Rice hull ash clears turbid water by acting as a nucleation site. Suspended colloidal impurities become attached to the site and are subsequently "swept out" of suspension as the ash particle settles.

Rubber Compounding.—Rice hull derivatives (silicates) can be effectively used in rubber compounding, particularly as a reinforcement and as a nonskid additive (Beagle 1974). For usage by the rubber industry it must be supported by an adequate engineering program to evaluate the advantages of utilizing rice hull silica. Initially, a very fine particle material must be prepared in the desired carbon-silica ratio. Then its characteristics must be determined for various rubber formulations. Reinforcement specifications will be required and advantages should be clearly determined. Thus, while the utilization of rice hull silica is highly valid, extensive applied studies are required to verify the reinforcement and antiskid usage. Colloidal silicas are used to reinforce nonblack rubbers and carbon-free rice hull ash also could be used similarly. Where particle size is too large or unsatisfactory properties are exhibited, it still offers a potential for replacing a portion of the expensive colloidal silica. Black ash or rice hull char can be used as a reinforcement in black rubbers. It is possible that mixtures of carbon black and rice hull ash could provide superior properties or lower cost when compared to utilizing carbon black by itself. Rice hull silica can be utilized to enhance antiskid and abrasive wearing qualities.

The silica ash produced from rice hull by the Mehta and Pitt process (1974) is in the form of a soft material, characterized by high surface area and cellular structure. The absence of any characteristic peaks of crystalline silica phases in the X-ray diffraction pattern shows that the silica is in a noncrystalline form.

Rubber compounds containing 50–100 parts of ash by weight of 100 parts of rubber show mechanical properties that are generally superior to those given by commercially available ground silica or clay fillers. Properties of rice hull ash-containing rubber compounds are comparable to those obtained by using medium thermal blacks as reinforcing agents. Based on a synthetic rubber (SBR 1502), the relative mechanical properties from different reinforcing fillers are compared in Table 22.5. It is

TABLE 22.5. EFFECT OF DIFFERENT REINFORCING AGENTS ON THE PROPERTIES OF RUBBER COMPOUNDS

	Commercial medium thermal black (100 parts[1])	Commercial medium thermal silica (100 parts[1])	Rice ash (RHA)[1] of the Mehta process			
			100 parts[1] RHA	50 parts[1] RHA	50 parts[1] RHA and 0.25 parts silane	30 parts[1] RHA and 25 parts HAF black
Tensile strength (kg/cm²)	79.4	54.8	106.8	68.2	95.6	130.8
Elongation (%)	480	510	610	500	460	550
Modulus at 100% strain (kg/cm²)	22.5	18.3	21.4	14.8	17.2	17.9
Modulus at 300% strain (kg/cm²)	63.3	28.8	45.3	37.3	52.0	51.3
Shore hardness number	65	63	62	54	55	60
Bashore rebound (%)	42	36	41	49	49	50
Compression set (%)	13.5	—	18.2	18.5	16.6	27.2

Source: Mehta and Pitt (1974).
[1]The proportion of the reinforcing agent is based on 100 parts of rubber by weight.

evident from the data that the strength, elastic and hardness characteristics of rubber compounds made with rice hull ash are further improved by slight modification of the compound compositions, such as by introducing silane or by using a combination of rice hull ash and a commercial carbon black (HAF Black). Advantages of using rice hull ash as a filler are that it incorporates readily into the rubber compounds and that the curing times are somewhat quicker when compared with carbon blacks.

The usefulness of rice hull ash is not limited to the Styrene-Butadiene type synthetic rubbers (SBR). Good quality rubber products based on several other types of synthetic rubbers have also been made. A surprising discovery was that when 60 parts of ash by weight were added to 100 parts of natural isoprene rubber, a product with unexpectedly superior mechanical properties was obtained. The resulting isoprene rubber can have 207 kg/cm^2 (2950 psi) tensile strength and 63 kg/cm^2 (900 psi) elastic modulus at 300% strain level (Mehta and Pitt 1974).

Silicon Carbide and Silicon Nitride.—A government funded project on the above subject has been under way at the University of Utah (Cutler et al. 1973). Silicon carbide of fine particle size can be produced from rice hull. This fine SiC can be used for making SiC-containing refractories.

Sodium Silicate.—The silica content of rice hulls varies from 13.2–29.0% (Houston 1972). The silica from rice hulls may serve better than bentonite and diatomaceous earth for the food industry because of its minimal amount of unwanted elements other than silica.

Based on the overall operation cost, a sodium silicate plant utilizing rice hulls is more feasible than a normal plant. This plant should be able to produce sodium silicate at an equivalent cost per metric ton slightly cheaper than a normal plant (Staackmann and Goodale 1970). However, the transportation cost and labor cost must be lowered and all of the available steam should be utilized. Two unconventional processes: the wet air oxidation process (Nemerow 1963) and the fluidized bed reactor process (Staackmann and Goodale 1970)—for making sodium silicate from rice hulls appear feasible, but some development work is necessary to achieve a desirable operation and to reduce the production cost.

Steel Industry.—Rice hull and its derivatives have been utilized by various segments of the steel industry (Beagle 1974).

Material responsive to particular use requirements must be available at competitive prices. Some of its current use stems from the remarkably high refractory capability and good insulation qualities offered by rice hulls.

Hydrolytic Products

Rice hulls have been hydrolyzed with hydrochloric acid, sodium hydroxide and sulfuric acid under various conditions of temperature and pressure. The resulting hydrolyzate can be used for the growth of microorganisms. Nakamura and Ichino (1948) extracted 94% of the pentosan fraction of rice hulls (16.1% of total dry wt.) by treating with 11% HC1 at 110°C (212°F) for 3 hr. A concentration of 20% NaOH extracted 72.8% of the pentosan at 80°C (176°F). The yield (96.2%) was improved by using 0.4% H_2SO_4 at 134°C (273.2°F) at 2.1 kg/cm² for 3.5 hr. More severe conditions resulted in decomposition of the sugars. Rice hulls have been hydrolyzed in a 2 step procedure by Dmitrenko (1976) who hydrolyzed wood chips with 0.4−0.6% H_2SO_4 in a continuous procedure at 170°−190°C (338°−375°F). Yeast was grown under nonasceptic conditions, yielding 210−221 kg of dry yeast per metric ton of bone-dry forestry waste. In earlier published reports (Savinykh et al. 1969), yields of 51.5% on the growth of Candida tropicalis were obtained on the basis of the sugar content.

A New Rice Hull Utilization Process.—Mehta and Pitt (1974, 1976) described a new, low cost process for large-scale disposal of rice hulls. The primary product of this process is a special silica material, suitable for making acid-resistant hydraulic cements and reinforcing agents for rubber. Heat energy is an important byproduct of this process.

The paddy milling operation yields hulls at the rate of about ⅕ of the weight of paddy. Typically, the hulls consist of about 40% cellulose, 30% lignin and 20% ash. The ash is derived mainly from the opaline silica which is present in the cellular structure of hulls. The combustion of organic matter released about 3200 kcals/kg of heat energy.

The new process described by Mehta and Pitt (1974,1976) is for large-scale, profitable disposal of rice hulls. The process consists of burning the organic matter to produce heat energy under conditions such that the residual ash, which is essentially silica, continues to be in an amorphous form. It has been possible to develop an industrial furnace, which can burn the hull continuously at relatively low temperatures. The burning time and temperature are controlled so that the cellular structure of the hull is not generally disrupted and the silica thus produced is in a reactive form. Burning the hull under various conditions changes the end characteristics of the material, but this new process is unique in that it removes most of the organic matter without significantly altering the silica structure.

A flow diagram of a typical plant equipped with a steam boiler is shown in Fig. 22.6. From the furnace the hot gases containing the rice hull ash are taken to a tube boiler, and finally to a multiclone type separator

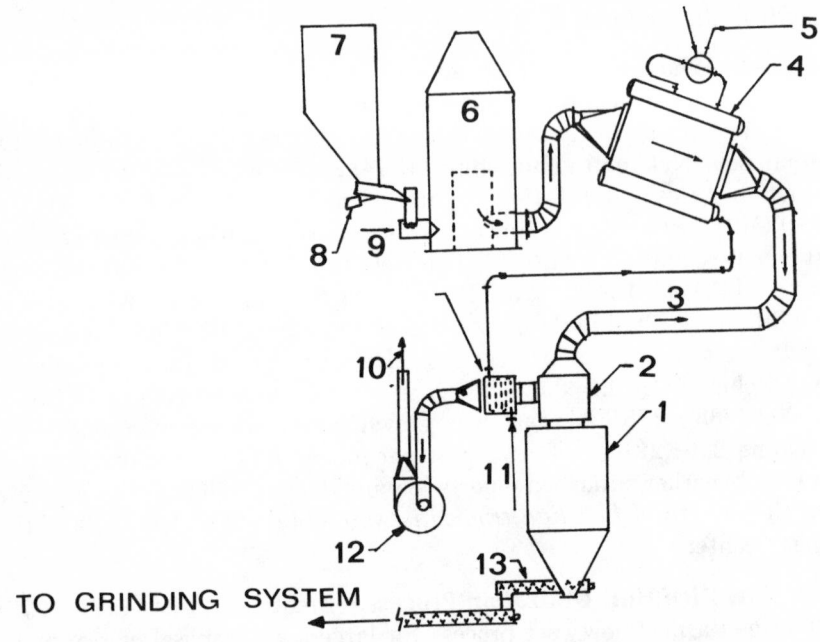

From Mehta and Pitt (1974)

FIG. 22.6. FLOW DIAGRAM OF A TYPICAL PLANT FOR PRODUCING RICE HULL ASH AND STEAM (MEHTA PROCESS)

(1) Underground ash silo, (2) dust collector, (3) preheated water, (4) boiler, (5) steam, (6) furnace, (7) surge bin, (8) feeder, (9) air, (10) exhaust or drying system, (11) water, (12) fan, and (13) screw feeder.

which separates the ash from the gases.

The processing equipment is simple and can be locally fabricated. The refractory-lined furnace looks like an inverted cone, into which hulls are sucked due to negative pressure conditions maintained by a fan. After the initial start and when the temperature in the furnace is high enough, the hulls burn by themselves.

Individual plants can burn hulls at the rate of ½ to 10 MT per hour. Bigger units can produce large quantities of hot air for crop drying and steam for the generation of electrical energy. After utilization for electrical generation, the low pressure steam can further be used for parboiling or other food processing operations. Smaller units may be more practical in areas where a large number of small rice mills are in existence. Such units may not be equipped with boilers, especially if the initial capital cost is to be kept low. In such cases, heat energy can be utilized in the form of by-product hot gases, which can be used for partial or complete drying of paddy.

The potential of the process can be judged from the world's supply of rice hulls. It would be possible to produce about 15 million MT per year of a hydraulic cement, and about 200×10^{12} kcals of heat energy, equivalent to about 30 million MT of coal each year.

REFERENCES

ARKANSAS RICE GROWERS' CO-OP. ASSOC. 1968. Rice hulls unground and special fractions. (mimeo data sheets)

BASU, P.K. 1974. Silicon tetrachloride. Ph.D. Dissertation. U.S. Dep. Agric. grant.

BEAGLE, E.C. 1974. Basic and applied research needs for optimizing utilization of rice husk. In Proc. of Rice By-Product Util. Int. Conf. Vol. 1. Valencia, Spain.

BEAGLE, E.C. and BEAGLE, C.A. 1971. Paddy husk utilization. Ppr. presented at U.N. Indus. Dev. Organ. Conf. on Rice Processing. Madras, India.

BROWNE, C.A. 1904. The chemical composition and feeding value of rice products. La. Agric. Exp. Sta. Bull. 77, 430.

BUSTON, H.W. 1947. The Methods of Cellulose Chemistry. 2nd Edition. E.R. Doree (Editor). Chapman and Hall, London.

CHOUNG, C.C. 1976. Utilization of rice hulls as ruminant feed. II. Metabolism studies on the alkali-treated rice hull in sheep. Hanguk Ch'uksan Hakhoe Chi. 18, 165-175. (Japanese)

CHOUNG, C.C. and McMANUS, W.R. 1976. Studies on forage cell walls, III. Effect of feeding alkali-treated rice hulls to sheep. J. Agric. Sci. 86, 517-530.

CUTLER, I.B. et al. 1973. Formation of silicon carbide and silicon nitride from rice hulls. Off. Res. and Develop.—Environ. Prot. Agency Project R. 801, 380.

DMITRENKO, L.A. 1976. Comprehensive processing of raw vegetable substances. Proc. from Conf. on Single Cell Protein of the U.S./USSR Working Group. March 24−26. MIT, Cambridge, Mass.

ENG, K.S. 1964. The value of ammoniated rice hulls as a ruminant feed. Feedstuff 36 (1) 4.

FAN, N.T. 1976. Effects of ammoniated rice hull on soil medium of horticultural seedling. Hsing Ta Yuen Yi. 1, 6−10.

FIEGER, B.A., CHOPPING, A.F. and TUCKER, P.W. 1947. Making use of rice hulls. Rice J. 50, 9−25.

FOOD AND AGRIC. ORGAN. U.N. (FAO) 1974−75. Rice bulletin. Crop forecast. 1974−75. FAO, Rome.

FAO-U.N. INDUS. DEV. ORGAN. 1973. Actual field test, Java Island, July 1973. FAO-UNIDO Rice Task Force Mission. Jakarta, Indonesia.

FRAPS, G.S. 1904. The composition of rice by-products. Texas Agric. Exp. Sta. Bull. 73, 14.

GRIST, D.H. 1959. Rice (3rd ed.). Longmans, Green and Co., London.

GRIST, D.H. 1975. Industrial uses of rice products. *In* Rice. 5th Edition. Longmans, London.

GUGGOLZ, J., SAUNDERS, R.M., KOHLER, C.O. and KLOPFENSTEIN, T.J. 1971. Enzymatic evaluation of processes for ruminant feeds. J. Animal Sci. *33*, 167–170.

HAMAD, M.A. *et al.* 1976. Effect of dry treatment of rice hulls with alkali and steam on their nutritive value. Indian J. Anim. Sci. *46*, 59–62.

HARTMAN, L. and LAGO, R.C.A. 1976. The composition of lipid from rice hulls and from the surface of rice caryopsis. J. Sci. Food Agric. *27*, 939–942.

HAWKEY, R. 1974. Rice husk utilization. *In* Proc. of the Rice By-Products Util. Int. Conf. Vol. I. Valencia, Spain.

HIROSHI, I., YOSHIAKI, T., NOBUHITO, T. and YUKIO, M. 1975. Improving the nutritive values of rice straw and rice hulls by ammonia treatment. Jpn. J. Zootech. Sci. *46*, 87–93.

HITCHCOCK, L.B. and DUFFEY, H.R. 1948. Commercial production of furfural in its twenty-fifth year. Chem. Eng. Progr. *44*, 669–702.

HOUGH, J.H. and BARR, H.T. 1956. Possible uses for waste rice hulls in building materials and other products. La. State Univ. and Agric. and Mech. College. Agric. Exp. Sta. Bull. No. *507*.

HOUSTON, D.F. 1972. Rice Chemistry & Technology. Amer. Assoc. Cereal Chem., St. Paul, Minnesota.

HOUSTON, D.F., ALLIS, M.E. and KOHLER, G.O. 1969. Amino acid composition of rice and rice by-products. Cereal Chem. *46*, 527–537.

HOUSTON, D.F., GARRETT, V.H., HILL, B. and KESTER, E.B. 1963. Rice quality measurement. Organic acids of rice and some other cereal seeds. J. Agric. Food Chem. *11*, 512–517.

HSU, W.H., LIN, K.C., WU, C.Y. and HUANG, F.M. 1976. Conversion of agro-cellulosic wastes to animal feed. I. Alkali treatment of rice hulls to improve nutritive value. Taiwan Food Ind. Res. and Dev. Inst. Res. Rept. No. *93*.

HUNTANUWATR, N., HINDS, F.C. and DAVIS, C.L. 1974. An evaluation of methods for improving the *in vitro* digestibility of rice hulls. J. Animal Sci. *38*, 140–148.

ISHIBASHI, H. 1954. Absorption of silica by rice seedlings from hull of rice. Yamaguti Univ. Agric. Fac. Bull. *5*, 31–32.

ISHIBASHI, H. 1956. On the effect of silica contained in carbonized rice hull on the growth of rice seedling. Yamaguti Univ. Agric. Fac. Bull. *7*, 333–336.

JONES, J.D. 1953. New refractory from vegetable sources. Can Metals *16*, 22.

JONES, J.D. 1954. Refractory insulators and porous media from vegetable sources. Can. Leram Soc. *23*, 99–101.

JULIANO, J.B. and ALDAMA, M.J. 1937. Morphology of *Oryza sativa* lineaeus. Philip. Agric. *26*, 1–3.

KAO, C.Y. 1977. The effect of ammoniated rice hulls on the yield and quality of leaf vegetable production. Chinese Soc. for Hortic. Sci. J. *23*, 39–44.

KARON, M.L. and ADAMS, M.E. 1949. Hygroscopic equilibrium of rice and rice fractions. Cereal Chem. *26*, 1–12.

KIK, M.C. 1943. Thiamine in products of commercial rice milling. Cereal Chem. *20*, 113.

KIK, M.C. 1945. Nicotinic acid (Niacin) in rice. Rice J. *48* (5) 18–22.

KIK, M.C. and VANLANDINGHAM, F.B. 1943A. Riboflavin in products of commercial rice milling and thiamine and riboflavin in rice varieties. Cereal Chem. *20*, 563–569.

KIK, M.C. and VANLANDINGHAM, F.B. 1943B. The influence of processing on the thiamine, riboflavin, and niacin content of rice. Cereal Chem. *20*, 569–572.

KIK, M.C. and VANLANDINGHAM, F.B. 1944. Nicotinic acid in products of commercial rice milling and in rice varieties. Cereal Chem. *21*, 154–159.

KIK, M.C. and WILLIAMS, R.R. 1945. The nutritional improvement of white rice. Natl. Acad. Sci. Natl. Res. Counc. Bull. *112.*

LATHROP, E.C. 1952. Industrial utilization of rice hulls. Rice J. Ann. Iss. *13.*

LEONZIO, M. 1966. The contents of lignin as a by-product during the elaboration of rice. Riso *15*, 219–223. (Italian)

LIMCANGO-LOPEZ, P.D., ACHACOSO, F.S. and CASTILLA, L.S. 1962. The chemical composition of rice and corn by-products. I. Proximate composition and minerals. Philip. Agric. *46*, 324–330.

LINDSEY, J.B. 1922. The effect of sodium hydrate upon the digestibility of grain hulls. Sci. *55*, 131–136.

MCMANUS, W.R. and CHOUNG, C.C. 1976. Studies on forage cell walls. II. Conditions for alkali treatment of rice straw and rice hulls. J. Agric. Sci. *86*, 453–470.

MCMANUS, W.R., MANTA, L., MCFARLANE, J.D. and GRAY, C. 1972B. The effects of diet supplements and gamma irradiation on dissimilation of low quality roughages by ruminants. II. Effect of gamma irradiation and urea supplementation on dissimilation in the rumen. J. Agric. Sci. *79*, 41–53.

MCMANUS, W.R., MANTA, L., MCFARLANE, J.D. and GRAY, C. 1972C. The effects of diet supplements and gamma irradiation on dissimilation of low quality roughage by ruminants. III. Effects of feeding gamma-irradiated base diets of wheat and straw to sheep. J. Agric. Sci. *79*, 55–56.

MEHTA, P.K. and PITT, N. 1974. A new process of rice husk utilization. *In* Proc. of the Rice By-Product Util. Int. Conf. Vol. I. Valencia, Spain.

MEHTA, P.K. and PITT, N. 1976. Energy and industrial materials from crop residues. Resource Recovery and Conservation *2*, 23–38.

MIKKELSEN, D.S. 1967. Germination inhibitors as a possible factor in rice dormancy. Int. Rice Comm. Newsl. (spec. iss.) *132.*

MIKKELSEN, D.S. and SINAH, M.N. 1961. Germination inhibition in *Oryza sativa* and control by preplanting soaking treatments. Crop Sci. *1*, 332.

MINSON, D.J. 1963. The effect of pelleting and wafering on the feeding value of roughage. British Grassland Soc. Rev. J. *18*, 39–44.

MORRISON, F.B. 1956. Feeds and Feeding: A Handbook for the Student and Stockman. 22nd Edition. Morrison Publishing Co., Ithaca, N.Y.

NAKAMURA, S. and ICHINO, K. 1948. Hydrolysis of fiber materials and their fermentation. Hakko Kogaku Zasshi *26*, 39, 78, 114, 151. (Japanese)

NATL. ACAD. OF SCI. 1971. U.S. and Canadian feeds. *In* Atlas of Nutritional Data. Natl. Acad. of Sci., Washington, D.C.

NEISH, A.C. 1965. Coumarins, phenylpropanes, and lignin. *In* Plant Biochemistry. J. Bonner and J.E. Varner (Editors). Academic Press, New York.

NELSON, G. H., TALLEY, L.E. and ARONOVSKY, S.I. 1950. Chemical composition of grain and seed hulls, nut shells, and fruit pits. Am. Assoc. Cereal Chem. Trans. *8*, 58–63.

NEMEROW, N. 1963. Theories and Practices of Industrial Waste Treatment. Addison-Wesley Publishing Co., Palo Alto, Calif.

NOLAND, P.R. and FORD, B.F. 1954. For wintering steers. Rice hulls and rice mill feed. Arkansas Farm Res. *3*.

NOLAND, P.R. and GAINER, J.H. 1953. Use of rice hulls as a roughage for wintering steer-calves and for gestating-lactating ewes. Arkansas Agric. Exp. Sta. Bull. *538*, 12.

PITT, N. 1976. Siliceous ashes. U.S. Pat. 276,134.

PRIMO, E. *et al.* 1970. Chemical composition of the by-products obtained in the different steps of the diagram of the elaboration of rice. J. Agric. Chem. and Food Technol. *10*, 244–248. (Spanish)

RAY, M.L. and CHILD, R.D. 1963. Rice hulls in wintering rations. Arkansas Farm Res. *12*, 2.

RUSOFF, C.L., FRYE, J.B. and EPPS, E.A. 1956. A look at uses for rice hulls. Hulls fed to cattle produce no ill effects. Rice J. *59* (4) 28.

SANTOS, J.K. 1933. Morphology of the flower and mature grain of Philippine rice. Philip. J. Sci. *52*, 475.

SAVINYKH, A.G. *et al.* 1969. Conversion of rice husk ligno-cellulose into yeast. Gidroliz. Lesokhim. Prom. *22*, 23–29. (Russian)

SCHULE, F.H. 1971. Steam power plants for husk. Schule Manufacturers, Hamburg, German Fed. Rep.

STAACKMANN, M. and GOODALE, T.C. 1970. Rice Hull Utilization Final Report. USR Research Co., San Mateo, California.

THOMAS, R.S., BASA, P.K. and JONES, F.T. 1972. Silicon tetrachloride synthesis from rice hulls: Transmisson and scanning electron microscope study. Proc. 30th Ann. Meeting Electron Microscopy Soc. of America, Aug. 9–13. Boston. Claitor's Publishing Div., Baton Rouge, La.

THOMAS, R.S. and JONES, F.R. 1970. Microincineration and scanning electron microscopy: Silica in rice hulls. Proc. 28th Ann. Meeting Electron Microscopy Soc. of America, Oct. 5–8. Houston. Claitor's Publishing Division, Baton Rouge, La.

ULREY, D.G. 1966. Rice hull products and method. U.S. Pat. 3,259,501. July 5.

U.S. FOOD AND DRUG ADMIN. (FDA). 1966. Ammoniated rice hulls. Federal Register *31*, 4, 955. FDA, U.S. Dep. Health, Education and Welfare, Washington, D.C.

VASISTH, R.C. 1971. Water resistant composite board. Paper presented at the joint UNIDO-FAO Econ. Com. for Asia and Far East Seminar. Madras, India.

WAHED, A.B.M.F. 1965. The influence of levels of low quality roughage upon feed lot performance and carcass characteristics of beef cattle fed high concentrate rations. M.S. Thesis. Texas A&M University, College Station, Texas.

WATSON, C.A. and DIKEMAN, E. 1977. Structure of the rice grain shown by scanning electron microscopy. Cereal Chem. *54*, 120–130.

WEATHERWAX, P. 1929. The morphology of the spikelets of six genera of oryzae. Amer. J. Bot. *16*, 547–549.

WESTON, R.H. 1967. Factors limiting the intake of feed by sheep. II. Studies with wheat and hay. Aust. J. Agric. Res. *18*, 983–1002.

WHITE, T.W., REYNOLD, W.L. and KLETT, R.H. 1969. Roughage sources and levels in steer rations. J. Anim. Sci. *29*, 1001–1002.

WILDMAN, W.E. and BRANDON, D.M. 1968. Rice hull soil incorporation studies, progress report. Univ. Calif. Agric. Ext. Ser. Pub., 44.

WINTON, A.C. and WINTON, K.B. 1932. The Structure and Composition of Foods. Vol. 1. Wiley, New York.

WOO, C.M. and LEE, S.R. 1972. Korea J. Food Sci. Tech. *4*, 300–308.

23

Rice Oil: Chemistry and Technology

Shiu-Chan Chang
Robin M. Saunders
Bor S. Luh

Rice oil is extracted from rice bran, which is produced in large quantities as a byproduct in rice-milling processing. High grade edible oil is extracted from fresh raw bran, followed by refining, bleaching and deodorization. It is extremely rich in nutrients such as vitamin E and has some advantageous effects in lowering blood cholesterol level.

The development of the rice bran oil industry has been studied and practiced for a relatively long period of time. This potentially important material has shown remarkable growth during the postwar period, but has still not been widely and profitably exploited in most underdeveloped rice producing countries. The main technical problem is rapid deterioration of the oil in the bran after separation from the brown rice and the subsequent deterioration in oil quality. Oil extracted from stored rice bran contains a high percentage of free fatty acids which are difficult to remove in refining owing to high refining losses. Stabilization of the bran or prevention of the lipase activity in it by heat treatment or other physicochemical methods has not been commercially applied to a wide extent.

A relatively new chemical processing technique of solvent extraction rice milling has been developed commercially by Riviana Foods, Houston, Texas. This process, carrying the trade name "X−M," starts from either fresh or stored brown rice and produces white rice, together with defatted rice bran containing high protein contents, a highly stable rice oil and a good quality wax. The products tend to be superior in quality to those made by the conventional extraction of bran. This technology has been patented both in the United States and abroad.

If conducted on a larger scale, the X−M process could make a valuable contribution to the nutrition and economy of the rice growing countries. Development of the edible rice bran oil industry through improving production techniques and facilities of the underdeveloped rice producing areas, will not only alleviate the shortage of edible oils and fats but will also improve the health and welfare of the population.

RICE OIL PROCESSING

Extraction Process

Conventional Process.—The general process as applied to receiving oil from rice bran by extraction consists of pretreatments, cleaning, thermal treatment, drying, extraction by pressure or solvent, postextraction treatments, and recovery of crude oil and meal products.

Pretreatments.—The first step in the processing is cleaning in order to separate broken rice, whole rice, germ, etc., from freshly milled rice bran; this is accomplished by screening. The next step is a thermal treatment; this operation is responsible for stabilizing the rice bran to avoid oil deterioration. The activity of a lipolytic enzyme may be destroyed in the bran by heat treatment at 95°–100°C (203°–212°F) for 3 min (Loeb *et al.* 1949; Yokochi 1974). Kopeikovskii and Arutyunyan (1971) wetted the bran to 19–21% moisture and kept the temperature at 99.4°–108°C (211°–227°F); this effectively inhibited the enzyme without increasing the oil acidity number. This thermal treatment also eliminated the fines problems by increasing the particle size and altering the physical characteristics of the particles with a "crisping" or "hardening" effect for better extractability and filtration rates. Gastrock *et al.* (1955) studied a process of mildly cooking the bran, which contained 14 to 26% moisture in the early stages and 6 to 18% in the final stages, using an overall time of 15 to 70 min and increasing the temperature from an initial 76.7°–99°C (170°–210°F) to a final temperature of nearly 113°C (235°F) and crisping the cooked particles by reducing temperature below 54.4°C (130°F) and moisture content by 2–4%. This cooking condition made the material easier to extract and gave a better yield of oil.

Other processes, with small differences in bran moisture content, heating time and temperature, and drying conditions, have been described elsewhere (Srimani *et al.* 1974; Bose and Srimani 1974; Barber *et al.* 1974; Fito *et al.* 1974).

Extraction.—Mechanical and solvent methods of oil extraction from rice bran have been practiced for many years, especially the latter since World War II, due to higher oil recovery.

The 2 important pressing methods currently employed are batch production by hydraulic presses (Fig. 23.1 and 23.2) and continuous production by the use of expellers (Fig. 23.3). The former operates with an open type press at a pressure of about 281 kg/cm^2; the latter is a closed type press with a pressure of 1406 to 2812 kg/cm^2. The amount of oil left in the residue approximates 6 to 12% and 4 to 8% under hydraulic pressing and expellers conditions, respectively.

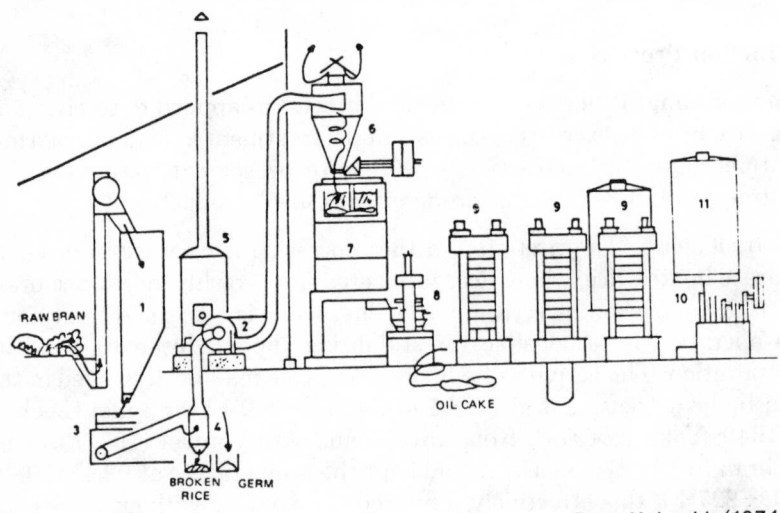

From Yokochi (1974)

FIG. 23.1. DIAGRAM OF A JAPANESE INSTALLATION FOR RICE BRAN OIL EXTRACTION BY THE PRESSING METHOD (HYDRAULIC PRESSES OF THE RING TYPE)

(1) Hopper bin, (2) fan, (3) sifter, (4) air separator, (5) steam boiler, (6) air cyclone, (7) dryer and cooker, (8) prepress, (9) hydraulic press, (10) hydraulic pump, and (11) accurator (low and high pressure).

Solvent extractions include batch type, battery type and continuous extraction. Batch type is the oldest system, using one or more extractors in which the pretreated raw material is placed; the hexane from the solvent tank is pumped to the extraction vessel and the solvent level is maintained to percolate the bran and dissolve the oil. The miscella pass through the filter to evaporator for desolventizing (Fig. 23.4). In battery extraction, sometimes called a semicontinuous or batch couer-current system, the fresh solvent is only applied to 1 batch, while the miscella obtained is used to treat the contents of all the other extraction vessels following in order in a countercurrent system. Continuous extraction (Fig. 23.5) is an ideal extraction system, achieving the highest economy of steam, power, labor and material. It is suitable and economical for a capacity of 50 to 200 MT a day or more. The latest models applied

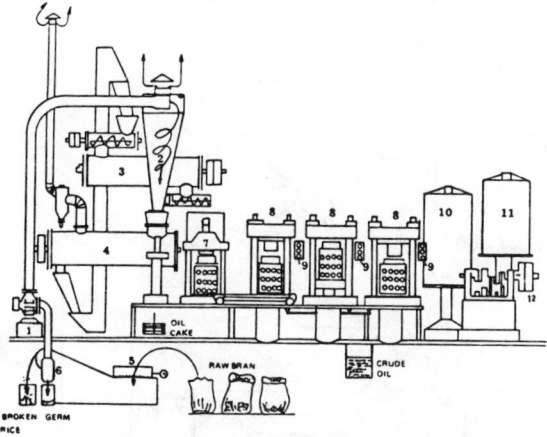

From Yokochi (1974)

FIG. 23.2.   DIAGRAM OF A JAPANESE INSTALLATION FOR RICE BRAN EXTRACTION
BY THE PRESSING METHOD (HYDRAULIC PRESSES OF THE CAGE TYPE)

1—Fan. 2—Cyclone. 3—Steam cooker. 4—Steam dryer. 5—Sifter. 6—Air separator.
7—Pre-press. 8—Hydraulic press. 9—Hydraulic valve (three ways). 10—Accurator
(low pressure). 11—Accurator (high pressure). 12—Hydraulic pump.

commonly in oil plants include the De-Smet type from Belgium, the
rotating O-W type from Japan, the Lurgi type from Germany and the
Rosedown type from England.

A filtration-extraction process for rice bran (Fig. 23.6) was revealed by
Graci *et al.* (1955). Its difference from the conventional solvent-extrac-
tion systems is that the major facility was designed with a series of the
unit operation of filtration through a continuous horizontal rotary vacuum filter to separate the concentrated miscella from the residual meal.
The advantages of this process include rapid filtration rate, low fines
content in the miscella, good oil extractability, high capacity, low solvent
to meal ratio, low solvent content in the mare (solvent-damp extracted
bran), a good quality of oil and meal products and low costs of investment.

X-M Extraction.—The X-M process, starting from either fresh or
stored brown rice, produces bran oil, wax and defatted bran of high
quality, all of which are nutritionally and economically important products. The oil, high in oleic and linoleic acids, can be used for human
consumption. The defatted bran has a well balanced amino acid profile
for animal feeding and potentially for human foods. The milled rice, low
in fat content, is whiter and more attractive in appearance than conventionally milled rice. Lower fat content also contributes to excellent
storage stability and improves beer brewing performance.

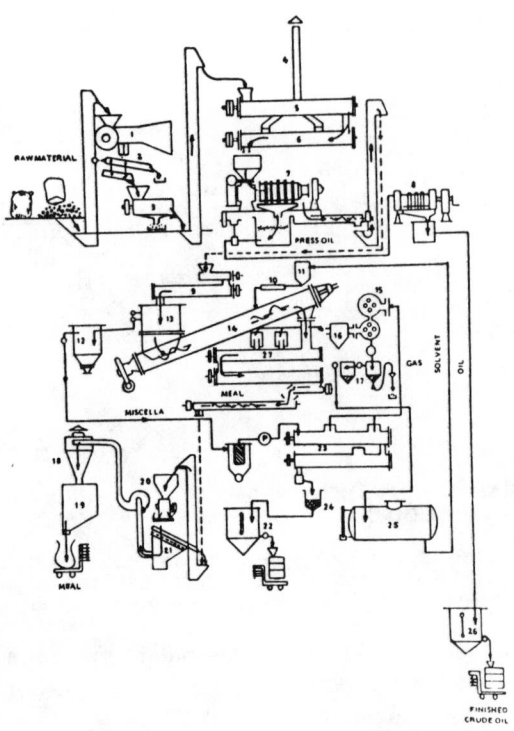

From Yokochi (1974)

FIG. 23.3. MODERN OIL MILLING PLANT WITH EXPELLER AND CONTINUOUS SOL-
VENT EXTRACTOR (FOR NUTS, OILSEEDS AND OTHER SEEDS)

(1) Air separator; (2) sieves; (3) polisher, (4) duct; (5) steam cooker; (6) dryer; (7)
expeller; (8) filter press; (9) preextractor; (10) heater; (11) solvent tank; (12) miscella settling tank; (13) miscella tank; (14) extractor; (15) condensor; (16) dust trap,
(17) water separator; (18) cyclone; (19) mill hopper bin; (20) hammer mill; (21) sieve;
(22) extracted oil settling tank; (23) miscella distillator; (24) filter; (25) solvent tank;
(26) expeller oil settling tank; (27) meal desolventizer.

A full scale commercialized production process (Fig. 23.7) may be divided into 3 main streams:

Rice Process Stream.—Brown rice, free from hull and other impurities, is
fed through an automatic scale *1* to the treater *2*, spray coated with
warm rice oil *3*, and followed to a holding bin *4*, for tempering. The ratio
of oil to brown rice is generally at a level of 0.5% and holding is generally
2 to 4 hr (each bin can provide 6 hr storage). During this treatment, the
bran is softened to facilitate removal during solvent extractive milling.
The treated grain is fed to the first break mills *5*, then to the second
break mills *6* via mass flow conveyors. Here, at controlled concentration

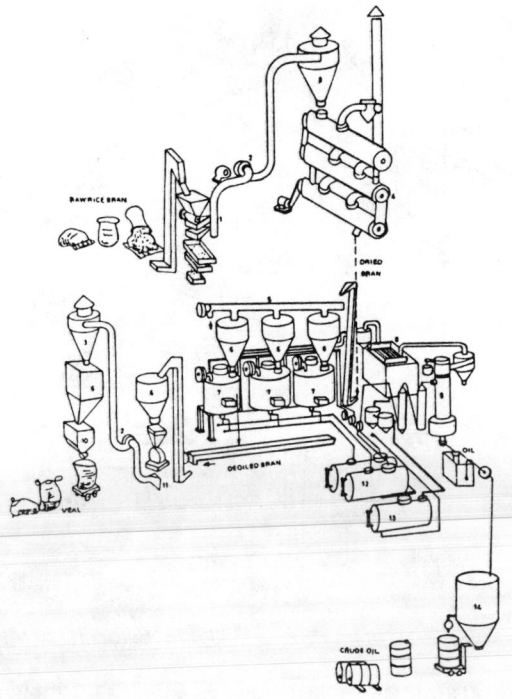

From Yokochi (1974)

FIG. 23.4. RICE BRAN EXTRACTION PLANT, BATCH TYPE

(1) Separation by air and sifting; (2) fan; (3) cyclone; (4) steam cooking & drying; (5) screw conveyor; (6) hopper; (7) extractor; (8) condenser; (9) miscella distillation; (10) measuring & packaging; (11) crusher; (12) miscella tank; (13) solvent tank; (14) settling tank.

and temperature, a recycle stream of clear rice oil-hexane miscella from the settler 7 flows into the mills, where the milling is done at low temperature to reduce whole grain breakage. The milled rice is transferred by a mixing conveyor to a series of vibratory screens 8 which are fixed with 8-mesh at the top and 16-mesh at the bottom to wash out the bran from the rice with fresh hexane. After draining, the product goes to a desolventizer 9 through a 2-stage unit in which the residual hexane of the rice kernel is picked up by superheated vapor and evaporated by the countercurrent flow of a warm mixture of solvent vapors and inert gases. Clean, solvent-free rice is conveyed to size grading, storage and packaging 10.

Bran Process Stream.—A bran slurry containing bran (15–20%), hexane and rice oil in the settler 7 is kept for about 1 hr, then discharged to the resettler 11 for further oil extraction from the bran. Then, the slurry is

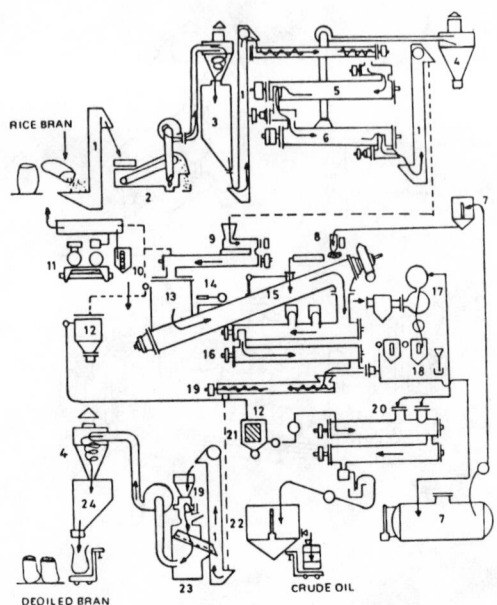

RICE BRAN

DEOILED BRAN

CRUDE OIL

From Yokochi (1974)

FIG. 23.5. FLOW DIAGRAM OF A CONTINUOUS COUNTERCURRENT TYPE SOLVENT
EXTRACTION PLANT (SUZUKI MODEL) FOR RICE BRAN

(1) Bucket elevator; (2) air separator; (3) hopper; (4) cyclone; (5) steam cooker;
(6) dryer; (7) solvent hold tank; (8) solvent flow meter & heater; (9) preextractor;
(10) recovered solvent tank; (11) solvent gas cold trap; (12) miscella settling tank;
(13) miscella tank; (14) miscella pump; (15) extractor; (16) extracted oil tank; (17)
dust trap & condenser; (18) solvent water separator; (19) meal desolventizer; (20)
miscella distiller; (21) miscella filter; (22) extracted oil tank; (23) hammer mill and
sifter; (24) meal hopper.

unloaded to the first of 2 bran mix tanks *12* equipped with an agitator to
keep a uniform consistency and to maintain extraction. Next, the slurry
is pumped into the first of 2 horizontal centrifuges *13* in a series to
separate the bran and miscella. The bran is washed with fresh hexane in
the second bran mix tank *14* for complete extraction and separated again
in the second horizontal centrifuge *15*. The bran dampened with hexane
is conveyed *16* to a desolventizer cyclone *17* where flash evaporation
removes the hexane. The last trace of organic solvent is removed in the
bran holder *18* by means of inert gas circulation. The dried bran is
conveyed to a cooling cyclone *19* and to storage *20*.

Oil Process Stream.—The oil rich miscella *21* recovered from the first
horizontal centrifuge *13* is carried to a surge tank *22*, then passed
through a liquid cyclone to remove larger solid particles prior to moving
to another surge tank *23*. The miscella is filtered *24* and held in the mis-

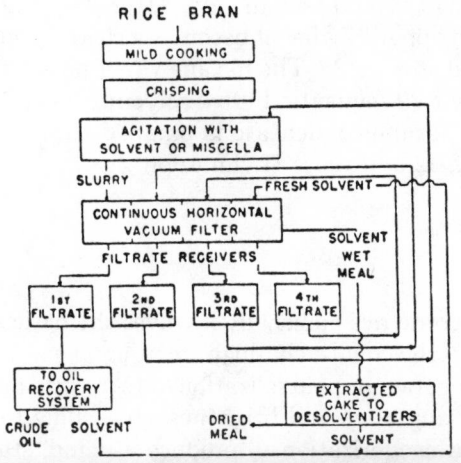

From Graci et al. (1955)

FIG. 23.6. SIMPLIFIED FLOW DIAGRAM OF THE FILTRATION-EXTRACTION PROCESS FOR RICE BRAN

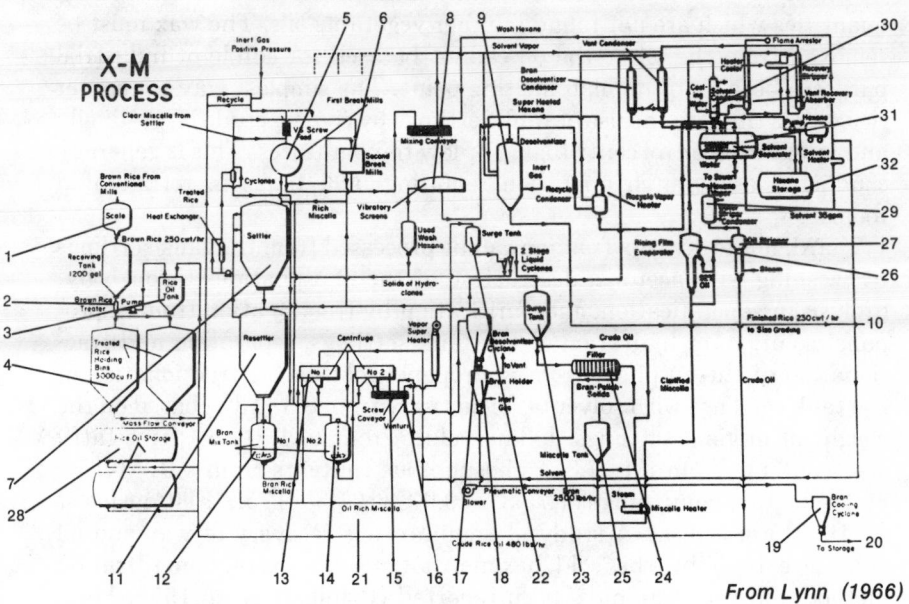

From Lynn (1966)

FIG. 23.7. FLOWSHEET OF X-M PROCESS

Shows rice, bran and oil phases of solvent milling and extraction.

cella tank *25* to obtain an oil concentration of about 11 to 14% by weight. The clarified miscella is transferred to a rising film evaporator *26* in which the oil content is increased to 92%. After stripping out all residual solvent in an oil stripper *27*, the oil becomes a clear medium-brown color and goes to rice oil storage *28*. The hexane vapor passes through a water stripping condensor *29* sequentially to a solvent condenser *30* to recover the solvent. The hexane condensate is pumped through a solvent separator *31* to store *32* and reuse (Lynn 1966).

Refining Process

The refining of crude rice oil usually involves the basic steps of dewaxing for removing the relatively high melting wax for consumption purposes, deacidification or neutralization of fatty acid and gums removal, thereby improving taste and brightness, bleaching to improve color, and finally, steam deodorization to produce a bland, stable edible oil.

Refining can be practiced as a batch process or preferably, as a continuous process.

Dewaxing.—Rice bran oil contains natural rice wax in substantial quantities which are not found in other vegetable oils. The wax must be removed from the oil or the oil cannot be used for edible or industrial purposes because of its high melting point. The simplest way to recover rice wax is through the use of tank settlings by cooling crude oil gradually and by filtering or by centrifuging at low temperatures. This is generally conducted as a batch operation requiring bulk tanks in refrigerated rooms.

A hard, nontacky wax fraction can be processed from the tank settlings by washing with acetone, destruction of the phosphatides through hydrolysis or saponification, and purification by fractionation from isopropanol solution, or hydration of the tank settlings, separation and fractionation of the oil phase from isopropanol solution, or fractionation of the tank settlings with solvents. Yields vary from 8.3–13%, based on the weight of original settlings. Iodine values from 11.1–17.6%, free fatty acid contents from 2.1–7.3%, phosphorous contents from 0.01–0.15% and a melting point range of 75.3°–79.9°C (167.5°–175.8°F) (Cousins *et al.* 1953) are typical. A study of simultaneous recovery of wax and oil from rice bran by the cold hexane; hot hexane extraction filtration method and the costs have been reported (Pominski *et al.* 1955; Pominski *et al.* 1955). Yields of rice wax vary from 1.29–1.82% of the extracted oil. Wax and other insoluble matter can be removed from rice oil by chilling the oil or miscella with a sodium silicate solution, and separating the wax flocculation from the oil by centrifugation and filtration (Kinsey and Hunnell 1969). A modern method for removing wax and pigments from rice oil by freezing-out refined oil, followed by filtration, has been studied by Fan *et al.* (1972). Koslowsky (1972) has disclosed a fractional crystallization method in isopropyl alcohol without using filters or centrifuges to obtain liquid fractions, with a chill stability of 25°C (77.0°F) for the first stage and solid fractions of 15°C (59°F) for the second stage.

Hard wax in the liquid state is almost black and difficult to bleach with activated clay or carbon. It can be treated with 0.5 part of 29% hydrogen peroxide in combination with 1 part of chromium trioxide per 1 part of wax, to produce practical white waxes (Cousin *et al.* 1953).

Deacidification.—

Conventional Alkali Refining Process.—In batch refining processes, a sufficient solution of sodium hydroxide (15–20%) with a 0.5–2% excess, depending on the calculation of the free fatty acids content, is added to a neutralizing kettle. The mixture is vigorously agitated for a short period to neutralize the free fatty acids, which transforms them into soap. Then they settle down for a while as "soapstock" or "foots." After separating the soapstock, which also absorbs impurities such as gums and part of the coloring constituents, the refined oil yields a lighter and more brilliant oil.

Continuous refining processes are only worthwhile for larger refineries. They consist of continuous mixers, heaters and centrifugal separators, which reduce the time of contact between oil and alkali to a minimum. Definite amounts of dilute, aqueous, caustic solution and oil are continuously combined in a small mixing unit in which the components are mixed for less than one minute at room temperature. Leaving the mixer, the charge is passed into a heating unit. Here, the temperature is rapidly raised to 55°–70°C (131°–158°F) to break the emulsion which invariably forms in the mixer. The charge is then passed through a battery of centrifuges which serve to separate the soapstock from the neutral oil. The time required for the total operation is approximately 2 min. The refined oil is washed continuously and the soapstock is separated in another series of centrifugal separators. This operation obtains considerably lower losses of neutral oil through saponification or occlusion in the soapstock.

In conventional alkali refining of rice oil by the procedure usually employed for cottonseed and similar oils, excessive losses occur. Losses in refining rice oil, which contain about 5% free fatty acids, may amount to at least 12% to more than 40% by the cup method. Losses approximating the absolute or Wesson Loss result from caustic soda refining in a hydrocarbon solvent as well as from refining with sodium carbonate (Reddi *et al.* 1948). Steam refining in conjunction with alkali refining have proven effective as a means of reducing the losses in refining (Swift *et al.* 1950).

It is indicated that refining losses of rice oil can be reduced 32—55% by adding certain organic compounds containing NH_2 and OH groups to the crude oil before the regular refining procedure. Blackstrap molasses and sucrose are the most effective and cheapest materials tried. In addition, ethanolamine and various glycols and alcohols may also be employed (Cousin *et al.* 1953). Two steps of neutralization by sodium carbonate in oil containing 2—25% free fatty acid have been found to cause a loss of only 2% oil per degree of acidity. The color of neutralized oils at less than 4 red Lovibond units can be achieved (Castro Ramos 1969). An optimum neutralization utilizes a ratio of at least 2 g/l of free lye in the soap-lye solution for oils with acidity number up to 30 mg KOH and with oil of more than 30 mg KOH the acidity number should be treated in at least 3 g/l solution. Neutralization in a soap-lye medium of rice oil with the acidity number up to 30 mg KOH helps to reduce waste by as much as 0.12% (Arutyunyan *et al.* 1971). Hartman and Dos Reis (1976) assume that the inordinately high refining losses of rice bran oil are due, along with the acidity, to the presence of hydroxylated compounds. It may be possible to diminish the refining losses by reducing the hydroxy number through treating with acetic anhydride at low acidity.

Miscella Alkali Refining Processes.—The basic idea in principle is to treat the oil as a miscella in solution with solvents and caustic soda to reduce the entrainment of neutral oil by the soapstock which is formed by the chemical action of caustic soda on the free fatty acids. The advantages of miscella refining compared to conventional caustic soda refining are lower refining losses, lighter colored oil without bleaching and eliminations of water washing the refined oil or miscella. However, the complete recovery of the solvent from the soapstock is still unsolved.

Miscella refining can be practiced batchwise or continuously, thus involving 2 main systems of refining and solvent recovery. In the batch miscella process, a refining vessel (Fig. 23.8) is available for small solvent extraction plants or where feed stocks change frequently. The crude oil, different solvents and caustic soda are fed from the respective tanks into a vessel fitted with a mixer. The concentration of the miscella is usually within the range of 40—60% by wt. of oil. Caustic soda is added in an amount equal to *c* 0.3% excess. When the soapstock begins to break free from the miscella, increased agitation and temperature to *c* 65°C (149°F) melt the soapstock. The soapstock is then separated by a centrifuge after the miscella is cooled rapidly to *c* 45°C (113°F). The refined miscella is filtered and hexane is stripped to produce a neutral yellow oil. Alternating this procedure between 2 tanks makes batch-continuous miscella refining possible (Cavanagh 1976).

Continuous miscella refining (Fig. 23.9) is usually the more efficient and preferred procedure. Filtered miscella from plant production is flowed

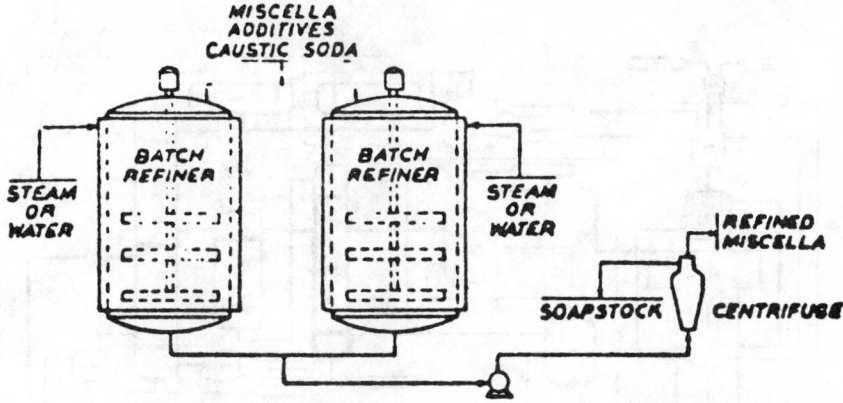

From Cavanagh (1976)

FIG. 23.8. BATCH MISCELLA REFINING

into 1 of 2 surge tanks in which precise concentration control and continuous operation are possible. Miscella is continuously agitated with a
high volume, low head pump which circulates crude miscella from the
cone bottom to near the top of the tank until it is filled to approximately
80% capacity. A moderate amount of caustic soda solution.is metered into the suction side of the refinery pump. This mixture is mixed immmed-
iately in a homogenizer, heated to *c* 65°C (149°F) to melt the soapstock,
cooled to *c* 45°C (113°F) and centrifugally separated (Cavanagh 1976).

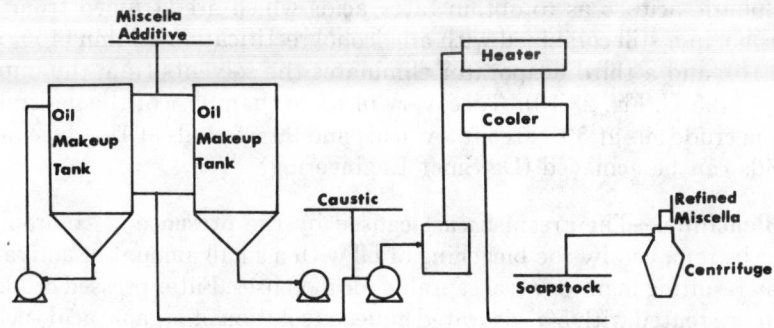

From Cavanagh (1976)

FIG. 23.9. CONTINUOUS MISCELLA REFINING

The De Smet "Neumi" miscella refining process (Fig. 23.10) provides a
unique purity and stability to oil refined from high acidity rice bran oil,
as well as complete automation to ensure uniform results and quality. In

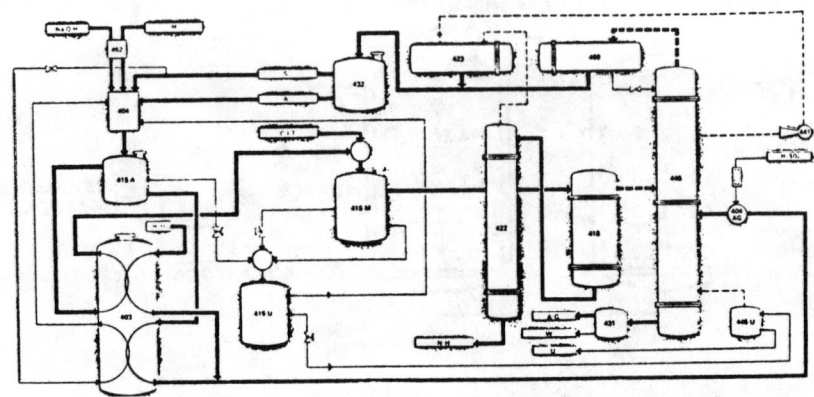

From De Smet Engineering

FIG. 23.10. DE SMET "NEUMI" MISCELLA REFINING PROCESS

addition, the fatty acids can be delivered with a purity of 95 to 98%, which can be used without further treatment for soapmaking. The mixture of neutral oil and soap in solution with hexane and alcohol overflows into a continuous decanter (415 A, Fig. 23.10) where it forms 3 well separated layers: refined oil in solution with hexane constitutes the upper layer, soap in solution with alcohol the lower layer and gums and other impurities the intermediate layer. A classical still (422, Fig. 23.10) strips the hexane from the refined oil, the lower layer being treated with sulphuric acid so as to obtain fatty acids which are stripped from the alcohol in a still combined with an alcohol rectification column (455, Fig. 23.10), and a third evaporator eliminates the solvents from the impurities (455 U, Fig. 23.10). A recovery of more than 95% of the neutral oil from crude oils at 50% free fatty acids and 99% for oils at 10% free fatty acids can be achieved (De Smet Engineering).

Bleaching.—The greenish cast caused by the presence of chlorophyll can be removed by the bleaching of oil with a small amount of activated clay resulting in products acceptable for food uses. Filter pressed crude oil can be treated with 5% saturated aqueous solution of organic acid such as oxalic, citric, etc., at room temperature for 30 min, then heated and allowed to settle down. The oily layer is washed with hot water until free of acid, dried in vacuum, and bleached with a mixture of activated earth (4%) and activated carbon (0.4%) at 110°C (230°F) for 20 min. After cooling and filtering, a light color with yellow 23 to 35, red 0 can be obtained. Alternatively, oil can be treated with mineral acid under anhydrous conditions and optionally bleached with a mixture of $C1O_2/C1_2$

inert gas. A clarified oil of 92–93% yield can be satisfactorily produced by adsorbent decolorizing with kaolin in an oil adsorbent ratio of 1:5, miscella concentration 20%, oil flow through the Kaolin of 4 m³/hr and adsorbent particle size of 0.25–0.5 mm (Nechaev *et al.* 1973).

Deodorizing.—Steam-vacuum deodorization is much more efficient and has found general application. The process involves blowing live steam through the heated oil at 220°–250°C (428°–482°F) under vacuum (3–5 mm Hg). This removes the objectionable odors such as peroxides, aldehydes and ketones, as well as the pleasant odors. It provides the additional advantage of destroying carotenoid and some other pigments, and reducing the free fatty acids. After deodorization, the oil is filtered and cooled down under vacuum to about 50°C (122°F). The deodorized oil is extremely tasteless and odorless.

The most common shape of the batch deodorizer is a vertical cylinder vessel 2 to 4 times as high as it is wide with dish or cone heads. The vacuum equipment consists of multistage steam ejectors with barometric condensers. Oil is heated internally or externally by steam, or direct firing. Stripping steam is arranged at the bottom of the vessel, consisting of perforated pipes to assure good steam distribution and provide effective agitation. The deodorization cycle usually lasts from 3 to 5 hr.

A semicontinuous deodorizer designed by the Girdler Corporation consists essentially of a large cylindrical vessel containing 5 trays stacked one above the other. At regular intervals a valve at the bottom of the trays is opened and closed automatically, allowing the oil to flow from one tray to the next tray below. The oil is preheated and deaerated in the top tray. In the second tray the temperature of the charge is increased to about 240°C (464°F). Deodorization is mainly carried out in the third and fourth trays. In the last tray the temperature is reduced by water cooling while the product is under vacuum. The total operating time is about 2½ hr.

The continuous deodorizer, in principle, consists essentially of a stainless steel column arranged with a number of shallow trays and bubble caps. The oil is usually first deaerated under vacuum, then preheated by a heat exchanger and heated to the required temperature by a heating system during the flow from top to bottom in countercurrent with the stripping steam. A new type of deodorization equipment developed in Japan is shown in Fig. 23.11. In the first stage, deodorizing is carried out rapidly using a vertical flushing evaporation equipment (5, Fig. 23.11); in the next stage, deodorizing is completed by a horizontal rotary type special finishing deodorizer located in the lower portion of the system (5, Fig. 23.11). This system permits efficient deodorizing in a short time and is very effective in minimizing the acid value. A more recent development consists of physical refining, which combines in one operation the deacid-

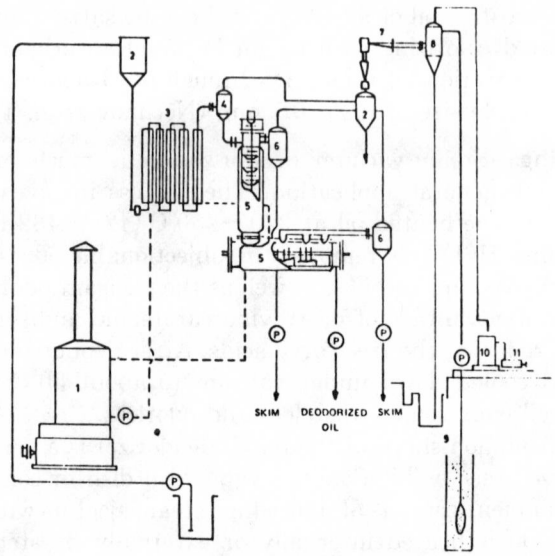

From Yokochi (1974)

FIG. 23.11. CONTINUOUS DEODORIZING EQUIPMENT FOR RICE BRAN OIL (Y.M. TYPE)

(1) Oil boiler; (2) oil reservoir; (3) heat exchanger; (4) degasing tank; (5) deodorizer; (6) oil trap; (7) steam ejector; (8) barometric condenser; (9) water seal; (10) vacuum reservoir; (10) vacuum pump.

ification by distillation and the deodorizing processes. This system is available for high acidity rice bran oil refining to obtain a far better yield of neutralized oil than the alkali refining method. A diagram of a plant for deacidification by distillation is given in Fig. 23.12.

Winterizing or Destearinization.—Winterizing is generally applied to remove saturated glycerides, commonly known as "stearine" which become soluble in the oil at medium and higher temperatures but precipitate in cold conditions resulting in a turbid oil.

The process consists of cooling the oil for a determined period and filtering by filterpress after approaching crystallization equilibrium. The cooling system uses cold water, brine or propylene glycol as the cooling agent. Oil is held at 30°–35°C (86°–95°F) and its temperature is lowered slowly at a uniform rate to 15°C (59°F) within about 12 hr with agitation, then to 4°–5°C (39.2°–41°F) without agitation, then maintained for 24 to 48 hr to enable the higher melting components to form crystals. It is essential to achieve a 1°C (1.8°F) differential between the temperature of the oil and that of the cooling agent. Otherwise small crystals of pre-

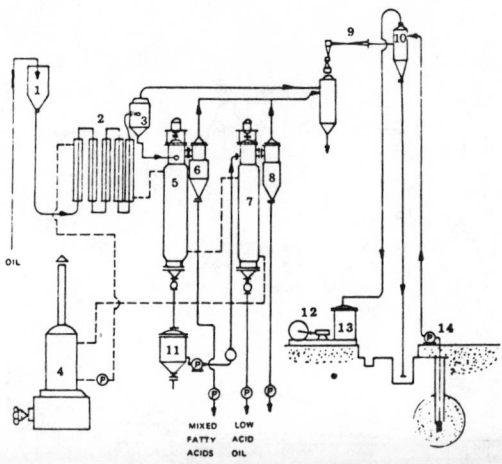

From Yokochi (1974)

FIG. 23.12. DEACIDIFICATION PLANT USING THE DISTILLATION SYSTEM FOR RICE
BRAN OIL (Y.M. TYPE)

(1) Oil reservoir; (2) heat exchanger (preheater); (3) degasing tank; (4) heating oil
boiler; (5) distillator No. 1 for deacidification; (6) condenser for fatty acids; (7) distillator No. 2 for finishing; (8) condenser; (9) steam ejector; (10) barometer condenser;
(11) deacidified oil reservoir; (12) vacuum pump; (13) vacuum reservoir; (14) cooling
water supply pump.

cipitate will be gelatinous and will retard filtration. The filtration is often
carried out in a cooling room or else the filterpress is provided with
internal channels for cooling in order to maintain a low temperature
during filtration. The addition of small amounts of filter aid, such as
diatomaceous earth, before cooling, may facilitate crystallization as well
as filtration.

Miscella winterizing more effectively separates high and low melting
glycerides at various temperatures than the older nonsolvent process.
Oil is mixed with solvents such as hexane, acetone, acetone/hexane,
methyl isobutyl ketone, isopropyl acetate, etc. The miscella is slowly
cooled to 15°C (59°F) with agitation during 12 hr, then cooled to 4°–5°C
(39.2°–41°F) without agitation and kept at this temperature for 24 to 48
hr. After winterization is completed, the low melting fraction is removed
from the miscella by use of a leaf filter. Subsequent heat and steam
treatment serve to recover the purified oil and solvent respectively.

A flowsheet describing a rice bran oil refining plant (alkali process) is
shown in Fig. 23.13. The yield of products obtainable, with their respective free fatty acid content, is listed in Table 23.1.

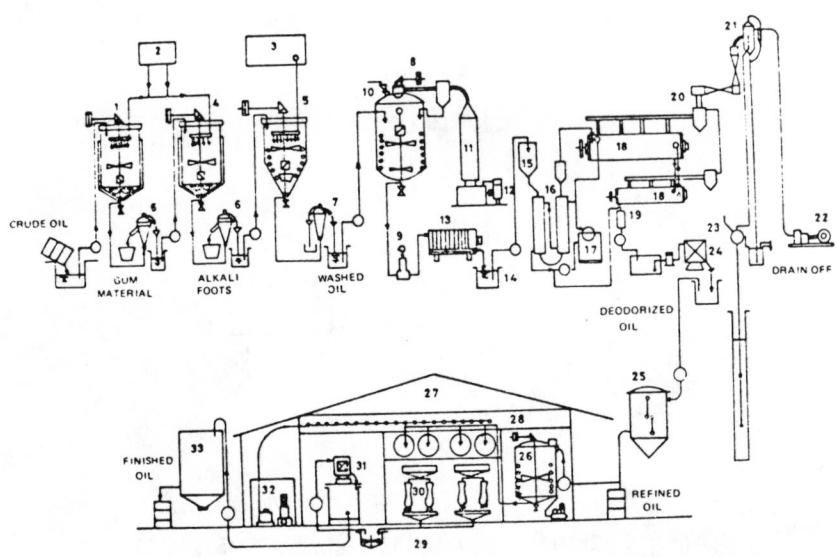

FIG. 23.13. DIAGRAM OF RICE BRAN OIL REFINING PLANT (ALKALI PROCESS)

(1) Degumming tank; (2) NaOH solution tank; (3) hot water tank; (4) neutralizer; (5) hot water washing vessel; (6) misco type sharples; (7) open type sharples; (8) vacuum bleacher; (9) plunger pump; (10) activated bleaching earth; (11) cooler; (12) vacuum pump; (13) filter; (14) bleached oil tank; (15) charge tank; (16) heat exchanger; (17) kanecrol boiler; (18) continuous deodorizer; (19) cooler; (20) steam ejector; (21) barometric condenser; (22) vacuum pump; (23) cooling water pump; (24) filter press; (25) deodorized oil storage; (26) chilling tank; (27) wintering apparatus; (28) cooling pipe; (29) cool room; (30) bag filter; (31) filter press for finishing; (32) refrigerator; (33) wintered oil tank.

COMPOSITION AND PROPERTIES

Chemical Composition

The fatty acids are the most important constituents of oils and fats. In rice oil, there are 3 major fatty acid components, palmitic, oleic, and linoleic acids and 5 minor constituents, myristic, palmitoleic, stearic, linolenic and arachidic acids. The data (Table 23.2) show that oil obtained by conventional methods has similar constituents to oil derived via the X-M process.

Izzo et al. (1972) has revealed that the variety of rice influences the amount of oil obtained but that the fatty acids of the oils had the same general composition. Resurreccion and Juliano (1975) indicate that oil

TABLE 23.1. YIELDS OF MATERIALS IN THE RICE BRAN OIL BY ALKALI REFINING PROCESS WITH VARIOUS FREE FATTY ACID CONTENT

Process	Products	Yields (%)	Acid value 10 (1000 kg)	20 (1000 kg)	30 (1000 kg)	40 (1000 kg)
	Crude oil					
Dewaxing	Dewaxed oil	86.5	865	865	865	865
	Crude wax	10.0	100	100	100	100
Alkali refining	Deacided oil	Y:97AV	752	665	580	493
	Soap stock	100-Y	113	200	285	372
Bleaching	Bleached oil	96.5	727	642	560	475
Deodorizing	Deodorized oil	97.3	707	625	545	462
	Scum oil	1.5	11	9.6	8.4	7.1
Winterizing	Salad oil	85.0	600	532	463	394
	Solid fat	15.0	107	93	82	68

Source: Tsuji (1971).

TABLE 23.2. FATTY ACID COMPOSITION OF RICE-BRAN OIL AND X-M RICE OIL

Fatty acid ester	Composition (%)	Rice bran oil[1] (%)	X-M rice oil[2] (%)
Myristic	(C14:0)	0.1– 1	0.49
Palmitic	(C16:0)	12 –18	13.80
Palmitoleic	(C16:1)	0.2– 0.6	–
Stearic	(C18:0)	1 – 3	2.01
Oleic	(C18:1)	40 –50	43.60
Linoleic	(C18:2)	20 –42	36.60
Linolenic	(C18:3)	0 – 1	1.77
Arachidic	(C20:0)	0 – 1	0.91

[1]Source: Lu and Williams (1965).
[2]Source: Hunnel and Nowlin (1972).

from bran-polish and oil from milled rice, extracted with petroleum ether, have similar fatty acid compositions. However, milled rice oil extracted with chloroform/methanol has a higher linoleic acid content and a lower oleic acid content than oil extracted with petroleum ether. The presence of phospholipids including phosphatidylethanolamine, phosphatidylcholine and phosphatidylinosital has been demonstrated in rice bran oil. The fatty acids of phospholipids contain linoleic 45.18%, oleic 34.02%, palmitic 16.62%, stearic 1.82%, palmitoleic 1.54% and myristic 0.82% acids (Babkhodzheava 1974).

There are also some unsaponifiable components other than fatty acids in the rice oil. The amount of this unsaponifiable matter is approximately 3 to 8% in raw oil. One of these components is δ-tocopherol (vitamin E) which comprises from 2 to 5% of crude oil. However, it is largely de-

TABLE 23.3. PHYSICAL AND CHEMICAL CHARACTERISTICS OF CRUDE RICE BRAN OIL AND X-M RICE OIL

	Rice bran oil[1]	X-M rice oil[2]
Classification	Semidry	Semidry
Rice wax (%)	1 – 4	2.5 – 3.5
Free fatty acid (%)	5 –120	2.5 – 5.0
Specific gravity (25°C)	0.916– 0.921	0.917– 0.920
Refractive index (40°C)	1.465– 1.467	—
Iodine value	92 –115	95 –102
Saponification value	175 –192	180 –190
Unsaponifiable (%)	3.0 – 8.0	2.5 – 4.0
Flash point (closed cup)	—	+300F
Moisture and volatile matter (%)	1.5	0.5 – 4.0
Insoluble impurities (%)	—	0.5 – 1.5
Hehner number	92.1 – 96.5	—
Reichert—Meissel value	0.59 – 1.75	—
Energy value (kcal/kg)		9438

[1]Source: Yokochi (1974).
[2]Source: Lynn and Anderson (1967).

stroyed in the finished oil after deodorization. Tocopherol is a natural antioxidant preventing fat deterioration. Another of the unsaponifiable components is oryzanol, which is obtained from alkali washed raw oil (Azuma 1973; Shimizu 1976) or from methyl esters (Park *et al.* 1971). Crude oil contains 2—3% oryzanol. It is claimed that this substance has an effect similar to that of vitamin E in expediting human growth, facilitating blood circulation and stimulating hormonal secretion. A medicine known as OZ has been developed from oryzanol in Japan (Yokochi 1974).

Sterols in rice oil have been found in amounts up to 5%, in which β-sitosterol is the most abundant sterol (Oka *et al.* 1972). Sterols are recoverable by means of esterification and saponification of the rice bran oil. Campesterol, stigmasterol and β-sitosterol in the sterol fractions and cycloartanol and 2,4-methylene-cycloartanol in the triterpene fractions have been the results of the action of lipases which give undesirable high free fatty acid levels in the edible oil produced (Viraktamath 1973; Sulimenko and Mitsenko 1974).

With regard to the storage of rice oil, it is important to know that unrefined oil will keep much longer than refined oil due to the former containing higher levels of tocopherols than the latter; these protect the oil against atmospheric oxidation and rancidity. Oil stored in tin cans shows a relatively higher rate of increase in free fatty acid concentration than does oil stored in PVC or polyethylene bags (Sidhom *et al.* 1975). Effects of cobalt-60 gamma radiation on rice bran oil show that BHT is more effective than caffeic acid in retarding peroxide formation during irradiation and storage, and is more effective at 2 Mrad than 7 Mrad. For storage of rice bran oil, the addition of antioxidants is preferably done after irradiation (Han *et al.* 1973). Properties of the processed oil are shown in Table 23.4.

UTILIZATION OF RICE OIL

The uses of rice bran oil in rice growing countries may be divided into 2 main categories, namely for food purposes or as a technical raw material. In general, the crude oils with high acid values are mostly used as raw material for industrial manufacturing products, while those of low acid values are refined for human consumption.

In Preparation of Foods

The main food use for rice oil with low free fatty acids, both extracted from the freshly milled rice bran and from the whole rice kernel during

TABLE 23.4. CHARACTERISTICS OF FINISHED OIL

	Dewax rice bran oil[1]	Refined oil[1]	Salad oil[1]	X-M cooking oil[2]
Color	35/5.0	30/3.5	30/3.5	1.5
Moisture and volatile (%)	0.2	0.1	0.1	—
Specific gravity	100/25°C	25/25°C	25/25°C	—
	0.841–0.869	0.913–0.919	0.913–0.919	—
Refractive index (25°C)	1.470–1.473	1.470–1.473	1.470–1.473	0.03
Free fatty acid (%)	0.25	0.10	0.07	190
Saponification value	180–195	180–195	180–195	100–105
Iodin value	92–112	92–112	92–112	2.0
Unsaponifiable (%)	5.0	4.5	3.0	36
Cold test hr/0°C	—	—	—	20
A.O.M. stability (hr)	—	—	—	17
Cloud point (°C)	—	—	—	

[1]Source: Tsuji (1971).
[2]Source: Hunnell and Nowlin (1972).

the milling process, is as cooking and frying oil after refining, bleaching and deodorization. It can be winterized easily and adapted for salads and mayonnaise. The refined, nonwinterized oil is more suitable than the winterized oil for the manufacture of margarine due to a higher content of saturated fatty acids. The hydrogenated oil can be used as specialty margarine and shortening-base materials or other emulsified products. A special use has also been as antioxidants and color stabilizer in foods by mixtures of tocopherol and unsaponified rice bran oil (Ohtaki 1972; Maruyama and Wakayama 1973). A dried granular form of baking yeast has been made by using compressed yeast or yeast concentrate liquors mixed with rice bran oil, monoglycerides, sorbitan fatty acid ester, soybean oil and alum. Some of the oil can be used as a pan-release agent in the baking industry or brominated as an essential oil stabilizer and soft drink clarifier, or sulfonated as emulsifiers, wetting agents and dispersing agents. A rice oil of 0.1−0.5% in pretreated rice is useful for sake brewing by fermentation (Sakamoto *et al.* 1973).

As Technical Raw Material

Crude rice bran oils of high acid value are usually used for producing solidified oil, stearic and oleic acids, glycerine and soap, especially powdered soap. Sometimes, they are combined with glycerine and converted into neutral oil, which is used as material for cooking oil, or it is combined with a drying agent for antirust and anticorrosive purposes. Some oil is sulfonated for lubricants used in textile or leather finishing. A good commercial grade of azelaic acid can be obtained as a major product in yields of 10−25% by ozonolysis of rice bran oil (Arida *et al.* 1970). A simple and effective method to recover ferulates from rice bran oil was made by Kato and Tanaka (1975). Ferulates are beneficial for reducing blood pressure and as blood vessel dilators. A product of gummy wastes from rice bran oil processing mixed with defatted rice bran or wheat bran or dried without additives is an excellent animal feed. It contains crude proteins 17.7% and crude lipids 1.9% and has a digestibility index of 49.2 (Tsuno *et al.* 1975). In regard to the fertility and growth of rats, rice oil is superior to that of sunflower, olive or peanut oils when given intragastrically at 1 g/rat/day during pregnancy, presumably due to high oleic and linoleic acids and vitamins, especially α-tocopherol (Luchetti 1975). Addition of 0.1−5.0% purified rice oil to basic foods containing mulberry leaf powder and other nutrients, increased the growth of silkworms and cocoon weight.

Rice Wax

Rice wax is useful for industrial, cosmetic and edible purposes, if extracted and refined to a pure bleached condition. The U.S. Food and Drug Administration has approved this material as an ingredient or coating agents for candy, chewing gum and fruits.

For further information on rice bran processing for production of rice bran oil, the reader is referred to the review by Yokochi (1974).

REFERENCES

ARIDA, V.P., BORLAGA, F.C. and SCHMITT, W.J. 1970. The ozonolysis of rice bran oil. Philip. J. Sci. 96 (3) 249–252.

ARUTYUNYAN, N.S. et al. 1971. Neutralization of high acidity number rice oil in soap-lye medium. Maslozhirovaya Promyshlennost' 37 (7) 15–17. (USSR)

AZUMA, M. 1973. Oryzanol from alkali washings of rice bran oil. Japanese Patent 7,304,617. January 20.

BABAKHODZHAEVA, S.A. 1974. Phospholipids and fatty acids of the phospholipids of rice bran oil. Russ. 119, 106–107. (USSR)

BARBER, S. et al. 1974. Process for the stabilization of rice bran. I. Basic research studies. Proc. Int. Conf. on Rice By-products Util. Valencia, Spain.

BOSE, A.N. and SRIMANI, B.N. 1974. Design of a moving bed continuous rice bran drier. Proc. Int. Conf. on Rice By-products Util. Valencia, Spain.

CASTRO RAMOS, R. DE 1969. Refining and properties of rice bran oil. J. Agric. Chem. and Food Technol. 9, 106–115. (Italian)

CAVANAGH, G.C. 1976. Miscella refining. J. Am. Oil Chem. Soc. 53 (6) 361–363.

COUSINS, E.R., FORE, S.P., JANSSEN, H.J. and FEUGE, R.O. 1953. Rice bran oil. VIII.: Tank settling from crude rice bran oil as a source of wax. J. Amer. Oil Chem. Soc. 30, 9–14.

ENDO, T. and INABA, Y. 1970. Ferulates in rice bran oil. VII.: Chemical structure of three new triterpenoids. J. Jpn. Oil Chem. Soc. (Yukagaku) 15 (5) 302–307.

FAN, TKHI-AN et al. 1972. A modern method for removing waxes and pigments from rice oil. Maslozhirovaga Promyshlennost' 38 (3) 19–21. (USSR)

FITO, P.J. et al. 1974. Process for the stabilization of rice bran. II. Development of pilot plant and industrial equipment. Proc. Int. Conf. on Rice By-products Util. Valencia, Spain.

GASTROCK, E.A. et al. 1955. Rice bran oil extraction process. U.S. Patent

2,727,914. December 20.

GRACI, A.V., Jr. *et al.* 1955. Pilot plant application of filtration-extraction to rice bran. J. Amer. Oil Chem. Soc. *30*, 139—143.

HAN, D.B., SUCK, H.G. and YOO, Y.J. 1973. Effect of cobalt-60 gamma radiation on rice bran oil. Korean J. of Food Sci. and Tech. *5* (2) 129—135.

HARTMAN, L. and DOS REIS, M.I.J. 1976. A study of rice bran oil refining. J. Am. Oil Chem. Soc. *53* (4) 149—151.

HUNNELL, J.W. and NOWLIN, J.F. 1972. Solvent extraction rice milling. *In* Rice: Chemistry and Technology. Amer. Assoc. of Cereal Chem., St. Paul, Minnesota.

IZZO, R., FABBRIN, F. and LOTTI, G. 1972. Studies of oils extracted from various parts of the rice grain. Riso *21* (2) 161—171. (Italian)

JEONG, T.M., TAMURA,T., ITOH, T. and MATSUMOTO, T. 1973. Cyclo-eucalenol in rice bran oil. Journal of Japan Oil Chem. Soc. (Yukagaku) *19* (5) 302—307.

KATO, A. and TANAKA, A. 1975. Recovering ferulates. Japanese Patent 7,530,686. October 3.

KINSEY, D.N. and HUNNELL, J.W. 1969. Dewaxing rice oil. U.S. Patent 3,481,960. December 2.

KOPEIKOVSKII, V.M. and ARUTYUNYAN, N.S. 1971. Effect of hydro-thermal treatment on enzymatic activity of rice bran. Izvestiya Vysshikh Uchebnykh Zavedenii, Pishchevaya Tekhnologiya *4*, 50—52. (Russian)

KOSLOWSKY, L. 1972. Rice bran oil dewaxing. Oleagineaux *27* (11) 557—560. (Israel)

LOEB, J.R., MORRIS, N.J. and DOLLEAR, F.G. 1949. Rice bran oil extraction process. J. Am. Oil Chem. Soc. *26*, 738—743.

LU, T.C. and WILLIAMS, V.R. 1965. The effect of length of storage on the fatty acid composition of rice lipids. Proc. 10th Rice Tech. Working Group. June 17—19. University of California, Davis.

LUCHETTI, G. 1975. Influence of dietary rice oil on fertility and growth of rats. Ernaehr.-Umsch. *22* (11) 332—335. (German)

LYNN, L. 1966. Revolutionizes rice milling. Food Eng. *38* (11) 68—73.

LYNN, L. and ANDERSON, R.M. 1967. X-M milling expands use of rice. Food Eng. *39*, 77—79.

MARUYAMA, S. and WAKAYAMA, T. 1970. Agents for stabilizing lipids against oxidation. Japan Patent 78,633. U.S. Patent 7,317,824. August 14.

MARUYAMA, S. and WAKAYAMA, T. 1973. Prevention of pigment discoloration. Japanese Patent 7,317,824. March 6.

NECHAEV, A.P. *et al.* 1973. The problem of refining rice oil with high acidity number. Maslozhirovaya Promyshlennost' *39* (12) 14—16. (USSR)

NOGUCHI, T. and INOUE, H. 1972. Components of free and ester type sterols and triterpenes in rice bran oil. Hokkaidoritsu Nogyo shikenjo Hokoku *189*, 1–5.

OHTAKI, G. 1972. Antirancidity agents for oils and fats. Japanese Patent 7,222,406. October 7.

OKA, Y., KIRIYAMA, S. and YOSHIDA, A. 1972. Sterol composition of Japanese foodstuffs. I.: Sterol composition of edible vegetable oils and investigation of conditions of sterol analysis. Jpn. Soc. Food and Nutr. J. *25* (2) 63– 67.

PARK, S.H., LEE, C.K. and NOH, C.S. 1971. Utilization of available ingredients in the rice bran oil. Composition of oryzanol from rice bran oil. Kungnip Kongop Yonguso Pago *21*, 185–189. (Korean)

POMINSKI, J., EAVES, P.H., VIX, H.L.E. and GASTROCK, E.A. 1955. Simultaneous recovery of wax and oil from rice bran by filtration extraction. J. Amer. Oil Chem. Soc. *31*, 451–455.

POMINSKI, J. *et al.* 1955. Preliminary cost study of rice wax filtration extraction. Ind. Eng. Chem. *47*, 2109–2111.

REDDI, P.B.V., MURTI, K.S. and FEUGE, R.O. 1948. Rice bran oil. I.: Oil obtained by solvent extraction. J. Amer. Oil Chem. Soc. *25*, 206–211.

RESURRECCION, A.P. and JULIANO, B.O. 1975. Fatty acid composition of rice oils. Sci. of Food and Agric. J. *26* (4) 437–439.

SAKAMOTO, M., KIMOTO, S. and MIYAGAWA, T. 1973. Sake from imported rice. Japan Patent 7,327,478. August 22.

SHIMIZU, H. 1976. Oryzanol-related steroids from rice bran oil. Japanese Patent 76,129,000. November 10.

SIDHOM, E.I., EL-TABEY, S.A.M. and MOHASSEB, Z.S. 1975. Effect of storage on rice bran oil and some of its constants. Alexandria J. Agric. Res. *23* (1) 109–111.

SRIMANI, B.N., CHATTOPADHYAY, P. and BOSE, A.N. 1974. Stabilization of rice bran. Proc. of Int. Conf. on Rice By-products Util., Valencia, Spain.

SULIMENKO, O.F. and MITSENKO, E.A. 1974. Industrial treatment of rice siftings. Maslo-Zhir Prom *1*, 37–38. (Russian)

SWIFT, C.E., FORE, S.P. and DOLLEAR, F.G. 1950. Rice bran oil. V.: The stability and processing characteristics of some rice bran oils. J. Amer. Oil Chem. Soc. *27*, 14–16.

TSUJI, S. 1971. Fats and oils processing products. *In* Foods Processing Technology Handbook. Kenpakusha Publishing Co., Japan.

TSUNO, M., TAKAHASHI, T. and HYAKUTA, Z. 1975. Rice bran oil waste treatment. Japanese Patent 7,574,602. June 19.

VIRAKTAMATH, C.S. 1973. Practical approach to the production of edible rice bran oil in India. Oils and Oilseeds J. *26* (4) 5–7.

YOKOCHI, K. 1974. Rice bran oil utilization. Vol. III. *In* Proc. of Int. Conf. on Rice By-products Util. Valencia, Spain.

YOKOCHI, K. 1975. Utilization of rice bran oil in South-East Asia. *In* The Assoc. of Jpn. Agric. Sci. Soc., Univ. of Tokyo.

24

Rice Bran: Chemistry and Technology

Salvador Barber
Carmen Benedito de Barber

Rice bran is a source of proteins, oil, nutrients and calories. Although most of the world's rice bran is produced in regions where optimum utilization of the food and feed potential resources is a must, paradoxically bran is underutilized and frequently wasted. This chapter summarizes the present knowledge on rice bran and the existing problems to be solved.

RICE MILLING TECHNOLOGIES AND RICE BRAN PRODUCTION

Milling methods vary from simple hand pounding, with a mortar and pestle, to large-scale processing in highly mechanized mill plants. Hand pounding, although declining in use, still accounts for a major percentage in many important areas—more than 75% of rice processing is done by this method in Indonesia. The remaining technologies can be classified under 2 major categories depending upon whether hulling and whitening are carried out in 1 or several steps. The Engleberg huller is widely used to remove hull and bran in a single operation. Hulls and bran are discharged together with the broken rice.

Bran from Shellers

When milling is carried out in 2 or more steps, the first step is hulling. The most common machines used in hulling paddy rice are the under-runner disk huller, which is the most widespread, and the rubber roll huller, which is slowly gaining popularity. Both machines produce what is known as sheller bran. It is made up fundamentally of hull fragments; it also may include a small proportion of pericarp, some germ and rachilla and small fragments of endosperm, as well as dust and soil.

The under-runner disk huller performs a coarser work. The paddy is forced between 2 horizontal cast iron discs which are lined with an abrasive coating. Pressure and friction are applied to remove the hull. Even though the clearance between the discs is adjusted to the optimum conditions, some damage of the pericarp and germ cannot be avoided. The rubber roll huller, when properly adjusted, does not damage the pericarp and rarely removes the germ from the caryopsis. The paddy grains are caught under pressure between the 2 rubber rolls and the hull is stripped off. Consequently, bran from the rubber roll huller basically consists of hull fragments and impuritites, and contains almost no germ. The under-runner disc huller produces 1—2.5% bran, with reference to paddy rice. The rubber roll huller gives productions lower than 0.5%. Due to the small quantities in which it is produced, coupled with its low feeding value, huller bran is a byproduct of little importance. It may be discharged as a waste product or alternatively, and more commonly, removed by the bran aspiration cyclone system.

The discharge from the huller is a mixture of brown rice, paddy, hulls, bran, dust, broken rice and immature paddy grains. Bran, dust and small broken rice may be separated by sieving. An airstream passing through it may separate the bran and dust from the brokens.

The industrial hullers do not hull all the paddy grains. Only 80—95% of them are actually hulled in a single pass. After separation of hulls, bran, dust and brokens in the hull separator, a mixture of brown rice and paddy remains; the paddy is then separated in a paddy separator. The presence of paddy in brown rice impairs the uniformity of the subsequent whitening. The hull, ground in the process of whitening, is incorporated in the bran. It may reach an important proportion—10 to 20 times the percentage of paddy in brown rice. Unless the return sheller is absent, the proportion of powdered hull in the bran is usually not large.

Bran from Whiteners

In the process of whitening the pericarp, the tegmen and the aleurone layer, together with the germ, are removed from brown rice. The bran is made up largely of fragments of the above coverings, plus some starchy endosperm, bits of the caryopsis and occasionally due to defective processing, some particles of hull.

In removing bran from brown rice 2 kinds of machines are widely used: (1) the abrasive type whitening machines, either vertical, like the European cono-mills, or horizontal—the Japanese design; and (2) the friction type whitening machines—like the Japanese horizontal jet pearler. In the abrasive machines, brown rice is whitened as it passes through the clearance between an abrasive roller and a wire screen (European mill) or perforated cylinder (Japanese mill). The abrasive roller—with the sharp edges of vitrified particles—acts as a blade to cut and remove small bits of the bran layer from the brown rice. The process is similar to cutting off an orange skin little by little in small pieces with a sharp razor blade in an effort to avoid bruising the fruit (Koga 1969). To this end, the abrasive type whitening machines have a high peripheral roll speed and a low pressure charge to grains. In the vertical whitening cone, the bran passes through the wire screen and drops into the bottom of the cone housing from which it is scraped by a rotating scraper and unloaded. Air sucked through the machine serves to cool the system but will remove some bran which is recovered by cylone separation. In the abrasive type horizontal whitening machine, jet air is blown through the hollow part of the main shaft into the milling chamber and, as it cools the rice grains, blows off the bran adhering to the whitened kernels. In the friction type whitening machines, Japanese jet pearler type, brown rice is whitened on a different principle. Pressure is applied to the grains as they pass through the clearance between the screen and the milling roller; the bran layer of the brown rice is loosened and peeled off by means of friction caused by the action of grains rubbing together under pressure. Unlike the abrasive type machines, bran here is removed in rather big flakes (Koga 1969). Jet air is blown through the hollow part of the main shaft through the rice grains. It blows off the adhered bran and removes the freed bran as it leaves the machine through the perforated screen.

Obviously the abrasive type and the friction type machines utilize both abrasion and friction. Each type of machine combines both actions, although one predominates. There is one Japanese machine which was designed to combine both principles. It has the abrasion and the friction zones on the main shaft. The rice is whitened by a precision pressure abrasive roller and then polished by a friction device. Bran from this 2-step process is removed altogether by a constant jet of air.

Because multipass whitening produces higher head rice output than single pass whitening, the former is gaining in popularity. Multipass whitening allows for serial removal of successive bran layers in separate machines. The processing line may include only 1 type of machine or a combination of abrasion type and friction types.

Although whitening is sometimes called polishing, the latter term has a distinct meaning. The process of polishing removes from the already whitened rice any fine particles adhering to the grain, just as the kernel acquires a glossy appearance. The 2 types of polishing machines more widely used—the vertical cone polisher and the horizontal polisher—are based on the same principle. A cone or a cylinder is mounted on a shaft and is housed in a screen casing. The space between the cone or cylinder and the screen forms the polishing chamber. The cone or cylinder is provided with leather strips on the surface and the polishing is accomplished by the rubbing effects produced by these leather strips when the shaft revolves. The detached loose flour—polish—is sucked out of the machine and recovered in a cyclone system.

In some countries calcium carbonate is used as a milling aid, usually 0.25% with reference to the paddy rice. When quality of parboiled bran is not a major concern, the rice hulls are incorporated as a milling aid with the double purpose of abrasion and antichoking. Other products such as diatomaceous earth and ground limestone have also been used. They may affect the characteristics of the bran produced.

No matter how well a whitener machine works, some breakage of the grain is unavoidable; the broken surface is exposed to abrasion and/or friction. Then more materials other than the bran are removed. Grain fragments of various sizes are produced. The finest fragments pass, along with the bran and the germ, through the perforations of the wire screen or cylinder of the milling chamber (openings of 1.30 × 1.25 mm are usual in Spanish cono mills). Large fragments may be found in the bran due to faulty closing of the wire screen frames or defects in the wire or perforated casing.

Germ and Bran Separation

The combined streams coming out from the whitener machines consist of a mixture of bran, germ, starchy endosperm particles and hull fragments. Sieving in combination with an air stream is used to accomplish separation of major fractions—bran, germ and brokens of various sizes. Although in the majority of countries the commercial bran includes the germ, in others like Italy and Spain, the separation of the germ is a generalized practice. In the diagrams of the Spanish mills the mixture of byproducts from the whitener machines pass through the rotating hexagonal reel and the germ separator. The former, a rotatory sieve, with a mesh of approximated 0.8 mm (No. 20 ASTM standard) allows bran, fragmented germ and germs of smaller size to pass through. The germ separator combines oscillating sieving and pneumatic force to separate particles differing in size and density. It comprises a first sieve, with openings of 0.8 mm, to remove residual bran; a second screen with openings of 1.5 mm to separate the germ, and a third one, with perforations of 2 mm, to separate small from large brokens and kernels of small size that might be present. An airstream is sucked through the germ fraction discharge, removing light impuritites such as bran and hull. Further purification of the germ fraction can be achieved by pneumatic separation.

HISTOLOGY AND HISTOCHEMISTRY OF COMMERCIAL RICE BRAN

The need for a large and detailed knowledge of commercial rice bran for the full utilization of its potential food value is recognized worldwide and it has recently been increased with the development of highly promising processes and products related to or based on this byproduct.

Available literature about the histology and histochemistry of rice bran as a commercial byproduct is limited. Papers dealing with outer layers of the caryopsis are relatively numerous (Barber *et al.* 1972 A,B, 1976; Borasio 1929; Bouharmont 1967; Cho 1938; Little and Dawson 1960; Santos 1933; Yasui 1936; Yung 1939). However, those on the morphology, the histological composition and the chemical nature of the individual particles composing commercial rice bran are scarce.

The gross structure of the rice grain and role of milling should be kept in mind when approaching the study of the morphology and the histological composition of commercial bran. The rice caryopsis is a fruit in which the seed occupies the greatest part. It is covered by 2 florescent glumes: the lemma, which is the largest and has an awn or beard, and the palea.

Two small sterile lemmas, the rachilla and the pedicel, can be present in the paddy grain to be milled. In the process of milling, the caryopsis becomes disclosed, showing the pericarp or outermost layer. Immediately under the pericarp, the seed covering, known as testa by some authors and as tegmen by others, the aleurone layer and the starchy endosperm are located. The germ remains in a cavity in the lower abdominal area of the caryopsis, stuck to the endosperm and outwardly covered by the aleurone layer, the seed coat and the pericarp, and externally by the lemma. The whitening process is intended to remove the pericarp, the tegmen, the aleurone layers and the germ, although parts of the starchy endosperm are also separated.

In a recent study[1,2] (Pineda 1976) 15 different kinds of discrete particles (Table 24.1) and a few small fragments of the abrasive coating of the cono mills were identified in commercial bran as described as follows.

Type a Particles: Fragments of Florescent Glumes

They have irregular polygonal shapes and are of variable size, usually ranging from 25 to 40 μ, although a few fragments can reach 2.5 × 1.5 mm; the thickness of cross-sections varies from 80 to 120 μ. There are

---

[1]Extensive use has been made in preparing this section of the doctoral thesis research work of Dr. J.A. Pineda, carried out at the author's laboratory.
[2]Milling diagram includes: aspirator-precleaner, rubber roller husker, husk aspirator, paddy separator, return sheller, abrasive type whitening machines (four successive vertical cone units), small brokens separator and germ separator. Processed rice varieties: Japonica type—Bahia and Balilla × Sollana.

TABLE 24.1. TYPES OF DISCRETE PARTICLES IDENTIFIED IN SPANISH COMMERCIAL RICE BRAN

Type	Simple particles Composition	Type	Compound particles Composition
a	Fragments of florescent glumes	i	Fragments of pericarp and seed coat
b	Fragments of sterile glumes	j	Fragments of seed coat and aleurone layer
c	Fragments of trichomes	k	Fragments of pericarp, seed coat and aleurone layer
d	Fragments of pedicel	l	Fragments of pericarp, seed coat, aleurone layer and starchy endosperm
e	Fragments of pericarp	m	Fragments of starchy endosperm and aleurone layer
f	Fragments of starchy endosperm	n	Fragments of germ, flattened cell layer and starchy endosperm
g	Germ, whole or fragments	o	Fragments of germ, aleurone layer, seed coat and pericarp
h	Fragments of fibers		

also particles made up of nervure and lemma awn. Most particles come from longitudinal fragmentation of the hull. The anatomohistological characteristics of the particle are similar to those of the hull (Santos 1933; Angladette 1966; Houston 1972A,B). Its outer surface is rough and wrinkled covered by entire and/or broken trichomes, whereas its inner surface is plain. In a cross section cut 4 layers can be seen: external epidermic layer, esclerenchymatic layer, parenchymatic layer and internal epidermic layer. An extracellular layer, the cuticle, is present outside the epidermis. The external epidermic layer is made up of thick-walled epithelial cells (from 30 to 80 by 20 to 10 μ, those of palea, and from 60 to 90 by 50 to 120 μ, those of lemma (Santos 1933)) which are radially elongated and axially oriented in rows. Among these cells, trichomes (see later on) are inserted. They cross the cutin-silica layer on the outside and inwardly reach the esclerenchymatic layer. The esclerenchymatic layer is composed of 2–4 layers of thick-walled (4–6μ) cells, 200–610 μ long, those of lemma, and 150–630 μ long, those of palea (Santos 1933).

The parenchymatic layer is made up of several layers of cells (Pineda 1976). Cells are of 2 types, tangentially elongated and short, with sizes varying from 20 to 30 by 10 to 15 μ, and thin cell walls (2 μ). Among the parenchymatic cells, tube and cross cells are placed (Angladette 1966; Houston 1972B). The internal epidermis (from 6 to 12 μ thick (Pineda 1976)) is formed by 3 thin-walled isodiametric cell layers, tangentially oriented. Between the esclerenchymatic and parenchymatic layers, the vascular bundles are located.

Carbohydrates[3] are present in the cell walls of type a particles (Pineda 1976). Cellulose is the major carbohydrate, along with hemicelluloses, mainly pentosans (Houston 1972B). Starch is reported to be absent (Houston 1972B). Lignin appears to be intermingled with the cellulose of the cell walls (Santos 1933; Pineda 1976) and is chemically combined with hemicelluloses (Neish 1965). Proteins are apparently absent (Pineda 1976) as shown by the mercuric chloride-bromophenol blue reagent (Mazia et al. 1953). Black Sudan stains it intensely on the outer surface, indicating the presence of wax and cutin (Pineda 1976). Fatty globules are absent (Pineda 1976). Silica is abundant in the extracellular cutin-silica layer (Houston 1972B; Yoshida et al. 1962). It is also plentiful in the outer epidermis and in the esclerenchymatic layer (Yoshida et al. 1962; Thomas and Jones 1970).

Silica also occurs among the cell walls of esclerenchymatic cells, par-

---

[3]Included in this group are the carbohydrates that give a positive acid-Schiff reaction (PAS) (Hotchkiss 1948; Pearce 1968), namely starch, cellulose, hemicellulose, pectic substances, glycoproteins and glycolipids.

enchymatic cells, outer and inner epidermis and vascular bundles (Pineda 1976).

Type b Particles: Fragments of Sterile Glumes

The sterile glumes are boat-like, with the forward end sharp-pointed and the rear end, where the rachilla is fixed, U-shaped. Fragments of sterile glumes, not numerous in bran, are of variable size and occasionally they are more than 2 mm long. The surface is plain. The trichomes are less frequent and are smaller than those in the lemma and palea. The thickness of the fragments varies from 60 to 120 μ (Pineda 1976). Their histological structure is simpler than that of florescent glumes (Santos 1933). In cross section there is an outer epidermis, an inner epidermis and, between them, a parenchymatic layer. The outer epidermis consists of a single layer of rectangular cells from 50 to 70 by 5 to 10 μ, tangentially elongated, with thick outer walls (3 μ), coated externally by a thick (5 μ) cuticle. The inner epidermis has similar characteristics. The intermediate parenchymatic tissue consists of 2–4 cell layers. Their cells are variable in size (20 to 70 by 7 μ).

Occurrence and distribution of major chemical constitutents in the tissues of particles of sterile glumes are similar to those described above for fluorescent glumes, as corresponds to their anatomic and histological characteristics.

Type c Particles: Trichome Fragments

The entire trichome is conic, transparent, uncolored and empty. In commercial bran (of short grained varieties, Bahia and Balilla × Sollana) entire trichomes and fragments, varying from 100 to 350 μ in length with a maximum diameter of 30 μ, have been detected. Size and number depend on variety; there are varieties which are almost awnless (Houston 1972B; Santos 1933).

In the cell walls of trichomes, carbohydrates were detected (Pineda 1976) with the PAS (periodic acid–Schiff) method (Hotchkiss 1948), lignin with Schiff's reagent (Davenport 1964) and lipids, only in cross sections (Pineda 1976), with Black Sudan (Jensen 1960). Staining with mercuric chloride-bromophenol blue reagent (Hg-BPB) (Mazia et al. 1953) and safranin-phenol (Yoshida et al. 1971) failed to detect proteins and silica respectively.

Type d Particles: Pedicel Fragments

Fragments of pedicel are scarce in bran (Pineda 1976; Santos 1933). They are cylindrical fragments of variable sizes, up to 2–3 mm, with grooved surfaces. Their cell walls react positively to stain tests for carbohydrates, lignin and silica, and negatively for proteins and silica.

Type e Particles: Pericarp Fragments

Type e particles have an elongated shape, variable size, generally no more than 800 μ and a thickness of 6–8 μ (Pineda 1976). The pericarp tissues can be strongly compressed, so it is difficult to distinguish their characteristic constitutive layers. They are visible in some varieties only (Esau 1965; Juliano 1972). Six layers, five of which are transversely elongated, can be distinguished (Winton and Winton 1932). Due to their high differentiation, these layers can be subdivided into epidermis, hypodermis, cross cells and tube cells, in that order, moving inward. The epidermis consists of a single layer of cells, transversely elongated, with sinuous end walls. Cells in noncompressed pericarp fragments vary in size from 5 × 3 to 20 × 10 μ. On the outer walls of the epidermal cells there is a thin cuticle. The hypodermis is composed of several layers of parenchymatic cells, partially crushed, transversely elongated, with dimensions varying from 15 × 80 to 5 × 12 μ. Two to five layers of elongated cross-cells of 6 μ width and 1–2 μ depth, arranged in files and rows (their long axes lie perpendicular to the long axis of the caryopsis), give the tissue along with the parenchymatic layers a net-like look. Finally, the tube cells, with a long and approximately cylindrical shape, of 3 to 5 μ in diameter, lie with their long axes parallel to the long axis of the caryopsis. The mean thickness of pericarp cell walls is 2 μ (Little and Dawson 1960).

Carbohydrates have been identified in the pericarp: cellulose (Little and Dawson 1960) and hemicellulose (Little and Dawson 1960; Pineda 1976). Also present are lignins (Pineda 1976), proteins (Little and Dawson 1960; Pineda 1976), lipids—cutin and/or wax (Pineda 1976), phytin (Santos 1933; Pineda 1976), anthocyanins (Chang and Bardenas 1965) and silica (Pineda 1976).

PAS reaction (Hotchkiss 1948) for carbohydrates is positive in pericarp cell walls (Pineda 1976); cross cell walls cannot be easily stained. Cellulose and hemicellulose are probably major components (Little and Dawson 1960). Pericarp cell walls, mainly those of epicarp, contribute notably, along with the seed coat, to higher fiber content of rice bran. Starch is apparently absent (Santos 1933; Pineda 1976). Lignin has been detected in pericarp cell walls. It is abundant in the epidermis. Schiff's staining (Davenport 1964) failed to detect it in tube cells (Pineda 1976) even though they are lignified (Santos 1933). Pericarp cell walls can be stained with typical protein reagents (Little and Dawson 1960). Staining with Hg-BPB (Mazia et al. 1953) is positive in hypodermis and tube cells and negative in epidermis (Pineda 1976); cross-cells are stained in yellow instead of blue (Pineda 1976). On the free surface of pericarp, lipids, cutin (Cho 1956) and probably wax, can be detected with Black Sudan (Pineda 1976).

Type f Particles: Starchy Endosperm Fragments

It is one of the most abundant types. Size varies amply; the most frequent vary from 1×1.5 to 30×15 μ. The largest particles usually are components of subaleurone endosperm. They are parenchymatic cells with lumen full of compound starch granules and protein bodies. The smallest particles come from the caryopsis nucleus and are cell fragments. Clusters of starch granules compressed by a protein matrix have been also identified. The cells are either polygonal, generally irregular pentagons or hexagons in cross-section, or slightly elongated. Those of the central core have isodiametric shapes ranging in size from 45×50 to 80×105 μ. The mean thickness of endosperm cell walls is about 0.25 μ (Little and Dawson 1960).

In the starchy endosperm, starch, cellulose, hemicellulose, proteins and lipids have been detected. The Schiff's reagent (Davenport 1964) failed to detect lignin. Cellulose and hemicellulose are located in cell walls and starch in the lumen. Starch granules are polygonal, usually 5-sided in shape and compound. Sizes of individual granules range from 0.3 to 13 μ (Yamazaki and Wilson 1964). Starch granules are small (2 to 4 μ) in subaleurone cells and larger (5 to 9 μ) in cells of the kernel nucleus (Little and Dawson 1960). Size of compound granules varies from 12×10 to 20×10 μ in the short grain Bahia variety (Pineda 1976). Proteins occur as protein bodies of about 1μ in size in the inner cells (Mitsuda et al. 1967); as cementing material between protein bodies and compound starch granules (Del Rosario et al. 1968; Little and Dawson 1960), and as a lipoprotein membrane surrounding individual starch granules. Spaces among starchy granules are stained with Black Sudan. This is most likely due to lipids associated with protein bodies (Mitsuda et al. 1967). Membrane surrounding starch granules also gets stained.

Type g Particles: Entire Germ or Fragments of It

Whether it is degermed or not at the mill, commercial rice bran always contains entire germ and fragments of it. These fragments can be distinguished from other particles easily. The most conspicuous fragments are (1) entire plumule, (2) coleoptile, (3) entire coleoptile and plumule, (4) primary root with root cap and part of coleorhiza, and (5) scutellum and epiblast (see also particles n and o).

The histology and histochemistry of rice germ have been the subject of

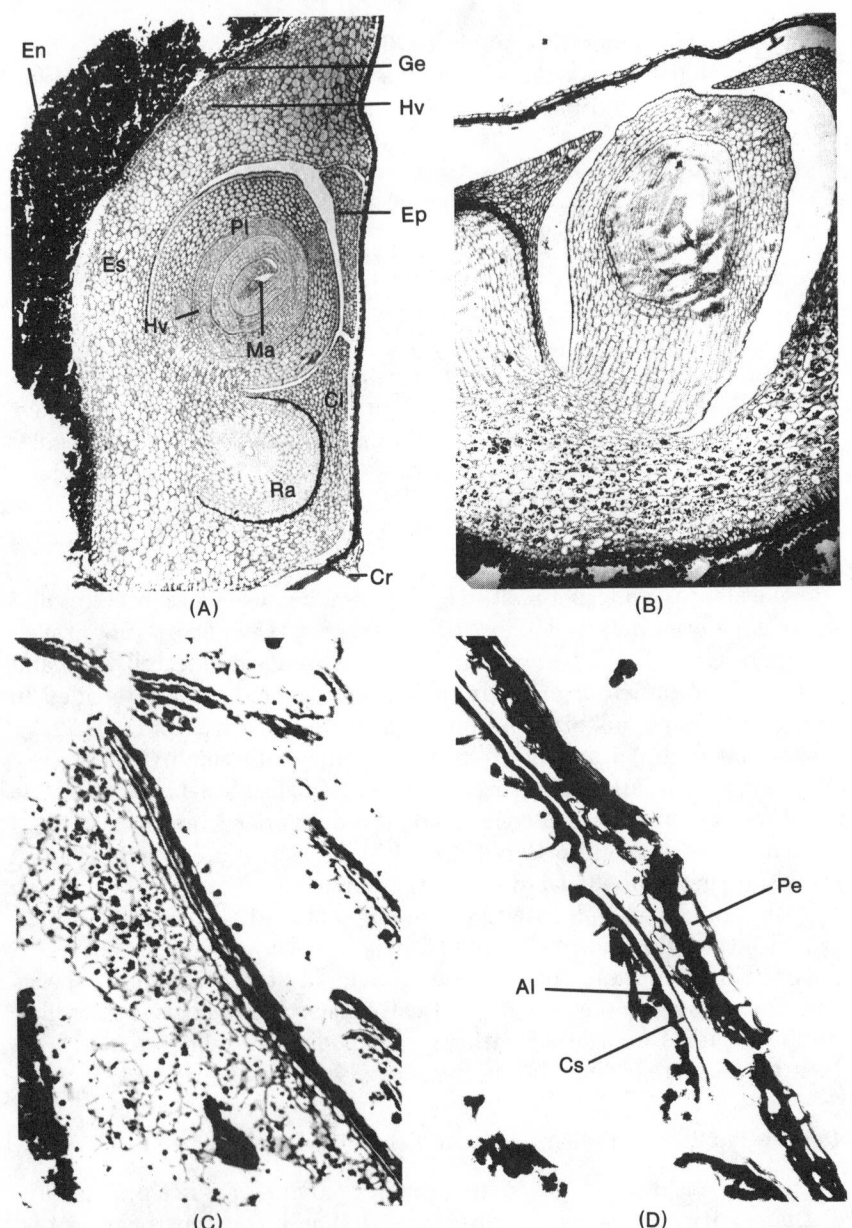

FIG. 24.1. HISTOLOGY AND HISTOCHEMISTRY OF COMMERCIAL RICE BRAN

(A) and (B) Longitudinal sections of germ, (C) Various types of bran particles, (D) Type *k* particle—composed of pericarp (Pe), seed coat (Te) and aleurone (Al). Al, aleurone layer; Cl, coleorhiza; Co, coleoptile, Cr, wrinkle; En, endosperm; Ep, epiblast; ES, scutellum, Ge, epitheliar gland; Hv, provascular bundle; Ma, apical meristem; Pe, pericarp; Pl, plumule; Ra, primary root; Te, seed coat.

recent detailed studies (Barber *et al.* 1972 A,B). The germ (Fig. 24.1) is composed of the embryonic axis (that includes the plumule, the coleoptile, the primary root and the hypocotyl), and the tissues surrounding it (the scutellum, the coleorhiza and the epiblast). The proportion of these parts in the grain is shown in Table 24.2. The germ is covered by the aleurone layer, the tegmen and the pericarp.

TABLE 24.2. WEIGHT DISTRIBUTION OF ANATOMICAL FRACTIONS OF BROWN RICE AND RICE GERM

Fraction	(%) Brown rice
Pericarp and aleurone	7.0
Germ	2.3
Plumule	0.34
Radicle	0.18
Scutellum	1.4
Epiblast	0.26
Coleorhiza	0.18
Endosperm	90.7

Source: Hinton (1948); Hinton and Shaw (1954).

The plumule has 2 to 3 embryonic leaves. It is igloo-like in shape, 0.3 mm in diameter and 0.4 mm in height (Balilla × Sollana variety). The embryonic leaves have polyhedral thin-walled parenchyma cells, isodiametric (7 μ) in cross-section and elongated in longitudinal section. They are covered by the upper and lower epidermis and coated by an 0.1–0.2 μ thick cuticle. The provascular bundles of the leaves have polygonal cells in cross-section of about 4 μ, and rectangular cells in longitudinal section of about 4 × 11 μ. The apical meristem has isodiametric cells (4 μ). The coleoptile covers the plumule and has a pore in its apex. It is 0.11–0.18 mm thick and is composed of 9–12 layers of polygonal cells. It is covered by an outer epidermis with a 1.1–1.3 μ thick cuticle and an inner epidermis with a 0.5 μ thick cuticle.

The primary root is cylindrical, about 300–350 μ in diameter and 0.45 mm in length. It is composed of several tissues arranged in radial symmetry: a 6–10 μ thick cuticle (thinner, 1.5 μ, in apical) meristem, epidermis and subepidermis (a single layer each of prismatic cells, of 11 × 23 μ and 7 × 14 μ respectively), exodermis (2 layers of cells of various sizes), cortex (70 μ thick, 5–7 layers of round cells of 10–20 μ in size), endodermis (a single layer of prismatic cells, of 6 × 5μ), pericycle (1–2 layers of cells) and the central cylinder (vascular cylinder) which consists of the protophloem, the protoxylem, the metaxylem and parenchymatic cells. The tip of the primary root is protected by the calyptra or root cap (about 40 μ high).

The hypocotyl joins the primary root and the plumule. It is made up of provascular bundles surrounded by parenchymatic cells.

The scutellum is located between the embryonic axis and the endosperm. The greatest part of the scutellum is composed of polygonal parenchymatic cells, of 14 to 28 μ in longitudinal section. The scutellum is surrounded externally by an epidermis which is modified to form the scutellar epithelium in the area next to the endosperm; this consists of radially elongated, thin-walled cells arranged in palisade and some epithelial glands. A provascular bundle extends into the scutellum from the plumule. The epiblast is fused to the scutellum and surrounds the coleoptile. Its thin-walled cells are parenchymatic, polygonal and isodiametric (about 6μ) in cross-section and elongated ($7-12\ \mu$) in longitudinal section. The epiblast is surrounded by an epithelium. The coleorhiza encloses the radicle. It consists of parenchymatic cells similar in size and shape to those of the epiblast, and is protected by an epidermis covered with a thin cuticle.

Starch, hemicellulose, cellulose, lipids, proteins, basic and sulfur amino acids, tryptophan, and ash have been identitifed histochemically in rice germ (Barber *et al.* 1972B, 1976). Pectic substances were apparently absent (Barber *et al.* 1972A; Pineda 1976), as determined either by the alkaline hydrolysis method (Jensen 1960) followed by PAS straining (Hotchkiss 1948) or by the hydroxylamine-ferric chloride test (Gee *et al.* 1959; Reeve 1959); lignin, by floroglucinol, chloride-sulphite and Schiff's reagents (Jensen 1960) and silica by safranin-phenol (Yoshida *et al.* 1971) were absent (Barber *et al.* 1972A; Pineda 1976).

Germ cell walls react positively to PAS reagent (Hotchkiss 1948; Pearse 1968) for total carbohydrates, and to cellulose and hemicellulose tests (Jensen 1960; Barber *et al.* 1972B); cell walls of the primary root and cuticles of the primary root, coleoptile and scutellum appear to be rich in hemicellulose. Cuticles of the epidermis of plumule, coleoptile, primary root and scutellum react positively with PAS (Pineda 1976). The cytoplasm of the cells of the embryonic axis and of surrounding tissues contains simple starch granules; they are not abundant (1 to 16 per cell), but are of round shape and variable size (0.5 to 5μ). Granules are more numerous in the scutellum and coleorhiza, and are apparently absent in the epidermic layers (Barber *et al.* 1972B; Pineda 1976). The protoplasms sometimes show the presence of dispersed PAS-straining material (Pineda 1976), presumably sugars (Santos 1933); glucose and sucrose have been reported in the scutellum (Akazawa 1972).

All the tissues of the germ were stained with Black Sudan (Barber *et al.* 1972B). Lipids occur in the cytoplasm as globules of diverse size (0.5 to 4 μ); they are abundant in the coleoptile and scutellum and scarce in the plumule, primary root and hypocotyl. The cuticle of the epidermis of plumule, coleoptile, primary root and scutellum reacts positively to Black Sudan (Pineda 1976).

The cytoplasm and the nucleus, including the plasmatic membrane, react positively to bromophenol blue test for proteins. In the cytoplasm of parenchyma cells of plumule, coleoptile, primary root, scutellum and coleorhiza, tiny protein granules (0.5–1 μ) and dispersed proteins are detected; the cytoplasm of the respective epidermis cells only shows dispersed proteins (Barber *et al.* 1972B; Pineda 1976). Provascular bundles are weakly stained with bromophenol blue. Proteins are apparently absent in the cuticles.

All the tissues containing proteins react positively to naphthol-yellow (Deitch 1955) for lysine, hydroxylysine, histidine, arginine, and terminal amino groups and to DDD (2,2′-dihydroxy-6,6′ dinaphthyl disulfide) (Barrnet and Seligman 1952) for tryptophan (Barber *et al.* 1974). Germ contains minerals (ash) distributed throughout. The plumule and the primary root are more intensely stained than the coleoptile; the scutellum is abundant in minerals, particularly the cell layers near the endosperm (Barber *et al.* 1974).

Recently the structure of rice scutellum has been examined by scanning electron microscopy (Tanaka *et al.* 1976). Scutellar cells contain many rounded particles, around 2–3 μ in diameter, apparently covered with a membraneous coat, which resembles aleurone particles of the aleurone layer. Particles in both tissues are characterized by a high magnesium content and potassium salts of phytic acid.

Type *h* Particles: Fibers

Fibers originate from knots of the panicula base, and are scarce. Some are very long (1 cm or even more), ribbon-like and twisted. They do not show a definite cellular structure; between 2 thick, almost parallel walls they show a cavity of about 30 μ wide. Fibers react positively to PAS stain for carbohydrates and negatively to stains for lignin, proteins, lipids and silica (Pineda 1976).

Type *i* Particles: Fragments of Pericarp and Seed Coat

These particles and those of type *k* are the major components of commercial bran. They are rolled or bent flakes with a smooth, shining surface and a round inner surface of different sizes, between 200–300 μ (Pineda 1976). In particles with compressed tissues the thickness ranges generally from 10 to 15 μ; with spongy tissues the thickness is much greater; most of the particles are removed from the caryopsis longitudinally. The histological and histochemical particularities of pericarp have been described above (see under Type *a* Particles). The seed coat (tegmen or testa) consists of (1) the fatty cuticle; (2) the spermoderm, which is located next to the pericarp; and (3) the perisperm, which is next to the aleurone layer. In commercial bran (Pineda 1976) only the cuticle and a layer (2 to 6 μ thick) containing the remains of the inner integument and nucella, have been detected. The cuticle, 1 to 3 μ thick, is occasionally discontinuous (Little and Dawson 1960). The tegmen in the germ area is thinner. Some commercial bran particles only consist of pericarp and cuticle (Pineda 1976).

The seed coat, except the cuticle, reacts positively to the PAS test for carbohydrates and to the Hg-BPB test for proteins (Pineda 1976). Cellulose is the major constituent of cell walls (Little and Dawson 1960). The cuticle absorbs Black Sudan but the remains of the inner integument and nucella are stained very weakly (Pineda 1976); the cell walls of the inner integument are cutinized (Cho 1956). The integument and nucella contain lignin. Neither the pericarp nor the seed coat react positively (Pineda 1976) to the silica test (Yoshida *et al.* 1971).

Type *j* Particles: Fragments of Seed Coat and Aleurone Layer

It is a type of particle not abundant in commercial bran. These particles contain only 1 layer of aleurone cells. Generally neither the seed coat nor its constituent layers are removed alone during milling. Although a part of endosperm, aleurone cells come out of the caryopsis preferably stuck to the seed coat (particles *j* and *k*) rather than with the starch endosperm (particle *m*).

The size of *j* particles ranges from 100 to 300 μ in length and from 15 to 20 μ in width. The seed coat is around 6 μ thick. The characteristics of the seed coat are described under particle type *i*. Aleurone cells are predominantly polygonal. Those from the ventral side are almost perfect rectangles and those from the dorsal side are irregular hexagons (Del Rosario *et al.* 1968; Little and Dawson 1960). They are nucleated parenchymatic cells, with cell walls around 1–3 μ thick, rich in protoplasmatic constituents. Aleurone cells are around 25–30 μ long and 6–10 μ thick. The aleurone layer on the germ is thinner and its cells are smaller.

The aleurone cells contain the so-called aleurone grains or aleurone particles. These have been recently isolated as a white ivory powder from rice bran using differential centrifugation in nonaqueous media (Tanaka *et al.* 1973). Its yield ranges between 2 and 3% by weight of the bran fraction used (2–4% in bran/unpolished grain). Subcellular aleurone particles are spherical grains about 1–3 μ in diameter. They exhibit no strata structure, and instead show embedded electron-dense bodies.

Cell walls and lumen of aleurone cells react positively to carbohydrates. Starch granules smaller than 5 μ have been detected. Santos (1933) reported that aleurone cells do not contain starch, but it may be found in more mature kernels (Juliano 1972).

Aleurone cell walls are stained by reagents for cellulose and hemicellulose or pectin (Little and Dawson 1960). Staining with Hg-BPB indicates the presence of proteins in the aleurone cells. A weak, pale staining suggests the presence of dispersed protein material in the lumen. The Hg-BPB reagent shows a thin cuticle along the inner border of aleurone cell walls (Pineda 1976). The nucleus of the aleurone cells is difficult to detect (Del Rosario *et al.* 1968; Pineda 1976). Aleurone particles are not distributed uniformly (Little and Dawson 1960). They occupy most of the lumen of cells in the dorsal side of the caryopsis whereas in other cells, starch granules and especially fat granules are numerous (Pineda 1976). A sheath of fat-staining material encloses the aleurone particles (Little and Dawson 1960). A positive weak reaction to lignin and negative reaction to silica have been reported (Pineda 1976).

The chemical composition of isolated aleurone particles has been recently reported (Table 24.3) (Tanaka *et al.* 1973). The soluble carbohydrates (Table 24.3) are not characterized further; however, isolated

TABLE 24.3. CHEMICAL COMPOSITION OF ALEURONE PARTICLES ISOLATED FROM RICE BRAN

Composition	Weight %
Nitrogen	1.95
Carbohydrate	7.93
Acid-soluble organic phosphorus	11.3
Inorganic phosphorus	0.20
Acid-insoluble organic phosphorus	0.10
Phospholipid phosphorus	0.012
RNA	0.23
DNA	0.11
Myoinositol	
Total	9.41
Free	0.026
Moisture	14.9
Metals	
Mg	8.30
K	9.45
Ca	0.42
Zn	0.0095
Fe	0.034
Cu	0.0025
Mn	0.046

Source: Adapted from Tanaka *et al.* (1973).

aleurone grains showed no iodine reaction and no starch granules were recognized under a transmission electron microscope. Protein content was 11.7%; protein solubility in water was 6.03%; 0.5 M NaCl, 0.60%; 0.05 M HCl, 0.25%; and 0.2 M NaOH 2.19%. γ_1 and γ_2 globulins are localized in the aleurone cells as shown by means of a fluorescent-antibody technique; the same γ globulins appear to exist in the scutellum (Horikoshi and Morita 1975). Acid-soluble organic phosphorus amounts to 97.3% of the total phosphorus and almost all of it exists as phytic acid. The aleurone particle is the accumulation site of phytic acid in rice grain. The distribution of ^{3}H-myoinositol administered to ripening grains of rice was followed by microradioautography and the ^{3}H was exclusively found in aleurone particles (Tanaka et al. 1974). The electron dense materials embedded in the aleurone particles of mature rice grains are potassium and magnesium salts of phytic acid (Ogawa et al. 1975).

The phosphorus and protein content of the protein bodies in the endosperm and the aleurone particles differ sharply. Protein bodies are rich in protein (about 60%) and very low in phosphorus (0.5%). About 6% of the total protein in the protein bodies is water soluble whereas in the aleurone particles, 70% of the total protein is water soluble (Tanaka et al. 1973).

Other Types of Bran Particles (k Through o)

Type k particles are composed of pericarp, seed coat and aleurone layers. Reported sizes range between 130 × 50 and 560 × 105 μ. Bran of such composition has been reported to be removed easier from the ventral than from the dorsal side (Cho 1956). Most particles have 1–2 aleurone cell layers; few have 5–7. Presumably this has a bearing on the fact that Japonica varieties have 1–2 aleurone cell layers in the ventral side of the caryopsis and 5–7 in the dorsal side (Cho 1956). Aleurone cells appear generally intact. In some particles, the lipid cuticle of the seed coat appears interrupted; then the pericarp and seed coat are around 20 μ thick. Particles of type l include pericarp, seed coat, aleurone layer and starchy endosperm, and they are scarce. Their size ranges generally between 400 × 260 and 735 × 200 μ. Particles of type m are composed of a starchy endosperm and aleurone layer. Their size varies; generally it is less than 95 × 60 μ. Particles of type n include germ, flattened cells layer and starchy endosperm. Usual sizes range between 690 × 320 and 410 × 225 μ. Germ, mainly scutellum, is the major component. The layer of flattened cells is around 5–20 μ thick. Particles of type o include germ, aleurone layer, seed coat and pericarp. They are scarce and of various sizes (Pineda 1976).

General histological and histochemical characteristics of anatomic layers constituting particles k through o (Table 24.1) coincide with those described in preceeding sections.

CHEMICAL COMPOSITION AND CHEMICAL CONSTITUENTS OF BRAN, POLISH AND GERM

Proximate Analysis of Bran and Polish

The composition of rice bran depends on a variety of factors associated with the rice grain itself and the milling process. Major factors associated with the rice grain are varietal and environmental variability in: (1) average chemical composition, (2) distribution of chemical constituents, (3) thickness of anatomical outer layers, (4) size and shape of grains, and (5) resistance of grains to breakage and abrasion.

Major factors associated with the milling process are: (1) processing methods and machines, and (2) milling conditions as they affect degree and uniformity of milling. Reported data can also vary because of differences in analytical techniques used by different authors. Lack of clear definition of byproducts (frequently, bran-polish-germ mixtures) for which analytical data are reported, contributes to some confusion and widens the range of variation of the chemical composition data.

Bran products obtainable from different alternatives of processing rice are: (1) those from 1 step milling, i.e., "huller bran," and (2) those from multistep milling such as (a) sheller bran from hulling and (b) bran partially degermed or with all of its accompanying germ, including or not including polishings, (c) polish and (d) germ from whitening.

Data published on the composition of bran vary within a wide range (Table 24.4). Some extreme values are generally associated with particular factors; e.g., very low protein and oil contents and very high fiber and ash contents are characteristic of bran from huller type mills (Tables 24.4 and 24.5). Because of frequent variations in bran composition, eventually resulting in commercial bran adulteration, it is a market practice to guarantee a limit composition (Barber and Benedito de Barber 1977).

TABLE 24.4. USUAL RANGE OF VARIATION OF THE GROSS CHEMICAL COMPOSITION OF RICE BRAN[1]

Constituent (% dry basis)	Minimum	Maximum
Protein	6.7[3]; 11.5	17.2
Fat	4.7[3]; 12.8	22.6
Fiber	6.2	14.4; 26.9[3]
Ash	8.0	17.7; 22.2[3]
NFE[2]	33.5	53.5

[1]Includes data from: India (Panda and Gupte 1965), Italy (Fossati *et al.* 1976; Leonzio 1965; Maymone *et al.* 1962), Japan (Yokochi 1977), Malaysia (Arnott and Lim 1966), Mexico (Yokochi 1977), Nepal (Yokochi 1977), Philippines (Limcangco-Lopez *et al.* 1962), Spain (Primo *et al.* 1970), Sri Lanka (Siriwardene 1969) and the United States (Betschart *et al.* 1977; Lew *et al.* 1975).
[2]Nitrogen-free extract.
[3]For huller type mills.

TABLE 24.5. PROXIMATE ANALYSIS OF RICE BRAN FROM HULLER TYPE AND CONE TYPE MILLS

Analysis[1]	Huller type mill	Cone type mill
Crude Protein	10.01	13.95
Oil	7.31	16.29
Crude Fiber	21.48	9.93
Ash	22.16	11.94
NFE	39.03	47.88

Source: Siriwardene (1969).
[1]Percentage dry basis.

Bran from the hulling step (Table 24.6) differs in composition from that produced in the whitening steps. It is lower in protein and fat content and higher in fiber and ash content. Bran from hulling can be close to that obtained from the huller type milling in gross chemical composition; nevertheless the high proportion of rice hull in the latter makes a major difference. The composition of "sheller bran" strongly depends on the type of sheller. The under-runner disk huller produces bran of higher protein and fat content and lower fiber content than the rubber roll huller (Table 24.6). The type of machine used in the hulling step also has an influence on the composition of the bran removed in subsequent whitening. The rubber roll huller produces undamaged brown rice from which a bran richer in fat will be removed. The under-runner disk huller removes some germ and outermost layers, rich in oil, which will not be incorporated into the bran.

TABLE 24.6. PROXIMATE ANALYSIS OF RICE BRAN FROM HULLING[1]

	Under-runner disk huller	Rubber roll huller
Moisture	9.2–13.5	11.7
Protein	8.1–11.6	3.8
Fat	6.5–10.4	0.8
Crude Fiber	14.8–22.6	41.5
Ash	11.2–20.4	13.2
NFE	31.0–40.3	28.9

Source: Borasio (1948); Leonzio (1965); Primo et al. (1970).
[1]Percentage wet basis.

The type of whitening machine used is another source of variation of the chemical composition of bran. The friction type machines produce bran of higher fat content than the abrasion type (Barber and Benedito de Barber 1977). The latter machines frequently scratch the grain, reaching into deeper layers of the starchy endosperm.

Kernel layers at various depths differ greatly in chemical composition (Barber 1972). Because the percentage removal of bran may range from about 4 to 14 (Barber and Benedito de Barber 1977) the degree of milling greatly influences the composition of commercial bran. In multi-step milling, the bran from latter cones is higher in NFE and lower in protein, fat, fiber and ash (Primo et al. 1970) (Table 24.7). Generally, differences are not great enough to justify the segregation at the mill site of individual bran fractions as distinct commercial products. In industrial practice, bran from successive whitening cones is combined into a single product; polishings are sometimes an exception. Rice polish composition is not too far from that of rice bran as concerns protein and fat, but it is lower in fiber and ash content and higher in NFE content (Tables 24.7 and 24.8). The unequal removal of bran layers due to the irregular geometrical shape of the caryopsis as well as to the rough and difficult-to-control abrading/friction action in the whitening machine, greatly contributes to minimum compositional differences. Large differences exist in the homogeneity of milling. These can be measured objectively and have been shown to occur in commercial milling (Barber and Benedito de Barber 1977).

The use of calcium carbonate or any other product as a milling aid can influence the chemical composition of bran. Since it is fully incorporated in this fraction, even low proportions will significantly increase the mineral content; e.g., 0.25 kg carbonate/100 kg paddy will represent about 5% carbonate in bran from rice milled to 5% degree of milling.

Chemical Constituents[1] of Bran and Polish

Carbohydrates.—Starch, which occurs abundantly only in the endosperm, has been identified in the germ and the aleurone layers and is apparently absent in the pericarp and seed coat. Commercial bran contains a fair amount of starch (McCall *et al.* 1953; Pascual and Primo 1955) due to the endosperm present. Reported values range from about 10 to 55% (dry basis). The starch content of bran increases from the first to the last whitening machine.

According to Santos (1933), free sugars are apparently absent in pericarp, tegmen and aleurone layers. Notwithstanding, reported values for total sugars content of rice bran range from about 3 to 5%, moisture-free basis; presumably the contribution of germ and endosperm is significant. Nonreducing sugars are more abundant than reducing sugars, the ratio being 3:1–4:1 (Pascual and Primo 1955). Glucose, fructose and sucrose have been reported to be present as free sugars in rice bran (Parihar 1955; Soldi 1946: Vadher and Chauhan 1964; Williams and Bevenue 1953); raffinose has been found in the 80% ethyl alchohol extract of bran and polish (Bevenue and Williams 1956).

Bran is rich in cellulose and hemicelluloses. Data for crude cellulose of 10 samples of Italian bran partially degermed at the mill site ranged from 9.64 to 12.80% (Leonzio 1966). Corresponding values for rice polish

---

[1]See Chapter 23 for lipids.

TABLE 24.7. VARIATION OF THE CHEMICAL COMPOSITION OF BRAN[1] WITH THE DEGREE OF MILLING OF RICE[2]

	Bran[1] from individual whiteners and polishers					
	1st cone	2nd cone	3rd cone	4th cone	1st polisher	2nd polisher
Increment of degree of milling (g bran/100 g brown rice)	0–3	3–6	6–9	9–10	10–10.5	10.5–11
Protein	17.03	17.63	16.97	16.74	16.37	15.27
Fat	17.65	17.11	16.45	14.23	14.96	11.64
Fiber	10.51	10.73	5.72	5.67	5.09	3.66
Ash	9.82	9.37	8.35	7.49	7.23	6.02
NFE	45.0	45.2	52.5	55.9	56.3	63.4

	Accumulated bran[1] fractions					
	1st cone	1st and 2nd cone	1st to 3rd cone	1st to 4th cone	1st cone to 1st polisher	1st cone to 2nd polisher
Degree of milling (g bran/100 g brown rice)	3	6	9	10	10.5	11
Protein	17.03	17.33	17.22	17.16	17.11	17.01
Fat	17.65	17.38	17.08	16.77	16.66	16.39
Fiber	10.51	10.62	9.03	8.67	8.18	8.20
Ash	9.82	9.59	9.19	9.01	8.90	8.40
NFE	45.0	45.10	47.52	48.40	48.84	49.62

Source: Adapted from Primo et al. (1970).
[1]Includes bran, polish (if any) and embryo.
[2]Percentage, dry basis.

TABLE 24.8. PROXIMATE ANALYSIS OF RICE POLISH[1]

Constituent (%, dry basis)	Range
Protein	10.5–16.5
Fat	10.4–19.0
Fiber	2.1– 8.0
Ash	5.0–13.2
NFE	51.8–71.3

[1]Includes data from: Italy (Fossati *et al.* 1976; Leonzio 1965; Maymone *et al.* 1962), Malaysia (Arnott and Lim 1966), Spain (Primo *et al.* 1970) and the United States (Kik 1956).

were lower: 2.10–5.25% (Leonzio 1966). Pentosans ranged from 8.59 to 10.87% in bran (Leonzio 1967) and from 3.15 to 6.01% in polish (Leonzio 1967). The pentosan content of bran decreases with successive steps in milling: 11.73, 9.11, 5.88 and 4.28% dry basis were reported for the bran from four successive whitening cones of an Italian mill (Leonzio 1967).

The hemicelluloses are a complex fraction not readily resolved into polysaccharides of single sugar units (Bevenue and Williams 1956). Hemicellulose B of bran has been reported to contain 67.9% reducing sugars, primarily pentoses (59.6%). Xylose, arabinose, galactose and uronic acid were identified as major components with the first 2 being predominant (Matsuo and Namba 1958). The hemicelluloses of bran polish have been found to contain about 0.1% water soluble fraction and 1% 0.5N sodium hydroxide extractable fraction (Cartano and Juliano 1970; Gremli and Juliano 1970). The former had an arabinose: xylose ratio of 1.8 and contained galactose and protein; the latter contained arabinose and xylose, in a 1:1 ratio, as major components, and galactose and minor amounts of glucose; uronic acid was present in fair amounts. As shown by paper chromatography (Bevenue and Williams 1956), the 4% and 24% KOH soluble fractions of hemicellulose, as well as the alpha-cellulose fraction (residue from alkaline extractions), of bran (including germ), and polish have a similar qualitative sugar pattern. They contain galactose, glucose, mannose (absent in the 4% KOH fraction), arabinose and xylose.

Lignin.—Lignin content ranged from 7.70 to 13.11% in the bran, and from 2.01 to 4.42% in the polish (Leonzio 1966). It decreased from the first to the fourth cone: 12.79, 8.41, 5.94 and 4.28%, dry basis, free of fat (De Rege 1964).

Proteins and Other Nitrogen Compounds.—The nitrogen content of bran varies within a wide range, about 1–3%, dry basis. The nitrogen content is usually multiplied by the factor 5.95 for its conversion into protein content. For commercial purposes, particularly in the feedstuff industry, the conversion factor 6.25 is widely used.

The largest part of rice bran nitrogen is protein nitrogen. Nonprotein nitrogen accounts for about 16% of the total nitrogen of rice bran and for about 11% of rice polish nitrogen (Baldi *et al.* 1976). Data reported for free amino N in bran range from 22 (Tamura and Kenmochi 1963) to 300 (Sugimura and Ebisawa 1955) mg/100 g bran. Major free amino acids in bran are glutamic acid (7–31%), alanine (11–16%) and serine (5–15%) (Ebisawa and Sugimura 1956; Sugimura and Ebisawa 1955; Tamura and Kenmochi 1963). Other nitrogen compounds reported are guanine, xanthine, adenine, hypoxanthine, histidine, ammonia, dimethylamine, trimethylamine, cytosine, nicotinic acid, guanidine, betaine, choline, uracil (van Veen 1931A), allantoin in rice polish (van Veen 1931B), flavine adenine dinucleotide (0.93 γ/g) and flavine mononucleotide (0.69 γ/g) (Mitsuda *et al.* 1958).

Reported data for amino acid composition of rice bran vary within wide limits (Table 24.9). The main source of variation appears to be the analytical procedure (Houston *et al.* 1969; Mauron 1973); rice variety and type of bran are of lesser significance. Conditions of hydrolysis affect the destruction of methionine, serine, threonine, tryptophan and cystine as well as defective liberation of the branched-chain amino acids.

Sheller (under-runner disc) bran, bran and polish were remarkably similar in amino acids content (Table 24.10). Differences among deeper

TABLE 24.9. AMINO ACID COMPOSITION OF RICE BRAN[1]

Amino acid	Average[2]	Standard deviation
Lysine	3.88	0.82
Histidine	2.11	0.57
Ammonia	1.72	0.96
Arginine	6.50	1.31
Aspartic acid	7.62	2.03
Threonine	3.06	0.69
Serine	4.24	0.73
Tryptophan	1.70	0.51
Glutamic acid	12.84	2.86
Proline	4.10	1.01
Glycine	4.52	0.86
Alanine	5.67	1.22
Cystine	1.63	0.58
Valine	5.45	0.43
Methionine	2.22	0.35
Isoleucine	3.94	0.54
Leucine	6.96	1.44
Tyrosine	3.65	1.28
Phenylalanine	4.47	0.83
% N recovered	85.6	11.3

Source: Barber and Benedito de Barber (1977).
[1]United States bran (Houston and Kohler 1970), Japanese bran (Nishihara and Tashiro 1960) (Tamura and Kenmochi 1963) and Spanish bran (Ronda and Soto 1965).
[2] g/16.0 gN

layers of the kernel were not important (Houston *et al.* 1969; Kik 1956; Normand *et al.* 1966). The proteins of commercial hulls contain more proline and less histidine, arginine, glutamic acid, methionine and tyrosine than those of bran (Table 24.11) (Houston *et al.* 1969).

TABLE 24.10. AMINO ACID COMPOSITION OF BRAN, POLISH AND GERM[1]

g AA/100 g P (N × 6.25)	Bran from under-runner disc huller	Bran from whitening (cones)	Polish	Germ
Lysine	3.82	4.11	4.22	4.27
Histidine	1.29	1.34	1.50	1.48
Ammonia	2.09	2.84	1.39	3.68
Arginine	5.81	5.83	6.87	6.28
Aspartic acid	9.20	9.15	8.98	9.01
Threonine	3.25	3.10	3.60	3.50
Serine	5.07	4.98	4.61	4.96
Tryptophan	1.30	1.34	1.42	1.47
Glutamic acid	16.23	16.03	16.38	14.56
Proline	4.89	5.25	5.61	7.70
Glycine	4.01	3.93	4.42	4.33
Alanine	6.59	6.72	5.67	5.27
Cystine	1.17	1.15	1.26	1.19
Valine	5.49	5.34	5.40	5.26
Methionine	2.34	2.55	2.52	2.88
Isoleucine	4.90	4.70	4.25	3.49
Leucine	8.29	8.72	7.86	7.70
Tyrosine	4.69	4.73	5.74	4.70
Phenylalanine	5.28	5.57	6.39	4.93
% N recovered	96.78	98.60	98.64	97.58

Source: Ronda and Soto (1965).
[1]Embryo separated at the mill site.

Information on protein fractions of bran and polish is meager. The protein contents of bran (including germ) and polish of high and low protein rice varieties are as follows: bran, 13.3–15.5% and 16.0–17.4%; and polish, 12.2–14.1% and 17.9–19.0%. The major protein fractions in bran are albumin and globulin. That in polish is glutelin. Prolamin is the minor fraction in both cases. The mean ratio of albumin:globulin:prolamin:glutelin for the 6 Philippine samples is 37: 36: 5: 22 for bran (including germ) and 30: 14: 5: 51 for polish. In contrast with these are the data reported for Italian samples: the mean ratio of albumin-globulin:prolamin:glutelin is 42: 3: 55 for bran and 34: 4: 62 for polish (Baldi *et al.* 1976).

The albumin-globulin fraction of bran has been resolved by gel filtration in Sephadex G-100 into 3 components with molecular weights of 2.0×10^5, 0.8×10^5 and 0.3×10^5 (Palmiano *et al.* 1968). The globulin fraction has also been fractionated by Sephadex G-200 chromatography into 3 components: γ globulin, with a molecular weight of 1.5×10^5 is the major one (Morita and Yoshida 1968).

Cytochrome C and a blue protein have been isolated and found as the

TABLE 24.11.   AMINO ACIDS OF RICE HULLS AND MILLFEEDS AS COMPARED WITH BRAN

Amino acid[1]	Hulls	Millfeed California	Texas	Commercial U.S. bran[2]
Lysine	3.82	4.67	4.51	4.69±0.42
Histidine	1.22	2.35	2.36	2.74±0.18
Ammonia	3.42	2.82	2.78	2.01±0.20
Arginine	4.30	7.35	7.35	8.47±0.35
Aspartic acid	8.60	9.32	8.96	8.67±0.75
Threonine	4.20	3.93	3.96	3.75±0.22
Serine	4.65	4.84	4.84	4.70±0.19
Triptophan	n.d.	n.d.	n.d.	n.d.
Glutamic acid	10.42	13.09	14.02	13.67±0.85
Proline	6.50	4.79	4.92	4.17±0.22
Glycine	5.43	4.91	5.54	5.38±0.14
Alanine	6.13	6.30	6.15	6.01±0.29
Cystine	1.90	2.24	2.25	2.27±0.11
Valine	5.69	5.89	6.10	6.11±0.34
Methionine	1.76	1.91	2.12	2.45±0.13
Isoleucine	3.66	3.96	3.98	3.97±0.22
Leucine	6.47	6.85	7.04	7.00±0.37
Tyrosine	2.16	2.72	2.80	3.28±0.26
Phenylalanine	4.40	4.58	4.60	4.54±0.21
% N recovered	82.9	91.6	92.0	90.1 ±3.63
% N in sample	0.32	1.02	0.98	2.42±0.21

Source: Houston et al. (1969).
[1]g/16.0 gN
[2]Data from 5 varieties.
n.d. = not determined.

major soluble basic proteins of bran (Morita and Ida 1968; Ida and Morita 1969; Morita et al. 1971). The blue protein is a copper containing glycoprotein, with an estimated molecular weight of approximately 18,300, which occurs predominantly in the aleurone layer but rarely in the embryo and other parts of the kernel. Its physicochemical properties and chemical constituents have been investigated recently (Morita et al. 1971).

Enzymes.—Rice bran is abundant in various enzyme systems. The following enzymes have been reported to occur: α-amylase and β-amylase (Sreenivasan 1939; Yamagishi 1935; Borasio 1932; Barber 1969), ascorbic acid oxidase (Tadokoro et al. 1941), catalase (Borasio 1929; Chikasuye 1953 A,B, 1954), cytochrome oxidase (Roberts 1964; Aleshin 1961; Aleshin and Filinkoldakov 1960), dehydrogenase (Delouche et al. 1962; Aleshin 1960), deoxyribonuclease I (Mukai 1966), esterase (Obara and Ogasawara 1958), flavin oxidase (Aleshin 1960; Aleshin and Filinkoldakov 1961), α and β glucosidase, ferredoxin NADP reductase (Ida and Morita 1970B), glutamic acid decarboxylase, glutamine synthetase, glutathion reductase (Ida and Morita 1971A, B), β-glycerophosphatase, invertase (Rosedale and Oliveiro 1928), lecithinase, lipase (Funatsu et al. 1971; El Hinnawy 1961; Aizono et al. 1971,1973,1976; Shastry and Rao

1971,1976; Shibuya *et al.* 1975), lipoxygenase (Shastry and Rao 1975), NADPH diaphorase (Ida and Morita 1970B), maltase, pectinase, peroxidase (Morita and Ida 1968; Ida *et al.* 1970,1972; Barber *et al.* 1977A, B), phosphodiesterase, phosphomonoesterase, phosphatases (Shastry and Rao 1972; Uzawa 1932), phytase (Emiliani and Fronticelli 1938), polyphenoloxidase (Aleshin 1961; Aleshin and Filinkoldakov 1961), protease (Barber 1969), piridine nucleotide transhydrogenase (Ida and Morita 1970B) and succinate dehydrogenase.

The germ and outer covers of the caryopsis are the site of higher enzymatic activities. Amylolytic activities (mg maltose/10 g) of rice milling products are reported to be brown rice, 39; milled rice, 15; bran, 320; polishings, 250; germ, 310 (Borasio 1932). Catalase activity ratio for brown rice: brown rice free of germ: milled rice was found to be about 40:16:1 (Borasio 1929). Proteolytic activities of degermed brown rice (I), bran (5.9% of I), polish (4.6% of I) and germ were 1.7, 17.4, 4.9 and 31.8 hemoglobin units/g rice, dry basis (Barber 1969). Enzymes in rice bran might also be of microbial origin. Among the enzymes, lipase has merited most attention because it affects the keeping quality and extent of industrial utilization of rice bran (see section on Processed Rice Bran and Rice Bran Products).

Minerals.—Phosphorus is one of the major mineral constituents of bran (Table 24.12). Also present in decreasing order are potassium, magnesium and silicon. The level of calcium, chlorine, manganese, iron and sodi-

TABLE 24.12. INORGANIC CONSTITUENTS OF RICE BRAN

Constituent	Content[1]
Aluminum	53.5 − 369[2]
Calcium	140 − 1,310
Chlorine	510 − 970[2]
Copper	0.37
Iodine	15
Iron	130 − 530
Magnesium	8,650 −12,300[2]
Manganese	110 − 877[2]
Mercury	0.3[3]
Phosphorus	14,800 −28,680[2]
Potassium	13,650 −23,960
Selenium	0.170
Silicon	1,700 −16,300[2]
Sodium	0 − 290
Sulfur	80
Tin	17.6 − 41.3
Titanium	26
Zinc	80

Source: Ferreti and Levander (1974); Fossati *et al.* (1976); Juliano (1972); McCall *et al.* (1953); Mendoza and Sutaria (1970).
[1] γ/g, dry basis.
[2] Polish and germ included.
[3] Unpublished results.

um in rice bran is low. Barium and boron are present in trace amounts. Cobalt, chromium, germanium, lead or vanadium have not been detected spectrochemically (McCall *et al.* 1953).

The concentration of mineral elements in bran varies with the degree of milling (Primo *et al.* 1970; Tanaka *et al.* 1973). Some elements (P, K, Mg) increase initially and decrease with deeper milling; others (Ca, Mn, Fe) exhibit an early sharp decrease (Tanaka *et al.* 1973). A decreasing concentration gradient occurs in subaleurone layers (Barber 1972; Kennedy and Schelstraete 1975). The distribution of mineral elements in endosperm, germ and pericarp plus aleurone has been reported (O'Dell *et al.* 1972).

Phosphorus in bran occurs as phytic acid, nucleic acid, inorganic phosphate, carbohydrate and phosphatide. The reported values, calculated as percentages of the total phosphorus, are 89.9, 4.4, 2.5, 2.3 and 1%, respectively (McCall *et al.* 1953). The largest part of phosphorus is linked to inositol as the calcium-magnesium salt of myo-inositol hexaphosphate or phytin. Bran may contain about 1.8% of phytin (Borasio 1944). Reported phytin phosphorus contents for bran-germ range between 2.2 to 2.6% (McCall *et al.* 1953).

TABLE 24.13. VITAMIN CONTENT OF RICE BRAN

Vitamin	Content[1]		
Vitamin A (carotenes)		4.2	
Thiamin	10.1	−	27.9
Riboflavin	1.7	−	3.4
Niacin (nicotinic acid)	236	−	590
Pyridoxine	10.3	−	32.1
Pantothenic acid	27.7	−	71.3
Biotin	0.16	−	0.60
Inositol	4,627	−	9,270
Choline	1,279	−	1,700
p−Amino benzoic acid		0.75	
Folic acid	0.50	−	1.46
Vitamin B_{12}		0.005	
Vitamin E (tocopherols)		149.2	

Source: Juliano (1972).
[1] γ/g, dry basis.

Vitamins.—Bran is abundant in vitamins of the B group and tocopherols and is poor in Vitamin A and C (Table 24.13). Carangian and Sutaria (1970) reported 0.003–0.007 μg ascorbic acid/g sample. No vitamin D has been reported in bran. Total loss of ascorbic acid after 1 month of storage at 27°C (80.6°F) has been reported (Carangian and Sutaria 1970). The rapid deterioration might be the reason for the absence of this vitamin in some rice bran samples.

Vitamins, as with other constituents, are not uniformly distributed within the grain. The greatest concentration is found in bran, where the major part of the B-vitamins in the grain is located (Hinton and Shaw

1954). However, the pattern of distribution is not identical for all the vitamins. For instance, nearly 80% of the nicotinic acid and only 35% of thiamin has been found to occur in the pericarp plus aleurone layer. In the germ (scutellum plus embryo), about 2.5% of the nicotinic acid and over 55% of the thiamin are located (Hinton and Shaw 1954). Concentration levels of some vitamins, like pantothenic acid and folic acid, in rice polish may be higher than in rice bran (Kik 1956).

Data reported in the literature for the vitamin content of rice bran vary within a wide range (Table 24.13). Although the vitamin content differs among rice varieties and to a lesser extent with location of growth (Kik 1945), major causes of variations are the analytical techniques employed, the uncontrolled proportion of germ and polish and the differences in the degree of milling.

Vitamins occur in bran in free or combined form: 75% of riboflavin was reported to be in esterified form (Sone 1958), 89% of the folic acid, 49% of the pantothenic acid (Kik 1956) and 86% of niacin in the bound form. Pyridoxine was also found in free and combined form (Scudi 1942).

Chemical Composition of Rice Germ

Literature on rice germ is less abundant than that on bran. However, present knowledge of the chemical composition of rice germ is quite extensive (see also sections on Histology and Histochemistry).

Rice germ is characteristically richer in proteins and fat, but lower in fiber than the bran (Table 24.14). Commercial germ available in Spain, however, always contains 20–30% impurities, most frequently endosperm and hull fragments. Ash is an important component. As rice germ may account for 20% or more by weight of rice bran (Japonica rice), germ removal affects bran composition, particularly decreasing the fat content.

TABLE 24.14. PROXIMATE ANALYSIS OF RICE GERM

Constituent (%,dry basis)	Range
Protein	17.3–26.4
Fat	16.6–39.8
Fiber	1.98–15.1
Ash	6.07–10.8
NFE	34.7–69.3

[1]Sources: Rice Germ from Italy (Fossati et al. 1976; Leonzio 1965), Spain (Primo et al. 1970) and India (Rao et al. 1960).

Sugar Content.—Data reported on total sugar content of germ differ widely. They vary from 8% (Pascual and Primo 1955; Soldi 1946) to 25% (Fossati et al. 1976). Reducing sugars range from 1% (Pascual and Primo

1955) to 11.63% (Yusta and Santos 1953). In one case (Fukui and Nikuni 1959), the main sugar of rice embryo was sucrose, together with small amounts of raffinose, glucose and fructose; in another (Soldi 1946) glucose, fructose and sucrose comprised about 50, 30 and 20% of total sugars, respectively.

Pentosans and Lignins.—Pentosans in germ amount to about 7% by weight (Leonzio 1967, Yusta and Santos 1953) and lignin about 3–6% (Leonzio 1966, 1967). The level of pentosan in the germ is intermediate between that of bran (9.65%) and polish (4.60%). Lignin content is much lower in the germ than in the bran (10.35%) but is similar to that in the polish (3.15%).

Nitrogenous Compounds.—Nonprotein nitrogen in rice germ is about 13% of the total nitrogen, intermediate between bran and polish (Baldi et al. 1976). Total free amino acids range from 24.3 (Yusta and Santos 1954) to 106.0 (Tamura and Kenmochi 1963) mg amino N/100 g germ. Rice embryo contains about 4 times more amino N than bran. The major free amino acids are alanine (14%), aspartic acid (12%), proline (29.5%) and serine (12%) (Tamur and Kenmochi 1963).

Polyamines occur in appreciable quantities in rice germ: cadaverine 133 γ/g; putrescine, 69 γ/g; spermidine, 153 γ/g, and spermine, 141 γ/g (Moruzzi and Caldarera 1964).

Adenine, xanthine, hypoxanthine and guanine (2-amino-hypoxanthine) have been isolated from rice embryo (Hirai 1925). Nucleic acid has also been prepared from the germ (Kimura 1934A, B). About 1% ribonucleic acid has been found in the germ (Matsushita 1958).

Amino acid composition of rice germ protein varies widely (Table 24.15) (Ronda and Soto 1965; Baldi et al. 1976; Tamura and Kenmochi 1963; Kik 1954). Rice germ contains more lysine than the bran. Differences in lysine and other amino acids between the germ, bran and polish are not significant even though they fluctuate somewhat (Table 24.10).

The protein fractions in Italian rice germ have been shown to be high in the brine soluble fraction, the mean albumin-globulin: prolamin: glutelin ratio being 71:1:28 (Baldi et al. 1976). In contrast, 17.4% albumin and 20.0% globulin have been reported for Japanese rice embryo protein (Morita and Yoshida 1968). Italian data also have shown the brine soluble fraction to be much higher in the germ than in the bran, and lowest in the polish (Baldi et al. 1976).

Similar to bran globulin, embryo globulin has been fractionated into three components by Sephadex G-200 chromatogrphy. The γ-fraction, with a molecular weight of 1.5 × 10^5, is the major one (Morita and

TABLE 24.15. AMINO ACID COMPOSITION OF RICE GERM

	AA[1] g/16.0 gN	AA[2] g/16.0 gN	AA[3] g/16.0 gN
Lysine	4.27	7.01	5.05
Histidine	1.48	3.44	2.76
Ammonia	3.68	4.62	1.24
Arginine	6.28	9.64	6.86
Aspartic acid	9.01	8.85	7.24
Threonine	3.50	4.18	3.24
Serine	4.96	4.58	3.62
Tryptophan	1.47	–	1.14
Glutamic acid	14.56	14.73	11.52
Proline	7.70	4.80	3.71
Glycine	4.33	5.84	4.86
Alanine	5.27	6.27	5.52
Cystine	1.19	2.67	2.10
Valine	5.26	4.94	4.48
Methionine	2.88	1.76	1.90
Isoleucine	3.49	3.13	1.33
Leucine	7.70	6.61	2.57
Tyrosine	4.70	3.14	0.86
Phenylalanine	4.93	3.77	1.62
% N recovered	97.58	95.3	74.6

[1]Source: Ronda and Soto (1965).
[2]Source: Baldi et al. (1976).
[3]Source: Tamura and Kenmochi (1963).

Yoshida 1968). γ-globulin has been in turn fractionated into 3 components by diethylaminoethyl Sephadex A-50 chromatography (Sawai and Morita 1970A). The molecular dimension and chemical characteristics of one of these γ fractions have been reported recently (Sawai and Morita 1970B,C; Horikoshi and Morita 1975). The histones of rice embryo have also been investigated (Tsuchiya et al. 1967).

Hemoproteins (Cytochrome C and peroxidase 556 as major constituents) and some other cromoproteins (flavoproteins and a blue protein) have been isolated from rice embryo, purified and partially characterized (Morita and Ida 1968, 1972; Ida and Morita 1970A, B, 1971A; Ida et al. 1970).

Minerals.—Rice germ is high in phosphorus and low in calcium; it contains somewhat more sodium and much less silicon (Table 24.16). In pure germ 75% of the total phosphorus is present as phytate phosphorus. In pericarp and aleurone, 91% of the total phosphorus is phytate P (O'Dell et al. 1972).

Vitamins.—The values reported for most of the vitamins of the rice germ (Table 24.17) fall within the ranges reported for bran (Table 24.13). The exceptions are thiamin, which is more abundant in the germ, and niacin, which is more abundant in the bran.

TABLE 24.16. INORGANIC CONSTITUENTS OF RICE GERM

Constituent (γ/g, dry basis)	Content	Constituent (γ/g, dry basis)	Content
Calcium	$250^2 - 2,750^4$		
Chlorine	$1,520^4$	Silicon	$455^3 - 1900^4$
Copper	$10^2 - 30^2$		
Iron	$110^2 - 489^4$		
Magnesium	$5,600^1 - 15,270^2$	Sodium	$161^3 - 1400^4$
Manganese	$110^1 - 140^2$		
Phosphorus	$13,000^1 - 27,280^3$	Zinc	$100^2 - 300^2$
Potassium	$3,850^4 - 21,466^3$		

[1]Source: O'Dell et al. (1972).
[2]Source: Primo et al. (1970).
[3]Source: Fossati et al. (1976).
[4]Source: Juliano (1972).

TABLE 24.17. VITAMIN CONTENT OF RICE GERM

Constituent (γ/g, dry basis)	Content
Vitamin A (carotenes)	1.3
Thiamin	45.3 — 76.0
Riboflavin	2.7 — 5.0
Niacin (nicotinic acid)	15.2 — 99.0
Pyridoxine	15.2 — 16.0
Pantothenic acid	13.2 — 30.0
Biotin	0.26— 0.58
Inositol	3725 —6400
Choline	2031 —3000
p-Aminobenzoic acid	1
Folic acid	0.9 — 4.3
Vitamin B_{12}	0.01
Vitamin E (tocopherols)	87.3

Source: Juliano (1972).

Processed Rice Brans

The composition and properties of various processed rice brans—parboiled, stabilized, defatted, dephytinized and adulterated—have been reviewed recently (Barber and Benedito de Barber 1977).

Parboiled rice bran is higher in oil content than the raw bran (Table 24.18), but is lower in thiamin and riboflavin. Differences in composition between raw and parboiled brans depend upon the degree of milling and parboiling conditions (Benedito de Barber et al. 1977).

Stabilization of bran by dry and moist heat treatment does not affect its gross chemical composition (see section on Processed Rice Bran and Rice Bran Products).

Removal of oil from regular bran by pressing or solvent extraction

TABLE 24.18. COMPOSITION OF RAW AND PARBOILED RICE BRAN FROM CEYLON

Analysis[1]	Bran from raw rice		Bran from parboiled rice	
	1st quality[2]	2nd quality[3]	1st quality[2]	2nd quality[3]
Dry matter	89.1	88.9	89.9	90.1
Crude protein	12.5	8.9	12.8	9.1
Oil	14.6	6.5	20.3	10.4
Crude fiber	8.9	19.1	10.0	17.1
Ash	10.7	19.7	10.7	21.2
NFE	42.9	34.7	36.1	32.9

Source: Siriwardene (1969).
[1]Percentage wet basis.
[2]From cone type mill.
[3]From huller type mill.

brings about proportional increases in the concentration of all major constituents except water. Bran from the solvent extraction milling process (X-M process, Hunnell and Nowlin 1972) has been reported to be higher in protein content and lower in fiber content than expected from the simple removal of fat.

Dephytinized bran has a lower ash and nitrogen free fraction, and higher protein, lipids and crude fiber than regular bran (Table 24.19).

TABLE 24.19. GROSS CHEMICAL COMPOSITION OF DEPHYTINIZED AND DEFATTED DEPHYTINIZED BRAN

Constituent[1]	Untreated rice bran	Dephytinized rice bran	Defatted dephytinized rice bran
Protein	16.6	17.2	21.3
Fat	19.7	25.4	2.7
Crude cellulose	7.5	14.1	17.0
Ash	10.7	3.2	4.0
NFE	45.6	40.1	55.0

Source: Maymone et al. (1961).
[1]Percentage, dry basis.

Some rice bran may be adulterated with hulls, corn cobs, peanut shells, wood sawdust and clay. Urea is used to increase the nitrogen content. Hulls, the most frequent adulterant, increase the lignin, silica and ash content of bran.

PHYSICAL PROPERTIES OF RICE BRAN

Information on the physical characteristics and properties of bran is necessary for appropriate processing of the byproduct in stabilizing, drying, handling and storage. Available information described by Houston (1972A) and Barber and Benedito de Barber (1977) is summarized as follows.

Particle size distribution of bran varies within a wide range. Data has been published for bran from friction vs abrasion type milling machines, different cones (Barber and Benedito de Barber 1977), raw vs heat stabilized rice (Barber and Maquieira 1977), raw vs parboiled rice (Rao et al. 1967A), and defatted bran and germ (Barber and Benedito de Barber 1977). They have shown that:

(1) Friction type machines produce a bran of larger particle size than the abrasion type.
(2) There is no clear relationship between fineness and cone number; notwithstanding this, the percentage of fine particles increases in the bran fraction of deeper layers. Distribution patterns for brans from different mills differ remarkably.
(3) Moist heat stabilization brings about agglomeration of rice bran particles (Table 24.20).

TABLE 24.20. PARTICLE SIZE DISTRIBUTION OF RAW AND HEAT STABILIZED BRANS[1]

Mesh[2]	μ	Raw	Moist heat stabilized[3]
>18	> 1000	0	0
18– 30	1000–595	2.4	18.6
30– 50	595–297	30.0	32.7
50– 80	297–177	12.2	18.5
80–100	177–149	8.5	10.8
< 100	< 149	46.7	19.4

Source: Barber and Maquieira (1977).
[1]Japonica rice type; cone mill.
[2]United States standard.
[3]Stabilized according to Barber et al. (1977): 3 min steaming, followed by flash drying and cooling.

(4) Bran from parboiled rice appears to be flakier, with larger flake size, than bran from raw rice (Table 24.21). Quantitative differences among varieties are large.
(5) Commercial rice germ comprises a wide range of sizes (Table 24.22).

The bulk density of rice bran varies from 0.2 to 0.4 g/cc (Table 24.23). That for purified commercial germ, Japonica type rice, is 0.51 g/cc. The density of first-break bran has been reported to be 1.29 and that of rice germ 0.9 (Kik 1957).

The angle of repose in piled material (the angle formed between the heap of bran and the horizontal surface) has been reported to be 38° (ARGCA 1968) which is within the range of variation of that for paddy and rice.

Bran exhibits moisture absorption and desorption properties. Data of equilibrium moisture content of bran, polish and moist-heat stabilized bran (Tables 24.24 and 24.25) are relevant since the stability of the

TABLE 24.21. SIEVE ANALYSIS OF RAW AND PARBOILED BRAN

Sieve[1]	Bangara Sanna Raw (%)	Bangara Sanna Parboiled (%)	Sonakathi Raw (%)	Sonakathi Parboiled (%)	Gurmatia Raw (%)	Gurmatia Parboiled (%)	Safri Raw (%)	Safri Parboiled (%)
−22	93.5	76.6	93.0	82.5	87.5	60.0	86.3	38.3
−18	5.7	10.6	6.3	15.0	12.0	32.5	12.7	49.3
−16	0.0	5.3	0.0	2.5	0.0	3.7	0.0	5.5
+16	0.0	6.8	0.0	0.0	0.0	3.1	0.0	6.2

Source: Rao et al. (1967A).
[1]Mesh size: British system standards.

TABLE 24.22. SIZE DISTRIBUTION OF RICE GERM

Commercial germ[1]		Purified germ[1,2]			
		Long axis		Short axis	
Size (μ)	% weight	Size (mm)	% weight	Size (mm)	% weight
< 149	0	< 1.65	9	< 0.90	9
149–420	0.3	1.65–1.80	32	0.90–1.05	34
420–595	0.6	1.80–1.95	35	1.05–1.20	48
595–840	23.6	1.95–2.10	10	1.20–1.35	8
> 840	75.7	2.10–2.25	14	1.35–1.50	1

Source: Barber and Benedito de Barber (1977).
[1] Japonica type rice. Different varieties.
[2] Sample purified by screening and pneumatic classification, defatted with petroleum ether (60°–70°C), (140°–158°F) fraction, and dried at room temperature.

TABLE 24.23. BULK DENSITY OF RICE BRAN AND GERM

Byproduct	g/cc
Bran[1]	
Raw[2]	0.32
Stabilized[2]	
powder	0.41
pelletized (10−20 × 7−8 mm)	0.50
Bran[3]	
First cone	0.19−0.21
Fourth cone	0.38−0.43
Germ[1]	
Purified, commercial, Japonica rice type	0.51
Impurifying endosperm fragments	0.83

[1]Source: Barber and Benedito de Barber (1977).
[2]Same lot of bran. Moist-heat stabilized.
[3]Source: Silvestre et al. (1960).

TABLE 24.24. EQUILIBRIUM MOISTURE CONTENT OF BRAN AND POLISH

Byproduct	Equilibrium moisture content at 25°C (kg water/100 kg bran)								
	10	20	30	40	50	60	70	80	90
Bran[1]	5.0	6.4	8.0	9.0	10.0	11.0	12.4	14.8	18.0
Bran[2]	4.6	5.8	6.6	7.4	8.3	9.2	10.6	—	—
Polish[1]	5.3	7.0	8.2	9.2	10.1	11.0	12.4	14.5	18.0

Source: Karon and Adams (1949).
[1]Field dried rough rice; initial moisture 16.8%.
[2]Rice artificially dried; initial moisture 12.8% From undermilled rice.

TABLE 24.25. EQUILIBRIUM MOISTURE CONTENT OF HEAT-STABILIZED RICE BRAN[1]

Relative humidity	Equilibrium moisture content (kg water/100 kg bran)			
	Temperature			
	20°C	25°C	30°C	35°C
0.11	5.6[1]	5.5	5.4	5.2
0.23	7.3	7.1	6.8	6.5
0.33	7.7	7.8	7.5[2]	7.1[2]
0.42	8.8[2]	8.7	8.3[1]	8.0[3]
0.52	9.3	9.6[1]	9.0	8.7[2]
0.57	10.2	9.8	9.6	9.3
0.67	11.2[1]	11.0	10.6	10.4
0.75	13.0	12.7	12.4	12.1
0.84	16.2[3]	15.8[3]	15.0	14.7
0.92	24.6[3]	21.3[2]	19.8	19.1[2]

Source: Adapted from Fito and Sanz (1974).
[1] + 0.1
[2] − 0.1
[3] + 0.2

byproducts is affected by their moisture content and this varies with environmental conditions and the gain or loss in moisture which takes place during storage.

PROCESSED RICE BRAN AND RICE BRAN PRODUCTS

The utilization of rice bran as food and feed is limited by its instability and high fiber content. In order to solve the problems, alternative processes and products have been developed: (1) rice bran stabilization, and (2) rice bran fractionation processes.

Rice bran oil extraction is dealt with in Chapter 23. Parboiled, defatted and dephytinized brans are included elsewhere (see section on Chemical Composition and Chemical Constituents of Bran Polish and Germ).

Rice Bran Stabilization

Fundamentals of Rice Bran Stabilization.—Bran contains valuable components such as oil, proteins, vitamins and essential minerals. It also contains enzymes, microorganisms, insects, natural toxicant constituents, harmful contaminants and adulterants. Some components have to be preserved; others must be removed or their activity arrested.

Enzymes, microorganisms and insects are major causes of deterioration of rice bran. Lipases and to a lesser extent oxidases, are responsible for the deteriorative changes. Lipases promote the hydrolysis of the bran oil into glycerol and free fatty acids (FFA). An extensive study on this reaction has been reported by De Rege (1953, 1954, 1955). The location of lipases in rice grain and its characteristics have been reviewed recently (Desikachar 1977). In the intact grain, lipases are dormant. The enzyme and the substrate are not together in the resting grain. Lipases are localized in the testa/cross layer of the rice grains (Shastry 1973) while the oil is in the aleurone and subaleurone layers and germ. A similar compartmentation should occur in the germ where 60% of the total lipase activity of the grain has been found (Shibuya et al. 1975). When bran is scoured during rice milling, the enzyme and substrate are brought together and oil deterioration starts. The rate of free fatty acid release is very high, depending on environmental conditions (Loeb et al. 1949; De Rege 1953, 1954, 1955). Rice bran lipase has been isolated and characterized (Funatsu et al. 1971; Aizono et al. 1971, 1973, 1976; Shastry and Rao 1971). It has a molecular weight of 40,000. The enzyme is activated by low concentration of Ca^{2+}, and inhibited by heavy metals. Its optimum pH is between 7.5 and 8.0, and the optimum temperature is about 37°C (98.6°F). Heating at 60°C (140°F) for 15 min inactivates the enzyme in solution. Under certain conditions, oxidases can cause deterioration of oil and other bran constituents. Thus, peroxidase causes oxidative spoilage of bran components (oil, tocopherols) at low moisture levels. Some enzymes, like peroxidases, can regenerate their activity after total deactivation if appropriate provisions are not made (Saunders *et al.* 1964). Other enzymes, like lipases whose activities have been suppressed by combining partial thermal inactivation and dehydration, are reactivated when moisture content increases. Commercial bran has a high microbial population, frequently exceeding several million microorganisms per gram (Barber *et al.* 1977A). Molds, including heat resistant spores, which are capable of producing active lipases and mycotoxins, are always present. Insects, whether adults, larvae or eggs, can cause spoilage, and are usual contaminants of commercial rice bran.

Bran contains toxic constituents which can cause inhibition of growth and/or a decrease in food efficiency (Liener and Kakade 1969). Trypsin inhibitors and hemagglutinins in bran have been investigated extensively. More than 90% of total antitrypsin and hemagglutinating activities of the caryopsis are located in the germ (Barber *et al.* 1978; Benedito de Barber 1978).

In order to process bran into a food grade product of good keeping quality and high industrial value, all the components causing deterioration must be removed or their activity arrested. Important in this respect is that enzyme inactivation must be complete and irreversible. At the same time, the valuable components must be preserved. Enzymes, microorganisms, insects and natural toxicants in bran are heat-labile. Although other inactivating agents are known, the application of heat is the only method that is safe and effective.

Recent studies have shown that *in situ* heat resistance of rice bran enzymes depends upon temperature and time of treatment as well as on moisture content, the latter being a critical parameter. The higher the moisture content, the lower the heat resistance (Fig. 24.2) (Barber *et al.* 1977A). Measurement of residual peroxidase activity has proven to be a reliable and convenient method to assess effectiveness of stabilization (Barber *et al.* 1977A). Peroxidase appears to be the most heat-stable enzyme in plants (Scott 1975) and is more resistant to heat than lipase and other enzymes in bran (Rothe 1967).

Processes and Equipment for Rice Bran Stabilization.—Most of the processes involve dry or moist heat-treatment. Use of chemicals (Arnott and Lim 1966; Loeb *et al.* 1949; Desikachar 1977; Rao *et al.* 1967B; Gomez and Primo 1953) or gamma irradiation as well as storage under low temperature (Loeb *et al.* 1949; David *et al.* 1965; Murthy 1966) and/or inert atmospheres (Loeb *et al.* 1949) have also been suggested, although they have never been accepted as reliable, practical procedures. Bran from the solvent extraction milling process (X-M process, Hunnell

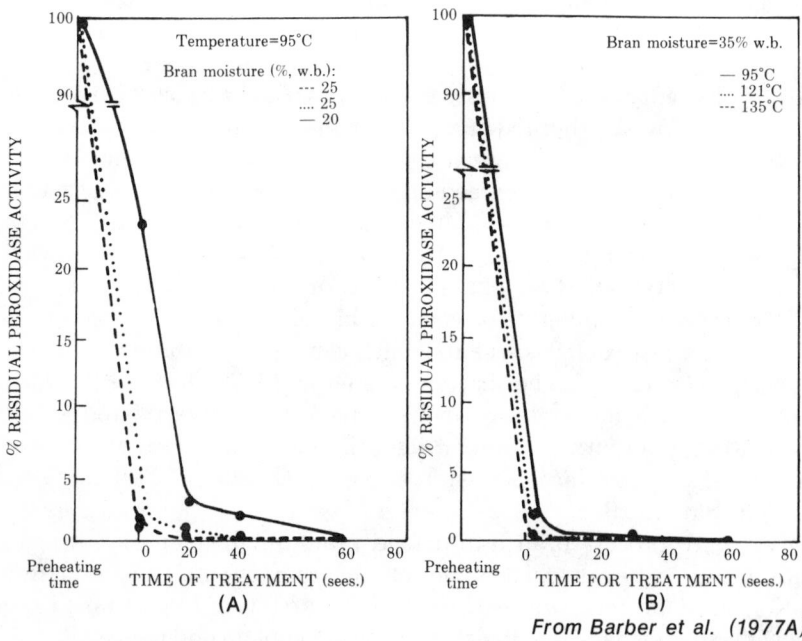

From Barber et al. (1977A)

FIG. 24.2. HEAT INACTIVATION OF RICE BRAN PEROXIDASE

(A) Influence of the treatment temperature. (B) Influence of the moisture content of bran.

and Nowlin 1972) still contains a small proportion of oil and some lipase activity (Lynn 1969).

The use of dry heat to stabilize rice bran has been investigated extensively and papers published on the subject are numerous. Recommended treatment conditions differ dramatically and results are very often discordant. In general, dry heat does not inactivate lipases totally (West and Cruz 1933; Yokochi 1977). Moisture levels of 3–6% are recommended to prevent an FFA rise (Yokochi 1977; West and Cruz 1933; Rao et al. 1967A; Loeb et al. 1952). Such conditions favor oxidative deterioration.

Heating the bran at 120°C (248°F) for 15 min in a pan with stirring did not arrest lipase activity; 20 min at 110°C (230°F) was needed for inactivation (Srimani et al. 1977). Experiments with heating bran in the oven at temperatures ranging from 70° to 200°C (158° to 392°F) and times from 5 min to 6 hr have been conducted (Arnott and Lim 1966; Gomez and Primo 1953; Loeb et al. 1949; West and Cruz 1933; Yokochi 1977). A definite increase in FFA in the bran was noted in most cases after a few weeks of storage. Heating treatments in a steam heated rotary dryer (Rao *et al.* 1967A) and jacketed drum shaped apparatus with paddle stirrer (West and Cruz 1933), ranging from 80° to 125°C (176° to 257°F) and from 10 min to 2 hr have been reported. An electrically heated, vertically rotating drum, intended to operate where facilities for raising steam are not available, has also been tested (Viraktamath and Desikachar 1971). After preheating and driving out the air, the temperature was raised and maintained 5 min at 110°C (230°F); moisture of the bran was reduced to about 7–8% by releasing the steam which evolved. The total processing time was 20–25 min. A pneumatic conveying dryer was used for stabilizing rice bran while bringing down its moisture content below 4% (Rao *et al.* 1965). Their recommended temperature was about 200°C with air rate around 200 cu m/hr. Above 250°C (392°F) the bran chars in 1 sec. Another process (Ramkrishniah *et al.* 1973) involves heat treatment of the bran in a fluidized bed, in which a stirrer rotating at low speed obviates channelling difficulties caused by the poor fluidization characteristics of rice bran; 4 min at 105°C (221°F) are reported to cause effective stabilization for 20 days. Although lowering of moisture content appears to be a major reason, the residual moisture was not reported. A bran stabilizer using the principles of combined phase fluidization—a combination of dense (sand) and dilute (bran) phase fluidized beds—has been designed to facilitate heating of all bran particles uniformly with a high heat transfer coefficient (Chand and Gupta 1975). Lighter particles (bran) are carried by the fluidizing air stream around the particles in dense phase (sand). A residence time of one second in the fluidizing column at 110°C (230°F) is reported to be sufficient to stabilize the bran. In addition, the possibility of applying frictional heat to accomplish the objective has been investigated, using a "Handler" type oil expeller unit (Viraktamath and Desikachar 1971), an Engleberg type rice mill and a plate grinder mill with minimum clearance (Desikachar 1977). All the units present serious operational difficulties involving compacting and hardening of the bran.

The drawbacks common in these methods are: (1) severe processing conditions capable of damaging the valuable components of bran, (2) substantial moisture removal with high calorie consumption, and (3) complete and irreversible inactivation of enzymes not achieved. The moisture content of the bran must be kept low which involves operational difficulties in industrial practice.

Moist heating processes generally involve steaming the bran for 1–30 min, drying the product to 3–12% moisture content and then cooling. Cooking and extrusion under high pressure is another alternative. It is generally recognized that moist heat is more effective than dry heat. Out of the many processes using steam, few have achieved satisfactory results. To achieve proper stabilization, every discrete particle of the bran must have a proper moisture content depending on the processing temperature and time. Bran agglomerates with moistened surfaces but dry cores are usually formed. When properly performed, the method of steaming bran for 3 min at 100°C (212°F), followed by drying to the initial moisture content and cooling, can yield satisfactory results (Barber et al. 1977A).

For moist heat treatment, several types of equipment have been used, including: steam cookers and blanchers (Viraktamath 1974; Roberts et al. 1949), autoclaves (Loeb and Mayne 1952), tempering and preconditioning units for oil expellers and parboiling steaming kettles (Viraktamath 1974), screw conveyors (Yokochi 1977; Burns and Cassidy 1949; Barber et al. 1977) and screw extrusion units with injected steam and/or water (Williams and Baer 1965). Treatment time ranges from 1 min to 3 hr and the temperature ranges from 95° to 135°C (203° to 275°F). A U.S. patent (Burns and Cassidy 1949) describes a stabilizer unit consisting of a 27.4 m screw conveyor, the first 6.1 m of which are subject to live steam injected through a perforated pipe in order to agitate the bran. The succeeding 18.3 m of the conveyor are heated with both live steam and external heat. The remaining 3 m of the conveyor serve to cool the material. Bran passing through the 3 sections is steamed at 100°–121°C (212°–250°F) for 1.5 min, its moisture content being raised by 3 to 5 percentage points. The product is then held for 3 min at 102°–104°C (215.6°–219.2°F) and finally cooled. A Japanese plant (Yokochi 1977) consists of 3 successive screw conveyors for the successive steaming (3 min at 95°C, 203°F) drying and cooling (3–4% moisture content) of the bran. A Spanish continuous stabilizer (Barber et al. 1977A; Fito et al. 1977) consists of a U-shaped trough mixer with live steam injected through its perforated bottom, an insulated screw conveyor, a flash dryer and a flash cooler. The bran is heated at about 100°C (212°F) for 3 min, during which its moisture content is raised by 4 percentage points. The product is then dried to its initial moisture content and cooled. Total processing time is about 4 min. Smaller units have a capacity of 50–125 kg bran/hr. In a plant using the extrusion method (Williams and Baer 1965), the bran is conveyed through the barrel by a continuous worm shaft toward the discharge plate, with water and steam being added, and the bran is finally extruded from the expander through the die plate where the expanded bran is dried and cooled. The bran is heated at 121°–163°C (249.8°– 325.4°F) and its moisture content raised by approximately 23% wet basis. After expansion, the product is dried to 8–10% moisture. Some plants (Expandolex) have capacities ranging from 10–300 MT/day.

Post-stabilization Technology.—"Stabilization" is generally understood to mean the heat processing only. However, "stabilization" in a broader sense includes the process which commences with heat treatment of the bran after milling and terminates with its use as an animal feed or as a raw material in further processing (UNIDO 1977). Heat treatment of bran is therefore only a part of the process. In fact, well stabilized bran has excellent keeping qualities with adequate protection from microbial, insect and other pest attacks. Like wheat flour and other food products, heat stabilized bran demands appropriate storage technology. Research and development work on rice bran "stabilization" has never gone beyond the heat stabilization stage. Basic and applied knowledge on transportation, packaging, storage, insecticidal treatments, mold growth prevention, rodent protection and their effects on the quality of the product are meager. Application of existing knowledge on other commodities will be most helpful; however, there is an urgent need to develop an appropriate post-stabilization technology for rice bran in regions where it is handled. Otherwise, the many valuable efforts made in developing a successful rice bran stabilization technology will be wasted.

Bran Fractionation Processes

Approaches to solving the problem of the high fiber content of rice bran have been oriented toward: (1) removal of more fibrous particles, and (2) isolation of proteins and/or protein-rich particles. Both dry and wet fractionation methods have been developed and they are dealt with below. A recent paper reviews this subject (Barber and Benedito de Barber 1977).

Dry Fractionation Processes.—The separate collection of the individual bran streams from the different whitening cones and the separation of germ from bran (see section on Rice Milling Technologies and Rice Bran Production) are probably the first attempts to fractionate bran. The differences among bran streams today do not justify the individual segregation. Only rice polish is sometimes collected separately.

In spite of the considerable range of particle sizes in bran, fractionation by dry sieving has not been proven successful (Barber and Benedito de Barber 1977). Grinding of defatted bran followed by air classification permits segregation of a high protein fraction containing 15% protein as compared with 10.6% in the original bran (Houston and Mohammad 1966). However, its fiber content is high (6.5%), the yield is low (25%) and the processing costs are high.

Wet Fractionation Processes.—Existing methods may be classified into 3 groups: (1) alkaline extraction (Chen and Houston 1970; Connor *et al.* 1977; Lew *et al.* 1971, 1975; Lynn 1969; Mitsuda *et al.* 1970, 1973, 1977A; Saunders *et al.* 1974; Youseff *et al.* 1974), (2) water extraction/sedimentation (Barber et al. 1977B; Desikachar and Parpia 1970; Mihara 1970) and (3) organic solvent sedimentation (Mitsuda et al. 1977B).

Alkaline Extraction Process.—Typical processes basically consist of alkaline extraction of the proteins, separation of the liquid phase and isoelectric precipitation of the proteins (Fig. 24.3, Chen and Houston 1970). Sodium hydroxide and hydrochloric acid have been found to be the most effective and economical agents for these steps (Lynn 1969). Addition of an anion exchange resin has been suggested to counteract the effect of phosphate and phytate, and to increase the final protein yield (Mitsuda et al. 1973). After isoelectric precipitation, a further amount of protein can be recovered by heat coagulation.

The effects of extraction conditions—pH (2 to 11), temperature (30° to 80°C, 86° to 176°F), solid-to-liquid ratio (1:30 to 1:3, weight/volume), time (2.5 min to 6 hr), particle size (<1000, 354, 210 and 149 μ) and ionic strength (0 to 0.06, using calcium sulfate)—on the nitrogen extractability of full-fat raw and heat-stabilized bran have been recently investigated (Barber and Maquieira 1977). Extractability increased with pH within the range 5–11 for raw bran and 7–11 for heat-stabilized bran (Fig. 24.4) and also with temperature of extraction within the pH range 9–11. The increase was greater in the case of heat-stabilized bran. At 80°C (176°F) and pH 9, the extractability of raw and stabilized brans were similar. Extractability was little affected by the solid-to-liquid ratio, time of extraction and particle size within the cited ranges, but it decreased with increasing ionic strenth. Grinding significantly increased N extraction at the pH of rice germ (Betschart et al. 1977). The extractability of defatted bran (Chen and Houston 1970) as well as the solubility of extracted rice bran protein (Connor et al. 1977) have also been studied.

Exposure of proteins to alkali may cause chemical changes, loss of nutritive value and formation of toxic compounds (DeGroot and Slump 1969). Cystine and serine residues may be converted to dehydroalanine and this may form lysinoalanine with the E-amino groups of lysine. The combination of heat and alkali results in severe degradation of the protein and undesirable flavor (Lynn 1969).

Although the protein content of concentrates is fairly good as compared with bran, the total yield is low (Table 24.26). The color of the rice protein concentrate varies from greenish at alkaline pH to a light tan at acid pH (Lynn 1969).

The amino acid composition of protein concentrates of rice bran does not differ substantially from that of the original bran (Connor et al. 1977). The protein efficiency ratio (PER) and nitrogen digestibility of acid precipitated (25.3% protein content) and heat precipitated (23.5%

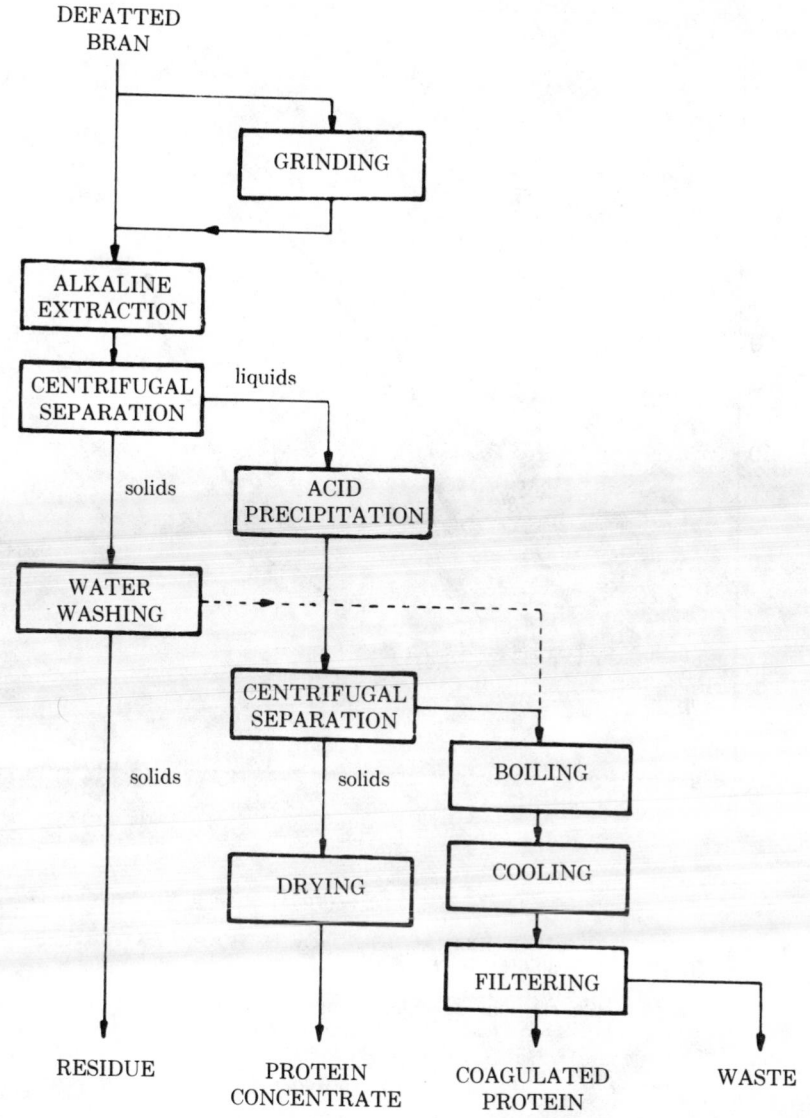

From Chen and Houston (1970)

FIG. 24.3. FLOW CHART OF WET–ALKALINE PROCESS FOR RICE BRAN PROTEIN CONCENTRATES

protein content) rice bran concentrates (1.99 and 2.19 vs 2.50 for casein and 89.6 and 83.4, respectively) compare favorably with values for other plant protein concentrates (Saunders *et al.* 1974; Connor *et al.* 1977). A protein concentrate (70% protein content) has been reported to have a

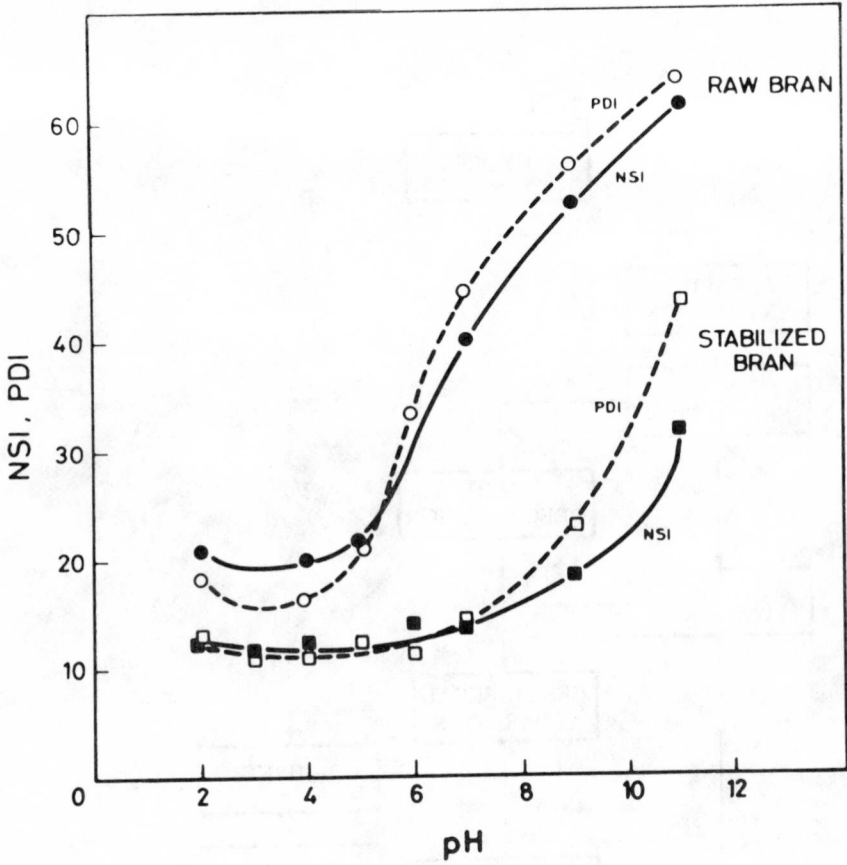

From Barber and Maquieira (1977)

FIG. 24.4. EFFECTS OF pH OF EXTRACTION UPON NITROGEN EXTRACTABILITY OF RAW AND HEAT STABILIZED RICE BRAN

NSI: Nitrogen solubility index

$$(NSI = \frac{\% \text{ Water soluble nitrogen} \times 100}{\% \text{ Total nitrogen}})$$

PDI: Protein dispersibility index

$$(PDI = \frac{\% \text{ Water dispersible protein} \times 100}{\% \text{ Total protein}})$$

ratio of essential amino acids above the Food and Agriculture Organization of the United Nations requirements and a PER of 2.6 vs 2.85 for the casein control (Lynn 1969).

Water Extraction/Sedimentation Processes.—In the Japanese process (Mi-

TABLE 24.26. YIELD AND COMPOSITION OF WET ALKALINE PROCESS RICE BRAN CONCENTRATES

Original bran and concentrate	Extraction pH	Yield		Composition[3]						Reference
		Weight[1]	Protein[2]	Protein[4]	Fat	Fiber	Ash	NFE	Starch	
Defatted rice bran	—	100	100	12.1	7.0	8.2	8.9	63.8	—	Youseff et al (1974).
Concentrate	5	45.5	65.1	17.3	5.8	6.8	5.8	64.3	—	
Full fat rice bran	—	100	100	12.0	13.7	14.4	12.1	47.8	25.4	Connor et al (1977).
Concentrate	9	20.5	38.8	22.7	32.7	0.7	11.7	32.2	22.9	
Full fat rice bran	—	100	100	14.9	17.8	9.6	8.6	49.1	—	Barber and Benedito de Barber (1978).
Concentrate	9	18.0	37.8	31.3	46.3	0.9	4.7	16.8	—	
Full fat rice bran	—	100	100	11.5	12.8	11.5	9.6	54.6	—	Lew et al. (1975).
Concentrate	11	6.6	43.7	76.1	14.1	—	3.5	—	—	

[1] g protein concentrate/100 g original bran, dry basis.
[2] % protein recovered.
[3] %, dry basis.
[4] N × 5.95
[5] 20 g bran extracted with 180 cc H_2O + 1.6 cc 19% NaOH sol.

hara 1970) the rice bran is ground in a suitable amount of water and the solids separated by centrifugation. The remainder is a colloidal solution containing protein oil compounds and also water insoluble phytin, proteins, vitamins and carbohydrates. The protein oil compound is chemically coagulated and can be separated by filtration. Dehydration and de-oiling produce an oil fraction and a white or grayish-white protein powder. The filtrate is further fractionated into phytin and a vitamin syrup by adding a phytin precipitant. The solid residue of the first centrifugation is further fractionated into a final residue and a mixture of starch and phytin, from which the latter is solubilized by acidic aqueous solution and recovered by precipitation. Yield and composition of products are shown in Table 24.27. A Spanish process (Barber *et al.* 1977B) is characterized by selective grinding in disk, rotating or colloidal mills, and fiber removal by means of sieving through vibrating or stationary meshes or screens and/or hydrocyclones, all in a water medium, with final segregation of 3 basic fractions: (1) a high protein, low fiber flour; (2) an aqueous concentrate of nutrients, mainly vitamins and minerals; and (3) a high fiber, medium protein meal (Fig. 24.5) (Table 24.28).

TABLE 24.27. NW PROCESS: YIELD AND ANALYSIS OF VARIOUS PRODUCTS

	Product					
	Protein	Starch	Phytin	Residue	Vitamin syrup concentrate	Rice oil
Yield (kg)	110	145	80	250	220	165
Water (%)	8.56	5.40	–	2.21	–	–
Protein (%)	78.5	1.61	0	13.1	11.2	–
Fats and Oils (%)	1.75	0.49	0	6.76	–	–
Cellulose (%)	0	1.67	0	26.3	0	–
Ashes (%)	2.9	0.67	41.2	2.60	–	–
Carbohydrates (%)	8.28	90.1	0	48.8	87.3	–

Source; Mihara (1970).

Fractional Sedimentation in n-Hexane.—The process (Mitsuda *et al.* 1977B) is intended to be attached to the conventional rice bran oil production system. It involves disintegration of the bran with 2 volumes of n-hexane in a high speed blender and subsequent sieving and centrifugation. A brown fraction, between 60- and 250-mesh screens, which contains the greatest amount of cellulose in rice bran, is finally separated by a basket centrifuge. A white fraction, passing through the 250-mesh screen, is obtained by sedimentation and continuous centrifugation of the supernatant. The white fraction amounts to 35–40% of original bran and contains around 22% protein, 50% carbohydrate, 4% fiber and 20% ash, dry basis. PER values of white and brown fractions are 1.80 and 1.31 as compared with 1.50 for the original bran.

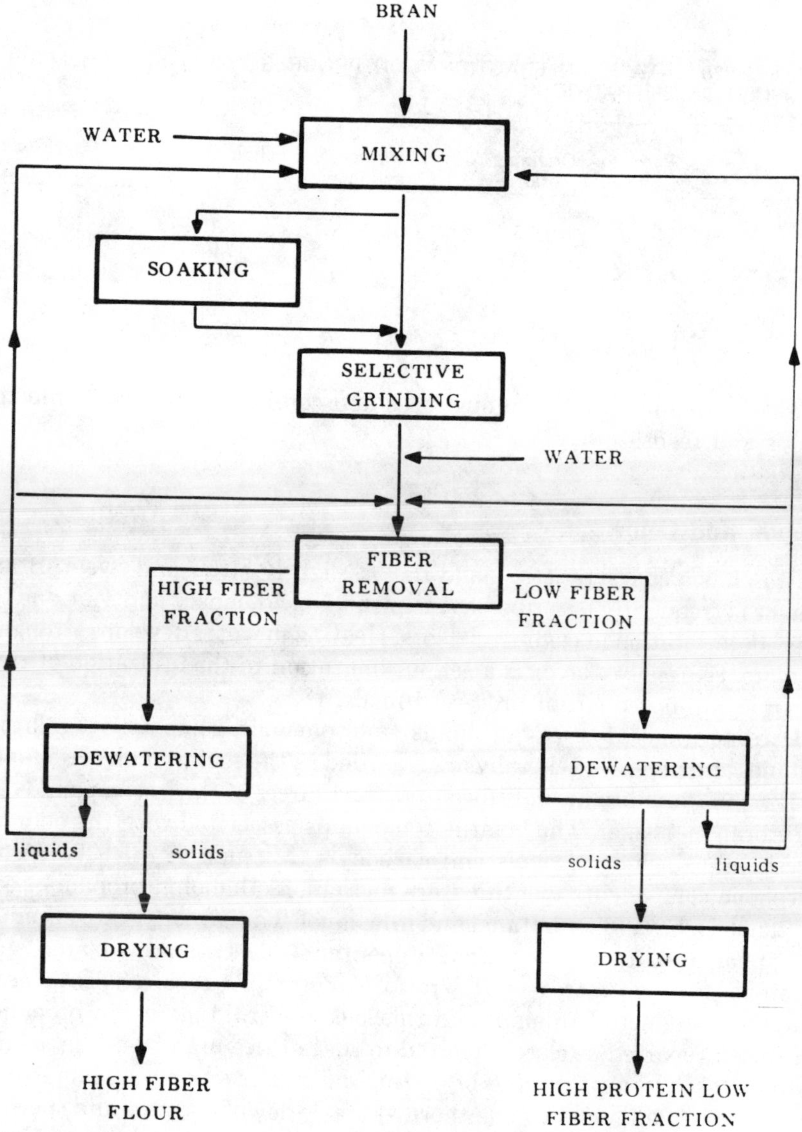

BRAN

WATER ——→ MIXING

SOAKING

SELECTIVE
GRINDING

WATER

FIBER
REMOVAL

HIGH FIBER
FRACTION

LOW FIBER
FRACTION

DEWATERING

DEWATERING

liquids   solids

solids   liquids

DRYING

DRYING

HIGH FIBER
FLOUR

HIGH PROTEIN LOW
FIBER FRACTION

From Barber et al. (1977B)

FIG. 24.5. FLOW CHART OF IATA RICE BRAN FRACTIONATION PROCESS

FACTORS AFFECTING FOOD/FEED USES OF RICE BRAN AND RICE BRAN PRODUCTS

Flavor and color, functional properties and nutritional properties are

TABLE 24.28. IATA BRAN FRACTIONATION PROCESS: YIELD AND CHEMICAL COMPOSITION OF FRACTIONS

		Product		
	Original bran	Low fiber fraction	High fiber fraction	Liquor
Yield[1]	100	57	32	11
Protein	14.5	16.9	13.1	6.1
Fat	14.6	17.1	13.0	6.3
Fiber	9.5	3.0	17.8	0.5

Source: Barber et al. (1977B).
[1]Percentage, dry basis.

the 3 major factors determining uses and consumption of rice bran in foods and feeds.

Flavor and Color

Bran has a characteristic bland flavor, slightly bitter and sweet. Bran flavor is frequently described as incipient, rancid, musty and sour due to easy deterioration in commercial lots. Heating in water develops stronger flavors. Generally, flavor is a serious limitation to the utilization of rice bran as an ingredient of processed foods.

Deterioration products of lipids and phenolic compounds, volatile amines and peptides generally are regarded as the source of objectionable flavors in grain products. Information on sources of flavor compounds in rice bran is meager. The volatile compounds associated with the odor of fresh bran include alcohols and carbonyls (Mitsuda et al. 1968). The alcoholic components identified are methanol, ethanol, n-propanol, sec-butanol, isobutanol, n-butanal and n-hexanol. The carbonyls are ethanal, propanal, isobutanal, n-butanal, isopentanal, n-pentanal, hexanal and acetone. Storage increases the ratio of carbonyls to alcohols. A degradation product of thiamin and thiazole, 3-aceto-3-mercapto-propanol has been reported as closely related to that of rice bran (Watanabe and Asahi 1957). In heat treated bran, the Maillard reaction is an important source of usual flavors by the thermal break-down of sugars and specific flavors by the decarboxylation of amino acids. Four substances, furfural, 5-methyl-furfural, furfurylalcohol and 4-vinylguaiacol seem to be the principal aroma constituents of roasted bran (Kasahara 1976).

Compounds responsible for the characteristic taste of rice bran are not known. The sweet taste is ascribed to the relative high sugars content of bran and germ (see section on Chemical Composition and Chemical Constituents of Bran Polish and Germ). The bitter taste is presumably associated, in part, with saponins which have been identified in rice bran

(Benedito de Barber and Tortosa 1978). Major sources of it are expected to be deterioration products, presumably from lipids and/or from proteins (the hydroperoxides of linoleic and linolenic acids (Kalbrener et al. 1974), and oxidized phosphatidylcholines (Sessa et al. 1974), as well as amino acids and peptides (Eriksen and Fagerson 1976) are known to have a bitter taste, as bran appropriately stabilized and kept under safe storage does not develop an objectionable bitter taste.

As concerns color, bran varies from light tan in ordinary samples to deep brown in parboiled samples. Moist heat treatment increases the red and yellow components of color, thus decreasing whiteness. Hexane extraction of stabilized bran removes most of the pigments and a white, creamy flour remains. Water/bran pastes are browner than the dry powder, especially after cooking (Table 24.29) (Tortosa and Benedito de Barber 1978).

Functional Properties

In addition to flavor, color (above par) and protein extractability and solubility of bran, other properties such as water and fat absorption, and emulsifying and foaming capacity, are important factors determining the potential use of bran in foods. Few preliminary data are available (Benedito de Barber et al. 1978). In experiments performed by Medcalf and Gilles (1965) and Sosulski (1962), water absorption was measured by slurrying bran in water and then centrifuging. Absorbed water, taken as the increase in bran weight, was close to 200 g water/100 g bran. Fat absorption was determined in a similar way using corn germ oil instead of water (Lin et al. 1973). Absorbed oil, taken as the decrease in volume of free oil, was around 150 g oil/100 g bran. Emulsifying capacity was measured by mixing water, bran and corn germ oil in a Virtis blender and then centrifuging (Yasumatsu et al. 1972; Puski 1975). The emulsified layer was 50% and emulsion stability after a 30 min heating was almost complete.

Data for water absorption, fat absorption and the emulsifying capacity of bran compared acceptably with those of commercially available soyabean 70% protein concentrate (270, 110 and 50, respectively). Bran exhibited very low foaming capacity. Results of the model tests above have not been evaluated in predicting performance of raw, stabilized and defatted brans in real food systems.

Nutritional Properties

The content of essential amino acids, the chemical score, protein efficiency ratio (PER) and nitrogen digestibility of rice byproducts have

TABLE 24.29. COLOR OF RAW, STABILIZED AND DEFATTED STABILIZED RICE BRAN. TRISTIMULUS COLOR FACTORS[1]

	Dry powder			3:1 Water:bran paste			6.5:1 Water-bran paste, cooked[2]		
	L	a	b	L	a	b	L	a	b
Raw bran	63.8	0.2	14.3	55.4	0.1	13.6	49.0	1.4	14.1
Moist-heat stabilized bran[3]	54.2	1.0	15.4	50.3	0.6	13.7	45.9	1.6	13.9
Defatted moist-heat stabilized bran[4]	70.1	0	11.5	51.4	0.9	13.6	42.3	2.1	13.6

Source: Tortosa and Benedito de Barber (1978).
[1]Hunter Color Difference Meter.
[2]Boiled for 5 min.
[3]Pilot-plant processed according to Barber et al. (1977A).
[4]Hexane extracted, desolventized at room conditions.

been reviewed recently (Barber and Benedito de Barber 1977). The content of essential amino acids of bran, polish and germ (Table 24.10) compares favorably with that of milled rice, and other cereals and food-stuffs. Differences in individual amino acids content between raw and parboiled and defatted brans, including X-M bran, are small—generally within a ± 10% range. Bran polish and germ have higher levels of lysine than milled rice. Notwithstanding this, lysine and threonine generally are the first and second limiting amino acids in regular, parboiled and defat-ted bran. Bran proteins are similar to those of other cereals and some oilseeds (cottonseed, safflowerseed, sunflowerseed) for their deficiency in lysine. The chemical score of rice bran is, in general, higher.

Levels of available lysine and methionine of raw bran (Benedito de Barber *et al.* 1977) are within the usual range of variation of each in-dividual total amino acid. Available tryptophan is extremely low. Levels of available amino acids in the bran of parboiled rices are lower than in bran of the parent rice; loss of availability increases with severity of treatment.

Regular bran, X-M bran, rice polish, parboiled rice polish and high protein rice flours have similar PER values (Table 24.30). Rice bran protein concentrates appear to have higher ones. Rice germ has the highest. Bran protein, although of lower nutritional value than egg and

TABLE 24.30. NUTRITIONAL VALUES AS GIVEN BY PER OF RICE BYPRODUCTS

Rice Byproduct	PER	Remarks	Reference
Bran	1.61	At 8.16% protein.	Kik (1957)
Bran	1.92	At 9.00% protein	Kik (1965)
Bran	1.83	—	Kik (1962)
X—M bran	1.7—1.9	—	Lynn (1969)
Polish	1.84	At 9.00% protein	Kik (1965)
Polish	1.88	—	Kik (1956)
Polish	1.92	—	Kik (1962)
High protein flour (deep milling)	1.84	—	Milner (1965)
3rd break commercial parboiled bran	1.70	—	Houston and Kohler (1970)
High-protein bran fraction:white	1.80	—	Mitsuda *et al.* (1977B)
High-protein bran fraction:brown	1.31	—	Mitsuda *et al.* (1977B)
Rice bran protein concentrate, acid ppt.	1.99		Saunders *et al.* (1974)
Rice bran protein concentrate, heat ppt.	2.19	—	Saunders *et al.* (1974)

animal proteins, compares favorably with that of soybeans (0.7–1.8), cottonseed, (1.3–2.1) and is much higher than those from corn (1.2) and wheat (1.0) (Liener 1972).

Data for weight gain, coefficient of apparent digestibility (CDA), coefficient of true digestibility (CDR), biological value (BV), net protein utilization (NPU) and productive protein value (PPV) (Table 24.31) show that the nutritional quality of the protein of raw bran is high and similar to the feeds of animal origin. Moist heat stabilization results in a slight increase in the weight gain and other parameters (Varela and Escriva 1977). A biological value of 78.1 for rice germ as compared with 61.4 for milled rice has been reported (Kik 1954).

TABLE 24.31. NUTRITIONAL QUALITY OF THE PROTEIN OF RAW AND MOIST HEAT STABILIZED RICE BRAN

Parameter	Raw bran	Moist heat stabilized bran[1]
Food intake[2]	9.84	12.37
Weight gain[3]	2.37	3.08
CDA[4]	62.6	64.7
CDR[4]	68.0	70.4
BV[4]	79.2	81.0
NPU[4]	53.9	57.1
PPV[4]	1.92	2.09
Nitrogen balance	+73.5	+96.1

Source: Varela and Escriva (1977).
[1]Pilot-plant processed according to Barber et al. (1977A).
[2]Dry substance ingested, g/rat/day.
[3]g/rat/day.
[4]CDA, coefficient of apparent digestibility; CDR, coefficient of true digestibility; BV, biological value; NPU, net protein utilization; PPV, productive protein value.

Information on digestible nutrients (protein, lipids, fiber and NFE) was obtained with sheep with maintenance level rations including 50% of rice byproducts. Digestibility of proteins was bran, 60.2; defatted bran, 70.0; polish 62.2; germ, 67.9; defatted germ, 67.8 and hulls, 11.0 (Maymone et al. 1962). Total digestible nutrients (TDN) calculated from above work are bran, 67; defatted bran, 65; polish, 73; defatted germ, 73; and hulls, 14 (Barber and Benedito de Barber 1977). Slightly higher values have been reported for bran in case of swine and for bran and polishings in the case of swine and cattle (Crampton and Harris 1969). Nitrogen digestibility of protein concentrate from rice bran by wet-processing at pH 8.6–9.0, was 89.6 (60.4 for parent bran), similar to protein concentrates from wheat and triticale brans—93.8 and 95.1, respectively—obtained by the same procedure (Saunders et al. 1974).

Feed energy values, digestible energy (DE) and metabolizable energy (ME), of rice bran are high for sheep and swine but relatively low for cattle. Values for rice polish and defatted rice germ are higher and those for defatted rice bran are lower. Rice hulls, a frequent contaminant of commercial bran, have very low values (Table 24.32).

TABLE 24.32. FEED ENERGY VALUES OF RICE BYPRODUCTS

Byproduct	Cattle		Energy[1] Sheep		Swine	
	DE	ME	DE	ME	DE	ME
Rice bran[2]	2648	2171	3210	2632	3256	3028
Defatted rice bran[3]	2167	1776	2247	1842	—	—
Rice polishings	3532	2896	3690	3026	3916	2658
Defatted rice germ[4]	—	—	3325	2727	—	—
Rice hulls	609	500	609	500	—	—

Source: Crampton and Harris (1969).
[1]DE: Digestible energy; ME: Metabolizable energy, kcal/kg.
[2]Rice bran with germ, dry-milled, maximum 13% fiber, $CaCO_3$ declared above 3% minimum.
[3]Rice bran with germ, solvent extracted, minimum 14% protein, maximum 14% fiber.
[4]Source: Barber and Benedito de Barber (1977).

In addition to proteins and calories, bran, germ and polish (to a lesser extent) are good sources of essential unsaturated fatty acids and tocopherols (see Chapter 23), vitamins B_1 and B_2 and minerals (see section on Chemical Composition and Chemical Constituents of Bran Polish and Germ). Sodium, potassium and chlorine are easily absorbed and easily excreted. Calcium, magnesium and phosphorus are imperfectly absorbed and the amount of each present in bran does not necessarily reflect its nutritive value. Other essential elements such as manganese, iron, copper and zinc are poorly absorbed and poorly excreted. As related to the daily requirements of humans the calcium content of bran is low (Natl. Acad. Sci. 1974). Vitamin D and proteins may assist its absorption. However, some bran components (phytic acid and to a lesser extent phosphate and fats) may interfere. The Ca/P ratio in food may be important in some diets, such as in that for infants needing to be fed with artificial milk. Although the recommendation is between 2:1 and 1:2, ratios outside this range can be satisfactory if vitamin D intake is adequate. The availability of iron in bran is low at 52% (Kik and Williams 1945). Phytic acid and phosphates form insoluble salts with this element. In addition to calcium and iron, phytic acid forms insoluble complexes with other mineral elements, rendering them biologically unavailable; zinc and magnesium are examples. Phytic acid also forms complexes with proteins, making them less soluble and more resistant to proteolytic digestion.

Many plants synthesize substances which are able to exert deleterious effects, such as inhibition of growth and a decrease in food efficiency, when ingested by man or animals. The occurrence in bran of substances of this type has been reported; trypsin inhibitor (Barber et al. 1978) (see section on Processed Rice Bran and Rice Bran Products), pepsin inhibitor (Mitsuda et al. 1977B), hemagglutinin (Benedito de Barber and Barber 1978), antithiamin factor (Chaudfuri 1964), and estrogenic activity, also in rice polish (Booth *et al.* 1960). Although activity is relatively low and generally can be arrested by heating, much research remains to be done if utilization of the full food and feed potential of bran is to be achieved.

UTILIZATION OF RICE BRAN, POLISH AND GERM

Uses of rice bran vary from fuel to food, including fertilizers, pharmaceuticals, soaps and feed. The rice bran oil production system (oil, soaps, waxes) is dealt with in Chapter 23. Bran as a source of materials for pharmaceuticals has been reviewed recently (Houston 1972A). Uses of bran as food and feed are dealt with below.

Food Uses

Rice bran has not yet been considered in its full potential for feeding other than animals. Rice polish is used in notable quantities in baby foods, in spite of its low keeping quality. However, consumption is far from being significant. In fact, only "potential" food uses of bran can be talked about. These are numerous: bread, muffins, pancakes, cookies, cakes, pies, made of regular bran (Wood 1967) and defatted bran (Lynn 1969); extruded snacks or breakfast cereals, coatings and crusts for finger foods or confections, spice carriers, deep fried preparations (Lynn 1969), puddings (Vanossi 1958), milk-like products (Lynn 1969; Desikachar and Parpia 1970) and pickles (Sakurai 1977). Koji malt obtained from de-oiled bran is suitable for the manufacture of miso (bean paste) and Shoyu (soy sauce) (Yokochi 1977).

Baked goods provide one of the most attractive possibilities. X-M bran increases dough yields, contributes to an attractive tan crumb and crust, does not disturb fermentation or mixing tolerances of the dough, causes baked products to remain fresher and more moist, and adds significant essential amino acids, minerals and vitamins to the baked goods (Lynn 1969). Nevertheless, the proteins of rice bran do not share with wheat the property of forming gluten. The incorporation of bran in the bread dough is limited by the final volume of the bread, which is less than normal with bran. X-M bran has been added to bread in amounts between 5 and 15% of flour weight, using a sponge and dough system. Greater proportions of bran may be incorporated using the bread making system developed with composite flours (Kim and De Ruiter 1969).

Protein concentrates from bran have been the object of extensive studies on the formulation of various products, such as paste products, beverages and confections (Lynn 1969).

The preparation of a milk-like protein beverage is another product of interest. A 3% protein drink made of X-M rice bran has been developed as a cow's milk substitute (Table 24.33). The inclusion of a suitable fat and an emulsifier gives a stable oil-water emulsion with the rice protein base. The base used is a light tan-colored and virtually odor- and flavor-free gel. It is well suited for flavoring and coloring with fruit and other flavors. Rice milk concentrate and rice milk made from this are both stable at retort times and temperatures.

TABLE 24.33. COMPOSITION OF MILK-LIKE BEVERAGE FROM RICE PROTEIN CONCENTRATE VS COW'S MILK

Component[1]	Bran protein milk	Cow's milk
Protein	3.60	3.42
Fat	4.00	3.67
Mineral	0.70	0.73
Carbohydrate	4.81	8.77
Water	87.10	87.30

Source: Lynn (1969).
[1]Percentage, wet basis.

Acceptable beverages of the type described promise to be a good outlet for the liquids resulting from the wet fractionation processes of bran. In fact, a milk-like emulsion with a dull creamy color can be obtained from full-fat bran by aqueous leaching or grinding in water and final decantation; it is an acceptable drink after flavoring, sweetening or spicing (Desikachar and Parpia 1970).

The use of bran in pickles (nukazuke) preparation is a Japanese alternative (Sakurai 1977). Takuanzuke (bran pickles of radishes) is an example; it consists of a mixture of sun dried radishes, rice bran, salt and pepper. Pickled radishes are edible after 2–6 months.

Feed Uses

Rice bran, mixed in adequate quantities with other ingredients, can be used to feed domestic animals. However, due to its composition and characteristics, it is subject to certain limitations. These are derived principally from: (1) high fiber content, (2) high content of fat rich in unsaturated fatty acids, (3) contents of calcium and phosphorus in proportions different from the required ones, (4) very variable composition from one lot to another, and (5) high instability.

An adequate ration for single-stomached species usually contain 6% or less of cellulose; that for ruminants may run as high as 35%. The high fiber content of rice bran limits its use in unlimited proportions in rations for monogastric animals. Defatted and huller brans require greater attention since they have a higher fiber content.

The high oil content of bran coupled with its high unsaturated fatty acids level makes it unsuitable for incorporation in unlimited proportions in certain rations. Excessive quantities of bran in diets for pigs may impair the quality of the carcass fat, which becomes soft.

Rice bran is a source of phosphorus, at least for certain animal diets; pigs apparently utilize phytin phosphorus (Morrison 1959). However, the Ca/P ratio in rice bran (1:10) compares unfavorably with recommended ratios for animals (1.5/1 to 3/1) (Piccioni 1970). The addition of calcium carbonate balances the ratio, but the calcium alone should not exceed certain limits. An excess of calcium in the diet will decrease gains and tend to cause parakeratosis (Morrison 1959).

In advanced animal husbandry, where balanced feed stuffs are used, the frequent variation of bran composition is a disadvantage, especially when bran is a major ingredient. The high chemical instability of bran is another factor. Hydrolysis and oxidation of bran oils develop easily. Off-flavors and loss of nutritive value are immediate consequences. Free fatty acids are more liable to oxidative deterioration than the corresponding esters, and oxidation brings about undesirable nutritional changes, the loss of tocopherols and available lysine, among other reactions. Parboiled rice bran is particularly susceptible to oxidative deterioration, since most of its natural antioxidants are destroyed. Although there is some controversy concerning the effects of bran-free fatty acids, the addition of free fatty acids to the diet is an industrial practice today. The possible occurrence of mycotoxins should be mentioned since mold-infested bran may contain mycotoxins, especially in warm and humid climates.

Bran is used in manufacturing the so-called "fish soluble adsorbed feed" (Takei 1977). Exhausted viscera and scrap from processing fish and the residual liquids from fish meal manufacturing are mixed with defatted rice bran and dried. The resulting feed contains around 45% crude protein, 90% of which is fish protein and 10% bran protein.

Rice bran has been used with satisfactory results in pig diets. The proportion of bran in the ratio varies according to the age and growth of the animal. Bran is not recommended for suckling pigs and pigs weighing less than 25 kg but can make up to 20% of the diet for fattening pigs (Piccioni 1970; Thrasher et al. 1967), and about 80% for pregnant sows (Thrasher et al. 1967). Other data (Noland et al. 1960) show that up to 40% rice bran can be used in growing and finishing swine rations; more than 40% bran results in reduced weight gains and feed efficiency, and poor quality fat in the carcasses.

Pilot and field feeding trials of growing-fattening pigs in which moist-heat stabilized bran formed 20% of the diet in partial substitution for sorghum and corn, conducted on 60 and 100 pigs, respectively, showed that daily weight gain, feed: weight gain and feed: carcass weight gain for diets containing stabilized bran were equal to or higher than those without bran. This, coupled with the lower cost of the diet containing stabilized bran and the good quality of the resulting carcasses allowed significant savings (Tortosa *et al.* 1977).

Up to 40% bran can be used in the total diet of poultry (Panda and Gupte 1965). The use of raw bran in poultry feeds is risky. Destruction of tocopherols may cause vitamin E deficiency, an increase in mortality rate and a decrease in masculine fecundity (Piccioni 1970). As concerns broilers, oxidized fats are less dangerous if the amount of vitamin E in the diet is increased (Tortuero 1966). In poultry feeding, the lack of carotenoids in bran, which give color to the skin and the eggs, should be corrected by the addition of xanthophylls to the feed. The possible effects of growth depressing factors is not well known. It has been reported (Kratzer *et al.* 1974) that when raw bran was used at 60% of the diet to replace corn in diets containing fish meal or soybean meal, the growth of chicks was depressed by approximately 30%, but when autoclaved or steamed rice bran was used instead of raw bran a marked improvement in growth was noted.

The incorporation of rice bran in diets for dairy cows has been reported to increase milk and butterfat production (Stallcup 1967). It should be mentioned, however, that diets containing too much bran affect the properties of the milk, resulting in poor conservation of cheeses and of the fatty acids in butter for which reason a proportion of 15–20% is recommended (Piccioni 1970). However, up to 60% bran has been used in the concentrate ration in diets of dairy cows (Panda and Gupte 1965). Good results have been obtained by feeding bran-containing diets to beef cattle (Piccioni 1970) and to working bullocks and buffaloes (Panda and Gupte 1965). Recommended proportions of bran in the diets vary.

REFERENCES

AIZONO, Y., FUNATSU, M., FUJIKI, Y. and WATANABE, M. 1976. Purification and characterization of rice bran lipase. Agric. Biol. Chem. *40* (2) 317–324.

AIZONO, Y. *et al.* 1971. Biochemical studies on rice bran lipase. Part II. Chemical properties. Agric. Biol. Chem. *35* (12) 1973–1979.

AIZONO, Y. *et al.* 1973. Enzymatic properties of rice bran lipase. Agric. Biol. Chem. *37* (9) 2031–2036.

AKAZAWA, T. 1972. Enzymes in rice. *In* Rice Chemistry and Technology. D.F. Houston (Editor). Amer. Assoc. of Cereal Chem., St. Paul, Minnesota.

ALESHIN, E.P. 1961. Physiological characteristics of germination of rice seeds. Plant Physiol. 7, 533–566.

ALESHIN, E.P. and FILINKOLDAKOV, B.V. 1961. Terminal oxidases of germinative rice grains. Rice J. *64*, (9) 28.

ANGLADETTE, A. 1966. The Rice. G.P. Maisonneuve and Larose, Paris, France. (French)

ARKANSAS RICE GROWERS CO-OPERATIVE ASSOCIATION (ARCG). 1968. Rice hulls—unground and special fractions. Mimeo data sheets. *In* Rice Chemistry and Technology. D.F. Houston (Editor). Amer. Assoc. of Cereal Chem., St. Paul, Minn.

ARNOTT, G.W. and LIM, H.K. 1966. Animal feeding stuffs in Malaya. 2. Quality of rice bran and polishing. Malaysian Agric. J. *45* (4) 387–403.

BALDI, G., FOSSATI, G. and FANTONE, G.C. 1976. Rice by-products. II. Protein soluble fraction and amino acids composition. Rice *25* (4) 347–356. (Italian)

BARBER, S. 1969. Basic studies on aging of milled rice and application to discriminating quality factors. U.S. Dep. Agric., Agric. Res. Ser., and For. Res. and Tech. Prog. Div., Project No. *E–25–AMS–(9)*, Final Report.

BARBER, S. 1972. Milled rice and changes during aging. *In* Rice, Chemistry and Technology. D.F. Houston (Editor). Amer. Assoc. of Cereal Chem., St. Paul, Minnesota.

BARBER, S. and BENEDITO DE BARBER, C. 1977. Basic and applied research needs for optimizing utilization of rice bran as food and feed. *In* Proc. of the Rice By-products Util. Int. Conf., 1974. Valencia, Spain. Vol. IV. Rice Bran Utilization: Food and Feed. S. Barber and E. Tortosa (Editors). Inst. Agric. Chem. and Food Technol., Valencia, Spain.

BARBER, S. and BENEDITO DE BARBER, C. 1978. Unpublished results. Inst. Agric. Chem. and Food Technol., Valencia, Spain.

BARBER, S. and MAQUIEIRA, A. 1977. Rice Bran Proteins. I. Extractability of proteins from raw and heat-stabilized bran. J. Agric. Chem. and Food Technol. *17* (2) 209–222. (English summary, Spanish)

BARBER, S., NAVARRO, L. and TORTOSA, E. 1972A. Histology of rice embryo. J. Agric. Chem. and Food Technol. *12* (2) 232–255. (Spanish)

BARBER, S., NAVARRO, L. and TORTOSA, E. 1972B. Histochemistry of the rice embryo. I. Carbohydrates. J. Agric. Chem. and Food Technol. *12* (4) 597–608. (English summary, Spanish)

BARBER, S., NAVARRO, L. and TORTOSA, E. 1976. Histochemistry of the rice embryo. II. Lipids, proteins, amino acids and minerals. J. Agric. Chem. and Food Technol. *16* (4) 516–530. (Spanish)

BARBER, S., BENEDITO DE BARBER, C., FLORES, M.J. and MONTES, J.J. 1978. Toxic constituents of rice bran. I. Trypsin inhibitor activity of raw and heat-treated bran. J. Agric. Chem. and Food Technol. *18* (1) 80–88. (English summary, Spanish)

BARBER, S., *et al.* 1977A. Process for the stabilization of rice bran. I. Basic research studies. *In* Proc. Rice By-products Util., Int. Conf., 1974. Valencia, Spain. Vol. II. Rice By-products Preservation. S. Barber and E. Tortosa (Editors). Inst. Agric. Chem. and Food Technol., Valencia, Spain.

BARBER, S. *et al.* 1977B. High protein flours from rice bran by wet fractionation process. *In* Proc. Rice By-products Utilization, Int. Conf., 1974. Valencia, Spain. Vol. IV. Rice Bran Utilization: Food and Feed. S. Barber and E. Tortosa (Editors). Inst. Agric. Chem. and Food Technol., Valencia, Spain.

BARRNET, R.J. and SELIGMAN, A.M. 1952. Histochemical demonstration of protein-bound sulphydril groups. Sci. *116*, 323–327.

BENEDITO DE BARBER, C. and BARBER, S. 1978. Toxic constituents of rice bran. II. Hemagglutinating activity of raw and heat-treated bran. J. Agric. Chem. and Food Technol. *18* (1) 89–94. (English summary, Spanish)

BENEDITO DE BARBER, C., MARTINEZ, J. and BARBER, S. 1977. Effects of parboiling processes on the chemical composition and nutritional characteristics of rice bran. *In* Proc. Rice By-products Util. Int. Conf., 1974. Valencia, Spain. Vol. IV. Rice Bran Utilization: Food and Feed. S. Barber and E. Tortosa (Editors). Inst. Agric. Chem. and Food Technol., Valencia, Spain.

BENEDITO DE BARBER, C., MARTINEZ, J. and BARBER, S. 1978. Unpublished results. Inst. Agric. Chem. and Food Technol., Valencia, Spain.

BENEDITO DE BARBER, C. and TORTOSA, E. 1978. Unpublished results. Inst. Agric. Chem. and Food Technol., Valencia, Spain.

BETSCHART, A.A., FONG, R.Y. and SAUNDERS, R.M. 1977. Rice by-products: comparative extraction and precipitation of nitrogen from U.S. and Spanish bran and germ. J. Food Sci. *42* (4) 1088–1093.

BEVENUE, A. and WILLIAMS, K.T. 1956. Hemicellulose components of rice. J. Agric. Food Chem. *4* (12) 1014–1017.

BOOTH, A.N., BICKOFF, E.M. and KOHLER, G.O. 1960. Estrogen-like activity in vegetable oils and mill by-products. Sci. *131*, 1807–1808.

BORASIO, L. 1929. Microchemical research and location of the chemical components in the rice grain. J. Rice Cultivation *19* (1) 69–77. (Italian)

BORASIO, L. 1932. Enzymes and baking. J. Rice Cultivation *22*, 137–149. (Italian)

BORASIO, L. 1944. By-products of rice cultivation and milling. Exp. Sta. on Rice Cultivation in Vercelli Bull. *26*. (Italian)

BORASIO, L. 1948. Proposal to the classification and commercial evaluation of rice milling by-products. Rice Cult. *36* (1) 4–8. (Italian)

BOUHARMONT, J. 1967. Anatomical study of the rice embryo and its germination. Cellule *46* (3) 273–298. (French)

BURNS, H.L. and CASSIDY, M.M. 1949. Method of treating rice bran and rice polish. U.S. Patent 2,563,798. June 29.

CARANGIAN, D.D. and SUTARIA, P.B. 1970. Analysis of seven varieties of rice bran and hull. Part II. Determination of some vitamins and effects of storage. Natr. Appl. Sci. Bull. *22*, 86–93.

CARTANO, A.V. and JULIANO, B.O. 1970. Hemicelluloses of milled rice. J. Agric. Food Chem. *18* (1) 40–42.

CHAND, K. and GUPTA, C.P. 1975. Design of rice bran stabilizer. Rice Process. Engin. Cntr. Rept. *1* (1) 39–40.

CHANG, T. and BARDENAS, E.A. 1965. Mutant Traits. *In* Morphology and Varietal Characteristics of the Rice Plant. Int. Rice Res. Inst. Tech. Bull. 4.

CHAUDFURI, D.K. 1964. Purification of antithiamin factor from rice bran. Sci. Cult. *30* (2) 97–98. (Indian)

CHEN, L. and HOUSTON, D.F. 1970. Solubilization and recovery of protein from defatted rice bran. Cereal Chem. *47* (1) 72–79.

CHIKASUYE, M. 1953A. Biochemical studies on cereal seeds. III. On the nature of catalase in rice seeds. Mie Univ. Agric. Fac. Bull. 7, 140–145.

CHIKASUYE, M. 1953B. Biochemical studies on cereal seed. IV. Minerals and catalase in cereal seeds and forms of catalases. Mie Univ. Agric. Fac. Bull. 7, 146–152.

CHIKASUYE, M. 1954. Biochemical studies on cereal seed. V. Changes of catalase activity during the germination and maturation of cereal seeds. J. Agric. Chem. Soc. Jpn. *28*, 105–111.

CHO, J. 1938. The anatomical observation of the embryo in the rice. Bot. Mag. *52* (622) 520–531.

CHO, J. 1956. Double fertilization in *Oryza sativa L.* and development of the endosperm with special reference to the aleurone layer. Natl. Inst. Agric. Sci. Bull. *D* (6) 61–101. (Japanese)

CONNOR, M.A., SAUNDERS, R.M. and KOHLER, G.O. 1977. Preparation and properties of protein concentrates obtained by wet alkaline processing of rice bran. *In* Proc. Rice By-products Util., Int. Conf., 1974. Valencia, Spain. Vol. IV. Rice Bran Utilization: Food and Feed. S. Barber and E. Tortosa (Editors). Inst. Agric. Chem. and Food Technol., Valencia, Spain.

CRAMPTON, E.W. and HARRIS, L.E. 1969. Applied Animal Nutrition. 2nd Edition. W.H. Freeman and Co., San Francisco.

DAVENPORT, H.A. 1964. Histological and Histochemical Techniques. W.B. Saunders Co., Philadelphia and London.

DAVID, J.S.K., RAO, S.K., RAO, S.D.T. and MURTI, K.S. 1965. Quality of bran oil as influenced by the conditions of storage of rice bran. J. Food Sci. Technol. 2 (3–4) 113–114.

DE GROOT, A.P. and SLUMPS, P. 1969. Effects of severe alkali treatment of proteins on amino acid composition and nutritive value. J. Nutr. *98*, 45–48.

DEITCH, A.D. 1955. Microspectrophotometric study of the binding of the anionic dye, naphthol yellow S, by tissue sections and by purified proteins. Lab. Invest. *4*, 324–351.

DELOUCHE, J.C., STILL, T.W., RASPET, M. and LIENHARD, M. 1962. The tetrazolium test for seed viability. Mississippi State Univ. Agric. Exp. Sta. Tech. Bull. *51*.

DEL ROSARIO, A.R., BRIONES, V.P., VIDAL, A.J. and JULIANO, B.O. 1968. Composition and endosperm structure of developing and mature rice kernel. Cereal Chem. *45* (3) 225–235.

DE REGE, F. 1953. Lipase fermentation of the rice lipids. Part I. Ann. Exp. Sta. on Rice Cultivation and Irrigation 1, 101–123. (Italian)

DE REGE, F. 1954. Lipase fermentation of the rice lipids. Part II. Ann. Exp. Sta. on Rice Cultivation and Irrigation 2, 181–201. (Italian)

DE REGE, F. 1955. Lipase fermentation of the rice lipids. Part III. Ann. Exp. Sta. on Rice Cultivation and Irrigation 3, 67–68. (Italian)

DE REGE, F. 1964. Quantitative method of determining adulteration of rice by-products. Riso 13 (4) 310–322 (English summary, Italian)

DESIKACHAR, H.S.R. 1977. Preservation of by-products of rice milling. In Proceedings Rice By-products Util., Int. Conf., 1974. Valencia, Spain. Vol. II. Rice By-products Preservation. S. Barber and E. Tortosa (Editors). Inst. Agric. Chem. and Food Technol., Valencia, Spain.

DESIKACHAR, H.S.R. and PARPIA, H.A.B. 1970. Processing and utilization of rice bran with special reference to its possibility for use in human food. In Proc. Fifth World Cereal and Bread Congr., Dresden, German Dem. Rep.

EBISAWA, H. and SUGIMURA, K. 1956. Studies on rice proteins. III. Non-protein amino acids of rice. Food Res. Inst. Bull. 11, 82–83 (English translation by T. Yoshida).

EL HINNAWY, S.I. 1961. Specifity and nature of lipolytic enzymes in rice bran. Oil and Soap 8 (1) 23–24. (Egyptian)

EMILIANI, E. and FRONTICELLI, F. 1938. Phytin extraction from the rice grain. J. Pharm. Chim. 27, 351–352. (French)

ERIKSEN, S. and FAGERSON, I.S. 1976. Flavours of amino acids and peptides. Int. Flavours 1, 13–16.

ESAU, K. 1965. Plant Anatomy. John Wiley and Sons, New York.

FERRETI, R.J. and LEVANDER, O.A. 1974. Effect of milling and processing on the selenium content of grains and cereal products. J. Agric. Food Chem. 22 (6) 1049–1051.

FITO, P. J. and SANZ, F.J. 1974. Equilibrium moisture contents in drying of rice bran. In Proc. 4th Int. Congr. Food Sci. and Technol. Vol. IV. Madrid.

FOSSATI, G., BALDI, G. and RANGHINO, F. 1976. Rice by-products. I. Chemical composition and inorganic constituents. Rice. 25 (4) 339–345. (Italian)

FUKUI, T. and NIKUNI, Z. 1959. Changes in sugar content during the germination of rice seeds. Application of ion-exchange resin chromatography. J. Agric. Chem. Soc. Jpn. 33, 72–78. (Japanese)

FUNATSU, M. et al. 1971. Biochemical studies on rice bran lipase. Part 1. Purification and physical properties. Agric. Biol. Chem. 35 (5) 734–742.

GEE, M., REEVE, R.M. and McCREADY, R.M. 1959. Reaction of hydroxylamine with pectinic acids. Chemical studies and histochemical estimation of the degree of sterification of pectic substances in fruits. J. Agric. Food Chem. 7 (1) 34–38.

GOMEZ, J.L. and PRIMO, E. 1953. Industrial utilization of the rice by-products. IX. Stabilization of rice germ and bran. Ann. of Span. Roy. Soc. of Phys. Chem. *49B* (12) 801–808. (English summary, Spanish)

GREMLI, H. and JULIANO, B.O. 1970. Studies on alkali-soluble, rice-bran hemicelluloses. Carbohydrate Res. *12* (2) 273–276.

HINTON, J.J.C. 1948. The distribution of vitamin B_1 in the rice grain. Brit. J. Nutr. *2*, 237–241.

HINTON, J.J.C. and SHAW, B. 1954. The distribution of nicotinic acid in rice grain. Brit. J. Nutr. *8* (1) 65–71.

HIRAI, S. 1925. On the purine bases of rice embryo. Kyoto Daigaku. Med. Sci. Bull. Med. School, Kyoto Univ. *7*, 459–462. (Japanese)

HORIKOSHI, M. and MORITA, Y. 1975. Localization of globulin in rice seed and changes in globulin content during seed development and germination. Agric. Biol. Chem. *39* (12) 2309–2314.

HOTCHKISS, R.D. 1948. A microchemical reaction resulting in the staining of polysaccharide structures in fixed tissues preparations. Arch. Biochem. *16*, 131–141.

HOUSTON, D.F. 1972A. Rice Bran and Polish. *In* Rice, Chemistry and Technology. D.F. Houston (Editor). Amer. Assoc. of Cereal Chem., St. Paul, Minnesota.

HOUSTON, D.F. 1972B. Rice Hulls. *In* Rice, Chemistry and Technology. D.F. Houston (Editor). Amer. Assoc. of Cereal Chem., St. Paul, Minnesota.

HOUSTON, D.F., ALLIS, M.E. and KOHLER, G.O. 1969. Amino acid composition of rice and rice by-products. Cereal Chem. *46* (5) 527–537.

HOUSTON, D.F. and KOHLER, G.O. 1970. Nutritional properties of rice. Natl. Acad. Sci., Washington, D.C.

HOUSTON, D.F. and MOHAMMAD, A. 1966. Air classification and sieving of rice bran and polish. Rice J. *69* (8) 20–21, 44.

HUNNELL, J.W. and NOWLIN, J.F. 1972. Solvent extractive rice milling. *In* Rice, Chemistry and Technology. D.F. Houston (Editor). Amer. Assoc. Cereal Chem., St. Paul, Minn

IDA, S., KITAMURA, I. and MORITA, Y. 1970. Studies on respiratory enzymes in rice kernel. Part III. Physicochemical and enzymatic properties of peroxidase 556 from rice embryo. Agric. Biol. Chem. *34* (5) 715–723.

IDA, S., KITAMURA, I., NIKAIDO, H. and MORITA, Y. 1972. Studies on respiratory enzymes in rice kernel. Part IX. Peroxidase isoenzymes of rice embryo. Agric. Biol. Chem. *36* (4) 611–620.

IDA, S. and MORITA, Y. 1969. Studies on respiratory enzymes in rice kernel. Part II. Isolation and purification of cytochrome c and a blue protein from rice bran. Agric. Biol. Chem. *33* (1) 10–17.

IDA, S. and MORITA, Y. 1970A. Studies on respiratory enzymes in rice kernel. Part IV. Purification and some properties of a flavoprotein from rice embryo. Agric. Biol. Chem. *34* (10) 1470–1476.

IDA, S. and MORITA, Y. 1970B. Studies on respiratory enzymes in rice kernel. Part V. NADPH Diaphorase and some other enzymatic activities of the flavoprotein from rice embryo. Agric. Biol. Chem. *34* (10) 1477–1483.

IDA, S. and MORITA, Y. 1971A. Studies on respiratory enzymes in rice kernel. Part VII. Purification of an acidic flavoprotein, glutathione reductase, from rice embryos. Agric. Biol. Chem. *35* (10) 1542–1549.

IDA, S. and MORITA, Y. 1971B. Studies on respiratory enzymes in rice kernel. Part VIII. Enzymatic properties and physical and chemical characterization of glutathione reductase from rice embryos. Agric. Biol. Chem. *35* (10) 1550–1557.

JENSEN, W.A. 1960. The composition of the developing primary wall in onion root tip cells. II. Cytochemical localization. Amer. J. Bot. *47*, 287–295.

JULIANO, B.O. 1972. The Rice Caryopsis and its Compositon. *In* Rice, Chemistry and Technology. D.F. Houston (Editor). Amer. Assoc. of Cereal Chem., St. Paul, Minnesota.

KALBRENER, J.E., WARNER, K. and ELDRIDGE, A.C. 1974. Flavors derived from linoleic and linolenic acid hydroperoxides. Cereal Chem. *51*, 406–416.

KARON, M.L. and ADAMS, M.E. 1949. Hygroscopic equilibrium of rice and rice fractions. Cereal Chem. *26* (1) 1–12.

KASAHARA, K. 1976. Identification of aroma-components of roasted rice bran. J. Jpn. Soc. Food and Nutr. *29* (3) 171–181.

KENNEDY, B.M. and SCHELSTRAETE, M. 1975. Chemical, physical, and nutritional properties of high-protein flours and residual kernel from the over-milling of uncoated milled rice. III. Iron, calcium, magnesium, phosphorus, sodium, potassium, and phytic acid. Cereal Chem. *52* (2) 173–182.

KIK, M.C. 1945. Effect of milling, processing, washing, cooking and storage on thiamine, riboflavin and niacin in rice. Arkansas Agric. Exp. Sta. Bull. *458*, Univ. Arkansas, Fayetteville, Ark.

KIK, M.C. 1954. Nutritive value of rice germ. J. Agric. Food Chem. *2*, 1179–1181.

KIK, M.C. 1956. Nutrients in rice bran and rice polish and improvement of protein quality with amino acids. J. Agric. Food Chem. *4* (2) 170–172.

KIK, M.C. 1957. The nutritive value of rice and its by-products. Arkansas Agric. Exp. Sta. Bull. *589*, Univ. Arkansas, Fayetteville, Ark.

KIK, M.C. 1962. Rice protein supplementation. Rice J. *65* (6) 20–22.

KIK, M.C. 1965. Nutritional improvement of rice diets and effect of rice on nutritive value of other foodstuffs. Arkansas Agric. Exp. Sta. Bull. *698*, Univ. Arkansas, Fayetteville, Ark.

KIK, M.C. and WILLIAMS, R.R. 1945. The nutritional improvement of white rice. Natl. Acad. Sci., Natl. Res. Counc., Bull. *112*. Washington, D.C. *In* Rice, Chemistry and Technology. D.F. Houston (Editor). Amer. Assoc. of Cereal Chem., St. Paul, Minn.

KIM, J.C. and DE RUITER, C. 1969. Bread from nonwheat flours. *In* Protein-enriched Cereal Foods for World Needs. M. Milner (Editor). Amer. Assoc. of Cereal Chem., St. Paul, Minn.

KIMURA, M. 1934A. Comparative studies on the nucleic acid in the sake pressed-cake and beer yeast. III. On the preparation of nucleic acid by Japanese acid clay. Soc. Chem. Ind. J. *37*, 8B–9B. (Japanese)

KIMURA, M. 1934B. On the nucleic acid of rice embryo (Oryza nucleic acid). I. On the separation of pyrimidine base (cytosine). Soc. Chem. Ind. J. *37*, 199B–200B. (Japanese)

KOGA, Y. 1969. Rice milling. *In* Drying, Husking and Milling in Japan. Jpn. Agric. Machinery Man. Assoc., Tokyo.

KRATZER, F.H., EARL, L. and CHIARAVANONT, C. 1974. Factors influencing the feeding value of rice bran for chickens. Poultry Sci. *53*, 1795–1800.

LEONZIO, M. 1965. Chemical analysis of the principal by-products of rice. Riso *14* (4) 331–345. (English summary, Italian)

LEONZIO, M. 1966. The lignin content of the principle by-products of rice. Riso *15* (3) 219–227. (English summary, Italian)

LEONZIO, M. 1967. The pentosans content of the rice caryopsis and principal by-products. Riso *16* (4) 313–319. (English summary, Italian)

LEW, E.J.L., HOUSTON, D.F. and FELLERS, D.A. 1971. Extraction of protein from full-fat rice bran. Proc. of Amer. Assoc. Cereal Chem. Meet., Oct. 10–14. Dallas, Texas. Cereal Sci. Today *16* (92).

LEW, E.J.L., HOUSTON, D.F. and FELLERS, D.A. 1975. A note on protein concentrate from full-fat rice bran. Cereal Chem. *52* (5) 748–750.

LIENER, I.E. 1972. Nutritional value of food protein products. *In* Soybeans: Chemistry and Technology, Vol. 1: Proteins. A.K. Smith and S.J. Circle (Editors). AVI Publishing Co., Westport, Conn.

LIENER, I.E. and KAKADE, M.L. 1969. Protease inhibitors. *In* Toxic Constituents of Plant Foodstuffs. I.E. Liener (Editor). Academic Press, New York and London.

LIMCANGCO-LOPEZ, P.D., ACHACOSO, F.S. and CASTILLO, L.S. 1962. The chemical composition of rice and corn by-products: I. Proximate composition and minerals. Philip. Agric. *46*, 324–342.

LIN, M.J.Y., HUMBERT, E.S. and SOSULSKI, F.W. 1973. Certain functional properties of sunflower protein. Presented at the 58th Ann. Amer. Assoc. Cereal Chem. Meeting, Nov. 4–8. St. Louis, Missouri.

LITTLE, R.R. and DAWSON, E.H. 1960. Histology and histochemistry of raw and cooked rice kernels. Food Res. *25* (5) 611–622.

LOEB, J.R. and MAYNE, R.Y. 1952. Effect of moisture on the microflora and formation of free fatty acids in rice bran. Cereal Chem. *29* (3) 163–175.

LOEB, J.R., MORRIS, N.J. and DOLLEAR, F.G. 1949. Rice bran oil. IV. Storage of the bran as it affects hydrolysis of the oil. J. Amer. Oil Chem. Soc. *26*, 738–743.

LYNN, J. 1969. Edible rice bran foods. *In* Protein-enriched Cereal Foods for World Needs. M. Milner (Editor). Amer. Assoc. Cereal Chem., St. Paul, Minn.

MATSUO, Y. and NAMBA, A. 1958. Hemicellulose in rice grain. J. Fermentation Engin. *36*, 190–193. (Japanese)

MATSUSHITA, S. 1958. The nucleic acids in plants. I. The nucleic acid contents of cereal and pulse seeds and their nucleic acid-containing fractions. Mem. Res. Inst., Food Sci., Kyoto Univ. *14*, 14–23.

MAURON, J. 1973. The analysis of food proteins, amino acid composition and nutritive value. *In* Proteins in Human Nutrition. Pg. J.W.G. Porter and B.A. Rolls (Editors). Academic Press, London and New York.

MAYMONE, B., BATTAGLINI, A. and TIBERIO, M. 1961. The chemical grading and nutritive value of dephytinized rice bran. Ann. of Exp. Agric. *15*, 751–763. (English summary, Italian)

MAYMONE, B., TIBERIO, M. and BATTAGLINI, A. 1962. Digestibility and nutritive value of the by-products of the rice processing. Ann. Sper. Agraria *16* (1–2) 477–513.

MAZIA, D., BREWER, P.A. and ALFERT, A. 1953. The cytochemical staining and measurement of protein with mercuric bromophenol blue. Biol. Bull. *104*, 57–67.

MCCALL, E.R. *et al.* 1953. Composition of rice. Influence of variety and environment on physical and chemical composition. J. Agric. Food Chem. *1* (16) 988–993.

MEDCALF, D.G. and GILLES, K.A. 1965. Wheat starches. Cereal Chem. *42* (6) 558–568.

MENDOZA, B.V. and SUTARIA, P.B. 1970. Analysis of seven varieties of rice bran and hull. III. Determination of some inorganic constituents. Natr. Appli. Aci. Bull. *22* (3–4) 94–102.

MIHARA, S. 1970. Nakataki water process separating rice bran components. Chem. Econ. Eng. Rev. 36–38, 42.

MILNER, M. 1965. FAO/WHO Protein Advisory Group News Bull. *5*, 39FA. *In* Nutritional properties of rice. Natl. Acad. Sci., Washington, D.C.

MITSUDA, H., IKEDA, K. and YASUMOTO, K. 1973. Studies on protein foods (Part 8). Protein extraction from rice bran by alkaline-anion exchange resin method. J. Jpn. Soc. Food and Nutr. *26* (3) 171–175.

MITSUDA, H., KAWAI, F. and MIYOSHI, T. 1958. Flavines 1. Distribution and forms of flavines in animal and plant tissue. J. Agric. Chem. Soc. Jpn. *32*, 847–851. (Japanese)

MITSUDA, H., KAWAI, F., SUZUKI, A. and HONJO, J. 1977B. Studies on the production of a protein-rich fraction from rice bran by means of fractional sedimentation in n-hexane. *In* Rice Report 1976. S. Barber, H. Mitsuda, H.S.R. Desikachar and E. Tortosa (Editors). Int. Union of Food Sci. and Technol. Working Party on Rice Utilization. Inst. Agric. Chem. and Food Technol., Valencia, Spain.

MITSUDA, H., MURAKAMI, K. and TAKAGI, S. 1970. Studies on protein foods. Part 7. Protein isolate from rice bran and its nutritive value. J. Jpn. Soc. Food 23, 80–84.

MITSUDA, H., YASUMOTO, K. and IWAMI, K. 1968. Analysis of volatile components in rice bran. Agric. Biol. Chem. 32 (4) 453–458. (Japanese)

MITSUDA, H., YASUMOTO, K. and YAMAMOTO, S. 1977A. Protein in rice bran and polish for human nutrition. In Proc. Rice By-products Util., Int. Conf., 1974. Valencia, Spain. Vol. IV. Rice Bran Utilization: Food and Feed. S. Barber and E. Tortosa (Editors). Inst. Agric. Chem. and Food Technol., Valencia, Spain.

MITSUDA, H. et al. 1967. Studies on the proteinaceous subcellular particles in rice endosperm: electron microscopy and isolation. Agric. Biol. Chem. 31 (3) 293–300. (Japanese)

MORITA, Y. and IDA, S. 1968. Studies on respiratory enzymes in rice kernel. Part I. Isolation and purification of cytochrome c. and peroxidase 556 from rice embryo. Agric. Biol. Chem. 32 (4) 441–447.

MORITA, Y. and IDA, S. 1972. A preliminary crystallographic investigation of rice cytochrome c. J. Mol. Biol. 71, 807–808.

MORITA, Y., WADANO, A. and IDA, S. 1971. Studies on respiratory enzymes in rice kernel. Part VI. Characterization of blue protein of rice bran. Agric. Biol. Chem. 35 (2) 255–260.

MORITA, Y. and YOSHIDA, C. 1968. Studies on γ-globulin of rice embryo. I. Isolation and purification of γ-globulin from rice embryo. Agric. Biol. Chem. 32 (5) 664–670. (Japanese)

MORRISON, S.H. 1959. Supplementation of cereal-based animal feed. In The Chemistry and Technology of Cereals as Food and Feed. S.A. Matz (Editor). AVI Publishing Co., Westport, Conn.

MORUZZI, G. and CALDARERA, C.M. 1964. Occurrence of polyamines in the germs of cereals. Arch. Biochem. Biophys. 105, 209–210.

MUKAI, J. 1965. A deoxyribonuclease I from rice bran. J. Fac. Agric., Kyushu Univ. 13 (3) 361–368.

MURTHY, K.S. 1966. Present status of rice bran oil industry in India. Indian Oil and Soap J. 31 (8) 217–233.

NATL. ACAD. SCI. 1974. National Academy of Sciences Recommended Dietary Allowances. 8th Revised Edition. Natl. Acad. Sci., Washington, D.C.

NEISH, A.C. 1965. Coumarins, phenylpropanes, and lignin. In Plant Biochemistry. J. Bonner and J.E. Varner (Editors). Academic Press, New York.

NISHIHARA, S. and TASHIRO, H. 1960. Amino acids in rice bran mash. I. Amino acids in rice bran. Domestic Sci. 5 (1) 1–6. (Japanese)

NOLAND, P.R., SCOTT, N.W. and NORTH, A.W. 1960. Rice bran can be used for growing-finishing swine. Arkansas Farm. Res. 9 (5) 9.

NORMAND, F.L., SOIGNET, D.M., HOGAN, J.T. and DEOBALD, H.J. 1966. Content of certain nutrients and amino acids pattern in high-protein rice flour. Rice J. 69 (9) 13–18.

OBARA, T. and OGASAWARA, Y. 1958. Studies on esterases from seeds. II. Preparation of esterase from rice embryo. J. Agric. Chem. Soc. Jpn. *32*, 867–871. (Japanese)

O'DELL, B.L., de BOLAND, A.R. and KOIRTYOHANN, S.R. 1972. Distribution of phytate and nutritionally important elements among the morphological components of cereal grains. J. Agric. Food Chem. *20* (3) 718–721.

OGAWA, M., TANAKA, K. and KASAI, Z. 1975. Isolation of high phytin containing particles from rice grains using an aqueous polymer two phase system. Agric. Biol. Chem. *39* (3) 695–700.

PALMIANO, E.P., ALMAZAN, A.M. and JULIANO, B.O. 1968. Physicochemical properties of protein of developing and mature rice grain. Cereal Chem. *45* (1) 1–12.

PANDA, B. and GUPTE, S.M. 1965. Utilization of rice by-products in animal industry. J. Food Sci. (Mysore) *2*, 120–123.

PARIHAR, D.B. 1955. Saccharides of different varieties of Indian rice. Nature *175*, 42–43.

PASCUAL, F. and PRIMO, E. 1955. Industrial utilization of the rice by-products. XIII. Nutritional value of defatted rice germ and bran. Ann. Span. Roy. Soc. Phys. Sci. *51B* (4) 301–304. (Spanish)

PEARSE, A.G.E. 1968. Histochemistry, Theoretical and Applied. J.E.A. Churchill, London.

PICCIONI, M. 1970. Animal nutrition dictionary. Acribia, Zaragoza, Spain. (Spanish)

PINEDA, J.A. 1976. Histology and histochemistry of rice bran and its application to IATA fractionation process. D.C. Thesis. University of Valencia, Spain. (Spanish)

PRIMO, E. *et al.* 1970. Chemical composition of the rice. V. By-products obtained in the different steps of the milling diagram. J. Agric. Chem. and Food Technol. *10* (2) 244–257. (English summary, Spanish)

PUSKI, G. 1975. Modification of functional properties of soy proteins by proteolytic enzyme treatment. Cereal Chem. *52*, 655–664.

RAMKRISHNIAH, P., SAWARKAR, S.K. and SEN, P. 1973. Stabilization of rice bran in a stirred fluid bed. Rice *22* (2) 185–190.

RAO, B.P., AHMED, S.A. and RAO, S.D.T. 1967B. Factors affecting the free fatty acid changes in rice bran. Indian Oil and Soap J. *32* (7) 203–210.

RAO, P.N.S., RAMANATHAN, P.K., RAJAGOPAL, P. and RAO, L.S.S. 1965. Pilot plant studies on the stabilization of bran-pneumatic conveying drying. Indian Chem. Eng. *7*, 72–76.

RAO, S.N.R., NARAYANA, M.N. and DESIKACHAR, H.S.R. 1967A. Studies on some comparative milling properties of raw and parboiled rice. J. Food Sci. Tech. *4*, 150–155. (Indian)

RAO, Y.K.R., RAMASWAMY, K.G., RAO, L.S.S. and GURUVENKATESH, A.S. 1960. Rice germ oil. Indian Oil Seed J. *4* (2) 53–56.

REEVE, R.M. 1959. A specific hydroxylamine–ferric chloride reaction for histochemical localization of pectin. Stain Tech. *34*, 209–211.

ROBERTS, E.H. 1964. The distribution of oxidation-reduction enzymes and the effects of respiratory inhibitors and oxidizing agents on dormancy in rice seed. Physiol. Plantarum *17* (1) 14–29.

ROBERTS, R.L. *et al.* 1949. Steam blanching of fresh rough rice curbs spoilage by fatty acids. Food Ind. *21*, 1041.

RONDA, E. and SOTO, E. 1965. Rice by-products. I. Content in immediate principles. II. Amino acids content. III. Proteinic quality. Indexes of Mitchell, Block and Oser. J. Animal Nutr. *3* (2) 92–98. (Spanish)

ROSEDALE, J.L. and OLIVEIRO, C.J. 1928. Studies on the antineuritic vitamin. II. The properties of the curative substance. Biochem. J. *22*, 1362-1367.

ROTHE, M. 1967. Heat inactivation of enzymes in the stabilization of rice. *In* Proc. of the 3rd Meeting Int. Problems of Modern Cereal Processing and Chemistry, Bergholz-Rehbrücke, German Dem. Rep. (German)

SAKURAI, J. 1977. Utilization of Rice By-products in Japan. *In* Proc. Rice By-products Util., Int. Conf., 1974. Valencia, Spain. Vol. IV. Rice Bran Utilization: Food and Feed. S. Barber and E. Tortosa (Editors). Inst. Agric. Chem. and Food Technol., Valencia, Spain. (Spanish)

SANTOS, J.K. 1933. Morphology of the flower and mature grain of Philippine rice. Philip. J. Sci. *52* (4) 475–508.

SAUNDERS, R.M., HOLMESSIEDLE, A.G. and STARK, B.P. 1964. Peroxidase. Butterworth, London.

SAUNDERS, R.M. *et al.* 1974. Preparation and properties of protein concentrates obtained by wet-processing of cereal grain milling by-products. *In* Proc. 4th Int. Cong. Food Sci. and Tech. 1974. Madrid, Spain. Vol. IV. Inst. Nat. Sci. and Food Technol. Spain. (Spanish)

SAWAI, H. and MORITA, Y. 1970A. Studies in γ-globulin of rice embryo. II. Separation of three components of γ-globulin by ion exchange chromatography. Agric. Biol. Chem. *34* (1) 53–60.

SAWAI, H. and MORITA, Y. 1970B. Studies on γ-globulin of rice embryo. III. Molecular dimension and chemical composition of γ₁-globulin. Agric. Biol. Chem. *34* (1) 61–67.

SAWAI, H. and MORITA, Y. 1970C. Studies on γ-globulin of rice embryo. IV. Some aspects of sub unit structure of γ₁-globulin. Agric. Biol. Chem. *34* (5) 771–779.

SCOTT, D. 1975. Oxidoreductases. *In* Enzymes in Food Processing. G. Reed (Editor). Academic Press, New York.

SCUDI, J.V. 1942. Conjugated pyridoxine in rice bran concentrates. J. Biol. Chem. *145*, 637–639.

SESSA, D.J., WARNER, K. and HONIG, D.H. 1974. Soybean phosphatidylcholine develops bitter taste on autoxidation. J. Food Sci. *39*, 69–72.

SHASTRY, B.S. 1973. Enzymes in rice bran. Ph.D. Dissertation. Univ. of Mysore, Mysore, India.

SHASTRY, B.S. and RAO, M.R.R. 1972. Phosphatases from rice bran. Indian J. Biochem. Biophys. 9 (4) 297–303.

SHASTRY, B.S. and RAO, M.R.R. 1975. Studies on lipoxygenase from rice bran. Cereal Chem. 52 (5) 597–603.

SHASTRY, B.S. and RAO, M.R.R. 1976. Chemical studies on rice bran lipase. Cereal Chem. 53 (2) 190–200.

SHASTRY, B.S. and RAO, R.M.R. 1971. Studies on rice bran lipase. Indian J. Biochem. Biophys. 8, 327–332.

SHIBUYA, N., IWASAKI, T. and CHIKUBU, S. 1975. Lipase activity in the rice kernel. Rep. Natl. Food Res. Inst. 30, 10–13.

SILVESTRE, P., BELEY, J. and QUET, R. 1960. Studies of the physical and chemical characteristics of rice milling by-products. Proc. of 5th Session of the Consultative Sub-committee on the Economic Problems of Rice, Food and Agric. Organ. U.N., Feb. Saigon. In Rice, Chemistry and Technology. D.F. Houston (Editor). St. Paul, Minn. (French)

SIRIWARDENE, J.A. DE S. 1969. Analytical data on rice bran processed in Ceylon. Ceylon Vet. J. 17 (3) 73–76.

SOLDI, A. 1946. Sugar content of rice germ. Sci. Tech. 1 (1) 39–40. (Italian)

SONE, K. 1958. The quality of plant food. I. Influence of variety and environment on the riboflavine content of hulled rice. I. The estimation of riboflavine and the selection of sample rice. Tohoku. J. Agric. Res. 9, 93–112.

SOSULSKI, F.W. 1962. The centrifuge method for determining flour absorption in hard red spring wheats. Cereal Chem. 39, 344–350.

SREENIVASAN, A. 1939. Studies on quality of rice. IV. Storage changes in rice after harvest. Ind. J. Agric. Sci. 9, 208–222.

SRIMANI, B.N., CHATTOPADHYAY, P. and BOSE, A.N. 1977. Stabilization of rice bran. I. Direct measurement of the lipase activity in rice bran and the methods for the inactivation of the same. In Proc. Rice By-products Util., Int. Conf. 1974. Valencia, Spain. Vol. II. Rice By-products preservation. S. Barber and E. Tortosa (Editors). Inst. Agric. Chem. and Food Technol., Valencia, Spain. (Spanish)

STALLCUP, O.T. 1967. Research on the use of rice products in dairy feeds at the Arkansas Agricultural Experiment Station. Rice J. 70 (7).

SUGIMURA, K. and EBISAWA, H. 1955. Amino acid composition of rice protein. II. A preliminary experiment on the non-protein amino acids. Inst. Food Res. Bull. 10, 183–184 (English translation by T. Yoshida). (Japanese)

TADOKORO, T., TAKASUGI, N. and SAITO, T. 1941. Chemical properties of ascorbic acid oxidase. XV. J. Chem. Soc. Jpn. 62, 119–122.

TAKEI, T. 1977. Utilization of defatted rice bran as fish soluble adsorbent material. In Proc. Rice By-products Util. Int. Conf., 1974. Valencia, Spain. Vol. IV. Rice Bran Utilization: Food and Feed. S. Barber and E. Tortosa (Editors). Inst. Agric. Chem. and Food Technol., Valencia, Spain.

TAMURA, S. and KENMOCHI, K. 1963. Studies on amino acid content of rice. III. Distribution of amino acid in rice grain. J. Agric. Chem. Soc. Jpn. *37* (12) 753–776 (English translation by H. Murakami).

TANAKA, K., OGAWA, M. and KASAI, Z. 1976. The rice scutellum: studies by scanning electron microscopy and electron microprobe X-ray analysis. Cereal Chem. *53* (5) 643–649.

TANAKA, K., YOSHIDA, T., ASADA, K. and KASAI, Z. 1973. Subcellular particles isolated from aleurone layer of rice seeds. Arch. Biochem. and Biophys. *155*, 136–143.

TANAKA, K., YOSHIDA, T. and KASAI, Z. 1974. Radioautographic demonstration of the accumulation site of phytic acid in rice and wheat grains. Plant and Cell Physiol. *15*, 147–151.

THOMAS, R. and JONES, F.T. 1970. Microincineration and scanning electron microscopy: silica in rice hulls. Proc. 28th Ann. Meeting, Electron Microscopy Soc. of America, Oct. 5–8. Houston, Texas. C.J. Arceneaux (Editor). Claitor's Publishing Division, Baton Rouge, La. *In* Rice, Chemistry and Technology. D.F. Houston (Editor). Amer. Assoc. Cereal Chem., St. Paul, Minn.

THRASHER, D.M., MULLINS, A.M. and SCOTT, V.B. 1967. Using rice bran in modern hog rations. La. Agric. *10* (3) 8–9, 13.

TORTOSA, E. *et al.* 1977. Process for the stabilization of rice bran. V. Pilot and field feeding trials of growing-fattening pigs with stabilized bran. *In* Proc. Rice By-products Util. Int. Conf., 1974. Valencia, Spain. Vol. IV. Rice Bran Utilization: Food and Feed. S. Barber and E. Tortosa (Editors). Inst. Agric. Chem. and Food Technol., Valencia, Spain.

TORTOSA, E. and BENEDITO DE BARBER, C. 1978. Unpublished results. Inst. Agric. Chem. and Food Technol., Valencia, Spain.

TORTUERO, F. 1966. Utilization of fats in broilers feeding. Library of Advances in Animal Nutr., Madrid, Spain. (Spanish)

TSUCHIYA, T., IWAI, K. and ANDO, T. 1967. The histones of rice embryo. Seikagaku *39* (2) 109–116.

U.N. INDUS. DEV. ORGAN. (UNIDO). 1977. Research and development of a small-scale, low cost rice bran stabilizing unit. *In* Proc. Sem. on Edible Rice Bran Oil. Solvent Extractor's Assoc. Bombay, India.

UZAWA, S. 1932. The phosphoesterases of bran. J. Biochem. Jpn. *15*, 1–10.

VADHER, C.C. and CHAUHAN, C.S. 1964. Chromatographic studies of rice bran constituents from Gurmatya variety of rice grown in Mandhya Pradesh, India. Indian J. Chem. *2* (2).

VANOSSI, L. 1958. The use of rice bran in cooking. Riso *5*, 23–24. (Italian)

VAN VEEN, A.G. 1931A. The constituent parts of the rice bran in reference to the isolation of the antineuritic vitamin. Communication, Serv. Public Health (Dutch East Indies) *20*, 80–96. (German)

VAN VEEN, A.G. 1931B. The antineuritic vitamin. III. The components of the active "acid clay" extract. J. Chem. Stud. (Netherlands) *50*, 208–220. (French)

VARELA, G. and ESCRIVA, J. 1977. Process for the stabilization of rice bran. IV. Nutritional quality of the protein of raw and stabilized bran. *In* Proc. Rice By-products Util. Int. Conf., 1974. Valencia, Spain. Vol. IV. Rice Bran Utilization: Food and Feed. S. Barber and E. Tortosa (Editors). Inst. Agric. Chem. and Food Technol. Valencia, Spain.

VIRAKTAMATH, C.S. 1974. Large scale trials for stabilization of rice bran with steam. J. Food Sci. Tech. (Mysore) *11* (4) 191–193.

VIRAKTAMATH, C.S. and DESIKACHAR, H.S.R. 1971. Inactivation of lipase in rice bran in Indian rice mills. J. Food Sci. Tech. (Mysore) *8* (2) 70–74.

WATANABE, A. and KASAHI, Y. 1957. Physicochemical studies on vitamin B_1 and its related compounds. V. Decomposition of the thiamine in alkaline solution. J. Pharm. Soc. Jpn. *77*, 153.

WEST, A.P. and CRUZ, A.O. 1933. Philippine rice-mill products with particular reference to the nutritive value and preservation of rice bran. Philip. J. Sci. *52* (1) 1–68.

WILLIAMS, K.T. and BEVENUE, A. 1953. A note on the sugars in rice. Cereal Chem. *30* (3) 267–269.

WILLIAMS, M. and BAER, S. 1965. The expansion and extraction of rice bran. J. Amer. Oil Chem. Soc. *42*, 151–155.

WINTON, A.L. and WINTON, K.B. 1932. Rice. *In* The Structure and Composition of Foods. John Wiley and Sons, New York.

WOOD, M.N. 1967. Gourmet food on a wheat-free diet. C.C. Thomas, Springfield, Ill. *In* Rice, Chemistry and Technology. D.F. Houston (Editor). Amer. Assoc. Cereal Chem., St. Paul, Minn.

YAMAGISHI, G. 1935. Chemical studies on grain enzymes. IV. Distribution of starch-liquefying enzyme in rice grain and effects of salts on the extraction of enzyme. J. Agric. Chem. Soc. Jpn. *11*, 964–970.

YAMAZAKI, W.T. and WILSON, J.T. 1964. *In* Methods in Carbohydrate Chemistry. R. L. Whistler (Editor). Academic Press, London.

YASUI, K. 1936. The anatomy of the embryo and seedling of *Oryza sativa* L. with special reference to the structure of cotyledon and mesocotyl in gramineae. Bot. Mag. *50*, 632–640.

YASUMATSU, K. *et al.* 1972. Whipping and emulsifying properties of soybean products. Agric. Biol. Chem. *36* (5) 719–727.

YOKOCHI, K. 1977. Rice bran processing for the production of rice bran oil and characteristics and uses of the oil and deoiled bran. *In* Proceedings Rice By-products Utilization, Int. Conf., 1974. Valencia, Spain. Vol. III. Rice Bran Utilization. S. Barber and E. Tortosa (Editors). Inst. Agric. Chem. and Food Technol., Valencia, Spain.

YOSHIDA, S., FORNO, D.A. and COCK, J.H. 1971. Safranin-phenol method for detection of silicified cell in rice tissue. Laboratory Manual for Physiological Studies of Rice, Int. Rice Res. Inst. Los Baños, Philippines.

YOSHIDA, S., OHNISHI, Y. and KITAGISHI, K. 1962. Histochemistry of silicon in rice plant. III. The presence of cutin-silica double layer in the epidermal tissue. Biol. Sci. and Plant Nutr. *8* (2) 1–5.

YOUSEFF, A.M., EL-FOULY, M.M. and EL-BAZ, F.K. 1974. Isolation and chemical composition of protein concentrates from soyabean, rice bran and protelan. Plant Foods for Human Nutr. *24* (1–2) 71–84.

YUNG, C.T. 1939. Developmental anatomy of the seeding of the rice plant. Bot. Gaz. *99*, 786–802.

YUSTA, A. and SANTOS, A. 1953. Chemical composition of the germ of *Oryza sativa L.* I. Gross composition and carbohydrates. Ann. Span. Roy. Soc. Phys. Sci. *49B*, 441–444. (English summary, Spanish)

YUSTA, A. and SANTOS, A. 1954. Chemical composition of the germ of *Oryza sativa L.* III. Proteins, mineral salts and vitamins. Ann. Span. Roy. Soc. Phys. Sci. *50B*, 95–98. (English summary, Spanish)

Appendix

APPENDIX 1. UNITED STATES STANDARDS FOR ROUGH RICE[1]
TERMS DEFINED

§ 68.201 Definition of Rough Rice.

Rice (*Oryza sativa L.*) which consists of 50.0 percent or more of paddy kernels [see § 68.202(i)] of rice.

§ 68.202 Definition of other terms.

For the purposes of these standards, the following terms shall have the meanings stated below:

(a) *Broken kernels.* Kernels of rice which are less than three-fourths of whole kernels.

(b) *Chalky kernels.* Whole or large broken kernels of rice which are one-half or more chalky.

(c) *Classes.* The following four classes:

Long Grain Rough Rice
Medium Grain Rough Rice
Short Grain Rough Rice
Mixed Rough Rice

Classes shall be based on the percentage of whole kernels, large broken kernels, and types of rice.

Source: USDA. 1977. United States Standards for Rough Rice, Brown Rice for Processing, Milled Rice. USDA, Federal Grain Inspection Service, Inspection Div., Washington, D.C.
[1]Compliance with the provisions of these standards does not excuse failure to comply with the provisions of the Federal Food, Drug, and Cosmetic Act, or other Federal laws.

(1) "Long grain rough rice" shall consist of rough rice which contains more than 25.0 percent of whole kernels and which, after milling to a well-milled degree, contains not more than 10.0 percent of whole or large broken kernels of medium- or short-grain rice.

(2) "Medium grain rough rice" shall consist of rough rice which contains more than 25.0 percent of whole kernels and which, after milling to a well-milled degree, contains not more than 10.0 percent of whole or large broken kernels of long-grain rice or whole kernels of short-grain rice.

(3) "Short grain rough rice" shall consist of rough rice which contains more than 25.0 percent of whole kernels and which, after milling to a well-milled degree, contains not more than 10.0 percent of whole or large broken kernels of long-grain rice or whole kernels of medium-grain rice.

(4) "Mixed rough rice" shall consist of rough rice which contains more than 25.0 percent of whole kernels and which, after milling to a well-milled degree, contains more than 10.0 percent of "other types" as defined in paragraph (h) of this section.

(d) *Damaged kernels.* Whole or large broken kernels of rice which are distinctly discolored or damaged by water, insects, heat, or any other means, and whole or large broken kernels of parboiled rice in nonparboiled rice. "Heat-damaged kernels" [see paragraph (e) of this section] shall not function as damaged kernels.

(e) *Heat-damaged kernels.* Whole or large broken kernels of rice which are materially discolored and damaged as a result of heating, and whole or large broken kernels of parboiled rice in nonparboiled rice which are as dark as, or darker in color than, the interpretive line for heat-damaged kernels.

(f) *Milling yield.* An estimate of the quantity of whole kernels and total milled rice (whole and broken kernels combined) that are produced in the milling of rough rice to a well-milled degree.

(g) *Objectionable seeds.* Seeds other than rice, except seeds of *Echinochloa crusgalli* (commonly known as barnyard grass, watergrass, and Japanese millet).

(h) *Other types.* (1) Whole kernels of: (i) Long-grain rice in medium- or short-grain rice, (ii) medium-grain rice in long- or short-grain rice,

(iii) short-grain rice in long- or medium-grain rice, and (2) Large broken kernels of long-grain rice in medium- or short-grain rice and large broken kernels of medium- or short-grain rice in long-grain rice.

NOTE—Large broken kernels of medium-grain rice in short-grain rice and large broken kernels of short-grain rice in medium-grain rice shall not be considered other types.

(i) *Paddy kernels.* Whole or broken unhulled kernels of rice.

(j) *Red rice.* Whole or large broken kernels of rice on which there is an appreciable amount of red bran.

(k) *Seeds.* Whole or broken seeds of any plant other than rice.

(l) *Smutty kernels.* Whole or broken kernels of rice which are distinctly infected by smut.

(m) *Types of rice.* The following three types:

> Long grain
> Medium grain
> Short grain

Types shall be based on the length/width ratio of kernels of rice that are unbroken and the width, thickness, and shape of kernels of rice that are broken as set forth in Inspection Handbook HB 918-11.[2]

(n) *Ungelatinized kernels.* Whole or large broken kernels of parboiled rice with distinct white or chalky areas due to incomplete gelatinization of the starch.

(o) *Whole and large broken kernels.* Rice (including seeds) that (1) passes over a 6 plate (for southern production), or (2) remains on top of a 6 sieve (for western production).

(p) *Whole kernels.* Unbroken kernels of rice and broken kernels of rice which are at least three-fourths of an unbroken kernel.

(q) *6 sieve.* A metal sieve 0.032-inch thick, perforated with rows of round holes 0.0938 ($\%_{64}$) inch in diameter.

(r) *6 plate.* A laminated metal plate 0.142-inch thick, with a top lamina 0.051-inch thick, perforated with rows of round holes 0.0938 ($\%_{64}$) inch in diameter, and a bottom lamina 0.091-inch thick, without perforations.

Principles Governing Application of Standards

§ 68.203 Basis of determination.

The determination of seeds, objectionable seeds, heat-damaged kernels, red rice and damaged kernels, chalky kernels, other types, color, and the special grade Parboiled rough rice shall be on the basis of the whole and large broken kernels of milled rice that are produced in the milling of rough rice to a well-milled degree. When determining class, the percentage of (a) whole kernels of rough rice shall be determined on the basis of the original sample, and (b) types of rice shall be determined on the basis of the whole and large broken kernels of milled rice that are produced in the milling of rough rice to a well-milled degree. Smutty kernels shall be determined on the basis of the rough rice after it has been cleaned and shelled as set forth in Inspection Handbook HB 918-11[3], or by any method that is approved by the Administrator as giving equivalent results[3]. All other determinations shall be on the basis of the original sample. Mechanical sizing of kernels shall be adjusted by handpicking as set forth in Inspection Handbook HB 918-11[2], or by any method that is approved by the Administrator as giving equivalent results[3].

§ 68.204 Temporary modifications in equipment and procedures.

The equipment and procedures referrred to in the rough rice standards are applicable to rice produced and harvested under normal environmental conditions. Abnormal environmental conditions during the production and harvest of rice may require minor temporary modifications in the equipment or procedures to obtain results expected under normal conditions. When these adjustments are necessary, Inspection Division Field Offices, official inspection agencies, and interested parties in the rice industry will be notified promptly in writing of the modification. These modifications shall not includes changes in interpretations of identity, class, quality, or condition.

---

[2]The following publications are referenced in these standards. Copies will be made available upon request to the Inspection Division, Federal Grain Inspection Service, U.S. Department of Agriculture, 1400 Independence Avenue, S.W., Washington, D.C. 20250.
  a. Equipment Handbook, effective September 25, 1968, as amended, U.S. Department of Agriculture, Federal Grain Inspection Service.
  b. Inspection Handbook HB 918-11, effective October 1976, as amended, U.S. Department of Agriculture, Federal Grain Inspection Service.

§ 68.205 Interpretive line samples.

Interpretive line samples showing the official scoring line for factors that are determined by visual examinations shall be maintained by the Standardization Division, Federal Grain Inspection Service, U.S. Department of Agriculture, and shall be available for reference in all inspection offices that inspect and grade rice.

§ 68.206 Milling requirements.

In determining milling yield [see § 68.202 (f)] in rough rice, the degree of milling shall be equal to or better than, that of the interpretive line sample for "well-milled" rice.

§ 68.207 Milling yield determination.

Milling yield shall be determined by the use of an approved device in accordance with procedures prescribed in Inspection Handbook HB 918-11, and the Equipment Handbook[2]. For the purpose of this paragraph, "approved device" shall include the McGill Miller No. 3 and any other equipment that is approved by the Administrator as giving equivalent results[3].

NOTE—Milling yield shall not be determined when the moisture content of the rough rice exceeds 18.0 percent.

§ 68.208 Moisture.

Water content in rough rice as determined by an approved device in accordance with procedures prescribed in the Inspection Handbook HB 918-11[2]. For the purpose of this paragraph "approved device" shall include the Motomco Moisture Meter and any other equipment that is approved by the Administrator as giving equivalent results[3].

§ 68.209 Percentages.

Percentages shall be determined on the basis of weight and shall be rounded off as follows:

(a) When the figure to be rounded is followed by a figure greater than 5, round to the next higher figure: e.g., 0.46, report as 0.5.

---

[3]Requests for information concerning approved devices and procedures, criteria for approved devices, and request for approval of devices should be directed to the Standardization Division, Federal Grain Inspection Service, U.S. Department of Agriculture, 1400 Independence Avenue, S.W., Washington, D.C. 20250.

§ 68.210 Grades and grade requirements for the classes of rough rice. (See also § 68.212.)

Grade	Seeds and heat-damaged kernels		Red rice and damaged kernels (singly or combined)	Maximum limits of — Chalky kernels[1]		Other types[2]	Color requirements[1]
	Total (singly or combined)	Heat-damaged kernels and objectionable seeds (singly or combined)		In long grain rice	In medium or short grain rice		
	Number in 500 grams	Number in 500 grams	Percent	Percent	Percent	Percent	
U.S. No. 1	4	3	0.5	1.0	2.0	1.0	Shall be white or creamy.
U.S. No. 2	7	5	1.5	2.0	4.0	2.0	May be slightly gray.
U.S. No. 3	10	8	2.5	4.0	6.0	3.0	May be light gray.
U.S. No. 4	27	22	4.0	6.0	8.0	5.0	May be gray or slightly rosy.
U.S. No. 5	37	32	6.0	10.0	10.0	10.0	May be dark gray or rosy.
U.S. No. 6	75	75	15.0[3]	15.0	15.0	10.0	May be dark gray or rosy.
U.S. Sample grade							

U.S. Sample grade shall be rough rice which: (a) does not meet the requirements for any of the grades from U.S. No. 1 to U.S. No. 6, inclusive; (b) contains more than 14.0 percent of moisture; (c) is musty, or sour, or heating; (d) has any commercially objectionable foreign odor; or (e) is otherwise of distinctly low quality.

[1] For the special grade Parboiled rough rice, see §68.212 (a).
[2] These limits do not apply to the class Mixed Rough Rice.
[3] Rice in grade U.S. No. 6 shall contain not more than 6.0 percent of damaged kernels.

(b) When the figure to be rounded is followed by a figure less than 5, round to the next lower figure; e.g., 0.54, report as 0.5.

(c) When the figure to be rounded is followed by the figure 5, round to the nearest even figure; e.g., 0.45, report as 0.4; 0.55, report as 0.6.

All percentages, except for milling yield, shall be stated in whole and tenth percent to the nearest tenth percent. Milling yield shall be stated to the nearest whole percent.

Grades, Grade Requirements and Grade Designations

§ 68.211 Grade designation.

(a) The grade designation for all classes of rough rice, except Mixed Rough Rice, shall include in the following order: (1) The letters "U.S."; (2) the number of the grade or the words "Sample grade," as warranted; (3) the words "or better" when applicable and requested by the applicant prior to inspection; (4) the class; (5) each applicable special grade (see § 68.213); and (6) a statement of the milling yield.

(b) The grade designation for the class Mixed Rough Rice shall include, in the following order: (1) The letters "U.S."; (2) the number of the grade or the words "Sample grade," as warranted; (3) the words "or better," when applicable and requested by the applicant prior to inspection; (4) the class; (5) each applicable special grade (see § 68.213); (6) the percentage of whole kernels of each type in the order of predominance; (7) the percentage of large broken kernels of each type in the order of predominance; (8) the percent of material removed by the No. 6 sieve or the No. 6 sizing plate; (9) when applicable, the percentage of seeds; and (10) a statement of the milling yield.

NOTE: Large broken kernels other than long grain, in Mixed Rough Rice, shall be certificated as "medium or short grain."

Special Grades, Special Grade Requirements and Special Grade Designations

§ 68.212 Special grades and requirements

A special grade, when applicable, is supplemental to the grade assigned under § 68.210. Such special grades for Rough Rice are established and determined as follows:

(a) *Parboiled rough rice.* Parboiled rough rice shall be rough rice in which the starch has been gelatinized by soaking, steaming, and drying. Grades U.S. No. 1 to U.S. No. 6 inclusive, shall contain not more than 10.0 percent of ungelatinized kernels. Grades U.S. No. 1 and U.S. No. 2 shall contain not more than 0.1 percent, grades U.S. No. 3 and U.S. No. 4 not more than 0.2 percent, and grades U.S. No. 5 and U.S. No. 6 not more than 0.5 percent of nonparboiled rice. If the rice is: (1) Not distinctly colored by the parboiling process, it shall be considered "Parboiled Light"; (2) distinctly but not materially colored by the parboiling process, it shall be considered "Parboiled"; (3) materially colored by the parboiling process, it shall be considered "Parboiled Dark." The color levels for "Parboiled Light," "Parboiled" and "Parboiled Dark" rice shall be in accordance with the interpretive line samples for parboiled rice.

NOTE: The maximum limits for "Chalky kernels," "Heat-damaged kernels," "Kernels damaged by heat," and the "Color requirements" shown in § 68.210 are not applicable to the special grade "Parboiled rough rice."

(b) *Smutty rough rice.* Smutty rough rice shall be rough rice which contains more than 3.0 percent of smutty kernels.

(c) *Weevily rough rice.* Weevily rough rice shall be rough rice which is infected with live weevils or other live insects injurious to stored rice.

§ 68.213 Special grade designation.

The grade designation for parboiled, smutty, or weevily rough rice shall include, following the class, the word(s) "Parboiled Light," "Parboiled," "Parboiled Dark," "Smutty," or "Weevily," as warranted, and all other information prescribed in § 68.211.

APPENDIX 2. UNITED STATES STANDARDS FOR BROWN RICE FOR PROCESSING[1] TERMS DEFINED

§ 68.251 Definition of brown rice for processing.

Rice (*Oryza sativa L.*) which consists of more than 50.0 percent of kernels of brown rice, and which is intended for processing to milled rice.

§ 68.252 Definition of other terms.

For the purposes of these standards, the following terms shall have the meanings stated below:

(a) *Broken kernels.* Kernels of rice which are less than three-fourths of whole kernels.

(b) *Brown rice.* Whole or broken kernels of rice from which the hulls have been removed.

(c) *Chalky kernels.* Whole or broken kernels of rice which are one-half or more chalky.

(d) *Classes.* There are four classes of brown rice for processing.

> Long Grain Brown Rice for Processing.
> Medium Grain Brown Rice for Processing.
> Short Grain Brown Rice for Processing.
> Mixed Brown Rice for Processing.

Classes shall be based on the percentage of whole kernels, broken kernels, and types of rice.

(1) "Long-grain brown rice for processing" shall consist of brown rice for processing which contains more than 25.0 percent of whole kernels of brown rice and not more than 10.0 percent of whole or broken kernels of medium- or short-grain rice.

(2) "Medium-grain brown rice for processing" shall consist of brown rice for processing which contains more than 25.0 percent of whole ker-

Source: USDA. 1977. United States Standards for Rough Rice, Brown Rice for Processing, Milled Rice. USDA, Federal Grain Inspection Services, Inspection Div., Washington, D.C.
[1]Compliance with the provisions of these standards does not excuse failure to comply with the provisions of the Federal Food, Drug, and Cosmetic Act, or other Federal laws.

nels of brown rice and not more than 10.0 percent of whole or broken kernels of long-grain rice or whole kernels of short-grain rice.

(3) "Short-grain brown rice for processing" shall consist of brown rice for processing which contains more than 25.0 percent of whole (kernels of brown rice and not more than 10.0 percent of whole) or broken kernels of long-grain rice or whole kernels of medium-grain rice.

(4) "Mixed brown rice for processing" shall be brown rice for processing which contains more than 25.0 percent of whole kernels of brown rice and more than 10.0 percent of "other types" as defined in paragraph (i) of this section.

(e) *Damaged kernels.* Whole or broken kernels of rice which are distinctly discolored or damaged by water, insects, heat, or any other means (including parboiled kernels in nonparboiled rice and smutty kernels). "Heat-damaged kernels" [see paragraph (f) of this section] shall not function as damaged kernels.

(f) *Heat-damaged kernels.* Whole or broken kernels of rice which are materially discolored and damaged as a result of heating and parboiled kernels in nonparboiled rice which are as dark as, or darker in color than, the interpretative line for heat-damaged kernels.

(g) *Milling yield.* An estimate of the quantity of whole kernels and total milled rice (whole and broken kernels combined) that is produced in the milling of brown rice for processing to a well-milled degree.

(h) *Objectionable seeds.* Whole or broken seeds other than rice, except seeds of *Echinochloa crusgalli* (commonly known as barnyard grass, watergrass, and Japanese millet).

(i) *Other types.* (1) Whole kernels of: (i) long-grain rice in medium- or short-grain rice, (ii) medium-grain rice in long- or short-grain rice, (iii) short-grain rice in long- or medium-grain rice, and (2) broken kernels of long-grain rice in medium- or short-grain rice and broken kernels of medium- or short-grain rice in long-grain rice.

NOTE: Broken kernels of medium-grain rice in short-grain rice and broken kernels of short-grain rice in medium-grain rice shall not be considered other types.

(j) *Paddy kernels.* Whole or broken unhulled kernels of rice.

(k) *Red rice.* Whole or broken kernels of rice on which the bran is distinctly red in color.

(l) *Related material.* All by-products of a paddy kernel, such as the outer glumes, lemma, palea, awn, embryo, and bran layers.

(m) *Seeds.* Whole or broken seeds of any plant other than rice.

(n) *Smutty kernels.* Whole or broken kernels of rice which are distinctly infected by smut.

(o) *Types of Rice.* There are three types of brown rice for processing:

Long grain
Medium grain
Short grain

Types shall be based on the length/width ratio of kernels of rice that are unbroken and the width, thickness, and shape of kernels of rice that are broken as set forth in Inspection Handbook HB 918-11[2].

(p) *Ungelatinized kernels.* Whole or broken kernels of parboiled rice with distinct white or chalky areas due to incomplete gelatinization of the starch.

(q) *Unrelated material.* All matter other than rice, related material, and seeds.

(r) *Well-milled kernels.* Whole or broken kernels of rice from which the hulls and practically all of the embryos and the bran layers have been removed.

(s) *Whole kernels.* Unbroken kernels of rice and broken kernels of rice which are at least three-fourths of an unbroken kernel.

(t) *6 plate.* A laminated metal plate 0.142-inch thick, with a top lamina 0.051-inch thick, perforated with rows of round holes 0.0938 (⁶⁄₆₄) inch in diameter, and a bottom lamina 0.091-inch thick, without perforations.

(u) *6 1/2 sieve.* A metal sieve 0.032-inch thick, perforated with rows of round holes 0.1016 (6 ½⁄₆₄) inch in diameter.

---

[2]The following publications are referenced in these standards. Copies will be made available upon request to the Inspection Division, Federal Grain Inspection Service, U.S. Department of Agriculture, 1400 Independence Avenue, S.W., Washington, D.C. 20250.
a. Equipment Handbook, effective September 25, 1968, as amended. U.S. Department of Agriculture, Federal Grain Inspection Service.
b. Inspection Handbook HB 918-11, effective October 1976, as amended U.S. Department of Agriculture, Federal Grain Inspection Service.

Principles Governing Application of Standards

§ 68.253 Basis of determination.

The determination of kernels damaged by heat, heat-damaged kernels, parboiled kernels in nonparboiled rice, and the special grade Parboiled brown rice for processing shall be on the basis of the brown rice for processing after it has been milled to a well-milled degree. All other determinations shall be on the basis of the original sample. Mechanical sizing of kernels shall be adjusted by handpicking as set forth in Inspection Handbook HB 918-11[2] or by any method which gives equivalent results.

§ 68.254 Temporary modifications in equipment and procedures.

The equipment and procedures referred to in the brown rice for processing standards are applicable to rice produced and harvested under normal environmental conditions. Abnormal environmental conditions during the production and harvest of rice may require minor temporary modifications in the equipment or procedures to obtain results expected under normal conditions. When these adjustments are necessary, Inspection Division Field Offices, official inspection agencies, and interested parties in the rice industry will be notified promptly in writing of the modification. These modifications shall not include changes in interpretations of identity, class, quality, or condition.

§ 68.255 Broken kernels determinations.

Broken kernels shall be determined by the use of equipment and procedures set forth in Inspection Handbook HB 918-11[2] or by any method which gives equivalent results.

§ 68.256 Interpretive line samples.

Interpretive line samples showing the official scoring line for factors that are determined by visual observation shall be maintained by the Standarization Division, Federal Grain Inspection Service, U.S. Department of Agriculture, and shall be available for reference in all inspection offices that inspect and grade rice.

§ 68.257 Milling requirements.

In determining milling yield [see § 68.252 (g)] in brown rice for processing, the degree of milling shall be equal to, or better than, that of the interpretive line sample for "well-milled" rice.

§ 68.258 Milling yield determination.

Milling yield shall be determined by the use of an approved device in accordance with procedures prescribed in Inspection Handbook HB 918-11[2] and the Equipment Handbook[2]. For the purpose of this paragraph, "approved device" shall include the McGill Miller No. 3 and any other equipment that is approved by the Administrator as giving equivalent results[3].

NOTE: Milling yield shall not be determined when the moisture content of the brown rice for processing exceeds 18.0 percent.

§ 68.259 Moisture.

Water content in brown rice for processing as determined by an approved device in accordance with procedures prescribed in the Inspection Handbook HB 918-11[2]. For the purpose of this paragrah, "approved device" shall include the Motomco Moisture Meter and any other equipment that is approved by the Administrator as giving equivalent results[3].

§ 68.260 Percentages.

Percentages shall be determined on the basis of weight and shall be rounded off as follows:

(a) When the figure to be rounded is followed by a figure greater than 5, round to the next higher figure; e.g., 0.46, report as 0.5.

(b) When the figure to be rounded is followed by a figure less than 5, round to the next lower figure; e.g., 0.54, report as 0.5.

(c) When the figure to be rounded is followed by the figure 5, round to the nearest even figure; e.g., 0.45, report as 0.4; 0.55, report as 0.6.

All percentages, except for milling yield, shall be stated in whole and tenth percent to the nearest tenth percent. Milling yield shall be stated to the nearest whole percent.

---

[3]Requests for information concerning approved devices and procedures, criteria for approved devices, and request for approval of devices should be directed to the Federal Grain Inspection Service, U.S. Department of Agriculture, 1400 Independence Avenue, S.W., Washington, D.C. 20250.

§ 68.261 Grade and grade requirements for the classes of brown rice for processing. (See also § 68.263.)

Maximum limits of —

Grade	Paddy Kernels		Seeds and heat-damaged kernels			Red rice and damaged kernels (singly or combined)	Chalky Kernels [1]	Broken Kernels Removed by a 6 plate or a 6½ sieve [2]	Other Types [3]	Well-Milled Kernels
			Total (singly or combined)	Heat-damaged kernels	Objectionable seeds					
	Percent	Number in 500 grams	Number in 500 grams	Number in 500 grams	Number in 500 grams	Percent	Percent	Percent	Percent	Percent
U.S. No. 1	—	20	10	1	2	1.0	2.0	1.0	1.0	1.0
U.S. No. 2	2.0	—	40	2	10	2.0	4.0	2.0	2.0	3.0
U.S. No. 3	2.0	—	70	4	20	4.0	6.0	3.0	5.0	10.0
U.S. No. 4	2.0	—	100	8	35	8.0	8.0	4.0	10.0	10.0
U.S. No. 5	2.0	—	150	15	50	15.0	15.0	6.0	10.0	10.0
U.S. Sample grade										

U.S. Sample grade shall be brown rice for processing which (a) does not meet the requirements for any of the grades from U.S. No. 1 to U.S. No. 5, inclusive; (b) contains more than 14.5 percent of moisture; (c) is musty, or sour, or heating; (d) has any commercially objectionable foreign odor; (e) contains more than 0.2 percent of related material or more than 0.1 percent of unrelated material; (f) contains live weevils or other live insects; or (g) is otherwise of distinctly low quality.

[1] For the special grade Parboiled brown rice for processing, see §68.263 (a).
[2] Plates should be used for southern production rice and sieves should be used for western production rice, but any device or method which gives equivalent results may be used.
[3] These limits do not apply to the class Mixed Brown Rice for Processing.

Grades, Grade Requirements and Grade Designations

§ 68.262 Grade designation.

(a) The grade designation for all classes of brown rice for processing, except Mixed Brown Rice for Processing, shall include in the following order: (1) The letters "U.S": (2) the number of the grade or the words "Sample grade," as warranted; (3) the words "or better," when applicable and requested by the applicant prior to inspection; (4) the class; and (5) each applicable special grade (see § 68.264).

(b) The grade designation for the class Mixed Brown Rice for Processing shall include in the following order: (1) The letters "U.S."; (2) the number of grade or the words "Sample grade," as warranted; (3) the words "or better," when applicable and requested by the applicant prior to inspection; (4) the class; (5) each applicable special grade (see § 68.264); (6) the percentage of whole kernels of each type in the order of predominance and when applicable; (7) the percentage of broken kernels of each type in the order of predominance; and (8) the percentage of seeds, related material, and unrelated material.

NOTE: Broken kernels other than long grain, in Mixed Brown Rice for Processing, shall be certificated as "medium or short grain."

Special Grades, Special Grade Requirements, and Special Grade Designations

§ 68.263 Special grades and special grade requirements.

A special grade, when applicable, is supplemental to the grade assigned under § 68.262. Such special grades for brown rice for processing are established and determined as follows:

(a) *Parboiled brown rice for processing.* Parboiled brown rice for processing shall be rice in which the starch has been gelatinized by soaking, steaming, and drying. Grades U.S. Nos. 1 to 5 inclusive shall contain not more than 10.0 percent of ungelatinized kernels. Grades U.S. No. 1 and U.S. No. 2 shall contain not more than 0.1 percent, grades U.S. No. 3 and U.S. No. 4 not more than 0.2 percent, and grade U.S. No. 5 not more than 0.5 percent of nonparboiled rice.

NOTE: The maximum limits for "chalky kernels," "Heat-damaged kernels," and "Kernels damaged by heat" shown in § 68.261 are not applicable to the special grade "Parboiled brown rice for processing."

(b) *Smutty brown rice for processing.* Smutty brown rice for processing shall be rice which contains more than 3.0 percent of smutty kernels.

§ 68.264 Special grade designation.

The grade designation for parboiled or smutty brown rice for processing shall include, following the class, the word(s) "Parboiled" or "Smutty," as warranted, and all other information prescribed in §68.262.

APPENDIX 3. UNITED STATES STANDARDS FOR MILLED RICE[1]
TERMS DEFINED

§ 68.301 Definition of milled rice.

Whole or broken kernels of rice (*Oryza sativa L.*) from which the hulls and at least the outer bran layers and a part of the germs have been removed and which contain not more than 10.0 percent of seeds, paddy kernels, or foreign material, either singly or combined.

§ 68.302 Definition of other terms.

For the purposes of these standards, the following terms shall have the meanings stated below:

(a) *Broken kernels.* Kernels of rice which are less than three-fourths of whole kernels.

(b) *Brown rice.* Whole or broken kernels of rice from which the hulls have been removed.

(c) *Chalky kernels.* Whole or broken kernels of rice which are one-half or more chalky.

(d) *Classes.* There are seven classes of milled rice. The following four classes shall be based on the percentage of whole kernels, broken kernels, and types of rice:

> Long-Grain Milled Rice
> Medium-Grain Milled Rice
> Short-Grain Milled Rice
> Mixed Milled Rice

The following three classes shall be based on the percentage of whole kernels and of broken kernels of different size:

> Second-Head Milled Rice
> Screenings Milled Rice
> Brewers Milled Rice

Source: USDA 1977. United States Standards for Rough Rice, Brown Rice for Processing, Milled Rice. USDA, Federal Grain Inspection Services, Inspection Div., Washington, D.C.
[1]Compliance with the provisions of these standards does not excuse failure to comply with the provisions of the Federal Food, Drug, and Cosmetic Act, or other Federal laws.

(1) "Long-grain milled rice" shall consist of milled rice which contains more than 25.0 percent of whole kernels of milled rice and U.S. Nos. 1 through 4 not more than 10.0 percent of whole or broken kernels of medium- or short-grain rice. U.S. No. 5 and U.S. No. 6 long-grain milled rice shall contain not more than 10.0 percent of whole kernels of medium- or short-grain milled rice (broken kernels do not apply).

(2) "Medium-grain milled rice" shall consist of milled rice which contains more than 25.0 percent of whole kernels of milled rice and in U.S. Nos. 1 through 4 not more than 10.0 percent of whole or broken kernels of long-grain rice or whole kernels of short-grain rice. U.S. No. 5 and U.S. No. 6 medium-grain milled rice shall contain not more than 10.0 percent of whole kernels of long- or short-grain milled rice (broken kernels do not apply).

(3) "Short-grain milled rice" shall consist of milled rice which contains more than 25.0 percent of whole kernels of milled rice and in U.S. Nos. 1 through 4 not more than 10.0 percent of whole or broken kernels of long-grain rice or whole kernels of medium-grain rice. U.S. No. 5 and U.S. No. 6 short-grain milled rice shall contain not more than 10.0 percent of whole kernels of long- or medium-grain milled rice (broken kernels do not apply).

(4) "Mixed milled rice" shall consist of milled rice which contains more than 25.0 percent of whole kernels of milled rice and more than 10.0 percent of "other types" as defined in paragraph (i) of this section. U.S. No. 5 and U.S. No. 6 mixed milled rice shall contain more than 10.0 percent of whole kernels of "other types" (broken kernels do not apply).

(5) "Second-head milled rice" shall consist of milled rice which, when determined in accordance with §§68.303 and 68.304, contains:

(i) Not more than (a) 25.0 percent of whole kernels, (b) 7.0 percent of broken kernels removed by a 6 plate, (c) 0.4 percent of broken kernels removed by a 5 plate, and (d) 0.05 percent of broken kernels passing through a 4 sieve (southern production); or

(ii) Not more than (a) 25.0 percent of whole kernels, (b) 50.0 percent of broken kernels passing through a 6½ sieve, and (c) 10.0 percent of broken kernels passing through a 6 sieve (western production).

(6) "Screenings milled rice" shall consist of milled rice which, when determined in accordance with §§68.303 and 68.304, contains:

(i) Not more than (a) 25.0 percent of whole kernels, (b) 10.0 percent of broken kernels removed by a 5 plate, and (c) 0.2 percent of broken kernels passing through a 4 sieve (southern production); or

(ii) Not more than (a) 25.0 percent of whole kernels, (b) 15.0 percent of broken kernels passing through a 5½ sieve; and more than (c) 50.0 percent of broken kernels passing through a 6½ sieve; and (d) 10.0 percent of broken kernels passing through a 6 sieve (western production).

(7) "Brewers milled rice" shall consist of milled rice which, when determined in accordance with §§ 68.303 and 68.304, contains not more than 25.0 percent of whole kernels and which does not meet the kernel-size requirements for the class Second Head Milled Rice or Screenings Milled Rice.

(e) *Damaged kernels.* Whole or broken kernels of rice which are distinctly discolored or damaged by water, insects, heat or any other means, and parboiled kernels in nonparboiled rice. "Heat-damaged kernels" [see paragraph (g) of this section] shall not function as damaged kernels.

(f) *Foreign material.* All matter other than rice and seeds. Hulls, germs, and bran which have separated from the kernels of rice shall be considered foreign material.

(g) *Heat-damaged kernels.* Whole or broken kernels of rice which are materially discolored and damaged as a result of heating and parboiled kernels in nonparboiled rice which are as dark as, or darker in color than, the interpretive line for heat-damaged kernels.

(h) *Objectionable seeds.* Seeds other than rice, except seeds of *Enchinochloa crusgalli* (commonly known as barnyard grass, watergrass, and Japanese Millet).

(i) *Other types.* (1) Whole kernels of: (i) Long-grain rice in medium- or short-grain rice, (ii) medium-grain rice in long- or short-grain rice, (iii) Short-grain rice in long- or medium-grain rice, and (2) broken kernels of long-grain rice in medium- or short-grain rice and broken kernels of medium- or short-grain rice in long-grain rice, except in U.S. No. 5 and U.S. No. 6 milled rice. In U.S. No. 5 and U.S. No. 6 milled rice, only whole kernels will apply.

NOTE: Broken kernels of medium grain rice in short-grain rice and broken kernels of short-grain rice in medium-grain rice shall not be considered other types.

(j) *Paddy kernels.* Whole or broken unhulled kernels of rice and whole or broken kernels or brown rice.

(k) *Red rice.* Whole or broken kernels of rice on which there is an appreciable amount of red bran.

(l) *Seeds.* Whole or broken seeds of any plant other than rice.

(m) *Types of rice.* There are three types of milled rice as follows:

> Long grain
> Medium grain
> Short grain

Types shall be based on the length-width ratio of kernels of rice that are unbroken and the width, thickness, and shape of kernels that are broken, as set forth in the Inspection Handbook HB 918-11[2].

(n) *Ungelatinized kernels.* Whole or broken kernels of parboiled rice with distinct white or chalky areas due to incomplete gelatinization of the starch.

(o) *Well-milled kernels.* Whole or broken kernels of rice from which the hulls and practically all of the germs and the bran layers have been removed.

NOTE: This factor is determined on an individual kernel basis and applies to the special grade Undermilled milled rice only.

(p) *Whole kernels.* Unbroken kernels of rice and broken kernels of rice which are at least three-fourths of an unbroken kernel.

(q) *5 plate.* A laminated metal plate 0.142-inch thick, with a top lamina 0.051-inch thick, perforated with rows of round holes 0.0781 ($5/64$) inch in diameter, $5/32$ inch from center to center, with each row staggered in relation to the adjacent rows, and a bottom lamina 0.091-inch thick, without perforations.

(r) *6 plate.* A laminated metal plate 0.142-inch thick, with a top lamina 0.051-inch thick, perforated with rows of round holes 0.0938 ($6/64$) inch in diameter, $5/32$ inch from center to center, with each row staggered in relation to the adjacent rows, and a bottom lamina 0.091-inch thick, without perforations.

---

[2]The following publications are referenced in these standards. Copies will be made available upon request to the Inspection Division, Federal Grain Inspection Service, U.S. Department of Agriculture, 1400 Independence Avenue, S.W., Washington, D.C. 20250.
a. Equipment Handbook, effective September 25, 1968, as amended, U.S. Department of Agriculture, Federal Grain Inspection Service.
b. Inspection Handbook HB 918-11, effective October 1976, as amended. U.S. Department of Agriculture, Federal Grain Inspection Service.

(s) 2½ sieve. A metal sieve 0.032-inch thick, perforated with rows of round holes 0.391 (2½/64) inch in diameter, 0.075-inch from center to center, with each row staggered in relation to the adjacent rows.

(t) *4 sieve.* A metal sieve 0.032-inch thick, perforated with rows of round holes 0.0625 (4/64) inch in diameter, ⅛ inch from center to center, with each row staggered in relation to the adjacent rows.

(u) *5 sieve.* A metal sieve 0.032-inch thick, perforated with rows of round holes 0.0781 (5/64) inch in diameter, 5/32 inch from center to center, with each row staggered in relation to the adjacent rows.

(v) *5½ sieve.* A metal sieve 0.032-inch thick, perforated with rows of round holes 0.0859 (5½/64) inch in diameter, 9/64 inch from center to center, with each row staggered in relation to the adjacent rows.

(w) *6 sieve.* A metal sieve 0.032-inch thick, perforated with rows of round holes 0.0938 (6/64) inch in diameter, 5/32 inch from center to center, with each row staggered in relation to the adjacent rows.

(x) *6½ sieve.* A metal sieve 0.032 inch thick, perforated with rows of round holes 0.1016 (6½/64) inch in diameter, 5/32 inch from center to center, with each row staggered in relation to the adjacent rows.

(y) *30 sieve.* A woven wire cloth sieve having 0.0234-inch openings, with a wire diameter of 0.0153 inch, and meeting the specifications of American Society for Testing and Materials Designation E-11-61, as set forth in the Equipment Handbook[2].

Principles Governing Application of Standards

§ 68.303 Basis of determination.

All determinations shall be on the basis of the original sample. Mechanical sizing of kernels shall be adjusted by handpicking, as set forth in the Inspection Handbook HB 918-11[2], or by any method which gives equivalent results.

§ 68.304. Temporary modifications in equipment and procedures.

The equipment and procedures referred to in the milled rice standards are applicable to rice produced and harvested under normal environmental conditions. Abnormal environmental conditions during the production and harvest of rice may require minor temporary modifications in the equipment or procedures to obtain results expected under normal conditions. When these adjustments are necessary, Federal Grain Inspection Service Field Offices, official inspection agencies, and interested parties in the rice industry will be notified promptly in writing of the modification. These modifications shall not include changes in interpretations of identity, class, quality, or condition.

§ 68.305 Broken kernels determination.

Broken kernels shall be determined by the use of equipment and procedures set forth in the Inspection Handbook HB 918-11[2], or by any method which gives equivalent results.

§ 68.306 Interpretive line samples.

Interpretive line samples showing the official scoring line for factors that are determined by visual observation shall be maintained by the Standardization Division, Federal Grain Inspection Service, U.S. Department of Agriculture, and shall be available for reference in all inspection offices that inspect and grade rice.

§ 68.307 Milling requirements.

The degree of milling for milled rice; i.e., "well milled," "reasonably well milled," and "lightly milled" shall be equal to, or better than, that of the interpretive line samples for such rice.

§ 68.308 Moisture.

Water content in milled rice as determined by an approved device in accordance with procedures prescribed in the Inspection Handbook HB 918-11[2]. For the purpose of this paragraph, "approved device" shall include the Montomco Moisture Meter and any other equipment that is approved by the Administrator as giving equivalent results[3].

§ 68.309 Percentages.

Percentages shall be determined on the basis of weight and shall be rounded off as follows:

(a) When the figure to be rounded is followed by a figure greater than 5, round to the next higher figure; e.g., 0.46, report as 0.5.

---

[3]Requests for information concerning approved devices and procedures, criteria for approved devices, and requests for approval of devices should be directed to the Standardization Division, Federal Grain Inspection Service, U.S. Department of Agriculture, 1400 Independence Avenue, S.W., Washington, D.C. 20250.

(b) When the figure to be rounded is followed by a figure less than 5, round to the next lowest figure; e.g., 0.54, report as 0.5.

(c) When the figure to be rounded is followed by the figure 5, round to the nearest even figure; e.g., 0.45, report as 0.4; 0.55, report as 0.6.

All percentages, except for milling yield, shall be stated in whole and tenth percent to the nearest tenth percent. Milling yield shall be stated to the nearest whole percent.

Grades, Grade Requirements and Grade Designations

§ 68.314 Grade designations.

(a) The grade designation for all classes of milled rice, except mixed Milled Rice, shall include in the following order: (1) The letters "U.S."; (2) the number of the grade or the words "Sample grade," as warranted; (3) the words "or better," when applicable and requested by the applicant prior to inspection; (4) the class; and (5) each applicable special grade (see § 68.316).

(b) The grade designation for the class Mixed Milled Rice shall include, in the following order: (1) The letters "U.S."; (2) the number of the grade or the words "Sample grade," as warranted; (3) the words "or better," when applicable and requested by the applicant prior to inspection; (4) the class; (5) each applicable special grade (see § 68.316); (6) the percentage of whole kernels of each type in the order of predominance and when applicable; (7) the percentage of broken kernels of each type in the order of predominance; and (8) the percentage of seeds and foreign material.

NOTE: Broken kernels other than long grain, in Mixed Milled Rice, shall be certificated as "medium or short grain."

Special Grades, Special Grade Requirements and Special Grade Designations

§ 68.315 Special grade and special grade requirements.

A special grade, when applicable, is supplemental to the grade assigned under § 68.314. Such special grades for milled rice are established and determined as follows:

§ 68.310 Grades and grade requirements for the classes Long-Grain Milled Rice, Medium-Grain Milled Rice, Short-Grain Milled Rice, and Mixed Milled Rice. (See also §68.315.)

Grade	Seeds, heat-damaged, and paddy kernels (singly or combined)		Red rice and damaged kernels (singly or combined) Percent	Chalky kernels [1]		Broken kernels				Other types [3]		Color Requirements [1]	Milling Requirements (Minimum) [4]
	Total Number in 500 grams	Heat-damaged kernels and objectionable seeds Number in 500 grams		In long-grain rice Percent	In medium- or short-grain rice Percent	Total Percent	Removed by a 5 plate [2] Percent	Removed by a 6 plate [2] Percent	Through a 6 sieve [2] Percent	Whole kernels Percent	Whole and broken kernels Percent		
U.S. No. 1	2	1	0.5	1.0	2.0	4.0	0.04	0.1	0.1	—	1.0	Shall be white or creamy	Well milled
U.S. No. 2	4	2	1.5	2.0	4.0	7.0	0.06	0.2	0.2	—	2.0	May be slightly gray	Well milled
U.S. No. 3	7	5	2.5	4.0	6.0	15.0	0.1	0.8	0.5	—	3.0	May be light gray	Reasonably well milled
U.S. No. 4	20	15	4.0	6.0	8.0	25.0	0.4	2.0	0.7	—	5.0	May be gray or slightly rosy	Reasonably well milled
U.S. No. 5	30	25	[4]6.0	10.0	10.0	35.0	0.7	3.0	1.0	10.0	—	May be dark gray or rosy	Lightly milled
U.S. No. 6	75	75	[5]15.0	15.0	15.0	50.0	1.0	4.0	2.0	10.0	—	May be dark gray or rosy	Lightly milled

U.S. Sample grade

U.S. Sample grade shall be milled rice of any of these classes which: (a) does not meet the requirements for any of the grades from U.S. No. 1 to U.S. No. 6, inclusive; (b) contains more than 15.0 percent of moisture; (c) is musty, or sour, or heating; (d) has any commercially objectionable foreign odor; (e) contains more than 0.1 percent of foreign material; (f) contains live or dead weevils or other insects, insect webbing, or insect refuse; or (g) is otherwise of distinctly low quality.

[1] For the special grade Parboiled milled rice, see § 68.315(c).
[2] Plates should be used for southern production rice, and sieves should be used for western production rice; but any device or method which gives equivalent results may be used.
[3] These limits do not apply to the class Mixed Milled Rice.
[4] For the special grade Undermilled milled rice, see § 68.315(d).
[5] Grade U.S. No. 6 shall contain not more than 6.0 percent of damaged kernels.

§ 68.311 Grades and Grade Requirements for the Class Second Head Milled Rice (See also § 68.315)

Grade	Maximum limits of —				Color Requirements [1]	Minimum Milling Requirements [2]
	Seeds, heat-damaged, and paddy kernels (singly or combined)		Red rice and damaged kernels (singly or combined)	Chalky kernels [1]		
	Total	Heat-damaged kernels and objectionable seeds				
	Number in 500 grams	Number in 500 grams	Percent	Percent		
U.S. No. 1	15	5	1.0	4.0	Shall be white or creamy	Well milled
U.S. No. 2	20	10	2.0	6.0	May be slightly gray	Well milled
U.S. No. 3	35	15	3.0	10.0	May be light gray	Reasonably well milled
U.S. No. 4	50	25	5.0	15.0	May be gray or slightly rosy	Reasonably well milled
U.S. No. 5	75	40	10.0	20.0	May be dark gray or rosy	Lightly milled
U.S. Sample Grade......	U.S. Sample Grade shall be milled rice of this class which: (a) does not meet the requirements for any of the grades from U.S. No. 1 to U.S. No. 5 inclusive; (b) contains more than 15.0 percent of moisture; (c) is musty, or sour, or heating; (d) has any commercially objectionably foreign odor; (e) contains more than 0.1 percent of foreign material; (f) contains live or dead weevils or other insects, insect webbing, or insect refuse; or (g) is otherwise of distinctly low quality.					

[1] For the special grade parboiled milled rice, see §68.315 (c).
[2] For the special grade Undermilled milled rice, see § 68.315. (d).

§ 68.312 Grades and Grade Requirements for the class Screenings Milled Rice (See also § 68.315)

Grade	Maximum limits of — Paddy kernels and seeds — Total (singly or combined) Number in 500 grams	Objectionable seeds Number in 500 grams	Chalky kernels [1] Percent	Color Requirements [1]	Minimum Milling Requirements [2]
U.S. No. 1 [3,4]	30	20	5.0	Shall be white or creamy	Well milled
U.S. No. 2 [3,4]	75	50	8.0	May be slightly gray	Well milled
U.S. No. 3 [3,4]	125	90	12.0	May be light gray or slightly rose	Reasonably well milled
U.S. No. 4 [3,4]	175	140	20.0	May be gray or rosy	Reasonably well milled
U.S. No. 5	250	200	30.0	May be dark gray or very rosy	Lightly milled
U.S. Sample Grade					

U.S. Sample Grade shall be milled rice of this class which: (a) does not meet the requirements for any of the grades from U.S. No. 1 to U.S. No. 5 inclusive; (b) contains more than 15.0 percent of moisture (c) is musty, or sour, or heating; (d) has any commercially objectionable foreign odor; (e) has a badly damaged or extremely red appearance; (f) contains more than 0.1 percent of foreign material; (g) contains live or dead weevils or other insects, insect webbing, or insect refuse; or (h) is otherwise of distinctly low quality.

[1] For the special grade Parboiled milled rice see § 68.315(c).
[2] For the special grade Undermilled milled rice see § 68.315(d).
[3] Grades U.S. No. 1 to U.S. No. 4, inclusive shall contain not more than 3.0 percent of heat-damaged kernels, kernels damaged by heat and/or parboiled kernels in nonparboiled rice.
[4] Grades U.S. No. 1 to U.S. No. 4, inclusive, shall contain not more than 1.0 percent of material passing through a 30 sieve.

§ 68.313 Grades and Grade Requirements for the Class Brewers Milled Rice. (See also § 68.315)

Grade	Maximum limits of —		Color Requirements [1]	Minimum Milling Requirements [2]
	Paddy kernels and seeds			
	Total (singly or combined)	Objectionable seeds		
	Percent	Percent		
U.S. No. 1 [3][4]...	0.5	0.05	Shall be white or creamy	Well milled
U.S. No. 2 [3][4]...	1.0	0.1	May be slightly gray	Well milled
U.S. No. 3 [3][4]...	1.5	0.2	May be light gray or slightly rosy	Reasonably well milled
U.S. No. 4 [3][4]...	3.0	0.4	May be gray or rosy	Reasonably well milled
U.S. No. 5	5.0	1.5	May be dark gray or very rosy	Lightly milled
U.S. Sample Grade ...	U.S. Sample Grade shall be milled rice of this class which: (a) does not meet the requirements for any of the grades from U.S. No. 1 to U.S. No. 5 inclusive; (b) contains more than 15.0 percent of moisture; (c) is musty, or sour, or heating; (d) has any commercially objectionable foreign odor; (e) has a badly damaged or extremely red appearance; (f) contains more than 0.1 percent of foreign material; (g) contains more than 15.0 percent of broken kernels that will pass through a 2½ sieve; (h) contains live or dead weevils or other insects, insect webbing, or insect refuse; or (i) is othewise of distinctly low quality.			

[1]For the special grade Parboiled milled rice see § 68.315(c).
[2]For the special grade Undermilled milled rice see § 68.315(d).
[3]Grades U.S. No. 1 to U.S. No. 4, inclusive, shall contain not more than 3.0 percent of heat-damaged kernels, kernels damaged by heat and/or parboiled kernels in nonparboiled rice.
[4]Grades U.S. No. 1 to U.S. No. 4, inclusive, shall contain not more than 1.0 percent of material passing through a 30 sieve. This limit does not apply to the special grade Granulated brewers milled rice.

(a) *Coated milled rice.* Coated milled rice shall be milled rice which is coated, in whole or in part, with glucose and talc.

(b) *Granulated brewers milled rice.* Granulated brewers milled rice shall be milled rice which has been crushed or granulated so that 95.0 percent or more will pass through a 5 sieve, 70.0 percent or more will pass through a 4 sieve, and not more than 15.0 percent will pass through a 2½ sieve.

(c) *Parboiled milled rice.* Parboiled milled rice shall be milled rice in which the starch has been gelatinized by soaking, steaming, and drying. Grades U.S. No. 1 to U.S. No. 6, inclusive, shall contain not more than 10.0 percent of ungelatinized kernels. Grades U.S. No. 1 and U.S. No. 2 shall contain not more than 0.1 percent, grades U.S. No. 3 and U.S. No. 4 not more than 0.2 percent, and grades U.S. No. 5 and U.S. No. 6 not more than 0.5 percent of nonparboiled rice. If the rice is: (1) Not distinctly colored by the parboiling process, it shall be considered "Parboiled Light"; (2) distinctly but not materially colored by the parboiled process, it shall be considered "Parboiled"; (3) materially colored by the parboiling process, it shall be considered "Parboiled Dark." The color levels for "Parboiled Light," "Parboiled," and "Parboiled Dark" shall be in accordance with the interpretive line samples for parboiled rice.

NOTE: The maximum limits for "Chalky kernels," "Heat-damaged kernels," "Kernels damaged by heat," and the "Color requirements" in §§ 68.310, 68.311, 68.312, and 68.313 are not applicable to the special grade "Parboiled milled rice."

(d) *Undermilled milled rice.* Undermilled milled rice shall be milled rice which is not equal to the milling requirements for "well-milled," "reasonably well milled," and "lightly milled" rice (see § 68.307). Grades U.S. No. 1 and U.S. No. 2 shall contain not more than 2.0 percent, grades U.S. No. 3 and U.S. No. 4 not more than 5.0 percent, grade U.S. No. 5 not more than 10.0 percent, and grade U.S. No. 6 not more than 15.0 percent of well-milled kernels. Grade U.S. No. 5 shall contain not more than 10.0 percent of red rice and damaged kernels (singly or combined) and in no case more than 6.0 percent of damaged kernels.

NOTE: The "Color and milling requirements" in §§ 68.310, 68.311, 68.312, and 68.313 are not applicable to the special grade "Undermilled milled rice."

§ 68.316 Special Grade designation.

The grade designation for coated, granulated brewers, parboiled, or undermilled milled rice shall include, following the class, the word(s) "Coated," "Granulated," "Parboiled Light," "Parboiled," "Parboiled Dark," or "Undermilled," as warranted, and all other information prescribed in § 68.314.

TEMPERATURE CONVERSION

The numbers in boldface type in the center column refer to the temperature, either in degree Celsius or Fahrenheit, which is to be converted to the other scale. If converting Fahrenheit to degree Celsius, the equivalent temperature will be found in the left column. If converting degree Celsius to Fahrenheit, the equivalent temperature will be found in the column on the right.

Temperature			Temperature			Temperature			Temperature		
Celsius	°C or F	Fahr	Celsius	°C or F	Fahr	Celsius	°C or F	Fahr	Celsius	°C or F	Fahr
-40.0	-40	-40.0	+1.7	+35	+95.0	+43.3	+110	+230.0	+85.0	+185	+365.0
-39.4	-39	-38.2	+2.2	+36	+96.8	+43.9	+111	+231.8	+85.6	+186	+366.8
-38.9	-38	-36.4	+2.8	+37	+98.6	+44.4	+112	+233.6	+86.1	+187	+368.6
-38.3	-37	-34.6	+3.3	+38	+100.4	+45.0	+113	+235.4	+86.7	+188	+370.4
-37.8	-36	-32.8	+3.9	+39	+102.2	+45.6	+114	+237.2	+87.2	+189	+372.2
-37.2	-35	-31.0	+4.4	+40	+104.0	+46.1	+115	+239.0	+87.8	+190	+374.0
-36.7	-34	-29.2	+5.0	+41	+105.8	+46.7	+116	+240.8	+88.3	+191	+375.8
-36.1	-33	-27.4	+5.5	+42	+107.6	+47.2	+117	+242.6	+88.9	+192	+377.6
-35.6	-32	-25.6	+6.1	+43	+109.4	+47.8	+118	+244.4	+89.4	+193	+379.4
-35.0	-31	-23.8	+6.7	+44	+111.2	+48.3	+119	+246.2	+90.0	+194	+381.2
-34.4	-30	-22.0	+7.2	+45	+113.0	+48.9	+120	+248.0	+90.6	+195	+383.0
-33.9	-29	-20.2	+7.8	+46	+114.8	+49.4	+121	+249.8	+91.1	+196	+384.8
-33.3	-28	-18.4	+8.3	+47	+116.6	+50.0	+122	+251.6	+91.7	+197	+386.6
-32.8	-27	-16.6	+8.9	+48	+118.4	+50.6	+123	+253.4	+92.2	+198	+388.4
-32.2	-26	-14.8	+9.4	+49	+120.2	+51.1	+124	+255.2	+92.8	+199	+390.2
-31.7	-25	-13.0	+10.0	+50	+122.0	+51.7	+125	+257.0	+93.3	+200	+392.0
-31.1	-24	-11.2	+10.6	+51	+123.8	+52.2	+126	+258.8	+93.9	+201	+393.8
-30.6	-23	-9.4	+11.1	+52	+125.6	+52.8	+127	+260.6	+94.4	+202	+395.6
-30.0	-22	-7.6	+11.7	+53	+127.4	+53.3	+128	+262.4	+95.0	+203	+397.4
-29.4	-21	-5.8	+12.2	+54	+129.2	+53.9	+129	+264.2	+95.6	+204	+399.2
-28.9	-20	-4.0	+12.8	+55	+131.0	+54.4	+130	+266.0	+96.1	+205	+401.0
-28.3	-19	-2.2	+13.3	+56	+132.8	+55.0	+131	+267.8	+96.7	+206	+402.8
-27.8	-18	-0.4	+13.9	+57	+134.6	+55.6	+132	+269.6	+97.2	+207	+404.6
-27.2	-17	+1.4	+14.4	+58	+136.4	+56.1	+133	+271.4	+97.8	+208	+406.4
-26.7	-16	+3.2	+15.0	+59	+138.2	+56.7	+134	+273.2	+98.3	+209	+408.2
-26.1	-15	+5.0	+15.6	+60	+140.0	+57.2	+135	+275.0	+98.9	+210	+410.0
-25.6	-14	+6.8	+16.1	+61	+141.8	+57.8	+136	+276.8	+99.4	+211	+411.8
-25.0	-13	+8.6	+16.7	+62	+143.6	+58.3	+137	+278.6	+100.0	+212	+413.6
-24.4	-12	+10.4	+17.2	+63	+145.4	+58.9	+138	+280.4	+100.6	+213	+415.4
-23.9	-11	+12.2	+17.8	+64	+147.2	+59.4	+139	+282.2	+101.1	+214	+417.2
-23.3	-10	+14.0	+18.3	+65	+149.0	+60.0	+140	+284.0	+101.7	+215	+419.0
-22.8	-9	+15.8	+18.9	+66	+150.8	+60.6	+141	+285.8	+102.2	+216	+420.8
-22.2	-8	+17.6	+19.4	+67	+152.6	+61.1	+142	+287.6	+102.8	+217	+422.6
-21.7	-7	+19.4	+20.0	+68	+154.4	+61.7	+143	+289.4	+103.3	+218	+424.4
-21.1	-6	+21.2	+20.6	+69	+156.2	+62.2	+144	+291.2	+103.9	+219	+426.2
-20.6	-5	+23.0	+21.1	+70	+158.0	+62.8	+145	+293.0	+104.4	+220	+428.0
-20.0	-4	+24.8	+21.7	+71	+159.8	+63.3	+146	+294.8	+105.6	+222	+431.6
-19.4	-3	+26.6	+22.2	+72	+161.6	+63.9	+147	+296.6	+106.7	+224	+435.2
-18.9	-2	+28.4	+22.8	+73	+163.4	+64.4	+148	+298.4	+107.8	+226	+438.8
-18.3	-1	+30.2	+23.3	+74	+165.2	+65.0	+149	+300.2	+108.9	+228	+442.4
-17.8	0	+32.0	+23.9	+75	+167.0	+65.6	+150	+302.0	+110.0	+230	+446.0
-17.2	+1	+33.8	+24.4	+76	+168.8	+66.1	+151	+303.8	+111.1	+232	+449.6
-16.7	+2	+35.6	+25.0	+77	+170.6	+66.7	+152	+305.6	+112.2	+234	+453.2
-16.1	+3	+37.4	+25.6	+78	+172.4	+67.2	+153	+307.4	+113.3	+236	+456.8
-15.6	+4	+39.2	+26.1	+79	+174.2	+67.8	+154	+309.2	+114.4	+238	+460.4
-15.0	+5	+41.0	+26.7	+80	+176.0	+68.3	+155	+311.0	+115.6	+240	+464.0
-14.4	+6	+42.8	+27.2	+81	+177.8	+68.9	+156	+312.8	+116.7	+242	+467.6
-13.9	+7	+44.6	+27.8	+82	+179.6	+69.4	+157	+314.6	+117.8	+244	+471.2
-13.3	+8	+46.4	+28.3	+83	+181.4	+70.0	+158	+316.4	+118.9	+246	+474.2
-12.8	+9	+48.2	+28.9	+84	+183.2	+70.6	+159	+318.2	+120.0	+248	+478.4
-12.2	+10	+50.0	+29.4	+85	+185.0	+71.1	+160	+320.0	+121.1	+250	+482.0
-11.7	+11	+51.8	+30.0	+86	+186.8	+71.7	+161	+321.8	+122.4	+252	+485.6
-11.1	+12	+53.6	+30.6	+87	+188.6	+72.2	+162	+323.6	+123.3	+254	+489.2
-10.6	+13	+55.4	+31.1	+88	+190.4	+72.8	+163	+325.4	+124.4	+256	+492.8
-10.0	+14	+57.2	+31.7	+89	+192.2	+73.3	+164	+327.2	+125.5	+258	+496.4
-9.4	+15	+59.0	+32.2	+90	+194.0	+73.9	+165	+329.0	+126.7	+260	+500.0
-8.9	+16	+60.8	+32.8	+91	+195.8	+74.4	+166	+330.8	+127.8	+262	+503.6
-8.3	+17	+62.6	+33.3	+92	+197.6	+75.0	+167	+332.6	+128.9	+264	+507.2
-7.8	+18	+64.4	+33.9	+93	+199.4	+75.6	+168	+334.4	+130.0	+266	+510.8
-7.2	+19	+66.2	+34.4	+94	+201.2	+76.1	+169	+336.2	+131.3	+268	+514.4
-6.7	+20	+68.0	+35.0	+95	+203.0	+76.7	+170	+338.0	+132.2	+270	+518.0
-6.1	+21	+69.8	+35.6	+96	+204.8	+77.2	+171	+339.8	+133.3	+272	+521.6
-5.5	+22	+71.6	+36.1	+97	+206.6	+77.8	+172	+341.6	+134.4	+274	+525.2
-5.0	+23	+73.4	+36.7	+98	+208.4	+78.3	+173	+343.4	+135.6	+276	+528.8
-4.4	+24	+75.2	+37.2	+99	+210.2	+78.9	+174	+345.2	+136.7	+278	+532.4
-3.9	+25	+77.0	+37.8	+100	+212.0	+79.4	+175	+347.0	+137.8	+280	+536.0
-3.3	+26	+78.8	+38.3	+101	+213.8	+80.0	+176	+348.8	+138.9	+282	+539.6
-2.8	+27	+80.6	+38.9	+102	+215.6	+80.6	+177	+350.6	+140.0	+284	+543.2
-2.2	+28	+82.4	+39.4	+103	+217.4	+81.1	+178	+352.4	+141.1	+286	+546.8
-1.7	+29	+84.2	+40.0	+104	+219.2	+81.7	+179	+354.2	+142.2	+288	+550.4
-1.1	+30	+86.0	+40.6	+105	+221.0	+82.2	+180	+356.0	+143.3	+290	+554.0
-0.6	+31	+87.8	+41.1	+106	+222.8	+82.8	+181	+357.8	+144.4	+292	+557.6
.0	+32	+89.6	+41.7	+107	+224.6	+83.3	+182	+359.6	+145.6	+294	+561.2
+0.6	+33	+91.4	+42.2	+108	+226.4	+83.9	+183	+361.4	+146.7	+296	+564.8
+1.1	+34	+93.2	+42.8	+109	+228.2	+84.4	+184	+363.2	+147.8	+298	+568.4

COMPARISON OF AVOIRDUPOIS AND METRIC UNITS OF WEIGHT

1 oz = 0.06 lb = 28.35 g	1 lb = 0.454 kg	1 g = 0.035 oz	1 kg = 2.205 lb
2 oz = 0.12 lb = 56.70 g	2 lb = 0.91 kg	2 g = 0.07 oz	2 kg = 4.41 lb
3 oz = 0.19 lb = 85.05 g	3 lb = 1.36 kg	3 g = 0.11 oz	3 kg = 6.61 lb
4 oz = 0.25 lb = 113.40 g	4 lb = 1.81 kg	4 g = 0.14 oz	4 kg = 8.82 lb
5 oz = 0.31 lb = 141.75 g	5 lb = 2.27 kg	5 g = 0.18 oz	5 kg = 11.02 lb
6 oz = 0.38 lb = 170.10 g	6 lb = 2.72 kg	6 g = 0.21 oz	6 kg = 13.23 lb
7 oz = 0.44 lb = 198.45 g	7 lb = 3.18 kg	7 g = 0.25 oz	7 kg = 15.43 lb
8 oz = 0.50 lb = 226.80 g	8 lb = 3.63 kg	8 g = 0.28 oz	8 kg = 17.64 lb
9 oz = 0.56 lb = 255.15 g	9 lb = 4.08 kg	9 g = 0.32 oz	9 kg = 19.84 lb
10 oz = 0.62 lb = 283.50 g	10 lb = 4.54 kg	10 g = 0.35 oz	10 kg = 22.05 lb
11 oz = 0.69 lb = 311.85 g	11 lb = 4.99 kg	11 g = 0.39 oz	11 kg = 24.26 lb
12 oz = 0.75 lb = 340.20 g	12 lb = 5.44 kg	12 g = 0.42 oz	12 kg = 26.46 lb
13 oz = 0.81 lb = 368.55 g	13 lb = 5.90 kg	13 g = 0.46 oz	13 kg = 28.67 lb
14 oz = 0.88 lb = 396.90 g	14 lb = 6.35 kg	14 g = 0.49 oz	14 kg = 30.87 lb
15 oz = 0.94 lb = 425.25 g	15 lb = 6.81 kg	15 g = 0.53 oz	15 kg = 33.08 lb
16 oz = 1.00 lb = 453.59 g	16 lb = 7.26 kg	16 g = 0.56 oz	16 kg = 35.28 lb

COMPARISON OF U.S. AND METRIC UNITS OF LIQUID MEASURE

1 fl oz = 29.573 ml	1 qt = 0.946 liter	1 gal. = 3.785 liters
2 fl oz = 59.15 ml	2 qt = 1.89 liters	2 gal. = 7.57 liters
3 fl oz = 88.72 ml	3 qt = 2.84 liters	3 gal. = 11.36 liters
4 fl oz = 118.30 ml	4 qt = 3.79 liters	4 gal. = 15.14 liters
5 fl oz = 147.87 ml	5 qt = 4.73 liters	5 gal. = 18.93 liters
6 fl oz = 177.44 ml	6 qt = 5.68 liters	6 gal. = 22.71 liters
7 fl oz = 207.02 ml	7 qt = 6.62 liters	7 gal. = 26.50 liters
8 fl oz = 236.59 ml	8 qt = 7.57 liters	8 gal. = 30.28 liters
9 fl oz = 266.16 ml	9 qt = 8.52 liters	9 gal. = 34.07 liters
10 fl oz = 295.73 ml	10 qt = 9.46 liters	10 gal. = 37.85 liters

1 ml = 0.034 fl oz	1 liter = 1.057 qt	1 liter = 0.264 gal.
2 ml = 0.07 fl oz	2 liters = 2.11 qt	2 liters = 0.53 gal.
3 ml = 0.10 fl oz	3 liters = 3.17 qt	3 liters = 0.79 gal.
4 ml = 0.14 fl oz	4 liters = 4.23 qt	4 liters = 1.06 gal.
5 ml = 0.17 fl oz	5 liters = 5.28 qt	5 liters = 1.32 gal.
6 ml = 0.20 fl oz	6 liters = 6.34 qt	6 liters = 1.59 gal.
7 ml = 0.24 fl oz	7 liters = 7.40 qt	7 liters = 1.85 gal.
8 ml = 0.27 fl oz	8 liters = 8.45 qt	8 liters = 2.11 gal.
9 ml = 0.30 fl oz	9 liters = 9.51 qt	9 liters = 2.38 gal.
10 ml = 0.34 fl oz	10 liters = 10.57 qt	10 liters = 2.64 gal.

CONVERSION OF OVEN TEMPERATURES

Conventional (Fahrenheit)		Metric (Celsius)
200 F		93 C
225 F		107 C
250 F	Very low	121 C
300 F	Low	149 C
325 F		163 C
350 F	Moderate	177 C
400 F	Hot	204 C
450 F	Very high	232 C
500 F	Extremely high	260 C

VOLUME CONVERSION DIFFERENCES
CONVENTIONAL VS. METRIC MEASUREMENTS

Utensil	Capacity (ml)	Tolerance (ml)
1 cup	236.6	11.8
½ cup	118.3	5.9
⅓ cup	78.9	3.9
¼ cup	59.2	3.0
1 tablespoon	14.79	0.73
1 teaspoon	4.93	0.24
½ teaspoon	2.46	0.12
¼ teaspoon	1.23	0.06

Index

deodorizing, 777−778
dewaxing, 772−773
winterizing or destearinization, 778−779
"Okaki,' 691
Orange leaf, symptoms, 240
Organic acids, in rice hulls, 738, 740
Organic manures, 188
Organic soils, 159
Organic solvents, solubility of vitamin B_1
    derivatives, 494
Oryza glaberrima, geographic distribution,
    17
Oryza perennis Moench, 8
Oryza sativa, extent of wild relatives and
    spread in Asia and Oceania, 2
  geographic dispersal, 14−17
    across the ocean, 16−17
    reaching Europe, 15−16
    spreading through Africa, 16
    spreading throughout Asia, 14−15
  mendelian genes, 88−91
Oven-puffed rice, 627−628

Paddy armyworm, 268
Paddy hispa, 270
Paddy kernels, definition by U.S.
    standards, 865, 872, 881
Paddy rice, characteristics before
    parboiling, 504−507
  awned rice and empty glumes, 505
  chalky, green and red grains, 506−507
  injuries caused by insects or threshing,
    506
  mold infestations, 505
  shelled grains, 505
Paddy root weevil, 264−265
  description of damage, 264
  losses due to damage, 265
Paddy separators, 366−368, 374−375
  compartment separator type, 366−367
  indented cylinder separator, 374
  indented disc separator, 375
Pakistan, rice export, 60−61
Panicle, 95−96, 116
Panicle initiation, 79−80
Parboiled brown rice, special grades and
    requirements by U.S. standards,
    877−878
Parboiled canned rice, effects of steaming
    conditions, 593
Parboiled milled rice, special grades and
    requirements by U.S. standards,
    890
Parboiled rice, 501−540
  characteristics of pady rice before
    parboiling, 504−507
  awned rice and empty glumes, 505
  chalky, green and red grains, 506−507
  injuries caused by insects or threshing,
    506
  mold infestations, 505

shelled grains, 505
cleaning, 507
color sorting, 522−523
drying, 517−520
  dryers, 518−520
  objectives and methods, 517−518
economics, 536−540
  comparative costs, 540
  costs of processing, 537−538
effect of epichlorohydrin treatment on
    solid loss and organoleptic
    evaluation, 598
grading, 507−509
milling, 520−521
parboiling process, 524−532
  affect on canned rice, 591−593
  changes in rice caryopsis, 424−426
  examples of parboiling systems, 525−
    532
    avorio process, 528
    CFTRI process, 526−527
    crystal rice process, 528
    CRGA process, 530−531
    Jadavpur Univ. process, 527−528
    malek process, 530
    rice conversion process, 528−530
    schule process, 525
    summary, 531−532
quick cooking parboiled rice, 533−535,
    584
recent developments in equipment, 535−
    536
  rubber rollers, 535−536
  transfer of materials, 535
stability of milled rice, 523−524
steaming, 513−516
steeping, 509−513
Parboiled bran, composition, 821
  nutritional values, 841
  sieve analysis, 823
Parboiled rough rice, special grades and
    requirements by U.S. standards,
    869−870
Parboiling systems, 525−532
  avorio process, 528
  CFTRI process, 526−527
  crystal rice process, 528
  CRGA process, 530−531
  Jadavpur Univ. process, 527−528
  malek process, 530
  rice conversion process, 528−530
  schule process, 525
  summary, 531−532
Pasteurization, rice vinegar, 726
Pearl rice, canned, effect of edible oil
    emulsions and surfactants on
    reduction of cohesion, 596
Pentosans, in rice germ, 818
Peru, rice planting procedures, 176
Philippines, rice import and export, 54
Phosphine gas, 308
Phosphorus, in endosperm, 452
Phosphorus deficiency, 186−187
Photoperiod, 85, 153−154

Other AVI Books

BREAD SCIENCE AND TECHNOLOGY
Pomeranz, Shellenberger

CARBOHYDRATES AND HEALTH
Hood, Wardrip, Bollenback

CEREAL TECHNOLOGY
Matz

CITRUS SCIENCE AND TECHNOLOGY
Nagy, Shaw, Veldhuis   Vol. 1 and 2

COCONUTS: PRODUCTION, PROCESSING, PRODUCTS
Woodroof

COMMERCIAL FRUIT PROCESSING
Woodroof, Luh

COMMERCIAL VEGETABLE PROCESSING
Luh, Woodroof

ENCYCLOPEDIA OF FOOD SCIENCE
Peterson, Johnson   Vol. 3

FOOD PRODUCTS FORMULARY: FRUIT, VEGETABLE AND NUT PRODUCTS
Tressler, Woodroof   Vol. 3

HANDLING, TRANSPORTATION AND STORAGE OF FRUITS AND VEGETABLES
Ryall, Lipton   Vol. 1 and 2

MICROBIOLOGY OF FOOD FERMENTATION
Pederson   Second Edition

PHYSIOENGINEERING PRINCIPLES
Merva

POTATO PROCESSING
Talburt, Smith   Third Edition

RHEOLOGY AND TEXTURE IN FOOD QUALITY
Deman, Voisey, Rasper, Stanley

SOILS AND OTHER GROWTH MEDIA
Flegmann, George

WHEAT: PRODUCTION AND UTILIZATION
Inglett

YEAST TECHNOLOGY
Reed, Peppler